Handboo

Mobile

Ubiquit

Comput

Status and Perspe

OTHE

Ad Hoc M
Principles
Subir Kum
and C. P
ISBN 978

Commur
Yang Xia

Decent
Interco
Magdi S
ISBN 9

Delay
Applic
Athan
Thras
ISBN

Eme
Tech
Chris
ISBN

Gar
Co
Ne
Jo
IS

G
E
T
H

Handbook on Mobile and Ubiquitous Computing

Status and Perspective

Edited by
Laurence T. Yang • Evi Syukur • Seng W. Loke

CRC Press
Taylor & Francis Group
Boca Raton London New York

CRC Press is an imprint of the
Taylor & Francis Group, an **informa** business

CRC Press
Taylor & Francis Group
6000 Broken Sound Parkway NW, Suite 300
Boca Raton, FL 33487-2742

© 2013 by Taylor & Francis Group, LLC
CRC Press is an imprint of Taylor & Francis Group, an Informa business

No claim to original U.S. Government works

Version Date: 20120822

International Standard Book Number: 978-1-4398-4811-1 (Hardback)

Library of Congress Cataloging-in-Publication Data

Handbook on mobile and ubiquitous computing : status and perspective / editors, Laurence T. Yang, Evi Syukur, Seng W. Loke.
    p. cm.
Includes bibliographical references and index.
ISBN 978-1-4398-4811-1 (hardback)
  1. Mobile computing--Handbooks, manuals, etc. 2. Ubiquitous computing--Handbooks, manuals, etc. I. Yang, Laurence Tianruo. II. Syukur, Evi. III. Loke, Seng.

QA76.59.H364 2012
004--dc23
                                          2012028459

Visit the Taylor & Francis Web site at
http://www.taylorandfrancis.com

and the CRC Press Web site at
http://www.crcpress.com

Contents

PART I   Mobile and Ubiquitous Computing

PART II   Smart Environments and Agent Systems

Preface

Mobile and ubiquitous computing is an essential part of computing today and will be so in the years to come. The late Mark Weiser's vision of ubiquitous computing, released to the world around 1990, has seized the imagination of many. The trend is that computing will get more mobile and more ubiquitous in the future. Perhaps 30 years (or earlier) from now, computing will be synonymous with ubiquitous computing, that is, the word "ubiquitous" will no longer need special mention; instead, it will be the norm.

To move forward, it is often useful, from time to time, to consolidate and review research in the area in order to gain an overarching perspective, and this book serves as a piece of the picture toward this end. Mobile and ubiquitous computing often involves a number of different areas in computing (this itself is a testimony of its impact), and this book testifies to that.

This book is intended for researchers, graduate students, and industry practitioners in computer science and engineering who are interested in recent developments in the area of mobile and ubiquitous computing. Folk wisdom has it that a good way to get acquainted with a particular area is to read a set of useful papers in the area. This book contains a useful set of papers/chapters, even if not exhaustive, for one to gain an understanding into the key issues and problems being tackled in this area. People familiar with the basics and active in the area will also find this book useful as a reference snapshot of recent research.

This book is divided into 6 parts and 27 chapters, as follows:

1. Part I: Mobile and Ubiquitous Computing
2. Part II: Smart Environments and Agent Systems
3. Part III: Human–Computer Interaction (HCI) and Multimedia Computing
4. Part IV: Security, Privacy, and Trust Management
5. Part V: Embedded Real-Time Systems
6. Part VI: Networking Sensing and Communications

In the following, we outline the main contributions of each chapter for the reader. Extensive references can be found in each chapter.

Part I: Mobile and Ubiquitous Computing

The availability of sensor devices, a convergence of advanced computing devices and connected software systems via wireless technology, and the Internet in the environment has opened up a research avenue known as "mobile and ubiquitous computing." This part illustrates the concepts, design, implementation, and deployment of mobile and ubiquitous systems, particularly in mobile and ubiquitous environments.

In location-aware computing systems, location information is considered an essential context, in order to deliver right information to users. However, more fine-grained information can be delivered by identifying the semantics of a location. Chapter 1 proposes a technique/mechanism to automatically identify the location semantics by taking into account the dynamic purpose of a space (e.g., activities happening in the space) and environmental entities such as users in the space and time of day.

Recent advances in mobile devices, wireless networking, communication, and software systems are offering new possibilities for Internet-based applications and commerce. This then opens up a new research field known as "wireless payment systems" that plays a crucial role in online commercial applications (e.g., online shopping from mobile devices, online trading, online banking, etc.). Chapter 2 gives a detailed overview of wireless payment systems. The chapter covers essential concepts, requirements, architecture, and prototype of secured wireless payment (known as P2P-Paid) between mobile clients and the payment server.

Chapter 3 presents a world model for ubiquitous computing environments. The model provides a unified view of the space that includes users, computing devices, and services. One of the advantages of the proposed model is that location-aware services can be managed without any database servers.

Nowadays, radio frequency identification (RFID) technology is being widely used in both educational and commercial environments. For example, RFID systems are used for tracking the location of deliveries, quantity of objects, etc. The deployment of RFID systems is currently targeting many fixed (predefined) locations (also known as static environments) in order to optimize accuracy and performance of the systems. Chapter 4 proposes novel concepts of developing RFID systems that target mobile environments. The chapter discusses technical issues such as reader collisions, frequency interference, flexibility, interoperability, mobility of users, accuracy, and performance for building such a system.

Part II: Smart Environments and Agent Systems

Through sensor devices and software systems (e.g., agent-based systems), which are invisibly installed in the environment of our everyday lives, embedded or non-embedded computing devices (e.g., sensor devices, RFID tags, and readers) communicate and exchange messages with each other. This part discusses a new trend toward intelligent systems that are completely connected, proactive, intuitive, and constantly available for users everywhere in the environment.

With the advances of computing devices, software systems, and communication technologies, requesting information services via the Internet has increased rapidly. Users can request and access multiple services at a time. However, in order to support this feature, there are several aspects that need to be considered such as the adaptability, timeliness, heterogeneity of users' behaviors, dynamically changing requirements, and system performance (i.e., reducing network traffic and minimizing users' wait time). Chapter 5 presents a demand-oriented architecture, called faded information field (FIF) that employs push and pull mechanisms in order to preserve the access time to all requested users, regardless of the trend of the demand and volume of the requested services.

Chapter 6 proposes a novel framework for detecting and improving potential collision between vehicles, especially at intersections. The framework employs ubiquitous data mining, multiagent systems, and context awareness in order to improve intersection safety.

Chapter 7 presents a new model for intelligent environments, where the proposed model is based on a motivated learning agent model, in order to allow environments to adapt to changes (e.g., addition) of new computing devices in the environment.

Chapter 8 proposes a framework that addresses the challenges of merging RFID and wireless sensor networks in smart computing environments. The idea is to provide users with smart objects, while moving around and providing the relevant services (e.g., tagged information) based on their current contexts.

Part III: Human–Computer Interaction (HCI) and Multimedia Computing

Two aspects that will continue to influence human–computer interaction today and in the near future will be, increasingly, sensory input and multimedia information.

Chapter 9 describes guidelines for designing multisensory input and output for mobile devices, covering spatial, direct (visual, auditory, and haptic), and temporal metaphors for interaction. The framework is comprehensive and provides fuel for thought about the next generation of mobile devices.

The new media would typically incorporate video and hyperlinking for interaction due to the influence of the web. Chapter 10 notes through experiments that hyperlinks are important while video transcriptions are not encouraged for multimedia information system interfaces.

Chapter 11 discusses a logic-based formalism for representing aspects of the world and reasoning with them, including attributes, sensors, actors, and spatial configurations. Multimedia information can also be input to a reasoning system for understanding integrated multimedia information. The sketch of the approach and examples suggest the utility of a knowledge representation formalism as the central basis for machine understanding of integrated sensory and multimedia information.

Part IV: Security, Privacy, and Trust Management

Pervasive computing systems can enable unprecedented flexibility and openness in interactions between devices and entities. However, there is a need to protect against unwanted behaviors and use of resources.

Trust between a system and its mobile client is imperative in pervasive computing environments, where servers might interact with clients in an ad hoc manner without previous agreements or a priori deep knowledge of each other. Chapter 12 presents an approach to dynamically establish such trust in a flexible manner using a multiagent negotiation mechanism.

Chapter 13 is another framework for establishing trust without a priori knowledge of the complete set of interacting entities and global trust registers. Fine-grained access control can be provided via context awareness that takes into account the contexts of clients and service providers in determining an authorization decision.

RFID systems will proliferate given their wide appeal, but not without privacy risks. RFID tags might have been used by a company to tag goods for its operations. However, such goods when sold must not be trackable as when they were company property, or RFID tags might be used to tag money, in which case the trackability they provide, while a convenience to some, becomes an intrusion of privacy for others. Information on such tags might be re-encrypted to prevent trackability in such cases. Chapter 14 discusses issues with re-encryption of such RFID tags.

Part V: Embedded Real-Time Systems

With the proliferation of mobile and ubiquitous systems in all aspects of human life, there is a great demand on embedded systems for real-time applications. The use of embedded computational artifacts, wireless networking, and sensor devices is an important instrument for facilitating invisible and seamless connectivity in mobile and ubiquitous environments. This part presents novel work on how mobile, ubiquitous, and intelligence computing can be realized in the field of real-time embedded systems world.

Chapter 15 proposes a new architecture for motion compensation (MC) of H.264/Advanced Video Coding (AVC) with power reduction, by avoiding redundant data transfers, whereas the data transfers dominate power consumption. The objective of this chapter is to employ a microarchitecture-level configurability, based on the value of the motion vectors. This allows the system to operate in different control flows during execution. This chapter demonstrates that it is possible to reduce up to 87% of power via data transfers in the conventional MC module.

Chapter 16 presents the architecture and design of a framework called verifiable embedded real-time application framework (VERTAF). VERTAF supports software component–based reuse, formal synthesis, and formal verification of embedded real-time applications with mobile and ubiquitous control access. The framework offers high-level reusability and extensibility of software components, which helps to increase productivity (during the design phase).

Chapter 17 presents an agent-based architecture on a reconfigurable system-on-chip (rSoC) platform in real-time systems. The author proposes an on-demand message passing (ODMP) protocol for the agent-to-agent communication. Two real-time systems have been employed by using P3-DX robot in order to evaluate the proposed agent-based model. The experimental results show that the proposed architecture is feasible, efficient, and extensible.

Part VI: Networking Sensing and Communications

Wireless sensor networks are installed everywhere in the environment. Sensor networks play a crucial part in helping humans accomplish their daily tasks—for example, from sensing a user's current location to reminding the user of the next tasks to do. Wireless sensor networks pretty much require no fixed infrastructure, such as self-organizing, rapid deployment, and dynamically reconfigurable. This then creates many ideas, challenges, designs, and prototype solutions on how to establish, manage, and maintain current working sensor networks in mobile and ubiquitous computing environments.

Chapter 18 proposes an efficient sensor node on an embedded ubiquitous computing platform. This is done by processing data locally at a lower layer and extracting particular information to be sent out to the upper layer in a wireless sensor network. In the proposed scheme, the software task scheduler and the hardware task scheduler are independent of each other. Regardless of their independence, an appropriate balance between computation time and energy consumption is maintained.

Chapter 19 proposes a model to support QoS-aware multicast routing targeting large-scale MANETs. The model uses the location information of mobile networks and takes into account several properties such as high fault tolerance, small size of diameter, regularity, and symmetry. The proposed model tackles the issue of node overload and bottleneck, which is more likely to occur in the traditional tree-based architectures.

Chapter 20 presents a comparative analysis of Mobile Ad hoc NETwork (MANET) routing protocols under each scenario by taking into account mobility, traffic, and network size parameters.

The protocol performance depends on network conditions and environment. The experimental results show that it is highly possible to obtain a throughput greater than 40 kbps for each transmitting node in a network composed of 20 mobile nodes, when 10 of the nodes are transmitting simultaneously.

Chapter 21 presents a geometrical probability approach to compute the connectivity of two adjacent blocks. The proposed approach is valid for further extension of probabilistic Quality of Service (QoS) in the ad hoc network with connectivity and real-time bandwidth being on demand. Based on the experimental results, the proposed approach can effectively determine the minimum transmission energy required for each node and the minimum number of nodes required for maintaining the expected connectivity probability.

Chapter 22 proposes an extended schema of weighted low-energy localized clustering (w-LLC) in order to cope with the situation of unequal events in a certain area of wireless networking. The simulation results show that w-LLC applies the concept of weight functions and achieves better throughput than the ordinary low-energy localized clustering (LLC). The proposed model helps to reduce the possibility of RFID reader collision problems in RFID networks.

Chapter 23 presents the coupled simulation approach in cross-layer design of mobile ad hoc networks that enables interaction between layers. The proposed approach is able to simulate the feedback loop between the network state and the application state. In addition, the emulation infrastructure can be easily extended and facilitates development of new cross-layer solutions and their interdependencies.

Chapter 24 proposes algorithms that support adaptivity of ant-based routing (ABR) and a hybrid ant-based routing (HABR) scheme in wavelength division multiplexing (WDM) networks. Based on the extensive simulation study, the proposed ABR algorithm performs better than the ordinary fixed-alternated routing algorithm in terms of blocking probability. Moreover, the performance of the ABR approach is further improved by a HABR approach by combining the mobile agent approach and the alternate routing methods.

Chapter 25 proposes a modular software architecture for mobile ad hoc networks using cross-layer design principles. The proposed model allows dynamic routing algorithm switches and employs adaptive parameter changes at runtime. The compiled modules can be run without modifications on the scalable MANET emulation test bed and the real Linux 2.6–based computer systems.

Chapter 26 proposes an efficient location update and paging scheme based on both the movement and distance directions of mobile hosts. The aim is to reduce the required managing location of mobile hosts by taking into account parameters such as location update cost, paging cost, and the total cost in various situations.

Chapter 27 proposes a Transmission Control Protocol (TCP) Mobile solution in order to improve TCP performance for data communication in the wireless network where packet losses occur mostly due to congestions, handoffs, and lousy links. The proposed approach includes connection establishment, data transfer, and handoff management phases. The extensive simulation proves that the proposed TCPMobile solution achieves better performance than Reno TCP and M-TCP in terms of throughput.

Acknowledgments

We would like to take this opportunity to thank all the authors for their excellent contributions. The authors have also been very punctual in the submission process as well as patient in all stages of the publication process. We also thank the referees for their valuable comments, which helped the authors to revise and improve the quality of their chapters. Help from Richard A. O'Hanley (the publisher of IT, Business, and Security, CRC Press), Stephanie Morkert (project coordinator, CRC Press), Robert Sims (project editor, CRC Press), and Saranyaa Moureharry (project manager, SPi Global) through the publication process is very much acknowledged and appreciated.

Evi Syukur
Laurence T. Yang
Seng W. Loke

Editors

Dr. Evi Syukur is the cofounder and CEO of HexaMobilIndo Consulting company. She served as a senior researcher at the University of New South Wales (UNSW), Sydney, New South Wales, Australia. She also served as a .NET developer at Yarra Trams Company, Melbourne, Australia. Syukur received her PhD in computer science from Monash University, Melbourne, Victoria, Australia. Her research interests include service-oriented systems/applications, context-aware service applications, and smart services/objects. She has also published widely in several international journals, conferences, and workshops. She has been on the PC cochairs, publicity cochairs, and program committees of numerous international conferences and workshops. She was the co-guest editor of selected *International Journal of Pervasive Computing and Communications* (*IJPCC*) special issue on ubiquitous intelligence in smart world.

Professor Laurence T. Yang is a professor in computer science in the Department of Computer Science at St. Francis Xavier University, Antigonish, Nova Scotia, Canada. His current research includes parallel and distributed computing as well as embedded and ubiquitous/pervasive computing.

He has published more than 350 papers in various refereed journals, conference proceedings, and book chapters in these areas, including around 120 international journal papers such as IEEE and ACM Transactions. He has been actively involved in conferences and workshops as a program/general/steering conference chair and numerous conference and workshops as a program committee member. He served as the vice chair of IEEE Technical Committee of Supercomputing Applications (2001–2004), the chair of IEEE Technical Committee of Scalable Computing (2008–2011), and the chair of IEEE Task Force on Ubiquitous Computing and Intelligence (2009–till present). He was in the steering committee of IEEE/ACM Supercomputing (SC-XY) conference series (2008–2011) and currently is in the National Resource Allocation Committee (NRAC) of Compute Canada (2009–present).

Professor Yang is also the editor in chief of several international journals. He has also authored/-coauthored or edited/coedited more than 25 books from well-known publishers. The book *Mobile Intelligence* (Wiley, 2010) received an honorable mention by the American Publishers Awards for Professional and Scholarly Excellence (The PROSE Awards). He has won several Best Paper Awards (including IEEE Best and Outstanding Conference Awards such as the IEEE 20th International Conference on Advanced Information Networking and Applications [IEEE AINA-06]); one Best Paper Nomination; Distinguished Achievement Award in 2005 and 2011; and Canada Foundation for Innovation Award in 2003. He has been invited to give around 20 keynote talks at various international conferences and symposia.

Dr. Seng W. Loke is a reader and associate professor in the Department of Computer Science and Computer Engineering at La Trobe University, Melbourne, Victoria, Australia. He leads the Pervasive Computing Interest Group at La Trobe. He has coauthored more than 220 research publications, including numerous works on context-aware computing and mobile and pervasive computing. He authored *Context-Aware Pervasive Systems: Architectures for a New Breed of Applications* (Auerbach, 2006). He has been actively involved on the program committee of numerous conferences/workshops in the area, including pervasive 2008. He received his PhD in 1998 from the University of Melbourne.

Contributors

Cláudia J.B. Abbas
Facultad de Informática
Departamento de Ingeniería del Software e
    Inteligencia Artificial
Grupo de Análisis, Seguridad y Sistemas
Universidad Complutense de Madrid
Madrid, Spain

and

Arab Academy for E-Business
Aleppo, Syria

Patroklos Argyroudis
Consus, Inc.
Athens, Greece

Oliver Battenfeld
Department of Mathematics and Computer
    Science
University of Marburg
Marburg, Germany

Jacky Cai
Department of Computer Engineering
San Jose State University
San Jose, California

Dempsey Chang
Shun Hwa Pharmaceutical Co., Ltd.
Taipei, Taiwan

Eric Y. Cheng
Department of Computer Information
    Systems
The State University of New York
Canton, New York

Jihoon Choi
Network Testing and Certification Department
IT Testing and Certification Laboratory
Telecommunications Technology Association
Seongnam-City, Gyconggi-do, Republic of
    Korea

Bernd Freisleben
Department of Mathematics and Computer
    Science
University of Marburg
Marburg, Germany

Jerry Gao
Department of Computer Engineering
San Jose State University
San Jose, California

Minyi Guo
Department of Computer Science
Shanghai Jiao Tong University
Shanghai, People's Republic of China

Sven Hanemann
Department of Mathematics and Computer
    Science
University of Marburg
Marburg, Germany

Vinh Dien Hoang
National Institute of Information and
    Communications Technology
Singapore, Singapore

Susumu Horiguchi
Graduate School of Information Sciences
Tohoku University
Sendai, Japan

Pao-Ann Hsiung
Department of Computer Science and
 Information Engineering
National Chung-Cheng University
Taiwan, Republic of China

Seok Joong Hwang
System Architecture Lab
Samsung Advanced Institute of Technology
Samsung Electronics Co., Ltd.
Yongin, South Korea

Yoshinori Isoda
R&D Centre
NTT DoCoMo, Inc.
Yokosuka, Japan

Dae-Young Jeong
Electronic Warfare R&D Laboratory
LIG Nex1
Gyeonggi-do, South Korea

Weijia Jia
Department of Computer Science
City University of Hong Kong
Hong Kong, China

Xiaohong Jiang
Graduate School of Information Sciences
Tohoku University
Sendai, Japan

Jaewon Jung
Equity Investment Department
IBK Asset Management Co., Ltd
Seoul, South Korea

Byung-Gil Kim
Supercomputing Laboratory
Department of Computer Science
Yonsei University
Seoul, South Korea

Cheong-Ghil Kim
Department of Computer Science
Namseoul University
CheonAnn, ChoongNam, Korea

Daeyoung Kim
Korean Advanced Institute of Science and
 Technology
Daejeon, South Korea

Dongshin Kim
Department of Computer Science and
 Engineering
Korea University
Seoul, South Korea

Eunkyo Kim
CTO Department
LG Electronics
Seoul, Korea

Joongheon Kim
CTO Department
LG Electronics
Seoul, Korea

Seon Wook Kim
Compiler and Microarchitecture Laboratory
School of Electrical Engineering
Korea University
Seoul, South Korea

Seung-Yeon Kim
Telecommunication R&D Center
Samsung Electronics Co., Ltd.
Paldal-gu, South Korea

Shin-Dug Kim
Supercomputing Laboratory
Department of Computer Science
Yonsei University
Seoul, South Korea

Woo-Jae Kim
Department of Computer Science and
 Engineering
Pohang University of Science and
 Technology
Pohang, South Korea

Shonali Krishnaswamy
Caulfield School of Information Technology
Monash University
Melbourne, Victoria, Australia

Goro Kunito
R&D Centre
NTT DoCoMo, Inc.
Yokosuka, Japan

Jin Kwak
Department of Information Security
    Engineering
Soonchunhyang University
Chungnam, Korea

Tyrone Tai-On Kwok
Department of Electrical and Electronic
    Engineering
The University of Hong Kong
Pokfulam, Hong Kong

Yu-Kwong Kwok
Department of Electrical and Electronic
    Engineering
The University of Hong Kong
Pokfulam, Hong Kong

Joon Goo Lee
Compiler and Microarchitecture Laboratory
School of Electrical Engineering
Korea University
Seoul, South Korea

Wonjun Lee
Department of Computer Science and
    Engineering
Korea University
Seoul, South Korea

Shang-Wei Lin
Temasek Laboratories
National University of Singapore
Singapore, Singapore

Zhaoyu Liu
Department of Software and Information
    Systems
University of North Carolina
Charlotte, North Carolina

Seng Wai Loke
Department of Computer Science and
    Computer Engineering
La Trobe University
Melbourne, Victoria, Australia

Tomas Sanchez Lopez
European Aeronautic Defence and Space
    Innovation Works
Newport, United Kingdom

Xiaodong Lu
Department of Computer Science
Tokyo Institute of Technology
Tokyo, Japan

Maode Ma
School of Electrical & Electronic Engineering
Nanyang Technological University
Singapore, Singapore

Owen Macindoe
Faculty of Architecture, Design and Planning
University of Sydney
Sydney, New South Wales, Australia

Mary Lou Maher
Faculty of Architecture, Design and Planning
University of Sydney
Sydney, New South Wales, Australia

Yan Meng
Department of Electrical and Computer
    Engineering
Stevens Institute of Technology
Hoboken, New Jersey

Kathryn Merrick
School of Information Technologies
University of Sydney
Sydney, New South Wales, Australia

Kinji Mori
Department of Computer Science
Tokyo Institute of Technology
Tokyo, Japan

Keith V. Nesbitt
School of Design, Communication and IT
The University of Newcastle
Newcastle, New South Wales, Australia

Son Hong Ngo
School of Information and Communication
Technology
Hanoi University of Technology
Hanoi, Vietnam

Donal O'Mahony
Department of Computer Science
Trinity College Dublin
Dublin, Ireland

Kyung Ho Park
GPU Research Center
Siliconarts, Inc.
Seoul, South Korea

Dichao Peng
Department of Mathematics
University of Zhejiang
Hangzhou, China

Andry Rakotonirainy
Centre for Accident Research and Road Safety
Queensland
Queensland University of Technology
Queensland, Brisbane, Australia

Patrick Reinhardt
Department of Mathematics and Computer
Science
University of Marburg
Marburg, Germany

Jae-Cheol Ryou
Division of Electrical and Computer
Engineering
Chungnam National University
Daejeon, South Korea

Junichiro Saito
Faculty of Information Science and Electrical
Engineering
Kyushu University
Fukuoka, Japan

Kenji Sakamoto
R&D Centre
NTT DoCoMo, Inc.
Yokosuka City, Japan

Kouichi Sakurai
Department of Computer Science and
Communication Engineering
Faculty of Information Science and Electrical
Engineering
Kyushu University
Fukuoka, Japan

Flora Dilys Salim
Spatial Information Architecture
Laboratory
The Royal Melbourne Institute
of Technology University
Melbourne, Victoria, Australia

Ichiro Satoh
National Institute of Informatics
Tokyo, Japan

Christian K. Shin
Department of Computer Science
The State University of New York
New York, New York

Matthew Smith
Department of Electrical Engineering and
Computer Science
Leibniz University Hannover
Hannover, Germany

Bala Srinivasan
Clayton School of Information
Technology
Monash University
Melbourne, Victoria, Australia

Young-Joo Suh
Department of Computer Science and
    Engineering
Pohang University of Science and
    Technology
Pohang, South Korea

Satoshi Tanaka
R&D Centre
NTT DoCoMo, Inc.
Yokosuka, Japan

L.J. García Villalba
Facultad de Informática
Grupo de Análisis, Seguridad y Sistemas
Departamento de Ingeniería del Software e
    Inteligencia Artificial
Universidad Complutense de Madrid
Madrid, Spain

Guojun Wang
School of Information Science and Engineering
Central South University
Changsha, People's Republic of China

Jianxin Wang
School of Information Science and Engineering
Central South University
Changsha, People's Republic of China

Daoxi Xiu
Department of Software and Information
    Systems
University of North Carolina
Charlotte, North Carolina

Naoharu Yamada
R&D Centre
NTT DoCoMo, Inc.
Yokosuka, Japan

Kenichi Yamazaki
R&D Centre
NTT DoCoMo, Inc.
Yokosuka, Japan

Sang-Soo Yeo
Faculty of Information Science and Electrical
    Engineering
Kyushu University
Fukuoka, Japan

Masao Yokota
Department of System Management
Faculty of Information Engineering
Fukuoka Institute of Technology
Fukuoka, Japan

Hyung Min Yoon
GPU Research Center
Siliconarts, Inc.
Seoul, South Korea

Lifan Zhang
School of Information Science and Engineering
Central South University
Changsha, People's Republic of China

Lili Zhang
Intelligent System Centre
Nanyang Technological University
Singapore, Singapore

Yan Zhang
Department of Network Systems
Simula Research Laboratory
Oslo, Norway

Yang Zhou
School of Information Science and Engineering
Central South University
Changsha, People's Republic of China

MOBILE AND UBIQUITOUS COMPUTING

Chapter 1

Automatic Identification of Location Semantics Based on Environmental Entities

Naoharu Yamada, Yoshinori Isoda, Kenji Sakamoto, Satoshi Tanaka, Goro Kunito, and Kenichi Yamazaki

Contents

1.1 Introduction

Acquiring context information is essential in achieving context-aware systems and applications. Since location is the most important context, many works have tried to identify the user's location [11,21,22,36,37]. By acquiring a user's location periodically, the system can provide man navigation and specify user's significant location and "hub of activity" [13,32]. Furthermore, by acquiring the location of multiple users, the system can identify friends in the neighborhood or how crowded a specific location is [1].

In order to implement context-aware applications, it is important to identify not only the location but also the location semantics [10]. Location semantics are public labels that indicate possible activities and social restrictions/permissions. For example, the semantics of a *restaurant* indicate that users can eat at the location. In some *restaurants*, smoking is prohibited. By acquiring the semantics of a user's location, the system can support the user by providing him with information or services. For example, if a location near a user can currently be labeled as *a flea market*, the system can provide information about *the flea market* to the user. Furthermore, the semantics of a user's location offer important clues for identifying the user's activities. For example, if the semantics of a user's location is *food stall*, we can imagine that the user is taking a meal. Therefore, by acquiring the semantics of a user's location, other users can imagine what the user is doing. For example, the user can be provided with a view of the life of an elderly relative by reviewing a list of the semantics of the locations visited by the relative. In addition, the system can provide a service even if the user does not make an explicit request. For example, when a user is shopping in a food court, the system can tell him what is in his refrigerator and what is missing.

Conventional research assumes that location semantics are entered by hand or via map references since they suppose that location semantics are static. Unfortunately, location semantics can change dynamically [4]. For example, a park should be labeled *a shopping place* when and only when it hosts *a flea market*. In addition, the dining room in most Japanese homes is used for studying, working, and socializing as well as eating. In this case, the semantics of dining room becomes *an eating place* or *a working place* only when the users have meal or work there. Conventional research cannot handle dynamically changing location semantics. Even if some location semantics are static with respect to location, manual identification does not scale well and map-based identification often fails because some location names such as shop names and company names do not indicate the semantics. For example, a map name, *NTT DoCoMo, Inc. Building*, a cell-phone company's building, may not indicate any particular activity. More detailed semantics such as *office* or *cell-phone shop* are required.

In this chapter, we propose a method to automatically identify the location semantics by handling dynamically changing semantics. We focus on environmental entities such as things, humans, and time of day to identify location semantics following the concept of affordance [9]. We explain related work in Section 1.2. Section 1.3 clarifies the definition, characteristics, and technical issues of location semantics. Section 1.4 describes our approach, which utilizes ontology and a multi-class naive Bayesian technique. Section 1.5 explains the simulations and experiments that were conducted using actual things to verify our proposal.

1.2 Related Work

Many papers have defined and categorized location semantics [2–4,12,14,29]. Jordan et al. [12] defines "place," which is very similar in its concept to location semantics, as something that provides a context for everyday action and a means for identification via the surrounding environment. Tuan

[33] defines "place" as space infused with human meaning. Entrikin [4] mentions that location semantics are identified by personal subjectivity created by personal experiences such as a memorial site. As for the categorization of location semantics, Curry et al. [3] classify location semantics into five categories: naming (e.g., geographic name), applying existing typologies (e.g., inner city), making/picking out a symbol (e.g., Pyramid in Egypt), telling stories (e.g., the place of heroic battles), and performing activities (e.g., eating place). Jordan et al. add the category of physical features such as things to Curry's classification. Matsuo [14] classifies location semantics as physical functions that indicate the enabling activities and social functions that indicate social restrictions/permissions. Therborn et al. [29] and Cresswell et al. [2] also mention social functions such as *in place* and *out of place*.

Though some works also mention that some location semantics dynamically change [4], few researchers have tackled automatic identification methods. Hightower [10] proposed human activity-based identification. With regard to human activity recognition, many works have focused on things that a user touches or grasps. Accordingly, they attempt to detect the series of things that the user has been interacting with using various sensors such as cameras [19], ultrasonic sensors [20], state change sensors [27], accelerometers [25], and RFID tags [7]. However, these approaches have implementation issues with regard to sensors and location semantics identification: As for the issue of sensors, cameras and ultrasonic sensors have high hardware costs and installation costs and the approaches that use these sensors have been shown to work only in limited areas such as laboratories. In addition, the approaches that require attaching wearable sensors to the user [7,29] face significant user resistance. An RFID tag is a low-cost sensor that consists of a microchip that stores a unique ID and a radio antenna [6]. Its is seen as replacing the barcode in the area of logistics. Some companies or governments now require suppliers to attach RFID tags to every item [30]. EPC Global [5] and Ubiquitous ID center [34] have proposed an ID scheme that makes it possible to put a unique serial number on every item. Considering these circumstances, we can assume that everything will have its own RFID tag in the near future. This means that RFID tags are the most promising approach to achieving human activity recognition. With regard to location semantics identification, location semantics exist even if no user is present. For example, *a time sale in a shop* exists even if no user is at the location and disappears once all goods are sold or the time limit is reached. Therefore, approaches based on user interaction such as touching and grasping are insufficient to automatically identify location semantics.

In this chapter, we focus on environmental entities such as things, humans, and time of day to identify dynamically changing location semantics based on the concept of affordance [9]. According to affordance, an environment offers particular activities to users. Based on the concept of affordance, the offered activities can be identified by the environmental entities in the environment. We detect things and people in a certain location by attaching RFID tags to them and placing RFID tag readers in each location. We identify the semantics of a location by assessing all environmental entities present. Since the number of such entities is likely to be enormous, our approach can identify location semantics even if some entities are not properly detected.

1.3 Location Semantics Identification Based on Environmental Entities

This section clarifies the definition, characteristics, and technical issues of location semantics. It also describes the difficulties of location semantics identification based on environmental entities.

1.3.1 Location Semantics: Definition, Characteristics, and Technical Issues

Location semantics are public labels that indicate possible activities and social restrictions/permissions. This definition is similar to Jordan's definition which states that location semantics are something that provides a context for everyday action and a means for identification via the surrounding environment. For example, locations with the semantics of *a shopping place* such as *supermarkets*, *flea markets*, and *stalls* indicate that a user can buy commodities. The locations with semantics of *eating place* such as *a dining room*, *a restaurant*, and *a cafeteria* indicate that a user can eat and drink. The locations with semantics of *a user's own territory* such as a user's own room in his/her house or hotel where he/she is staying and a user's desk at his/her office indicate that a user can store his/her possessions. Each physical location can have multiple location semantics. In the above examples, the user's desk at the office is labeled as both *a working place* and *the user's own territory*.

This chapter classifies location semantics into physical location semantics and social location semantics. The classification of location semantics is shown in Figure 1.1

> *Physical location semantics*: Physical location semantics are specified by context provided by the environment. According to the concept of affordance [9], users recognize the activities enabled by the environment. Based on this concept, users can acquire the semantics of a location even if they are visiting the location for the first time. For example, if we look at a kitchen for the first time, we can recognize that cooking is possible at this location. If people look at a market stall in a foreign country for the first time, they can recognize that they can buy products at the location. In general, an environment is formed by the key environmental entities of things, humans, and time of day. Therefore, physical location semantics is further classified into these environmental entities as follows.

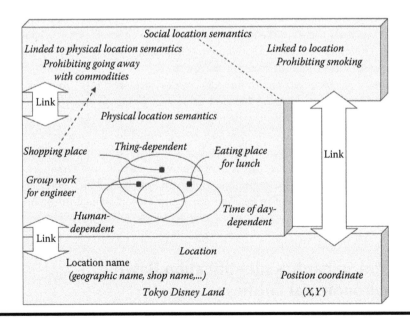

Figure 1.1   Classification of location semantics.

Thing-dependent: This category covers activities performed with things such as *a flea market* and *a kitchen*. The semantics are divided into two types: the semantics that depend on movable things and the semantics that depend on fixed things. The former covers *flea markets* and *working places in a dining room*. The latter covers *kitchens*, *bathrooms*, and *dressing rooms*. The latter are relatively static since the things comprising the semantics are fixed and be seldom moved.

Human-dependent: This category covers the activities associated with humans. *A board meeting with senior executives* falls into this category. Even if the system cannot identify what activities the humans are performing, the system can identify the enabling activities from the semantics of the attendees such as *business title* and *type of business*. This type of semantics can dynamically change since humans move.

Time of day-dependent: This category covers activities that are time sensitive. Performing activities within some time constraint falls into this category. This type of semantics can dynamically change as is obvious. This category has fewer entries and most are associated with the semantics of other categories as shown below.

Some semantics depend on multiple environmental entities. For example, *an eating place for lunch* depends on things and time. *Group work place of engineers* depends on things and humans.

Social location semantics: Social location semantics are determined by people's determination, agreement, or consensus. Examples of social location semantics include *prohibition on smoking* and *prohibition on phone calls*. Social location semantics have two types: social location semantics linked to location and social location semantics linked to physical location semantics.

Social location semantics linked to location: This type of location semantics is determined by explicit or implicit assignment. It can be difficult for users to identify the location semantics of a location just by looking at it. Therefore, users need to acquire knowledge of the location semantics from other people or official sources. For example, we cannot tell whether a river is class A or class B by just looking at it. Solutions include checking a map or signs along the river. People cannot tell if smoking is prohibited at a location just by looking at the location. They need refer to signs such as "No smoking." Once people reach agreement on the social semantics of a location, it seldom changes. Therefore, social location semantics is static with respect to location.

Social location semantics linked to physical location semantics: This type of location semantics are universally recognized as the restrictions/permissions set for specific physical location semantics. For example, the social location semantics of *prohibiting phone calls* in the physical location semantics of *theaters* is universally accepted. The social location semantics of *prohibiting the removal of items* in the physical location semantics of *museums*, *exhibitions*, and *shopping place* such as *a super market* and *a flea market* is universally accepted. In the last example, the social location semantics of *prohibiting the removal of items* exists only when an exhibition or a flea market is held. Therefore, these social location semantics dynamically change.

In this classification of physical location semantics, the dynamically changing semantics must be automatically identified. Even if the physical location semantics are fixed, automatic identification is needed to reduce the cost of manually inputting location semantics.

The social location semantics associated with physical location semantics also dynamically change. We preliminarily link these semantics to physical location semantics, which enables the system to automatically identify these semantics by automatically identifying physical location semantics.

1.3.2 Physical Location Semantics Detection Using Environmental Entities and the Difficulties

This chapter utilizes the semantics of environmental entities such as things, humans, and time of day to identify physical location semantics. However, identifying physical location semantics from environmental entities has several difficulties. We list them below based on the characteristics of physical location semantics and environmental entities.

Multiple physical location semantics: There are spatial relationships among physical location semantics such as inclusion, overlap, and adjacency. Figure 1.2 shows an example. Therefore, multiple semantics can be assigned to the same spatial position. This requires multiple semantics identification (P1).

Multiple semantics and representations of environmental entities: Each environmental entity has multiple semantics that also have multiple representations. For example, the thing with the semantics of *pencil* can represent *a writing tool* and *stationery*, and at the store, it can represent *a discount product* as a sub concept of *a product*. The human *Michel* with semantics of *computer engineer* can represent *engineer* in his office and *a student* in a private English school. The time of day *8 o'clock* with semantics of *morning* can represent *A.M.* This requires multiple semantics and representations of environmental entities (P2).

Variability of constituent environmental entities: Two locations may have identical location semantics but, of course, contain different things and people. For example, each house has a kitchen, a cooking place, and each kitchen contains different things. This means that the manual creation of detection rules is difficult (P3). Furthermore, even if some learning approaches are utilized to automatically extract inference rules, the system cannot deal with environmental entities that have not been learned (P4).

Mobility of environmental entities: Things and people can be moved or move for different reasons. The things that are moved due to the user's intention such as food or dishes for preparing meals are important in identifying physical location semantics. People who move to a specific location for a certain purpose are also important. Since things and people in a target location can change, near real-time physical location semantics identification is required (P5).

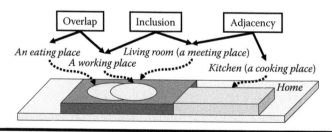

Figure 1.2 Spatial relationships among location semantics.

Environmental entities irrelevant to physical location semantics: While some environmental entities are effective in identifying physical location semantics, others are not useful since people do not use these things to perform particular activities or people do not perform particular activities with these people. Examples of the former are *lamps, trash,* and *users' dressing.* Those of the latter are *strangers passing through.* Therefore, tolerance to the noise of irrelevant things is required to correctly identify physical location semantics (P6).

Massiveness of things: People are surrounded by a huge number of things. Therefore, processing load may be excessive when surrounding things are used in a naive manner to identify physical location semantics (P7).

Other problems arise from the characteristics of RFID systems: RFID tag detection is not completely reliable because of collision and different ID transmission intervals [8].

1.4 Ontology and PMM for Identifying Physical Location Semantics via Environmental Entities

This section describes our approach, which uses ontology and the parametric mixture model (PMM). To solve the difficulties caused by multiple semantics/representation of entities (P2), unlearned entities (P4), and some part of entity mobility (P5), we utilize ontology. To counter the difficulties caused by multiple physical location semantics (P1), constituent variability (P3), entity mobility (P5), irrelevant things (P6), and massiveness of things (P7), we employ the parametric mixture model (PMM) [35], a text classification method, because we draw an analogy between documents composed of words and physical location semantics composed of environmental entities. Figure 1.3 shows the overall process flow of the proposed system. Though our description of the proposed system assumes the use of EPC Global, this assumption is not a definitive requirement. In the preprocess stage, the system aggregates detected RFID tags and extracts distinct things and people. For example, the system may extract only those things and people that appeared recently to detect newly generated physical location semantics. In the representation stage, the system acquires terms that represent each environmental entity. We acquire the attribute information of each detected environmental entity from Physical Markup Language servers (PML servers) [23]. Utilizing this information, all terms representing the environmental entities are acquired through ontology. In the learning stage, the probability of an environmental entity being an indicator of physical location semantics and an appropriate term representing each environmental entity is specified by utilizing the terms and supervised physical location semantics data. At the classification stage, the system uses PMM to classify a set of terms into physical location semantics.

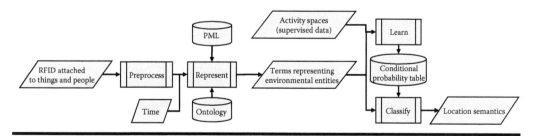

Figure 1.3   Overall process flow of physical location semantics identification.

1.4.1 Ontology to Manage Representations

Ontology has a long history in philosophy; it refers to the subject of existence. Although there is no agreed definition of ontology, one definition involves the specification of terms in each domain and the relations among them. Ontology sets the following concepts of terms and relations between terms [16–18].

> *Basic concept*: The underlying concept of an environmental entity such as *pencil* and *morning*.
> *Role concept*: The concept to represent a role that an environmental entity plays in a particular domain such as *product* and *engineer*.
> *Is-a relation*: The relation to represent the sub-concept between two terms. For example, "A pen is-a writing tool" means *a pen* is a sub-concept of *a writing tool*. "A computer engineer is-a an engineer" means *a computer engineer* is a sub-concept of *an engineer*.

We call the hierarchy based on these is-a relations the abstraction level. In addition, relations among environmental entities are needed. The following relation represents the owner of each thing.

> *Is-owned-by relation*: The relation that represents ownership between two terms. (e.g., "pencil is-owned-by Michel" means *a pencil* is owned by *Michel*.)

Utilizing these concepts and relations makes it possible to acquire all terms related to environmental entities by tracing the relations, which resolves the difficulty caused by multiple semantics/representation of entities (P2). Figure 1.4 shows an example. Acquired terms of the thing whose ID is 1 are *pencil, writing tool* and *stationery* by the is-a relation through its basic concept, and *discount product* and *product* by the is-a relation through its role concept. Here, the terms at the lowest abstraction level in each concept are preliminarily linked to the ID of each environmental entity in PML.

Among all terms related to an environmental entity, we need to identify the appropriate abstraction level to identify physical location semantics. Since it is difficult to manually identify a proper abstraction level, we manually choose only the appropriate "concept" such as basic concept and role concept in the learning phase of PMM. For example, the basic concept is selected for a working place, and role concept is selected for a shopping place in the learning phase.

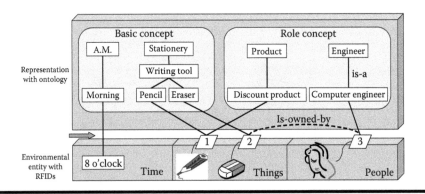

Figure 1.4 Representing environmental entities by utilizing ontology.

This approach, however, leaves unanswered how to select the proper term or abstraction level in the selected concept; this is solved by Section 1.4.2. After selecting a proper abstraction level, we utilize the terms in the proper abstraction level to identify physical location semantics in the classifying phase. To solve the difficulty caused by unlearned entities (P4), we transform the terms that have not been learned into the terms that have been learned by raising the abstraction level. For example, in Figure 1.4, if *eraser* has not been learned but *pencil* has; we can treat both as *stationery*, which has already been learned. In addition, raising the abstraction level reduces the number of kinds of terms. Therefore, the processing load of PMM can be reduced, which resolves the difficulty caused by entity mobility (P5).

1.4.2 Identifying Physical Location Semantics via Topic Detection

Many schemes for tackling the identification of the topics of documents or web contents have been proposed. The characteristics of their target are very similar to those of our objective: a document consists of a set of words that includes noise such as stop words [26]; each document on the same topic consists of different words, but people can identify the topic of a document at a glance. Among the many approaches proposed for topic detection, most assume that a document has only one topic. The parametric mixture model (PMM), however, allows one document to have multiple topics. It employs a probabilistic approach, which is efficient and robust against noise; it offers the highest accuracy in detecting multiple topics [35]. Since it is highly likely that multiple physical location semantics (P1) will be identified from one set of environmental entities, we employ PMM. PMM represents documents as a set of words, known as "Bag of Words" (BOW). It is based on the naive Bayes model [15]. While the naive Bayes model only provides binary classification, PMM extends this model to provide multi-topic detection. PMM assumes that a multi-topic document is composed of a mixture of words typical of each topic. Based on this assumption, a multi-topic document can be represented as the linear summation of the word occurrence probability vector of each topic as shown in Equation 1.1. Here, conditional probability $p(t_i|c_l)$ is calculated with MAP estimation. By replacing words, topics with environmental entities, physical location semantics, we can use Equation 1.1 to detect multiple physical location semantics (P1) from a set of environmental entities.

$$p(d|c) = p(t_1, \cdots, t_n|c) = \prod_{j=1}^{n} \left(\sum_{l=1}^{L} h_l(y) p(t_i|c_l) \right)^{x_i} \qquad (1.1)$$

where

$h_l(y) = \frac{y_l}{\sum_{l'=1}^{L} y_{l'}}, \ l = 1, \cdots, L$

$y_l = 1 (y_l$ belongs$)$ or $y_l = 0 (y_l$ does not belong$)$

d is document

c is topic

x_i is frequency of word t_i

L is # of topics

n is # of word kinds

To select the appropriate abstraction level of is-a relation in the manually selected concept, conditional probability $p(t_i|PLS_l)$ (t: thing, PLS: Physical Location Semantics) is acquired from the environmental entity lists of each abstraction level in the learning phase. PMM then acquires the classification accuracy of location semantics through the learned conditional probability of each abstraction level. Finally, the abstraction level with the highest classification accuracy is employed to classify test sets of environmental entities.

1.5 Simulations and Experiments

This chapter conducted simulations and experiments using only things. The simulations, which used actual but manually collected things, evaluated the performance of the proposed method in terms of physical location semantics detection; we focused on the difficulties caused by multiple semantics/representation of entities (P2), constituent variability (P3), unlearned entities (P4), and massiveness of things (P7). In the experiments in an actual environment, we attached RFID tags to each thing and confirmed the feasibility of the proposed method in the face of the difficulties presented by RFID characteristics. To create thing data at each abstraction level, we surveyed existing ontology bases. We decided to employ WordNet [39] since it contains the largest number of kinds of terms at each abstraction level and it is easily handled by external programs. We set "artifact" in WordNet as abstraction level 1, the highest abstraction level. Instead of utilizing the attribute information stored in PML, we manually set the terms of abstraction level 6, the lowest abstraction level, representing each thing. The terms on abstraction levels 2–6 can be acquired by utilizing the is-a relations specified in WordNet (Figure 1.5).

We implemented PMM in Java and ran the programs on a Pentium 4 3 GHz, 2 GB RAM PC. F-measure was used to evaluate the detection accuracy of physical location semantics, which is defined as the harmonic mean of precision and recall. Precision is the ratio of the number of physical location semantics correctly identified to the total number of physical location semantics identified. Recall is the ratio of the number of physical location semantics correctly detected to the number of correct physical location semantics.

1.5.1 Simulation

The simulations examined two cases: (sim.1) detection of frequently changing physical location semantics to address the difficulties caused by multiple semantics/representation of entities (P2), constituent's variability (P3), and unlearned entities (P4), and (sim.2) physical location semantics detection with a large number of things to address the difficulty caused by massiveness of things (P7).

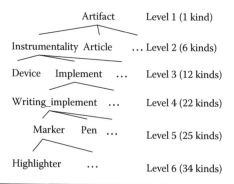

Figure 1.5   Term example and the number of kinds of terms at each abstraction level in WordNet.

Sim.1 considered a dining table in a dining room since it supports several activities. Based on the observation of activities at a dining table in one author's house, we found that there were three activities: having meeting, eating, and working. Therefore, we set the physical location semantics as meeting place, eating place, and working place. Observations showed meetings involved only fixed things, while movable things were associated with eating and working. This means that meeting place always exists regardless of whether working place or eating place is active. Finally, we set three physical location semantics: a meeting place, both a meeting place and a working place, and both a meeting place and an eating place.

Since relatively few things are associated with the physical location semantics (26 kinds of 96 things), simulation #2 (sim.2) focused on rooms in a home: each room of the home has many things (totally 472 kinds of 836 things). We examined four physical location semantics: a dining room that has the semantics of meeting place; a kitchen that has the semantics of cooking place; a bathroom that has the semantics of bathing place; and a study room that has the semantics of working place.

1.5.1.1 Thing Dataset Acquisition

For simulation #1 (sim.1), we manually identified all things present on an actual dining table. PMM must be informed of the things $thing_i$ associated with each physical location semantics PLS_l to learn the conditional probability $p(thing_i|PLS_l)$. However, acquired things for eating place or working place include the things associated with meeting place since meeting place always exists regardless of whether working place or eating place is active. Therefore, we acquired the things specific to eating place and working place by eliminating the things specific to meeting place from those associated with eating place and working place. The things associated with each physical location semantics are shown in Table 1.1. As for sim.2, we used the sets of things in an actual Korean family's house as collected by the National Museum of Ethnology [31]. They listed up all the possessions of an actual family. Since the Korean family's house did not have a study room, we manually identified all the things in and on a typical office desk from photos taken at various angles (Figure 1.6)

Table 1.1 Things Associated with Each Physical Location Semantics (PLS) for sim.1

PLS	Things
Meeting place	1 table, 4 chairs and cushions, 4 newspapers, 1 vase, 5 window-type envelops, 5 ballpoints, 1 in-basket, 1 wastepaper basket, 2 coasters, 1 jotter
Eating place (breakfast)	6 dishes, 2 chopsticks, 2 table spoons, 2 mugs, 2 table linens
Eating place (lunch)	6 dishes, 2 chopsticks, 2 forks, 2 table knives, 2 glasses, 2 table linens
Eating place (dinner)	6 dishes, 2 chopsticks, 2 forks, 2 table knives, 2 glasses, 2 beer cans, 2 table linens
Working place	2 ballpoints, 4 highlighters, 1 commonplace book, 1 digital computer, 1 power cord, 1 mouse, 7 files

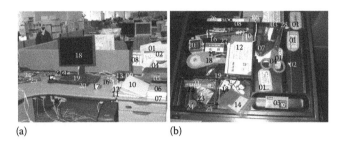

(a)   (b)

Figure 1.6   Manual collection: a list of things in an actual office desk for sim.2. (a) Numbering the things on an office desk. (b) Numbering the things in the drawer of an officer desk.

We then added noise to the abstracted data sets using noise ratios of 0%, 12.5%, and 25%. Noise was created by adding and deleting things in equal measure. In detail, we added the things of another physical location semantics to reflect the characteristic that some existing things are not related to the physical location semantics. In addition, we randomly eliminated things from the data sets to reflect RFID detection errors and the presence of things without RFID tags. By randomly adding noise, we created 1000 data sets for each physical location semantics. To include unlearned things in the test data for evaluating the difficulty caused by unlearned entities (P4), we used the eating places of breakfast as learning data and those of lunch and dinner as test data in sim.1.

1.5.2 Experiments in Actual Environment

To check the feasibility of the proposed method, we implemented the system and repeated sim.1 in the actual environment. Each thing listed in Table 1.1 was tagged with an active RFID tag (RF CODE Spider III [24]). Transmission interval of the RFID is 0.4 s. The RFID tags were detected by TAVIS Concentrator [28]. Figure 1.7 shows the things equipped with RFID tags. As the learning data and test data, we acquired RFID datasets for 1 h each in the various combinations of physical location semantics. We then split the datasets into 1 or 6 s intervals to create learning and test datasets, which means that the system performed the learning and estimating function every 1 or 6 s.

1.5.3 Results

1.5.3.1 Simulations

Table 1.2 shows the F-measure of each of the three types of physical location semantics in sim.1. The proposed method successfully detected each physical location semantics with a high degree of accuracy. The degree of accuracy achieved in detecting working places was slightly lower than the other physical location semantics since some of the things indicative of working place are also the indicative of meeting place such as ballpoints and jotters. On the other hand, meeting place and eating place have more discriminative things. Therefore, while the multiple physical location semantics of working and meeting can be successfully detected, the single meeting place was classified as the multiple physical location semantics of working and meeting.

This result also demonstrates the noise tolerance of the proposed method since the accuracy of physical location semantics detection did not drop as the noise ratio was raised. Furthermore, the accuracy of working place detection increased when the abstraction level was raised. Raising the

Figure 1.7   Things equipped with RFID tags.

Table 1.2   F-Measure of Detecting Each Physical Location Semantics (PLS) at Each Abstraction Level: sim.1

PLS	Noise (%)	Lev1	Lev2	Lev3	Lev4	Lev5	Lev6
Meet	0	100	100	100	100	100	100
	12.5	100	99.9	100	100	100	100
	25	100	98.3	100	100	100	100
Eat	0	0.0	100	100	100	100	100
	12.5	0.0	100	100	100	100	100
	25	0.0	100	100	100	99.8	100
Work	0	0.0	100	100	100	100	100
	12.5	0.0	75.9	93.5	94.2	96.4	90.5
	25	0.0	62.9	81.0	79.3	86.1	78.0

abstraction level decreased the number of kinds of terms: 1 kind in level 1, 6 kinds in level 2, and 34 kinds in level 6 (Figure 1.5). This means that the information amount decreased and the detection accuracy of physical location semantics generally fell when the abstraction level rose. Ontology can provide an explanation: each physical location semantics has many kinds of indicative terms but only a few instances of each; the use of ontology raised the abstraction level which yielded fewer kinds of indicative terms but a larger number of each. Note that it makes sense that the F-measure is 0 at level 1 in most physical location semantics since the term of abstraction level 1 is just "Artifact." As for unlearned things data, we did not learn forks, table knives, and glasses. WordNet transformed both forks and table knives, which were not learned, and tablespoons, which were learned, into cutlery at abstraction level 5. It also transformed both glasses, which were not learned, and mugs, which were learned, into containers at abstraction level 3. Therefore, ontology can utilize unlearned things in detecting physical location semantics by raising the abstraction level. The appropriate abstraction level to identify physical location semantics depends on the physical location semantics being targeted. The results of sim.1 show that abstraction level 5 yielded the highest accuracy while for sim.2, abstraction level 4 offered the highest accuracy. Therefore, the appropriate abstraction level needs to be specified in the learning phase.

Table 1.3 shows the F-measure of each physical location semantics as determined in sim.2. This result also demonstrates the feasibility of the proposed method. Figure 1.8 shows the processing time for learning and estimating 4000 datasets of things and the number of kinds of terms at each abstraction level. This demonstrates that the proposed method can rapidly handle large sets of things and that increasing the abstraction level makes it possible to reduce the processing time. Furthermore, though 472 kinds of things were aggregated into 17 kinds at abstraction level 2, the

Table 1.3   F-Measure of Detecting Each Physical Location Semantics at Each Abstraction Level: sim.2

PLS	Noise (%)	Lev1	Lev2	Lev3	Lev4	Lev5	Lev6
Bath	0	0.0	100	100	100	100	100
	12.5	0.0	97.1	95.6	99.8	98.9	100
	25	0.0	86.6	92.2	99.4	97.8	100
Cook	0	40	100	100	100	100	100
	12.5	40	91.1	96.5	98.7	99.4	100
	25	40	82.5	92.5	97.6	96.9	100
Meet	0	100	100	100	100	100	100
	12.5	100	94.6	97.2	98.1	98.8	100
	25	100	89.9	94.0	97.5	97.7	100
Work	0	0.0	100	100	100	100	100
	12.5	0.0	90.6	97.5	99.3	99.4	100
	25	0.0	83.9	95.1	98.4	97.4	100

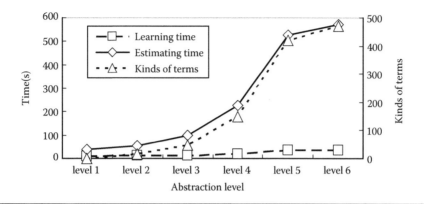

Figure 1.8 Processing time for learning and estimating 4000 datasets of things, and average number of kinds of terms contained in the datasets of things with regard to each abstraction level.

F-measure of each physical location semantics did not decrease, which clearly demonstrates the effectiveness of ontology.

1.5.3.2 Experiments

Figure 1.9 shows the detection rate of things in each combination of physical location semantics. Though each RFID sent its ID every 0.4 s, it took 6 s for the detection ratio to stabilize. Furthermore, the detection rate stayed under 60% when over 60 things were present on the table. The reason for this is that some RFID tags were attached to metallic things such as beer cans while others were attached to things that were covered up. The detection rate was 9%–15% (3–7 things) and 54%–100% (18–42 things) for the intervals of 1 and 6 s, respectively. Table 1.4 shows

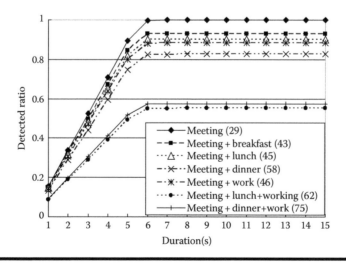

Figure 1.9   Detection rate of things in each combination of physical location semantics (numbers in parentheses indicate the number of things).

Table 1.4   F-Measure of Detecting Each Physical Location Semantics at Each Abstraction Level and Each Interval in the Experiment

Intvl	PLS	Lev1	Lev2	Lev3	Lev4	Lev5	Lev6
	Meet	99.7	75.8	92.7	96.5	96.7	95.7
1s	Eat	1.2	90.7	90.7	91.1	86.0	86.7
	Work	0.0	65.0	85.0	90.5	90.3	90.8
	Meet	100	100	100	100	100	100
6s	Eat	0.0	100	100	100	100	100
	Work	0.0	91.4	100	100	100	100

the F-measure of each physical location semantics in detecting combinations of physical location semantics for the intervals of 1s and 6s. This result demonstrates that the proposed method can identify physical location semantics in actual environments. As in the case of sim.1, the F-measure of working place was slightly lower than those of the other physical location semantics. In addition, the abstraction offered by ontology raised the F-measure of eating place.

At the interval of 6 s, most physical location semantics were successfully detected. At the interval of 1 s, the F-measure was still high even though the system detected only 3 – 7 things. The reason is that the things used in this experiment are quite indicative of the particular physical location semantics targeted. To prove this consideration, we calculated the cosine similarity of feature vectors among the physical location semantics. The feature vector is provided by the term and its conditional probability for each physical location semantics. In Table 1.5, the similarity is quite low for abstraction levels 3, 4, 5, and 6, which demonstrates that the physical location semantics have completely different indicative things. Therefore, even if just one thing indicative of meeting place was detected, the system output meeting place. Since the total number of things was few (3 – 7 things), the influence of each thing on physical location semantics identification was relatively high.

1.5.4 Discussion

The above simulations and experiments demonstrate the ability of our proposal to rapidly detect multiple physical location semantics and to resist noise. Though the findings of this experiment are meaningful and interesting, the following issues still remain.

Table 1.5   Cosine Similarity of the Terms' Conditional Probabilities between Two Physical Location Semantics

PLSs	Lev1	Lev2	Lev3	Lev4	Lev5	Lev6
M–E	0.9993	0.1833	0.0798	0.0048	0.0002	0.0050
M–W	0.9994	0.7995	0.1238	0.0382	0.0144	0.0218
E–W	0.9994	0.2966	0.0003	0.0003	0.0003	0.0003

M, Meeting places; E, eating places; W, working places.

Target physical location semantics: While this experiment focused on three or four physical location semantics, the variety of target physical location semantics must be expanded and refined. With regard to expansion, we need to acquire as many physical location semantics as the types of human activities. As for refinement, we need to split each initial physical location semantics, if possible, into more detailed physical location semantics. For example, a meeting place has the sub-concepts of a director's meeting place and a group meeting place. Ontology would be helpful in achieving this.

Concepts of ontology: In the experiment, we utilized WordNet which defines only the basic concept of terms. Though it is useful in identifying many physical location semantics, it fails to help in identifying other physical location semantics such as *shopping place* and *user's territory*. Since no existing ontology base defines role concept such as product, we need to build the ontology of role concept.

PMM for detection of physical location semantics: When PMM is employed for topic detection of texts, stop word elimination is an efficient technique to improve performance since so many words remain. When PMM is employed for detection of physical location semantics, especially where movable entities are involved, the amount of information available may be much less. For example, an eating place can be created anywhere the user brings takeout food. This issue can be solved by weighting things in a preprocessing step.

1.6 Conclusion

This chapter proposed a novel approach to automatically detect physical location semantics. Inspired by the concept of affordance, physical location semantics are identified through the surrounding environmental entities. We utilize ontology to specify terms representing environmental entities and a multi-class naive Bayesian approach to identify physical location semantics from the terms. In order to evaluate the proposed method, we established an environment where everything (26 kinds of 94 things) was tagged with an RFID tag. Experiments on the environment demonstrated that the proposed system has excellent noise tolerance and high accuracy in the detection of physical location semantics. In addition, we manually collected all things in a home (472 kinds of 836 things). Simulations using these things demonstrated that the proposed system has the ability to rapidly handle large amounts of data. Future work includes identifying human activities by utilizing physical location semantics as well as tackling the remaining issues described in Section 1.5.4. Since physical location semantics focus on the enabling activities offered by a location to users, the system needs to specify one or multiple activities that the user is actually performing. One approach is identifying the thing that the user is interacting with and selecting the physical location semantics that has the highest conditional probability.

References

1. Ashbrook D., Starner T.: Using GPS to learn significant locations and predict movement across multiple users. *Personal Ubiquitous Computing*, 7(5), 275–286, 2003.
2. Cresswell T.: *In Place/Out of Place*. University of Minnesota Press, Minneapolis, MN, 1996.
3. Curry M.: *The Work in the World—Geographical Practice and the Written Word*. University of Minnesota Press, Minneapolis, MN, 1996.
4. Entrikin J.: *The Betweenness of Place—Towards a Geography of Modernity*. Johns Hopkins University Press, Baltimore, MD, 1973.

5. EPC Global: http://www.epcglobalinc.org/ (accessed on June 8, 2012).

6. Finkenzeller K.: *RFID Handbook: Radio-Frequency Identification Fundamentals and Applications.* John Wiley & Sons, Chichester, U.K., 2000.

7. Fishkin K., Jiang B., Philipose M., Roy S.: I sense a disturbance in the force: Unobtrusive detection of interactions with RFID-tagged objects. *Proceedings of sixth International Conference on Ubiquitous Computing (UbiComp2004)*, Nottingham, U.K., pp. 268–282, 2004.

8. Floerkemeier C., Lampe M.: Issues with RFID usage in ubiquitous computing applications. *Proceedings of Pervasive Computing 2004(PERVASIVE), Volume 3001 of Lecture Notes in Computer Science*, Vienna, Austria, pp. 188–193, 2004.

9. Gibson J.: *The Ecological Approach to Visual Perception.* Lawrence Erlbaum Assoc Inc., Hillsdale, NJ, 1979.

10. Hightower J.: From position to place. *Proceedings of International Workshop on LoCA 2003*, pp. 10–12, 2003.

11. Hightower J., Borriello G.: Location systems for ubiquitous computing. *IEEE Computer*, 34(8), 57–66, 2001.

12. Jordan T., Raubal M., Gartrell B., Egenhofer M.: An affordance-based model of place in GIS. *Proceedings of Eighth International Symposium on Spatial Data Handling (SDH'98)*, Vancouver, British Columbia, Canada, 1998.

13. Kang J., Welbourne W., Stewart B., Borriello G.: Extracting places from traces of locations. *Proceedings of International Workshop on Wireless Mobile Applications and Services on WLAN Hotspots (WMASH)*, Philadelphia, PA, pp. 110–118, 2004.

14. Matsuo Y.: Social knowledge for ubiquitous environment—Human network and spatial semantics. *Proceedings of International Symposium on Life-World Semantics and Digital City Design*, Kyoto, Japan, 2004.

15. McCallum A., Nigam K.: A comparison of event models for naive Bayes text classification. *Proceedings of International Workshop on Learning for Text Categorization in AAAI-98*, Madison, Wisconsin, 1998.

16. Mizoguchi R.: Tutorial on Ontological engineering part1: Introduction to ontological engineering. *New Generation Computing, OhmSha&Springer*, 21(4), 365–384, 2003.

17. Mizoguchi R.: Tutorial on ontological engineering part2: Ontology development, tools and languages. *New Generation Computing, OhmSha&Springer*, 22(1), 61–96, 2004.

18. Mizoguchi R.: Tutorial on ontological engineering part3: Advanced course of ontological engineering. *New Generation Computing, OhmSha&Springer*, 22(2), 2004.

19. Moore D., Essa I., Hayes M.: Exploiting human actions and object context for recognition tasks. *Proceedings of Fourth International Conference on Computer Vision (ICCV'99)*, Corfu, Greece, 1999.

20. Nishida Y., Kitamura K., Hori T., Nishitani A., Kanade T., Mizoguchi H.: Quick realization of function for detecting human activity events by ultrasonic 3D tag and stereo vision. *Proceedings of Second IEEE International Conference on Pervasive Computing and Communications (PerCom2004)*, Orlando, FL, pp. 43–54, 2004.

21. Nissanka P., Anit C., Hari B.: The cricket compass for context-aware mobile applications. *Proceedings of Seventh ACM MOBICOM*, Rome, Italy, July 2001.

22. Ogawa T., Yoshino S., Shimizu M., Suda H.: A new in-door location detection method adopting learning algorithms. *Proceedings of IEEE International Conference on Pervasive Computing and Communications (PerCom2003)*, Fort Worth, TX, March 2003.

23. Brock D.: The Physical Markup Language. http://xml.coverpages.org/PML-MIT-AUTOID-WH-003.pdf (accessed on June 8, 2012).

24. RF CODE: http://www.rfcode.com/ (accessed on June 8, 2012).

25. Seon-Woo L., Mase K.: Activity and location recognition using wearable sensors. *IEEE Pervasive Computing* 1(3), 10–18, 2002.

26. Stop list: ftp://ftp.cs.cornell.edu/pub/smart/english.stop (accessed on June 8, 2012).

27. Tapia E., Intille S., Larson K.: Activity recognition in the home using simple and ubiquitous sensors. *Proceedings of Second International Conference on Pervasive Computing 2004 (Pervasive2004)*, Vienna, Austria, pp. 158–175, 2004.

28. RF CODE, Tavis Concentrator: http://www.egomexico.com/images/Productos/RFCode/tavismod.pdf (accessed on June 8, 2012).

29. Terborn G.: *The Ideology of Power and the Power of Ideology*. Routledge, London, U.K., 1980.

30. The American Department of Defense, DoD Announces Radio Frequency Identification Policy. http://www.defense.gov/releases/release.aspx?releaseid=5725 (accessed on June 8, 2012).

31. The National Museum of Ethnology: *Seoul Style 2002*. Osaka, Japan, Senri Foundation, 2002.

32. Toyama N., Ota T., Kato F., Toyota Y., Hattori T., Hagino T.: Exploiting multiple radii to learn significant locations. *Proceedings of First International Workshop on LoCA 2005*, Oberpfaffenhofen, Germany, pp. 7–168, 2005.

33. Tuan Y.: *Space and Place: The Perspective of Experience*. University of Minnesota Press, Minneapolis, London, U.K., 1977.

34. Ubiquitous ID Center: http://www.uidcenter.org/

35. Ueda N., Saito K.: Singleshot detection of multi-category text using parametric mixture models. *Proceedings of Eighth International Conference on Knowledge Discovery and Data Mining (SIGKDD2002)*, Edmonton, Alberta, Canada, pp. 626–631, 2002.

36. Ward A., Jones A, Hopper A.: A new location technique for the active office. *IEEE Personal Communications*, 4 (5), 42–47, 1997.

37. Want R., Hopper A., Falcao V., and Gibbons J.: The active badge location system. *ACM Transactions on Information Systems*, 10, 91–102, 1992.

38. Weiser M.: The computer for the 21st century. *Scientific American*, 265, 94–104, 1991.

39. Princeton University, WordNet: http://wordnet.princeton.edu/ (accessed on June 8, 2012).

Chapter 2

Wireless Payment and Systems

Jerry Gao and Jacky Cai

Contents

2.1 Introduction

The fast advance of wireless networking, communication, and mobile technology is making a big impact on daily life. As there is a significant increase of mobile device users, more wireless information services and commerce applications are needed [1–3]. Since wireless payment is an essential part of mobile commerce applications (such as mobile banking, wireless trades, and mobile shopping), how to build secured wireless payment systems to support mobile payment transactions becomes a hot research topic. According to the Wireless World Forum, mobile payment on wireless devices will provide excellent business opportunities in the coming years. Building secure, practical, and cost-effective wireless payment solutions to support mobile device users not only provides good business opportunities but also brings new technical challenges and issues to engineers. Although there are a number of types of electronic payment solutions for Internet-based applications and commerce, we are still faced with new issues and challenges in wireless payment because of lack of study and experience in wireless payment. Although there are a number of papers discussing businesses markets, payment process, payment methods, and standards in wireless payment [4–7], there are a very few papers discussing how to build wireless payment systems, including protocols, design issues, and security solutions [8–12].

This chapter provides a tutorial overview of wireless payment concepts and wireless payment systems. It first covers basic concepts, requirements, payment processes, schemes, and solutions as well as major players. Then, it classifies and discusses different types of wireless payment solutions and systems. Finally, it presents an example of wireless payment system, known as P2P-Paid. This chapter uses this example to explain the design and implementation of a peer-to-peer wireless payment system, which allows two mobile users to conduct wireless payment transactions over the Bluetooth communications and supports related secured transactions between the payment server and mobile clients.

This chapter is structured as follows. The next section introduces some basic concepts of wireless payment. Section 2.3 introduces different types of mobile payment systems. Some major players in wireless payment are introduced in Section 2.4. Mobile payment models, challenges, and issues are discussed in Section 2.5. Section 2.6 introduces the P2P-Paid wireless payment system. Its system design, including system architecture, functional components, as well as used technologies and other features, and security solutions are detailed presented. The current P2P-Paid system implementation examples are also given. Finally, conclusions are given in Section 2.7.

2.2 Wireless Payment

2.2.1 Basic Concepts

What is wireless payment? According to [11], wireless payment refers to any transaction with a monetary value that is conducted via a mobile telecommunications network. More precisely, wireless payment refers to wireless-based electronic payment for mobile commerce to support point-of-sale (POS) and/or point-of-service payment transactions on mobile devices, such as cellular telephones, smart phones, and personal digital assistants (PDAs), or mobile terminals.

What is a wireless payment system? A wireless payment system is a mobile commerce system that processes electronic commerce payment transactions supporting mobile commerce applications in wireless networks and wireless Internet infrastructures [11].

In general, wireless payment systems can be used by wireless-based merchants, mobile content vendors, and wireless information and commerce service providers to process and support payment transactions driven from wireless-based commerce applications, including wireless-based trading systems, mobile portals, wireless information, and commerce service applications.

Wireless-based payment systems can be viewed as one type of electronic payment systems. Similar to current online payment systems, wireless-based payment systems also provide automatic payment transaction processing. Unlike online payment systems, where payment transactions are driven by web user accesses and processed by Internet-based payment systems, wireless-based payment systems process transaction requests from mobile devices and location-oriented terminals. Wireless-based payment systems have the following common features:

- *Mobile accessibility*: This enables mobile users to conduct payment by using mobile devices anywhere.
- *Mobile-driven payment processing*: This enables wireless-based payment transactions, including POS transactions, point-of-service transactions, or location-oriented transactions.
- *Secure wireless payment protocols*: This supports secure payment transactions over wireless networks (or wireless Internet infrastructures) between mobile devices and their payment system.

2.2.2 Basic Requirements

The basic requirements of mobile payment transactions are listed in the following:

- *Simple and convenient*: The payment method must be easy for mobile users to make payments through mobile devices anytime and anywhere.
- *Fast and efficient*: Users must get fast response and quick payment processing.
- *Secure*: The system must protect buyers, sellers, and all involved parties. Consumers should have the assurance that their payment accounts will be protected from unauthorized use and their transactions are secured.
- *Universal acceptance*: This allows consumers to shop and pay anyone, anywhere, and anyhow using mobile payment solutions and underlying wireless networking.

2.2.3 Parties Involved in Wireless Payment

According to the Telecom Media Networks [13], the identified key roles to be managed in mobile payment are content provider, authentication provider, payment authorization and settlement provider, and consumer (see Figure 2.1).

- *Consumer*: The consumer is the person who owns a mobile device and buys content or services from the content provider through his or her mobile device.
- *Content provider*: A content provider is an individual (merchant) or organization that sells either electronic or physical content (products or services) to consumers.
- *Trusted third party*: The trusted third party (TTP) is the company that checks the authentication and the authorization of transaction parties and makes the settlements among companies involved. TTP could be a telephone company, a bank, or a credit card company.
- *Payment service provider*: The payment service provider (PSP) is the central entity responsible for the payment process. It enables the payment message initiated from the mobile device to be routed to and cleared by the TTP. This service generally includes an "e-wallet" application that enables payers to store their payment details, such as credit card account numbers and shipping addresses, on a provider's secure server so that they do not need to enter in all

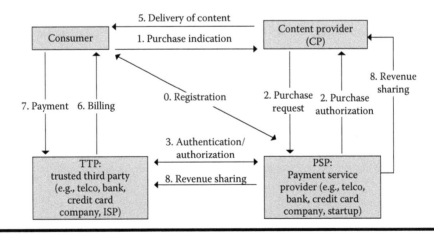

Figure 2.1   Steps involved in M-payment transactions.

the pertinent information required for each sale on small and difficult-to-use mobile keypad devices. The PSP may also act as a clearinghouse to share the revenues among all the partners involved in the payment process. It could be a Telco, a bank, a credit card company, or a start-up. A Telco could be positioned at the same time as PSP, TTP, and content provider.

2.2.4 Wireless Payment Schemes

There are three popular wireless payment schemes: cash payment, prepayment, and postpayment. In the cash payment scheme, consumers make an instant payment using digital cash to exchange a good (or service) with the merchant. In the prepay scheme, the charge is made before purchasing a product (or service). The ATM card–based payments use this scheme. In the postpay scheme, consumers make actual payments after purchasing. Current credit card–based electronic payment is a typical example. To support the three payment schemes, three types of online electronic payment solutions are developed in the past years:

1. Credit card–based electronic payment solutions, in which consumers use credit cards to make automatic payments
2. Electronic check–based solutions, in which consumers use electronic checks to make their payments in automatic manner
3. Electronic cash–based systems, in which consumers use digital cashes to make payments in a systematic way

Intuitively, the three types of online electronic payment solutions can be extended or adopted in wireless Internet payment. According to Telecom Media Networks [13], potential mobile payment falls into several distinct categories: content type, content value, transaction type, and transaction settlement method.

- *Content type*: digital goods, hard goods, voting, and ticketing.
- *Content value*: micropayments and macropayments. Micropayments are small order of smallest valued coins. Macropayments are the higher valued payments such as anything above $25.
- *Transaction type*: pay per view (PPV), pay per unit (PPU), and recurrent subscription.
- *Transaction settlement method*: prepaid (debit) and postpaid (credit).

2.2.5 Mobile Payment Process

The Telecom Media Networks identified four key roles in mobile payment as mentioned earlier: content provider, authentication provider, payment authorization and settlement provider, and consumer. The main process of m-payment can be divided into the following phases:

- *Registration*: Firstly the consumer needs to open an account with the PSP to enable the payment service through a particular payment method. During this phase, the PSP requires confirmation from the TTP that handles the relationship with the customer.
- *Transaction*: Consumer indicates the desire to purchase some content. Content provider forwards the purchase request to the PSP. PSP then requests authentication and authorization from the TTP. PSP informs the content provider about the success of the purchase demand. Content provider then delivers the purchased content.

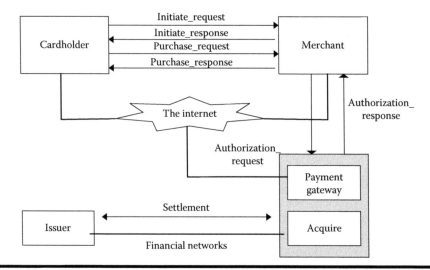

Figure 2.2 Processing flows for purchase request and authorization in SET.

- *Clearing and settlement*: Settlement can take place in real time during the purchase or in a postpaid mode. Real-time settlement can be conducted via a prepaid account if the TTP is a Telco or directly through a bank account if the TTP is a bank. In postpaid mode, the PSP sends the billing information to the TTP. The TTP sends the bill to the consumers, gets the money back, and forwards it to the TTP. The PSP is then responsible for computing the revenues of each entity and distributing the funds accordingly (Figure 2.2).

2.2.6 Security Solutions for M-Payment Systems

There are a number of security solutions for mobile payment systems. They are listed in the following:

- *Mobile telephone encryption*: The encryption of the payment system is established on all major mobile phone system providers. Encryption is built into the system and is invisible to users. However, the system is beyond mobile payment system control.
- *PIN or password*: PIN and password are one of the easiest ways for mobile payment authorization. PIN uses numbers and password uses a combination of letters and numbers. User can enter PIN when making payment via mobile phone and password when browsing personal profile online. Password is more secure because it is harder to be cracked. However, the user enters a PIN not a password because it is easier to input on a mobile phone with a small numeric keypad.
- *Caller ID*: Caller identification (caller ID) is a service provided by many mobile telephone service providers. It allows a receiver to see the number of the caller. If the mobile number of the user does not match the mobile number in user account repository, the transaction is canceled. A caller ID is not easy to fake or change because it is only generated at the caller party's exchange and not the mobile phone.
- *Callback*: In conjunction with PIN, callback is another way to authenticate and authorize the user for mobile payment. It authenticates the person who is making a mobile payment by asking the person to enter his PIN when he is called back.

- *SMS*: Short message service (SMS) is a mechanism for transmitting "short" messages to and from wireless handsets over the encrypted GSM network, CDMA, or TDMA. SMS can be used by a mobile payment system to send a receipt containing the transaction information to the sender and receiver of the transaction. It is also a good way for a mobile payment system to ensure nonrepudiation by notifying the user that a transaction occurred in their account.
- *WAP-WTLS*: The Wireless Transport Layer Security (WTLS) is the security layer protocol in the WAP architecture. WTLS operates above the transport protocol layer and provides the upper-level layer of WAP with a secure transport service interface that preserves the transport service interface below it. In addition, WTLS provides an interface for managing secure connections. The primary goal of the WTLS is to provide privacy, data integrity, and authentication between two communicating applications.
- *Biometric authentication*: As most mobile devices today are multimedia enabled, biometric security is a compelling solution to m-commerce system. Biometric security solutions such as voice authentication and facial recognition are practical solutions for mobile payment security [14–17]. Later in this chapter, we will propose a security solution that especially addresses security needs of m-commerce system. The secured solution is designed and implemented for the application level, which eliminates the need of changing the existing security standards and technologies in underline layers. The proposed voice authentication on top of traditional e-key security solution guarantees the most secure mobile payment system. The integration of the secured solution to the payment system does not have any detrimental impact on user experience. Instead, the outcome of the mobile payment system is easy to use, more secured, and more compelling to average users.

2.3 Different Types of Wireless Payment Systems

Current wireless-based payment systems could be classified into the following types: account-based payment system, mobile POS payment system, and mobile wallet. They are described in the following paragraphs.

2.3.1 Account-Based Payment Systems

Account-based payment systems can be further classified into three categories: (1) mobile phone–based payment systems, (2) smart card payment systems, and (3) credit card mobile payment systems:

1. *Mobile phone–based payment systems*: Mobile phone–based payment system enables customers to purchase and pay for goods or services via mobile phones. Some systems also allow users to send or receive money with their account via the mobile phone. The mobile phone is used as the personal payment tool in connection with the remote sales. A phone card–based payment system has the advantage over the traditional card-based payment in that the mobile phone replaces both the physical card and the card terminal as well. Payments can take place anywhere far away from both the recipient and the bank. The only condition being to have is a mobile phone and be within the reach of a GSM network.
2. *Smart card payment systems*: This type of payment systems uses a smart card; an embedded microcircuit, which contains memory; and a microprocessor together with an operating

system for memory control. The smart card is a secure storage location for secret information. It can be used for electronic identification, electronic signature, encryption, payment, and data storage. However, a smart card requires a reader to get required computing energy and can only perform calculations when connected to one. The security is based on the public key cryptography, and any attempt to access its data directly is more likely to destroy the card than let secret data leak out because it only can be accessed through its reader. Today, widely used SIM cards are the typical examples of smart cards. SIM cards can be used to customize mobile phones regardless of the communication standards, such as GSM, personal communications service (PCS), satellite, digital cellular system (DCS), and so on. Although there are a number of applications of smart cards, mobile payment is one of them. A SIM card could potentially be used to store payment information such as card numbers or used to authenticate a customer making a payment. Alternatively "dual-slot" phones are available, and these contain a slot into which the customer can insert their credit card (smart card). This communicates with the SIM card to obtain a card authorization. Other "dual-chip" phones have a second small slot for taking just the chip element of a payment card, which can work independently of the SIM card. SIM cards will continue to provide enhanced functionality beyond the basic communications link. Payments and authentication are going to become increasingly important applications, and the mobile operators, with their large customer base and existing SIM/smart card infrastructure, are going to become a key player in the marketplace.

3. *Credit card mobile payment systems*: This type of mobile payment systems allows customers to make payments on mobile devices using their credit cards. These payment systems are developed based on the existing credit card–based financial infrastructure by adding wireless payment capability for consumers on mobile devices. A credit card–based mobile payment system allows consumers to make payments during mobile commerce trading sessions using their mobile devices. The existing SET secure protocol, developed by Visa and MasterCard for secure transfer of credit card transactions, has been extended and known as three dimensional (3D)-SET to support mobile payment for mobile device users [10,14]. Current SET-based payment systems have not been widely adopted because it was inconvenient to both the cardholder and the merchant. To use SET protocol, the merchant has to issue each cardholder a software digital certificate installed on his transaction terminal. Therefore, the cardholder would not be able to use SET at his colleagues' PC or at public terminals. On the merchant's end, SET used a hefty set of algorithms that cost a lot of computing power to process. This dissuaded many smaller merchants from using SET. To solve these problems, Visa starts a new initiative, known as "3D mo-del" (three domains), for secure mobile credit card transactions. It covers the different areas of Visa transaction flow. It looks at the activity between the following domains: (a) the acquirer domain (including merchants and their bank), (b) the issuer domain (including cardholders and their bank), and (c) the interoperability domain (including cardholders' banks and merchants' banks), which is covered by the original SET protocol. The 3D-SET protocol is developed to meet the needs of the three domains. It has two major advantages over the SET protocol. First, it reduces the effort of performing a SET payment on behalf of the cardholder, and it allows the cardholder to use their certificate from any mobile device access. A payment authentication takes place during a mobile shopping transaction when buyers and sellers are identified proving cardholder verification, card verification, and merchant authentication. Second, the 3D-SET protocol frees merchants from the charge back problem because it does not require the cardholder to download software. However, according to Berlecon Research [18], a dark cloud above

the future of 3D-SET is the danger that issuing banks might want to shift the costs to cardholders.

2.3.2 Mobile Point-of-Sale Payment

Mobile POS payment system enables customers to purchase products on vending machines or in retail stores with their mobile phones. Many companies equip mobile devices with POS payment system. It is designed to complement existing credit and debit card systems by enabling all mobile phone users to turn their phones into the payment instruments of their choice.

Mobile POS systems can be applied to two different payment applications. The first type is known as automated POS payments. They are frequently used over ATM machines, retail vending machines, parking meters or toll collectors, and ticket machines to allow mobile users to purchase goods (such as snacks, parking permits, and movie tickets) through mobile devices. The other type is known as attended POS payments (shop counters, taxis), which allow mobile users to make payments using mobile devices with the assistance from a service party, such as a taxi driver or a counter clerk.

There are four leading industry groups developing open standards for local mobile transactions in mobile POS payments [19]. They are Infrared Data Association's (IrDA) Infrared Financial Messaging Special Interest Group (IrFM SIG), Mobile Electronic Transaction Forum (MeT Forum), Bluetooth Special Interest Group's (Bluetooth SIG) Short Range Financial Transaction Study Group (SRFT SG), and National Retail Federation (NRF).

2.3.3 Mobile Wallets

M-wallets are the most popular type of mobile payment option for transactions. Like e-wallets, they allow a user to store billing and shipping information that the user can recall with one click while shopping from a mobile device [20].

The main types of mobile wallet schemes in the market are client wallet and hosted wallet [21]. Client wallets are stored on a user's device in the form of a SIM Application Toolkit card that resides in a mobile phone. Since the wallet is based on hardware, it is difficult to update, and potentially the user's sensitive financial information is compromised if the device is lost or stolen. Hosted wallets are hosted on a server. This gives the service provider much greater control over the functionality it delivers and the security of the data and transactions. Hosted wallets can be self-hosted wallets or third party hosted wallets.

MasterCard Global Mobile Commerce Working Group proposed "Remote Wallet Server Architecture." In the server-based m-wallet scheme, the payment account information is stored on a financial institution's remote server. The cardholder identifies himself/herself via an ID and password through the mobile device and approves the transaction. The merchant then charges purchase to the buyer's payment account and processes the transaction through their acquiring bank and MasterCard network. The payment account information itself is never transmitted from the mobile device, making the transaction more secure than schemes in which the payment account information is transmitted over the airwaves (Figure 2.3).

Server-based mobile e-wallets using SET technology are already being used, providing secure transaction capability for merchants and cardholders. We can distinguish three phases in the payment architecture shown earlier. The first phase is the initiation phase. In this phase, the merchant server sends a payment initiation message to a cardholder device. The second phase is the cardholder device, the SET wallet server interaction phase. In this phase, the cardholder device(s)

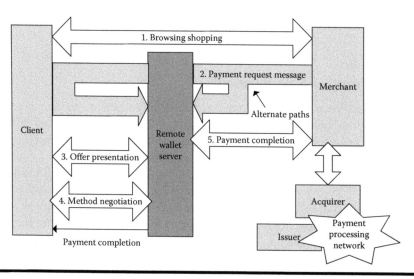

Figure 2.3   Mobile wallet server architectural overview.

forward the merchant's initiation message such that the wallet server either receives or is able to retrieve the SET wake-up message. The cardholder approves the transaction, and the wallet server authenticates the cardholder. The third phase is the SET transaction phase. In this phase the SET wallet server and the merchant SET server conduct a SET transaction. Eventually SET will support multiple cardholder authentication schemes, including SET certificates, PIN, chip cryptograms, digital IDs, and non-SET certificates (Figure 2.4).

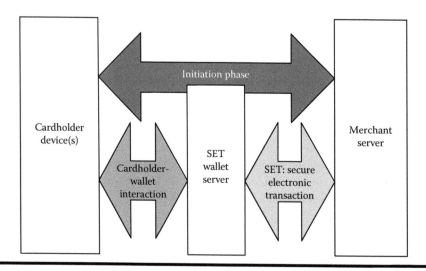

Figure 2.4   SET wallet server–based architecture.

2.4 Major Players in Wireless Payment

Up to now, there are a number of major players in wireless payment. In this section, we discuss two groups of major players: major players in account-based wireless payment and major players in mobile wallet.

2.4.1 Major Players in Account-Based Wireless Payment

2.4.1.1 PhonePaid

PhonePaid is a UK-based company that allows its customers to use GSM mobile phones to pay for goods and services or transfer money using their PhonePaid accounts. PhonePaid's mobile payment system provides the following personal services:

- *Send money*: To send money, a mobile user calls a PhonePaid transaction number first using a mobile phone and then enters the recipient party's mobile phone number for sending the money. Next, the user must enter the amount of money and invoke the sending transaction. When the transaction is completed, both the sender and the recipient will receive SMS receipts within seconds.
- *Receive money*: To receive money, a mobile user needs to call a PhonePaid transaction number first using a mobile phone and enter the other party's mobile phone number and the amount of money for requesting money. In seconds later, the other party will receive the money request.
- *Mobile purchase*: To make mobile purchase, a mobile user calls a PhonePaid transaction number first with a registered mobile and then enters the mobile pin. Next, the user follows the voice prompts and enters the Merchant ID and Product Code, as well as the payment amount of the purchased product (or service). Then, the user will receive voice prompts to request a confirmation for the purchase. Once the user accepts it, the purchase transaction is completed. The money is transferred from the user's PhonePaid account to the merchant's PhonePaid account immediately. Within a minute both the user and the merchant will receive SMS-based e-mail notification for the purchase.

2.4.1.2 Paybox

Paybox was raised by Paybox.net AG (belongs to Deutsche Bank) in May 2000. It works like a debit card. Each payment is debited from user bank account only after users have authorized the transaction by entering his/her Paybox PIN on the mobile phone. It enables users to purchase goods, services, and make bank transactions with their mobile phones. All transactions are conducted over the existing GSM mobile phone network. Figure 2.5 shows a procedure of payment transactions in Paybox.

Paybox uses EBPP (electronic bill presentment and payment) protocol that helps companies to optimize the payment process costs. The traditional invoice is replaced with a link, which leads the recipient to a prefilled electronic remittance form. After entering the mobile phone number or the Paybox security number, the invoice recipient is called and can authorize the settlement of the bill by entering his or her four-digit Paybox PIN. The exact sum is then debited from the bill recipient's account and transferred to the account of the invoice issuer. Electronic bill presentment will be particularly useful to delivery service, telecommunications, utility, publishing, and medical companies who would like to deploy a more efficient billing system.

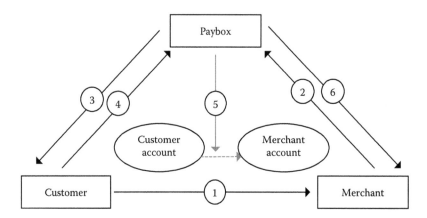

Payment transaction (communication via mobile phone):
1. Customer's mobile phone number
2. Merchant's mobile phone number, customer's mobile phone
   number and payment amount x
3. Payment of amount x to merchant.
4. Confirmation of payment by providing the customer PIN
5. Settle customer/merchant accounts
6. Confirmation

Figure 2.5 Paybox payment transaction steps. (From Paybox.net, Mobile payment delivery made simple, http://www.paybox.net/publicrelations/public_relations_whitepapers.html, retrieved on February 12, 2004.)

The risk of abuse can be greatly reduced for traders and customers by means of payment confirmation via a second, independent channel (mobile phone). As customers must register in advance as a Paybox user, he is clearly authenticated by entering his PIN. When confirming the payment, Paybox informs the customer of the trader's name so that there is additional security.

2.4.1.3 Ultra's M-Pay

M-Pay is a mobile payment system, enabling customers to purchase products on vending machines or in retail stores with their mobile phones. M-Pay is based on Ultra's patented payment terminal using voice to transfer authorization data. It is network and network operator independent (the same terminal works in GSM/GPRS/UMTS terminals). The user's phone is used to transfer data that simplifies the terminal design and leaves the terminal to focus on safe payment authorization.

M-Pay system enables the mobile operators and financial institutions enter the m-commerce without having to invest heavily: the existing mobile phone network is being used to carry financial data and the processing centers are being used for credit card authorization processing. The customers are required to have a relation with at least one mobile operator and an open account with at least one financial institution. To access the service, the customer calls the M-Pay Access Point, which is typically installed within mobile operator's premises. The payment process of the M-Pay system involves the following parties:

■ M-Pay Access Points enables telecommunication operators to ensure reliable and secure multiple connections to the M-Pay Center, customer identification, and sufficient data throughput.

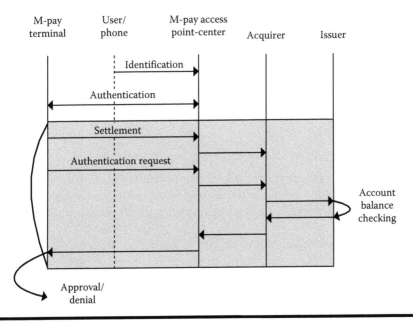

Figure 2.6 Sequence diagram of payment flow.

- Billing operators are banks and card authentication systems. They provide monetary transaction control (debt collection from customers). They consist of issuers, acquirers, and merchants.
- M-Pay Center is connected with M-Pay Access Point for transaction authorization.

M-Pay uses the security protocol PIN to authenticate users. The user's identity is defined on a SIM card in the mobile phone and is further secured by entering a special PIN either on a phone or payment terminal (Figure 2.6).

The payment terminal and payment center authenticate them with a digital signature based on the ECC cryptographic system (elliptic curve cryptography—public key cryptography using elliptic curves). Data encryption is performed according to a validated digital signature. The systems protect all of the parties involved in a transaction—customers, merchants, and processing centers. M-Pay can be applied to two different payment applications: automated POS payments (e.g., vending machines, parking meters, and ticket machines) and attended POS payments (e.g., shop counters, taxis).

2.4.2 Major Players in M-Wallet

This section introduces the current major players in mobile wallet, including Trintech, PayWare, Encorus PaymentWorks, and SNAZ.

2.4.2.1 Trintech

Trintech provides payment infrastructure solutions to financial institutions, payment processors, enterprise retailers, and network operators globally. Trintech deployed a secure SIM toolkit application in client/server architecture with some of the functionality on the phone and some on a server. This enables transactions to be executed in an encrypted form over bearers such as SMS and WAP. Users' payment methods are preregistered in a server wallet, allowing the user to quickly complete a transaction with a minimum amount of traffic—especially important over the SMS channel. In order to preserve the integrity of SMS and WAP-based transactions, Trintech has developed a suite of secure payment solutions.

2.4.2.2 PayWare

PayWare Issuer is a secure wallet server for multiple payment instruments. Virtual credit cards allow card issuers to offer cardholders a secure and convenient mechanism for online shopping. Shoppers simply drag and drop these virtual cards from their desktops or cell phones to automatically fill out online merchant forms.

PayWare Acquirer provides real-time routing of authorization and settlement to financial systems. It enables merchants to process transactions over the Internet, provide credits or refunds, make adjustments, or void sales. PayWare's mAccess provides a front-end interface, so that transactions can be executed in a multichannel environment. Consumers can use a variety of access channels, such as cell phones, PDAs, or other wireless access devices, to pay for goods and services with convenience and PIN-protected security.

2.4.2.3 Encorus PaymentWorks

According to Encorus Technologies, PaymentWorks Mobile is a software application for enabling payment transactions from cellular phones, the Internet, WAP-enabled mobile devices, and PDAs. PaymentWorks Mobile is specifically designed to address the needs of mobile network operators and other payment providers who process millions of transactions daily. PaymentWorks was designed to be secure and scalable. It consists of several components that can be combined to meet the individual needs of payment and PSP/mobile wallet providers: mobile wallet server, merchant component and modules for payment routing, virtual consumer cards, and merchant billing.

The wallet server is the central component of PaymentWorks Mobile. Installed and hosted by the mobile network operator, it processes and administers customer data and handles payment transactions. All relevant customer data needed to process payments is stored in the wallet server, like customer addresses, billing information, and the customer's preferred payment methods (i.e., credit card, telephone bill, direct debit). The data is stored in a secure customer specific wallet that the customer can modify from an Internet PC or mobile device.

2.4.2.4 SNAZ

The SNAZ mobile wallet (m-wallet) makes single click wireless shopping a reality. In addition to storing the customer's username and password, the m-wallet retains, on the server side, multiple billing and shipping details provided by the user during the one-time registration process. SNAZ's mobile wallet can easily interact with any transaction system to execute transactions. Only one username and password is needed to transact with multiple merchants in the SNAZ network. Unlike traditional wallets that force users to buy or abandon, the SNAZ mobile wallet allows users to have total control of their baskets. Users can save multiple items from multiple merchants into lists. With the mobile wallet, users can e-mail their lists to family and friends. Items are removed only when the user decides to remove them and not when they log off for the day. The SNAZ m-wallet also lets users search for products within the SNAZ merchant network and use the price compare option to save on the products they want.

2.5 Payment Models, Challengers, and Issues

According to the Telecom Media Networks [13], there are many business challenges, technical challenges, and interoperability challenges in mobile payment that current wireless vendors and mobile commerce service businesses must cope with to meet the expectations of the mobile commerce market.

2.5.1 Payment Model

A number of possible payment models can be set up for wireless payment in mobile commerce. We classify them into the following groups:

- Service-oriented payment models drive mobile payment transactions for the wireless information and commerce services.
- Merchant-oriented payment models drive mobile payment transactions for product purchases made by mobile users.
- Peer-to-peer payment models drive mobile payment transactions by two mobile device users (e.g., a taxi driver and a customer) at a mobile location anywhere and anytime.

2.5.2 Business Challenges

2.5.2.1 Business Model

One important business issue that the telecom industry must face is: Should Telcos collaborate with banks to address this business opportunity? Studies from Forrester Research show that most retailers would favor a joint venture including a financial company as a payment provider. But previous joint experiences between Telcos and banks have not really been a success.

2.5.2.2 Cost

Cost is another issue that could slow the m-payment development process. The questions from consumers, content providers, and service providers need to be concerned for m-payment to gain popularity. A return on the m-payment investment is not likely to be achieved within the first 2 years. This does not necessarily mean that one should wait for a more mature market—successful early adopters will gain significant competitive market advantages that may be impossible to reach.

2.5.2.3 Customer Apathy

One of the main reasons for mobile commerce's slow start is customer apathy. According to Forrester Research [1], most consumers over the world are uncomfortable with the idea of mobile payment even though different wireless security solutions are used. The result indicated the consumers' fear of an unknown medium due to their lack of understanding about wireless security and current wireless payment solutions. Many of them are not even willing to try paying with their mobile devices. Applications enable people to make a payment more efficiently and quickly than what they are used to will be critical.

2.5.3 Technical Challenges

2.5.3.1 Security

Security is a very crucial issue for m-payment method to gain widespread acceptance. It is the consumer's top concern, because they will have little confidence if the payment method cannot provide guarantees on authenticity, confidentiality, and integrity. Security can be viewed from five aspects: confidentiality, authentication, integrity, authorization, and nonrepudiation. Confidentiality protects the payment details (e.g., a consumer's personal particulars, password) against passive monitoring. Authentication ensures that the consumer and content provider are the ones they claim to be. Integrity protects payment details from being modified from the time they are sent to the time they are received. Authorization ensures that only authorized consumers are allowed to participate in the payment transaction. Nonrepudiation guarantees that consumers cannot falsely claim that they did not participate in the transaction. This provides the benefits for merchants and PSPs.

2.5.3.2 Accessibility

Accessibility is considered as a combination of convenience, speed, and ease of use. Convenience indicates the capability of the payment method to pay for any type of content, from any location in the world, using any device. Some payment methods might require consumers to upgrade their existing handsets or be preregistered with a company. Speed is the amount of time spent on payment. Consumers care more if they have to pay for the access. Ease of use is especially important for micropayments. Accessibility also strongly depends on the devices' capabilities and the quality of the network.

2.5.3.3 Interoperability Challenges

Interoperability means that applications will work on different networks. Interoperability challenge underpins any global payment system, ensuring that any participating payment product can be used at any participating merchant location. Mobile operators' principal concerns revolve around standardization and interoperability. Operators want payment to be seamless, allowing them to compete on services and applications.

2.5.3.4 Standardization

According to Telecom Media Networks, a wide variety of technologies for mobile payments exist today, ranging from simple premium-charged SMS solutions for mobile content to advanced dual-slot phone technology for real-world technology. Payment across networks is not standardized today, and the issue is complicated because different players have different needs and concerns, and many different solutions exist. There are a few of existing standards nowadays for the mobile payment:

- *UMTS*: A revolutionary new world standard for mobility, Universal Mobile Telecommunication System (UMTS) is not just a new, pioneering standard for mobile communications and payments. It also represents a new market situation following a different set of rules. Being aware of the market and revenue opportunities for UMTS, existing network operators, carrier consortiums, and even new entrants are vying to obtain operating licenses. Because of the massive investment required for UMTS, however, speed to market may become the deciding factor for commercial success.
- *EMV Standard*: In 1995, Europay, MasterCard, and Visa International developed an industry-wide chip-based debit/credit payment specification in order to provide global interoperability and reduce rising fraud levels. The EMV specification ensures that all Europay-, MasterCard-, and Visa-branded smart cards will operate with all chip-reading devices, regardless of location, financial institution, or manufacturer.
- *Jalda*: Ericsson and Hewlett Packard formed a joint venture called Ericsson Hewlett Packard Telecom AB (EHPT). EHPT has developed a potential standard Jalda for enabling micropayment transactions. Jalda supports various merchant billing options. When buyers make purchases, they receive SMS messages containing purchase information and costs via SMS or WAP cell-phone interfaces. If the buyers agree to the charges, they send the agreements back to the Internet payment providers (IPP) to authorize the purchases.
- *GMCIG standard for m-wallet*: The Global Mobile Commerce Interoperability Group (GMCIG), a nonprofit organization, has developed specifications for the remote m-wallet. GMCIG members include credit card companies, telecommunications network operators, wireless-device manufacturers, mobile-technology developers, content providers, and financial institutions. There are four leading industry groups developing open standards for local transaction solutions: IrDA's IrFM SIG, MeT Forum, Bluetooth SIG's SRFT SG, and NRF.

2.6 P2P-Paid Payment System

This section is dedicated to present the design and implementation of an example wireless payment application known as P2P-Paid. The application uses a two-dimensional (2D) secured protocol, which not only supports the peer-to-peer payment transactions between two mobile clients over the Bluetooth communications but also supports the related secured transactions between the payment server and mobile clients over wide area networks (WAN). This type of mobile payments can be used in a dynamic mobile environment to allow a payer and payee to conduct wireless payment transactions for mobile commerce. Typical application scenarios can be as follows [23]:

- Mobile payments between a taxi passenger and taxi driver
- Mobile payments between a merchant in flea market and his or her customers
- Mobile payments for parking fees or subways

2.6.1 P2P-Paid System Overview

P2P-Paid is a mobile payment system, which has been developed at San Jose State University as a research prototype system since 2004. Its objective is to develop a wireless-based payment system to assist mobile users to perform mobile payment transactions over mobile phones. The P2P-Paid system provides the following major functions to mobile phone users over its mobile client interface:

- *P2P peer discovery*: Only valid payers can discover valid payees who have published their IDs within the Bluetooth network.
- *P2P session management*: Mobile client sessions are established and kept track of by the system. P2P session management allows a user to establish or drop a P2P payment session over the Bluetooth network.
- *P2P payment management*: This allows a user to conduct a P2P mobile payment transaction over the Bluetooth network based on an established P2P payment session. The P2P payment transactions are carried out using the 2D payment transaction protocols defined in the previous section.
- *Account management*: This allows a user to perform basic account management functions, such as checking account balance and displaying an account summary.
- *Payment scheduling*: This allows a user to set up, update, delete, and view payment schedules using mobile phones.
- *Payee management*: This allows a user to perform payee management functions on a mobile phone, such as adding, editing, deleting, and viewing the payee information.

To assist mobile users, the P2P-Paid system also provides them with essential web-based functions through an online user interface. The basic functions include

- User registration and service registration to set up a P2P ID and password and to update user information
- User online security and authentication checking
- P2P-Paid account management such as balance checking and transaction history reporting
- Bank account management, which allows users to connect their P2P-Paid account to their bank account and to transfer balance from or to their bank account
- Mobile payment scheduling, which allows users to schedule their payments
- Payment management, which allows users to perform online payment transactions
- Mobile payee management, which allows users to perform payee management online

2.6.2 P2P-Paid System Architecture

As shown in Figure 2.7, P2P-Paid system has a four-tier architecture, which includes mobile client, middleware, P2P-Paid server, and database server.

P2P-Paid Client: The P2P-Paid client includes (a) J2ME-based P2P-Paid mobile client software for mobile phone users and (b) HTML-based online client for Internet accesses. The P2P-Paid mobile client software includes the following functional parts:

- *User interface*: It supports the interactions with a mobile user to accept and process user requests and displays the system responses.
- *P2P service module*: It supports peer-to-peer interactions between two mobile phone users (payer and payee) to conduct the payment party discovery based on the Service Discovery Protocol. Both the payer and payee must be registered with the system through online registration.

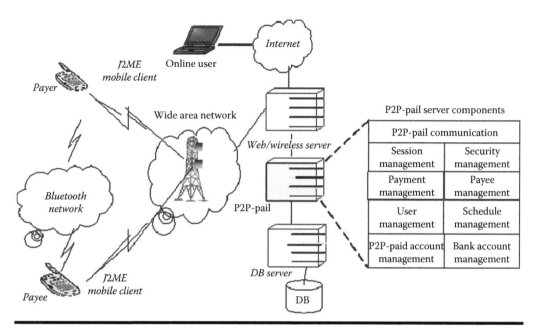

Figure 2.7 P2P-Paid system architecture.

- *Mobile security module*: It performs basic security functions with the support of the security management in the P2P-Paid server, including mobile user authentication and voice verification. In the current version, the voice verification has not been completely implemented.
- *Mobile payment module*: It supports P2P mobile payment communications between a mobile client and the P2P-Paid server over a wireless Internet infrastructure. A 2D wireless payment protocol is implemented here.

Middleware: The system middleware includes a Tomcat application server with the wireless Internet support and other middleware software, such as JSP and Servlets.

Payment database server: The system payment database server is a database server (MySQL Server) that works with the database access programs to maintain necessary user and account information and payment transactions.

P2P-Paid server: The P2P-Paid server works with middleware by communicating with mobile client software to support wireless payment functions. As shown in Figure 2.7, the P2P-Paid server consists of the following functional components to provide user-supporting features:

- *P2P-Paid communication*: It implements the peer-to-peer 2D payment protocol to support all mobile payment communications between a mobile client and the server.
- *User management*: It supports user registration and maintains two types of user information, including registered mobile users and administration users.
- *P2P-Paid account management*: It supports user account creation and updates.
- *Bank account management*: It supports bank account connection, update, and money transfer.
- *Payment management*: It manages and maintains all mobile payment transaction records.
- *Schedule management*: It manages and maintains mobile payment schedules, such as adding, updating, or deleting a payment schedule. Whenever a scheduled payment is due, a transaction is invoked.

- *Payee management*: It manages and maintains payee records for a mobile user, such as adding, updating, and deleting a payee.
- *Session management*: It establishes and controls mobile payment sessions for mobile payment transactions.
- *Security management*: It supports several security functions such as user authentication, access control, key management, data encryption/decryption, data digest, voice verification, etc.

2.6.3 P2P-Paid Payment Protocol

The goal of designing the P2P-Paid payment protocol is to provide a convenient, secured, and lightweight protocol built on top of HTTP and Bluetooth communication protocol for supporting monetary transactions. P2P-Paid provides various services for mobile users to make monetary transaction securely. Those services include the following: send instant money, schedule payment, payee management, Bluetooth device/service discovery, etc. P2P-Paid payment protocol has two parts: mobile client to P2P-Paid server communication protocol and mobile client to mobile client communication protocol.

2.6.3.1 P2P-Paid Client–Server Protocol

Figure 2.8 encapsulates the client–server communication sequences in the P2P-Paid payment system. A service a user selects in step 1.1 can be any service provided by the P2P-Paid payment

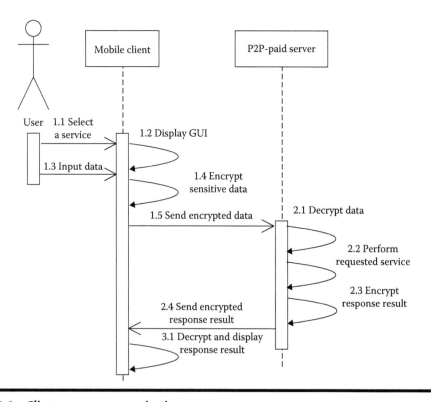

Figure 2.8   Client–server communications.

system. The key used for encrypting and decrypting data is an authentication key generated from the mobile user's secret key. The security solution section will explain more on key generation.

2.6.3.2 P2P-Paid Client–Client Protocol

Mobile client to mobile client over the Bluetooth communication mainly involve five steps as shown on Figure 2.9. In step 1, a mobile user needs to choose a role either as a payer or a payee before entering the Bluetooth network. In step 2, a payee needs to publish his or her information (P2PID) onto the network to become visible to payers. Step 3 is for the payer to search for payees within the Bluetooth network. Bluetooth's Service Discovery Protocol (SDP) provides standard means for a Bluetooth device to query and discover services supported by a peer Bluetooth device. Figure 2.10 shows a JSR-82 capable MIDlet using the DiscoveryAgent, which provides methods to perform

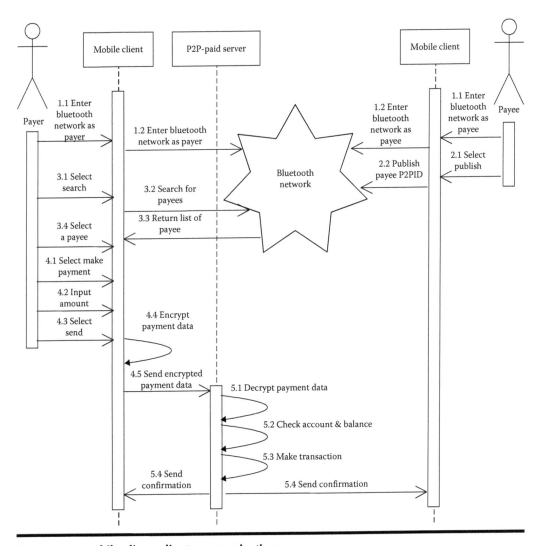

Figure 2.9   Mobile client–client communications.

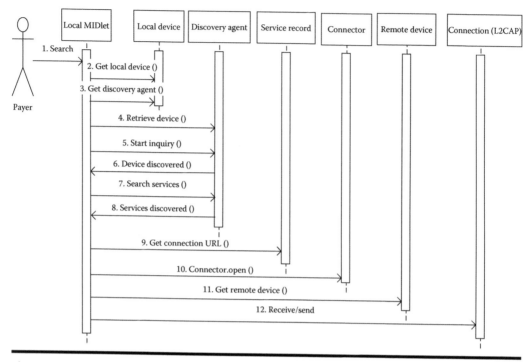

Figure 2.10   Using SDP API sequence.

device and service discovery. In step 4, a payer selects to make a payment to the selected payee, then he or she inputs the amount and a simple description of the payment. The payer's mobile client encrypts the payment data and sends it to the P2P-Paid server. On step 5, the P2P-Paid server decrypts the payment data, checks for validation of the payer and the payee's accounts, checks for the sufficiency of the payer's balance, and then performs the transaction. The server then sends a confirmation to both the payer and the payee to finish the transaction.

2.6.3.3 P2P-Paid Message Format

P2P-Paid client–server message format: Mobile and web clients communicate with the P2P-Paid server by using HTTP/HTTPS requests and responses to send or receive messages. The formats of the request and response messages are shown in Figure 2.11. A security header is attached for each message sent across the Internet or the air. The security header is described separately in the security solution section.

P2P-Paid Bluetooth message format: P2P-Paid Bluetooth Protocol is used to establish a connection and send data between two MIDlet clients, where these clients are considered as peers. This protocol carries the L2CAP packets in small units and uses the HCI interface to organize the data between software and hardware. Figure 2.12 shows the P2P Bluetooth message format.

- *Packet boundary*: Packet boundary flag identifies whether the packet data carries the start of the L2CAP packet. The L2CAP packets are divided into small units as mentioned earlier. The first packet is always set to START flag and the remaining packets are set to CONTINUE flag.

Request

Request-Line			Request Header			Entity Header				Entity Body		SE-Hdr
Method	REQ-URI	HTTP-VERS	From	User-AG	Host	CONT-LEN	CONT-Type	Expires	Last-UPD	SES-ID	Body	

Response

Status-Line			Response Header		Entity Header				Entity Body			SE-Hdr
HTTP-VERS	Status-Code	REAS-Phrase	LOC	Server	CONT-LEN	CONT-Type	Expires	Last-UPD	SES-ID	Type	Body	

Figure 2.11   P2P-Paid client–server message format.

Packet Boundary	Broadcast Flag	L2CAP Header	Data	SE-Hdr

RequestData

sessionID	SendFrom	sendTo	sendAmt	sendDesc

ResponseData

typeError	errorMessage	typeSuccess	successMessage

Figure 2.12   P2P-Paid Bluetooth message format.

- *Broadcast flag*: Broadcast flag distinguishes the data between point-to-point and broadcast communication.
- *L2CAP header*: L2CAP header contains two bytes of channel identifier (CID), two bytes of total length of L2CAP.
- *Data*: The data section is the XML document wrapped inside the L2CAP packet. It is either request or response message type, and the elements in the request and response message are different as shown in the figure.

2.6.4 P2P-Paid Security Solution

The security solution of the P2P-Paid payment system is an integration of the secured payment protocol, a biometric verification, and optimized conventional security methods. It provides the following security features: service registration, access control, security code attachment, and speaker verification.

2.6.4.1 Service Registration

Before using the P2P-Paid payment services, a user must first register with the system. There are two types of registration a user must go through. They are online registration and mobile registration. On completion of online registration, a user account is created, and a key pair is associated with the web client and the system server (see Figure 2.13). On completion of mobile

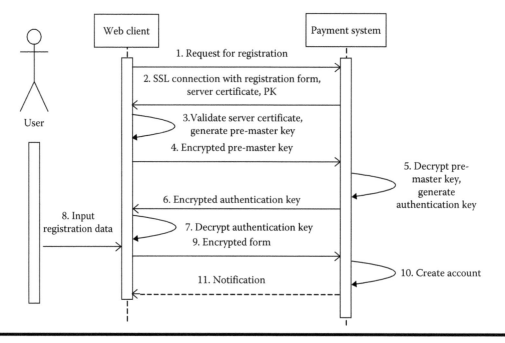

Figure 2.13 Secured web registration sequence.

registration, a voiceprint is created for the user for future mobile authentication, and a symmetric key is also associated with the mobile device and the system server. Symmetric key encryption and decryption is chosen here because of its better performance. The symmetric key is protected with secure connection (HTTPS) when it is sent to the mobile client, and the key is stored on the mobile device in an encoded form (see Figure 2.14). Therefore, only the P2P-Paid server recognizes the key, and the mobile device is also authenticated when it successfully communicates with the P2P-Paid server. The sensitive data transmitted between a mobile client and the server is encrypted by using an authentication key that is generated based on the symmetric key of the mobile user for each communication session. Another alternative is to use the public key encryption and decryption approach. The related papers can be found in [25–26].

2.6.4.2 Access Control

Authorization comes after authentication. In order to use the P2P-Paid payment service, a mobile user has to log in the system first. Only valid user receives the access to the system. There are two types of login: web login and mobile login. Web login requires the user to enter the valid P2PID and password. Mobile login requires the user to enter P2PID, PIN, and voiceprint. Figure 2.15 presents the P2P-Paid mobile login procedures.

2.6.4.3 Security Code Attachment

As mentioned earlier, any sensitive data is encrypted before transmitting over the network. A security header is attached on each piece of message sent across the Internet or the air. The security header

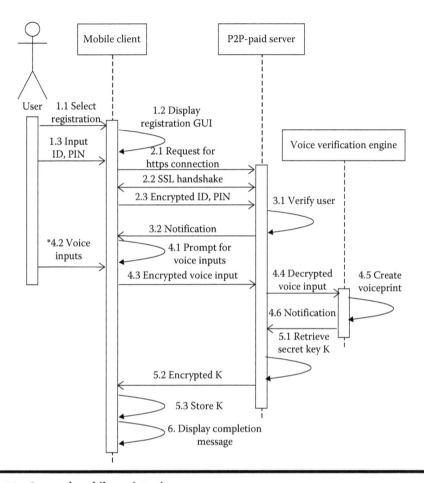

Figure 2.14 Secured mobile registration sequence.

carries the information about the secured protocol. Each entity in the security header can be empty or contain value depending on what type of information is sent. The format of the security header is shown on Figure 2.16.

- *MAC*: MAC is the message authentication code. MAC is the digest of authentication key and the message to be sent. MAC allows data authentication, which allows the receiver to verify the integrity of a message.
- *Key used*: For the first time when two parties communicate with each other, they have to negotiate what key is used for establishing a secure connection.
- *Key length*: Due to the ability of different devices for handling different key lengths, an appropriate key length must be negotiated before two devices can start encrypting the traffic between them.
- *Encrypted fields*: It indicates what fields or which pieces on the message are encoded. In general, all sensitive data such as user ID, password, PIN, account number, etc., should be encrypted before transmitting across the network.

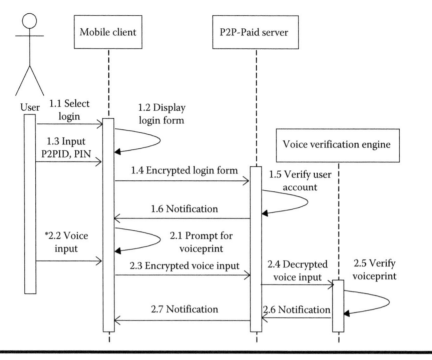

Figure 2.15 Secured mobile login sequence.

MAC	Key used	Key length	Encrypted fields (e.g., userID, PIN, account No.)

Figure 2.16 Security header format.

2.6.4.4 Speaker Verification

Speaker verification, as a subclass of voice recognition, is the process of automatically recognizing who is speaking on the basis of individual information included in speech waves. Its task is a hypothesis-testing problem where the system has to accept or reject a claimed identity associated with an utterance [24]. There are two types of speaker verification approaches: text dependent and text independent. We use speaker verification as a part of user authentication in P2P-Paid payment system.

Based on our research, there are a number of existing biometric verification systems [24,14–16] in which the feature vectors and trained model of a speech are based on a set of world models. To obtain a more accurate model that represents better speech characteristics, a large set of world models is needed. This can be time consuming. Since most existing solutions are not designed to address mobile client limitations (such as dynamic mobile environment and limited computing power), we are working on to develop a refined text-dependent speaker verification module to focus on optimizing feature extracting. Before using the P2P-Paid services with a mobile device, a user has to be authenticated by verifying his or her voiceprint. With this approach, the system is able to perform the authentication for the mobile device and the mobile user who is using the device.

The speaker verification model implements a new algorithm to choose better feature vectors for each mobile user. This algorithm is designed based on the fact that only those stable features in a voice signal can better be used for the feature vector set. For example, during enrolment procedures, some Mel-frequency cepstral coefficients (MFCC) and linear prediction cepstral coefficients (LPCC) may not change or change slightly from time to time. However, some of those coefficients could change significantly in different utterances. Those coefficients that don't change or change slightly can be said to be more stable and better represent the speech characteristics than those that change. Therefore, only those stable coefficients should be selected to build the reference models during enrolment and used to compare those coefficients during verification. Since the selection of those coefficients is different for each mobile user, the system is able to compare them dynamically for each user during verification. In theory, this optimized feature extracting approach should be more accurate in speaker verification because each voice model is built based on individual speaker. In addition, the approach should be faster and simpler because a smaller set of coefficients is used for feature extracting and a smaller set world model is used for training. The detailed implementation and experimental results of the speaker verification will be reported in the future publications.

2.6.5 Implementation Status and Used Technologies

The first version of the P2P-Paid system only implemented the P2P payment protocol with basic security support. The voice verification feature will be added in the later version. The current server is deployed on a Tomcat application server (Version 5.0.28). On the mobile client side, NetBeans 4.0 IED with the J2ME Wireless Toolkit Emulator integrated is used to develop and test the MIDP client. JSR-82 is used to test the Bluetooth communications. In addition, the kXML parser is used to parse XML data on MIDP client and Xercer parser is used on the P2P-Paid server.

2.6.6 Application Examples

In this section, a couple of payment scenarios are presented to demonstrate the P2P-Paid payment system prototype.

2.6.6.1 Web Examples

Figure 2.17a through g shows some web interfaces for new user registration and for registered users to perform various types of management:

a. P2P-Paid home page shows some general information about the system.
b. Registration page provides a form for entering registration information. Note that all the registration information, transaction, and related sensitive information are protected by SSL.
c. P2P-Paid account summary provides quick summary about the user account.
d. Bank management enables user to change banking information and to transfer money between user's bank account and P2P-Paid account.
e. Payment history page enables user to view and search his account transaction history.
f. Schedule payment management page allows user to schedule and manage scheduled transactions.
g. Payee management page enables user to add payee and manage payee easily.

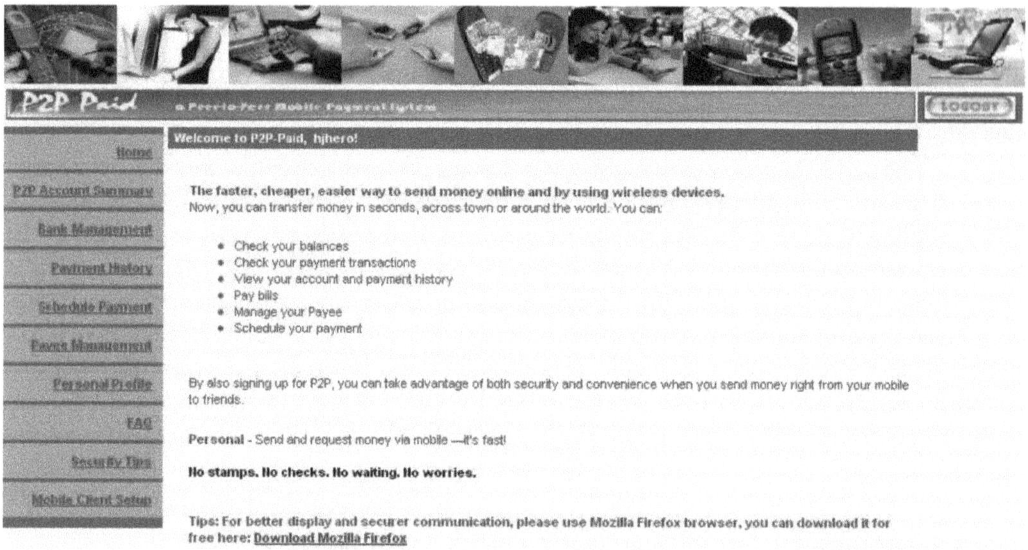

(a)

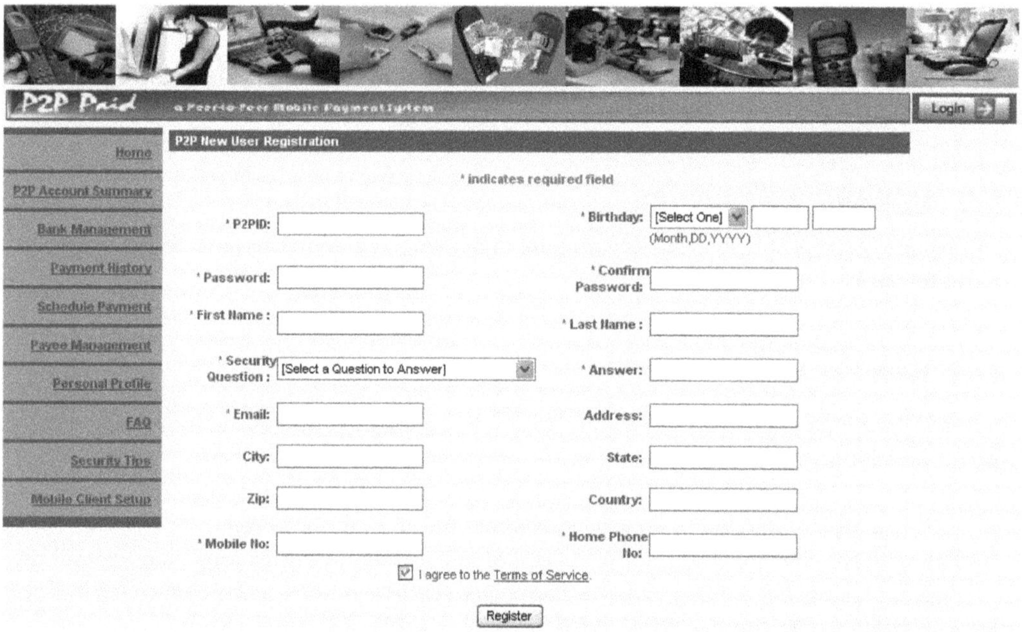

(b)

Figure 2.17   P2P-Paid web examples. (a) P2P-Paid home page, (b) registration page, (c) P2P-Paid account summary, (d) bank management, (e) payment history, (f) schedule payment management, and (g) payee management.

(c)

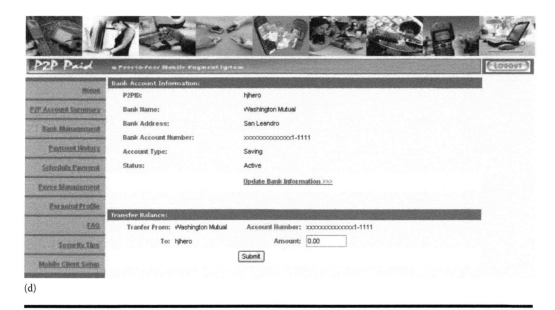

(d)

Figure 2.17   (continued)

2.6.6.2 Mobile Examples

Figure 2.18 shows some mobile interfaces that mobile client provides for mobile user to perform mobile payment transactions and management. Some scenarios are described in the following:

    a. After successfully logging in, mobile user (JackyCai) is presented with the main menu.
    b. JackyCai chooses to view account summary.

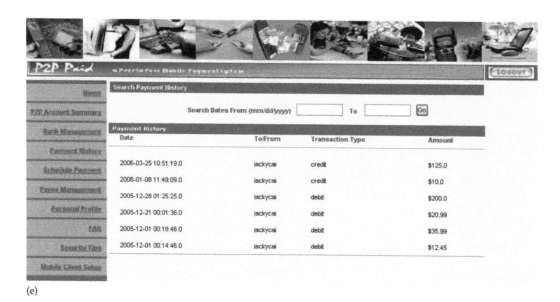

(e)

(f)

Figure 2.17   (continued)

c. JackyCai chooses to send money now. JackyCai needs to input payee's (Kiranp's) P2PID, amount, and payment description.

d. JackyCai chooses to view last payment summary from the main menu in (a), and the summary of the payment just made in (c) is presented.

e. The payee, Kiranp, can log in with her mobile device and view the summary of the payment made by JackyCai in (c).

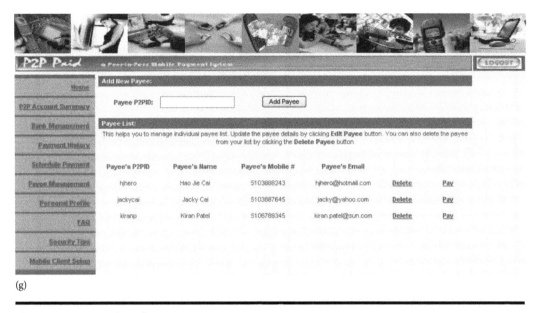

(g)

Figure 2.17   (continued)

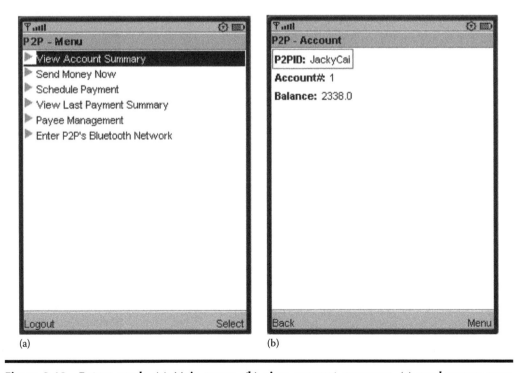

(a)   (b)

Figure 2.18   Future work. (a) Main menu, (b) view account summary, (c) send money now, (d) last payment summary, (e) last payment summary, (f) schedule payment menu, (g) add schedule payment (h) schedule payment menu, (i) payment management, (j) select a role as payee, (k) payee's menu, (l) published P2PID, (m) select a role as payer, (n) peer search result, (o) make a payment, (p) JackyCai's receipt, and (q) Kiranp's receipt.

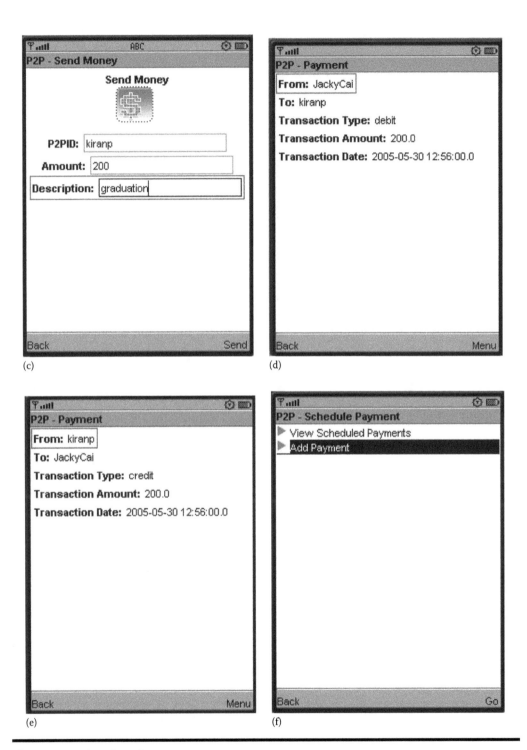

(c)

(d)

(e)

(f)

Figure 2.18 (continued)

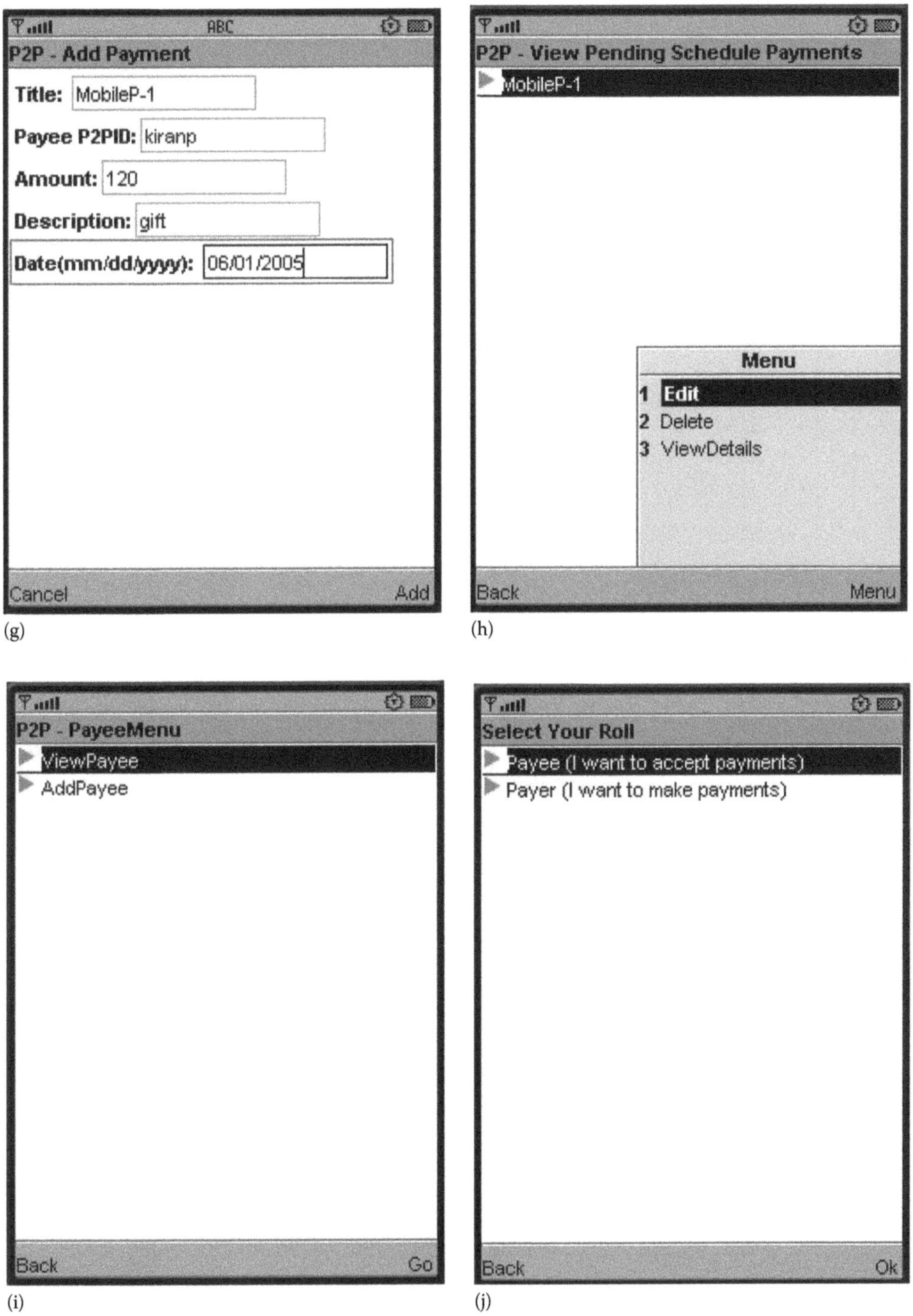

(g)

(h)

(i)

(j)

Figure 2.18   (continued)

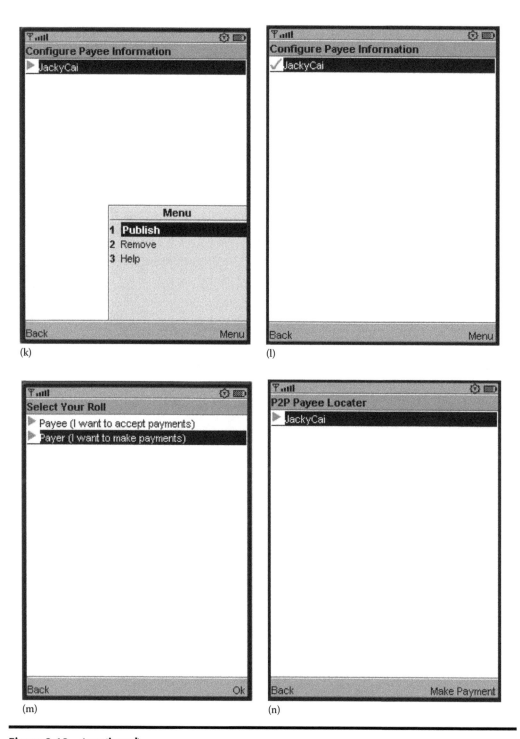

(k)

(l)

(m)

(n)

Figure 2.18   (continued)

(o)

(p)

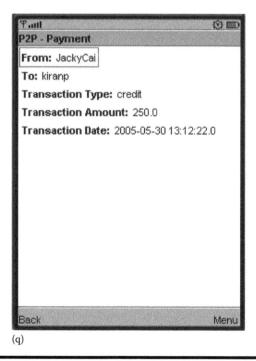

(q)

Figure 2.18 (continued)

f. User (JackyCai) can select schedule payment from the main menu in (a), and the user is presented with the schedule payment menu, from which the user can choose to view scheduled payments and add schedule payment.

g. JackyCai selects to add a schedule payment. His mobile device presents a form for inputting schedule payment data. The schedule payment data includes schedule payment title, payee's P2PID, amount, description, and schedule date.

h. After sending the schedule payment, JackyCai selects to view schedule payments, and he is presented with a menu to edit, delete, and view detail of the scheduled payment.

i. User can select Payee Management from the main menu in (a) to perform payee management. Payee Management has similar features as schedule payment.

j. From (j) to (o) are the P2P interactions within Bluetooth network. After JackyCai selects Enter P2P's Bluetooth Network from the main menu in (a), he select a role as payee to enter the Bluetooth network.

k. Payee (JackyCai) is represented with a menu for publishing/removing his information (P2PID) onto the network; payee also can select Help to see some helpful tips.

l. Payee selects to publish his information, and his P2PID is visible in the network.

m. Another user (Kiranp) selects a role as payer to enter into the Bluetooth network.

n. Payer performs a search to discover payees in the network. One payee (JackyCai) is found. Note that if multiple payees have published their information onto the network, a payer is able to discover them and select them from the list to make payment.

o. Payer (Kiranp) selects to make payment. The payment form with the selected payee's P2PID attached is presented. The payer only needs to enter the amount and a simple description to make a transaction.

p. Payee can view his receipt of the transaction made in (o).

q. Payer also can view her receipt of the transaction made in (o).

The next phase for this P2P-Paid project is to load the application onto some mobile devices to test how it performs. The future works also include developing and implementing a lightweight speaker verification solution to support mobile user authentication and authorization for mobile payment transactions. The study about its performance, scalability, usability, etc., is also needed for the completion of the project.

2.7 Conclusions

The fast increase of the number of mobile device users is creating many new demands and business opportunities in wireless information and mobile commerce services. To provide various mobile contents, diverse information services, and mobile commerce applications, we need cost-effective mobile payment solutions. Wireless service vendors and mobile content providers need to study different business models in mobile payment and develop and/or deploy new wireless payment systems to support mobile devices to make payments anywhere and anytime.

The economic viability of business models based on mobile data remains suspect due to the lack of clear business models for mobile payments and revenue sharing. Today wireless payment vendors need more time to study and compare different payment models and schemes through actual deployment experience in wireless service and mobile commerce applications because the business models for wireless payments are still on trial. With the fast advance of diverse wireless information services and commerce applications, charging based on the airtime of wireless communication is no longer an appropriate payment method in the wireless world. Therefore, new wireless payment methods and electronic solutions are strongly on demand for wireless information services and mobile commerce. Making a successful wireless payment system has four needs: (a) easy mobile payment interfaces with good interoperability and strong accessibility to diverse mobile devices; (b) secure and reliable payment transactions based on a standardized wireless payment protocol; (c) effective security solutions to provide authentication, authorization, and integrity of involved parties and encryption of wireless communication messages; and (d) cost-effective and efficient payment processes.

Successful payment methods and solutions will be those that can continue to meet the mentioned challenges, particularly in mobile accessibility, security, standardization, and interoperability. Mobile consumers need more time to understand the advantages of mobile commerce and mobile payment system. Finally, the success of mobile payment will be driven by the success of mobile applications and services.

References

1. J.Z. Gao, S. Shim, H. Mei, and X. Su, *Engineering Wireless Based Software Systems and Applications*, Artech House Publishers, Norwood, MA, 2006, Retrieved on June 6, 2012 from http://www.amazon.com/Engineering-Wireless-Based-Software-Systems-Applications/dp/1580538207

2. J. Ondrus and Y. Pigneur, A disruption analysis in the mobile payment market, in *Proceedings of the 38th Hawaii International Conference on System Sciences (HICSS-38'05)*, Big Island, HI, 2005.

3. N.M. Sadeh, *M-Commerce: Technologies, Services, and Business Models*, John Wiley & Sons, Inc., New York, 2002.

4. L. Antovski and M. Gusev, M-payments, in *Proceedings of the 25th International Conference Information Technology Interfaces (ITI'03)*, Cavtat, Croatia, 2003.

5. K. Pousttchi and M. Zhenker, Current mobile payment procedures on the German market from the view of customer requirements, in *Proceedings of the 14th International Workshop on Database and Expert Systems Application (DEXA'03)*, Prague, Czech Republic, 2003.

6. S. Nambiar and T.L. Chang, M-payment solutions and M-commerce fraud management, Retrieved September 9, 2004 from http://europa.nvc.cs.vt.edu/~ctlu/Publication/M-Payment-Solutions.pdf

7. X. Zheng and D. Chen, Study of mobile payments system, in *Proceedings of the IEEE International Conference on E-Commerce (CEC'03)*, Newport Beach, CA, 2003.

8. S. Kungpisdan, B. Srinivasan, and P.D. Le, A secure account-based mobile payment protocol, in *Proceedings of the International Conference on Information Technology: Coding and Computing (ITCC'04)*, Las Vegas, NV, 2004.

9. Y. Lin, M. Chang, and H. Rao, Mobile prepaid phone services, *IEEE Personal Communications*, 7(3): 6–14, 2000.

10. A. Fourati, H.K.B. Ayed, F. Kamoun, and A. Benzekri, A SET based approach to secure the payment in mobile commerce, in *Proceedings of 27th Annual IEEE Conference on Local Computer Networks (LCN'02)*, Tampa, FL, November 6–8, 2002.

11. J.Z. Gao, J. Cai, M. Li, and S.M. Venkateshi, Wireless payment—Opportunities, challenges, and solutions, *High Technology Letter*, May Issue, 2006.

12. Z. Huang and K. Chen, Electronic payment in mobile environment, in *Proceedings of 13th International Workshop on Database and Expert Systems Applications (DEXA'02)*, Aix-en-Provence, France, September 2–6, 2002.

13. Telecom Media Networks, Mobile payments in M-commerce, September 2002, Retrieved February 12, 2004 from http://www.cgey.com/tmn/pdf/MobilePaymentsinMCommerce.pdf

14. D.A. Reynolds, T.F. Quatieri, and R.B. Dunn, Speaker verification using adapted Gaussian mixture models, Retrieved January 5, 2005 from http://www.cse.ohio-state.edu/~dwang/teaching/cis788/papers/Reynolds-dsp00.pdf

15. D. Neiberg, Text independent speaker verification using adapted Gaussian mixture models, Retrieved January 5, 2005 from http://www.speech.kth.se/~neiberg/neiberg02mst.pdf

16. T.B. Nordstrom, H. Melin, and J. Lindberg, A comparative study of speaker verification using the polycost database, Retrieved January 11, 2005 from http://www.speech.kth.se/ctt/publications/papers/icslp98_1359.pdf

17. S. Marinov, Text dependent and text independent speaker verification system: Technology and application, Retrieved January 19, 2005 from http://citeseerx.ist.psu.edu/viewdoc/summary?doi=10.1.1.134.1529

18. Berlecon Research, Getting your Bills paid in E-Commerce—An analysis of relevant payment systems from the Distributor's Point of View, Retrieved on Feburary, 2001 from http://www.berlecon.de/research/en/reports.php?we_objectID=34

19. N.A. Thomas, Infrared and bluetooth transactions at the point of sale, Retrieved on February 12, 2004 from http://www.tdap.co.uk/uk/archive/mobile/mob(in2m_0109).html

20. H.M. Deitel, P.J. Deitel, T.R. Nieto, and K. Steinbuhler, *Wireless Internet & Mobile Businesses—How to Program*, Prentice Hall, Upper Saddle River, NJ, 2002.

21. D. Hennessy (a white paper from Valista^{TM}), The value of the mobile wallet, Retrieved on February 10, 2004 from http://www.valista.com/downloads/whitepaper/mobile_wallet.pdf

22. Paybox.net, Mobile payment delivery made simple, Retrieved on February 12, 2004 from http://www.paybox.net/publicrelations/public_relations_whitepapers.html

23. J. Gao, J. Cai, K. Patel, and S. Shim, A wireless payment system, in *Proceedings of the Second International Conference on Embedded Software and Systems (ICESS'05)*, IEEE Computer Society Press, Xian, China, December 2005.

24. J. Olsson, Text dependent speaker verification with a hybrid HMM/ANN system, Retrieved January 5, 2005 from http://www.speech.kth.se/ctt/publications/exjobb/exjobb_jolsson.pdf

25. N. Potlapally, S. Ravi, A. Raghunathan, and G. Lakshminarayana, Algorithm exploration for efficient public-key security processing in wireless handsets, Retrieved on June 9, 2012 from http://scholar.googleusercontent.com/scholar?q=cache:t643DXFrgHgJ:scholar.google.com/&hl=en&as_sdt=0,5

26. N.R. Potlapally, S. Ravi, A. Raghunathan, and G. Lakshminarayan, Optimizing public-key encryption for wireless clients, in *IEEE International Conference on Communications (ICC)*, New York, May 2002.

Chapter 3

A World Model for Smart Spaces

Ichiro Satoh

Contents

3.1 Introduction

Recent technological advances have enabled parts of the real world to be turned into so-called *smart spaces*. In fact, computers are embedded into everyday objects, including appliances, chairs, and walls and sensing devices, and are in fact already present in almost every room of modern buildings or houses and in many of the public facilities with cities. As a result, spaces are becoming perceptual and smart. For example, location-sensing technologies, e.g., RFID, computer vision, and GPS, have been used to identify physical objects and track the positions of objects. These sensors have made it possible to detect and track the presence and location of people, computers, and practically any other object we want to monitor. There have been several attempts for narrowing gaps between the physical world and cyberspaces, but most existing approaches or infrastructures inherently depend on particular sensing systems and have inherently been designed for their initial applications.

A solution to this problem would be to provide a general world model for representing the physical world in cyberspaces. There have been several world models. However, existing models have aimed at maintaining the locations of only people and objects in the physical world, but we often require the locations of computing devices that provide location-based or personalized services and the software that defines the services in cyberspace. Furthermore, most existing models are not available in smart spaces, because these need to be maintained in centralized database systems, whereas smart spaces are often managed in an ad hoc manner without any database servers. Therefore, we need a new world model that can be used in smart spaces, which may not have database servers, and that can specify logical and physical entities in a unified manner. This chapter is aimed at discussing the construction of such a model, called *M-Spaces*, as a programming interface between location sensors and application-specific services in smart spaces. We presented an earlier version of the model presented in this chapter in our previous paper [1]. The previous paper was aimed at proposing the model, whereas this chapter is aimed at presenting a prototype implementation of the model, its programming interfaces, some advanced applications, in addition to the model itself. This chapter describes our experience with the model, which the previous paper lacked.

In the remainder of this chapter, we outline an approach to building and managing location-based and personalized information services in smart spaces (Section 3.2), the design of our model (Section 3.3), and an implementation of the framework and its programming interfaces (Section 3.4). We describe some experience we have had with four applications, which we used the framework to develop (Section 3.5). We also outline related work (Section 3.5), provide a summary, and discuss some future issues (Section 3.6).

3.2 Basic Approach

This chapter proposes a world model for location-based and personalized services in indoor settings, e.g., buildings and houses, rather than outdoor ones.

3.2.1 Requirements

Smart spaces, which our model is targeted at, have several unique requirements as follows:

- Not only entities, such as physical objects and people, but also computing devices, which provide application-specific services, can be moved from location to location by users. A world model for smart spaces is required to be able to represent mobile computing devices and spaces as well as mobile entities.

- A smart space consists of multiple computing devices, e.g., embedded computers, PDAs, and public terminals. Such devices are often too heterogeneous to mask the differences between them. The model is required to maintain the capabilities of computing devices as well as their locations.
- Computing devices in smart spaces may also have limited memories and non-powerful processors, so they cannot support all the services that they need to provide. To conserve their limited resources, service-provider software must be able to be deployed at computing devices that are located at appropriate locations with device capabilities that can satisfy the requirements of the software. The model should be able to manage the location and deployment of service-provider software.
- Since computing devices in smart spaces are dynamically connected to and occasionally disconnected from networks, they are required to be dynamically organized in an ad hoc and peer-to-peer manner. As a result, they cannot always access database servers to maintain world models. The model should be available without database servers enabling computing devices to be organized without centralized management servers.

3.2.2 Background

Many researchers have explored world models for smart spaces. Most existing models have been aimed at identifying and locating entities, e.g., people and physical objects and computing devices in the physical world. These existing models can be classified into two types: physical-location and symbolic-location models. The former represents the position of people and objects as geometric information, e.g., NEXUS [2,3] and Cooltown [4]. A few applications like moving-map navigation can easily be constructed on a physical-location model with GPS systems. However, most emerging applications require a more symbolic notion: place. Generically, place is the human-readable labeling of positions. The latter represent the position of entities as labels for potentially overlapping geometric volumes, e.g., names of rooms, and buildings, e.g., Sentient Computing [5], and RAUM [6]. Existing approaches assume that their models are maintained in centralized database servers, which may not always be used in smart spaces. Therefore, our model should be managed in a decentralized manner and be dynamically organized in an ad hoc and peer-to-peer manner. Virtual Counterpart [7] supports RFID-based tracking systems and provides objects attached to RFID-tags with Jini-based services. Since it enables objects attached to RFID-tags to have their counterparts, it is similar to our model. However, it only supports physical entities except for computing devices and places. Our model should not distinguish between physical entities, places, and software-based services so that it can provide a unified view of smart spaces, where not only physical entities are mobile but also computing devices and spaces.

The framework presented in this chapter was inspired by our previous work, called SpatialAgents [8], which is an infrastructure that enables services to be dynamically deployed at computing devices according to the positions of people, objects, and places that are attached to RFID tags. The previous framework lacked any general-purpose world model and specified the positions of physical entities according to just the coverage areas of the RFID readers so that it could not represent any containment relationship of physical spaces, e.g., rooms and buildings. Moreover, we presented another world model, called *M-Space* [1,9] and the previous model aimed at integrating between software-based services running on introducing computing devices and service-provider computing devices whereas the model presented in this chapter aims at modeling containment relationship between physical and logical entities, including computing devices and software for defining services.

3.2.3 Design Principles

Existing location models can be classified into two types: physical location and symbolic location [6,10,11]. The former represents the position of people and objects as geometric information. A few outdoor applications like moving-map navigation can easily be constructed on such physical-location models. Most emerging applications, on the other hand, require a more symbolic notion: place. Generically, place is the human-readable labeling of positions. A more rigorous definition is an evolving set of both communal and personal labels for potentially overlapping geometric volumes, e.g., the names of rooms and buildings. An object contained in a volume is reported to be in that place.

Therefore, the approach presented in this chapter addresses symbolic location as a programming model that directly maps to event-driven application programming. For example, when a person enters a place, services should be provided from his or her portable terminal, or his or her stationary terminals should provide personalized services to assist him or her. Our model also introduces the containment relationship between spaces. This is because physical spaces are often organized in a containment relationship, because each space is often composed of more than one subspace. For example, each floor is contained within at most one building and each room is contained within at most one floor. Therefore, our world model is constructed as a tree, based on geographical containment. Such a tree-based model may not be unique but our model has the following features that existing models do not.

a. *Virtual counterparts*: No physical objects or spaces may specify their attributes or interact with one another, because of resource limitations. The model introduces the notion of counterparts. These are digital representations of physical entities or spaces. An application does not directly interact with physical objects or places, but with their virtual counterparts. The model spatially binds the positions of entities and spaces with the locations of their virtual counterparts and, when they move in the physical world, it deploys their counterparts at proper locations in it.

b. *Unified view*: Location-based and personalized services must be executed at computing devices whose capabilities can satisfy the requirements of the services and that are at the locations that the services should be provided. The model can maintain the locations and capabilities of computing devices as well as those of physical entities and services. It also manages the deployment of application-specific services according to changes in the locations of physical entities, spaces, and computing devices. That is, the model does not distinguish between physical entities, spaces, computing devices, including the computers that maintains themselves, or application-specific services.

c. *Sensor independence*: There have been a variety of location-sensing systems. They can be classified into two types: tracking and positioning. The former, including RFID tags, measures the location of other objects. The latter, including GPS, measures its own location. Since it is almost impossible to support all kinds of sensors, the model aims at supporting various kinds of tracking sensors, e.g., RFID, infrared, or ultrasonic tags and computer vision, as much as possible. The model can have a mechanism for managing location sensors outside itself so that it is designed independently of sensors. The model transforms geometric information about the positions of objects into corresponding containment relations, but it allows application-specific services to explicitly know the geometric-information locations measured by sensors.

d. *Extensibility*: The model can be managed by one or more computers, which may also offer application-specific services, as well as database servers. It enables the computers that maintain it to be managed within it so that we can easily configure the system structure. It provides a demand-driven mechanism, which was inspired by ad hoc mobile networking technology, that discovers the computing devices and services that are required. It also enables service-provider software to be dynamically deployed at computing devices, but only when the software is needed.

e. *Local interaction*: In human communication, a person wants to communicate with someone in front of him or her rather than with people in another room. Smart spaces are equipped with computing devices that can provide services to users. Services running on a device should be valid within a bounding region surrounding the device and limiting their presence in space. An entity, including a person, or the computing device's surrounding scope can be a medium, e.g., visual, audio, or hand-manipulation, which enables it to interact with other entities or devices. The model enables each entity to specify the scope within which it can receive services running on the device according to the media between the entity and the device. It enables each computing device and service to specify the scope within which each is available depending on the capabilities of the device and the requirements of the service.

3.3 World Model

This section describes the world model presented in this chapter. The model manages the locations of physical entities and spaces through symbolic names.

3.3.1 Hierarchical World Model

Our model consists of elements, called components, which are just computing devices or software, or which are implemented as virtual counterpart objects of physical entities or places. The model represents facts about entities or places in terms of the semantic or spatial containment relationships between components associated with these entities or places (Figure 3.1).

- *Virtual counterpart*: Each component is a virtual counterpart of a physical entity or place, including the coverage area of the sensor, computing device, or service-provider software.
- *Component structure*: Each component can be contained within at most one component according to containment relationships in the physical world and cyberspace.
- *Inter-component movement*: Each component can move between components as a whole with all its inner components.

When a component contains other components, the former component is called a *parent* and the latter *children*, like the MobileSpaces model [12]. When physical entities, spaces, and computing devices move from location to location in the physical world, the model detects their movements through location-sensing systems and changes the containment relationships of components corresponding to moving entities, their source and destination. Each component is a

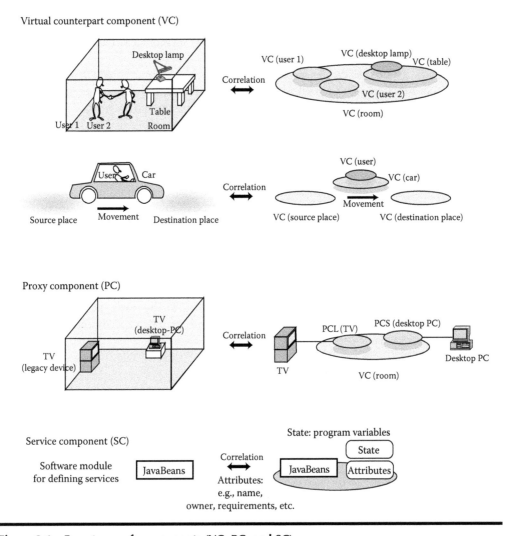

Figure 3.1 Four types of components (VC, PC, and SC).

virtual counterpart of its target in the world model and maintains the target's attributes. Figure 3.2 shows the correlation between spaces and entities in the physical world and their counterpart components. Containment relationships between components should reflect on the structural containment relationships between entities or places as we can see in Figure 3.2. The model monitors changes in the locations of entities, places, and computing devices through location-sensing systems and it dynamically configures the containment relationships between corresponding components.

3.3.2 Components

The model is unique to existing world models because it not only maintains the location of physical entities, such as people and objects, but also the locations of computing devices and services in a unified manner. Components can be classified into three types.

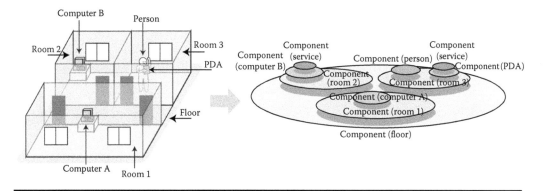

Figure 3.2 **Rooms on floor in physical world and counterpart components in world model.**

1. *Virtual Counterpart Component* (*VC*) is a digital representation of a physical entity, such as a person or object, except for a computing device, or a physical place, such as a building or room.
2. *Proxy Component* (*PC*) is a proxy component that bridges the world model and computing device, and maintains a subtree of the model or executes services located in a VC.
3. *Service Component* (*SC*) is software that defines application-specific services dependent on physical entities or places.

For example, when a person moves from the coverage area of a sensor to the coverage area of another sensor, the model detects and moves the VC corresponding to the moving person from the VC corresponding to the source location, to the VC corresponding to the destination location. If the person has a computing device, the VC corresponding to the person carries the PC corresponding to the device. The model also offers at least two basic events, entering and leaving, which enable application-specific services to react to actions in the physical world. Since each component in the model is treated as an autonomous programmable entity, it can define behaviors with some intelligence. Furthermore, the model also classifies PCs into three subtypes, PCM (PC for Model manager), PCS (PC for Service provider), and PCL (PC for Legacy device), according to the functions of the devices. Our model can be maintained by not only the server but also multiple computing devices in smart spaces.

- The first component, i.e., PCM, is a proxy of a computing device maintaining a subtree of the components in the world model (Figure 3.3a). It attaches the subtree of its target device to a tree maintained by another computing device. Some computing devices can provide runtime systems to execute services defined as SCs.
- The second component, i.e., PCS, is a proxy of the computing device that can execute SCs (Figure 3.3b). If such a device is in a space, its proxy is contained by the VC corresponding to the space. When a PCS receives SCs, it forwards these to the device that it refers to.
- The third component, called PCL (PC for Legacy device), is a proxy of the computing device that cannot execute SCs (Figure 3.3c). If such a device is in a space, its proxy is contained by the VC corresponding to the space and it communicates with the device through the device's favorite protocols.

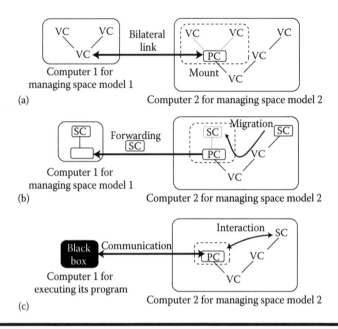

Figure 3.3   Three types of proxy components. (a) Virtual counterpart component. (b) Service component. (c) Proxy component.

For example, a television, which does not have any computing capabilities, can have an SC in the VC corresponding to the physical space that it is contained in and can be controlled in, and the SC can send infrared signals to it. A computing device can have different PCs whereby it can provide the capabilities to them.

3.4 Implementation

To evaluate the model described in Section 3.4, we implemented a prototype system that builds on this model. The model itself is independent of programming languages but the current implementation uses Java (J2SE or later versions) as an implementation language for components.

3.4.1 Components

We describe APIs that components provide.

3.4.1.1 Virtual Counterpart Component

Each VC is defined from an abstract class, which has some built-in methods that are used to control its mobility and life cycle. It can explicitly defines its own identifier and attributes.

```
class VirtualCounterComponent extends Component {
  void setIdentity(String name) { ... }
  void setAttribute(String attribute, String value){ ... }
```

```
  String getAttribute(String attribute) {..}
  ComponentInfo getParentComponent() { ... }
  ComponentInfo[] getChildren() { ... }
  ServiceInfo[] getParentServices(String name) { ... }
  ServiceInfo[] getAncestorServices(String name) { ... }
  Object execService(ServiceInfo si,
    Message m) throws NoSuchServiceException { ... }
  ....
}
```

3.4.1.2 Proxy Component (PC)

PCs are key elements in the model. According to the types of computing devices, PCs can be classified into two classes, i.e., PCS and PCL. Note that a computing device can have one or more and different PCs.

a. *Proxy component for service provider* (*PCS*): Each PCS is a representation of a computing device that can execute software modules for services. When it receives a software module, it automatically forwards its visiting modules to its target device, which supports the Java virtual machine, e.g., J2ME, J2SE, and Personal Java. If the device supports Java's object serialization mechanism, the device's PCS can forward both the classes and state of the module to the device by using the component migration mechanism. The modules can continue processing at the computing devices that PCSs refer to. A PCS can also extract Java classes from the modules and deploy only the classes at the device that it refers to and support a J2ME-based virtual machine through an HTTP or SMTP-based communication protocol. The devices provide runtime systems for executing components under the protection of Java's security manager. Each PCS allows other components to fetch modules and access the methods of the modules as if they were in them.

b. *Proxy component for legacy device* (*PCL*): Each PCL supports a legacy computing device that cannot execute SCs due to limitations with computational resources. Each is located at a VC corresponding to the space that contains its target device. Each establishes communication with its target device through its favorite approach, e.g., serial communication and infrared signals. For example, a television, which does not have any computing capabilities, can have an SC in the VC corresponding to the physical space that it is contained in and can be controlled in, and the SC can send infrared signals to it.

3.4.1.3 Service Component (SC)

Many computing devices in smart spaces only have a small amount of memory and slower processors. They cannot always support all services. Here, we introduce an approach to dynamically installing upgraded software that is immediately required in computing devices that may be running. SCs are mobile software that can travel from computing device to computing device achieved by using mobile agent technology. The current implementation assumes SCs to be Java programs. It can be dynamically deployed at computing devices. Each SC consists of service methods and is defined as a subclass of abstract class `ServiceComponent`. Most serializable JavaBeans can be used as SCs.

```
class ServiceComponent extends Component {
  void setName(String name)
  Host getCurrentHost() { ... }
  void setComponentProfile(ComponentProfile cpf) { ... }
  ....
}
```

When an SC migrates to another computer, not only the program code but also its state are transferred to the destination. For example, if an SC is included in a VC corresponding to a user, when the user moves to another location, it is migrated with the VC to a VC corresponding to the location. The model allows each SC to specify the minimal (and preferable) capabilities of PCSs that it may visit, e.g., vendor and model class of the device (i.e., PC, PDA, or phone), its screen size, number of colors, CPU, memory, input devices, and secondary storage, in CC/PP (composite capability/preference profiles) form [13]. Each SC can register such capabilities by invoking the `setComponentProfile()` method.

3.4.2 Component Management System

Our model can manage the computing devices that maintain it. This is because a PCM is a proxy for a subtree that its target computing device maintains and is located in the subtree that another computing device maintains. As a result, it can attach the former subtree to the latter. When it receives other components and control messages, it automatically forwards the visiting components or messages to the device that it refers to (and vice versa) by using a component migration mechanism, like PCSs. Therefore, even when the model consists of subtrees that multiple computing devices maintain, it can be treated as a single tree. Note that a computing device can maintain more than one subtree. Since the model does not distinguish between computing devices that maintain subtrees and computing devices that can execute services, the former can be the latter.

Component migration in a component hierarchy is done merely as a transformation of the tree structure of the hierarchy. When a component is moved to another component, a subtree, whose root corresponds to the component and branches correspond to its descendent component is moved to a subtree representing the destination. When a component is transferred over a network, the runtime system stores the state and the code of the component, including the components embedded within it, into a bit stream formed in Java's JAR file format that can support digital signatures for authentication. The system has a built-in mechanism for transmitting the bit stream over the network through an extension of the HTTP protocol. The current system basically uses the Java object serialization package for marshaling components. The package does not support the stack frames of threads being captured. Instead, when a component is serialized, the system propagates certain events within its embedded components to instruct the agent to stop its active threads.

People should only be able to access location-bound services, e.g., printers and lights, that are installed in a space, when they enter it carrying their own terminals or using public terminals located in the space. Therefore, this model introduces a component as a service provider for its inner components. That is, each VC can access its neighboring components, e.g., SCs and PCs located in the parent (or an ancestor) of the VC. For example, when a person is in the room of a building, the VC corresponding to the person can access SCs (or SCs on PCs) in the VC corresponding to the room or the VC corresponding to the building. In contrast, it has no direct access over other components, which do not contain it, for reasons of security. Furthermore, like Unix's file directory, the model enables each VC to specify its owner and group. For example, a component can explicitly permit descendent components that belong to a specified group or are owned by its user to access its services, e.g., PCs or SCs.

3.4.3 Location-Sensor Management System

The model permits us to manually customize the structure of components. Nevertheless, the model can dynamically configure containment relationships among components according to the physical world by using location sensing systems. To bridge location sensors and devices that maintain subtrees, the model introduces location-management systems, called LSMs, outside the component management systems. Each LSM manages location sensors and exchanges information between other LSMs in a peer-to-peer manner. This is a lightweight system because it can be operated on embedded computers initially designed to manage sensors, or computers that can maintain the model or execute application services. Hereafter, we will assume that entities and computing devices have been attached with RFID, infrared, or ultrasonic tags and LSMs manage the tracking sensors for those tags.

3.4.3.1 Monitoring Location Sensors

An LSM monitors more than one RFID reader for such tags. When an LSM detects changes in the position or presence of a tag in the coverage area of the sensor that it manages, there are two possible scenarios: the tag may be attached to an entity, or the tag may be attached to a computing device. The LSM next tries to discover VCs or PCs corresponding to the visiting entity or device as follows:

- *Step 1*: The LSM detects VCs or PCs bound to the entity or device in the subtree that contains the VC bound to the area in a breadth-first-search (BFS), because such new tags often emanate from one of its neighboring spaces or its surrounding space.
- *Step 2*: If the LSM cannot discover any VCs or PCs in the first step, it multicasts a query message with the identifier of the tag to other LSMs and computing devices that are maintaining their subtrees and it receives reply messages from LSMs or devices that know where these the VCs or PCs are located.

When the LSM detects the absence of a tag, the VC bound to the tag is contained in the VC bound to the area. If LSMs manage sensors that can measure geometric locations, they can define one or more virtual spaces within the coverage areas of their sensors and transform the geometric positions of entities within the areas into qualitative information concerning the presence or absence of entities in the spaces.

3.4.3.2 Location-Based Deployment of Components

After an LSM knows where the VC bound to the visiting entity or device is located, it instructs the model that is maintaining the VC bound to the area to configure the location of the VC at suitable components.

- When a new tag is bound to an entity, the VC bound to the entity is deployed at the VC bound to the area that physically contains the entity. The LSM also sends events to the area's VC to notify that the entity has arrived and informs the entity's VC about computing devices that can execute services within the area. If the VC has services and the devices can satisfy their requirements, it deploys the services at the devices.

■ When a new tag is bound to a computing device, the PCS or PCL bound to the device is deployed at the VC bound to the area that physically contains the device. The LSM also sends events to the PCS or PCL and informs the area's VC about the device's capabilities. If the VC has services and the device can satisfy the requirements for these, it migrates them to the PCS.

3.4.4 Current Status

A prototype implementation of this model was built with Sun's J2SE version 1.4. It uses the MobileSpaces mobile agent system to provide mobile components. The system is responsible for maintaining, detecting, executing, and migrating components. The implementation supports three commercial locating systems: Elpas's system (infrared tag sensing system), RF Code's Spider (active RF-tag system), and Alien Technology's 915 or 950 MHz RFID-tag (passive RFID system). Moreover, the current implementation supports a real-time locating system, which locates the geometric positions of WiFi-tags by using the Time Difference of Arrival (TDoA), and maps the positions into symbolic spaces.

Although the current implementation was not built for performance, we measured the cost of migrating a 4-Kb component (zip-compressed) from the source component to the destination component recommended by the LSM, where the source and destination components were maintained by different computers. The latency of component migration to the destination after the LSM had detected the presence of the component's tag was 420 ms and the cost of component migration between two hosts over a TCP connection was 42 ms. This experiment was performed with two LSMs and two computing devices that maintained a component tree, each of which was running on one of four computers (Pentium M-1.4 GHz with Windows XP and JDK 1.4) connected through a Fast Ethernet network. The latency included the costs of the following processes: multicasting the tags' identifiers from the LSM to the source host with UDP, transmitting the component's requirements from the source host to the LSM with TCP, transmitting a candidate destination from the LSM to the source host with TCP, marshaling the component, migrating the component from the source host to the destination host, unmarshaling the component, and verifying security. We believe that this latency is acceptable for a location-aware system used in a room or building. In fact, the migration of components tends to be inside individual sub-models maintained by different computers rather than between sub-models. As a result, we have no problem in its scalability with our early experiments.

3.5 Applications

This section briefly discusses how the model represents and implements typical applications and what advantages the model has.

3.5.1 Follow-Me Applications

Follow-me services are a typical application in smart spaces. For example, Cambridge University's Sentient Computing project [5] enabled applications to provide a location-aware platform using

Step 1

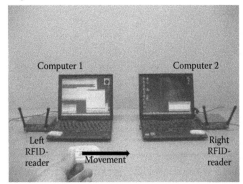

Step 2

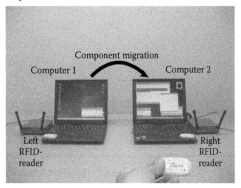

Figure 3.4   Follow-me desktop applications between two computers.

infrared-based or ultrasonic-based locating systems in a building.* While a user is moving around, the platform can track his or her movement so that the graphical user interfaces of the user's applications follow the user. The model presented in this chapter, on the other hand, enables moving users to be naturally represented independently of location-sensing systems. Unlike previous studies on the applications, it can also migrate such applications themselves to computers near the moving users. That is, the model provides each user with more than one VC and can migrate this VC to a VC corresponding to the destination. For example, we developed a mobile window manager, which is a mobile agent and could carry its desktop applications as a whole to another computer and control the size, position, and overlap in the windows of the applications. Using the model presented in this chapter, the window manager could be easily and naturally implemented as a VC bound to the user and desktop applications as SCs. They could be automatically moved to a VC corresponding to the computer that was in the current location of the user by an LCM and could then continue processing at the computer, as outlined in Figure 3.4.

3.5.2 *Location-Based Navigation Systems*

The next example is a user navigation system application running on portable computing devices, e.g., PDAs, tablet-PCs, and notebook PCs. The initial result on the system was presented in a previous paper [8]. There has been a lot of research on commercial systems for similar navigation, e.g., CyberGuide [14] and NEXUS [2]. Most of those have assumed that portable computing devices are equipped with GPSs and are used outdoors. Our system is aimed at use in a building. As a PDA enters rooms, it displays a map on its current position. We have assumed that each room in a building has a coverage of more than one RFID reader managed by an LSM, the room is bound to a VC that has a service module for location-based navigation, and each PDA can execute service modules and is attached to an RFID tag. When a PDA enters a room, the RFID reader for the room detects the presence of the tag and the LSM tries to discover the component bound to the PDA through the procedure presented in the previous section. After it has information about the component, i.e., a PCS bound to a PDA, it informs to the VC corresponding to the room

---

* The project does not report their world model but their systems seem to model the position of people and things through lower-level results from underlying location-sensing systems.

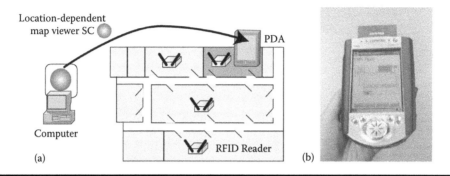

Figure 3.5 **(a) RFID-based location-aware map-viewer service and (b) location-aware map-viewer service running on PDA.**

about the capabilities of the visiting PDA . Next, the VC deploys a copy of its service module at the PCS and then the PCS forwards the module to the PDA to which it refers to display a map of the room. When the PDA leaves from the room, the model issues events to the PCS and VC and instructs the PCS to returns to the VC. Figure 3.5a outlines the architecture for the system. Figure 3.5b shows a service module running on a visiting PDA displaying a map on the PDA's screen.

3.5.3 Software Testing for Location-Based Services

To test software for location-based services running on a portable device, the developer often has to carry the device to locations that a user's device may move to and test whether software can connect to appropriate services provided in the locations. We developed a novel approach to test location-aware software running on portable computing devices [15]. The approach involves a mobile emulator for portable computing devices that can travel between computers, and emulates the physical mobility and reconnection of a device to subnetworks by the logical mobility of the emulator between subnetworks. In this model, such an emulator can be naturally implemented as a PC, which provides application-level software, with the internal execution environments of its target portable computing devices and target software as SCs. The emulator carries the software from a VC that is running on a computer on the source-side subnetwork to another VC that is running on another computer on the destination-side subnetwork. After migrating to the destination VC, it enables its inner SCs to access network resources provided within the destination-side subnetwork. Furthermore, SCs, which were tested successfully in the emulator, can run on target computing devices without modifying or recompiling the SCs. This is because this model provides a unified view of computing devices and software and enables SCs to be executed in both VCs and PCs.

3.6 Conclusion

We presented a world model for context-aware services, e.g., location-aware and personalized information services, in smart spaces. Like existing related models, it can be dynamically organized like a tree based on geographical containment, such as a user-room-floor-building hierarchy and each node in the tree can be constructed as an executable software component. It also has several advantages in that it can be used to model not only stationary but also moving spaces, e.g., cars. It enables context-aware services to be managed without databases and can be managed by multiple computers. It can provide a unified view of the locations of not only physical entities and spaces, including users and objects, but also computing devices and services. We also designed and implemented a prototype system based on the model and demonstrated its effectiveness in several practical applications.

Finally, we would like to identify further issues that need to be resolved. Since the model presented in this chapter is general purpose, in future work, we need to apply it to a variety of services. The prototype implementation presented in this chapter was constructed on Java but the model itself is independent of programming languages. We are therefore interested in developing it with other languages. We plan to design more elegant and flexible APIs for the model by incorporating existing spatial database technologies.

References

1. I. Satoh, A location model for pervasive computing environments, in *Proceedings of IEEE 3rd International Conference on Pervasive Computing and Communications (PerCom'05)* (IEEE Computer Society, Washington, DC, 2005), Kavai, HI pp. 215–224.
2. F. Hohl, U. Kubach, A. Leonhardi, K. Rothermel, and M. Schwehm, Next century challenges: Nexus—An open global infrastructure for spatial-aware applications, in *Proceedings of Conference on Mobile Computing and Networking (MOBICOM'99)* (ACM Press, NewYork 1999), Seattle, WA, pp. 249–255.
3. C. B. M. Bauer and K. Rothermel, Location models from the perspective of context-aware applications and mobile ad hoc networks, *Personal and Ubiquitous Computing* **6**, pp. 322–328, (2002).
4. E. A. T. Kindberg, People, places, things: Web presence for the real world, Technical Report HPL-2000-16, Internet and Mobile Systems Laboratory, HP Laboratories Palo Alto, CA (2000).
5. P. S. A. W. A. Harter, A. Hopper and P. Webster, The anatomy of a context-aware application, in *Proceedings of Conference on Mobile Computing and Networking (MOBICOM'99)* (ACM Press, New York 1999), Seattle, WA, pp. 59–68.
6. M. Beigl, T. Zimmer, C. Decker: A location model for communicating and processing of context, *Personal and Ubiquitous Computing*, 6(5–6), 341–357, (2002).
7. K. Romer, T. Schoch, F. Mattern, and T. Dubendorfer, Smart identification frameworks for ubiquitous computing applications, in *IEEE International Conference on Pervasive Computing and Communications (PerCom'03)* (IEEE Computer Society, Los Alamitos, CA, 2003), Fort Worth, TX, pp. 253–262.
8. I. Satoh, Linking physical worlds to logical worlds with mobile agents, in *Proceedings of International Conference on Mobile Data Management (MDM'2004)* (IEEE Computer Society, 2004), Berkeley, CA.
9. I. Satoh, A world model for smart spaces, in *1st International Symposium on Ubiquitous Intelligence and Smart Worlds (UISW2005)* LNCS, vol. 3823, pp. 31–40 (2005).
10. C. Becker, Context-Aware Computing, *Tutorial Text in IEEE International Conference on Mobile Data Management, (MDM'2004)*, (2004).
11. U. Leonhardt and J. Magee, Towards a general location service for mobile environments, in *Proceedings of IEEE Workshop on Services in Distributed and Networked Environments* (IEEE Computer Society, Macau, China 1996), Los Alamitos, CA, pp. 43–50.
12. I. Satoh, MobileSpaces: A framework for building adaptive distributed applications using a hierarchical mobile agent system, in *Proceedings of IEEE International Conference on Distributed Computing Systems (ICDCS'2000)*, pp. 161–168 (2000).

13. World Wide Web Consortium (W3C), Composite capability/preference profiles (CC/PP), World Wide Web Consortium (W3C) (1999).

14. G.D. Abowd, C. G. Atkeson, J. Hong, S. Long, R. Kooper, and M. Pinkerton, Cyberguide: A mobile context-aware tour guide, *ACM Wireless Networks* **3**, pp. 421–433, (1991).

15. I. Satoh, A testing framework for mobile computing software, *IEEE Transactions on Software Engineering*, **29**(12), 1112–1121, (2003).

Chapter 4

Software Architecture for a Multi-Protocol RFID Reader on Mobile Devices

Joon Goo Lee, Seok Joong Hwang, Seon Wook Kim, Hyung Min Yoon, and Kyung Ho Park

Contents

4.1 Introduction

For a ubiquitous environment, we try to assign a unique identification to each object. With the object identification, we would access its associated information through an information infrastructure such as LAN, WAN, Internet, sensor network, and so on. There are many currently available and acceptable solutions in data delivery and management on the network infrastructure [1–7], but there are still many constraints to capture data from versatile objects. One of currently existing solutions is manual data recording or manual scanning of bar-codes, and more advanced solutions are automatic scanning of bar-codes and sophisticated machine vision/sensing systems [8–10]. All of these solutions, however, are very expensive, time-consuming, and sometimes inaccurate. Even they have more significant constraints in real-work surroundings, for example, due to intensity of illumination and dirty spots on bar-codes. But in this case, we cost expensively even for getting a simple information.

One of the potential and attractive solutions to capture data easily from objects is to use RFID (Radio Frequency IDentification) systems [11]. In RFID systems, we attach a tag onto an object, and an identification code is stored in the memory of the tag. A tag reader or a tag interrogator communicates with the tag through RF technology, reads the identification code from the tag, and accesses its related database through a network infrastructure for getting more information.

Nowadays, RFID systems are widely used in delivery systems to track the location of deliveries [12,13]. In general, an industrial reader system in delivery and manufacturing services is implemented with many big and heavy RFID readers at fixed, that is, predefined, locations, and the system is highly optimized for accuracy and speed performance. Therefore, many technical issues such as reader collisions, frequency interference, and so on would be resolved by fully integrating and optimizing the whole system. So, the delivery application has become one of the better alternatives than traditional identification systems like bar-codes.

The current focus of research and development in RFID systems has shifted gradually toward the adaptation of RFID into the mobile environment. When an RFID reader is offered with mobile devices, every individual is capable of carrying an RFID reader, and it makes RFID readers ubiquitous. The RFID system in a mobile environment, however, has some different scenarios from an industrial reader system. Figure 4.1 shows a typical environment for mobile-based RFID systems.

First, since a reader can be carried anywhere and anytime under a ubiquitous computing/networking environment, it is very difficult to implement a *good* system in terms of speed and accuracy. For example, imagine that many people try to use their RFID readers to access the tag on a movie poster at the same time. In this environment, a reader collision problem becomes severe, such as RF interference between readers [14,15]. Many solutions have been proposed, such as coloring [16] and colorwave [17] methods, but these methods cannot be easily adapted to the real mobile world due to limited performance of passive tags. For example, passive tags cannot recognize more than one channel at a time. The tags only respond at the same carrier frequency that a reader uses. This gets a tag confused when multiple readers try to access it simultaneously.

Second, we should be concerned about the flexibility of an interface with mobile devices and RF modules more than a reader's performance. A manufacturer of an industrial reader develops and offers its own and fully customized system including readers, middlewares, and so on, but such a dedicated system does not provide an interoperability with other manufacturers' systems in general. However, an interoperability and an interface problem are quite important issues for ubiquitous service. When we apply a system for a u-world, manufacturers and service providers need to be distinguished to support new and manifold services, and follow standard specifications

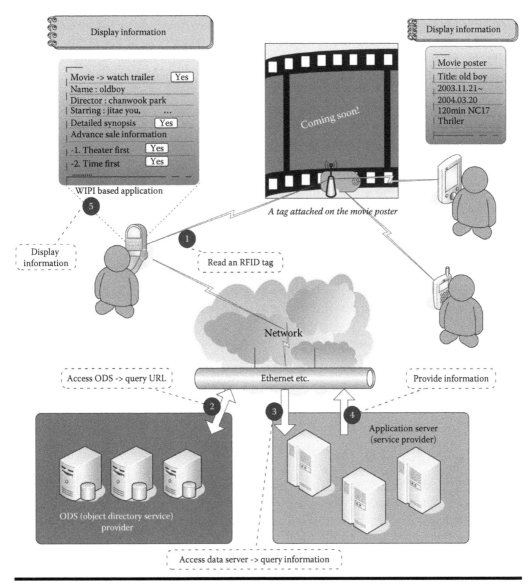

Figure 4.1   A typical example of mobile-based RFID systems.

for easy interface. If a reader is developed with its own interface, an application to use the reader needs to be implemented differently depending on each reader's interface. The application can be made by various service providers, so a common interface is needed in order to reduce extra work. In order to provide an easy interface, our firmware interacts with a Handset Adaptation Layer (HAL) to support WIPI (Wireless Internet Platform for Interoperability) [18]. Any WIPI application can access an RFID reader and query tags' information from the Internet through HAL interfaces without code adaption to hardware.

How about the performance issue? In a delivery system, an RFID reader should identify all tags in their reading zones. In a mobile environment, a user does not want to read all tags around him/her. A user wants to read only a few tags near the user's reader at one time. For an industrial use, efficiency and accuracy are important to optimize the whole system and to reduce the cost. In mobile circumstance, other issues such as interaction between objects and human beings [19,20], security and privacy issues [21–23], and so on, may be more critical than the performance. All these issues depend on the RFID service space [21].

In this chapter, we focus on the second issue and present the detailed software architecture for a multi-protocol RFID reader on mobile devices. There are many research activities in RFID hardware systems, but there are few in software infrastructure, especially on mobile devices. We have designed and implemented most functionalities of an RFID reader in software except for code modulation and demodulation to support multi-protocols and the next coming standards easily. For code modulation and demodulation, we made our own hardware logics on FPGA to process incoming and outgoing data in real time to meet the RFID standard specification [24–27], and RF parts were assembled using commercial chips. The implemented software components for a mobile RFID reader are mainly categorized into two components: baseband modem APIs, and a firmware including anticollision engines, read/write functions, packetizer/unpacketizer of air interface commands/responses for each protocol, interface functions for HAL, and miscellaneous functionalities.

For our architecture verification, we have prototyped our system on the ARM-based Excalibur FPGA [28] with iPAQ PDA. It was also fabricated into one chip with 0.18 μm technology and embedded into a PDA phone.

This chapter is organized as follows: Section 4.2 presents a background of RFID, and Section 4.3 presents an overall architecture of our mobile multi-protocol RFID reader. Section 4.4 describes the hardware parts briefly, and Section 4.5 explains the software architecture in detail. Section 4.6 presents implementation and verification of our prototype system and its chip version with evaluation in Section 4.7. Finally, Section 4.8 concludes and gives some issues to be solved.

4.2 Background of RFID

Including JTC 1/SC 31 and JTC1/SC 17, many working groups of ISO/IEC (International Organization for Standardization/International Electrotechnical Commission) for automatic identification techniques have proposed an RFID architecture and its related several standards for an application, a reader, and a tag, such as ISO/IEC 14443 series, 15693, 15961, 15962, 15963, 18000 series, and so on [27,29–41]. EPCglobal and AutoID Labs, one of famous leading groups for industry-driven technologies, also proposed and published EPC Class-0, Class-1, and Class-1 Generation2 in 860–960 MHz frequency band [24–26].

We can classify an RFID system according to an operating frequency of a reader, a physical coupling method, a range of the system, and a memory capacity and usage.

An RFID system is operated at different frequencies from 135 kHz to 2.4 GHz, and each frequency has its own physical characteristics and usages as follows, and they are summarized in Table 4.1.

- Low frequency (LF) band, under 125 kHz: This band has a short reading range and a slow data rate. The tag at this band is big, but it stands against environment changes well. Therefore, it is usually used for animal identification [38,42,43].
- High frequency (HF) band, 13.56 MHz band: This band is commonly used for contact-less smart cards. It can easily penetrate a textile fabric and a leather, so the tag in this band is used for a transport card, an ID-Card, a security card, and so on [29–34].

Table 4.1 Comparisons of Each Frequency Band for an RFID System

Frequency	LF (135 kHz)	HF (13.56 MHz)	UHF (433 MHz Active)	UHF (860–960 MHz)	Microwave (2.4 GHz)
Standard	18000-2	18000-3	18000-7	18000-6	18000-4
Reading range	Less than 0.5 m	Less than 1 m	Over 50 m	Less than 10 m	Less than 1 m
Purpose	Animal identification antitheft	Smart cards library	Container management sensor network	Logistics supply chains	Logistics antiforgery
Data rate	500 bps	Up to 106 kbps	27.7 kbps	Up to 640 kbps	Up to 384 kbps
Power supplying to tags	Inductive coupling	Inductive/ capacitive coupling	Battery	Backscatter coupling	Backscatter coupling or battery
Tag size	Large	Palm size	Large	Finger size	Coin size

- Ultrahigh frequency (UHF) band, 433 MHz band: This band is for an active air interface communication. The tag used at this band uses a power source such as a battery, so it can send data longer than any other passive-typed tags [41].
- Ultrahigh frequency (UHF) band, between 860 and 960 MHz: This band has a long identification range, a fast data transmission rate, and a small size tag. It has been used widely for logistics and supply chains [24–27,44].
- Microwave band, 2.4 GHz: This band has the smallest tag, but less reading range than the UHF band. For instance, the size of a μ-chip is only 0.4 mm by 0.4 mm, but its reading range is less than 30 cm [40,45].

We can select a physical coupling method among electric, magnetic, and electromagnetic fields; however, many documents about an RFID system tend to use only the terms *inductively* or *capacitively* coupled system, and *microwave* or *backscatter* system for the RFID classification. Actually, most remote-coupled systems are based on an inductive (magnetic) coupling, so many people recognize a coupling system as an inductive coupling system. The backscatter system is used at 860–960 MHz UHF band, and it has a longer range of an identification than any other frequency bands. The range of the system depends on its physical coupling method and an operating frequency due to its physical limitations. In general, the short-range system, up to 1 m, uses an inductively or a capacitively coupling method and a longwave operating frequency. The long-range system used in logistics and industrial automations uses a backscattering method at 860–960 MHz band.

A tag with ROM (Read Only Memory), called a WORM tag (Write Once Read Many), is only readable, and it contains only a tag ID. The memory can be read many times, but cannot be written. This kind of tag is used in cost-sensitive mass applications. If a system wants to write data into a tag, the tag includes EEPROM (Electrical Erasable Programmable Read Only Memory) or RAM (Random Access Memory) memory. A tag with RAM may need a battery backup for refreshment. Currently, the memory capacity is upgrading up to 64 kb.

Each classification affects other characteristics. For example, operating frequency may influence a range of reading/writing distance, and a physical coupling method may limit an operating frequency. In general, a system with a lower frequency is more stable and less affected against surrounding environments. The choice of a suitable RFID system heavily depends on its service models. For security and credit card services, long-range system is useless, but the memory capacity of a tag becomes important. For logistics or supply chains, the reading/writing range of an identification and fast data transmission are the most important. A short-range system may have no advantage than a traditional bar-code system.

In the next section, we discuss about the selection of RFID standards for a mobile RFID system and present an overall architecture of our system.

4.3 Multi-Protocol RFID Reader for Mobile Devices

4.3.1 Consideration of a Mobile RFID System

What standards should be considered for mobile RFID services? As you saw the typical example in Figure 4.1, we need a long-range system and should use a sufficiently small size of tags. The tag attached on an object should contain not only a tag ID, but also additional basic information of an object for which the tag has to embed rewritable memory. From these constrains, a 860–960 MHz UHF RFID system is a good candidate standard for mobile RFID system. Of course, this frequency band has a little disadvantage, that is, weak adaptation against the influence of environment. In spite of this problem, we argue that this band is the best choice for mobile RFID services. Considering all advantages and disadvantages of RF bands, a 860–960 MHz band is sufficient for our purpose, mobile RFID systems, since it has enough reading range of up to 10 m, smaller size tag than LF and HF, and a fast data rate. A tag price is still expensive, but it will be down if mass production is accomplished. In Korea, the Korean Ministry of Information and Communication is promoting mobile RFID services named Mobion (Mobile identification on) using this band actively. For the use of a 860–960 MHz range, several protocols have been used in commercial purposes. Basically, they have similar implementation. They use electromagnetic backscatter coupling for supplying power, modulated backscatter method for tag replies, and half duplex procedure for communication.

Table 4.2 shows forward and return link parameters and anticollision algorithms of each RFID protocol in a 860–960 MHz band. It shows that ASK modulation is common to all the protocols, since it is easy to implement ASK modulation in a passive tag that has cost and power concerns; however, there are many versatile baseband coding schemes [11]. Binary 1 and 0 can be represented in various line coding methods such as NRZ, Unipolar RZ, Manchester, Miller, FM0, and pulse time modulation (PTM) such as pulse width modulation (PWM) and pulse position modulation (PPM). RFID systems use one of these baseband coding schemes with consideration of easy bit synchronization, probability of error, bandwidth, and receiver complexity. The details about coding schemes can be found in [24–27].

In order to distinguish each tag from a tag pool, a party of tags in the reading zone, an algorithm is required. A reader without the algorithm cannot differentiate tag responses. If more than two tags respond at the same time, tag responses are mixed in the air, so a reader receives nonrecognizable signals. We call this situation a collision. To avoid the collision, each protocol offers an anticollision method. The anticollision algorithm of each RFID protocol is based on time division multiple access (TDMA) [11], since RFID systems are constrained by low computational ability, low power consumption, and little amount of memory. These constraints allow a simple anticollision

Table 4.2   Link Parameters and Anticollision Algorithms in Standard RFID Protocols

	Forward Link		Return Link		
Protocol	Modulation	Baseband Coding	Modulation	Baseband Coding	Anticollision
ISO 18000-6 Type A	ASK	PIE	ASK	FM0	Dynamic framed slotted ALOHA
ISO 18000-6 Type B		Manchester		FM0	Binary tree
EPC Class-0		PWM		Subcarrier FSK	Binary tree (bit-by-bit)
EPC Class-1		PWM		BitCell encoding	Binary tree (bin slot)
EPC Class-1 Generation2	ASK or PR-ASK	PIE	ASK or PSK	FM0 or miller	Dynamic framed slotted ALOHA

Source: Auto-ID Center, 860 MHz–960 MHz Class I radio frequency identification tag radio frequency & logical communication interface specification recommended standard, version 1.0.0, Technical Report MIT-AUTOID-TROOT, November 2002; Class 1 Generation 2 UHF Air Interface Protocol Standard version 1.0.9: "Gen 2," EPC global standard documents, http://www.epcglobalinc.org/standards; Auto-ID center, 860 MHz–930 MHz class 0 radio frequency identification tag protocol specification candidate recommendation, version 1.0.0, Technical Report MIT-AUTOID-TR016, June 2003; Information Technology-Radio frequency identification for item management-part 6: Parameters for air interface communications at 860 MHz to 960 MHz, ISO/IEC 18000-6 standard document, http://www.iso.org.

algorithm sufficient for our purpose. There are two major methods in implementation. One is ALOHA and the other is Binary Tree schemes. The original ALOHA is the simplest probabilistic method, but it is inefficient. For this reason, RFID systems use dynamic framed slotted ALOHA for efficiency. The Binary Tree scheme, a deterministic method of anticollision, has many varieties such as Bit-by-Bit in EPC Class-0 and Bin slot in EPC Class-1 for fast identification rate.

For identifying tags simultaneously, we need to consider a state transition of a tag, and the relationship of each command with its corresponding response. All protocols use a half duplex procedure, and a reader talks first. It implies that we can guess a tag state from a corresponding reply of the tag. If a reader knows how it can reduce the number of responses, it can finally recognize the tag response by reducing participant tags in reply gradually. Finding good algorithm for each protocol is one of the challenges for system efficiency.

4.3.2 Overall Architecture

In our target frequency range, there are several different kinds of air interface protocols. ISO/IEC published ISO 18000-6 Type A and Type B. EPCglobal has used its own protocols such as EPC Class-0, Class-1, and Class-1 Generation2. EPC Class-1 Generation2 became an ISO standard in 2006, and ISO/IEC named it ISO 18000-6 Type C. Since there are at least three standard protocols, we should develop an RFID reader to support several protocols.

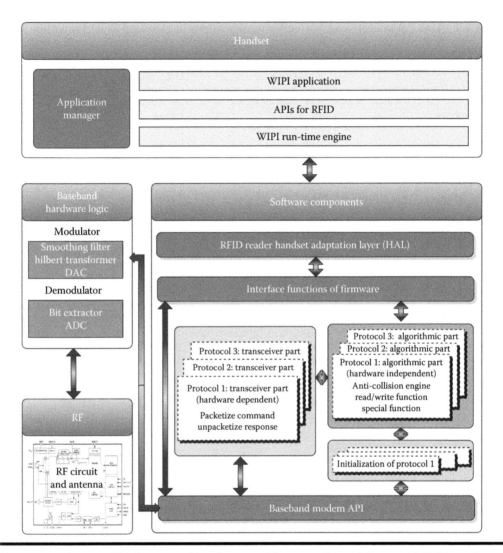

Figure 4.2   An overall software and hardware architecture of a multi-protocol RFID reader on mobile devices.

We investigated all available protocols, and then found out that each protocol uses a different encoding scheme and an anticollision algorithm, but it uses a similar or the same modulation/demodulation scheme, that is, ASK. Since we choose an ASK scheme, we are able to implement modulation and demodulation in hardware logics, and other components in software for efficiency and flexibility.

Figure 4.2 shows an overall software and hardware architecture of a multi-protocol RFID reader for our prototype system. The hardware component includes RF circuits and baseband hardware logics, and the software component does RFID software components and a handset.

The RF circuits consist of up/down mixers, PLL (Phase Locked Loop), PA (Power Amplifier), filters, and so on. The RF parts were assembled using commercial chips, and hardware components for controlling RF and modulation/demodulation were implemented on FPGA. The major modem hardware logic consists of a modulator and a demodulator. A modem controls a digital analog converter (DAC) for transmission, an analog digital converter (ADC) for reception, and RF circuits for analog signals. While the modem is modulating, the transmitted data on a forward link are converted to rectangular pulses, and then filtered by a smoothing filter. In a double side band (DSB) transmission, a pulse-shaped signal is transmitted to I and Q channels, which is tied to zero. In a single side band (SSB) transmission, I and Q channel signals are generated by a Hilbert transformer from the pulse-shaped signal. The signal is changed to an analog in DAC, and mixed with a carrier frequency in an up mixer, and then finally transmitted to tags. The carrier frequency is generated in PLL. In general, a reader may need PA to amplify the final signals for enough powering to tags optionally. While the modem is demodulating, the received analog signals from an antenna go through a down mixer. The down mixer gets rid of carrier frequency and extracts baseband signals. These signals are quantized in ADC, and then passed digital signals go to a bit extractor. These extracted bit symbols are decoded to logical bits, and then used in a firmware. Usually, a passive-typed tag response is very small and weak, so we can amplify the response signals by using an LNA (Low Noise Amplifier) optionally.

The RFID software components consist of baseband modem APIs, a firmware, HAL interfaces, and WIPI-based applications. The baseband modem APIs provide interfaces to access baseband modem hardware logics and RF circuits for controlling modem hardware. APIs should be carefully designed for software programmers.

The firmware includes anticollision algorithms, read/write functions, reader collision algorithms, interfaces with handset adaptation layer (HAL), and so on. For a multi-protocol reader, the firmware should be organized well. If not, maintenance will be irksome work. Since our selected protocols have common/distinguished parts and process similar sequences, we divide each protocol into two parts. One is hardware-independent parts, and the other is a hardware-dependent parts. The former deals with logical parts with air interface commands and responses, and the latter processes packetizing commands and unpacketizing responses. The HAL is an abstract layer to provide hardware independence to an application [18]. Any WIPI-based application and any application that follows HAL can access and control an RFID reader through HAL interfaces in a uniform manner.

The hardware architecture will be described more in Section 4.4, and the software architecture will be discussed in detail in Section 4.5.

4.4 Hardware Considerations

In an RFID system, especially in a passive-typed RFID system, we should develop a delicate receiver. The signal of a tag is very weak, but detectable if there is no noise. The tag is powered by a reader using a backscatter coupling, and the maximum power of the transmission signal is less than 0.5–4 W erp (Effective Radiated Power) in general [46]. The power of backscattered signal at the reader's receiver side is extremely less. The transmitter of a backscatter reader supplies a carrier frequency permanently in order to activate a tag, and induces a significant amount of additional noise. The greater part of this noise is a phase noise, and it comes from an oscillator in the transmitter. The noise dramatically reduces the sensitivity of a receiver in a reader. For a tag detection, as a rule of thumb in practice, the tag's signal must lie no more than 100 dB below the level of the transmitter's carrier frequency [47]. If we consider more conditions such as a modulation index and a power splitting effect, the receiver will be downgraded more. We need to design the receiver parts with these harsh conditions. A matched filter method has been widely

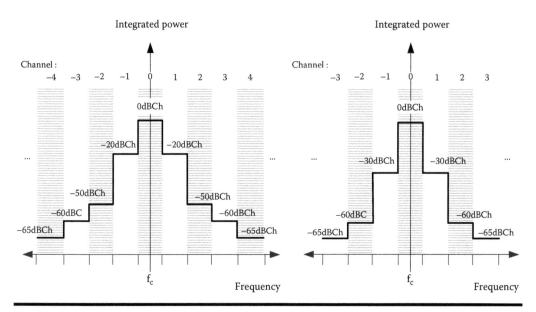

Figure 4.3   An example of a transmit mask. This mask is different with each regional regulation.

used for the receiver of communication devices; however, it has poor performance when DC offset fluctuates. This situation often appears in RFID systems since a tag performance is unstable. Our research team made a solution for this problem using a transition triggering method [48].

In contrast, a transmitter implementation is easier than a receiver. Making digital signal for each command is not complex. In a transmitter, the most important part is keeping the radio regulation of each region. In order to do that, the spectrum of a transmitter needs to satisfy the strict transmission mask in Figure 4.3. Due to this reason, the transmitter should include filters. A smoothing filter is required for DSB and SSB modulations, and an additional transformer is also needed for SSB modulation. These filters can be implemented by various methods: RF circuits, hardware FIR, or a software manner. Since each country has different radio regulations, we need flexible filters. For this reason, we made hardware reconfigurable FIR filters to allow us to configure a number of taps and their filter coefficients freely. We can also make them by using a software manner for flexibility, but a software-coded filter has lots of multiplications and adds due to convolution calculations. The software method needs more powerful processor to meet time constraint specified in the standard documents, which is not acceptable for mobile devices due to power consumption.

4.4.1 Time Constraints

In a procedure of communication between a reader and tags, four kinds of time constraints should be considered. Figure 4.4 shows them graphically, and Table 4.3 summarizes them in ISO 18000-6 Type A, B and C. In these constrains, we should consider about T_1 and T_2 time for the least performance.

T_1 and T_2 are related to reader performance, and T_1 is involved in tag performance. If the reader cannot prepare for the reception of a tag response before the minimum time of T_1, then the reader misses a part of the response. In this case, even though there is a response from a tag, the reader cannot catch the whole data. Therefore, the reader would fail to identify the tag. When a reader's transition time between an end of transmission and a start of reception takes longer than T_1,

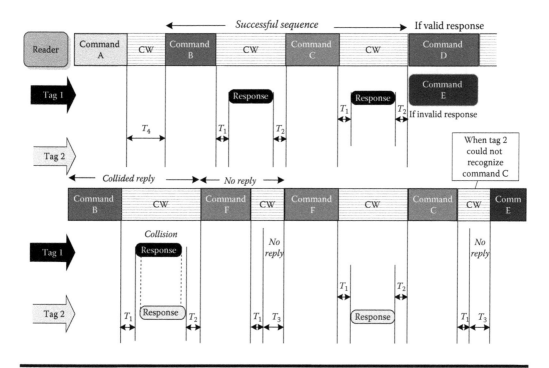

Figure 4.4 An example of a link timing and its time constraints. (Class 1 Generation 2 UHF Air Interface Protocol Standard version 1.09: "Gen 2", EPC global standard documents, http://www.epcglobalinc.org/standards; Information Technology-Radio frequency identification for item management-part 6: Parameters for air interface communications at 860 MHz to 960 MHz, ISO/IEC 18000-6 standard document, http://www.iso.org.)

this hazard appears. Let us assume the following scenario: There is no constraint like T_1. A reader sends a command to two tags and the tags are responding, but one takes a short time and the other a very long time due to tags' performances. Fortunately, the reader could recognize the first reception. After the first recognition of a tag, the reader sends another command, but at that time, the other tag's response arrives. We cannot resolve this situation easily. In order to avoid this situation, T_1 time is needed, and tag manufacturers should keep this T_1 time.

T_2 is a time constraint for avoiding an interruption of other readers. A tag can know whether the next coming command is the continuing one or not with T_2. A tag runs a timer from an end of transmission of the response. If the next command comes within T_2, the tag considers it as a continuous one that follows the prior communication procedure. If T_2 expires, the tag regards the incoming command as a new one from another reader or the start of new communication. A reader manufacturer should follow T_2 time. In order to do it, the reader needs to process the tag's reply, conform the next command that should be issued, and prepare packet for it within the maximum time of T_2 at least. The reader needs to be optimized for this time constraint.

T_3 is the time that a reader waits for the decision whether there is a reply or not. The reader should continue to listen to the received signal for this time. If it could not detect any tag response within T_3, the reader decides this situation as "No Reply." T_4 is another time interval for reader commands. When a reader sends a command that has no response, the reader should issue the next command after this time.

Table 4.3   Examples of Link Timing Parameters

Time Constraint	ISO 18000-6 Type A	ISO 18000-6 Type B	ISO 18000-6 Type C	Description
T_1	150 ~ 1150 μs The reader transmit/ receive setting time shall not exceed 85 μs	85 ~ 460 μs $16\,T_{return\ data\ rate}$ ~ $0.75\,T_{forward\ data\ rate}$	MAX(RTcal,10Tpri) × $(1 - 4FT) - 2$ μs ~MAX(RTcal,10Tpri) × $(1 + FT) + 2$ μs $RTcal = 0_{length} +$ 1_{length} $Tpri = 1/LF$ FT is frequency tolerance	Time from reader transmission to tag response, measured at the tag's antenna terminals
T_2	50 μs	400 μs	3.0Tpri ~ 20.0Tpri $Tpri = 1/LF$	Time required if a tag is to demodulate the reader signal, measured from the last falling edge of the last bit of the tag response to the first falling edge of the reader transmission
T_3	≥300 μs	≥400 μs	$\geq T_1$	Time a reader waits for the decision of No Reply
T_4	Not specified	Not specified	2.0RTcal $RTcal =$ $0_{length} + 1_{length}$	Minimum time between reader commands

Source: Class 1 Generation 2 UHF Air Interface Protocol Standard version 1.09: "Gen 2," EPC global standard documents, http://www.epcglobalinc.org/standards; Information Technology-Radio frequency identification for item management-part 6: Parameters for air interface communications at 860 MHz to 960 MHz, ISO/IEC 18000-6 standard document, http://www.iso.org.

4.4.2 Necessary Parts

In order to construct an RFID reader, we need to have some knowledge about various hardware parts for RF. The following show each part and its function in RFID systems briefly.

PLL: A phase locked loop (PLL) generates a fixed frequency for a carrier frequency. The generated carrier frequency is mixed with a baseband signal, which is a command signal with a smoothed edge from filters, in a mixer. It is also used for hopping frequency channel for FHSS (Frequency Hopping Spread Spectrum). PLL consists of a phase detector, a loop filter, a voltage-controlled oscillator, and so on. LF and HF RFID systems can use an accurate clock source such as a X-tal, but it is almost impossible to be used in UHF or microwave bands because of the accuracy, so they use a complex circuit like PLL to use an accurate UHF. An accurate clock generator is very important to operate an RFID reader and tags.

DAC: A digital analog converter (DAC) converts a digital signal into an analog signal. Digital circuits use a digital signal, that is, a baseband signal for managing bit streams, that is, square waves. However, we need an analog signal for RF communication. In order to make an analog signal from a digital signal, DAC is needed.

Mixer: For up conversion, a mixer mixes a carrier frequency generated at PLL and an analog-converted baseband signal for transmission in an RFID system. An up-converted signal preserves data as an envelope, and is sent to tags with a carrier frequency. For down conversion, the mixer gets rid of the carrier frequency from a backscattered signal, and extracts a baseband signal containing data.

PA: A power amplifier (PA) is needed optionally when a signal of reader's transmitter has not enough power to operate tags or when we want to control a transmission power. PA may compensate for loss of power in a mixer, an antenna, and so on.

Antenna: RF systems regard it as a component or a structure for the radiation or the reception of electromagnetic waves. An antenna is the end terminal for communicating between a reader and tags. There are manifold antennas, and each antenna has been optimized for certain frequency ranges. The behavior of an antenna can be predicted mathematically.

LNA: For the reception of a tag response, a low noise amplifier (LNA) may be needed. The signal of the tag response is significantly weak, and contains lots of noise components. If a receiver uses a general amplifier, it will amplify noise components together. Due to this reason, a dedicated receiver uses LNA, which amplifies a signal while it minimizes noises.

ADC: An extracted baseband signal from a mixer goes through ADC for quantization. The quantized signal can be used in digital circuits. A baseband demodulator extracts bit symbols from the signal.

4.4.3 Commercial IC and IP Selection

Selecting each part to be used in RFID systems is quite important for the performance of the systems. We need to select suitable commercial chips or IPs (Intellectual Property) for constructing an RFID reader. The parts supporting specific characteristics of UHF band are required, and each part should be chosen with consideration of low power consumption, a small size, and a low cost for a mobile RFID reader. We also have to check operating voltages and frequencies. If one hardware component uses a different operating voltage or frequency from others, we may need to add an additional power regulator or a clock generator only for this part. In the prototype RFID, we used one clock source of 25 MHz for operating clock frequency in RF circuits and digital parts including a general processor (we used ARM9), so we could narrow our target parts.

For hardware implementation, we can use commercial chips for a fast prototype development and IPs for SoC-based implementation. If we use a commercially verified one-chip RFIC (Radio Frequency Integrated Circuits) such as FSK/ASK transceiver IC for ISM band from Analog Device Inc. [49] or single-chip narrow band RF transceiver from Texas Instruments Inc. [50], we do not need to worry about implementation of most RF parts such as PLL, mixer, LNA, PA, DAC, and ADC. These chips already include all or parts of them, which allows developers to control RF parts easily.

If you want to integrate all your work into SoC (System-on-Chip), you need to get verified IPs with your fabrication environment. Since IPs could be verified with different technologies, confirming each technology is also important.

4.5 Software Architecture

Figure 4.5 shows an overall software architecture for a mobile-based multi-protocol RFID reader. We have implemented all RFID reader functions in software except for modulation and demodulation. It makes easy to support multiple protocols and to develop new standards in the near future. Our software architecture consists of four layers: baseband modem APIs, a firmware, a handset adaption layer, and a WIPI-based application [18,51,52].

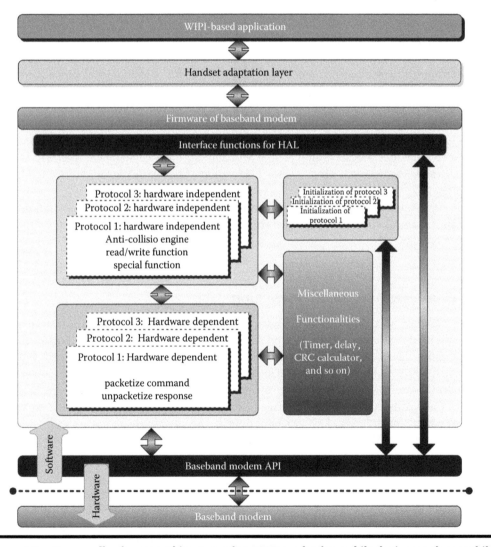

Figure 4.5 **An overall software architecture of an RFID reader for mobile devices such as mobile phones and PDAs.**

For easy understanding of our software architecture in a mobile RFID service, we assume a service scenario as in Figure 4.1 and then follow the whole procedure of our software. The scenario and the sequence will be as follows:

1. There is a movie poster on a billboard. The movie poster has a tag that contains a unique ID and basic information of the movie. One man is interested in it, and he stops to look at it for more information. He has a cell phone that embeds a mobile RFID reader, so he starts scanning the tag by pushing a button to enable an RFID service.
   a. The pushed button runs a WIPI-based application for RFID services.
   b. The application sends a HAL command for activating an embedded RFID reader.
   c. The HAL command decoder, one of interface functions for HAL, decodes the command and invokes the proper function in the firmware to control hardware.
   d. The baseband modem API function controls a baseband modem and returns the processed result.
   e. The HAL response encoder, one of interface functions for HAL, encodes a response from the result and sends it back to the application.
   f. The application knows the reader's condition from the response. If the reader is ready, the application sends the command for reading tags' ID through HAL.
   g. The HAL command decoder decodes the command and invokes a function for reading tags' ID, and then the firmware processes it. The firmware selects an appropriate air interface command, packetizes the command, and transmits it to the baseband modem using the baseband modem APIs.
   h. The baseband modem controls RF circuits and itself, and issues the command to tags. If a tag responds, the modem processes the air interface response and sends extracted bit streams to the firmware.
   i. The firmware extracts the tag ID from the bit streams, and the HAL response encoder encodes valuable data and replies to the application.
   j. The application may send another command for more information of the tag. In this case, the procedure for getting additional data is very similar to Steps f–i.
2. Through a tag's ID, he can get the basic information related to the movie such as the title, showtimes, names of the director, actors and actresses, and so on. If he is interested in the movie, he would get more information by using another RFID service.
   a. He activates an Internet-based service by pushing a button.
   b. The application accesses an ODS (Object Directory Server) and queries an URL of the movie information with the tag ID. The ODS gives the URL to the phone.
   c. The application accesses a data server with the URL, and queries for information of the movie. The data server provides information such as the movie trailer, the detailed synopsis, information about the director, actors and actresses, the nearest cinema, a ticketing information, and so on.
   d. The application displays those information and waits for a user's next actions.
3. He reads the information of the movie, watches the movie trailer, and finally purchases the movie ticket in advance on the Internet using his phone.

This kind of a mobile service is instinctive and already available using a current Internet infrastructure. The only two things we need to do are attach tags to each useful object and link each tag's ID with the associated URLs.

How about the privacy and security problem? Basically a tag has an ID and little amount of information. These are harmless because they do not have any personally sensitive information. For getting more data about some objects, we need to use a network resource. Typically, we need to be authenticated to access a network infrastructure. If we use a mobile RFID reader, a cell phone or a PDA helps authentication for data accessing. We are using Internet services on mobile devices with 2D bar-code [53], but in order to use the bar-code system, we need to run a camera program, focus on a bar-code, take a picture of it, and finally access the information. This sequence is irksome. The RFID technologies just help gathering information about objects and services more easily and instinctively without severe privacy or security problems.

In our architecture, each software layer works for its own task. A software of an upper layer sends a command to the next level software, which processes the command and responds a result as a reply. Any WIPI-based application can be programmed using a HAL interface. There is no necessity to concern about an RFID reader on a handset application developer's side, and similarly there is no problem to use any application if an RFID reader supports a HAL interface. This layered architecture has great flexibility and interoperability.

Details about each software component are described in the next sections from the bottom to the top.

4.5.1 Baseband Modem API

The software component for an RFID reader should allow users to control the hardware logics such as a modulator, a demodulator, and an RF circuit, and it is provided in forms of baseband modem APIs. They provide interfaces to access baseband modem hardware logics and RF circuits for controlling a modem hardware.

As we discussed in Section 4.4, an interface between hardware and software components are very important. An RFID reader can include various RF parts, and a different RF circuit may have different timing characteristics. When a hardware part is changed, a code segment or a hardware logic to interact with the hardware part needs to be modified also. Therefore, we need to develop well-predefined interfaces for easy maintenance of the system. In a software's view, many functionalities are needed to be implemented for hardware control, such as a center frequency, a transmission power gain, a receiver gain, a forward/return link frequency, selection of a modulation scheme, a modulation index, and so on. For low power consumption, we may need functions to turn on/off each part also. The software does not care about how we can activate PLL, and how we can provide data to DAC or get data from ADC. These can be hidden to the software. Full details are unnecessary. Even if we use a united IC, we may distinguish and divide each control part for easy maintenance. Table 4.4 shows some examples of our APIs. All APIs are used inside the firmware, and each interface function accesses its related logic or an RF part and modifies the configuration of each part.

Additionally, the baseband modem APIs offer functions for transmitting/receiving bit streams from air interface commands and tag responses. There are various baseband coding schemes as shown in Table 4.2. If hardware does not support a specific encode/decode function, a software engineer may be able to implement the function in this layer or in transceiver part of each protocol by programming. Since a software only deals with bit streams, providing APIs for transmitting/receiving bit streams is important for easy processing of reader commands and tag responses. Each interface function was implemented by using a memory mapped method. The baseband modem defined the memory map for each part, and we configured the memory values for each purpose of function.

Table 4.4   Implemented Baseband Modem APIs

Function Name	Description
BBM_init	Initialize a baseband modem. Configure a modulator, a demodulator, and RF parts as default setting
BBM_PLL_init	Initialize PLL
BBM_PLL_enable/disable	Enable/disable PLL for controlling of power consumption
BBM_PLL_set	Set a center frequency for a carrier frequency
BBM_TX_ask_modulation_rate	Set ASK modulation rate (index)
BBM_TX_mode	Set a transmission mode to OOK, ASK, or PR-ASK
BBM_TX_gain	Set a transmitter gain
BBM_RX_gain	Set a receiver gain
BBM_TX_link_frequency	Set a forward link frequency. It may set a sampling frequency of DAC and a timer, etc.
BBM_RX_link_frequency	Set a return link frequency. It may set a sampling frequency of ADC, etc.
BBM_TX_transmit_bitstream	Transmit baseband-coded bit streams to a transmitter. In our case, bit streams are sent to FIFO
BBM_RX_receive_bitstream	Receive ADC signal from a receiver and make NRZ bit streams
BBM_RX_RSSI	Listen to an incoming signal to know whether a channel is used or not
BBM_TX_filter_init	Initialize FIR filters as default number of taps and coefficients. In order to set filter coefficients, we used macro functions
BBM_TX_flush	Flush FIFO for a modulator
BBM_RX_flush	Flush FIFO for a demodulator
BBM_PA_turn_on/off	Turn on/off a power amplifier
BBM_RF_turn_on/off	Turn on/off all RF parts
BBM_ADC_turn_on/off	Turn on/off ADC
BBM_DAC_turn_on/off	Turn on/off DAC
BBM_TX_turn_on/off	Turn on/off transmitter parts such as PA, DAC, etc.

4.5.2 Firmware

The firmware includes four parts: an algorithmic part, a transceiver part, interface functions for HAL, and miscellaneous functionalities.

We separated an algorithmic and a transceiver part for tag control due to easy programming and debugging. The algorithmic part is hardware independent, and it includes an anticollision engine, read/write and special function support. The algorithmic part deals with air interface commands and responses logically, and uses functions provided by a relevant transceiver part. When it wants to operate the hardware units, the algorithmic part calls only functions from the transceiver part and an initialization part. The transceiver part includes packetizing/unpacketizing functionalities for each protocol. It uses baseband modem APIs freely to encode a command and to transmit bit streams to a modem. It also receives a response from the modem by using baseband modem APIs, decodes the response, and decides whether the response is successful or not. The receiver part notifies one of four return types, *collide response, no reply, response with error*, and *successful response*, to the algorithmic part, so the algorithmic part can process its procedure of collision arbitrations and other tasks easily.

There are two kinds of anticollisions: a tag collision and a reader collision.

1. A tag collision problem is solved by an anticollision algorithm, as shown in Table 4.2. We considered air interface commands/responses and a tag's state transitions for a tag collision arbitration. Each protocol has its own air interface commands/responses, states, and a sequence for an inventory. For example, ISO 18000-6 Type A reader can arbitrate a collision using `Init_round_all`, `Next_slot` mandatory commands, their corresponding responses, and three states (*Ready, Round_active, Quiet*) for the inventory. Type B reader uses `GROUP_SELECT/UNSELECT`, `FAIL`, `SUCCESS` commands, relevant responses, and two states of *READY, ID* for the tag inventory. Type C (EPC Class-1 Generation2) reader inventories tag IDs use `Select`, `Query`, `QueryRep`, `QueryAdjust`, `ACK`, `NAK` commands, their related replies, and three states of *Arbitrate, Reply, Acknowledged*.

   A general sequence of inventorying is as follows: (1) A reader sends a command for selecting a particular tag population. Typically, all tags in the reading zone of the reader become participants, and send their response to the reader; (2) If there is a collision, the reader sends a command that can reduce a tag population such as `Next_slot` of Type A, `FAIL` of Type B, and `Query` series of Type C; and (3) The reader repeats (2) until there is no collision. In real implementation, the sequence of inventory is a little more complicated. Each protocol provides a detailed sequence for inventory and also offers a state transition table in general. We can make these procedures for an anticollision from standard documents [25,27].

2. We also meditated a reader collision problem. When multiple readers try to access one tag simultaneously, the tag may not answer because the tag accepts multiple channels. The tag may consider some or all of the received signals as a noise. This implies that only one reader should access the tag at one time. In this situation, a coloring method using different channels may not work anymore [16,17]. Another reader collision occurs when a tag is replying to a correct reader and another reader attempts to send a command to the tag. In this case, the correct reader receives a tag response and an unexpected command together. If readers use different channels from each other, the correct reader might extract the desiring tag signal from mixed signals.

   For these reader collision problems, we apply LBT (Listen Before Talk) and a frequency hopping to minimize them. LBT is a method to avoid channel contention while communicating. A reader listens to a channel signal before it occupies the channel. If there is any signal, it tries again after random back-off time. If the channel is empty, the reader occupies it and sends an air interface command. Frequency hopping can help avoid channel collision. A reader can hold a channel for only limited dwell time. Before the dwell time expires, the reader should hop to other channels and continue their communication. Using these methods, we can prevent permanent occupancy and inefficiency. We can apply two methods together. In this case, a reader listens before communication, and then if a channel is safe for communication, it starts to transmit a command and activates a timer to keep a dwell time. Before the dwell time expires, it listens to other channels and hops to a void channel.

Our firmware supports read/write functions and special commands. If a tag has a user-writable memory, we need to provide read/write functionalities to access the memory. In this case, the firmware processes read/write commands with different parameters and methods for each supported protocol in one interface function. The AutoID center, one of the famous RFID laboratories, also proposed special commands, such as `Kill` command for protection of user privacy. EPC tags include this function essentially. If a tag is killed by this command with a password, the tag is deactivated permanently. This feature allows us to keep our privacy by stopping the tag's function before others get it.

The firmware also provides HAL interfaces. These interfaces are necessary parts of whole HAL interface functions in the WIPI platform. You can find detailed information about HAL in Section 4.5.3. The firmware processes predefined commands with HAL. These commands are of two types: a tag control and a reader control. The tag control commands include functionalities to read a tag's ID and read/write data from/to a user memory of tags. The reader control commands handle and examine the hardware including an RF circuit and a baseband hardware logic. The firmware packetizes and unpacketizes air interface commands/responses for each protocol. While an RFID reader is running, the firmware is waiting for a message from HAL. The received messages are decoded, and the firmware enables a proper function with the extracted parameters. We categorized HAL interface functions into tag-associated functions, RF-associated functions, reader-associated functions, and reader environment–associated functions. Each category and its example functions are shown in Table 4.5.

Finally, the firmware gives miscellaneous functionalities. There are many common functions to all protocols, such as CRC calculation of commands and responses, timer and delay, and so on. The functions are necessary to algorithmic and/or transceiver parts for most protocols, and they are in one functional block for efficiency and prevention of waste.

4.5.3 HAL

An RFID modem offers a firmware to applications through HAL, and HAL can access modem logics and functions with this firmware. In WIPI, HAL is an abstract layer for hardware independence to be transplanted to any platform easily [18]. The WIPI-based applications access hardware resources indirectly by calling HAL API functions. In this manner, any platform can be constructed independently regardless of hardware implementation. The communication between HAL and the firmware is asynchronous. At first, HAL sends a command, and a firmware accepts and executes this command. After that, the firmware sends a message back to HAL, but in this manner, a program may wait forever until the response comes while the channel is disconnected. Since this problem can occur, an asynchronous method is implemented in forms of Listeners in WIPI JAVA API and registered Callback functions in WIPI C API.

HAL is elaborately designed to offer a common interface for every RFID reader and application. Therefore, HAL basically defines intrinsic and primitive functions for a mobile RFID reader as mandatory commands, while the rest of them are defined as optional commands. HAL commands are classified into five types: a reader control, a tag access, a buffer control, a filter control, and a

Table 4.5   Examples of HAL Interface Functions in Each Category

Category	Function	Description
Tag-associated functions	readOIDs	Read tags' identification codes in a reading zone
	writeOID	Write OID into a tag memory
	readUserData	Read data from user data memory
	killTag	Let a tag deactivate
RF-associated functions	set/getRFStrength	Set/get RF signal strength
	set/getRegion	Set/get a region and its regulation. It may set a carrier frequency and configure filters
Reader-associated functions	set/getReadCycles	Set/get a reader's read cycles (try times)
	open/closeReader	Open/close a reader's communication channel
	resetReader	Reset a reader as a default state
Reader environment–associated functions	busyReader	Check whether a reader is busy or not
	getManufacturer	Get a manufacturer's name and number
	getModel	Get a model name and a number

reader status report command. Brief descriptions of each command type are Table 4.6 as follows and explains each command:

- Reader control commands are used for setting a reader and getting its information. In this type, mandatory commands provide basic functionalities similar to the available one in other wireless communication devices. On the other hand, optional commands offer some special functions related to the RFID reader performance.
- Tag access commands are for a tag inventory and a tag memory access. Mandatory commands provide these functionalities except for tag memory writing ones. Memory writing is available in optional commands, because most of RFID standards define tag memory writing–related functions as optional ones.
- We can have inventory tags and access them using tag control commands. Since an asynchronous method is used between a command issue and its final result, some buffers should be provided to store the results. Due to the fact, HAL defines a specific buffer structure to store tag information as shown in Table 4.7. Finally, it is managed by buffer control commands in a FIFO manner.
- Filter control commands provide filtering functions to omit unwanted tag information from the buffer.
- A reader status report command is used for getting information of a reader status.

Table 4.6   Description of HAL Commands

Type	Command	Description
Reader control	Mandatory commands	
	MH_rfidPowerOn	Power on an RFID reader.
	MH_rfidPowerOff	Power off the reader.
	MH_rfidOpenReader	Arrange a logical channel to communicate with a reader, and make the reader ready to get tag control commands
	MH_rfidCloseReader	Close a logical channel and terminate all the performing operation
	MH_rfidResetReader	Clear all the registers in the reader, and reconstruct a logical channel
	MH_rfidisOpenReader	Check whether a logical channel is opened or not
	MH_rfidisBusyReader	Check whether a reader is operating for tag control commands or not
	MH_rfidSetRegion	Set a carrier frequency range for each region to obey its regulation
	MH_rfidGetRegion	Get current region settings
	MH_rfidSetRFStrength	Set RF emission power level
	MH_rfidGetRFStrength	Get current RF emission power level
	MH_rfidGetManufacturer	Get a reader manufacturer
	MH_rfidGetModel	Get a reader model
	Optional commands	
	MH_rfidSetReadCycle	Set the number of read operations per tag control command
	MH_rfidGetReadCycle	Get current read cycle setting
	MH_rfidSetReadDelayTime	Set a delay time between successive read operations
	MH_rfidGetReadDelayTime	Get current read delay time setting
Tag control	Mandatory commands	
	MH_rfidReadTagOIDs	Read all the surrounding tags, and then store their OIDs in a buffer (as shown in Table 4.7)
	MH_rfidReadTagTID	Read the surrounding Type C tag, and then store its TIDs in a buffer
		(continued)

Table 4.6   (continued) Description of HAL Commands

Type	Command	Description
	MH_rfidReadUserData	Read a user memory data of a tag having given OID. A start address and data size are also given as its parameters
	MH_rfidKillTag	Make a specified Type C tag dormant permanently
		Optional commands
	MH_rfidWriteOID	Write a new OID in a specified tag
	MH_rfidWriteUserData	Write a user data in a specified tag with given address and size
	MH_rfidLockTagByBit	Lock the memory bits on a given address of a specified tag to avoid unwanted accesses
	MH_rfidLockTagByField	Lock a memory block on a given field of a specified tag
	MH_rfidUnlockTagByField	Unlock a memory block on a given field of a specified tag
Buffer control	MH_rfidCreateBuffer	Create a buffer for storing tag information
	MH_rfidDestoryBuffer	Destroy an existing buffer
	MH_rfidReadBuffer	Read a tag information entry that is stored in a given buffer
	MH_rfidWriteBuffer	Write a tag information entry in a buffer
	MH_rfidDeleteBuffer	Delete an entry in a buffer
	MH_rfidClearBuffer	Delete all the entries in a buffer
	MH_rfidGetNumBuffer	Get the number of entries stored in a buffer
	MH_rfidGetMaxNumBuffer	Get the capacity of a buffer to store the maximum number of entries
	MH_rfidSortBuffer	Sort buffer entries in specified orders: entry index, OID, and stored time
	MH_rfidValidateBuffer	Delete invalid entries with given policies: entries duplicated, stored before a given time, and kept during a given duration
Filter Control	MH_rfidAddFilter	Add a filter to prevent from storing entries having a specified pattern in a buffer
	MH_rfidDeleteFileter	Delete an existing filter
	MH_rfidEnableFilter	Enable a filter to carry out the filtering operation
	MH_rfidDisableFilter	Disable a filter to not operate

Table 4.6 (continued) Description of HAL Commands

Type	Command	Description
Reader Status Report	MH_rfidReportReaderStatus	Report a composite reader status. It includes most of the information provided by reader control commands. In addition, a serial number of a reader and supporting tag types by a reader is provided.

Table 4.7 A Buffer Structure of HAL

Field	Description
index	A sequence number of buffer entry
tagtype	A type of a corresponding tag
ID	An OID of the tag
udata	A data which reader reads in a user memory of a tag
addr	A start address of a user memory of a tag
readcycle	A consecutive number of read cycles at which the tag has been read
time	The time of the read cycle

4.5.4 WIPI

Wireless Internet Platform for Interoperability (WIPI) is the common platform envisioned by the Korean Ministry of Information and Communication for running mobile applications on any handset regardless of vendors [18]. WIPI serves as a backbone for content providers to develop applications that run seamlessly on any mobile platform. WIPI supports multiple programming languages such as Java and C, and downloads and runs content provider applications in a binary format from application servers. Using WIPI, any WIPI application can access an RFID reader through the Handset Adaptation Layer (HAL) and query tags from Internet for additional information with network APIs extended for RFID systems.

As shown in Figure 4.6a, the WIPI architecture is comprised of several layers: a handset hardware, Native System Software for the hardware, HAL for providing abstraction for WIPI, the WIPI run-time engine for providing run-time environment, basic and extended APIs and dynamic components for applications, WIPI applications, and an application manager. Based on the WIPI architecture, a mobile RFID reader is constructed as Figure 4.6b.

Each component of Figure 4.6b is explained in detail as follows:

- RFID Reader Module/Chip is an RFID reader device, a module or a chip type, especially embedded or installed in a mobile phone.
- RFID Reader Device Driver is a system software for providing control functions on RFID reader Module/Chip.
- RFID Reader Control API provides control functions on an RFID reader to applications.
- RFID Code-Support API provides tag data–related functions; it interprets information read from a tag and translates information to be written.

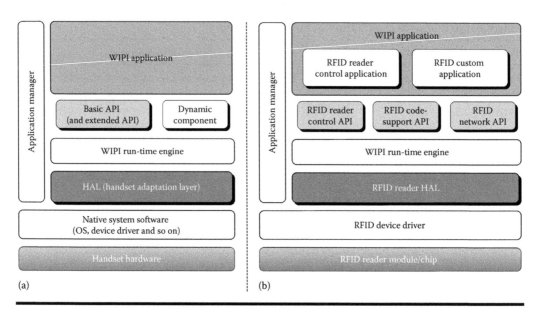

Figure 4.6 Overall WIPI architectures (a) Basic WIPI architecture (b) The WIPI architecture using an RFID reader.

- Network API provides networking functions, extended for RFID, to get related information of a tag by connecting with ODS (Object Directory Service), OIS (Object Information Service), and contents server. OIS offers information about an object while ODS searches URIs (Uniform Resource Identifier) of corresponding OISs using the tag's ID. The contents server provides contents related to the object.
- RFID Reader Control APP is a default application installed in a mobile phone embedding an RFID reader. It controls the reader, and then invokes a tag-related service using WAP or ME browser.
- RFID Custom APP is a custom application that is downloadable from a specific service provider through an existing mobile phone network infrastructure. With this application, a user can get custom service specific to his/her interest. For example, in a market a shopper can download a custom application of the market in place. Then the downloaded application helps the shopper search tags easily according to their interest, and the application provides the shopper specific information on products by using their own database as well as using their user-friendly interface. However, if there is no common platform like WIPI, service providers should develop their custom applications for each different platform. WIPI takes very important roles in this point.

With the WIPI architecture using an RFID reader, RFID applications provide RFID-related services to its user with the curtain flow as depicted in Figure 4.7 through methods of using OIS servers. As already mentioned, there are two kinds of RFID applications: RFID Reader Control and RFID Custom Applications. The operation flow of the RFID-related services are as follows.

(1) A user invokes an RFID reader application by pressing a button or something like that; (2) The application calls the RFID Reader Control API to inventory the surrounding tags; (3) The API sends corresponding HAL commands to do that; (4) HAL calls a device driver of an RFID

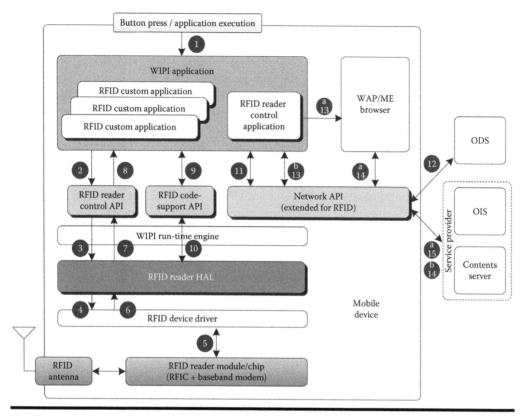

Figure 4.7   The RFID service operation flow in the WIPI architecture.

reader; (5) The device driver controls the RFID reader, and the reader returns its inventory result (i.e., tag data bit streams, tag types, and so on); (6) The device driver returns the result to HAL; (7) HAL also returns the result to the API; (8) The API finally returns the result to the application; (9) The application calls the RFID Code Support API to interpret the result tag data; (10) The API interprets the tag data by accessing a buffer storing tag information; (11) The application extracts actually needed codes from the interpreting result, then creates ODS query messages for the codes; (12) Using the Network API, the application connects with ODS servers and sends the query message, then gets URIs for the corresponding codes; (a13) The RFID reader control application invokes a WAP or ME browser to connect a service provider (OIS or contents server); (a14) The browser calls the Network API to connect the service provider; and (a15) The browser obtains and displays information of the object from a database of the service provider. On the other hand, (b13) the RFID custom application directly calls the Network API to connect the service provider, and then (b14) the application obtains and displays information of the object from a database by itself using their own specific interface.

As described, WIPI serves as a backbone for content providers to develop applications that run seamlessly on any mobile platform. Therefore, this concept works well even for an RFID reader by providing a uniform manner to application developers with plentiful degrees of freedom for various applications.

4.6 Implementation

Figure 4.8 shows a picture of our prototyped system. We implemented our baseband modem logic on the ARM-based Excalibur FPGA [28], ultrahigh frequency (UHF) RF circuits with filters, direct conversion up/down mixers, PLL, PA, and palm-sized antennas. The RFID firmware is running on the ARM9 processor on the Excalibur, and embodied baseband modem APIs control forward/return link frequencies, transmitting/receiving bit streams, PLL, TX/RX gains, and turning on/off each RF part. The firmware has functions of inventory using anticollision algorithms, and also has read/write functions if a protocol supports. Special functionalities for security such as kill or lock at EPC protocols are also accomplished.

The HAL application was coded on a personal digital assistant (PDA) using GUI (Graphic User Interface) based on PocketPC 2002. Using a UART (Universal Asynchronous Receiver/-Transmitter), the HAL application on PDA and the RFID baseband modem were connected. When a user chooses a HAL command on GUI, the HAL application sends a message to the UART with a predefined format and waits for a response from the firmware. The firmware of the baseband modem receives this message from the HAL application and decodes the message. After decoding the message, the firmware calls and executes a suitable function in an algorithmic part or a baseband modem API. If the firmware executes the function successfully, it sends back a response message, also in a predefined format, to the HAL application. The HAL application displays proper information according to the responded message.

After having prototyped and verified the reader system, we developed the system as SoC. The hardware logic of a baseband modem was designed with many IPs such as ADC for a demodulator, DAC for a modulator, SPI (Serial Peripheral Interface) for RFIC controlling, the ARM9 processor for our firmware, and cache memories for ARM9. After having verified the new system with enough simulations with many harsh bench codes, we fabricated it into one chip with 0.18 μm technology of Samsung Electronics. Figure 4.9 shows the first version of our baseband modem module and

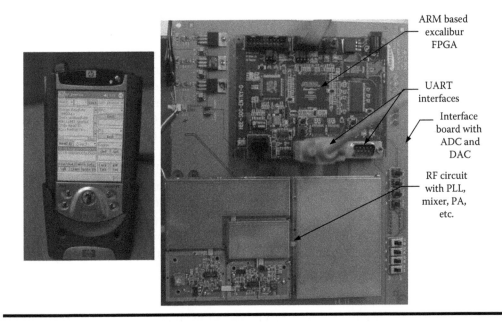

ARM based
excalibur
FPGA

UART
interfaces

Interface
board with
ADC and
DAC

RF circuit
with PLL,
mixer, PA,
etc.

Figure 4.8   Our prototyped multi-protocol RFID reader for mobile devices.

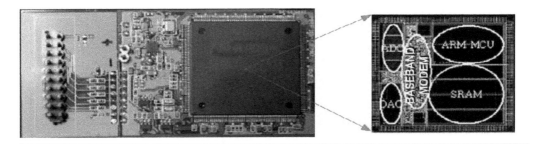

Figure 4.9   Our prototyped multi-protocol UHF band RFID baseband modem processor module for mobile devices and the layout of the processor.

Figure 4.10   Demonstration of a mobile RFID service using PDA phone to embed an RFID baseband modem processor module at the 2005 USN/RFID Conference & Exhibition held in Seoul, Korea.

the layout of the chip. An RF R&D team of S.A.I.T supported RFIC for our research. This chip has been revised many times from the first version, and the most recent version is the unified PoC (Package on Chip) to embed the baseband modem and the RFIC all together.

We verified the baseband modem module with many commercial tags, and succeeded in reading tags of ISO 18000-6 Type B, Type C, and EPC class-1. At last, we demonstrated our chip embedding to the PDA phone at the 2005 USN/RFID Conference & Exhibition held successfully in Seoul, Korea, as shown in Figure 4.10. The demonstrated scenario was exactly the same as in Figure 4.1, and including this, various service scenarios have been developed to promote mobile RFID services.

4.7 Evaluation

The baseband modem processor was implemented with about 80K gate baseband logic, the ARM9 processor, ADC/DAC, and SRAM. The baseband logic includes a modulator, a demodulator, a SPI controller, and two FIR filters. The die size of the chip is 5 mm by 5.27 mm. The power consumption of the baseband logic is estimated to be 21.705 mW with 25 MHz operating frequency, 30% flip-flop ratio, and 10% switching activity. Since the necessary clock frequency is

Table 4.8   Code and Data Sizes for Each Firmware Part

Part Name	Text	Data	bss	Data sum(= Data + bss)	Total
Main	1204	—	—	—	1, 204
Interface functions for HAL	6168	—	—	—	6, 168
EPC Class-1: Algorithmic part	3280	28	8740	8768	12, 048
EPC Class-1: Transceiver part	1720	100	7124	7224	8, 944
ISO18000-6 Type B: Algorithmic part	2104	—	—	—	2, 104
ISO18000-6 Type B: Transceiver part	3520	4	5048	5052	8, 572
ISO18000-6 Type C: Algorithmic part	8192	—	148	148	8, 340
ISO18000-6 Type C: Transceiver part	2752	—	4104	4104	6, 856
Baseband modem APIs	4400	32	—	32	4, 432
Miscellaneous functions	2224	—	12	12	1, 864
Total size (kB)	34.7	0.2	24.6	24.7	59.5

The result is estimated from our prototyped RFID reader firmware, and these were coded without consideration of resource constraint.

lower than 25 MHz, actually it is just about 4 MHz and can be reduced even more. The reason we used 25 MHz clock frequency is to eliminate an additional PLL by using a unified clock for MCU, the baseband logic, and RFIC.

The total code size of the prototyped RFID reader is less than 35 kB including three different protocols, baseband modem APIs, interface functions for HAL, and the miscellaneous parts. The total data size is about 25 kB. The estimated code size and data size of each part are summarized in Table 4.8. It is not an absolute size, and it can be various. For optimization, we could reduce the total code size under 24 kB without redundant parts for the target service. The data size for transceiver parts is quite big commonly since it stores lots of bit values in arrays. Unlike other two protocols, EPC Class-1 used one byte for one bit logical value in the algorithmic part. In the others, they used only one bit for one bit. Due to this reason, EPC Class-1 spent much more memory. ISO 18000-6 Type C is more complicated than the other protocols. Its commands and responses have different lengths, and some commands and responses are not byte ordered. For example, `Query` command has 22 bits. When we develop this kind of a command in software manner, it requires extra codes. For this reason, Type C requires more codes than any other protocol.

4.8 Conclusions and Discussions

In this chapter, we presented a software architecture and an implementation of a multi-protocol RFID reader on mobile devices such as mobile phones and PDAs. There are many research activities in RFID hardware systems, but there are few in software infrastructure, especially on mobile devices. We have implemented most functionalities of an RFID reader in software except for code modulation and demodulation to support multi-protocols and the next coming standards easily.

Our software architecture consists of four layers, which are baseband modem APIs, a firmware, a handset adaption layer, and the WIPI-based application. The baseband modem APIs allow us to control hardware logics such as a modulator, a demodulator, and RF circuits, and the firmware includes RFID protocol engines. In order to embed an RFID reader's functionalities on mobile devices, our firmware also interfaces with HAL on the WIPI platform. Through HAL interfaces, any WIPI application can access an RFID reader and query tags' information from the Internet. We have prototyped our system on the ARM-based Excalibur FPGA with iPAQ PDA for verification, and finally fabricated it with 0.18 μm technology. We demonstrated the PDA phone to embed our developed chip for a real mobile service. It read multiple commercial tags' IDs, and requested the URI of an object with the tag's ID to OIS. An application server of a service provider offered information with the URI to the PDA phone, and then the PDA phone displayed the information. A user could access additional information using a hyperlinked text.

The architecture we proposed gives great interoperability, flexibility, and expandability with enough performance for mobile use. It has a hierarchical architecture for maintainability and provides a uniform manner for accessing hardware. Any WIPI-based application and other applications that follow HAL can also use any RFID reader freely with minimal efforts.

In spite of these advantages, there still are obstacles for mobile use. Unlike industrial readers, mobile devices are moved anywhere. Therefore, many people may try to read the tag at the same time in the small area. It incurs a reader collision problem severely. Since a tag has limited ability, the problem is difficult to solve. In order to resolve this problem, readers use LBT and/or a frequency hopping; however, even using these methods limitations still exist. Lower power consumption is one of the most important issues in embedded systems. Someone may select low power parts, operate each part with a low frequency, and optimize software. In the half duplex procedure, while a reader receives signals, many parts of a transmitter can be turned off. Even the passive-typed RFID reader needs to radiate a carrier frequency, and we may turn off DAC, filters, and a modulator. In contrast, when the reader sends a command, most parts of a receiver can be turned off. This simple approach will help reduce power consumption. When a programmer applies this idea, he/she considers about time constraints reviewed in Section 4.4.1. A power control may be required for good system performance in terms of power consumption and receiver sensitivity. If a tag is far from a reader, the reader should radiate a strong signal to the tag of course; however if the tag is near, the reader has to lower the transmission signal. Since a strong signal amplifies noise components, it may saturate the receiver's signal and tend to hide a response signal. Due to this characteristic, a good power control for passive-typed RFID system and also a good noise compensating method are needed. A tag price is one of the obstacles for spreading mobile services. It is still expensive for broad use; however, this will be settled by mass production. Security problem also is a big issue for people. Many researchers have tried to find good solutions. Security for mobile RFID services is one of the challenges. Radio regulations may restrict the use of mobile RFID. Each country has different radio usage, transmit mask and radiation power for preventing channel interference, a bandwidth that can limit communication speed, and so on. These kinds of restrictions cannot be solved without the support of the government. However, a developer should keep these restrictions in mind while he/she constructs a mobile RFID system.

Our research team has two plans, one is to replace all software parts with hardware logics, and the other is to replace all hardware parts except essential RF parts with software using Java platform.

After that, we can see the trade-ffs of each system in terms of power consumption, size, efficiency, and flexibility. Overall, our hardware and software co-design methodology has many advantages, but a complete hardware design may help mass production easily, and a complete software design may help any device use an RFID reader with minimal cost.

Acknowledgment

We wish to acknowledge the support of the Samsung Advanced Institute of Technology, Korea.

References

1. R. Khare. Telnet: The mother of all (application) protocols. *IEEE Internet Computing*, 2(3):88–91, 1998.
2. R. Khare. I want my FTP: Bits on demand. *IEEE Internet Computing*, 2(4):88–91, 1998.
3. M. J. Donahoo and K. L. Calvert. *The Pocket Guide to TCP/IP Sockets*. Morgan Kaufmann, San Francisco, CA, 2001.
4. B. Goode. Voice over internet protocol (VoIP). *Proceedings of the IEEE*, 90(9):1495–1517, 2002.
5. B. Totty, D. Gourley, M. Sayer, A. Aggarwal, and S. Reddy. *Http: The Definitive Guide*. O'Reilly & Associates, Inc., Sebastopol, CA, 2002.
6. S. Saroiu, K. P. Gummadi, R. J. Dunn, S. D. Gribble, and H. M. Levy. An analysis of internet content delivery systems. In *OSDI '02: Proceedings of the 5th symposium on Operating systems design and implementation*, Boston, MA, pp. 315–327, 2002. ACM Press, New York.
7. S.-Q. Lee, N. Park, C. Cho, H. Lee, and S. Ryu. The wireless broadband (Wibro) system for broadband wireless internet services. *Vehicular Technology Magazine*, 44:106–112, 2006.
8. A. D. Smith and F. Offodile. Information management of automatic data capture: An overview of technical developments. *Information Management & Computer Security*, 10:109–118, 2002.
9. N. Arica and F. T. Yarman-Vural. Optical character recognition for cursive handwriting. *IEEE Transactions on Pattern Analysis and Machine Intelligence*, 24(6):801–813, 2002.
10. T. Acharya and A. K. Ray. *Image Processing—Principles and Applications*. Wiley-Interscience, Hoboken, NJ, 2005.
11. K. Finkenzeller; translated by R. Waddington. *RFID Handbook*. John Wiley & Sons, Chichester, West Sussex, England, 2003.
12. P. Ibach and M. Horbank. Highly available location-based services in mobile environments. *Lecture Notes in Computer Science*, 3335:134–147, 2005.
13. V. Stanford. Pervasive computing goes the last 100 feet with RFID systems. *IEEE Pervasive Computing*, 2(2):9–14, 2003.
14. K. S. Leong, M. L. Ng, and P. H. Cole. The reader collision problem in RFID systems. In *Proceedings of Microwave, Antenna, Propagation and EMC Technologies for Wireless Communications (MAPE 2005)*, Beijing, China, vol. 1, pp. 658–661, August 2005.
15. D. W. Engels. White paper of the reader collision problem. Auto-ID Center Massachusetts Institute of Technology, Cambridge, MA, November 2001.
16. D. W. Engels and S. E. Sarma. The reader collision problem. *IEEE Transactions on Systems, Mans, and Cybernetics*, vol. 3, Hammamet, Tunisia, October 2002.
17. J. Waldrop, D. W. Engels, and S. E. Sarma. Colorwave: An anticollision algorithm for the reader collision problem. In *International Conference on Communication*, vol. 2, Anchorage, AK, pp. 1206–1210, May 2003.
18. Wireless Internet Platform for Interoperability (WIPI). http://wipi.or.kr

19. M. Lampe, S. Hinske, and S. Brockmann. Mobile device based interaction patterns in augmented toy environments. In *Proceedings of Pervasive 2006 Workshop (Third International Workshop on Pervasive Gaming Applications, PerGames 2006)*, Dublin, Ireland, pp. 109–118, May 2006.

20. S. Puglia and A. Vitaletti. Alternative RFID based architectures for mobile HCI with physical objects. In *Mobile Interaction with the Real World (MIRW 2006)*, Espoo, Finland, pp. 34–38, September 2006.

21. D. M. Konidala and K. Kim. Mobile RFID applications and security challenges. *Information Security and CryptologyŮICISC 2006*, Busan, South Korea, vol. 4296 of LNCS, pp. 194–205, Springer, Berlin/Heidelberg, Germany, 2006.

22. M. R. Rieback, B. Crispo, and A. S. Tanenbaum. RFID guardian: A battery-powered mobile device for RFID privacy management. In *Proceedings of 10th Australasian Conference on Information Security and Privacy (ACISP 2005)*, Brisbane, Queensland, Australia, vol. 3574 of LNCS, pp. 184–194, Springer, Berlin/Heidelberg, Germany, July 2005.

23. M. R. Rieback, B. Crispo, and A. S. Tanenbaum. The evolution of RFID security. *Pervasive Computing, IEEE*, 5(1):62–69, 2006.

24. Auto-ID Center. 860MHzŮ960MHz Class I radio frequency identification tag radio frequency & logical communication interface specification recommended standard, Version 1.0.0. Technical Report from Auto-ID Center, Massachusetts Institute of Technology, Cambridge, MA, November 2002. http://www.autoidlabs.org/uploads/media/MIT-AUTOID-TR007.pdf

25. EPC global. GS1 US. EPC^{TM} radio-frequency identity protocols class-1 generation-2 UHF RFID protocol for communications at 860 MHz–960 MHz, Version 1.0.9. EPCglobal standard document. EPCglobal Inc., Corporate Headquaters: Princeton Pike Corporate Center, Lawrenceville, NJ, January 2005. http://www.epcglobalus.org/dnn_epcus/KnowledgeBase/Browse/tabid/277/DMXModule/706/Command/Core_Download/Default.aspx?EntryId=292

26. Auto-ID Center. 860MHzŮ930MHz Class 0 radio frequency identification tag protocol specification candidate recommendation, Version 1.0.0. Technical Report from Auto-ID Center Massachusetts Institute of Technology, Cambridge, MA, June 2003. http://www.autoidlabs.org/uploads/media/MIT-AUTOID-TR016.pdf

27. JTC 1/SC 31 (Committee for Automatic Identification and Data Capture Techniques). Information technology—Radio frequency identification for item management—Part 6: Parameters for air interface communications at 860 MHz to 960 MHz. ISO/IEC 18000-6 standard document. International Organization for Standardization, 2004.

28. Altera Excalibur. EPXA10 Development Board—Hardware Reference Manual. Altera, San Jose, CA, April 2002. http://www.altera.com/literature/manual/mnl_epxa10devbd.pdf?GSA_pos=1&WT.oss_r=1&WT.oss=excalibur

29. JTC 1/SC 17 (Committee for Cards and Personal Identification). Identification cards—Contactless integrated circuit cards—Proximity cards—Part 1: Physical characteristics. ISO/IEC 14443-1:2000 standard document. International Organization for Standardization, 2000.

30. JTC 1/SC 17 (Committee for Cards and Personal Identification). Identification cards—Contactless integrated circuit cards—Proximity cards—Part 2: Radio frequency power and signal interface. ISO/IEC 14443-2:2001/Amd 1:2005 standard document. International Organization for Standardization, 2005.

31. JTC 1/SC 17 (Committee for Cards and Personal Identification). Identification cards—Contactless integrated circuit cards—Proximity cards—Part 3: Initialization and anticollision. ISO/IEC 14443-2:2001/Amd 3:2006 standard document. International Organization for Standardization, 2006.

32. JTC 1/SC 17 (Committee for Cards and Personal Identification). Identification cards—Contactless integrated circuit cards—Vicinity cards—Part 1: Physical characteristics. ISO/IEC 15693-1:2000 standard document. International Organization for Standardization, 2000.

33. JTC 1/SC 17 (Committee for Cards and Personal Identification). Identification cards—Contactless integrated circuit cards—Vicinity cards—Part 2: Air interface and initialization. ISO/IEC 15693-2:2006 standard document. International Organization for Standardization, 2006.

34. JTC 1/SC 17 (Committee for Cards and Personal Identification). Identification cards—Contactless integrated circuit cards—Vicinity cards—Part 3: Anticollision and transmission protocol. ISO/IEC 15693-3:2001 standard document. International Organization for Standardization, 2001.

35. JTC 1/SC 31 (Committee for Automatic Identification and Data Capture Techniques). Information technology—Radio frequency identification (RFID) for item management—Data protocol: Application interface. ISO/IEC 15961:2004 standard document. International Organization for Standardization, 2004.

36. JTC 1/SC 31 (Committee for Automatic Identification and Data Capture Techniques). Information Technology—Radio frequency identification (RFID) for item management—Data protocol: Data encoding rules and logical memory functions. ISO/IEC 15962:2004 standard document. International Organization for Standardization, 2004.

37. JTC 1/SC 31 (Committee for Automatic Identification and Data Capture Techniques). Information technology—Radio frequency identification for item management—Unique identification for RF tags. ISO/IEC 15963:2004 standard document. International Organization for Standardization, 2004.

38. JTC 1/SC 31 (Committee for Automatic Identification and Data Capture Techniques). Information technology—Radio frequency identification for item management—Part 2: Parameters for air interface communications below 135 kHz. ISO/IEC 18000-2:2004 standard document. International Organization for Standardization, 2004.

39. JTC 1/SC 31 (Committee for Automatic Identification and Data Capture Techniques). Information technology—Radio frequency identification for item management—Part 3: Parameters for air interface communications at 13,56 MHz. ISO/IEC 18000-3:2004 standard document. International Organization for Standardization, 2004.

40. JTC 1/SC 31 (Committee for Automatic Identification and Data Capture Techniques). Information technology—Radio frequency identification for item management—Part 4: Parameters for air interface communications at 2,45 GHz. ISO/IEC 18000-4:2004 standard document. International Organization for Standardization, 2004.

41. JTC 1/SC 31 (Committee for Automatic Identification and Data Capture Techniques). Information technology—Radio frequency identification for item management—Part 7: Parameters for active air interface communications at 433 MHz. ISO/IEC 18000-7:2004 standard document. International Organization for Standardization, 2004.

42. TC 23/SC 19 (Committee for Agriculture Electronics). Radio frequency identification of animals—Code structure. ISO 11784:1996/Amd 1:2004 standard document. International Organization for Standardization, 2004.

43. TC 23/SC 19 (Committee for Agriculture Electronics). Radio frequency identification of animals—Technical concept. ISO 11785:1996 standard document. International Organization for Standardization, 1996.

44. GS1 EPCglobal. Improved efficiencies with EPC Gen2. Case study document. GS1 EPCglobal, 2006. http://www.gs1.org/docs/transportlogistics/8-Case-study-Exel.pdf

45. Hitach Chemical. RFID products (IC Card & Tags) for 2.45GHz microwave frequency (μ-chip). Official website of Hitachi Chemical Co., Ltd. http://www.hitachi-chem.co.jp/english/products/ppcm/014.html

46. H. Barthel. Regulatory status for using RFID in the UHF spectrum. GS1 EPC global, August 2006. http://www.gs1.org/docs/epcglobal/UHF_Regulations.pdf

47. K. Finkenzeller; translated by R. Waddington. *RFID Handbook*. John Wiley & Sons, Chichester, West Sussex, England, p. 142, 2003.

48. S. J. Hwang, J. G. Lee, S. W. Kim, S. Ahn, S.-G. Koo, J. H. Koo, K. H. Park, and W. S. Kang. A multi-protocol baseband modem processor for a mobile RFID reader. *Lecture Note in Computer Science*, 4096:785–794, 2006.

49. Analog Devices, Inc. High performance, ISM band, FSK/ASK transceiver IC. Datasheet for ADF7020: ISM band transceiver IC. Analog Devices, Inc., Norwood, MA, June 2005. http://www.analog.com/static/imported-files/data_sheets/ADF7020.pdf

50. Texas Instruments Inc. CC1020 Low-power RF transceiver for narrowband systems. Datasheet for CC1020: Single-chip FSK/OOK CMOS RF transceiver for narrowband apps in 402-470 and 804-940 MHz range Rev.1.8. Texas Instruments Inc., Dallas, TX, January 2006. http://www.ti.com/lit/ds/symlink/cc1020.pdf

51. Mobile RFID Forum. WIPI network APIs for mobile RFID services. MRFS-2-02 TTA (Telecommunications Technology Association) standard document. Mobile RFID forum, Seoul, South Korea, August 2005.

52. Mobile RFID Forum. HAL API standard for mobile RFID reader. MRFS-1-03/R1 TTA standard document. Mobile RFID Forum, Seoul, South Korea, August 2005.

53. iconlab. History. Company history about ICONLAB Co., Ltd. on its official website. Seoul, Korea, 2006. http://www.iconlab.co.kr/EN/2d_history.html

SMART ENVIRONMENTS AND AGENT SYSTEMS

Chapter 5

Autonomous Timeliness and Reliability Oriented Information Services Integration and Access in Multi-Agent Systems

Xiaodong Lu and Kinji Mori

Contents

5.1 Introduction

With the advances in communication technologies and the decreasing costs of computers, global information systems such as Internet has become an attractive alternative for satisfying the information service needs of global users. Currently, most information service systems enable one-to-one interactions, which take place between one user and one provider. But sometimes the user requests multi-service, for example, a complete travel package made up of several flights and hotel information. In this case, the adaptability and timeliness have to be assured by the system. Conventional information service systems based on client/server model cannot meet users' heterogeneous and dynamically changing requirements.

The demand for information services is increasing at an explosive rate. Due to large geographical region of business operations and external competitive pressures, information services are expected to collect, store, retrieve, integrate, and distribute timely at remote and dispersed nodes. In order to reduce network traffic and improve users' access time, we are developing a demand-oriented architecture called Faded Information Field (FIF), sustained by push/pull mobile agents [1]. The characteristic of the FIF is to balance the cost of information allocation performed by push mobile agents and the cost of the access to the information performed by the pull mobile agents [2]. The information structure consequently permits to preserve the same access time to all unspecified users, whatever their current demand trend and volume.

Nowadays, most requests on the Internet involve two or more services at the same time. However, the basis of the current information systems is to provide the integrated information through the middle process in a passive and noncooperative way. For complex interdependent requests, the timeliness cannot be satisfied. It is necessary to find a trade-off between the heterogeneous requirements for services provision and the need for an efficient utilization of those correlated services. Under FIF architecture, services available in a common replication area can be integrated when they can be demonstrated with high correlation. In this chapter, the relationship that exists between the timeliness and reliability of correlated services allocation and access is clarified. Based on these factors, we propose the autonomous network-based integration technology to achieve timeliness for users multi-service access in rapidly evolving situations.

However, the increase in the amount of requests will cause disproportional increase in the load of servicing nodes. In addition, when the users requirements and preferences change, how to guarantee the load balancing and avoid the local overload is the main problem we should consider. Conventional technologies make many efforts to improve query planning process in information integration [3–5]. But these dynamic schedules algorithms applied on the query server or the integration server are centralized schemes that are associated with problem of reliability and low responsiveness. Under constantly changing environment, neither the requested correlated information volume, nor its arrival distribution on the network and the entire system structure can be predicted so that the efficient correlated services access cannot be achieved through current techniques without undesirable time overheads due to the centralized process. The autonomous correlated services access technology implemented above the network-based integration architecture is proposed. The autonomous processing and adjusting mechanisms permit to dynamically balance the load in the locality and offload some congested nodes to achieve timeliness and assure the fairness under rapidly changing environments.

The structure of this chapter is organized as follows. In the next section, the system architecture of FIF is presented. Autonomous network-based integration of information services is presented in Section 5.3. Autonomous access for correlated services is presented in Section 5.4. The simulation results in Section 5.5 show the improvement and effectiveness of proposed technology. We discuss related work in Section 5.6 and conclude in Section 5.7.

5.2 System Architecture

The main goal of the faded information field is to guarantee the assurance of autonomous information service provision and utilization [6]. Service providers trace the demand trend for information, and allocate to accepting storing nodes the most accessed segment of their information services. The storing nodes then, in a recursive pruning process, further allocate the information services to adjacent nodes. As a result, the multi-level distributed information services area is created. Nodes adjacent to the SP contain more information but are also far from users, while outmost nodes storing the minimum information amount are located at the edge of the users' network. Users with different requirements for information can be satisfied at different levels in the FIF. Consequently, the cost of service utilization (access time) and cost of service provision (update) are balanced by allocating closer to the majority of the users the most accessed part of the information services. Figure 5.1 shows the FIF architecture and the information volume on each level.

In the FIF system, information contents are uniquely defined by Content Code (CC). The information contents are further specified by its Characteristic Codes (CHs). For instance, the CC identified by CC1 standing for Flight information can be further specified by the CHs: Flight No., Origin/Destination, Fare, and Class. Service providers transmit information services through Push-MAs by specifying content codes (CCs) of information to the nodes. The nodes receive information and select to store autonomously based on CCs. Users search information based on CCs by exploiting Pull-MAs.

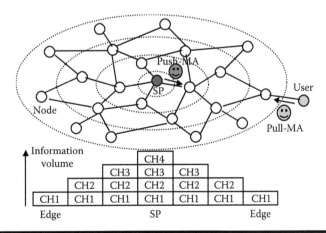

Figure 5.1 FIF architecture.

The FIF is defined as a multilevel information structure. The relationship of available information between layers can be expressed as: $I_{l+1} \subset I_l$, where I is the information volume which is defined as the number of CHs stored in each node, and l is the layer. SP indicates the top layer 0, where the maximum volume of information is stored. Each node only knows its upper and lower neighbor nodes and coordinates among them to form a gradient information structure.

But how to find the free nodes on the network to allocate information services autonomously for SPs and how to discover the requested service on the network and access it quickly for users are main difficulties for the FIF construction. To solve these problems, we proposed the Content Code Lookup Service (CCLUS) not only for the SPs to make their information services automatically available, but also for users transparently discovery the desired service within their local domains. The CCLUS placed in each domain is the major point of contact between the SPs, nodes, and users in the system. Only the edge nodes, which is defined as the node located in the edge layer with minimum volume of information register the information with CCLUS, and autonomously navigate the Pull-MAs to upper nodes with required information service [7].

In the FIF system, the mobile agent bridges the autonomous entities to achieve coordination. A reliable mobile agent platform is available on each node, providing an execution and routing environment for mobile agents. In one word, node, Push-MA and Pull-MA are three autonomous subsystems, mainly responsible for information storing, allocation, and utilization, respectively. Every subsystem has autonomous controllability over its own operations and autonomous coordinability with the other subsystems to continue its operations even under evolving situations [8].

5.3 Autonomous Network-Based Integration

In order to meet users' multi-service requests, various SPs will make temporary alliance to achieve common goals. The travel agency application is a classical case study in the agent and information integration research area. The user asks for a bundle of services comprising flight and hotel information. Services are tied together by requirements expressed by the user. Currently the user has to cope with this situation by searching information from different SPs and combining these services by himself or a certain application server, which is difficult to meet timeliness required by the users. Autonomy of services integration is necessary to cope with evolving environment to make alliance possible among service providers to provide one-stop service.

The proposed technology consists of two parts: (1) How to detect the correlated services autonomously based on the users' joint requests? (2) How to determine the information volume of correlated services in the integrated area to trade-off between timeliness and reliability? We describe these techniques separately in the next sections.

5.3.1 Service-Oriented Detection and Integration

In the FIF, a service is described by content code and related characteristic codes dependent on the type. We consider two service providers, SP1 and SP2, which offer online flight and hotel information services. Flight information service, represented with content codes CC1, consists of characteristic codes CH11, CH12, CH13, and CH14, for flight number, origin/destination, fare, and class, respectively. Hotel information service, represented with content codes CC2, consists of characteristic codes CH21, CH22, CH23, and CH24, for name, location, fare, and class, respectively.

Two services are correlated or not depending on the number of users who request simultaneously. The correlated degree $\text{Cor}_{ij}(t_k)$ at time t_k and changing tendency $T_{ij}(t)$ during certain period t are defined to measure the correlation between two services.

$$\text{Cor}_{ij}(t_k) = \frac{N_{ij}(CCi, CCj, t_k)}{N_i(CCi, t_k) + N_{ij}(CCi, CCj, t_k)}$$

$$T_{ij}(t) = \sum \frac{\text{Cor}_{ij}(t_{k+1}) - \text{Cor}_{ij}(t_k)}{t_{k+1} - t_k}$$

where

$N_i(CCi, t_k)$ is the number of users who only request service CCi

$N_{ij}(CCi, CCj, t_k)$ is the number of users who request both services CCi and CCj simultaneously at time t_k

In the system each CCLUS records the users' joint requests, and after certain timeout send the message to one edge node with the small CC, like CC1. After received the message, the edge node forwards it to the upper node until SP1. If the Correlated degree and Changing tendency are more than certain threshold, SP1 will send the Push-MAs with the correlated table of CC2. The threshold is used in order to provide stability in the integration process in a highly dynamic environment. After receiving the Push-MA, the node in the common area registers the CC1 and CC2 with the CCLUS in its own domain and the available message is forwarded to other CCLUS. As a result, the knowledge about the content, source, and relationship of correlated services is provided. And then the node modifies the Push-MA's fading behavior and forwards it into CC2 service area to allocate CC1 service into FIF2.

Gradually, each node in the FIF2 can get relevant information volume of CC1. Reciprocally, Push-MAs from SP2 allocate its information in the area covered by SP1. The information level of each FIF is gradually updated to reflect the new service consumption tendency of users with joint request and an area of integrated services is created. The integration is maximal in the area where all the correlated CHs can be found.

5.3.2 Timeliness and Reliability-Oriented Allocation

Through the service-oriented integration, amount of correlated information stored by each node can satisfy certain users' joint requests simultaneously. But how much information volume of correlated CHs should be allocated in the integrated area is still a main problem to assure both timeliness and reliability of correlated services utilization.

5.3.2.1 Timeliness of Correlated Services Utilization

The total time of correlated services utilization, $T(CC1, CC2)$, is composed of the push time of correlated services provision, $T_{push}(CC1, CC2)$, and the pull time of correlated services access, $T_{pull}(CC1, CC2)$.

Both $T_{push}(CC1, CC2)$ and $T_{pull}(CC1, CC2)$ depend on the amount of correlated information allocated and requested at each node. The more correlated CHs have been allocated in the integrated area, the longer push time of correlated services provision becomes. On the other hand, the less correlated CHs have been allocated, the longer pull time of correlated services access becomes. Through the trade-off between the push time and pull time, appropriate amount of correlated information can be determined.

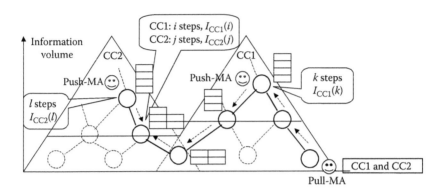

Figure 5.2 Push time of correlated services provision.

Push-MA traverses from the upper node to lower node, and stores certain information, while fading others in coordination with the node, and moves to the next node. Figure 5.2 shows the correlated services provision model of Push-MAs to allocate CC1 and CC2 services in the integrated area. If $I_{CC1}(i)$ is the information volume of CC1 at step i from the SP1, I_{push} is the code size of Push-MA, the time to transmit unit information to adjacent node is λ, and the time to process unit information by each node is α, then the CC1 service provision time $T_{push}(CC1)$ at the node N can be given by:

$$T_{push}(CC1) = \sum_{i=1}^{m}((I_{push} + I_{CC1}(i))\lambda + I_{CC1}(i)\alpha)$$

Accordingly, the CC2 service provision time $T_{push}(CC2)$ at the node N is expressed:

$$T_{push}(CC2) = \sum_{j=1}^{n}((I_{push} + I_{CC2}(j))\lambda + I_{CC2}(j)\alpha)$$

For the Pull-MA with joint request of CC1 and CC2, only both correlated services to be allocated, the request can be satisfied. Therefore, the push time of correlated services provision is defined as the larger one of $T_{push}(CC1)$ and $T_{push}(CC2)$.

$$T_{push}(CC1, CC2) = \max\{T_{push}(CC1), T_{push}(CC2)\}$$

Figure 5.2 shows the access model of a Pull-MA with joint request of CC1 and CC2. First, the Pull-MA traverses from the edge node to the upper node, and retrieves the required CHs of CC1 at a certain node. And then through the integrated service area to search a corresponding node of CC2 to get the correlated CHs. Finally, the Pull-MA carries the retrieved integrated information to the user. The correlated services access time $T_{pull}(CC1, CC2)$ of a Pull-MA is therefore composed of the transmission, process, and return time.

T_{proc} depends on the requested amount of information. T_{tran} and T_{retn} depend not only on the amount of information, but also on the size of Pull-MA, I_{pull}, and the number of steps, k and l, moving to the relevant CC1 and CC2 nodes. Then access time $T_{\text{pull}}(\text{CC1}, \text{CC2})$ of a Pull-MA with joint request is the sum of these three values.

$$
\begin{aligned}
T_{\text{pull}}(\text{CC1}, \text{CC2}) &= T_{\text{tran}} + T_{\text{proc}} + T_{\text{retn}} \\
&= kI_{\text{pull}}\lambda + I_{\text{CC1}}(k)\alpha + (l-k)(I_{\text{pull}} + I_{\text{CC1}}(k))\lambda \\
&\quad + I_{\text{CC2}}(l)\alpha + l(I_{\text{pull}} + I_{\text{CC1}}(k) + I_{\text{CC2}}(l))\lambda
\end{aligned}
$$

where

$I_{\text{CC1}}(k)$ and $I_{\text{CC2}}(l)$ is the information volume of CC1 and CC2 at step k and l from the user
λ is the time to transmit unit information to the adjacent node
α is the time to process unit information for each node

5.3.2.2 Reliability of Correlated Services Utilization

The reliability gives the probability of successful information provision and access. The more correlated CHs have been allocated, the higher the reliability of Pull-MAs with joint requests is. On the contrary, the more correlated CHs have been allocated, the greater the risk that the data have been corrupted. The reliability of correlated services utilization, $R(\text{CC1}, \text{CC2})$, is defined as the product of push reliability, $R_{\text{push}}(\text{CC1}, \text{CC2})$, and pull reliability, $R_{\text{pull}}(\text{CC1}, \text{CC2})$.

The reliability of correlated services allocation at each node depends on the information volume transmitted and processed on each step. Let us consider that the node agents and links between two nodes offer the same probability for transmitting and processing unit information as β and τ. Therefore, as represented in Figure 5.2, considering certain allocation paths of correlated services CC1 and CC2, the push reliability of correlated services at a node agent is expressed:

$$
\begin{aligned}
R_{\text{push}}(\text{CC1}, \text{CC2}) &= R_{\text{push}}(\text{CC1}) \cdot R_{\text{push}}(\text{CC2}) \\
&= (\beta\tau)^{\sum_{i=1}^{m} I_{\text{CC1}}(i)} \cdot (\beta\tau)^{\sum_{j=1}^{n} I_{\text{CC2}}(j)}
\end{aligned}
$$

where

$I_{\text{CC1}}(i)$ is the information volume of CC1 at step i
$I_{\text{CC2}}(j)$ is the information volume of CC2 at step j

Let us consider the same access model represented in Figure 5.2 to get correlated CHs of CC1 and CC2. The pull reliability of correlated services access is composed of the reliability to get correlated CHs of CC1 and CC2 and return reliability, and is then the product of these three values:

$$
R_{\text{pull}}(\text{CC1}, \text{CC2}) = R_{\text{pull}}(\text{CC1}) \cdot R_{\text{pull}}(\text{CC2}) \cdot R_{\text{retn}}
$$

If I_{pull} is the size of Pull-MA, $I_{\text{CC1}}(k)$ and $I_{\text{CC2}}(l)$ is the information volume of CC1 and CC2 at step k and l from the user, then the pull reliability of correlated services access is further written by:

$$
R_{\text{pull}}(\text{CC1}) = \beta^{\sum_{i=1}^{k} I_{\text{pull}}} \cdot \tau^{I_{\text{CC1}}(k)}
$$

$$
R_{\text{pull}}(\text{CC2}) = \beta^{\sum_{i=1}^{l-k}(I_{\text{pull}}+I_{\text{CC1}}(k))} \cdot \tau^{I_{\text{CC2}}(l)}
$$

$$
R_{\text{retn}} = \beta^{\sum_{i=1}^{l}(I_{\text{pull}}+I_{\text{CC1}}(k)+I_{\text{CC2}}(l))}
$$

By trade-off between the push reliability and pull reliability of correlated services provision and access, we can determine the optimal amount of correlated information that should be allocated in the integrated area.

From this discussion we can see that increasing the correlated information volume of the integrated service area, Pull-MAs with joint request can get high response, but the push reliability of correlated services provision will decrease, vice versa. To determine the correlated information volume allocated in the integrated area we must consider these two parameters at the same time.

5.3.2.3 Dynamic Correlated CHs Allocation

Each node in the system monitors local situation and status brought by a Push-MA or Pull-MA and autonomously adjusts the amount of information. The push time and reliability of CC1 and CC2 of upper paths are brought by Push-MAs. When a Push-MA arrives at a node, the transmission time and the process time of current node is detected and the sum of current value and the push time of upper paths is written into the Push-MA. And then, the reliability can be calculated based on the given parameters. The node agent calculates the pull time and reliability of joint Pull-MAs that can be satisfied at current node based on the moving time and steps in the system recorded by each Pull-MA.

To make the trade-off relationship clearly, we define the access performance, $P(t_i, k)$, as the reciprocal of the time of correlated services utilization. And $R(t_i, k)$ is the access reliability at time t_i detected at node k. Considering at time t_i Pull-MAs requests for CC1 and CC2 as amount of a and b at node k in CC1 service area, the amount of information CC1 can satisfy Pull-MAs' joint requests. In order to monitor users preferences and decide information volume of CC2 to be stored, the degree of satisfaction and access ratio are defined as follows:

$$S_l(t_i, k) = \frac{N_{12}(b > I_2(t_i, k))}{N_{12}(a \leq I_1(t_i, k))}$$

$$S_m(t_i, k) = \frac{N_{12}(b < I_2(t_i, k))}{N_{12}(a \leq I_1(t_i, k))}$$

$$A_{2i} = \frac{N(\text{CH2}i)}{N(\text{CC1}, \text{CC2})}$$

where $I_1(t_i, k)$ and $I_2(t_i, k)$ are information volume of CC1 and CC2 stored at node k at time t_i. $S_l(t_i, k)$ and $S_m(t_i, k)$ reflect that correlated CHs of CC2 stored at node k at time t_i are shortage or not. According to the access ratio $A_{2i}(i = 1, \ldots, n)$, we can get the most or least accessed correlated information of CH2i.

With correlated information volume increase in the integrated area, the access performance and reliability at current node will increase, but in lower path will decrease. The change of access performance and reliability can be obtained step by step through the monitoring information of Push/Pull-MAs. In the system, each node must keep the balance between the reliability and the timeliness of service utilization. Initially, the nodes store certain information volume, and monitor information of Push/Pull-MAs, which gives actual measurement of the access performance and reliability depending on the information amount stored on the node. The node agent detects $S_l(t_i, k)$ and $S_m(t_i, k)$, and adjusts information amount autonomously. If stored correlated CHs of CC2 are in shortage, and the increase of access performance and reliability at current node are more than the decrease in the lower path, the node agent increases information amount of CC2 upon the arrival of Push-MA, and vice versa. A node agent determines information amount $I_2(t_{i+1}, k)$ at time t_{i+1} at the node k as follows:

if $(S_l(t_i, k) > 0)$ and $(S_m(t_i, k) < 1)$ and
$\quad (P(t_i, k) > P(t_{i-1}, k))$ or $(R(t_i, k) > R(t_{i-1}, k))$
$\quad$ then
$$I_2(t_{i+1}, k) = I_2(t_i, k) + CH2i(\max(A_{2i}))$$
$\quad\quad$ else
$$I_2(t_{i+1}, k) = I_2(t_{i-1}, k)$$

According to this process, the node can detect the correlated information volume that makes access performance and reliability to get the maximum value respectively. And then, the correlated information volume that should be allocated in the integrated area is the point that the difference between the access performance and reliability is minimum. As a result, the appropriate correlated CHs allocated in the integrated area can be determined dynamically.

5.4 Autonomous Correlated Services Access

Through the autonomous integration and dynamic online allocation, majority of users, multi-service requests can be satisfied on one node at the same time. But to still have some Pull-MAs with joint requests we must get the correlated CHs from the separated node in each service area. In this case, the access time of the Pull-MA with joint request is composed of the time to move inside integrated service area to the corresponding nodes which store the requested correlated CHs, the time of processing at each node, and the time of returning to the user.

In a study on the performance of mobile agent systems, it is validated that the transmission time and the processing time strongly depends on the amount of information volume retrieved by the Pull-MA. For multi-service access, in order to obtain customized information, the users have to specify their personal conditions about the correlated services and the correlated CHs of each service.

As an example shown in Figure 5.3, the Pull-MA is emitted with joint request about CH(11, 12, 14) of CC1 and CH(21, 22, 24) of CC2. But the most correlated CHs in the integrated area only consist of CH(11, 12, 21, 22). The Pull-MA cannot be satisfied on one node for multi-service request. As a result, it must move to the upper node to get the required CHs of CC1 at a certain node. And then through the integrated service area to move to a node to get the correlated CHs of CC2. If the Pull-MA gets all matching CHs from node C and carries these CHs to move to the node G, the transmission time between node C and G will be high. As CH(11, 12, 21, 22) allocated in the integrated area are common correlated CHs for all nodes, the Pull-MA can also get them at node G. It is enough that the Pull-MA just gets the lacking CHs of CC1 from node C. For instance, if the Pull-MA only takes CH14 moving from node C to node G and gets other correlated CHs from node G, the response time can be improved while transmission time decreases.

The information structure of integrated service area permits to efficiently distribute Pull-MAs with joint requests between the levels. However, under rapidly changing environment, local overload in certain nodes might arise depending upon changing arrival rate. As the gradient of the information structure, any request bounded for a lower level can be processed in a higher

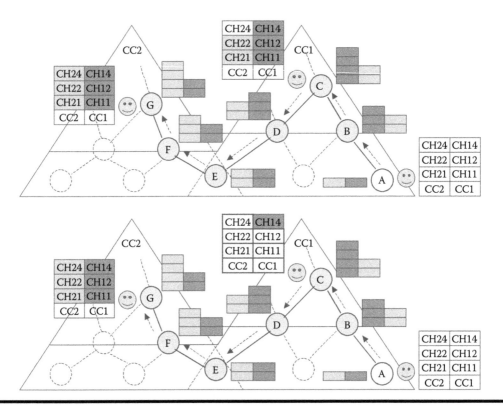

Figure 5.3 Autonomous moving with correlated CHs.

level of information if some free resources are available. Similarly, the processes of higher-level requests for multi-service access can be executed at the lower nodes with the lacking CHs. But extra access overhead are induced and impair the overall response time in the case that access processes are executed in more different levels. To distribute the multi-service access in separated process is benefit only if all upper nodes able to satisfy the request are currently congested. In other conditions, to satisfy the multi-service request at appropriate nodes is more efficient.

To analyze the load distribution of Pull-MA with joint request, the simple information structure is considered, as shown in Figure 5.4, and the M/M/k queuing model is applied. There are two kinds of Pull-MA's request in this model:

1. *Single service request*: The Pull-MA only requests CHs of CC2 information service.
2. *Multiple services request*: The Pull-MA requests correlated CHs of CC1 and CC2 at the same time.

If only the average processing time of each node is considered, two process methods exist for Pull-MAs with joint requests that cannot be satisfied at one node.

1. *Forward to process*: The Pull-MAs with joint requests are processed in the appropriate level, i.e., the node B always forwards the unsatisfied Pull-MAs to the upper level.
2. *Get and process*: The unsatisfied Pull-MAs with joint requests just get the lacking CH from the upper level and return back to process at the lower node.

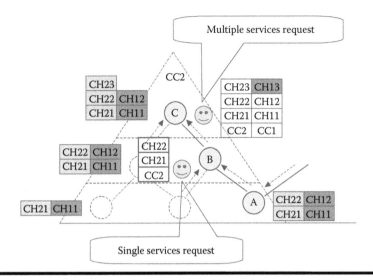

Figure 5.4   The model of the load distribution analysis.

Table 5.1   Illustration of Parameters

λ_{01}	Arrival rate of Pull-MAs to request CH(21, 22)
λ_{02}	Arrival rate of Pull-MAs to request CH(21, 22, 23)
λ_{11}	Arrival rate of Pull-MAs with joint request for CH(11, 12) and CH(21, 22)
λ_{12}	Arrival rate of Pull-MAs with joint request for CH(11, 12, 13) and CH(21, 22, 23)
μ_0	Service rate of Pull-MAs with single request
μ_1	Service rate of Pull-MAs with joint request
μ_2	Service rate of Pull-MAs for just getting the lacking CH

These two methods are compared in this analysis. Therefore, the average response time of the node B and node C in two models can be expressed as follows, and Table 5.1 gives the illustration of parameters used in the formulas.

Forward to process:

$$\begin{cases} T_B = \dfrac{1}{\mu_0 - \lambda_{01}} + \dfrac{1}{\mu_1 - \lambda_{11}} \\ T_C = \dfrac{1}{\mu_0 - \lambda_{02}} + \dfrac{1}{\mu_1 - \lambda_{12}} \end{cases}$$

Get and process:

$$\begin{cases} T_B' = \dfrac{1}{\mu_0 - \lambda_{01}} + \dfrac{1}{\mu_1 - (\lambda_{11} + \lambda_{12})} + \dfrac{1}{\mu_2 - \lambda_{12}} \\ T_C' = \dfrac{1}{\mu_0 - \lambda_{02}} + \dfrac{1}{\mu_2 - \lambda_{12}} \end{cases}$$

From above analysis, we can see that the variation of the average response time of upper and lower nodes for the Pull-MAs relates to the changes in the users preferences. If the sum of $\lambda_{11} + \lambda_{12}$ is a constant and $\mu_1 < \mu_2$, the average processing time of T_C increases rapidly in the case of overload with the λ_{12} increase. But in such case, the lower node becomes underload. It is efficient to improve the average response time of the upper node by using *Get and process* method. Both waiting time and processing time at the upper node are reduced. On the other hand, this approach has the disadvantage to burden the Pull-MAs that can be satisfied at the lower node with the increasing waiting time.

5.4.1 Autonomous Processing

On improving the average response time of the upper node blindly, some Pull-MAs might only improve their response time at the expense of a drastic impeding the processing time of other Pull-MAs. The access load distribution must provide both efficiency in response time and fairness among the Pull-MAs with heterogeneous requests. Therefore, the autonomous access distribution mechanism is proposed to fairly reduce congestion under dynamically changing users preferences conditions. According to the current condition, Pull-MAs with joint request execute two process methods adaptively.

1. If the current load is more than that of the upper node, it forwards Pull-MAs directly to upper level and executes processes on upper node.
2. If the heavy load in upper level is detected, it gets the lacking CH from the upper node and executes processes on the lower node to reduce the processing overhead of upper level.

The *forward to process* is reactive to a current overload situation making a reduction of the waiting time of Pull-MAs. The *get and process* proactive to reduce the servicing time on the upper node with heave load. Both processes assure the dynamic adaptation to unbalance of workload in the locality. If T_{up} and T_{low} are the current processing time of the upper and lower nodes, a new process behavior leads to new processing times as T'_{up} and T'_{low}. The fairness for the process behavior of Pull-MAs with joint request permits to the total decrease in the response time of the latter is equivalent or improved compared to the increase in the response time of the former.

$$(T_{up} - T'_{up}) + (T_{low} - T'_{low}) \geq 0, (T_{up} > T_{low} \ or \ T_{up} < T_{low})$$

5.4.2 Autonomous Monitoring

Under rapidly changing conditions, neither the demand correlated CHs nor the required conditions of each user can be predicted in the system. The workload unbalance hence cannot be solved through the centralized approach without incurring large time overheads. It is necessary that the decision of the process behaviors for Pull-MAs with joint request at each node be only based on the node's local information. The autonomous process behavior of each node assures to keep the fairness of the changes of the processing times in current and upper nodes. The local processing time of each node is mainly due to the node utilization, which is defined as the ration between the arrival rate of satisfied Pull-MAs over the service rate. To estimate the load in the locality, each node monitors the node utilization of current and upper nodes. As shown in Figure 5.4, the node utilization of the current node B and the upper node C can be given by:

$$\rho_{cu} = \frac{1}{2}\left(\frac{\lambda_{01}}{\mu_0} + \frac{\lambda_p}{\mu_1}\right)$$

$$\rho_{up} = \frac{1}{3}\left(\frac{\lambda_{02}}{\mu_0} + \frac{\lambda_f}{\mu_1} + \frac{\lambda_p}{\mu_2}\right)$$

- λ_p: Process rate of Pull-MAs with joint request that cannot be satisfied at the current node, i.e., number of Pull-MAs undertaking the *get and process* behavior.
- λ_f: Forward rate of unsatisfied Pull-MAs with joint request to the upper node, i.e., number of Pull-MAs undertaking the *forward to process* behavior.

5.4.3 Autonomous Adjusting

Any change of the access behavior has the effect on the processing times in current and upper nodes. The average response time of each node is mainly due to the length of the waiting queue. Based on the monitored value of the node utilization, each node estimates the average length of queue in current and upper nodes. Meanwhile, the forward ratio r_f which is defined as the ratio of the forward rate to the total number of unsatisfied Pull-MAs with joint request is dynamically adjusted according to the changes of node utilization. The more the processing time of the upper node decreases, the less the forward ratio is adjusted. Consequently, we can modify the forward ration with a new value, by comparing the average length of queue in current and upper nodes.

$$r_f' = \begin{cases} r_f \cdot \dfrac{\rho_{cu}}{\rho_{up}}, & \dfrac{\rho_{cu}}{1-\rho_{cu}} - \dfrac{\rho_{up}}{1-\rho_{up}} > \lambda_f \\[2ex] r_f \cdot \dfrac{\rho_{cu}}{\rho_{up}}, & \dfrac{\rho_{up}}{1-\rho_{up}} - \dfrac{\rho_{cu}}{1-\rho_{cu}} > \lambda_f \end{cases}$$

Through the autonomous adjusting mechanism, the access distribution of Pull-MAs with joint request is gradually refined to improve the response time and assure the fairness of multi-service access. This step-by-step approach permits to achieve timeliness under rapidly changing environments with high autonomy of information services.

5.5 Performance Evaluation

5.5.1 Autonomous Network-Based Integration

5.5.1.1 Simulation Model

To evaluate the performance of proposed technology, we consider two FIF systems containing two SPs. As shown in Figure 5.5, each FIF is composed of 20 nodes dispatched on 4 levels with varying degrees of connectivity from one level to another. From SP to the edge level the information volume is 10CHs, 8CHs, 6CHs, and 3CHs respectively, and each CH is 1kBytes.

Users have heterogeneous demands for information services, majority of the users need small portion of information. Consequently, we consider only multi-service requests by Pull-MAs on this basis. The distribution of users who require information volume is shown in Table 5.2.

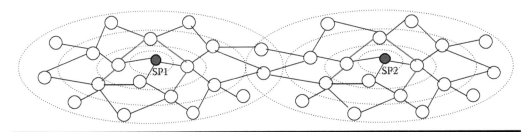

Figure 5.5   The simulation model of integration.

5.5.1.2 Correlated Information Volume Decision

How much correlated information should be allocated in the integrated service area is very important for timeliness and reliability of correlated services utilization. Pull-MAs with joint request are dispatched to the edge nodes following the trends described in Table 5.2. The data size of each Push/Pull-MA is 2 kB. We consider the system with the following parameters: $\lambda = 1$ ms/kB, $\alpha = 10$ ms/kB, $\beta = \tau = 0.99$.

In order to make the relationship between the timeliness and the reliability of correlated services utilization clear, the normalization method is applied that each value is divided by the maximum of each set. From Figure 5.6, we can see that the performance of correlated services utilization gets the highest value with 10 correlated CHs allocated at the integrated area. When six correlated CHs are allocated at the integrated area, the total reliability of correlated services utilization is highest. As shown in Figure 5.6, the close points of P_{access} and R_{access} means that the amount of correlated information at this point can satisfy both the timeliness and the reliability requirements of correlated services utilization. The numeric evaluation shows that the eight correlated CHs allocated at the integrated area is optimal information volume under given system structure and users' distribution.

5.5.1.3 Results of Comparison

The first experiment is to show the effectiveness of proposed technique compared to server-based integration model. For comparison, the mirror system with server-based integration model was adopted. As shown in Figure 5.5, there are 20 nodes in each service area and the total volume of information stored in each service area is 100 CHs. Assuming the node in the mirror system stores all 10 CHs and the total information volume is set to be the same for the fair comparison, the equivalent mirror model of each service should consist of 10 nodes allocated at the boundary layer. In the server-based integration model, users send requests to the broker. Then the broker gets the information services from each mirror system and returns the integrated information to the users.

We consider only joint requests of Pull-MAs with the distribution described in the Table 5.2. The result shown in Figure 5.7 reveals that the proposed autonomous integration technology permits to significantly leverage the average response time of correlated services utilization even under complicated joint requests. The average improvement of response time is 80% compared with server-based integration. For a small number of users, the average response time of server-based integration is little better than that of proposed network-based integration. Because some Pull-MAs with joint requests cannot be satisfied at one node, they must get the correlated CHs separately from the corresponding nodes in the system. However, when accessing users are getting

Table 5.2 The Users Distribution with Joint Request

SP1	Info. ratio	0.1–0.3				0.4–0.6				0.7–0.8				0.9–1.0			
	users%	50				30				15				5			
SP2	Info. ratio	0.3	0.6	0.8	1.0	0.3	0.6	0.8	1.0	0.3	0.6	0.8	1.0	0.3	0.6	0.8	1.0
	users%	50	30	15	5	30	50	15	5	15	30	50	5	5	15	30	50

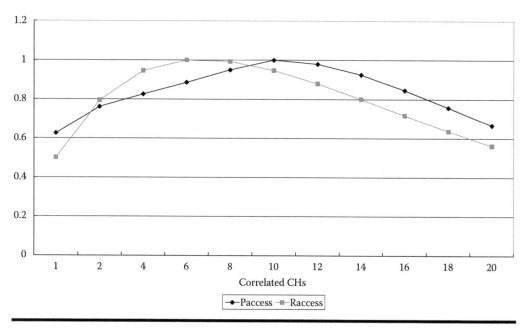

Figure 5.6   Access performance and reliability.

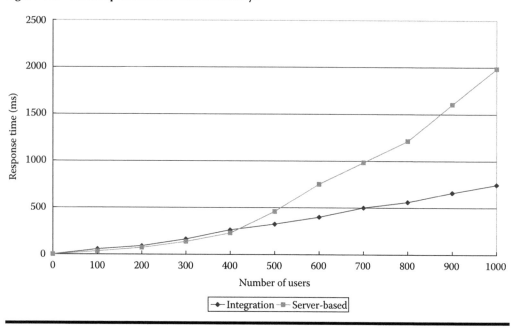

Figure 5.7   Comparison with server-based model.

more, nodes in server-based model get much more congested. Therefore, the average response time increases but exponentially. In proposed system, as the integrated service area is constructed taking users preferences into consideration and the joint requests are distributed into each level, in such case the average response time increases but not as quickly as the situation in the server-based model does.

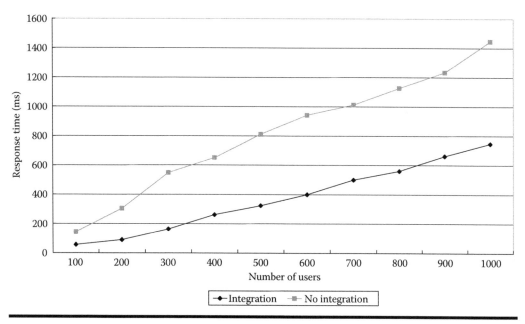

Figure 5.8   Comparison with separated service model.

In the second experiment, two separate service areas of CC1 and CC2, which means system without integration, and a hop-by-hop routing protocol were adopted. Each service area provides only one service and Pull-MAs are dynamically guided towards the relevant service going from node to node. And the same model of users' distribution is realized in the experiment. This simulation shows that the autonomous integration technology improves access time of Pull-MAs with joint requests. We computed the average response time for number of requests from 100 to 1000 per second. The simulation result showed that the response time decreases more than 70%. First reason for such improvement is that the integration increases the size of the overlapping area, where both services are provided. Pull-MAs with join requests find more easily the corresponding service. Second, the information levels adaptation in the common area leads to the satisfaction of a larger number of users' multi-service accesses as shown in Figure 5.8.

5.5.2 Autonomous Correlated Services Access

In this subsection, two simulations are illustrated to show the efficiency of the proposed autonomous correlated services access technology.

1. *Load reduction*: To show that the autonomous correlated services structure and access techniques are not only to improve the response time of multi-service access, but also to reduce the total load of the system.
2. *Access distribution*: To show that the autonomous process adjusting technique is efficient to reduce the local overload in rapidly changing user demand.

5.5.2.1 Load Reduction Evaluation

To evaluate the performance of load distribution, we consider two SPs to provide CC1 and CC2 services and heterogeneous user distributions. Figure 5.9 shows the fixed information structure of

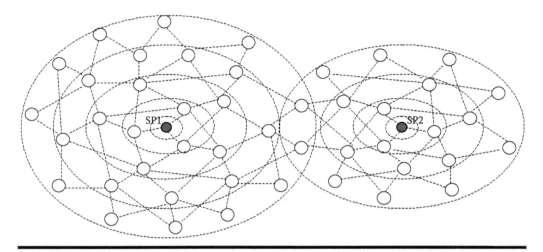

Figure 5.9   The simulation model of load distribution.

Table 5.3   The User Distributions

CC1	Info. ratio	0.2	0.4	0.6	0.8	1.0
	users%	40	30	17	10	3
CC2	Info. ratio	0.3	0.6		0.8	1.0
	users%	50	30		15	5
CC1 and CC2	Info. ratio	0.2	0.4	0.6	0.8	1.0
	users%	45	25	15	10	5

each service that has been constructed to map a certain volume of user requests. The CC1 service area is composed of 30 nodes dispatched on 5 levels and from SP to the edge level the information volume is 10CHs, 8CHs, 6CHs, 4CHs, and 2CHs respectively. The CC2 service area is set up by 20 nodes with 4 levels and 10CHs, 8CHs, 6CHs, and 3CHs are allocated from SP to the edge level.

The distributions of user demands for separated service and integrated service are shown in Table 5.3. Total number of CHs is 10 for each SP and the data size of each CH is 1 kB.

For comparison, three experiments are executed on the simulation model.

1. The separated access method is applied for Pull-MAs with joint request on the separated service areas. The average response time of users who request only one service is observed.
2. The integrated access method is applied for Pull-MAs with joint request on the separated service areas. The average response time of multi-service accesses is observed.
3. The integrated access method is applied for Pull-MAs with joint request on the integrated service area. The average response times of not only separated requests but also joint requests are recorded.

The users preferences realized in the experiments are same, and among the number of users 80% for separated service and 20% for integrated service. We computed the average response time

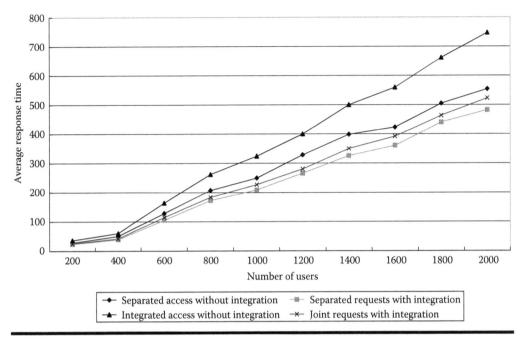

Figure 5.10 Comparison of the average response time.

of each case for number of users from 200 to 2000 per second. The average response time obtained in the condition without integration is compared with the proposed technique in Figure 5.10. The improvement for users' separated requests of each service and joint requests of integrated service is about 20% and 40% in average. The main reason is that the integrated access method reduces the total arrival rate of Pull-MAs to the system in one unit time. In addition, the information structure of integrated service area is effective to improve the ratio of the satisfaction of Pull-MAs with joint request on one node and the service rate of the system is therefore improved. As a result, the total load of the system is reduced autonomously thanks to the proposed integrated information service structure and access method.

5.5.2.2 Access Distribution Evaluation

The performance of the autonomous access distribution is evaluated on the model that contains two SPs, as shown in Figure 5.11. Each service area is composed of 10 nodes dispatched on 3 levels. From SP to the edge level the information volume is 3CHs, 2CHs, and 1CH, respectively. The integrated service area has been constructed to map a certain multi-service requests. The nodes in the system are assumed to be uniform in processing. Each node has three parallel queues for the different requests of Pull-MAs. The processing time depends on the information volume requested by the Pull-MA.

The arrival rate of Pull-MAs for each edge node is assumed to be constant and equal to 20 MA/s. Two kinds of Pull-MAs are considered in this simulation.

1. Among of them, 10 Pull-MAs request only one service and percentages of 1, 2, and 3 CHs requests are 60%, 30%, and 10%, respectively.

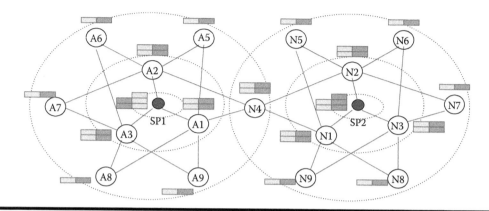

Figure 5.11 **The simulation model of access distribution.**

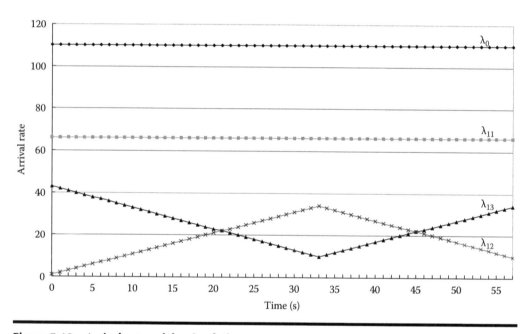

Figure 5.12 **Arrival rates of the simulation.**

2. Other 10 Pull-MAs are joint requests and the percentage of CH(11, 21) requests is 60%. The proportion of users who want to get correlated CHs of CH(11, 12, 21, 22) and CH(11, 12, 13, 21, 22, 23) changes dynamically (see Figure 5.12).

The proposed autonomous access distribution mechanism is measured and compared with the conventional process that Pull-MAs with joint request are forwarded to the corresponding level according to their required correlated CHs. The initial setting to perform the simulation is given in Table 5.4.

In this experiment, each Pull-MA is dynamically guided toward the relevant service going from node to node. And the changing arrival rates of users' preferences are realized in the experiment.

Table 5.4 Initial Setting

Parameter	Means	Value
λ_0	Arrival rate of Pull-MAs to request one service	110
λ_{11}	Arrival rate of Pull-MAs with joint request for CH(11, 21)	66
λ_{12}	Arrival rate of Pull-MAs with joint request for CH(11, 12, 21, 22)	Dynamic
λ_{13}	Arrival rate of Pull-MAs with joint request for all CHs	Dynamic
μ_0	Service rate of Pull-MAs with single request	40 MA/s
μ_1	Service rate of Pull-MAs with joint request	20 MA/s
μ_2	Service rate of Pull-MAs for just getting the lacking CH	60 MA/s

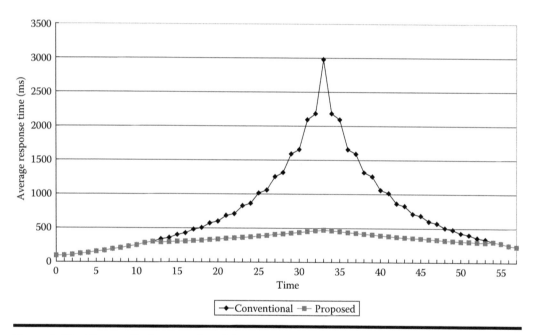

Figure 5.13 The result of comparison (a).

Figure 5.13 shows the average response time of users who want to get all correlated CHs. The autonomous processing and adjusting mechanisms clearly permit to maintain the access time of the Pull-MAs in an acceptable range. The proposed technique makes a 200% improvement in average compared to the *forward to process* case. On the other hand, the average response time of lower level is shown in Figure 5.14. In the case of *get and process*, the average response time of lower level is increased. But taking into account the improvement of the access time, the time cost in lower level is acceptable. As a result, with the changing arrival rates, the autonomous access distribution technique permits much improvement of the access time of joint requests when compared to the conventional process case.

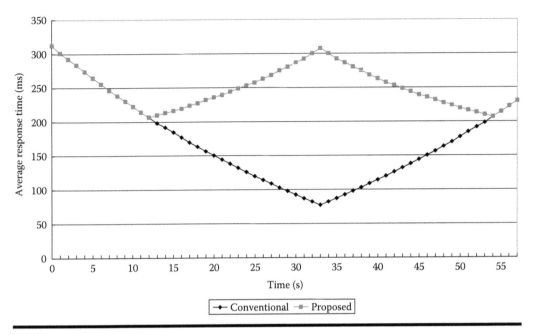

Figure 5.14   The result of comparison (b).

5.6 Related Work

Several standards have been offered to solve the problem of enterprise application and information integration such as CORBA [9] and UDDI [10]. The CORBA standard allows distributed objects to communicate with each other using a commonly defined interface language. However, it is not efficient in large decentralized information systems. Web services standard UDDI is a specification for distributed Web-based information registries of Web services. It focuses more on information registry and discovery issues than integration issue.

With the continued emergence of agent technologies, multi-agent systems have been introduced in information services integration. The Carnot [11] project developed semantic modeling techniques that enable the integration of static information resources and pioneered the use of agents to provide interoperation among autonomous systems. However, Carnot was not designed to operate in a dynamic environment where information sources change over time, and where new information can be added autonomously and without formal control. The InfoSleuth [12] project extends the capabilities of the Carnot system into dynamically changing environments, where resource agents may join and leave the system dynamically. This information is kept by the broker agent which enables the task planning agent to reformulate its plan to access the relevant information sources. In the SIMS [13] project, a model of the application domain is created using a knowledge representation system to establish a fixed vocabulary describing objects in the domain, their attributes, and relationships. For each information source a model is constructed that indicates the data model used, query language, network location, and size estimates, and describes the contents of its fields in relation to the domain model.

All these projects are server-based information services integration system, and centralized knowledge of information services is necessary. The unique aspect of our work is the concept of network-based information services integration. Under the FIF system architecture, every autonomous entity controls over its own operations and coordinates with other entities autonomously and locally just based on the local information.

5.7 Conclusion and Future Work

The Faded Information Field has been designed to satisfy the heterogeneous requirements of users and SPs in rapidly changing environments. In this architecture, the most popular information services are autonomously allocated near to the users to reduce users' access time and increase SP reliability. For mono-service requests, the FIF succeeds in greatly reducing access time while maintaining data integrity [14,15]. However, service provision has to be adapted to satisfy complex multi-service requests. A new approach for network-based integration and timeliness and reliability oriented allocation of information services that share a common replication area in the FIF environment has been proposed. Moreover, a new approach of autonomous correlated services access is proposed to reduce the load of the system and achieve the timeliness of correlated services utilization.

By this means, SPs can increase their service consumption by attracting new types of users with complex multi-service demands. An area of multi-service provision is dynamically created by autonomous allocation of Push-MAs. Users are able to access correlated services simultaneously through Pull-MAs from a number of nodes. The efficiency of the autonomous network-based integration and access for correlated information services has also been confirmed by the results of simulation.

The question of whether the multi-level information structure and autonomous mobile agent technology is superior to contemporary multimedia information services integration remains open. The next stage of validation for proposed system would be to undertake a course of research on the multimedia information services provision and utilization. We have built a simple prototype based on the autonomous agents model to provide on demand streaming information with different qualities [16]. But the current model still has many areas where it can be expanded. Increasing the size and complexity of the system would allow us to reapply the scenarios for change. A comparative study with current application models would be a valuable exercise to ascertain how much of an effect size and complexity has on system adaptability.

The continuous introduction of new applications, the criticality of information services, and the unpredictable demands require for taking a high-assurance approach in global information service systems. In the future ubiquitous computing system, autonomy and decentralization are fundamental virtues to assure the continuous operation. In the multiple services provision and utilization domain, autonomous network-based integration is a valid alternative to the centralized application server that is implemented in current systems.

References

1. K. Mori, Autonomous fading and navigation for information allocation and search under evolving service system, *Proceedings of IEEE Conference on APSITT*, Ulaanbaatar, Mongolia, pp. 326–330, 1999.
2. X.D. Lu, H. Arfaoui, and K. Mori, Autonomous information fading and provision to achieve high response time in distributed information systems, *IEEJ EIS*, 125–C(4), 645–652, 2005.
3. O. Etzioni, K. Golden, and D. Weld, Sound and efficient close-world reasoning for planning, *Journal of Artificial Intelligence*, 89(1–2), 113–148, 1997.

4. O.M. Duschka and A.Y. Levy, Recursive plans for information gathering, *Proceedings of 15th International Joint Conference on Artificial Intelligence*, San Fransisco, CA, 1997.

5. O.M. Duschka, Query optimization using local completeness, *Proceedings of 14th National Conference on Artificial Intelligence*, Providence, RI, 1997.

6. H.F. Ahmad and K. Mori, Autonomous information service system: Basic concepts for evaluation, *IEICE Transactions on Fundamentals*, E83-A(11), 2228–2235, 2000.

7. X.D. Lu and K. Mori et al., Timeliness and reliability oriented autonomous network-based information services integration in multi-agent systems, *IEICE Transactions on Information and Systems*, E86-D(9), 2089–2097, 2005.

8. K. Mori, Autonomous decentralized systems: Concepts, data field architecture and future trends, *Proceedings of IEEE Conference on ISADS*, Kawasaki, Japan, pp. 28–34, 1993.

9. OMG, The common object request broker: Architecture and specification, http://www.omg.org/cgi-bin/doc?1991/91-08-01.pdf

10. UDDI.ORG, UDDI technical white paper, http://www.uddi.org/pubs/Iru_UDDI_Technical_White_Paper.pdf

11. M. Huhns, N. Jacobs, T. Ksiezyk, W.M. Shen, M. Singh, P. Canata, Enterprise information modeling and model integration in carnot, *Proceedings of Enterprise Integration Modeling*, MIT Press, Cambridge, MA, 1992.

12. R. Bayardo Jr. et al., InfoSleuth: Agent-based semantic integration of information in open and dynamic environments, *Proceedings of ACM SIGMOD*, Tucson, AZ, pp. 195–206, 1997.

13. Y. Arens, C.A. Knoblock, and W. Shen, Query reformulation for dynamic information integration, *Journal of Intelligent Information Systems*, 6(2–3), 1996.

14. H.F. Ahmad and K. Mori, Autonomous information provision to achieve reliability for users and providers, *Proceedings of IEEE Conference on ISADS*, Dallas, TX, pp. 65–72, 2001.

15. X.D. Lu and K. Mori, Autonomous information provision for high response time in distributed information system, *Proceedings of IEEE Conference on IWADS*, Beijing, China, pp. 22–27, 2002.

16. M. Kanda, M. Tasaka, X.D. Lu, I. Luque, Y. Jiang, K. Moriyama, R. Takanuki, Y. Kuba, and K. Mori, Autonomous video-on-demand system for heterogeneous quality levels to achieve high assurance, *Proceedings of IEEE Conference on ISADS*, Chengdu, China, pp. 133–140, 2005.

Chapter 6

Ubiquitous Intersection Awareness (U&I Aware)

A Framework for Intersection Safety

Flora Dilys Salim, Seng Wai Loke, Shonali Krishnaswamy, Andry Rakotonirainy, and Bala Srinivasan

Contents

6.1 Background

Every minute, on average, no less than one person dies in a crash worldwide [1]. According to the International Road Traffic Accident Database, globally, there are likely to be 10 million road crashes every year, which claim $1\frac{1}{2}$ million fatalities [2]. The figure of fatal motor vehicle crashes at traffic signals is increasing more rapidly than any other type of fatal crash in the United States. Each year, more than 1,700 people die on Australian roads and more than 60,000 are injured. In Victoria, around 400 people have died on the roads each year from the year 1997 to 2002. Road crashes cost Australia $17 billion a year [3].

Intelligent transportation systems (ITS) aims to create safer and more efficient transport systems [4]. One of the main focuses of ITS is to improve intersection safety, which is a complex issue that requires support from all areas of ITS [IVI02]. The need for enhancing intersection safety is supported by the fact that the figure of the annual toll of human loss caused by intersection crashes has not significantly changed, regardless of improved intersection design and more sophisticated ITS technology over the years [1]. Intersections are among the most dangerous locations on the U.S. roads [2]. In 2002, in the United States, approximately 3.2 million intersection-related crashes occurred, corresponding to 50% of all reported crashes. The 9612 fatalities (22% of total fatalities) and roughly 1.5 million injuries and 3 million crashes took place at or within an intersection [1]. Yearly, 27% of the crashes in the United States take place at intersections [2]. In Japan, intersection collision figures are even more devastating; more than 58% of all traffic crashes occur at intersections. Intersection-related fatalities in Japan are about 30% of all Japanese traffic accidents, and those fatal crashes mainly happen at intersections without traffic signals [2].

The high accident and fatality rate at intersections is due mainly to the complexity of each intersection. Each intersection is unique because of the variety of intersections' characteristics [2,3], such as different intersection shapes, number of intersection legs, signalized/unsignalized, traffic volume, rural/urban setting, types of vehicles using the intersection, various average traffic speed, median width, road turn types, and number of lanes [3]. An intersection is safer if the combination of intersection attributes is less complex. For example, "T" intersections are generally safer than cross-intersections [3]. A decrease in the levels of perception of an intersection will increase accident rates [3]. Therefore, the complex nature of intersection collisions requires systems that warn drivers about possible collisions, based on an awareness of situations at the intersection. In addition, given the uniqueness of each intersection, rather than manually fine-tuning a system for each intersection, an intelligent system for intersection safety should be able to adapt automatically to different types of intersections [10].

The first collision avoidance technology such as adaptive cruise control (ACC) is available on modern vehicles. ACC systems use laser beams or radar to measure the distance of the vehicle from the car ahead and compare both cars' relative speeds. ACC maintains the car's speed on a given

Table 6.1   List of Sensors That Can Be Used to Capture Traffic Data

Data	In-Vehicle Sensors	Roadside Sensors
Speed	Speedometer, GPS	Camera, inductive loop detector
Vehicle size	Built-in information	Vision/camera, radar, lidar
Travel direction	Camera, compass, GPS	Vision/camera
Vehicle position	GPS, GIS	Vision/camera, inductive loop detector, radar, lidar
Angle	Camera, steering wheel, GPS	Vision/camera, inductive loop detector
Vehicle registration number	Built-in information	Vision/camera, automatic number plate recognition (ANPR)
Vehicle manoeuver	Eye and gaze sensors, GPS	Vision/camera

value and distance between itself and the other cars that are ahead. Unfortunately, ACC is mainly effective for driving on sparsely populated roads, such as highway and rural roads. Along with ACC technology, there are many sensors that could enhance car safety [5,6]. Table 6.1 lists type of sensors that are currently available and their usage.

Long before in-vehicle sensors existed, many roadside sensors have been implemented and used for traffic monitoring, such as inductive loop detectors and vision-based sensors. For example, to sense a vehicle's speed at a point, conventional inductive loop detectors, self-powered vehicle detectors, optical sensors, or radar sensors are employed [7]. Inductive loop detectors are used to detect presence of vehicles in a certain road segment. They are also used to measure traffic flow and estimate vehicle speed. In the past few years, inductive loop detectors have been proven effective for detecting incidents, such as road blockage or traffic jam. Vision-based sensors utilize video cameras to monitor certain traffic conditions, for example, speed camera and red light cameras are used to detect traffic violations. Traffic-Dot [8] is able to detect the presence, speed, length, and size of vehicles with up to 97% accuracy, which is better than inductive loop detectors.

Despite of the encouraging development of in-vehicle sensors, in-vehicle collision sensors still have limited visibility and range of detection; therefore, they are effective when used for detecting short-range obstacles such as when parking, but not for longer range problems. Consequently, in-vehicle collision sensors alone cannot guarantee that a car be free from collision.

Pervasive computing research, which has been developing rapidly in recent years, has introduced the notion of context awareness and amplified the use of artificial intelligence. Branches of artificial intelligence such as intelligent agents, machine learning, and data mining are said to be very appropriate to provide enhancement to ITS, because they take into account the human factors aspect of computer systems, including human–computer interaction, distributed problem solving, and simulation of social systems [9]. This has motivated the application of such intelligent systems in ITS. The next section discusses the emerging ubiquitous computing techniques and intelligent systems that have been applied in ITS.

6.2 Emerging Intelligent Systems in ITS

According to traffic engineers, human factor specialists, and others in this research area, emerging intelligent systems provide significant potential for improving intersection safety [10]. Safety of intersection environments can be improved by utilizing and integrating the advances in sensor development, ITS, and pervasive computing research. Safety of intersections can be improved by integrating the advances in sensor development, ITS, and situation-based reasoning.

Multiagent technology is very fitting for coordination of entities on the road. The abstraction of independent and autonomous entities, which are able to communicate with other entities and make independent decisions, maps eminently to the situation of an on-road scenario, where each entity, such as a vehicle or a traffic light, can be represented by an intelligent agent [10]. According to Wang [11], agent-based techniques are very appropriate for traffic and transportation management due to the distributed nature of traffic entities and the variation between busy and idle periods of traffic. The need for autonomy in managing and coordinating traffic in the rapidly changing environment of road transportation systems can be satisfied by multiagent systems, as each intelligent agent is able to learn, evolve, and react independently in the face of various road and traffic conditions. Agents are typically situated in an environment and able to sense changes in environment. Agents are capable of learning from their experiences[12] and also capable of autonomous actions [13]. Therefore, agents are, in fact, software models for decision-making systems embedded in an environment and acting autonomously in order to meet its design objectives [12,13]. According to France [14], agents are required to work autonomously within a real-time persistent environment. Moreover, agents have the capabilities to manage events that dynamically arise, recognize traffic changes, and deal with a vast array of traffic patterns [14]. The autonomy of agents is to the extent of the ability to compute its decision based on input sensory data, messages, or commands received to achieve the goals [15]. The autonomous nature of agents can either be proactive, reactive, or a combination of both. The problem-solving element in an agent, which mainly constitutes its autonomy, can be a rule-based algorithm, fuzzy rule, neural network, or any other types of learning algorithm [15]. Therefore, the necessity of integrating autonomous behavior within ITS has motivated the usage of intelligent agents to perform autonomous actions on behalf of the user in the traffic and transportation systems, such as for autonomous vehicle navigation, traffic optimization and control, traffic modeling and simulation, and intersection safety.

However, apart from the societal aspect of the traffic and transportation systems, there is a considerable amount of data from in-vehicle and roadside sensors. Hence, it is essential to make sense of the sensors data and act accordingly. Given the huge amount of data, a question arises whether computer systems can learn and improve automatically from past experience. Effective algorithms have been formulated to facilitate better understanding of data, better ways tasks are being executed, or performance improvement through experience [16]. Machine learning has been implemented widely in Intelligent Transportation System. The ALVINN system [17] applied machine learning to drive unaided at 70 miles/h for the distance of 90 miles on public highways among other vehicles. With the huge amount of data available at present in databases, spreadsheets, data from sensors, and many other organizational data, data mining has become popular over the last decade. According to Fayyad, data mining is the development of methods and techniques for making sense of data by pattern discovery and extraction [18]. There have been a number of research projects on data mining in the area of ITS, such as for driver's behavior recognition, traffic optimization, and incident detection.

Although data mining has been effectively used for extracting useful knowledge from data storage, however, the advances in sensor technology have resulted in a huge amount of sensor data to be understood. Therefore, it is not efficient to store real-time sensor data for later processing. Preferably, data processing should be done on the streams of sensor data, not on sensors' data storage. The analysis of data streams to discover useful knowledge such as patterns and association rules on mobile, embedded, and ubiquitous devices is called ubiquitous data mining (UDM) [19]. There are only a limited number of research projects that have applied UDM for traffic and transportation systems, which include drunk-driving behavior monitoring [20], vehicle health monitoring, and driving pattern recognition [21].

Data analysis techniques have the potential to facilitate better understanding of the vehicle, the driver, and the road environment for different purposes. Hence, information about the vehicles, infrastructures, and environment (road, traffic) extracted from sensors and further data analysis of sensor data can be utilized for better situation recognition and management. The notion of context awareness has been adopted in ITS, since a context-aware application has the capability to adapt to situation changes informed by the sensors. Furthermore, this is also supported by the advancing wireless technology, which facilitates communication among vehicles, road infrastructures, and traffic authorities.

Context is "any information that can be used to characterize the situation of an entity" [22]. An entity is "a person, place, or object that is considered relevant to the interaction between a user and an application, including the user and applications themselves" [22]. The availability of context information may influence the behaviors of the application or device [23,24]. The most important types of context are identity, location, time, and activity [24]. Therefore, context-aware applications observe the "who's, where's, when's, and what's" of entities and use this information to find out why the situation is happening [22]. An application can then use available context information to adapt to environment changes. Hence, context awareness is a necessity in ubiquitous and mobile computing paradigm to provide faultless and adaptive computing infrastructure, so that applications or devices can discover, use, and adapt to contextual information [23,25]. It is essential to incorporate knowledge about context to properly make decisions in complex dynamic environments such as driving [26]. Context awareness has been applied in ITS for driver's behavior recognition, cooperative and autonomous vehicle navigation, traffic modeling and monitoring, and environment mapping and monitoring.

The next section discusses existing approaches to intersection collision warning and/or avoidance systems and presents the current gaps for context-aware intersection collision avoidance systems.

6.3 Intersection Collision Warning and/or Avoidance Systems

There have been a number of initiatives in developing intersection collision warning systems and/or avoidance systems. Currently, no existing intersection collision warning and avoidance systems can tackle intersection collision problems entirely. Intersection collision warning and avoidance systems can be categorized as either *vehicle-based*, *infrastructure-only*, or as *infrastructure-vehicle cooperative* [7]:

1. *Vehicle-based systems* rely only on in-vehicle sensors, processors, and interface to detect threats and produce warnings.
2. *Infrastructure-only systems* utilize "roadside sensors, processors, and warning devices; roadside-vehicle communication devices; other roadside informational or warning devices, and traffic signals to provide driving assistance to motorists" [7]. Infrastructure-only systems rely only on roadside warning devices to inform drivers.

3. *Cooperative systems* communicate information straight to vehicles and drivers. The main advantage of cooperative systems rests in their potential to improve the interface to the driver and thus to almost guarantee that a warning is received. Cooperative systems include vehicle-to-vehicle communication and also infrastructure-to-vehicle communication.

6.3.1 Infrastructure-Only Intersection Collision Warning and Avoidance Systems

The Intelligent Vehicle Initiative of the U.S. Department of Transportation proposed initial concepts for intersection collision avoidance systems, which include traffic signal violation warning, stop sign violation warning, traffic signal left turn assistance, and stop sign movement assistance. They aimed to install sensors on the roadside to detect speed, acceleration rate, deceleration rate, stopping, and movement of each vehicle approaching the intersection from all directions. Warnings will be issued to drivers as violations or potential conflicts are detected. Warnings are given by (1) activating warning lights to notify a need for caution and possibly to point out the source of the conflict, (2) activating intelligent rumble strips to notify the other motorist to slow down and advance carefully at the intersection, and (3) using a variable message sign (VMS) or graphic display sign used to notify drivers of the potential conflict with the signal violator. All the four systems are still categorized into infrastructure-only systems, because there is no direct infrastructure-to-vehicle communication. As warning messages are given from the roadside, warning can be distractive and less effective. In the near future, vehicles will also be equipped to receive intersection collision warnings from a driver interface with the development of in-vehicle sensor and communication technologies [27].

A vision-based sensing system for monitoring an intersection and predicting vehicle collisions is currently under development [28,29]. It uses a single camera arbitrarily positioned at an intersection to observe the traffic flows. The system classifies moving objects (such as vehicles and pedestrians), tracks each of their movements [29], and collects traffic data such as vehicle speeds, positions, routes, accelerations/decelerations, vehicle sizes, and signal status [28]. It is able to compute promptly the potential collisions and near misses by applying algorithms that analyze the speeds and routes of the moving objects collection. This ongoing project still has unsolved issues, such as obstructions of the image (such as trees) and shadows in the vehicle image. In addition, vehicles that are not moving cannot be tracked by this vision-based system [28].

In summary, existing intersection collision warning systems are still infrastructure-only systems and are limited in certain aspects [10], such as the following:

1. In delivery of a warning message, it is distractive and less effective as warnings are only displayed on the roadside.
2. There is no communication means that exists between road infrastructure and vehicles, and therefore, there is no exchange of useful information between them.
3. Information about the intersection might not be comprehensive as the only data source is roadside sensors.
4. The systems are mostly reactive. Reactive behavior is required for such a real-time solution; however, deliberative reasoning can supplement and enhance these systems.
5. Each system is built for a particular intersection and cannot be generalized for other types of intersections, and therefore, each application requires a field study on that intersection.

6.3.2 Vehicle-Based Intersection Collision Warning and Avoidance Systems

Safety countermeasures for a single car have been developed by the National Highway Traffic Safety Administration (part of U.S. Department of Transportation) to cope with four different cross-intersection crash scenarios. Nonetheless, the system does not include any means of communication between the infrastructure and the vehicle. Investigating usage of communication is listed as one of their future works.

A research project by Miller and Huang [30] uses a peer-to-peer concept where information and messages are communicated between a pair of vehicles and so, falls into the category of vehicle-based collision detection systems. Threat detection relies on location, velocity, and acceleration information shared by other vehicles that use the system. Their proposed collision detection algorithms consist of algorithm to predict future collision point, predict collision time, and issue timely warning. Their algorithms were implemented using a multiagent system approach for intersection collision warning system. Each vehicle has a multiagent-based software architecture and hardware architecture installed to detect potential dangers. The software architecture consists of three layers [30]:

1. Sensory agents (i.e., global positioning system (GPS) agent, brake sensor agent)
2. Decision/control agents (i.e., collision warning system agent)
3. Presentation agents (e.g., speaker agents) that deliver warnings to driver

However, the limitations of this system are as follows:

1. The agent architecture is reactive, without learning new knowledge, such as driving behaviors and crash patterns of the intersection, which can enhance the system to react better.
2. Useful information about the infrastructure and environment is not incorporated here as there is no communication between infrastructure and vehicles, and between vehicles and external parties.
3. Changes in velocity in terms of acceleration and deceleration of vehicles in calculating collision time prediction are not considered.
4. Miller and Huang algorithms require very frequently updated information due to split-second velocity and location changes, thereby incurring high communication costs. However, with known acceleration or deceleration to predict future velocities, communication costs can be reduced.

In summary, most of ongoing research for in-vehicle collision warning system (e.g., forward collision, rear-end collision, and side collision warning system) enable the system to work in all road types, either in rural or urban areas, on highways or small streets, and also at intersections. These vehicle-based collision warning systems are fairly effective for a single vehicle. However, in an intersection, a potential danger normally impacts more than one vehicle, and therefore, a cooperative system is a preferred solution. Communication between infrastructure and vehicles is sought to be implemented as it will improve the effectiveness of vehicle-based systems.

6.3.3 Cooperative Intersection Collision Warning and Avoidance Systems

Research projects on cooperative intersection warning and/or collision avoidance systems have been initiated to improve intersection safety. One of the recently started initiatives that are developing cooperative safety system for intersection is the cooperative intersection collision avoidance systems (CICAS) by the U. S. Department of Transportation, which seeks to develop vehicle-based systems, infrastructure-only systems, and finally, infrastructure-vehicle cooperative systems. Vehicle-based systems include sensors, processors, and interfaces for the driver inside each vehicle. Infrastructure-only systems depend on roadside sensors to identify vehicles and threats and then generate signals or other means to warn motorists of potential collisions. Infrastructure-only operations also necessitate data processing techniques, an essential evolutionary move toward deployment of subsequent cooperative systems. Infrastructure-vehicle cooperative systems will use infrastructure-only systems and will also have a communications system, for example, dedicated short range communications (DSRC) to exchange warnings and data directly with drivers in vehicles capable of accepting and displaying the warnings within the vehicle. It has been stated that data processing and analysis techniques are also required to assess situations in such contexts [31].

The INTERSAFE project employs two different methods in parallel [32]. The first method is to develop the basic intersection safety system that is implemented on a VW test vehicle with two laser scanners for object detection, one video camera for road marking detection, and vehicle-to-infrastructure communication. Communication units will be installed at selected intersections to enable communication between the vehicle and traffic lights. A static world model is constructed from object detection, road marking detection, landmark navigation, GPS, and map [32]. The second method is to develop advanced intersection safety system that is implemented on a BMW driving simulator. This driving simulator will examine dangerous states beyond the limitation of sensors in detecting the environment. In this second methodology, a dynamic risk assessment is executed based on object tracking and classification, communication with traffic management, and driver intention. Hence, potential threats to other road users and conflicts with traffic controls can be detected. As a result, the system by INTERSAFE will be able to provide stop sign assistance, traffic light assistance, turning assistance, and right of way assistance [32]. INTERSAFE also identifies the need for analyzing the situation and collision risks at an intersection, but specifics of how to learn and what techniques are appropriate have not been investigated or addressed.

In summary, research initiatives in developing cooperative systems for intersection safety such as INTERSAFE [32] and CICAS [31] have commenced. To our knowledge, these projects do not mention techniques to discover crash patterns and precrash behavior associations, which are essential to detecting and reacting to potential threats. A generic framework that can automatically adapt to different intersections is required for efficient deployment; however, these projects have not addressed this issue [10].

6.3.4 Summary of Gaps and Challenges

From our review of current work, it is evident that there is a need for a cooperative intersection collision warning and avoidance system that addresses the following challenges [10]:

1. There is a requirement for an intersection safety model to detect high-risk situations and foresee threats in particular intersections. Given that sensor techniques have been deployed in cars and on infrastructures and there are considerable amount of sensor data, there is an opportunity to reason and use these data to develop patterns and associations that can help in better understanding of high-risk situations and behaviors that lead to crashes. While current systems tend to be reactive to situations, there is increased recognition [31–33] that reasoning and learning can be integrated to supplement reactivity.

2. As each intersection is unique, the profile of high-risk situations in one intersection is different from another; therefore, a generic model that is able to adapt to particular intersections over a period of time is required. Each system in different intersections should have a localized knowledge that is applicable only within its defined area, and this knowledge is gained through reasoning and learning. Hence, this approach alleviates the inefficiency of the current method of developing a different intersection collision warning and avoidance system for different intersections [4,6,28,34].

3. There is a necessity for a model that calculates the cost of real-time communication for delivering collision avoidance warning messages. A comprehensive contextual understanding of a particular intersection is required so that the system is able to act or respond better to a hazardous situation. The system should also be capable of dealing with high volumes of data in a highly dynamic environment and with mobile entities.

An intersection safety system should be able to detect collision in real time, since collision warning must be delivered in time before collision occurs. An early and accurate detection should allow time for the system to warn a potential collision, for drivers to respond to warnings, and for avoidance systems or drivers to steer clear of the potential collision. Firstly, the collision detection algorithm should be simple and optimized. Secondly, reducing the number of vehicle pairs to be calculated in real time can reduce the computational time, because calculating each possible pair of vehicles located at an intersection for a potential collision is not prudent due to time constraints. In order to address these gaps, we propose the ubiquitous intersection awareness (U&I Aware) framework, which is presented in the next section.

6.4 U&I Aware Framework

In order to improve the safety and design of an intersection, one of the first procedures is to execute a field observation and statistical analysis of collision patterns. Understanding patterns of collisions in an intersection can assist in planning for countermeasures. The process of learning patterns of collisions is mainly done manually and repeated for each intersection. Results of those studies cannot be applied for all types of intersections due to uniqueness of each intersection. An intersection safety system should be able to adjust to different types of intersection through computer-based pattern acquisition, not manual field observation. It is necessary to have a comprehensive collision pattern in an intersection safety system in order not to miss detecting a potential collision, since the system can only detect and warn vehicles that match those patterns kept in the knowledge base.

We implemented the U&I Aware framework (Figure 6.1), which aims to achieve holistic situation recognition at road intersections. We incorporate a means of filtering and matching vehicle pairs that have the potential of colliding with each other to reduce the number of collision detection computation. We suggest that patterns of collisions that are accurate can be used as selection criteria for finding and matching a pair of vehicles and therefore reduce the number of vehicle pairs to be calculated by the collision detection algorithm. Therefore, we store collision patterns in a dynamic knowledge base, which is populated through data mining of historical traffic and collision data. The usage of collision patterns has been customary in intersection safety studies, although it is not for the purpose of improving the performance of detecting potential collisions.

Currently, collision warning systems mostly react to events that might cause collision. Intersection collision warning systems should also evolve by adapting to information gained from analysis of sensor and historical data in the intersection. By learning from historical data of collision and

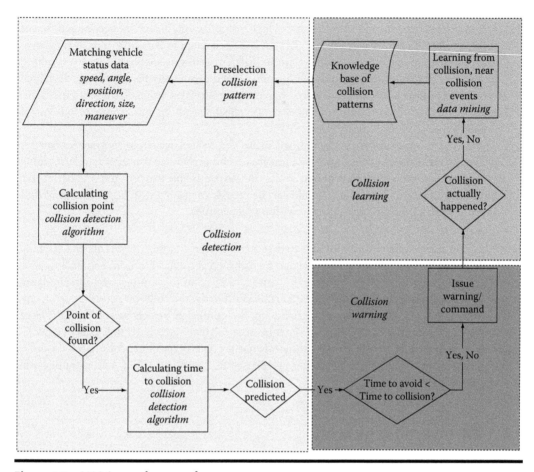

Figure 6.1   U&I Aware framework.

near-collision events, improved detection and reactive behavior can be achieved since the knowledge base of the intersection is evolving in the U&I Aware framework. Thus, the system can gain better knowledge of any intersection where it is installed for better crash prediction.

Each of the components is described as follows:

1. Collision learning: The collision learning component consists of the following elements:
    a. Data collection: Historical collisions as well as online real-time vehicular and traffic data from the intersection's vicinity are collected to be analyzed. Since collision data are rare, "near collision" or "near miss" events are also captured to support data collection. The U&I Aware framework only consumes sensor data; it does not perform any sensor data fusion or processing of raw sensor data from a particular sensor or a sensor network. Learning can start as soon as data are collected. A minimum quantity of data required is not specified since patterns (such as collision patterns) can be extracted once there are data. However, as a general rule of thumb, the more data are acquired, the higher are the support and confidence of the patterns and rules extracted from it.
    b. Data mining: Due to the need for a generic intersection collision warning and avoidance system, learning of specific collision patterns that are relevant for each particular

intersection needs to be performed using data mining techniques. Once data are collected, data mining is applied on the collected data.

c. Knowledge base integration: The results of learning that are relevant only for that particular intersection are integrated into the knowledge base of the framework for that intersection. Hence, the knowledge base is specific to an intersection and the situations that occur at the intersection. The knowledge base is used as the basis for preselection, which is an algorithm to match the vehicles that pass through the intersection with the collision patterns in the knowledge base. This is the key to reducing time to collision (TTC).

2. Collision detection: In this article, the terms "collision detection" and "collision prediction" are used interchangeably as both refer to recognizing potential collisions (i.e., future threats). These terms do not refer to identifying past or existing collision events. The collision detection component contains the following elements:

   a. Preselection: Based on the status data of a vehicle and key collision patterns in the knowledge base of the intersection, the preselection algorithm identifies vehicle pair combinations that have possibilities to collide.

   b. Calculate future collision point: The potentially colliding vehicles provide data to the collision detection algorithm. Each vehicle pair selected by preselection is assessed to see if a future collision point exists.

   c. Calculate TTC: If a future collision point is detected, then the TTC of each vehicle in the pair to the future collision point is calculated and compared. When the TTCs of both vehicles are almost equivalent, then a future collision is imminent.

3. Collision warning: The elements of the collision warning component are listed in the following:

   a. Calculate TTA: TTAs (time to avoidance) of both vehicles are calculated using the TTA cost model. We present our proposed TTA cost model that addresses the need for a real-time communication protocol.

   b. Issue warning or command: Depending on the TTA of each vehicle, either warning messages are issued to drivers of the relevant vehicles or command messages are generated and sent directly to the vehicle systems to avoid or minimize impact of an impending collision.

6.4.1 Novelty of the U&I Aware Framework

Currently, existing collision warning and avoidance systems only have detection and warning components. Consequently, they can merely react and respond to certain events as preprogrammed. However, the U&I Aware framework contains a learning component. The novelty of collision learning enables new intersection collision warning and avoidance systems (that can suit to various intersections) to be developed on the basis of the U&I Aware framework as the governing principle. This is because the adaptation of new knowledge and information gained from mining of sensor and historical data at the intersection is performed as an integral part of the framework. By learning from historical data of collision and near-collision events, improved detection and reactive behavior can be achieved since the knowledge base of the intersection continues to evolve. Thus, the system can operate in any intersection where it is installed and learns of collisions that are specific to that intersection.

The U&I Aware framework, as a basis for a cooperative and generic intersection collision warning and avoidance system that works on various intersections, is inspired by the notion of context awareness, since a context-aware application is capable of being conscious of the changes in its environment and adjusting its behavior accordingly. A context-aware application consists of a set of context attributes that become the basis for recognizing a situation, adjusting the behavior of the application, and issuing a specific response.

As the U&I Aware framework is generic and adaptable to different locations, it can be considered as a context-aware application (or to be more specific, a location-aware application). The framework can be aware of changes in the location context and able to use the context information (e.g., collision patterns, traffic patterns, road user behaviors) as stored in the knowledge base to adapt to location changes by learning from sensor data. This knowledge base has the ability to grow over a period of time if incremental learning from current events is incorporated into the system. In fact, collision patterns are the main context attributes that are used in the U&I Aware framework that makes it a context-aware application. There are multiple context attributes that can determine the behavior of the application, which are applicable to this application domain. Examples of context attributes that can be used in an intersection safety system might be speed profile of a driver, acceleration behavior of a driver, speed limit of the intersection, traffic patterns during different times of the day or different days of the week, etc. In this thesis, collision patterns are the context attributes that determine the circumstances in which collision detection is performed. Collision detection is only performed when matching vehicle status data with the context attributes (i.e., collision patterns) are found.

The key to the context awareness of the U&I Aware framework lies in the integration of data mining techniques and a knowledge base to facilitate the framework to learn from its environment (*and accumulate context attributes of a specific intersection location*), be aware of the occurrence of learnt events or incoming threats in the environment (*monitor the intersection for events that can be identified with the context attributes*), and respond to the incoming threats contextually (*based on a given context attribute, the system yields a certain action, e.g., issuing a specific warning to the relevant drivers*).

6.4.2 Scope of Implementation

For implementation, the U&I Aware framework is mapped to agents in intersections and vehicles. The notion of an agent is used to signify a piece of software that can act autonomously on behalf of the user. Ideally, each agent needs to be capable of learning from sensory and historical data, detecting threats, and issuing warning to one another. Learning needs to be enabled in each agent depending on the context. For example, an intersection agent (IA) can learn patterns of collisions and traffic at the intersection's vicinity. A vehicle agent (VA) can learn driver behavioral patterns in driving context as well as dangerous driving behaviors (such as drink driving and drowsiness). Threat detection can also be enabled in every vehicle and IA based on the patterns learnt on the agent. For example, a VA is able to detect drunk-driving behavior and threats that are faced by the driver when such behaviors are learnt. However, collision detection based on collision patterns can only be done by the IA, since collision patterns are learnt by each respective IA and not by VAs. When a threat is detected, the agent can then issue warning messages to other agents that may be impacted.

The communication between the IA and VAs in the U&I Aware framework is regulated inside the *administration zone* (Figure 6.2). An *administration zone* is the spatial domain that determines the region of authority of an IA to coordinate VAs in the approaching and passing vehicles. A wireless infrastructure is required for the administration zone and messaging of U&I Aware framework to operate. Each vehicle needs to be equipped with a wireless device, or at least with

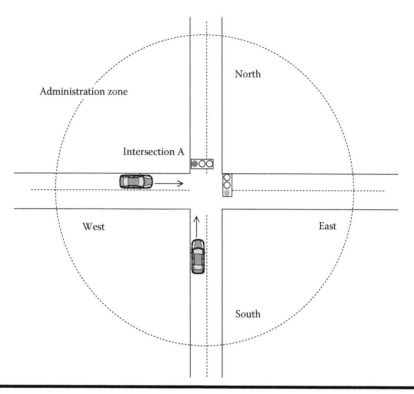

Figure 6.2 Intersection administration zone.

Bluetooth. The size of an administration zone is dependent on the effective wireless signal strength. The maximum radius of an administration zone depends on the network protocol being used. The current vehicular access network (VANET) protocol is being revised into a new proposal of the IEEE 802.11/p network protocol for vehicle-to-vehicle and vehicle-to-infrastructure communication. As soon as the IEEE 802.11/p hits the road worldwide, this scenario is going to be feasible for implementation.

An IA acts as a central traffic authority to learn collision patterns, detect threats, and warn possibly affected vehicles of incoming hazards. The VA of each car at the intersection should always report to the IA of its entry into and exit from a designated area in the vicinity of the intersection and also send its status periodically. Vehicle information and driver's behavior information, such as driving maneuvers, are retrieved from in-vehicle sensors. An IA manages the tasks of communication, learning, detection, and warning. The protocol of the communication is described further in the later section.

6.5 Collision Learning

Ubiquitous computing research provides a significant opportunity to develop novel ways of improving road intersection safety. In-vehicle sensors have received considerable research and development focus and are now a reality in today's roads. The increased proliferation of such sensors has brought with it the question of how the sensory data can be leveraged for effective and efficient road safety enhancement. First, given the large amount of sensor data that are obtained from intersections and sensor-equipped cars, analysis and learning from these data can help detect intersection accident patterns. Second, such patterns can be incorporated in accident detection systems. These patterns can be learnt through the historical collision and traffic data, which are collected from roadside sensors. We can also incorporate positive/negative results of the past collision warning ("collision actually happened?"), which can be communicated by the system in the vehicle, for refinement of the collision patterns. Data mining is proven to be effective for extracting traffic patterns and trends [35].

In opposite to static knowledge base, dynamic knowledge base involves learning to accumulate and refine rules in knowledge base to adapt to situational changes. The dynamic knowledge base in an intersection collision detection system should contain valid and comprehensive collision patterns. Collision pattern learning is performed by using classification rules of data mining. New events are matched with the existing classes in the patterns repository of the intersection central agent or the car agent, depending on where learning happens. If a collision happens outside a known pattern, a learning process can detect and add a new collision pattern. There are a number of improvements and enhancements that can be added to the plain collision warning system that is based only on trajectory calculations. These are done via mining of data from our simulation, assumed to be obtained from on-the-road sensors in order to characterize collision patterns. The sensor data simulated in our system resemble the real-world data gathered from sensors installed on the freeway by The Pantheon Gateway Project [35].

The simulated sensor data have six attributes, three of which (i.e., *direction, maneuver, and angle*) are from colliding vehicle pairs. Whenever there is a collision or near-collision event in our intersection simulation, data from the colliding (or near-colliding) pair of vehicles are collected and mined. In the real world, such data can be collected with conventional sensors such as inductive loop detectors on the road or speedometer in the vehicle. We have successfully classified types of side collisions or perpendicular crashes in a cross-intersection using the C4.5 decision tree (J48 classifier from Weka [36]), and the second vehicle direction (*Veh2_Direction*) attribute is nominated as the class. The implementation results also exhibit the most common crash patterns that exist within the particular intersection where the traffic data are acquired. Then, to realize all the possible crash patterns that involve a specific driving maneuver (e.g., straight) in an intersection, a Bayesian network classifier [36] is used to classify the same data. The crash patterns enumerate four possible straight driving directions in a four-leg cross-intersection, which are left, right, up, and down. The classification shows all the possible collision patterns that might happen with the probability rate of each crash pattern (see Figure 6.3). The highest probability of a crash pattern in each direction is circled in red in Figure 6.3. Out of all the collisions that occur to vehicles that travel from the right leg to the left leg (i.e., "LEFT" direction), 93.1% of the collisions occur with vehicles from the lower leg to the upper leg (i.e., "UP" direction). Based on the result, we can also deduce that vehicles that travel with a straight maneuver from the left leg to the right leg of the intersection

Probability Distribution Table For Veh1_Direction				✕
Veh2_Direction UP		DOWN	RIGHT	LEFT
UP	0.014	0.014	0.042	0.931
DOWN	0.022	0.065	0.891	0.022
LEFT	0.583	0.25	0.083	0.083
RIGHT	0.611	0.278	0.056	0.056

Figure 6.3   Collision patterns based on vehicle direction classified with Bayesian network.

Probability Distribution Table For Coll_Type		[X]
Veh1_Direction	SideCollision	RearEndCollision
DOWN	0.411	0.589
UP	0.023	0.977
RIGHT	0.044	0.956
LEFT	0.145	0.855

Figure 6.4   Collision patterns based on collision types as classified by Bayesian network.

("RIGHT" direction) tend to collide with vehicles that travel with a straight maneuver from the upper leg to the lower leg ("DOWN" direction). Note that these results were obtained from our simulated data for one intersection. Applying the same technique to a different intersection (with different data) could lead to different likely situations for collisions—the point is that applying such learning techniques would enable such collision situations to be recognized automatically and identified as "dangerous" patterns.

We also included data of rear collision events that occur in the simulation in a latter experiment. The test data contain seven attributes, that is, *direction, maneuver, and angle* from each vehicle in a colliding pair, and *collision type* (side collision or rear-end collision) and 20–30 rows in a file. In this particular intersection, when Bayesian network classification is applied with *collision type* nominated as the class, the result shows that rear-end collision occurs much more often than side collisions in this particular intersection (Figure 6.4). Using the same set of data, when the expectation maximization (EM) [36] is applied, it also exhibits the same highest probability of side collision patterns as in Figure 6.4.

In order to find trends in maneuver involved in certain collisions, we use EM clustering and the C4.5 decision tree. Visualization of EM results shows clusters of side collision with stopped maneuver, rear-end collision with straight maneuver, and rear-end collision with stopping maneuver. This is confirmed by C4.5 result (Figure 6.5). We conclude that in this particular intersection, most side collisions occur when one of the vehicle pair is stopped and rear-end collisions happen mostly when both vehicles are on the move with straight maneuvers and secondly when both vehicles are stopping.

Based on the results of knowledge acquisition, there are two types of collisions in this particular intersection, which are rear-end collisions and side collisions. Each of this collision type consists of a number of subtypes, for example, rear-end collision with stopping maneuvers and rear-end collision with straight maneuvers. The collision patterns are stored in the system's knowledge base. The knowledge base in our system has a hash table of collision patterns. Each pattern stores information about a pair of colliding trajectory, which is represented in vehicle's direction, maneuver, and leg position. We use the collision patterns in the knowledge base to improve the speed of detection by preselection technique, which is discussed in the next section.

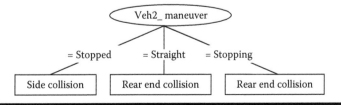

Figure 6.5   Classification of collision types based on vehicle maneuvers.

6.6 Collision Detection

The basic of calculating collision detection is the well-known speed formula, which is calculated by

$$v = \frac{s}{t} \qquad (6.1)$$

where
  v is speed
  s is distance
  t is travel time within the distance

Based on the formula (6.1), collision detection can be calculated by the following steps:

- Calculate future collision point, which is by finding route contention of a pair of vehicles.
- Calculate time for each vehicle to reach future collision point (TTC) based on the aforementioned speed formula.
- If TTC of one vehicle is equal or nearly equal with TTC of another vehicle to reach the same collision point, then collision is detected.

The peer-to-peer collision warning system by Miller and Huang [14], as discussed previously, consists of a pair-wise collision detection algorithm that computes the point of collision, TTC, and TTA. Their proposed algorithms to calculate the future collision point (x_+, y_+) are stated in (6.2) and (6.3) and the symbols used in those formulas are illustrated in the Figure 6.6.

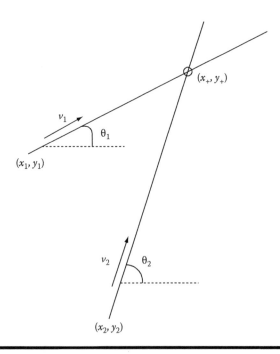

Figure 6.6 Collision detection algorithm. (From Miller, R. and Huang, Q., An adaptive peer-to-peer collision warning system, *Proceedings of Vehicular Technology Conference (VTC)*, Birmingham, AL, Spring 2002.)

The algorithm to calculate future collision point (x_+, y_+) is as follows:

$$x_+ = \frac{(y_2 - y_1) - (x_2 \tan \theta_2 - x_1 \tan \theta_1)}{\tan \theta_1 - \tan \theta_2} \tag{6.2}$$

$$y_+ = \frac{(x_2 - x_1) - (y_2 \cot \theta_2 - y_1 \cot \theta_1)}{\cot \theta_1 - \cot \theta_2} \tag{6.3}$$

The time for each car to reach the future collision point (TTX) [14] is calculated by

$$TTX_1 = \frac{|\vec{r}_+ - \vec{r}_1|}{|\vec{v}_1|} \mathrm{sign}((\vec{r}_+ - \vec{r}_1).\vec{v}_1) \tag{6.4}$$

$$TTX_2 = \frac{|\vec{r}_+ - \vec{r}_2|}{|\vec{v}_2|} \mathrm{sign}((\vec{r}_+ - \vec{r}_2).\vec{v}_2) \tag{6.5}$$

where
v is velocity of each car
r is the vector of the coordinate (x, y) [14]

As vehicles have variation in size, collision can no longer be expressed as a point; instead as a region. The α parameter, which is used to represent the size of the region, depends on the vehicle size. Therefore, a future collision is detected if time for both vehicles to reach the collision point is the equal or nearly equal, that is, $|TTX_1 - TTX_2| < \alpha$ [14].

This algorithm incurs high computational cost because the algorithm requires calculation for each possible pair of vehicles in the intersection (brute force). Therefore, real-time detection is challenging when the number of vehicles increases at the intersection. As centralized approach for collision detection computation is adopted in the U&I Aware framework, the formula to calculate the number of vehicle pairs to be monitored for collision detection is as follows:

$$\sum_{i=1}^{n} (i - 1) \tag{6.6}$$

where n is the number of vehicles. Hence, the number of vehicle pairs grows in a linear square as the number of vehicles in the intersection grows. In order to sustain the performance and scalability of collision detection of vehicle pairs in an intersection, there is a need for reducing the number of vehicle pairs for which collision detection points need to be calculated.

Furthermore, mere application of the algorithm only enables the system to react to threat. There is a need for analyzing collision, near-collision, or near-miss data to enhance collision detection. Therefore, applying data mining techniques, as discussed previously, along with implementation of the pair-wise collision detection algorithm helps better situation recognition. In addition, with the results gained from mining collision patterns, the number of vehicle pairs to be calculated for collision detection can be reduced by applying the preselection method.

In dealing with the issue of the high computational cost of a conventional collision detection algorithm, we propose a preselection strategy. Preselection is a method to improve the performance of the conventional collision detection by reducing the number of vehicle pairs in the intersection to be calculated for collision detection. Every subject vehicle (SV) is paired up with the potential principal other vehicle (POV) based on the collision patterns learnt at the intersection. This pair is then added into the pool of matching vehicle pairs. Other vehicles that do not match with the SV based on the characteristics of any collision pattern are not included in the pool. An SV is paired up with a POV based on the direction pair, maneuver pair, and/or location pair in an existing collision pattern. Whenever a pair of vehicles for potential collision is found, the matching collision pattern yields the collision type (i.e., side collision or rear-end collision) and the relevant collision detection computation based on the collision type is applied to assess the possibility of an imminent collision. Therefore, pair-wise collision detection is only performed on pairs of vehicles that have the possibility of collisions based on the known intersection collision patterns.

6.6.1 Implementation

As an implementation test bed, we use a computer-based simulation of two different scenarios: intersection with traffic lights (Figure 6.7) and without traffic lights. At this stage, computer-based simulation is an acceptable proof of concept, since the scenarios that we implement involve collisions that are difficult to be simulated in the real world due to the constraint of resources and technology. The simulation parameters are as follows:

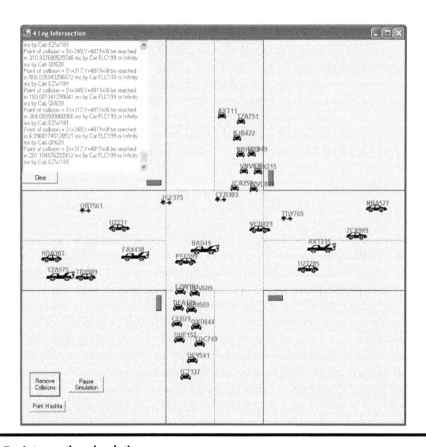

Figure 6.7   Intersection simulation.

1. Intersection module: intersection type, leg, lane, lane group, traffic control
2. Vehicle: speed, acceleration, size, type, position, angle, maneuver
3. Driver: profile, intended destination, choices of maneuver

The vehicles are randomly generated at a fixed time period (deterministic traffic flow/distribution) with different speeds, maneuvers, position, and trajectory at the end of each intersection leg. Each vehicle should observe the traffic light signals, safe following distance (3 s), safe stopping distance (2 s), and the speed limit. Random "naughty" vehicles (that will violate speed limit or perform red light running) are generated in the simulation to test the ability of the collision detection and learning algorithms. The probability of naughty vehicles in the intersection is 1:5. When a naughty vehicle is generated, its speed will be a random number up to 40 km/h above the speed limit.

6.6.2 Evaluation

We evaluate our approach using the following methods:

1. Speed of detection
2. Performance/accuracy: precision and coverage

Each method is performed in our system in two ways: first, the side collision detection is performed without knowledge base and preselection (i.e., pure implementation of pair-wise collision algorithm [14] where each possible pair of all the vehicles in the intersection is calculated) and second, the side collision detection is performed after applying preselection criteria from the knowledge base. Those methods are further discussed in the following subsections.

6.6.2.1 Speed of Detection

Whenever a future collision event is detected for the first time, it is recorded in a log file, with attributes as follows: *registration number of both vehicles, collision point, TTC, leg location of both vehicles, and collision type.* Afterward, the average of detection time (*TTC*) for each run is calculated. In each execution, the average TTC is calculated. At the evening peak vehicle distribution model (average traffic volume 37–42 vehicles): if preselection is ignored in collision detection, the average TTC is 5.6 s; however, when preselection is used, the average TTC is 10.7 s, which is around 5 s earlier than the previous method. In each distribution model, preselection yields faster detection result. Therefore, preselection is proven to speed up the process of collision detection. The greater the number of vehicles in an intersection, the more preselection is useful and effective.

6.6.2.2 Accuracy: Precision and Coverage

Whenever a prediction of a future collision event is issued, it is evaluated on whether the collision really happens. If it does, it is counted as a *true positive* (valid detection). However, when a predicted collision does not happen, it is counted as a *false positive* (invalid detection). When a collision occurs, and it is not previously predicted, then it is counted as *false negative* (undetected collision). The terms are described in Figure 6.8.

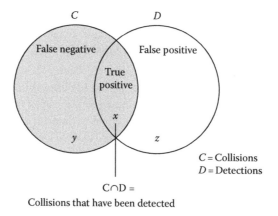

$C \cap D =$
Collisions that have been detected

Figure 6.8 Evaluation terms.

We determine performance based on the terms of *precision* (of all the detections) and *coverage* (of all the collisions), respectively:

$$\text{Precision} = \frac{\text{No. of valid detections}}{\text{Total collision detections}}$$
$$= \frac{\text{True positive}}{(\text{True positive} + \text{false positive})}$$
$$= \frac{x}{(x + z)} \tag{6.7}$$

$$\text{Coverage} = \frac{\text{No. of valid detections}}{\text{Total collisions}}$$
$$= \frac{\text{True positive}}{(\text{True positive} + \text{false positive})}$$
$$= \frac{x}{(x + z)} \tag{6.8}$$

Based on the accuracy evaluation on side collision detection in our simulation, we achieve 100% precision when side collision detections are present and 100% coverage when side collisions are present. A 100% precision and coverage can be realistically achieved when the collision detection algorithm is correct and effective. Besides proving the effectiveness of side collision detection in our system, this evaluation method helps to find parts of the collision detection that need improvement.

6.7 Collision Warning

As the time needed to avoid collision (TTA) should be less than the elapsed time to the predicted collision point (TTC) in order to avoid a collision, we need to increase the speed of detection (thus increasing the TTC value) and reduce the communication cost (thus lessening TTA value). In order to achieve a generic and real-time framework for intersection safety, it is apparent that

1. To speed up collision detection (increasing TTC), we need centralized location of computations, reduction of number of vehicle pairs for collision detection by preselection method, and knowledge base and learning of collision patterns, as discussed in the previous section.
2. To reduce the time needed to avoid collision (reducing TTA), we need accurate cost models of TTA to achieve timely warning or command message and real-time communication protocol. This is the purpose of the discussion in this section.

6.7.1 TTA Cost Model

When a collision is detected, it is important that we only send messages to affected vehicles. For collision warning particularly, point-to-point messaging is used between vehicle and IAs instead of broadcasting. As there is a need for real-time warning, the messages sent are short and, thus, only require a short processing time. However, when TTA is not enough to issue a warning to notify the driver, it is better to send a command message to the VA directly to brake. Therefore we propose two types of avoidance messages with two types of TTA accordingly (Figure 6.9), which are

1. Warning message, intended for driver, measured by *TTA warning*
2. Command message, intended for vehicle braking system, measured by *TTA command*

The choice of generating a warning or command message depends on the following rule of thumb:

- If $TTC > TTA_{warning}$, send Collision Warning message
- If $TTC <= TTA_{warning}$, send Command message

We propose the cost model for $TTA_{warning}$ (6.9), where driver initiates the avoidance, which is as follows:

$$TTA_{warning} = t_{message} + t_{receive} + t_{response} + t_{brake} + \frac{v}{a} \qquad (6.9)$$

where $t_{message}$ is the required time to generate, transmit and read a warning message by the software, $t_{receive}$ is the time for a driver to receive the message, $t_{response}$ is the response time for a driver to take

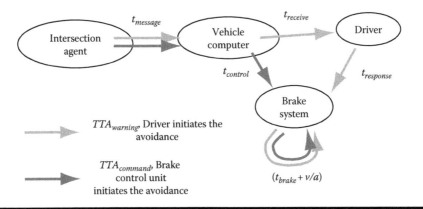

Figure 6.9   TTA cost model diagram.

an action, t_{brake} is the response time of braking system, and v/a is the time to full stop (v is velocity and a is acceleration). The cost model for $TTA_{command}$ (6.10), where the brake control unit initiates the avoidance, is

$$TTA_{command} = t_{message} + t_{control} + t_{brake} + \frac{v}{a} \qquad (6.10)$$

where $t_{control}$ is the response time of brake control unit. The cost of issuing a warning (in time units) after notification of new information or event ($t_{message}$) is computed by (6.11):

$$t_{message} = t_{generate} + t_{transmit} + t_{read} \qquad (6.11)$$

where $t_{generate}$ is time to generate the message, $t_{transmit}$ is time for message transmission, and t_{read} is time for the vehicle's computer to read the message. $t_{transmit}$ can be calculated by (6.12):

$$t_{transmit} = \frac{Message\_size}{Bandwidth} \qquad (6.12)$$

where message_size is the size of the message in bits, and bandwidth is the capacity of the communication channel in bits per second.

6.7.2 Message Protocol

We propose three types of messages transmitted within the administration zone (Figure 6.2), which are status report, registration, and warning report (Figure 6.10).

6.7.2.1 Status Message

When a VA receives an IA's signal, it sends a status message to the IA to report its status every specified interval time, for example, 6 ms. This message includes the vehicle's dynamic information:

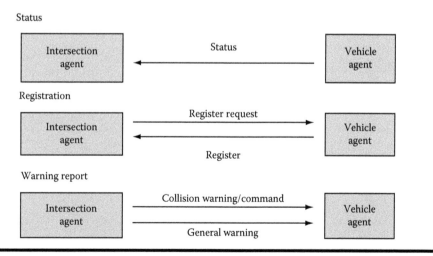

Figure 6.10 Message protocol.

vehicle ID, speed, position, angle, and maneuver. When all the required information can be retrieved from roadside sensors, status report from the VA is no longer necessary, as the IA is getting the data straight from the roadside sensors. The message structure is

status |<vehicle ID>|<x>|<y>|<speed>|<acceleration>|<direction>|<angle>|<maneuver>

The word "status" is to indicate the message type. The *vehicle ID* is the registration number of the vehicle, for example, "VICABC001." The *x, y* values are the coordinate values of the vehicle's position in our simulated environment, for example, 213, 320. The *speed* is the velocity of the vehicle, for example, 16.666. Its unit is m/s. The *acceleration* is the acceleration of the vehicle, for example, 1.471. Its unit is m/s^2. The *direction* is the travel direction of the vehicle, for example, 0.00 (toward north). If the vehicle travels toward east, the direction value is 90.00; toward south, 180.00; toward west, 270.00. The *angle* is the steering angle of the vehicle, for example, 0.00 for going straight. If the vehicle turns 5° to the left, the value is −5.00. If it turns 5° to the right, the value is 5.00. The *maneuver* is the intended driving maneuver that is predicted by in-vehicle devices 1 s before it occurs [34]. The value of *maneuver* includes Passing, TurnLeft, TurnRight, ChangeLaneLeft, ChangeLaneRight, Starting, and Stopping. Each parameter is separated by a vertical bar "|".

6.7.2.2 Registration Message

After the IA receives a VA status message, the IA checks its own database for the static information of the vehicle, such as the size of the vehicle. If it does not have this information, it sends a Registration Request message to the vehicle. When a vehicle receives a Registration Request message from an IA, it replies it with a Registration message. The message includes the vehicle's static information: vehicle ID, size.

The content of the Registration Request message is very simple. Its structure is

regreq |<vehicle ID>

The word "*regreq*" is to indicate the message type.

The Registration message includes the vehicle's static information, such as vehicle ID and size. The message structure is

regist |<vehicle ID>|<length>|<width>

The word "*regist*" indicates the message type. The *length* is the length of the vehicle in meter. The *width* is the width of the vehicle in meter.

6.7.2.3 Warning Report

There are two types of warning reports.

Firstly, *General Warning message*, which is broadcast to all vehicles including information for speed limit and drunk driver warning. The message structure can be one type of the following:

spdlmt |<value>,

for example, "spdlmt|60.000" means that speed limit is 60 km/h

$$drkdrv \mid <vehicle\ ID> \mid <x\ value> \mid <y\ value>,$$

for example, "drkdrv |VICPAD123|221|578" means that a drunk driver is driving vehicle "PAD-123" at the position (221, 578).

Secondly, *Collision Avoidance message*, which can either be a *Collision Warning* or *Command* message. If an intersection system detects that a collision will happen, its IA sends *Collision Warning* to notify the driver of the pair of involved vehicles. This message includes the following data: vehicle ID, TTC, collision position, and collision type. The message structure is

$$collwn \mid <vehicle\ ID> \mid <TTC> \mid <x> \mid <y> \mid <type>$$

The word "collwn" is used to indicate the message type. The *TTC* is the TTC for the particular vehicle. The *x*, *y* are the positions of the collision point in our simulated environment. The *type* is the collision type, for example, *Side* or *RearEnd*. The VA receives it, processes it, and warns the driver.

However, if the TTA is less than the TTC, the IA sends a *Command* message to the VA so that the vehicle takes an action automatically without the driver's intervention. This message includes the following data: vehicle ID and action. The message structure is

$$commnd \mid <vehicle\ ID> \mid <acceleration>$$

The word "*commnd*" indicates the message type. If *acceleration* is negative, the vehicle needs to slow down. Otherwise, the vehicle needs to speed up. All the protocols have been implemented on the simulated IA and VA for the evaluation of the communication cost.

6.7.3 Message Protocol Simulation

In order to evaluate the communication cost, the protocols are implemented on a simulated IA and VA. The implementation prototype is described as follows:

- The IA is simulated on a powerful computer or server. Since the IA is stationary and needs to perform learning, predict collision, communicate with numerous VAs, and calculate the TTA, it needs to run on a powerful and stable machine. The IA is implemented on Java Virtual Machine (see Figure 6.11). It is developed by using NetBeans 5.5.1 and Java 2 Standard Edition 1.6.0.02.
- The VA is simulated on a small device. Since this agent only needs to communicate with one IA, it does not need much computing power. Furthermore, because the device needs to sit in a vehicle, it is easier if the VA is installed on a small device rather than a huge full-size computer. The VA is implemented on Java Kilobyte Virtual Machine (see Figure 6.12). It is developed by using NetBeans 5.5.1 with the Connected Limited Device Configuration (CLDC) 1.1 and the Mobile Information Device Profile (MIDP) 2.0, which together provide a standard Java runtime environment for mobile device such as cell phones and personal digital assistants (PDA). It is important to note that a VA does not have to run on cell phone

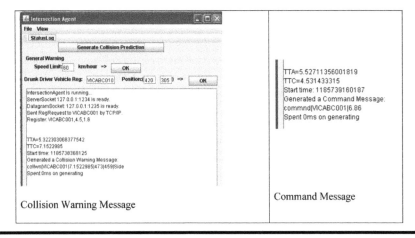

Figure 6.11   Simulation of an IA.

Figure 6.12   Simulation of a VA.

or PDA. Although it is implemented with a cell phone interface, it demonstrates that VA can run on a small device.

■ Since our proposed communication protocol is an application layer protocol in the ISO-OSI Reference Model, it needs to work with the protocols in the lower layer. In our implemented prototype, we employ TCP and UDP for the transmission layer protocol, IP for the network layer protocol, and IEEE 802.3 for the data-link layer and physical layer protocol. The status message should be sent through the UDP/IP protocol. Although UDP is not reliable protocol, it is faster than TCP. Since status message is sent frequently, transmission speed is more important than the reliability of the connection protocol. Other message types should be sent through the TCP/IP protocol because TCP provides a reliable connection. The performance would differ if the underlying network protocol is changed in the future, such as when the IEEE 802.11/p wireless VANET for vehicle-to-vehicle and vehicle-to-infrastructure communication has been introduced and implemented on the road.

6.8 Conclusion

This chapter has presented the challenges faced by intersection collision detection systems. We have investigated the features required in a cooperative intersection collision warning and avoidance systems that are able to adapt to various characteristics of intersections.

We have proposed the U&I Aware framework, which is a generic and real-time context-aware framework for collision detection and warning at road intersections. The following qualities have been incorporated into the U&I Aware framework: *adaptability* of the framework to various intersections, improvement of *performance and scalability* of the collision detection (or prediction) process, usage of appropriate *real-time data sources*, and a *real-time communication model and protocol* between vehicles and the system infrastructure with an *effective warning delivery* based on the available time before collision is predicted to occur.

The ubiquitous computing techniques—data mining, knowledge-based systems, multiagent systems, and context awareness—which enable learning and adaptability have inspired this research project and are integrated as components of the framework, which are collision learning, collision detection, and collision warning.

In the collision learning component, the collision patterns that are acquired from mining traffic and collision data are stored in the knowledge base of the collision detection system. These patterns are used for matching vehicle pairs to be calculated for the possibility of route contention and future collision events.

In the collision detection component, the patterns learnt at the intersection can be used as the basis for preselection, which identifies vehicle pairs that are likely to collide. This approach improves the performance of collision detection in the U&I Aware framework. The speed of the detection is evaluated by calculating the average of TTC in the first detection of a future collision event. The accuracy of collision detection is evaluated using precision and coverage measurements. An evaluation has been carried out using our custom-built intersection traffic simulation.

In order to have real-time collision avoidance, TTA should be less than TTC; therefore, we need to increase TTC by quickening the collision detection process (so available time before collision is greater) and decrease TTA by shortening communication and warning process. In the collision warning component, a real-time message protocol for collision avoidance has been proposed along with the cost model of TTA. As a result, impending collisions can be avoided in real-world situations.

References

1. U.S. Department of Transportation—Federal Highway Administration, Institute of Transportation Engineers, *Intersection Safety Briefing Sheet*, April 2004. http://www.ite.org/library/IntersectionSafety/BreifingSheets.pdf
2. C. Frye, International cooperation to prevent collisions at intersections, *Public Roads Magazine*, 65(1), Federal Highway Administration, USA, July–August 2001. http://www.tfhrc.gov/pubrds/julaug01/preventcollisions.htm
3. O. K. Arndt, Relationship between unsignalised intersection geometry and accident rates, School of Civil Engineering, Queensland University of Technology, Brisbane, Queensland, Australia, PhD Thesis, March 2003.
4. ITS Australia, *Handbook on Intelligent Transport Systems*, Darwin, NT, Australia, 2003.
5. P. Sharke, Smart cars, *Mechanical Engineering*, March 2003. http://www.memagazine.org/backissues/mar03/features/smartcar/smartcar.html

6. T. Strobel, A. Servel, C. Coue, and T. Tatschke, Compendium on sensors—State-of-the-art of sensors and sensor data fusion for automotive preventive safety applications, ProFusion IP Deliverable, PREVENT project, ERTICO EU, Brussels Belgium, July 19, 2004.

7. R. A. Ferlis, Infrastructure intersection collision avoidance, *Proceedings of Intersection Safety Conference*, Milwaukee, WI, November 2001. http://www.ite.org/library/IntersectionSafety/Ferlis.pdf

8. S. Coleri, S. Y. Cheung, and P. Varaiya, Sensor networks for monitoring traffic, invited paper to *Allerton Conference*, Monticello, IL, September 2004. http://www.eecs.berkeley.edu/~csinem/academic/publications/allerton_traffic_2005.pdf

9. R. Schleiffer, Intelligent agents in traffic and transportation, *Transportation Research Part C: Emerging Technologies (Special Issue)*, 10(5), 325–329, 2002.

10. F. D. Salim, S. Krishnaswamy, S. W. Loke, and A. Rakotonirainy, Context-aware ubiquitous data mining based agent model for intersection safety, *Proceedings of the Second International Symposium on Ubiquitous Intelligence and Smart Worlds (UISW 2005)*, in conjunction with the *2005 IFIP International Conference on Embedded and Ubiquitous Computing (EUC 2005)*, December 6–7, 2005, Lecture Notes in Computer Science, Springer-Verlag, EUC Workshops 2005, Nagasaki, Japan, pp. 61–70.

11. F.-Y. Wang, Agent-based control for networked traffic management systems, *IEEE Intelligent Systems*, 20(5), 92–96, September/October 2005.

12. M. Wooldridge, *An Introduction to Multiagent Systems*, John Wiley & Sons, Chichester, U.K., February 2002.

13. M. Wooldridge and N. R. Jennings, Intelligent agents: Theory and practice, *Knowledge Engineering Review* 10(2), 115–152, 1995.

14. J. France and A. A. Ghorbani, A multi-agent system for optimizing urban traffic, *Proceedings of IEEE/WIC International Conference on Intelligent Agent Technology (IAT 2003)*, Halifax, Nova Scotia, Canada, October 13–17, 2003.

15. D. A. Roozemond, Using intelligent agents for pro-active, real-time urban intersection control, *European Journal of Operational Research*, 131(2), 293–301, 2001.

16. T. Mitchell, *Machine Learning*, McGraw Hill, New York, 1997.

17. D. Pomerleau, Alvinn: An autonomous land vehicle in an neural network, in D. Touretzky, ed., *Advances in Neural Information Processing Systems 1*, Morgan Kaufmann, Denver, CO, 1989.

18. U. Fayyad, G. Piatetsky-Shapiro, and P. Smyth, From data mining to knowledge discovery in databases, *AI Magazine*, 17(3), 37–54, Fall 1996.

19. M. M. Gaber, S. Krishnaswamy, and A. Zaslavsky, Ubiquitous data stream mining, *Proceedings of Current Research and Future Directions Workshop*, in conjunction with *The Eighth Pacific-Asia Conference on Knowledge Discovery and Data Mining*, Sydney, New South Wales, Australia, May 26, 2004.

20. O. Horovitz, M. M. Gaber, and S. Krishnaswamy, Making sense of ubiquitous data streams: A fuzzy logic approach, *Proceedings of the Ninth International Conference on Knowledge-based Intelligent Information & Engineering Systems 2005 (KES 2005)*, Melbourne, Victoria, Australia, September 14–16, 2005.

21. H. Kargupta, R. Bhargava, K. Liu, M. Powers, P. Blair, S. Bushra, J. Dull, K. Sarkar, M. Klein, M. Vasa, and D. Handy, VEDAS: A mobile and distributed data stream mining system for real-time vehicle monitoring, *Proceedings of the SIAM International Data Mining Conference*, Orlando, FL, 2004.

22. K. Dey and G. D. Abowd, Towards a better understanding of context and context-awareness, GVU Technical Report GIT-GVU-99-22, Atlanta, GA, *First International Symposium on Handheld and Ubiquitous Computing*, June 1999.

23. G. Chen and D. Kotz, A survey of context aware mobile computing research, Technical Report TR2000-381, Department of Computer Science, Dartmouth College, Hanover, NH, November 2000.

24. T. P. Moran and P. Dourish, Context-aware computing, *Human-Computer Interaction*, 16(2–4), 87–95, 2001. http://hci-journal.com/editorial/si-context-aware-intro.pdf

25. J. Indulska, T. McFadden, M. Kind, and K. Henricksen, Scalable location management for context-aware systems, *Proceedings of the Fourth IFIP WG 6.1 International Conference on Distributed Applications and Interoperable Systems*, Springer Verlag, Paris, France, Lecture Notes in Computer Science 2893, November 2003, pp. 224–235.

26. N. Oliver and A. Pentland, Graphical models for driver behavior recognition in a smart car, *Proceedings of IEEE International Conference on Intelligent Vehicles 2000*, Detroit, MI, October 2000.

27. K. A. Funderburg, Update on intelligent vehicles and intersections, *Public Roads Magazine*, 67(4), January–February 2004. http://www.tfhrc.gov/pubrds/04jan/08.htm

28. K. Stubbs, H. Arumugam, O. Masoud, C. McMillen, H. Veeraraghavan, R. Janardan, and N. Papanikolopoulos, A real-time collision warning system for intersections, *Proceedings of Intelligent Transportation Systems America*, Minneapolis, MN, May 2003.

29. H. Veeraraghavan, O. Masoud, and N. Papanikolopoulos, Vision-based monitoring of intersections, *Proceedings of Intelligent Transportation Systems Conference*, Boston, MA, September 2002.

30. R. Miller and Q. Huang, An adaptive peer-to-peer collision warning system, *Proceedings of Vehicular Technology Conference (VTC)*, Birmingham, AL, Spring 2002.

31. U. S. Department of Transportation, Cooperative intersection collision avoidance systems, May 2006. http://www.its.dot.gov/cicas/index.htm

32. INTERSAFE, D40.4 Requirements for intersection safety applications, October 28, 2005. http://www.prevent-ip.org/en/public_documents/deliverables/d40d4_intersafe.htm

33. C.-Y. Chan and D. Marco, Traffic monitoring at signal-controlled intersections and data-mining for safety applications, *IEEE Intelligent Transportation System Conference*, Washington, DC, October 2004.

34. K. W. Ogden and S. V. Newstead, Analysis of crash patterns at Victorian signalised intersections, Monash University Accident Research Centre, Melbourne, VIC, Australia, Report No. 60, Australian Road Research Board, February 1994.

35. R. L. Grossman, M. Sabala, J. Alimohideen, A. Aanand, J. Chaves, J. Dillenburg, S. Eick et al., Real time change detection and alerts from highway traffic data, *Proceedings of ACM/IEEE Supercomputing*, Seattle, WA, 2005, pp. 62–69.

36. I. H. Witten and E. Frank, *Data Mining: Practical Machine Learning Tools and Techniques*, 2nd edn., Morgan Kaufmann, San Francisco, CA, 2005.

Chapter 7

Agent-Based Intrinsically Motivated Intelligent Environments

Owen Macindoe, Mary Lou Maher, and Kathryn Merrick

Contents

7.1 Introduction

In their seminal papers on intelligent environment (IE) design, Brookes [1] and Coen [2] argued that a key design goal for developing IEs is to enable them to adapt to and be useful for everyday activities. However, research in IEs has been dominated by the development of sensor and effector configurations and software architectures that specify protocols for interpreting and responding to sensor data. The configuration of new sensor and effector systems to allow IEs to produce useful behaviors is time consuming and labor intensive. The ability of IEs to adapt their behaviors autonomously to changes in activity patterns remains an open research area.

This chapter draws on recent advances in motivated learning agent research to develop new models for adaptive, responsive IEs. Motivated learning agents use computational models of motivation to direct machine learning algorithms. Self-motivated learning as a basis for IEs creates physical spaces that can adapt to new sensors and effectors and changing usage patterns. This chapter presents two IE applications experimenting with motivation and learning agent models. In the first application, an MRL agent changes the structure and content of a curious information display in response to patterns including human movement in a physical space. The second application experiments with unsupervised learning (UL) agents as a precursor to MUL agents for controlling the hardware devices in a simulated intelligent room to facilitate activities in the space. These agents are evaluated in terms of their adaptive, emergent behavior. We show that such agents can exhibit adaptive, learned behaviors to facilitate or augment the human activities in an IE.

7.2 Current Models for Intelligent Environments

An IE is a physical space for living or working that can bring embedded computational power to bear in a manner that facilitates or augments the actions of users of the environment as they perform their daily tasks. IEs can monitor the activities that take place within them using sensors and respond to sensations using effectors in order to exhibit intelligent behavior and assist users. Sensors may include devices such as motion detectors or pressure pads while effectors may include devices such as lights, projectors, or doors.

IE research can be regarded as a subfield of ubiquitous computing that aims to integrate computers seamlessly into everyday living. IEs have several specific design requirements. Brooks [1] and Coen [2] have argued that IEs should

- Adapt to and be useful for ordinary everyday activities
- Assist the user without requiring the user to attend to them

■ Have a high degree of interactivity
■ Be able to understand the context in which people are trying to use them and behave appropriately

An IE is essentially, as Kulkarni suggests [3], an immobile robot. However, the design requirements of an IE differ from those of normal robots in that they are oriented toward maintaining an internal space rather than exploring or manipulating an external environment.

Existing agent-based approaches to IE design include MIT's intelligent room prototype e21, which facilitates activities via a system called ReBa [4]. ReBa is a context handling system that observes a user's actions via the reports of other agents connected to sensors in the room's multiagent society and uses them to build a higher level representation of the user's activity. Each activity, such as watching a movie or giving a presentation, has an associated software agent, called a behavior agent, which responds to a user action and performs a reaction, such as turning on the lights when a user enters the room. Behaviors can form layers based on the order of user actions, acknowledging differences in context such as showing a presentation in a lecture setting versus showing one in an informal meeting. Although ReBa can infer context in this way, it cannot adapt to patterns of usage. In order for an entirely new context to be created, ReBa's behavior agents must be preprogrammed to recognize the actions of the user and take an appropriate action. It does not self-adapt to new usage patterns. Furthermore, when new sensors are added to the room, the existing rules must be modified manually if they are to take advantage of the new sensor data.

Other researchers have taken approaches to designing IEs that are not explicitly agent based. Both the University of Illinois' Gaia Project [5] and Stanford University's Interactive Workspace Project [6] have taken an operating systems approach, developing Active Spaces and Interactive Workspaces, respectively, which focus on the role of the physical space as a platform for running applications rather than as a proactive facilitator. The specification of an action in these systems is triggered by the user and the behavior is programmed by an applications developer. Gaia's context service provides the tools for applications developers to create agent-based facilitating applications so the overall model is reactive rather than adaptive.

7.3 Existing Approaches to Motivated Agents

An agent is a system that can perceive its environment through sensors and use some characteristic reasoning process to generate actions using effectors. Agent models correspond naturally to IEs as both are described as having sensors for monitoring their environment and effectors for making changes to the environment. A variety of agent models have been developed with different characteristic reasoning processes for mapping sensor input to effector output. These range from simple rule-based reactive agents to complex cognitive agents that try to maintain and reason about an internal model of the world using planning or machine learning algorithms. This raises the question of what kind of agent model is most suitable as a basis for an IE.

An IE needs to be driven to assist users, adapt to changes in its configuration, adapt to changing usage patterns, and understand context. Drives of this kind have been modeled by the concept of motivation in agent research, leading to several different motivated agent models. Norman and Long [7] developed a motivated agent model for a warehouse environment in which motivation is modeled by the temporal urgency of tasks to be completed. They use task-oriented motives such as the need to keep the warehouse tidy or fill orders in a timely manner. These motives are mitigated using planning algorithms to generate sequences of actions to fulfill the goals triggered by strong motivational drives. In a different approach, Beaudoin and Sloman [8] developed a simulation of a robot nursery in which a robot nursemaid implementing a motivated agent model was shown to effectively prioritize tasks using a sophisticated model of motivation that included logical propositions, temporal urgency, and levels of insistence. In another application, Aylett et al. [9] design an agent-controlled water filtration plant, which shows promise despite the relatively simplistic motivation model used.

Other researchers have experimented with more general cognitive models of motivation that promise greater adaptability. The requirement for adaptation in an IE can be satisfied with a model of learning new behaviors through the interpretation of sensor data. Rather than specifying a specific set of competencies or goals with an external reward, we look for computational models of novelty and curiosity that allow the agent to respond to unexpected changes in the kinds of activities in the room. Saunders and Gero [10] modeled curiosity computationally as a model of interest. Interest is partly determined by novelty but is also related to how well an agent can learn the information gained from novel experiences. Interest in a situation is aroused when its novelty is at a moderate level, meaning that the most interesting experiences are those that are similar yet different to previously encountered experiences as shown in Figure 7.1.

Saunders and Gero demonstrate the utility of this model by using it to simulate the formation of cliques in artistic communities [11], to explore the design space of a simple architectural problem [12], and to provide a richer social force model of human crowds in museums [10]. Merrick and Maher [13] experimented further with this model of interest by using it as a trigger for RL in nonplayer characters in computer games. They showed that it is possible for task-oriented behavior to emerge as a result of a motivation process based on task-independent concepts such as interest. Later work by Maher et al. [14] proposes three motivated learning agent models corresponding to the three traditional classes of machine learning algorithm: motivated reinforcement learning

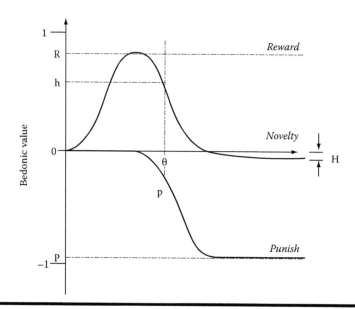

Figure 7.1 The Wundt curve. Interest rises and then falls as novelty increases. (From Saunders, R. and Gero, J.S., Designing for interest and novelty: Motivating design agents, presented at *CAAD Futures 2001*, Kluwer, Dordrecht, the Netherlands, 2001.)

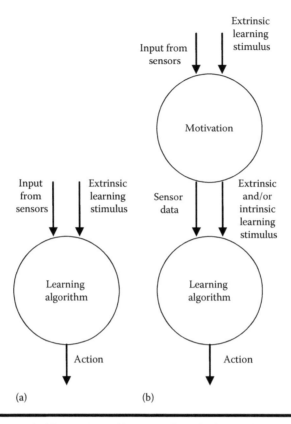

Figure 7.2   RL (a) is extended by motivated learning (b), which introduces an intrinsic motivation signal in addition to or instead of the extrinsic reward signal from the environment.

(MRL), motivated supervised learning (MSL), and motivated unsupervised learning (MUL). These intrinsically motivated learning agent models introduce an intrinsic learning stimulus in addition to or instead of the traditional extrinsic learning stimulus. The intrinsic motivation signal is computed by a motivation process. Figure 7.2 compares intrinsically motivated learning algorithms to traditional learning algorithms diagrammatically. In the following sections, we review each of the three motivated learning agent models with reference to their potential applications in IEs.

7.3.1 Motivated Reinforcement Learning

Reinforcement learning (RL) [15] uses rewards to guide agents to learn a function that represents the value of taking a given action in a given state with respect to some task. An RL agent is connected to its environment via perception and action. On each step of interaction with the environment, the agent receives an input that contains some indication of the current state of the environment and an extrinsic reward signal indicating the value of that state to the agent. The agent then chooses an action as output. This action changes the state of the environment. The agent's behavior should choose actions that tend to increase the long-run sum of values of the extrinsic reward signal. This behavior is learnt over time by systematic trial and error. In contrast, MRL agents [13] aim to use systematic trial and error to maximize the long-term sum of rewards produced by their motivation process in addition to any extrinsic reward from their environment. In MRL agents, the reward received in each state may change over time, resulting in the emergence, stabilization, and disappearance of multiple mappings from states to actions within the learned behavior. Over time, MRL agents are able to learn to solve multiple tasks.

Where standard RL requires the user to specify a reward signal for training, MRL is less obtrusive as it does not. However, as all RL requires trial and error to learn, MRL may still be too disruptive to consider as a means of controlling equipment such as projectors or lights. MSL and MUL models become useful in such situations.

7.3.2 Motivated Supervised Learning

Supervised learning (SL) uses examples of correct behavior to guide agents to learn a function that represents a mapping between observations and correct actions with respect to some task. An SL agent is connected to its environment via perception and action. On each step of interaction with the environment, the agent receives an input that contains some indication of the current state of the environment and, optionally, an example of the correct action to take when in that state. When an example is not provided, the agent chooses an action as output.

Unlike standard SL agents, MSL agents can choose to act on or ignore observed states that are not accompanied by example actions. The motivation process can then elect not to act in situations where it is not confident that it knows what to do. In MSL agents, the motivation process acts as a filter for observations and examples, so attention can be autonomously focused on learning specific tasks. With the development of appropriate motivation functions, MSL agents have the potential to derive contextual information from observations and examples rather than requiring a separate perception process dedicated to this task. In an IE, this may be useful for distinguishing task boundaries when there are several people in a room, performing multiple tasks. This remains an open research area.

7.3.3 Motivated Unsupervised Learning

Unlike SL and RL algorithms that learn functions mapping observed states to actions, UL algorithms [16] aim to identify patterns or important features in observed data. Thus, unlike MRL and MSL, MUL does not build behaviors that represent mappings from input to predefined output values such as actions. Rather, the output values produced by UL represent important patterns or features in the input data such as clusters, principal components, or repeated patterns in temporal data. The activation process uses a set of predefined behavioral rules to act on the features identified by the learning process. The chosen action triggers a corresponding effector that makes a change to the agent's environment.

In IEs, predefined behavioral rules might be represented as a mapping from sensors to actions. Maher et al. [17,18] proposed such a technique in which agents use sequential pattern mining to identify repeated patterns in sensor data and then act by associating actions with these observed states. While behavioral rules are fixed, the actual actions performed by the agent change over time as new observed states cause the learning process to identify different features or patterns as being important. This use of the MUL architecture is essentially quite similar to the SL architecture but opens the way for the use of UL such as data mining. Because behavioral rules in the MUL architecture are fixed and only fire when triggered, not necessarily at every time step, this architecture can also be used to create agents in which the sensed world and the effectible world are mutually exclusive.

7.4 The Sentient: A Sensed Environment for Supporting IE Applications

In order to experiment with IE applications, we are developing the Sentient, a physical space in the Key Centre for Design Computing and Cognition at the University of Sydney equipped with a range of sensor and effector hardware and an agent package for controlling these systems. The layout of the Sentient is shown in Figure 7.3. The sensor and effector architecture of the Sentient currently consists of four sensor systems: the Teleo system for sensing movement via pressure pads embedded in the floor, the Bluetooth system for sensing and controlling Bluetooth devices, the room booking system for sensing and modifying bookings of the room, and the camera system, which is an independent system that provides a video stream via network cameras. Each of these systems has a device monitor that records the sensor inputs arriving from the systems into tables in a MySQL context database as shown in Figure 7.4. These tables can be read by software agents in order to provide them with information about the state of the environment. The agent is also able to command the systems in the room by writing requests for actions into an action queue in the context database. These actions are carried out by device controllers, software demons that poll the action queue and pass on the requested action to the appropriate hardware system. A key feature of the Sentient's sensor and effector architecture is its minimalist approach. The focus of our research is on the agent algorithms that sample the sensor data and send actions to the effectors rather than the sensor and effector systems themselves.

Agents have a set $\mathbf{N} = \{N_1, N_2, \ldots N_{|N|}\}$ of several different sensors that return sensations from the data tables the agent is monitoring. These sensations are combined using a context-free grammar (CFG) [19]. Strings from a CFG have variable length, making them particularly suitable for real-world environments for which it is difficult to predict the exact contents. For example, in the Sentient, it is known that Bluetooth devices may be present but it is not known to whom they may belong or how many may be present at any particular time.

The process by which individual sensations are combined to form a complete representation of the current sensed state of the environment is shown in. Each sensor N_n polls a database table at some rate appropriate for the information stored in that table and returns one or more sensations. The sensations retrieved by the sensor N_n overwrite the previous sensations from that sensor stored in iconic memory [20] and are combined with the most recent sensations from all other sensors in order to form the sensed state $S_{(t)}$. Each sensor N_n assigns a label L to each sensation $s_{n(t)}$ such that

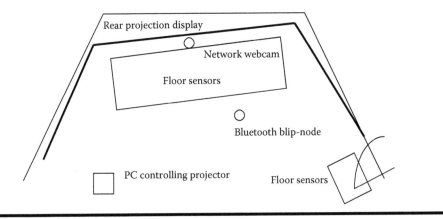

Figure 7.3 Layout of the Sentient.

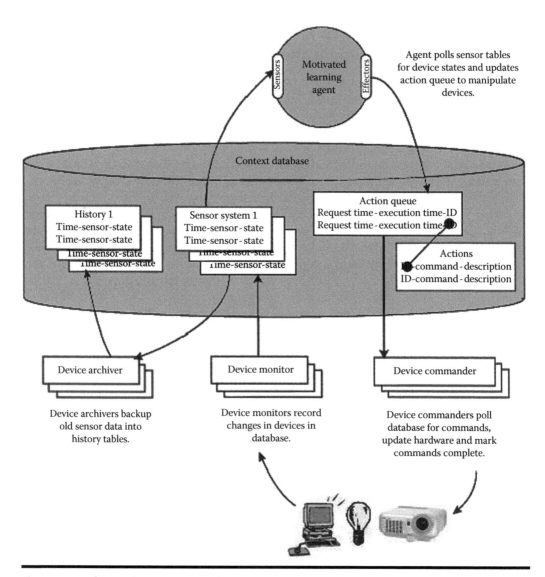

Figure 7.4   The Sentient's sensor and effector architecture.

that two sensed states can be compared using the values of elements with the same label. Thus, $S_{(t)}$ is a tuple of labeled sensations $S_{(t)} = (s_{1(t)}, s_{2(t)}, \ldots s_{L(t)}, \ldots)$ (Figure 7.5).

7.5 Curious Information Displays: A Motivated Reinforcement Learning IE Application

A curious information display is a system that transforms a physical space into an IE. Traditional information displays such as posters and billboards present a fixed image to observers. More recently, with the advances in large screen display technologies, digital displays are becoming more prevalent as an alternative means of presenting information. Digital displays have allowed the amount of

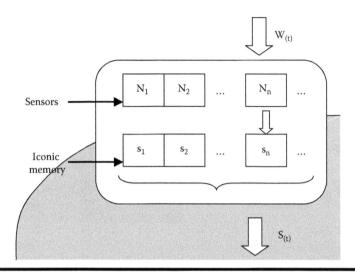

Figure 7.5   Sensor model for learning agents in curious places. The sensor N_n has just registered information about the world state $W_{(t)}$ at time t. The new sensation s_n is combined with sensations from iconic memory for the other sensors to produce the sensed state $S_{(t)}$.

information being presented to be increased by attaching the display to a computer that executes software to change the contents of the display automatically. However, as yet, the full power of a digital display has not been realized, with the most common scenario using a database of images and displaying images at random or in a predefined order.

In scenarios where the display of digital information is more familiar, such as web browsers for the display of information from the World Wide Web, novel interaction algorithms have been developed to automatically personalize the digital space [21]. Similarly, intelligent tutoring systems use artificial intelligence algorithms to tailor learning material to the individual needs of students [22]. Large digital information displays in public spaces have the same capacity for the use of novel techniques to improve the usefulness of the displays; however the public, multiuser nature of these displays calls for new algorithms to improve the ability of such displays to impart information. Curious information displays are a means of creating digital displays that can augment physical places by attracting the interest of observers and imparting information by being curious about and learning about the structure and content of the information they display.

This section introduces two models of curious information displays that display information about research work undertaken in the Key Centre for Design Computing and Cognition at the University of Sydney. The broad themes of the display are design, computing, agents, and curiosity. Information on these themes is obtained from a research image database, the World Wide Web, and live webcam images. The display is located in the Sentient and uses an MRL agent model to detect and learn about interesting events or observations.

Our curious information display comprises a matrix of displayed information items (IIs). Each II may be a definition or an image from a database, an image from the web, or video from a webcam. Each item can be displayed in a 1×1, 2×2, or 4×4 cell as shown in Figure 7.6.

The layout of the curious information display is implemented using a tree data structure in which non-leaf nodes represent 1×1, 2×2, and 4×4 resolution displays and leaf nodes represent IIs. For example, the internal data structure for the layout in Figure 7.6 is shown in Figure 7.7.

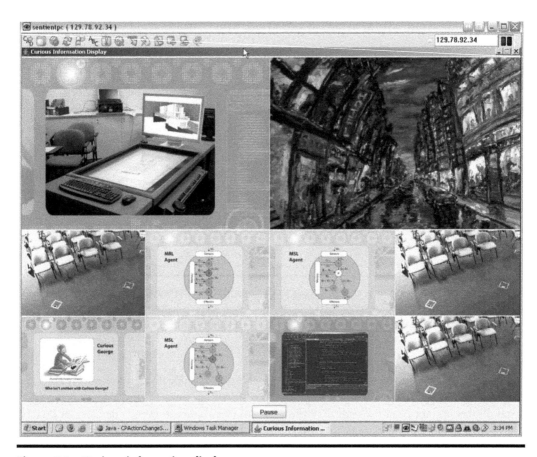

Figure 7.6   Curious information display.

Each leaf node has a number of properties describing the II it holds. These properties describe the information content of the display and comprise strings from the following grammar:

```
<leaf-node>          ◇   <id><source>
<id>                 ◇   [1, 999]
<source>       ◇        <simple-source> |
<parameterised-source>
<simple-source>      ◇   webcam
<parameterised-source> ◇      web <keyword> | database <keyword>
<keyword>            ◇   curious | design | agent | computing
```

Each non-leaf node also has a number of properties summarizing the layout of the IIs it holds. These properties describe the structural layout of the information being displayed. In particular, these properties recognize the 11 different ways in which two, three, or four IIs with the same source and keyword can be configured within a non-leaf node. Non-leaf node properties are described by strings from the grammar in the following. Recognized layout patterns are labeled according to the numbering in Figure 7.6.

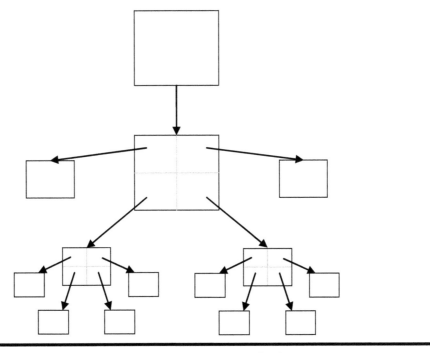

Figure 7.7   Internal data structure of the curious information display.

```
<non-leaf node>        ◇   <id><pattern><ii-summary>
<id>                   ◇   [1, 999]
<pattern>              ◇   <pair> | <triple> | square
<pair>                 ◇   p12 | p34 | p13 | p24 | p14 | p23
<triple>               ◇   t123 | t124 | t134 | t234
<ii-summary>           ◇   <common-source>
<common-source>        ◇   <simple-source> | <parameterised-source>
<simple-source>        ◇   webcam
<parameterised-source> ◇       web <common-keyword> | database
                               <common-keyword>
<common-keyword>       ◇   curious | design | agent | computing
```

Both non-leaf and leaf node properties can be sensed by a software agent that reasons about ways in which to modify the display. These properties are sensed according to the following grammar, which combines the properties of leaf nodes and non-leaf nodes:

```
DisplayState           ◇   <non-leaf nodes><leaf node><leaf nodes>
<non-leaf nodes>       ◇   <non-leaf node><non-leaf nodes> | ε
<leaf nodes>           ◇   <leaf-node><leaf nodes> | ε
<leaf node>            ◇   ...
<non-leaf node>        ◇   ...
                       ◇
```

Sensation values are enumerated by the sensors that produce them to provide the agent with numerical data for reasoning. Thus, a sensed state displaying a single 4 × 4 image from the web might look like the following:

```
S((pic1source:1)(pic1keyword:2))
```

A sensed state containing a combination of 1 × 1 and 2 × 2 images might look like the following:

```
S((pic141keyword:2)(pic144source:3)(pic11source:3)(pic13source:3)
  (pic142source:2)(pic120source:4)(pic120colour:1)(pic141source:2)
  (pic142keyword:2)(pic143source:3))
```

7.5.1 Motivated Reinforcement Learning Agent Model

The MRL agent model used in this application is shown in Figure 7.8. In this model, $W_{(t)}$ represents the state of the agent's environment at time t, while $S_{(t)}$ represents the state of the environment as sensed by the agent at time t. The agent's reasoning process can be broken into four subprocesses: sensation, motivation, learning, and activation. The sensation process S transforms raw data $S_{(t)}$ from the agent's sensors into structures that facilitate further reasoning. This includes

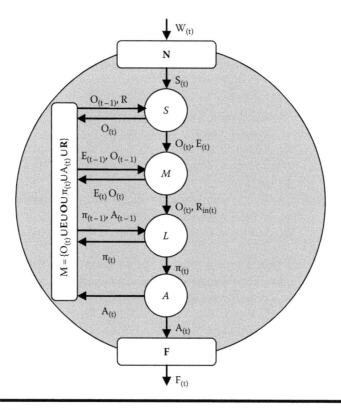

Figure 7.8   An MRL agent.

the observed state of the environment $O_{(t)}$ and the change or event $E_{(t)}$ between the current and previous observed states. The motivation process $\mathcal{M}$ reasons about the current observed state $O_{(t)}$, a representation of the set **E** of all events encountered so far, and/or a representation of the set **O** of all observed states encountered so far to produce an intrinsic reward signal $R_{in(t)}$. In this application, there is no extrinsic reward signal. The learning process $\mathcal{L}$ performs an RL update to incorporate the previous observed state, action, and current extrinsic and intrinsic rewards into the policy $\pi_{(t-1)}$ in memory M. Finally, the activation process $\mathcal{A}$ uses some exploration function of the RL action selection rule to select an action $A_{(t)}$ to perform from the updated behavior $\pi_{(t)}$.

7.5.1.1 Sensation Process

The sensation process transforms raw data $S_{(t)}$ from the agent's sensors into structures that facilitate further reasoning. This includes the observed state of the environment $O_{(t)}$ and the change or event $E_{(t)}$ between the current and previous observed states. Events allow agents to represent and reason about changes in their environment in addition to information about the current observed state. Events are computed as the difference between two observed states $O_{(t)} = (o_{1(t)}, o_{2(t)}, \ldots o_{L(t)} \ldots)$ and $O_{(t+1)} = (o_{1(t+1)}, o_{2(t+1)}, \ldots o_{L(t+1)} \ldots)$ using a difference function Δ:

$$E_{(t)} = O_{(t)} - O_{(t')}$$
$$= (\Delta(o_{1(t)}, o_{1(t')}), \Delta(o_{2(t)}, o_{2(t')}), \ldots \Delta(o_{L(t)}, o_{L(t')}), \ldots)$$

Either events or observations may be used as the basis for further reasoning in subsequent processes.

7.5.1.2 Motivation Process

The role of the motivation process in MRL is to provide an intrinsic reward signal to direct the learning process. Figure 7.9 summarizes diagrammatically our framework for modeling intrinsic motivation in this scenario. Firstly, events and observed states are received from the sensation process. Next, a focus of attention structure is used to distinguish between different input stimuli. One or more characteristic motivation functions are then used to compute motivation values. These motivation values are combined using a reward function that computes a single intrinsic reward signal. This value and the observed state are then passed to the RL process.

In order to create displays that are both interesting to their viewers and also interesting in the structure and content of the information they display, we modify the Saunders and Gero model of interest to create an intrinsic motivation signal for MRL. The Saunders and Gero model first computes the novelty of a stimulus from the environment using a Habituated Self-Organizing Map (HSOM) as a focus of attention mechanism. The characteristic motivation functions are Stanley's model of habituation for computing novelty [23] and the Wundt curve for computing interest [24]. The computed value of the interest function is used directly as the reward signal.

In this application, we replace the SOM component of the HSOM with K-means clustering as the attention focus mechanism. In environments such as the curious information display application where sequential stimuli share a large number of common features, SOMs can be dragged toward

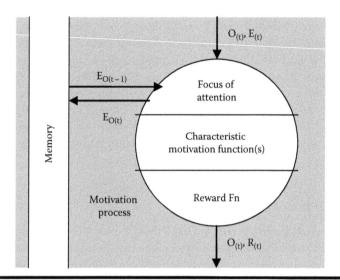

Figure 7.9   Framework for intrinsic motivation for RL.

a corner of the state space when the same neuron is repeatedly selected as the winner. In K-means clustering, a single neuron can move without deforming the entire network, making it a more appropriate attention focus mechanism in this application.

7.5.1.3 Learning Process

While the MRL agent model is designed to be independent of any specific RL algorithm, certain classes of RL are more appropriate in applications such as IE applications where learning is continuous rather than episodic. Temporal difference RL algorithms such as SARSA [25] and Q-learning [26] are the most appropriate in these settings as they do not require a model of the environment from which to learn and learning occurs after each action that is performed by the agent. As the curious information display has a large problem space, we use Q-learning combined with neural network function approximation in the learning process in this application. For environments that use an attribute-based state representation $O_{(t)} = (o_{(t)1}, o_{(t)2}, o_{(t)3}, \ldots)$ and have $|\mathbf{A}|$ actions, a network of perceptrons of the form shown in Figure 7.10 can be used to represent the utility of each action with respect to each observed state in the learned policy π.

7.5.1.4 Actions and the Activation Process

The agent modifies the display area using the following actions to manipulate the leaf nodes in the underlying data structure:

```
A₁=Change source of <leaf node> to <source>
A₂=Change keyword of <leaf node> to <keywd>
A₂ = Change image in <leaf node>
A₄ = Change resolution of <non-leaf node>
```

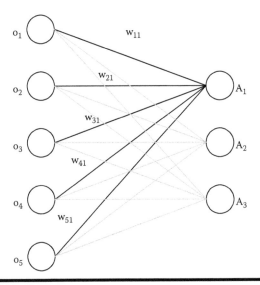

Figure 7.10   Network structure for perceptrons for function approximation in RL.

The agent cannot directly sense changes to the resolution of a node but it can indirectly sense these changes as a change in the configuration of nodes and leaf nodes.

7.5.2 Experimental Results

We implemented two types of curious information display using different aspects of our model of motivation that offer different capabilities. The two displays are one-way displays, that is, they are able to reason about structure and content of the IIs they show. The first uses events to trigger intrinsic motivation while the second uses observed states.

7.5.2.1 Curious Information Display Using Interesting Events

This curious information display extends static information displays by reasoning about the changes it can make in the structure and content of the IIs it displays and finding interesting patterns of behavior to modify the structure and content. The aim of this type of display is to achieve sequences of actions that make interesting changes to the structure and content of IIs being displayed. To facilitate this, the motivation process reasons about events in order to identify interesting changes in the display.

Figure 7.11 shows the change in behavioral variety over time for this curious information display agent. It shows that the agent progressively develops new behaviors at a steady rate over the course of its life. These behaviors represent several different types of events including keyword changes:

```
E((pic1keyword_3.0:1.0)(pic1keyword_2.0:-1.0))
```

source changes:

```
E((pic1source_2.0:1.0)(pic1keyword_3.0:1.0)(pic1source_3.0:-1.0))
```

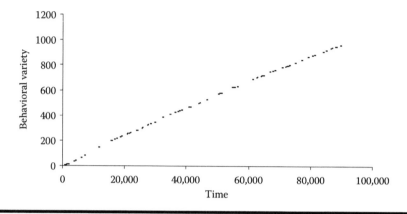

Figure 7.11   Increasing behavioral variety over time by the curious information display agent.

the creation of blocks of two (pairs p), three (an L shape l), or four (a square s) IIs with the same source at various levels of resolution and different parts of the display:

```
E((pic140keyword_5.0:1.0)(pic140source_2.0:1.0)(s140source_3.0:-1.0))
E((pic1source_2.0:-1.0)(pic1keyword_2.0:-1.0)(s10source_3.0:1.0))
E((pic133source_1.0:1.0)(12_130source_3.0:1.0)(pic133keyword_4.0:1.0)
   (pic130source_3.0:-1.0))
E((pic114keyword_5.0:1.0)(pic111keyword_2.0:1.0)(pic114source_2.0:1.0)
   (p4_110source_3.0:1.0)(pic111source_2.0:1.0)(pic110source_3.0:-1.0))
```

and resolution changes at various levels of resolution:

```
E((pic144keyword_2.0:-1.0)(pic112source_3.0:-1.0)
   (pic132keyword_4.0:-1.0)(pic120keyword_5.0:-1.0)
   (pic1keyword_5.0: 1.0)(pic1source_2.0:1.0)
   (pic120source_2.0:-1.0)(p2_130source_3.0:-1.0)
   (pic142keyword_5.0:-1.0)(pic113keyword_5.0:-1.0)
   (pic111keyword_2.0:-1.0)(pic133source_1.0:-1.0)
   (p3_140source_3.0:-1.0)(pic114keyword_3.0:-1.0)
   (pic133keyword_3.0:-1.0)(pic113source_2.0:-1.0)
   (pic111source_1.0:-1.0)(pic132source_2.0:-1.0)
   (pic144source_2.0:-1.0)(pic114source_2.0:-1.0)
   (pic142 source_2.0:-1.0))
```

One weakness of the agent from a visual perspective is its tendency to favor simple behaviors of only one or two actions as shown in Figure 7.12. This is because the simplest technique for repeating most of the events in the environment is to cause the event, undo it, then repeat it. This "shortest path" is naturally favored by RL. This phenomenon is a result of a state space with a moderate level of structure and complexity. There is enough complexity to continually stimulate the agent's motivation process to produce high reward and focus learning on new two-step changes; however, there is not enough structure to motivate the emergence of more complex behaviors as

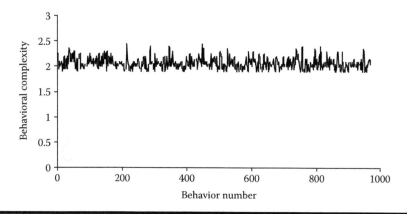

Figure 7.12 Similar behavioral complexity of behaviors learned by the curious information display agent.

has previously been possible in MRL agents in other environments [13]. Further work is required to understand the impact of the state space structure on learning.

The screen shots in Figure 7.13 show the key characteristic of this type of display is its ability to change rapidly between different configurations and different information content. This is because the MRL agent controlling the display is reasoning about events or changes in the display and is thus motivated to continue to change the layout and content of the display either to focus on an interesting change or to search for new changes that might be interesting. This type of curious information display could be useful as an ambient display device for information in pictorial or diagrammatic form, which can be understood at a glance rather than requiring reading. Changes in the display are eye-catching, and the movement between different displays holds the viewers' attention by displaying related information.

7.5.2.2 Curious Information Display Using Interesting Observed States

This curious information display extends static information displays by reasoning about the structure and content of the IIs it displays and learning to maintain interesting displays. The aim of this type of display is to maintain interesting display states. To facilitate this, the motivation process reasons about observed states in order to identify interesting display states.

Figures 7.14 and 7.15 compare the behavioral variety and complexity for this curious information display agent to the previous agent. It shows that both agents progressively develop new behaviors at a steady rate over the course of their life; however, this agent develops simpler behaviors at a slower rate. This is because the types of behaviors developed by this agent are maintenance behaviors to maintain interesting states. Such behaviors usually consist of only one or two actions to form and maintain the interesting state. The behaviors are learned at a slower rate as they tend to be maintained for a longer period of time. Observations show that this display appears to react much more quickly to high reward than the previous display. When high reward is encountered, the agent freezes the display within an action or two and focuses on that configuration until reward is reduced and boredom triggers exploration.

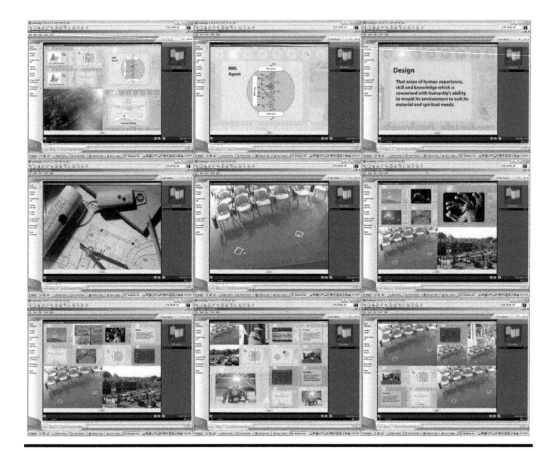

Figure 7.13   Screen shots from a 2 min video clip show many different configurations, and IIs are displayed.

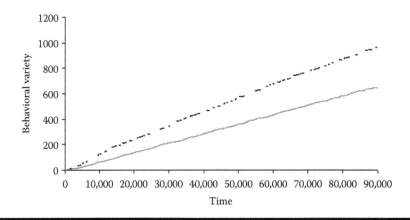

Figure 7.14   Comparison of change in behavioral variety over time by the first (blue) and second (purple) curious information display agents.

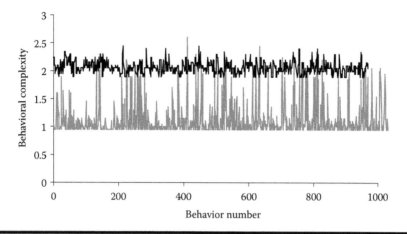

Figure 7.15 **Comparison of behavioral complexity of behaviors learned by the first (black) and second (grey) curious information display agents.**

In contrast to the previous display, this type of curious information display tends to maintain specific configurations of the display for longer periods of time. For example, the screen shots in Figure 7.16 show that there are many more common IIs and configurations maintained throughout the 2 min period. This is because the MRL agent controlling the display is reasoning about observed states and thus is motivated to find and maintain interesting observed states. This results in a display that changes more slowly over time.

This type of curious information display could be useful for both diagrammatic and textual information that requires more time for the viewer to understand. Because this display naturally changes more slowly, sudden changes caused by change in the reward signal are highly noticeable.

7.6 Intelligent Rooms: An Unsupervised Learning IE Simulation

Existing implementations of IEs generally try to anticipate the needs of their users by following preprogrammed rules. For example, this is the case of MIT's ReBa system in their e21 prototype [4]. However, preprogrammed rules do not take into account that the use of a room may change over time. In order to avoid the need to manually redefine the rule set that an IE follows whenever new sensor and effector systems are installed or whenever users wish to use the room in a different way, we have introduced the idea of using an MUL agent to observe changing usage patterns and use data mining techniques to infer appropriate rules for behavior (Macindoe and Maher, 2005).

An MUL agent would share control over the IE's effector system with the humans that use the IE, observe the way that humans use the room, and then attempt to use the effector system in the same manner to complete common patterns of usage for users. For instance, the agent may learn that when the pressure pads around the room's podium area are active and a user dims the lights, then the projector ought to be activated and the presentation software started. The agent's motivation process would help decide what action to take where several mutually exclusive actions might be applicable, such as whether to turn on the lights when a user enters the room, which could be the correct action when the room is not in use but incorrect when a presentation is in progress. Motivation could also help focus the learning process toward an interesting subset of sensor data or select an appropriate time to perform computationally intensive offline learning.

Figure 7.16   Screen shots from a 2 min video clip show many common IIs and configurations.

As a precursor to the development of a full MUL agent-based system, a simulation was performed in order to explore the potential pitfalls for such a system and to investigate appropriate parameters for the UL process. The simulation made use of a UL agent with no motivation process to be used as a baseline against which to compare future motivated agent-based approaches. The simulation was designed to investigate how a UL agent would cope with changes to the patterns of usage in an IE, particularly when potentially ambiguous situations arise where the agent has to distinguish between two or more similar usage patterns.

The world state was represented in the simulated data as a fixed length tuple of 732 attribute-value pairs, $\{(s_1, v_1), \ldots, (s_{732}, v_{732})\}$, where $s \in S$, the set of all sensor input types, and v_n is its corresponding value. The following table shows a breakdown of the structure of this tuple.

This simulated world state assumes the existence of a number of systems not currently installed in the Sentient, including motion sensors, a file sensor that detects the presence of files in a directory for presentations, a light sensor, a lighting control system, and a program detection and launching system. The value type column indicates the kinds of values the given sensor systems can generate. All integer values were normalized to the range 0–255. The effectible column states whether the simulated Sentient has an effector linked to the device sensed by the sensor system, so, for instance, the Sentient could not cause an activity booking to become active, that would have to be done by a human through the booking system, but it could cause the projectors to turn on and off.

7.6.1 *Unsupervised Learning Agent Model*

The UL agent model used in this simulation is shown in Figure 7.17. The model is similar to the MRL agent model presented in Section 7.5.1 but with several key differences: Firstly the motivation process is absent. Secondly the learning process is disconnected from the sensation and action processes. This is because learning is performed as a batch process driven by an implementation of the WINEPI algorithm (Manilla et al. 1997) and the events generated by the sensation process are used directly by the activation process without passing though any intermediate filtering process.

As in the MRL model, $W_{(t)}$ represents the state of the agent's environment at time t, while $S_{(t)}$ represents the state of the environment as sensed by the agent at time t. The sensation, learning, and activation processes constitute the agent's reasoning process. The sensation process S transforms raw data $S_{(t)}$ from the agent's sensors into structures used for reasoning in a manner similar to the MRL agent model. This includes the observed state of the environment $O_{(t)}$ and the change or event $E_{(t)}$ between the current and previous observed states. Only the most recently observed state is stored in the agent's memory, M, because its learning process only looks for patterns of changes that the agent can itself bring about. By contrast, a representation of the set **E** of all events encountered so far is maintained in the agent's memory for data mining. The activation process A selects action $A_{(t)}$ to perform on the basis of the most recent event $E_{(t)}$ and the most recent behavior $\pi_{(t)}$. The behavior $\pi_{(t)}$ is generated as a result of running the learning process $\mathcal{L}$ as a batch process that uses sequential patterns of events in **E** that meet certain requirements for frequency and confidence and that the agent is capable of undertaking itself.

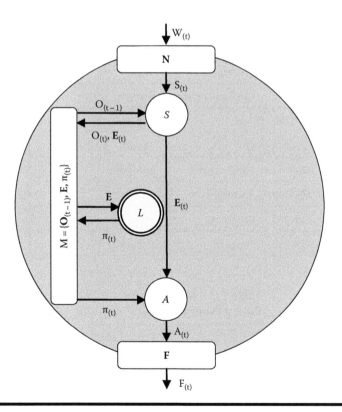

Figure 7.17   UL agent.

7.6.1.1 Sensation Process

The sensation process transforms raw data $S_{(t)}$ from the agent's sensors into structures that facilitate further reasoning. This includes the observed state of the environment $O_{(t)}$ and the change or event $E_{(t)}$ between the current and previous observed states. Events allow agents to represent and reason about changes in their environment in addition to information about the current observed state. Events are computed as the difference between two observed states $O_{(t)} = (o_{1(t)}, o_{2(t)}, \ldots o_{L(t)} \ldots)$ and $O_{(t+1)} = (o_{1(t+1)}, o_{2(t+1)}, \ldots o_{L(t+1)} \ldots)$ using a difference function Δ:

$$E_{(t)} = O_{(t)} - O_{(t')}$$
$$= (\Delta(o_{1(t)}, o_{1(t')}), \Delta(o_{2(t)}, o_{2(t')}), \ldots \Delta(o_{L(t)}, o_{L(t')}), \ldots)$$

The difference function Δ maps two state variable values to an event label prefixed by an identifier O_L based on the sensor from which the observation came. For instance, considering Table 7.1 O_{430} maps to the prefix PROJECTOR_0 because the 430th position in the world state tuple refers to the first projector. When the state variables are binary valued, the mapping is as in Table 7.2.

Because pattern mining using WINEPI requires discrete labels for events, Δ maps changes in integer-valued attributes to discrete labels as shown in Table 7.3.

Table 7.1 Structure of the World State Tuple

Position	Attribute Type	Value Type	Effectible?
1–22	Pressure pad	Binary	No
23–24	Motion sensor	Binary	No
25–124	Activity	Binary	No
125–424	File	Binary	No
425	Light sensor	Integer	No
426–429	Spotlight	Integer	Yes
430–432	Projector	Binary	Yes
433–732	Program	Binary	Yes

Table 7.2 Mappings for Binary Event Labels

$(\Delta(o_{L(t)}, o_{L(t')}))$	Event Label
0	No event
+1	O_L_ON
−1	O_L_OFF

Table 7.3   Mappings for Continuous Event Labels

Difference	Label
+0 to 31	No event
+32 to 95	O_L_PLUSONEQUARTER
+96 to 159	O_L_PLUSONEHALF
+160 to 223	O_L_PLUSTHREEQUARTERS
+224 to 255	O_L_PLUSFULL
+0 to 31	No event
−32 to 95	O_L_MINUSONEQUARTER
−96 to 159	O_L_MINUSONEHALF
−160 to 223	O_L_MINUSTHREEQUARTERS
−224 to 255	O_L_MINUSFULL

7.6.1.2 Learning Process

The UL approach to designing a learning IE assumes that it is inappropriate for an intelligent room to experiment with changes in the state of the room, as an RL agent might, or to require labeled examples from a human instructor. Instead learning must rely upon drawing inferences from previously experienced world states via data mining techniques without being able to affect the environment during the learning process or refer to any set of behaviors known to be correct beforehand. The aim of the learning process of the agent model is to infer behavior $\pi_{(t)}$ from the set of stored event data **E** and then store $\pi_{(t)}$ in memory for the activation process to utilize.

Behaviors will be rules of the form $(A_1, A_2, \ldots, A_{n-1}) \rightarrow (A_1, A_2, \ldots, A_{n-1}, A_n)$, where $A_i \in V$, the set of all possible event types, occurring within a given window of time and A_n is required to be an event type that is affectable. Such rules are formed by mining the frequent episodes of events that occur within a given time window and satisfy a minimum level of frequency, then taking the set of these episodes such that they satisfy the criteria of A_n being affectable. After data mining is complete, the rules are clustered into two sets using K-means clustering [27] on the confidence of the rules with $K = 2$. The cluster of high confidence rules is then stored as the behavior $\pi_{(t)}$.

The WINEPI algorithm [28] was used for the implementation of the UL agent in this simulation, an algorithm based loosely upon the a priori algorithm introduced by Agrawal et al. [29]. The WINEPI algorithm uses a sliding window approach to count the frequency of patterns and an a priori-like candidate generation strategy, which takes advantage of the fact that any subepisode of a frequent episode must itself be frequent. An event sequence s on E, the set of all events, is a triple (s, T_s, T_e) where s = {$(A_1, t_1), (A_2, t_2) \ldots (A_n, t_n)$} of events where $t_i < t_{i+1}$ for all i = 1 … n − 1, and T_s and T_e are start and end times such that $T_s <= t_i < T_e$ for all i = 1 … n. That is, the events must all occur within the specified time window. An episode is a partially ordered collection of events occurring together. There are two main classes of episode, serial episodes and parallel episodes, which can be combined into mixed episodes. In Figure 7.18 α, β, and γ represent examples of serial, parallel, and mixed episode classes, respectively.

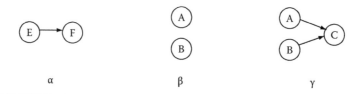

Figure 7.18 Serial, parallel, and mixed episode classes. (From Mannila, H. et al., *Data Min. Knowl. Discov.*, 1, 259, 1997.)

Informally, a serial episode is characterized by events following one another in a strict temporal sequence; for instance, α represents a type F event following a type E event. A parallel episode is characterized by the interchangeable occurrence of one or more types of events, for instance, β represents the occurrence of a type A event or a type B event. A mixed episode combines these types of episodes, so γ represents a type A event or a type B event followed by a type C event. For the purposes of the simulation, only serial episodes were considered for the formation of behaviors.

7.6.1.3 Actions and the Activation Process

The UL agent did not perform actions in the simulation. In principle, the activation process keeps track of all the events that have occurred within a fixed time window, and when there is a match between recent events and the left-hand side of one of the rules in the current behavior $\pi_{(t)}$, the agent performs the action $A_{(t)}$ corresponding to the event label appearing on the right-hand side of the rule. For instance, one of the rules in $\pi_{(t)}$ is

```
(SPOT2_PLUSFULL, PC0_PROG1_ON) -> (SPOT2_PLUSFULL, PC0_PROG1_ON, PROJ1_ON)
```

and the agent observes that spotlight 2 has recently been turned on followed by program 1 on PC0 being launched, then the agent will turn on projector 1.

7.6.2 Simulation Setup

Two weeks of simulated sensor data were generated based upon a timetable of room activities that was a reasonable extrapolation of the activities in the Sentient. For each hour of the 2 week period, an activity type was specified. The activity types were the following: Idle, representing times in which nobody was active in the room such as over the weekend and in evenings and mornings; Noisy, representing times when the room was not officially in use, but people wandered in and out, turned on lights, loaded programs on computers, and generally used the room in less predictable ways; Meeting, representing groups of less than 20 people using the room for meetings; Seminar, representing groups of around 20 people seeing a formal presentation followed by a question and answer session; and Class, representing as many as 40 people attending a formal lecture.

In the 1st week of simulated data, there were no class activities, whereas they did occur in the 2nd week. This was in order to demonstrate the agent's ability to adapt to a new kind of use for the room. Timetables showing a breakdown of the time allocations in the simulation are shown in Tables 7.4 and 7.5. The sensor data characteristic of the simulated activities is shown in Table 7.6.

Table 7.4 Timetable for Simulated Week 1 Data without Classes

Day	0000–0900	0900–1000	1000–1100	1100–1200	1200–1300	1300–1400	1400–1500	1500–1600	1600–1700	1700–2400
Mon	Idle	Noise	Noise	Noise	Seminar	Noise	Meeting	Seminar	Idle	Idle
Tues	Idle	Noise	Noise	Meeting	Noise	Noise	Meeting	Seminar	Idle	Idle
Wed	Idle	Idle	Meeting	Noise	Seminar	Noise	Meeting	Noise	Seminar	Idle
Thurs	Idle	Noise	Noise	Noise	Noise	Noise	Meeting	Noise	Seminar	Idle
Fri	Idle	Noise	Noise	Meeting	Seminar	Noise	Meeting	Seminar	Idle	Idle
Sat	Idle	Idle	Idle	Idle	Idle	Idle	Idle	Idle	Idle	Idle
Sun	Idle	Idle	Idle	Idle	Idle	Idle	Idle	Idle	Idle	Idle

Table 7.5   Timetable for Simulated Week 2 Data with Classes

Day	0000–0900	0900–1000	1000–1100	1100–1200	1200–1300	1300–1400	1400–1500	1500–1600	1600–1700	1700–2400
Mon	Idle	Noise	Noise	Class	Seminar	Noise	Meeting	Class	Idle	Idle
Tues	Idle	Noise	Noise	Class	Noise	Noise	Class	Class	Idle	Idle
Wed	Idle	Idle	Class	Class	Seminar	Noise	Meeting	Noise	Meeting	Idle
Thurs	Idle	Noise	Noise	Noise	Noise	Noise	Class	Class	Seminar	Idle
Fri	Idle	Noise	Noise	Meeting	Seminar	Noise	Meeting	Noise	Idle	Idle
Sat	Idle	Idle	Idle	Idle	Idle	Idle	Idle	Idle	Idle	Idle
Sun	Idle	Idle	Idle	Idle	Idle	Idle	Idle	Idle	Idle	Idle

Table 7.6   Summary of the Sensor Data Generated for Simulated Activities

Activity	Sensor Data Generated
Idle	The world state stays constant throughout the idle period. No sensors activate or deactivate
Noisy	Changes in the world state are randomly generated. The world state changes an average of around once every 2 min
Meeting	The activity sensor triggers for the meeting activity. There is motion sensor activity and the lights are turned on. For 10 min motion sensor activity occurs around 5–10 times. Pressure pads randomly activate. After 10 min there is a 60% chance of a file being placed on PC0, the meeting application being launched, and some of the projectors turning on. The motion sensors trigger occasionally as late-comers arrive and people drift in and out in the background. Pressure pads trigger randomly as people walk around. After 40 min the projectors turn off, the application ends, and the file is removed if PC0 was used. The pressure pads gradually deactivate and the motion sensors trigger around 5–10 times. Then the lights turn out and the activity sensor deactivates
Seminar	The activity sensor triggers for the seminar activity. There is motion sensor activity and the lights are turned on. For 10 min motion sensor activity occurs around 10–20 times. Pressure pads randomly activate. After 10 min a file is placed on PC0, the seminar application is launched, the spotlights turn on, and some of the projectors activate. The motion sensors trigger occasionally as late-comers arrive and people drift in and out in the background. Pressure pads trigger randomly as people walk around. After 40 min of presentation the application ends, the projectors are turned off, the file is removed, and the spotlights turn off. There is a quick burst of motion sensor and pressure pad activity as some people leave before the question and answer session. After 10 min the pressure pads gradually deactivate and the motion sensors trigger around 10–20 times. Then the lights turn out and the activity sensor deactivates
Class	The activity sensor triggers for the class activity. There is motion sensor activity and the lights are turned on. For 5 min motion sensor activity occurs around 20–40 times. Pressure pads randomly activate. After 5 min a file is placed on PC0, the class application is launched, the spotlights turn on, and some of the projectors activate. The motion sensors trigger occasionally as late-comers arrive and people drift in and out in the background. Pressure pads trigger randomly as people walk around. After 50 min of presentation the application ends, the projectors are turned off, the file is removed, and the spotlights turn off. The pressure pads gradually deactivate and the motion sensors trigger around 20–40 times. Then the lights turn out and the activity sensor deactivates

7.6.3 Experimental Results

This section examines the implementation of the learning process from two perspectives. Section 7.6.3.1 looks at the effects of varying the size of the simulated data set and the frequency threshold parameter of the WINEPI algorithm in terms of run time costs, the number of rules produced, and the confidence of those rules. From comparing the performance of the learning process on the two simulated data sets, it is shown that the algorithm performed as theoretically predicted and raises questions about the algorithm's scalability without a motivation process to help time learning sessions or developing an online version of the WINEPI algorithm. Section 7.6.3.2 looks qualitatively at the differences in rules produced firstly when different frequency thresholds are used for the same simulated data set and secondly when the same frequency threshold is used for two different simulated data sets. By considering the meaning of the rules in the context of the IE and the simulations, the capacity of the learning algorithm to support adaptability is shown and areas where additional development of the IE agent is need are revealed. These two perspectives are brought together in Section 7.6.3.3.

7.6.3.1 Learner Metrics

The results of the learning process running on the simulated data for the single week without classes and the double week whose 2nd week contained classes are shown in Tables 7.7 and 7.8. In these tables, minimum occurrences refer to the number of windows in which an episode must appear to be considered frequent; frequency threshold is the proportion of total windows that this minimum frequency requirement represents; rules generated is the total number of rules extracted from the simulated data by the WINEPI algorithm; rules selected is the number of rules placed in the "high confidence" cluster by the learning process; and run time is the total time taken for the learning process to extract the rules measured from the activation of the learning process by the sensation process to the storing of the extracted rules by the memory component.

From these results it can be seen that a doubling in the size of the data set resulted in the production of around a third of the number of rules selected. This is to be expected since a new activity with affectable events was added to the data set, which already contained two such activities. As the frequency threshold was reduced, the quality of the rules, measured by the minimum confidence bound, decreased. This is to be expected as rules with low frequencies are expected to be idiosyncratic, and the lower the frequency threshold, the more likely it becomes that patterns in the random noise are mistaken for usage patterns by the learning process. As the number of rules increased, the minimum confidence bound of the clustering also decreased; this is due to the phenomenon just mentioned and the relationship between the frequency threshold and number of rules generated shown in Figure 7.19, which reveals a steep increase in the number of rules generated as the frequency threshold drops.

Table 7.7 Rules Generation and Run Time for Single Week Data with a 60 s Window

Minimum Occurrences	Frequency Threshold	Rules Generated	Rules Selected	Run Time (ms)	Minimum Confidence	Maximum Confidence
600	0.0001	0	0	1,219	0.0	0.0
60	0.00001	15	11	6,031	0.69	0.92
50	0.000008	16	12	6,688	0.69	0.92
30	0.000005	22	17	16,876	0.61	0.92
25	0.000004	28	22	22,047	0.61	0.92

Table 7.8   Rules Generation and Run Time for Double Week Data with a 60 s Window

Minimum Occurrences	Frequency Threshold	Rules Generated	Rules Selected	Run Time (ms)	Minimum Confidence	Maximum Confidence
600	0.00005	0	0	2,735	0.0	0.0
60	0.000005	33	26	119,984	0.53	0.92
50	0.000004	58	35	178,420	0.47	0.92
30	0.0000025	120	59	836,593	0.45	0.92
25	0.0000002	175	80	126,3655	0.45	0.92

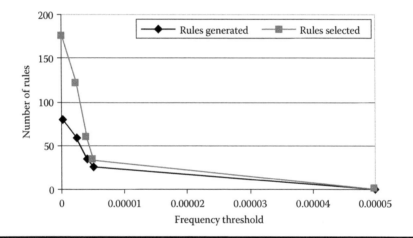

Figure 7.19   Frequency threshold versus rules generated for the double week data set.

Figure 7.20 shows the steep rise in run time resulting from decreasing the frequency threshold for the double week data set, a trend imitated by the single week data set as can be seen from the increase in run time from frequency thresholds 0.000005–0.000004 in Table 7.7. This increase in run time cost is due to the substantial increase in the number of rules generated as a result of the lowering threshold, a trend demonstrated in Figure 7.19.

Figure 7.21 demonstrates that there is a near linear relationship between the total number of rules mined and the run time cost for the learning process. The slight dip around at the 58 rule mark is possibly due to there being a slightly lower proportion of longer rules generated at that point, since longer rules take longer to mine as will be seen in the next set of results.

These results make it clear that the learning process is highly sensitive to changes in frequency threshold at the lower end of the scale for these data sets with strong increases in the number of rules mined as the frequency threshold drops below 0.000005, but that the learning process's performance as the number of rules mined increases is favorable, with close to a linear relationship between rules produced and run time.

A doubling in the size of the set of events considered by the algorithm increased the run time by almost a factor of 10 when comparing the results for the thresholds 0.000005 and 0.000004.

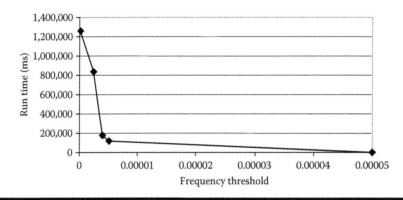

Figure 7.20   Frequency threshold versus run time for the double week data set.

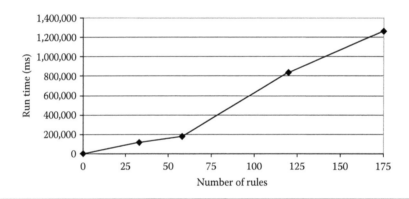

Figure 7.21   Rules generated versus run time for the double week data set.

Tables 7.9 and 7.10 show a breakdown of the run time costs for the threshold 0.000005 case for the single and double week data sets.

The difference in the size of the initial candidates is partly due to the increase in event types from the introduction of a new type of activity and partly due to the larger number of random noise events, some of which produced events in the 2nd week that had not been seen in the 1st week. The overall number of initial frequent episodes discovered decreased slightly because the relative proportions of events were disturbed by the introduction of events related to classes, pushing the proportions of some events below the threshold. The increase in the maximum candidate length suggests that the proportion of simultaneously occurring frequent events increased with the introduction of classes and this meant that episodes could be longer than the window size. These simultaneous events were most likely related to motion sensor activity since class activities include a large amount of door noise in a relatively small time period. The overall increase in the number of frequent episodes found and the subsequent increase in the number of candidates after length 3 were also most likely due to the set of episodes related to the new class activity and the associated simultaneous event episodes mentioned previously.

According to Mannila et al. (1997), the time complexity for the WINEPI to perform a scan for frequent episodes is $O(n \, |C| \, l)$, where n is the number of events in the database, $|C|$ is the size

Table 7.9   Candidate Generation and Frequency Testing Breakdown for Single Week Data, Window Size 60 s, and Frequency Threshold 0.000005

Candidate Length	Candidates Generated	Generation Time (ms)	Frequent Episodes Found	Scan Time (ms)
1	728	32	115	250
2	13, 225	234	302	7547
3	1, 578	359	275	6026
4	356	141	135	1687
5	78	16	56	375

Table 7.10   Candidate Generation and Frequency Testing Breakdown for Double Week Data, Window Size 60 s, and Frequency Threshold 0.000005

Candidate Length	Candidates Generated	Generation Time (ms)	Frequent Episodes Found	Scan Time (ms)
1	997	63	93	500
2	8649	94	451	9, 562
3	2411	922	667	27, 687
4	1237	953	400	19, 875
5	993	187	499	47, 328
6	415	297	330	10, 984
7	66	63	58	969

of the candidate set, and l is the length of the candidates. Since both |C| and n were increasing simultaneously in the comparison between the two data sets for reasons explained earlier, quadratic worst case increases in time costs are to be expected from length 3 candidates onward. This accounts for the dramatic increase in run time costs seen previously in Tables 7.7 and 7.8. Taking this scalability behavior into account, use of the WINEPI algorithm may prove to be problematic for analyzing longer periods, such as a month or a year of data.

7.6.3.2 Qualitative Rule Assessment

Since the goal of adaptation to changing patterns of usage is quite a specific requirement, some qualitative assessment of the rules and the degree to which they meet this goal is necessary in addition to the quantitative metrics discussed previously.

Since each run of the learning process results in a large number of rules being generated, especially in the 2 week case, an exhaustive listing of rules is infeasible, so two illustrative comparisons will be made in this section: firstly a comparison between two runs of the learning process on the single week data set at two different frequency thresholds and secondly a comparison between two runs of the learning process with the same frequency threshold on the single week and double week data sets.

7.6.3.2.1 Rule Differences due to Frequency Thresholds

A comparison of the rules mined for frequency threshold 0.00001 and 0.000005 with window size 60 s on the single week data set shows 11 rules in common and 6 new rules mined for the lower frequency threshold. An example of a typical high confidence rule mined in both cases is

```
A.(PC0_PROG1_OFF) -> (PC0_PROG1_OFF,PROJ1_OFF) (confidence: 0.92)
```

This indicates that when the seminar program ends, the main projector should be deactivated. Intuitively, this is a good rule since the pattern of usage defined for seminars always shows this behavior. An example of a borderline low confidence rule held in common is

```
B.(PC0_PROG1_ON) -> (PC0_PROG1_ON,PROJ1_ON) (confidence: 0.6875)
```

The lower confidence for rule B is due to the fact that sometimes random noise causes the main projector to be left on as a seminar starts, so the projector does not always need to be turned on. The six rules added by the halving of the frequency threshold are as follows:

```
1. (PC0_PROG2_OFF) -> (PC0_PROG2_OFF, PROJ2_OFF) (confidence: 0.76)
2. (PC0_PROG2_ON) -> (PC0_PROG2_ON, PROJ1_ON) (confidence: 0.61)
3. (SPOT1_PLUSFULL) -> (SPOT1_PLUSFULL, PROJ1_ON) (confidence: 0.75)
4. (SPOT2_PLUSFUL1) -> (SPOT2_PLUSFULL, PROJ1_ON) (confidence: 0.75)
5. (SPOT2_PLUSFULL) -> (SPOT2_PLUSFULL, PC0_PROG1_ON) (confidence: 0.83)
6. (SPOT2_PLUSFULL, PC0_PROG1_ON) -> (SPOT2_PLUSFULL, PC0_PROG1_ON,
                                        PROJ1_ON) (confidence: 0.9)
```

Rule 1 relates to deactivating projectors after meetings. This is always done and is correct, but has a low frequency due to the scarcity of presentations in meetings. Halving the frequency threshold reduced the lower bound of the high confidence cluster by 8% due to the introduction of rule 2. Rule 2 is functionally similar to rule B, but for meetings that less frequently have presentations made on the computer so the rule frequency is lower. It has a low confidence threshold for similar reasons to rule B. Rules 3 and 4 follow from the activation of spotlights during seminars being a precursor to a presentation being made. Rule 5, activating the seminar software when spotlights activate at the start of a presentation is beginning, also follows from this pattern, but as discussed in the following section will become a problematic rule when classes are introduced that follow a similar usage pattern. These three rules have relatively low frequencies due to interference by noise in a similar way to rule 2. Rule 6 is a redundant rule, since rule B is sufficient to satisfy it, but it has a higher confidence due to its specificity. Such redundant rules are fairly typical of longer rules that are mined from the data, since their subepisodes would have to be frequent in order for them to be frequent.

This example illustrates the two characteristic problems associated with decreasing the learning process's frequency threshold, namely, the drop in the lower confidence bound for the "high confidence" rule cluster and the introduction of very specific and often redundant rules.

7.6.3.2.2 Rule Differences between Data Sets

Comparing the rules generated by the learning process from the single week data set without classes against the rules generated by the learning process from the double week data set with classes, both with a window size of 60 s and a frequency threshold of 0.000005, shows 16 rules in common, 16 additional rules, and 1 rule eliminated.

The eliminated rule is rule 5 from the set earlier. The most likely reason for its elimination is that with the introduction of the class activity, there is a conflict over which program is meant to be run after the pattern of events that indicates the start of a formal presentation, that is, one involving lights dimming and the use of spotlights, since there are examples of both the class program and the seminar program being run. This clash probably caused the rule's confidence to lower so that it fell into the "low confidence" cluster and was discarded. The following rules are characteristic of the rules that were added as a result of the extra data being mined:

```
7.  (SPOT4_PLUSFULL) -> (SPOT4_PLUSFULL, PROJ1_ON) (confidence: 0.63)
8.  (PC0_PROG3_OFF) -> (PC0_PROG3_OFF, SPOT1_MINUSFULL) (confidence: 0.75)
9.  (PC0_PROG3_ON) -> (PC0_PROG3_ON, PROJ1_ON) (confidence: 0.71)
10. (PROJ1_OFF) -> (PROJ1_OFF, SPOT1_MINUSFULL) (confidence: 0.53)
11. (PC0_PROG3_OFF, PROJ1_OFF) -> (PC0_PROG3_OFF, PROJ1_OFF,
                                   SPOT1_MINUSFULL) (confidence: 0.82)
12. (SPOT1_PLUSFULL, PC0_PROG1_ON) -> (SPOT1_PLUSFULL, PC0_PROG1_ON,
                                        PROJ1_ON) (confidence: 0.9)
```

Rules like 7 are an extension of the set of rules like 3, which were on the border of being frequent enough in the single week data set and have been completely brought into the set of frequent episodes by the inclusion of extra presentations due to the addition of classes. Rules 8, 9, and 11 are equivalent to rules A and B for seminars, but they apply in classes. Rules like 10 are now included in the rule set because a higher proportion of uses of the main projector are now done in a presentation style due to the inclusion of classes. The low confidence reflects the fact that meetings generally have projectors on without spotlights being on and therefore not needing to be turned off. Rule 12 is the same as rule 6, suggesting that it was on the border of being frequent and the expanded timetable made it frequent.

Rules such as 8, 9, and 11 strongly demonstrate the implementation's capacity for adaptability. The agent's learning process was able to discern and adopt behavioral rules appropriate to classes from the changing usage patterns of the room. Furthermore, the learning process avoided adopting any rules that might result in a clash between behaviors appropriate for classes and behaviors appropriate for seminars and discarded one such rule learnt from the previous week's data that would have resulted in such an inappropriate behavior. Admittedly, clustering by confidence does not guarantee that no such behaviors could possibly be developed, but for this test it appears to have been a good heuristic.

7.6.3.3 Discussion

From the results of running the learning process on the simulated data, it is clear that the agent implementation met the basic goal of the simulation, namely, to show that it could adapt to and be useful for typical activities in a university meeting room. The implementation demonstrated its ability to adapt to the new patterns of usage implicit in class activities while at the same time avoiding the obvious pitfall of introducing rule clashes between class and seminar behaviors, which have similar characteristic usage patterns. In no cases for the 0.000005 frequency threshold runs were rules generated that were prima facie undesirable. Redundant rules were generated, but there is no evidence of any rules that would produce frustrating or contradictory behaviors. This bodes well for the future development of the implementation as the Sentient's sensor and effector systems are installed. There is no guarantee, however, that such rules cannot be generated, which raises the issue of the need for a system to deal with problematic rules. Furthermore, when rules are mined that contradict or clash with one another, it may be inappropriate to simply abandon them with the rest of the low confidence cluster rules. Such rules often indicate that there are more subtle rules that require disambiguation.

The quantitative analysis of the learning process on the simulated data shows that the learning process is heavily dependent upon both the size of the data set and the frequency threshold chosen for the learning process. This raises concerns about the scalability of the learning process to larger data sets, such as a month's worth of data, and also the question of how to determine an appropriate frequency threshold for the learning process. There are two key approaches to the scalability issue, firstly improving the performance of the WINEPI algorithm and secondly devising a strategy for directing the learning process as to when to learn and thereby avoid the scalability issues.

Although the results of the simulation are promising, testing with live sensor data is still required. From the qualitative analysis presented in Section 7.6.3.2, it seems that a threshold of 0.000005 is sufficient to produce meaningful rules for the simulated data set that are not present in the rule set when the frequency threshold is doubled. These results are only for simulated data, however, so they only provide an estimate of the kind of effect that different frequency thresholds would have on mining live data and so would need to be tuned once live testing is done.

Two directions for the adaptable IE agent model take advantage of our experience with MRL agents:

1. Use a motivation model to focus the agent's attention on a specific learning task or data set when appropriate.
2. Use a motivation model to direct learning by specifying the parameters of the learning algorithm.

In the case of focusing attention, what is needed is an arousal model appropriate to the domain, such that when sequences of events that could potentially form rules arise, the agent is driven to activate its learning process. Attention focusing in motivated reinforcement learners can be driven by event frequency, where the frequency with which an event occurs is taken as the difficulty of causing that event to occur and the reinforcement learner is driven to attempt to recreate the event that occurs most frequently in the cluster of infrequent events, the easiest "hard" event. The difficulty in adapting this model lies in the fact that a reinforcement learner's actions cause the events that it is trying to learn behaviors to bring about, which in turn raises the frequency of those events, making the agent focus on something else. In the case of an IE where RL is inappropriate, the agent has no influence over the frequency of events occurring within it so it would be driven to learn the same task repeatedly. However, changing proportions of events may be sufficient to indicate that a new pattern of behavior is occurring and therefore trigger the agent's learning process in a way similar to MRL. For instance, an increase in class-related events, as a proportion of events, may be sufficient to trigger learning. The challenge, however, would be to go further than this and to find a way to efficiently detect changes in the proportion of whole patterns of events and use these as a trigger for learning.

An alternative approach to focusing attention on learning could be taken from Norman and Long's (1996) work on using alarms as heuristics to model motivation. The idea would be that, building upon the idea in the previous section of learning after a fixed time interval, the motivational intensity of the agent toward learning would increase with the time since it last mined rules until it reached a certain intensity threshold, and then learning would be triggered. In addition to this, the agent's motivational intensity could increase faster as time passes in which no events occur, since this would represent the agent taking notice of the fact that the room is not in use at the moment so it is appropriate to spend its computational resources on learning. Likewise the ongoing occurrence of events could decrease the motivational intensity of the agent so that it is disinclined to trigger learning while its computational resources are needed for attending to its action process. This follows Norman and Long's idea that a motivated agent ought to increase its motivational intensity when the opportunity to fulfill one of its goals arises. In this case the room's goal is to learn without inconveniencing its users by spending computational resources when they are needed for other things.

A key limitation of the current implementation of the adaptable IE model is that it lacks awareness of context. A motivational model may go some way to helping solve this problem. Currently where several usage patterns exist such that rules with contradictory outcomes could be inferred, these rules either tend to fail to reach the required threshold to be clustered as "high confidence," becoming discarded, or one or all are adopted inappropriately. This contrasts unfavorably with current manually configured IEs, such as MIT's ReBa (Hassens et al. 2002) that can resolve such clashes. ReBa, for instance, has a layered structure to its behavioral rules, allowing it to modify its behavior depending on context so that a user entering the IE when the lights are off will result in the lights being turned on unless a film is being played since ReBa's rules specify that the lighting rule for film activities overrides the regular rule.

Rules that conflict in this way have a particular characteristic; they are not merely rules that have low confidence values, they are similar but different in some key ways to the other rules with which they conflict. The computational model of curiosity used in the curious information display develops this idea in its use of the HSOM. In the case of the IE, the arousal value generated by the motivation process could be used to indicate whether a contextual discovery mechanism may need to be triggered in order to learn the set of rules that triggered it due to their similarity to existing rules that the map has been exposed to before. As a result of this trigger, a user may be called upon to disambiguate the rule, or a search may be performed for contextual clues for disambiguation within the data, such as the current active scheduled event taking place during each episode that lead to the conflicting rules being mined. A key challenge that would need to be overcome in the use of a curiosity model using a self-organizing map-based approach would be to find an appropriate mapping between the events comprising an episode and the space of the map, since the dimensionality of the events may be extremely high due to the number of possible event types and the number of events occurring in an episode is variable, potentially making events very distant from one another in the space.

7.7 Conclusion

This chapter presents general models for intrinsically motivated learning in IEs and demonstrates these models for RL and UL in two applications. In the first application, an MRL agent controls the structure and content of a curious information display. Curious information displays are a means of creating digital displays that can augment physical places to attract the interest of observers by being curious about and learning about the structure and content of information being displayed. This application has experimented with the use of both events and observed states as the input for the motivation process and shown that different types of emergent behavior results from these alternatives. Reasoning about events produces behaviors that change rapidly between different states while reasoning about observed states produces behaviors that seek to maintain interesting states. This application has also experimented with MRL in conjunction with a function approximation RL algorithm. It has shown that task-oriented emergent behavior using a general intrinsic reward signal is still possible in this setting.

In the second application, a UL agent controls the hardware devices in a simulated intelligent room. This application facilitates usage of a physical space. We showed that UL agents can adapt to and be useful for typical activities in a university meeting room. The implementation demonstrated this ability to adapt to the new patterns of usage implicit in class activities while at the same time avoiding the obvious pitfall of introducing rule clashes between class and seminar behaviors, which have similar characteristic usage patterns.

Acknowledgments

This work was supported by a grant from the Australian Research Council for the project "Curious Place: Agent-Mediated Self-Aware Worlds." This chapter was written while Mary Lou Maher was working at the National Science Foundation in the United States. Any opinion, findings, and conclusions or recommendations expressed in this chapter are those of the authors and do not necessarily reflect the views of the National Science Foundation.

References

1. R. A. Brooks, M. Coen, D. Dang, J. DeBonet, J. Kramer, T. Lozano-Perez, J. Mellor et al., The intelligent room project, presented at *The Second International Cognitive Technology Conference (CT'97)*, Aizu, Japan, 1997.
2. M. Coen, Design principles for intelligent environments, presented at *The 15th National/10th Conference on Artificial Intelligence/Innovative Applications of Artificial Intelligence*, Madison, WI, 1998.
3. A. Kulkarni, Design principles of a reactive behavioural system for the intelligent room, *Bitstream: The MIT Journal of EECS Student Research*, 1–5, April 2002.
4. N. Hanssens, A. Kulkarni, R. Tuchinda, and T. Horton, Building agent-based intelligent workspaces, presented at *The Third International Conference on Internet Computing*, Las Vegas, NV, 2002.
5. M. Roman, C. K. Hess, R. Cerqueira, A. Ranganathan, R. H. Campbell, and K. N. Nahrstedt, Gaia: A middleware infrastructure to enable active spaces, *IEEE Pervasive Computing*, 1, 74–83, 2002.
6. B. Johanson, A. Fox, and T. Winograd, The interactive workspaces project: Experiences with ubiquitous computing rooms, *IEEE Pervasive Computing*, 1, 67–74, 2002.
7. T. J. Norman and D. Long, Goal creation in motivated agents, presented at *Intelligent Agents: Theories, Architectures and Languages*, 1995.
8. L. P. Beaudoin and A. Sloman, A study of motive processing and attention, in *Prospects for Artificial Intelligence*, A. Sloman, D. Hogg, G. Humphreys, D. Partridge, and A. Ramsay, Eds., IOS Press, Amsterdam, the Netherlands, 1993, pp. 229–238.
9. R. Aylett, A. Coddington, and G. Petley, Agent-based continuous planning, presented at *The 19th Workshop of the UK Planning and Scheduling Special Interest Group (PLANSIG 2000)*, Milton Keynes, U.K., 2000.

10. R. Saunders and J. S. Gero, Designing for interest and novelty: Motivating design agents, presented at *CAAD Futures 2001*, Kluwer, Dordrecht, the Netherlands, 2001.

11. R. Saunders and J. S. Gero, Curious agents and situated design evaluations, in *Agents in Design*, J. S. Gero and F. M. T. Brazier, Eds., Key Centre of Design Computing and Cognition, University of Sydney, Sydney, New South Wales, Australia, 2002, pp. 133–149.

12. R. Saunders and J. S. Gero, The digital clockwork muse: A computational model of aesthetic evolution, presented at *The AISB'01 Symposium on AI and Creativity in Arts and Science, SSAISB*, York, U.K., 2001.

13. K. Merrick and M.-L. Maher, Motivated reinforcement learning for non-player characters in persistent computer game worlds, presented at *International Conference on Advances in Computer Entertainment Technology*, Hollywood, CA, 2006.

14. M.-L. Maher, K. Merrick, and O. Macindoe, Intrinsically motivated intelligent sensed environments, presented at *EGICE 2006*, Ascona, Switzerland, 2006.

15. R. S. Sutton and A. G. Barto, *Reinforcement Learning: An Introduction*, The MIT Press, Cambridge, MA, 2000.

16. G. Zoubin, Unsupervised learning, presented at *Advanced Lectures on Machine Learning*, Canberra, Australian Capital Territory, Australia, 2003.

17. M.-L. Maher, K. Merrick, and O. Macindoe, Can designs themselves be creative? presented at *Computational and Cognitive Models of Creative Design*, Heron Island, Queensland, Australia, 2005.

18. O. Macindoe and M.-L. Maher, Intrinsically motivated intelligent rooms, presented at *Embedded and Ubiquitous Computing*, Nagasaki, Japan, 2005.

19. A. Merceron, *Languages and Logic*: Pearson Education Australia, Sydney, New South Wales, Australia, 2001.

20. C. Ware, *Information Visualisation: Perception for Design*, Morgan Kaufmann, San Francisco, CA, 2000.

21. H. Dieterich, U. Malinowski, T. Khme, and M. Schneider-Hufschmidt, State of the art in adaptive user interfaces, presented at *Adaptive User Interfaces: Principle and Practice*, North Holland, the Netherlands, 1993.

22. A. Graesser, K. VanLehn, C. Rose, P. Jordan, and D. Harter, Intelligent tutoring systems with conversational dialogue, *AI Magazine*, 22, 39–52, 2001.

23. J. C. Stanley, Computer simulation of a model of habituation, *Nature*, 261, 146–148, 1976.

24. W. Wundt, *Principles of Physiological Psychology*, Macmillan, New York, 1910.

25. G. A. Rummery and M. Niranjan, On-line q-learning using connectionist systems, Technical Report CUED/F-INFENG/TR 166, Engineering Department, Cambridge University, Cambridge, U.K., 1994.

26. C. Watkins and P. Dayan, Q-learning, *Machine Learning*, 8, 279–292, 1992.

27. J. B. MacQueen, Some methods for classification and analysis of multivariate observations, presented at *Proceedings of the Fifth Berkeley Symposium on Mathematical Statistics and Probability*, Berkeley, CA, 1967.

28. H. Mannila, H. Toivonen, and A. Verkamo, Discovery of frequent episodes in event sequences, *Data Mining and Knowledge Discovery*, 1, 259–289, 1997.

29. R. Agrawal, T. Imielinski, and A. Swami, Mining association rules between sets of items in large databases, presented at *The ACM SIGMOD Conference on Management of Data*, Washington, DC, 1993.

Chapter 8

Building Smart Spaces with Intelligent Objects

Tomas Sanchez Lopez and Daeyoung Kim

Contents

8.1 Introduction

The vision of a future populated with intelligent objects to react and adapt to the environment for better serving the human being originated over a decade ago. Many new concepts have arisen from this idea, including terms such as "smart environments," "ubiquitous computing," and "pervasive computing." However, the underlying concept is simple: If computing and sensing units become small enough, they might be integrated everywhere, providing a rich flow of anytime–anywhere services.

Together with the maturing of the concept, technology has evolved to produce new and more feasible pervasive computing units. Among all, RFID (radio frequency IDentification) and WSN (wireless sensor networks) have attracted special interest for their promising applications. On one hand, RFID tags are finally becoming mature enough for their massive deployment. On the other hand, WSN are now on the verge of producing practical applications, not only for industrial use but also to support people's daily life.

RFID technology is already a de facto standard in wireless electronic identification whose main applications reside in integrating RFID chips in real-world objects. Envisioned as a substitute for the bar code, the number of RFID applications get a boost when considering new advances in supply chain management and the integration of additional processing and sensory capabilities in the tags. RFID tags are specified in a variety of classes featuring different capabilities. Those proven most relevant until now are the so-called passive tags, which lack battery and are powered by reader devices situated in close range.

WSN are networks of small, cost-effective devices that can cooperate to gather environment information via simple integrated sensors. Sensor nodes are low-power, low-memory, low-range wireless devices that can operate at various radio frequencies. Additionally, sensor nodes may include simple actuators or be attached to more complex ones. WSN have gained popularity for their social implications through concepts such as "smart home," "smart classroom," "smart objects," or, in general, "smart spaces." These concepts involve sensor networks that gather users' *context*, meaning circumstances or conditions that surround users' behavior. Services are then offered to the context source, involving most of the time automatic actions that the user may desire in a particular situation.

Many believers in WSN picture a future of *ubiquitous* (anywhere at any time) sensor and actuator networks, referring to infrastructures such as houses, farms, or cities. However, we believe that the meaning of "ubiquitous" should not just be restricted to place sensor devices everywhere, but rather to make those WSN also part of real-world objects. In the same way as RFID technology aims to provide object information (simple identification) transparent to the user, wireless sensor devices installed in those same objects could *augment* their information. Merging RFID and sensor networks into the same objects would result in truly *smart* objects, moving around while providing their sensed context with a unique identification number. Furthermore, if at some point in time several objects are related to produce information about the same context, they could interact and decide to produce joint information under a single identification.

In our research, we analyze the challenges that arise from merging RFID and WSN to use them in smart spaces. We design a framework that considers all the processes involving users receiving services according to their context. This includes RFID tag memory design, WSN association protocols, routing and addressing schemes, RFID-sensor-actuator data integration and management, service definition and delivery, context and service matching, distributed middleware, etc. We also consider the best ways for integrating our ideas with existing infrastructures and provide implementation scenarios that prove our concepts. The rest of this chapter is organized as follows.

Section 8.2 gives an overview of the system. Section 8.3 describes the *entities* and the algorithms involved in the association processes. Section 8.4 presents the concept of *virtual entity* (VE) and Section 8.5 shows how they fit into the WISSE (wireless sensors and RFID for smart environments) middleware. Section 8.6 lists the different types of WISSE services. The potential application of WISSE are overviewed in Section 8.7, while Section 8.8 describes our implementation of the framework. Section 8.9 presents our future work in this area. Finally Section 8.10 compares our work with existing work in the area, and Section 8.11 concludes this chapter.

8.2 Wireless Sensors and RFID for Smart Environments

As mentioned in the introduction, in our work we envision a future of smart environments in which mobile objects and users carrying RFID and WSN will receive ubiquitous services according to their identity and real-time sensor and actuator information. We call this framework WISSE [1]. Consider the following example: Anne finishes her work and goes back home on a hot summer day. Her combination of today's clothes carries RFID tags and temperature sensors. Moreover, her RFID-enabled digital ID card and her GPS cell phone complete Anne's usual outfit. On her way home, her shirt sensor detects unusual high temperature due to the atypical hot weather. Fortunately for Anne, her profile in the WISSE network has noticed this fact as well, and realizing that she is walking home (GPS) sends an actuation command on Anne's behalf to her home server, turning on the AC just once. Moreover, Anne's cell phone receives a map with the location of a new ice cream shop on her way home, just in case she feels like making a refreshing stop. To make this example possible, WISSE defines a framework from the service receivers to the service providers and the logic for automatic, optimum service delivery. Following the separation between service providers and receivers, WISSE defines two layers and the border that separates them:

- *Context layer:* Contains the real-world objects carrying RFID/WSN. These objects, which we call *entities*, may associate forming groups of meaningful information (Section 8.3.1). Examples of entities could be clothes, vehicle parts, tableware, cell phones, etc. From now on, we will use the term *entity* to refer to equally grouped or ungrouped objects in general, whereas if necessary we will use *entity group* and *single entity* to specifically name grouped and ungrouped objects, respectively.
- *Service layer:* Contains the service providers registered to the WISSE network. As we will see later in more detail, the service providers specify a series of requirements in terms of sensors and actuators that clients must meet in order to be eligible to receive their services. Additionally, service providers hold the logic and processing power needed for executing the services they offer, plus a standard way of specifying how the clients should access their services once they have been granted access.
- *Service network:* As we have seen so far, mobile *entities* receive services according to their context. A key issue is, however, how to match the context of the entities in the context layer and the service requirements of the service layer when that context is deeply dynamic (as entities are mobile and thus their associations are spontaneous and temporal). Furthermore, according to our premises, those entities may be anywhere, anytime, and potentially in a big number. Hence, we need a distributed infrastructure that will connect clients and servers while tracking their context and requirements, respectively. The service network edge holds the tools for receiving, storing, and matching user and service information in a distributed manner.

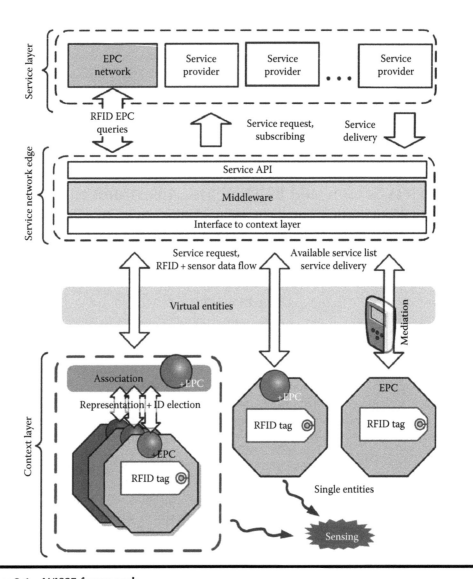

Figure 8.1 WISSE framework.

Figure 8.1 depicts the WISSE framework. Starting from the bottom, the context layer holds several entities that may associate around a *grouped entity*, which will immediately begin to produce joint context information. When associating, the entities choose a common identification (EPC) and a representative for sending out their context to the WISSE service network edge. The main goal of the framework is how to interpret that context information in order to offer the context entities meaningful services in an automatic, transparent manner. Throughout the following sections, we detail each one of the building blocks of WISSE and the interactions between them.

8.3 Entities

WISSE entities are no physically different from any object that may carry an RFID tag. The difference is functional: While normal RFID-tagged objects are just individually read by some

RFID reader, WISSE entities may associate and then transmit their information jointly. In order to achieve this, entities use the wireless communication capabilities of their sensor nodes. It is also possible for entities to lack sensor nodes, although this requires the mediation of an RFID reader that will implement WISSE protocols.

From the broad concept of radio frequency identification (RFID), we choose the EPC Global tag data standards [2]. The electronic product code (EPC) is an identification scheme designed to support the needs of various industries by accommodating both existing and new coding schemes. As we will see later, the decision of using EPCs as identification tokens for our entities goes beyond the tag data specifications. Rather, the EPC Global standards specify a whole architecture framework in a collection of interrelated standards for hardware, software, and core services. This architecture enables numerous advantages when managing information related to EPCs. A detailed description of the EPC Global architecture and the advantages that it brings to WISSE can be found on Section 8.6.2.

The EPC Global standards consider various kinds of tags distributed in *classes*. A total of four classes are defined, from the well-known *passive* tags (class 1 and 2) to the battery powered *active* tags. In December 2004, EPC Global ratified the second generation of part of the tag standards, known as the Class-1 Generation 2 UHF RFID, or air interface [3]. This generation involves noticeable improvements, including better compatibility, security, and extended tag memories. In WISSE, we want to emphasize the advantages of combining RFID and WSN to provide context information, but with a dose of reality. Therefore, we focus our research on current available technology, such as Class-1 Gen2 RFID tags and existing wireless sensor networks. However, our current research does not exclude the use of future technology. In fact, when Class 3 and 4 tags become available, the integration issues among RFID and sensor information will be solved by the standards themselves, and part of our present considerations regarding issues such as reader mediation and information capture will not be necessary.

In accordance with what was argued before, WISSE entities should carry one (and only one) passive RFID tag. The tag's EPC will uniquely identify the entity, the same way today's tagged objects are identified by one EPC. In addition, each entity may carry several sensor nodes, which may contain any combination of sensors and actuators. Ideally, each sensor node has a map of the RFID tag's memory of the entity it belongs. In this way, each node has a sense of identity by holding an EPC and has also access to other parts of the memory needed by WISSE protocols. Unfortunately, we cannot assume WSN to be able to read RFID tags. It would be desirable, though, that the same manufacturers that tag the objects and install the sensor nodes should also transfer the RFID memory map to the node's memory. This is in fact a reasonable assumption for factory production lines using RFID gates holding also WSN transceivers, for example, installed in conveyor belts. In this way, while RFID tags would still maintain their functionality if read by regular RFID readers, the sensor nodes could participate in the WISSE network on behalf of the same objects. Nevertheless, it is also possible for those nodes to be added after object manufacturing and, besides, we would like to consider the participation of objects without sensor nodes. The rest of this chapter assumes the RFID tag's memory is mapped into the sensor nodes memory belonging to their object. A discussion of how to overcome the problem of nodes without this map is provided in Section 8.3.1.2.

8.3.1 Grouping

Single entity data is normally poor in the sense that it cannot offer enough context information for getting relevant services. Providing as much data as possible for the same client may extend its context information and hence the possible available services. WISSE entities sense not only the environment but also the presence of other entities, and so may associate with them if they have the same interest. *Grouping* is the process by which two or more entities decide to collaborate by sharing their context information. Entities periodically advertise their presence by sending advertisement packets and listening for other entities in periodic, unsynchronized intervals. In order for an entity to decide which other entities it should associate with, we divide the grouping process in two phases. The first phase involves the information stored in the RFID tags to make basic decisions and prioritize the association process. The second phase involves choosing representatives, invoke the addressing process, and distribute the results to all the group members. Only when the grouping process is completed, the entities will be aware of their new membership and the result will be communicated to the rest of the WISSE infrastructure.

In order to deal properly with dynamic wireless networks, WISSE organizes any combination of associated entities in a double-clustered architecture. On one hand, each individual entity chooses a cluster head, which will communicate with other cluster heads from other entities. On the other hand, a group of associated entities chooses a correspondent cluster head to communicate with the service network. Both election processes are aperiodic and based on remaining node energy. Election processes are triggered by a number of reasons, including assigned time expiration, node battery exhaustion, or node discovery. Finally, each entity group chooses a meaningful identifier (EPC) among all entities, which will represent the group for the rest of the system. An entity will report its interactions dynamically, getting services according to the context gained through them. The following sections detail the processes that take place in the context layer, from choosing which entities shall form a group to which identifier should be chosen.

8.3.1.1 Representative Election

8.3.1.1.1 Entity Cluster Head Election

Entity cluster heads (ECH) are used by *single entities* to communicate with other entities. Hence, the first task of the entity sensor nodes is to undertake an ECH election process. This process is as follows.

When a node marked as "ECH capable" is unable to find its entity cluster head, it sends a proposal for being ECH for a time T_{entity}, which is a function of the node's remaining energy. Only nodes that compute higher T_{entity} will respond to the proposal. In order to avoid collisions when more than one node tries to answer, a delay in the response is introduced, also function of the remaining power plus a random component. We can compute the proposed time and delay as follows:

$$T_{entity} = C_1 \times \text{Remaining energy}$$

$$Delay = \frac{C_2}{\text{Remaining energy}} + \text{Random}$$

where $0 \leq \text{Random} \leq \frac{C_2}{2 \times \text{Max energy}}$ and C_1, C_2 are constants

The ECH election process starts when

1. Time T_{entity} expires
2. No cluster head is selected in current entity
3. A node cannot communicate with its ECH before T_{entity} expires

Once the process is completed, the ECH election result will be broadcasted to the rest of the entity nodes, which from that moment on will use the ECH as a rely node to communicate with the outside.

8.3.1.1.2 Correspondent Election

WISSE *correspondents* are defined as ECHs that are elected to communicate the entity with the service network edge. Only one *correspondent* can exist per group of associated entities. If an entity does not belong to any group, its ECH is automatically promoted to correspondent.

Not any ECH is eligible at any time to become a *correspondent*. It is possible for certain nodes of an entity to be in range with some sensor network gateway (edge devices) while some others remain "hidden" or out of range. Apart from an efficient energy use, the correspondent election procedure should not choose a correspondent which is hidden while some other ECH from the group is in range with an edge device. To address this issue, edge devices send advertisement packets (holding their edge device ID and subsystem ID (Section 8.5)) to announce their presence. Only ECHs that receive an advertisement (ADV packet) will start the correspondent election process.

Correspondent election procedure follows the same algorithm exposed before for the ECH election, using $T_{corresp}$ as the proposed time. It starts when

1. No correspondent is currently selected in the group
2. $T_{corresp}$ expires
3. Current correspondent loses range with all the edge devices
4. An ECH who did not participate in the previous election and has more remaining power than current correspondent now receives an ADV packet
5. An ECH cannot communicate with current correspondent before $T_{corresp}$ expires
6. A new entity is added to the correspondent's group

According to this algorithm, if an entity loses its correspondent and no ECH receives an ADV packet, the election process will not start again to choose a new correspondent. This situation is undesirable because the context information of the grouped entities may still be useful locally. Moreover, we cannot allow new groupings to be discarded due to temporal disconnection. To avoid this problem, the correspondent election procedure will be started by any node which runs more than a certain time T_{No-ADV} without being able to communicate with its correspondent. When connection with the service network is reestablished, entities holding a correspondent selected in this way will start a regular correspondent election procedure again.

In our research, we focus on groups of entities that can associate dynamically and spontaneously. This consideration not only involves that the information produced by those entities might evolve, but also that those associations might break as easily as they started. What is more, although during the association period we had time to exchange data between involved parties, disassociation will be typically triggered when there is an interruption on the communication: leaving entities will not have time to say "good bye." For this reason, we should avoid storing vital grouping information in a centralized manner, but rather distribute this information. This is especially true for the elected *correspondent*, as the entity it is contained in may ungroup without prior notice. Unlike correspondents, however, EHCs are unlikely to die without warning as they remain attached to their entities. In fact, apart from some rare HW failure, the major reason for the disappearance of an ECH node would be due to battery exhaustion. Therefore, while correspondent nodes should not store irreplaceable information, ECHs might store some data to ease the sensor data collection process. WISSE defines that each ECH shall store a table with a list of its entity's nodes and the sensors they hold (*node table*). Recall that one entity might have one or more sensor nodes and each sensor node can hold one or more sensors and actuators. Additionally, the ECH will store its address from the Sequence Chain addressing scheme (Section 8.3.2) and may store some other additional information such as its level in the association hierarchy (Section 8.4). When an ECH finishes its representation period or is about to exhaust its battery, its entity table and node table will be transferred to the next elected ECH.

8.3.1.2 Grouping Procedure

The grouping procedure concept behind WISSE associations is rather simple: Entity correspondents send periodic broadcast of *grouping-request* packets looking for other entities to associate with. Entities that receive *grouping-request* packets may process the packet information contained in them and decide to associate sending back a *grouping-response*. The responding entity makes decisions on which will be the new representative EPC of the resulting entity. Results are communicated to all entity members and a new correspondent election process begins. This process may involve several entities at the same time.

However, a too simplistic approach may lead to an unrealistic design. In general, we need to monitor the grouping procedure to avoid uncontrollable chain associations that will exhaust the nodes' batteries. In particular, if the objective is to provide shared context relevant to the clients, association nature and priorities should be considered carefully before using the devices scarce resources. In our work, we try to meet these requirements by proposing a two-phase grouping procedure. The first phase, or pregrouping, aims to organize and filter the entities that should undertake a full association procedure, which takes place in the second phase.

One of the most important principles of WISSE is that entities are represented in the system by one EPC. When entities associate, one of their EPCs is chosen as representative of the group. We could say that the group of associated entities form a wireless sensor network and that the EPC is its network ID (or NetID, for short). This design allows the addressing of meaningful groups of entities globally under a unique identifier. Nonetheless, WISSE still maintains the possibility of accessing every individual EPC through a database of associations called the virtual entity database (VED). The VED design is described on Section 8.5.

The pregrouping phase uses the user memory bank from the EPC Global Class-1 Gen2 tags to store simple logical information about the entity that carries the tag. Entities will send this information in their *grouping-request* packets, and the receiving entities will compare it with their own. The output of the pregrouping phase is a filtered list of *grouping-request* packets received by this entity in a single *grouping period* organized in priorities. Phase 2 receives as input the set of requests with higher priority. The main goal of Phase 2 is to process the *grouping-requests* coming from Phase 1 and choose a NetID among all the entities that will associate. The chosen NetID must be notified to all the members of the group.

In the case of grouping entities without sensor nodes, RFID reader devices present in any of the grouping entities will be used as mediators in the process. Readers will issue *grouping-request* packets on behalf of the tags they read and will keep a list of read entities to avoid repeating the processes. This list will also be used to avoid mediating with tags that have been detected to belong to entities that have sensor nodes.

As we mentioned at the beginning of this section, the ideal scenario for our framework is for the RFID tag's memory to be mapped into the sensor node's memory. Of course, this map should be kept in a place where it would survive power-off and rebooting, as, for example, the microcontroller's ROM. Nevertheless, we also argued that it is possible for this assumption to be invalidated by scenarios where manufacturers do not provide the mapping or where new nodes are added to the entities by the users. In order to read RFID tags inside the entities without using sensor nodes, we need some kind of standard RFID reader. As the entities in WISSE framework are mobile, we should consider mobile RFID readers. Several commercial products exist that already combine RFID reader and wireless communication technologies, such as the Symbol RD5000 (WLAN) [4] or the IDBlueTM (Bluetooth) [5].

8.3.2 Addressing

As nodes belonging to a network, WISSE entities need to have a network address to communicate. A simple and rather naive approach its to assume one hop communication and a flat network topology. A centralized agent, for example, the *correspondent*, could assign unique addresses to the entity's nodes when new entities join. However, this approach contradicts the principle of dynamic operation: Should the correspondent's entity leave the group without delegating its functions, the group state will become unstable. As a result, it is probable that stability could only be reached by reassigning all addresses again. Furthermore, if a group of entities becomes big enough, entities will not be able to communicate with each other within one hop. Hence, this approach is static and nonscalable.

In order to design an appropriate addressing scheme for WISSE, let us first analyze our requirements:

- Address must be unique inside of a network.
- Addresses must be reused when entities leave.
- Addressing must be dynamic. Address assignation should be fully distributed.
- Addressing mush must be scalable.
- Support for network merge should be provided.
- Overhead must be minimized.

To meet these requirements, we propose an addressing scheme with the following properties:

- *Hierarchical:* The nodes involved in this addressing scheme are entity's ECHs. Nodes receive addresses organized in a tree structure. In the grouping algorithm, the ECH that computes an address becomes a parent. Hence, senders of *grouping-request* packets become children. Inside of a *single entity*, the ECH assigns each sensor node a network address unique only inside that entity.
- *Distributed and unique:* Each node is responsible for assigning addresses only to its children. The address a child receives is derived from its parent address in a way that makes that address unique for the network.
- *Scalable:* If a node leaves a group, its address becomes automatically available for any other node joining with the same parent. Moreover, the addresses are not limited in size by the scheme, but rather increase in size as the network becomes bigger. Network merges are also supported by reassigning the addresses of the network with the smallest number of nodes.
- *Low overhead:* A parent only needs to know its immediate children to assign addresses in a unique manner. Moreover, our addressing scheme provides routing along the tree with no cost. A node can know how many hops it is away from any destination by just analyzing the address, and parents can route packets following the tree by just comparing its address with

the packet destination address. Additionally, parents can provide shortcuts to the destination in a equally simple way by routing packets to neighbors that are closer to the destination than following the tree.

8.4 Virtual Entity

WISSE context information is provided directly by the users and objects of the context layer. For each entity, a dynamically elected representative plays the role of a bridge with upper layers to transfer the entity's context periodically. In order to do so, representatives route information through gateways that we call "edge devices." The information transferred by each entity is then stored in a database for further processing. This information is organized around the entities' unique identifier (EPC), which assures a global data search key directly linked with the real entities. The piece of information associated with a particular EPC representing an entity is called virtual entity.

Virtual entities (VE) represent entities for the rest of the WISSE architecture. A VE is a table that stores information of the entities associated with a certain NetID, including the entities EPCs, each entity's sensor nodes, and which sensors/actuators each of those nodes are carrying. The virtual entity table (VET) is hierarchical. This design decision is taken from the perception that real-world objects are usually organized in a containment relationship. For example, a shirt is stored in a wardrobe that is located in a bedroom from the user's house. The wheels belong to the car that the user is driving, which may be contained at a certain moment in the garage of the user's house. Furthermore, the hierarchical design eases the process of choosing representative EPCs (NetID) when performing associations. The higher the entities are located in the hierarchy, the more representative they are in the group and the more meaningful their selection would be. Figure 8.2 shows a logical representation of the VET. The fields *NetID* and *Associations* represent the NetID of the registered entity and the number of grouped entities, respectively. Every entity is stored at a level pointing to its parent. The NetID will be, thus, one of the entities stored for the first level. The field "inflated entity" is used in cases where an RFID readers mediator reads more than one tag at the same time and cannot differentiate which entity they belong to.

Having discarded correspondents to store the entity's VET (Section 8.3.1.1), the question is which part of the upper layers should hold the database. Edge devices are not a good choice since this would also incur traffic overhead each time an entity roams between them. Moreover, handoff between edge devices suggests the idea of some centralized controller to manage the process. Edge device controllers (EDC) perform the role of managers of all the edge devices in their *subsystem*. The amount of edge devices in one subsystem will depend on several factors such as the average traffic load (average number of VETs to manage) or geographical reasons (i.e., a shopping mall area might want to control their own subsystem for localized services). EDCs will thus hold a database with the VETs of the entities present in their subsystem, so entities roaming between edge devices belonging to the same subsystem will not require any VET-related data transmission. Inter-subsystem handoff, however, will require EDCs to perform a VET transfer in order to avoid entities to carry out VE registration again. Figure 8.3 summarizes the process.

The procedure for an entity to upload for the first time its information to the EDC VET is called VE *registration*. VE registration takes place when an entity is first provided with an ECH and correspondent. From that moment, any grouping process will trigger a VE *update*. When a grouping occurs and, thus the VE update, the grouping party whose NetID was discarded requests the removal of its VET and the transfer of its contents to the VET of the chosen NetID's entity. This process involves the copy of the entities and node tables and the updates of the fields *Associations*

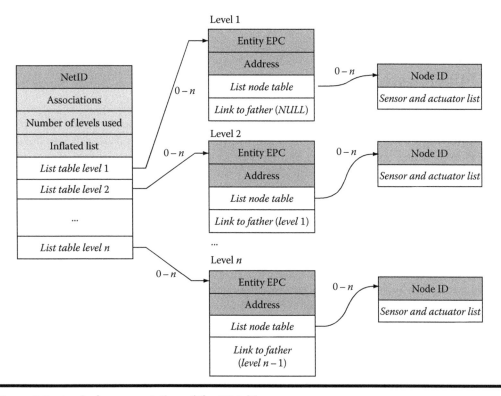

Figure 8.2   Logical representation of the VE table.

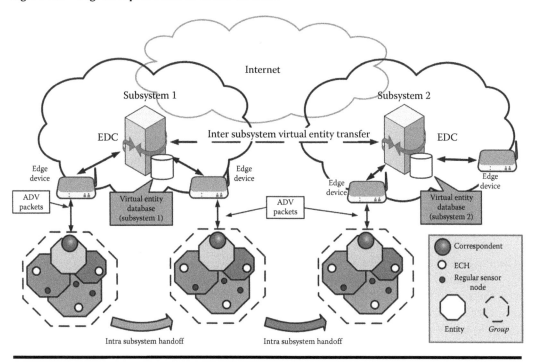

Figure 8.3   WISSE subsystem infrastructure.

and *number of levels*. When the update is finished, the VETs of the grouping entities in the EDC will have merged into one VET representing the new *group*.

8.5 Middleware

In the Introduction, we described WISSE as a framework where mobile users receive services according to their context. In Section 8.3, we explained how the context is provided by entities carrying RFID tags and sensor nodes. We also mentioned that entities may associate building a shared context and how this process can be done in an intelligent way by following a set of algorithms and communication protocols. However, so far we have only described those WISSE elements sitting on the context layer. How does that mix of entities with various capabilities receive services according to their context?

Apart from the tasks of managing the edge devices and store the database of VEs, the EDCs have the responsibility of acting as mediators between the service layer and the context layer in order to build the logic for meaningful service delivery. The part of the EDC that performs this process is called *EDC middleware*. The middleware from the EDCs will thus manage the available services for the VEs and will be in charge of processing the requests from the SSL and contact the SRL on behalf the entities. A block diagram of the EDC middleware is shown in Figure 8.4.

8.5.1 Service Manager

The main part of the EDC middleware logic is implemented in the *service manager*. The service manager maintains a *service pool* of available services and which requirements are needed for them. Requirements are expressed in terms of sensor and actuator types necessary to execute the service. It is also possible to express requirements in terms of the EPCs of the entities requesting the service. EPC requirements must be satisfied by contacting the EPC network and querying its repositories. The *updater* receives information from new services and updates the requirements of existing ones. With this information, a pool with services is built and maintained in the service manager. The VED transfers VE information to the service coordination. These updates are triggered by new registrations or by VET updates. To compute the available services for a certain VE, the service coordinator performs a cross-checking between the VET information extractor and the service pool and stores the matches in a database of *available services*.

Once the available services for each VE are computed, the service list is delivered to the actual entities through the *context request agent*. Entities are then free to choose which of those services they wish to receive. There are two ways in which a service may be requested:

1. *Direct request:* A formatted list of available services is delivered to the SSL waiting for manual service selection. For example, a user carrying a PDA is displayed a list of matching services from which she will choose if needed.
2. *Profile based request:* The service layer contains a list of profiles that have been registered in advance. This list is kept as a WISSE network infrastructure service and registered with the service manager in the same way regular services are. A profile lookup module will check available profile repositories and inform the service coordinator if some match with a managed VE is found. For the example in Section 8.3, a profile similar to the following could have been used: "If Anne is going home and her temperature is over X, send 'Turn On AC' command to Anne's home server." Note that temperature and location refer to temperature and GPS sensors, while Anne refers to Anne's signature (i.e., her digital ID RFID EPC).

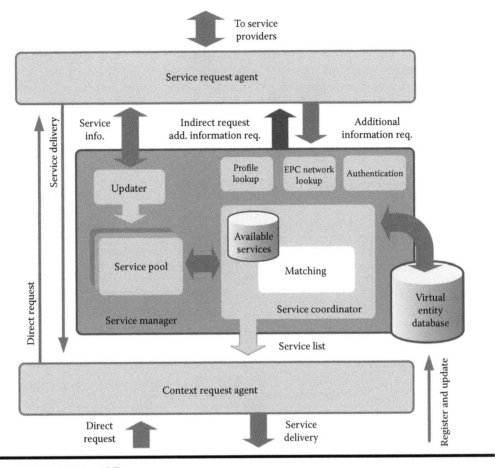

Figure 8.4 WISSE middleware.

8.5.2 Request Agents

The *service request agent* (SRA) and the *context request agent* (CRA) serve as intermediaries between the context layer, the Service Manager, and the service layer. Once a user performs a *direct request*, fruit of his choice from a previously delivered service list, the CRA will receive the request, check it for consistency, and hand it over to the SRA for actual service request. After the service request is processed in the service layer and a response is received, the SRA will transfer it to the CRA and then the response will be delivered to the requester entity. If the service request comes from a profile match rather than from a user direct choice, an *indirect request* will be delivered from the service manager directly to the SRA. The service response will follow the same path as the direct request.

8.6 Service Definitions

The various services that the service layer may provide are innumerable and depend on the service providers. In general, service providers will make available server equipment attached to the WISSE network following a standardized interface. This interface should include directives such as handlers for EDC *service request* and *service subscription* or issuing commands such as *service response* or *service announce.*

WISSE classifies the type of services into four different groups. *indirect services, personal services,* and *object services* are categorized according to the service receiver. *public services* and *subscriber services* classify them by their access policy. Local and remote services refer to the location of the service provider. Finally, we also consider *infrastructure services* and the *EPC network service* as special services. Table 8.1 summarizes this classification. Figure 8.5 shows a diagram of the relationship between the first group, together with a list of examples.

Table 8.1   WISSE Service Classification

Service Classification	Service Types	Description
According to service receiver	Indirect services	Delivered to an intermediate server for further processing
	Personal services	Requested and delivered to a user
	Object services	Delivered to a different client than the one that requested the service
According to access policy	Public services	Publicly available for any client
	Subscriber services	Available upon subscription
According to the location of the service provider	Local services	Belonging to the owner of the access network
	Remote services	Belonging to third-party service providers
Special services	Infrastructure services	Related with the middleware operation
	EPC network services	Manages requests to the EPC network

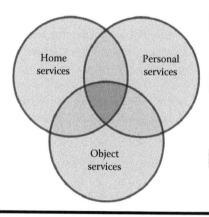

Indirect services:

Remote appliance control, home status reception (i.e., TV on/off, home video surveillance streaming, etc.)

Personal services:

Map delivery, shopping and transportation assistance, e-banking, automatic payment (i.e., theater, restaurants, etc.), multimedia delivery (i.e., music, movie, etc.), health monitoring.

Object services:

Object tracking, software downloads, status monitoring, remote actuation, etc.

Figure 8.5   Service classification according to the service receiver.

8.6.1 Service Types

Personal services are those services requested by a user directly or indirectly (i.e., by manual selection or served through some profile). *Indirect services* refer to any service requested by an entity and delivered to some intermediate server for further processing (i.e., a home server). *Object services* are those that refer to inanimate entities and that cannot be manually requested and delivered to the same entity. The overlapping in the classification obeys to combinations of the service types.

Public services refer to those services that are publicly available in the network and need no registration or subscription. *Subscriber services*, on the other hand, are those services that require an agreement with the provider (i.e., subscription and fee payment) to be used. The owner of a particular implementation of the WISSE network will normally provide a set of services for the clients using their subsystems. For example, customers using the supermarket's access network will have access to services locally offered by the supermarket, such as location of products, special offers, etc. These kinds of services are called *local services*. On the other hand, clients accessing services not attached to the access network they are currently using will be accessing *remote services*.

Finally, *infrastructure services* provide high-level resources for the WISSE middleware and decision-making process, such as profile repositories, authentication servers, etc.

8.6.2 EPC Network Service

The *EPC network* provides a standard infrastructure to discover and share information about the RFID tags' EPCs . The EPC network is a set of services that is built on top of the current Internet and whose main goal is to provide full product visibility along the supply chain. By enabling real-time sharing of information from RFID-tagged products, business partners can benefit from the extremely efficient and effective supply chain management.

The basic operation of the EPC network is as follows: Manufacturers provide the products with RFID tags that accompany them from the factories to the retail shops. At every point in the supply chain, *RFID readers* identify those products by reading automatically the EPCs stored in their RFID tags [6]. The EPC information obtained by the RFID readers may be filtered at a middleware layer called ALE (application level events [7]). The final destination of the information are database repositories owned by the business partners involved in the specific product supply chain. These repositories are part of the EPC information services, or EPCIS [8]. EPCIS are business entities that offer a set of standard interfaces, not only for *querying* the repositories, but also for *capturing* the EPC data. The EPCIS capturing application, which is part of the EPCIS, manages the EPC data capture by coordinating the rest of the EPC network components, delivering events to the ALE. Parties interested in accessing certain product information may provide the *object name service* (ONS) [9] with EPCs, obtaining in return records pointing to EPCIS repositories where the product information is stored. The EPC network also defines several other elements and interfaces such as tag protocols and a reader management component.

One of the main advantages of using EPC information as the NetID for WISSE entities is that those entities can benefit from the use of the EPC network infrastructure. When the information about a VE is retrieved from its VET, its NetID (or any of the entity's ID in its entities list) may be encapsulated inside a query packet and sent to the EPC ONS. The ONS will return the location of the manufacturer's EPC information service, which contains information about the concrete EPC of the associated entity.

The exposed infrastructure coupling provides significant service enhancement. On one hand, real-world objects and users are able to get ubiquitous services according to the real-time sensor information and its abstract ID. On the other hand, providing EPC network connectivity converts those abstract identifications into real universal identifiers, managed by their product manufactures and even able to take part in the supply chain. This could allow, for example, to automatically adjust the service delivery to certain models of objects (i.e., cell phones) whose specification may be retrieved from their manufacturer's EPCIS.

8.7 Applications and Infrastructure

In the previous sections, we have discussed many examples and applications of the WISSE service framework. In this section, these have been correctly discussed. Figure 8.6 represents this example graphically, including several of the concepts explained in the previous sections.

From a practical point of view, developing from scratch the infrastructure to support WISSE would be a rather complex and costly task. Although the context layer can scale steadily with the production of new objects carrying RFID tags and sensor devices, the service layer and its service network edge would need a bigger investment in equipments and network infrastructure before they start offering services. To alleviate this problem, as WISSE underlying communication protocols are not restricted, it is possible to integrate part of the WISSE infrastructure with existing telecommunication networks. For example, in the case of mobile phone telecommunication networks, edge devices can be deployed that connect to the operator's network through the existing cellular infrastructure. EDCs would be located in the *base station subsystems* (BSS) connected to the *mobile switching centers* (MSC) and *base station controllers* (BSC) of the cellular networks. The reuse of part of the existing network would not only ease common procedures such as handover between EDCs, but also provide an affordable way for the telecommunication operators to expand

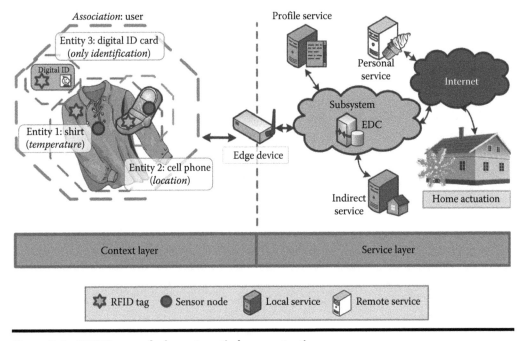

Figure 8.6 WISSE example for automatic home actuation.

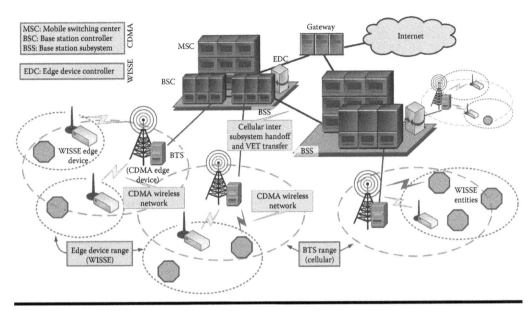

Figure 8.7   WISSE implementation on top of the CDMA infrastructure.

their services to mobile sensor networks. Figure 8.7 depicts a hypothetical implementation of the WISSE network on top of the CDMA cellular network.

8.8 Implementation

In order to provide a prove of concept of the WISSE framework, we implemented a simple scenario using real wireless sensor nodes. We used our own WSN technology being developed under the name of ANTS [10]. ANTS is a platform for the development of WSN and is based on the concept of evolvability: hardware and core protocols evolve with the environment producing adaptable networks. The evolvable operating system (EOS) is the heart of the software architecture and delivers a multithreaded OS with features such as dynamic memory allocation, small footprint, low power consumption, and upgrading capability. The EOS also provides the base for several MAC and network stack modules. Regarding the hardware support, ANTS hardware family comprises four different categories and seven different members, growing in complexity depending on the application needs. The platform supports up to four kinds of controller units, five types of RF transceivers, and a variety of sensors.

Figure 8.8 presents the elements of our scenario and the interaction between them. A sensor relays user's body temperature through its correspondent (PDA) to a *water supply* service on the service layer. The user receives a list of available services on its PDA, and once he selects the *water supply* service, water supply points (i.e., fountain) will appear on the screen if the temperature gets too high. GPS is used for locating the user and the closest water supply points.

For this implementation, we used the H20 nodes belonging to the H2 hardware family. The main features of this sensor node are an Atmega 128 microcontroller (128 KB of Flash, 4 KB of RAM), CC2420 as RF transceiver (2.4 GHz), and various sensors included in a separate pluggable board (Figure 8.9). For the PDA, we used an IPAQ hx470. In order to connect the PDA with the sensor node, we attached a Tellord Zigbee module in the IPAQ expansion slot. The tag information

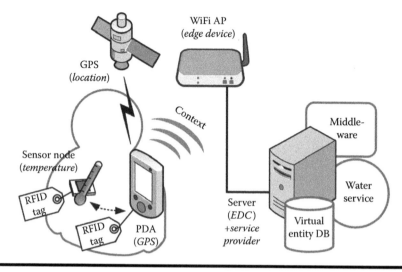

Figure 8.8 WISSE implementation scenario.

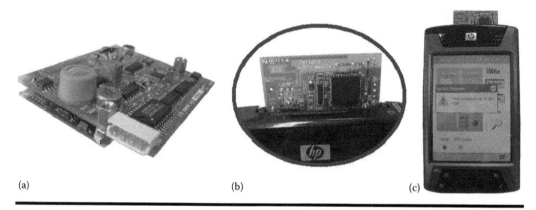

(a)   (b)   (c)

Figure 8.9 (a) ANTS platform H20 sensor node, (b) Tellord Zigbee module, and (c) IPAQ hx470 running the WISSE implementation.

was recoded in both the node and the PDA memory. For the GPS, a Bluetooth module was also used. In order to simulate the WISSE service network, we used on campus WiFi access points as edge devices. Finally, a computer connected to the campus wired network contained the rest of the software necessary for the demonstration, such as the VED and a simplified version of the WISSE middleware.

The sensor's node software was programmed using the C language and the driver libraries from the EOS. The first time the system is powered on, both the sensor node and the PDA associate in a group making the PDA correspondent. The information regarding the association and the sensory capabilities of the group is passed through to the middleware using the WiFi network. We used simple socket programming between the embedded C++ of the PDA and the Java from the middleware. Once the data from the group arrives to the middleware, a data base record is created indexed by the EPC of the correspondent, that is, the EPC that is stored in the PDA's memory. Note that since the sensor node cannot communicate directly with the middleware, its entity was not previously registered in the database. Otherwise its record would have been rearranged rather than created by just pointing a "parent" field to the EPC of the PDA.

The *water service* is also stored in a database of WISSE services. Its requirements are "localization" and "temperature" sensors. The middleware periodically checks the services from the service database against the entities of the VE DB. When the match of the *water service* and the PDA-sensor node group is detected, a message is sent to the PDA authorizing the use of the "find water supply" program. Originally, the software would have been downloaded from the service provider and installed in the PDA. However, to simplify matters, in this demo the program was already installed and it was only triggered by the information from the middleware.

Once the software in the PDA is running, the sensor node sends periodic temperature readings in the 2.4 GHz band to the Tellord module installed in the PDA. This module, embedded in a compact flash card, communicates via a serial interface with the software running in the Windows Mobile 2003 OS. GPS position is obtained through the GPS receiver of the PDA, and the coordinates are adapted to a simple campus map displayed on the LCD. When the temperature sent from the sensor node reaches a certain threshold, a message appears on the screen and several water supply spots on the campus appear on the map. According to the current location, the closest spot is indicated as the optimal choice.

We believe that the implementation scenario described here is a valid proof of concept of what kind of functionality can be achieved by the WISSE framework. In this case, there was only one service and one client, but note that the client could only access the service when the middleware layer decided that such client was eligible for the service. Should another node with different sensors have been associated in the first place, the service would have never been granted. This scenario shows that spontaneous groups with various capabilities on gathering context from a user can receive different services automatically. As a natural extension to our implementation, imagine the middleware contacting the EPC network and asking specific information about the entities, for example, the model of the PDA, to find out if the program can be successfully installed and run.

8.9 Future Work

The contents of this chapter constitute our initial work to make WISSE framework a reality. Many issues need to be studied in order to solve the technical and research challenges that an integration among RFID and WSN bring. First, the algorithms that perform the associations of entities according to their EPCs and context need to be studied to provide optimal performance and coherence results. For example, a too naive approach on handling advertisements of entities could lead to incontrollable chains of associations in dense areas. To solve this problem, we study the inclusion of some extra information inside the RFID tags exploiting the flexibility of the new EPC standards in *Class 1 Generation 2 (Gen2)* tags. This information could be used during the grouping procedure to differentiate which kind of entities participate and infer some grouping priorities.

Although, as we explained before, the WISSE framework is designed to run on top of existing communication networks, its distributed nature and integration with the EPC network also poses some challenges. In this respect, we investigate the best ways of integrating our ideas side by side with the Internet-based and distributed EPC network. By using EPC network's interfaces, we can transform our software in a kind of "accessing application" that can subscribe and benefit from the information related to EPCs. Furthermore, we also consider merging VET and EPC databases in order to leverage the EPC network infrastructure by extending the EPC concepts with sensor and actuator information. Part of our work is also related to the *EPC sensor network* project being carried out under the Auto-ID Lab Korea.

8.10 Related Work

In general, the difference between the WISSE approach and other works that build ubiquitous smart spaces lies in a difference of concept. Rather than enabling smart spaces by building external, context-aware infrastructures that sense, interact, and react to the users, we build context-aware users that gather their context while they move around. In this sense, the external infrastructures needed are similar to current ubiquitous networks such as the cellular phone network: users just require communication gateways to some service network. The key difference is that clients are no longer "dumb" clients, but they also provide information (context) that influences the services they receive. This smart client concept, compared to the traditional smart space concept, brings a series of fundamental differences. We highlight the most important ones in the remainder of this section.

Many works on ubiquitous environments build the users' or objects' context entirely on the middleware from independent sensor readings [11–15]. Context information is thus interpreted to extract an overall meaning, and actions (services) according to that meaning are taken. This approach is clearly a limitation because it is very difficult to interpret a context from a system's point of view (as opposed to a user's point of view) and based on independent information. In some works such as [16], users are augmented with portable devices and sensors, but the service infrastructure is inexistent and all the middleware is included on the portable devices. Other works such as [17] build agents represent the sources of context and perform grouping of related agents. However, entities with various capabilities should communicate with the system independently, which makes middleware inflexible and nonscalable. WISSE middleware is based on context information obtained from the user's point of view, but still maintains a distributed infrastructure that analyzes the context and offers services. Furthermore, WISSE transfers part of the context reasoning to the users' level by allowing conditional associations of entities before the context is transmitted. We believe that this framework balances the context processing tasks, providing an optimum combination of preprocessing and postprocessing of information. In this way, WISSE simplifies the middleware decisions to a requirements matching, although it also increases the complexity on the user side that has to deal with entity associations and information updates.

Another key difference of the WISSE framework is the complete independence between users, services, and middleware. Independent service providers publish their services with a standardized interface to a distributed repository. Users are offered a subset of services without the need to know anything about the service providers, and the middleware will just match the services from the repositories with the users' context at each moment in time. Projects such as [18–20] focus on augmenting everyday objects the same way we do. However, they lack a general infrastructure to provide spontaneous services which follow the independent pattern that our work follows.

But maybe the most important difference and key contribution of our work is that WISSE provides a real integration of RFID and sensor networks in the sense that RFID data and sensor data refer to the same objects and users. On one hand, assigning unique identification to context allows an efficient and powerful middleware management. On the other hand, the use of EPCs allows us to leverage the advantages of the EPC network, offering potentially unlimited entity information beyond the WISSE infrastructure itself while inheriting RFID services such as track and trace and b2b (business-to-business) relationships. There is little related work on this area, and those research efforts that embrace together both RFID and SN technologies just reflect static SN such as those inside buildings or "smart homes" [21].

Other parts of our research are also related to existing works in the area. In Section 8.3.1, the part of the representative election process where nodes are chosen according to their remaining energy is based on [22], although we adapt the algorithms for mobile WSN. The addressing scheme exposed in Section 8.3.2 resembles the one used by the Zigbee Networking [23], although our work minimizes the overhead and provides dynamic address spaces, support from merging networks, and a truly decentralized strategy.

We believe that our work and other related work in the area are complementary. Indeed, starting from a difference of concept, applications can be quite different. However, we also believe that our approach is a more realistic approach to true smart environments, while still offering a great deal of flexibility if necessary. For example, as suggested in Section 8.7, the gateway infrastructure represented by the edge devices could be easily adapted to that of cellular networks, and the middleware instances installed at the BSS that are already distributed and connected to the Internet.

8.11 Conclusion

WISSE framework describes a future in which mobile entities will actively participate in ubiquitous services both providing and receiving information from the network. Our architecture defines "smart" entities that provide sensor data stamped with their unique ID (RFID's EPC) and that can spontaneously and dynamically interact, enabling an infinite number of possible services. Additionally, our architecture might be merged into existing telecommunication infrastructure for fast deployment and low cost, offering a realistic approach to the present infrastructure status.

References

1. T.L. Sánchez, D. Kim, and T. Park, A service framework for mobile ubiquitous sensor networks and RFID, *ISWPC'06*, Phuket, Thailand, 2006.
2. EPC Global Ratified Standard, EPC^{TM} tag data standards, Version 1.3, July 2005.
3. Specification for RFID Air Interface. EPC^{TM} radio-frequency identity protocols class-1 generation-2 UHF RFID, Protocol for Communications at 860 MHz - 960 Mhz, Version 1.0.9, January 2005.
4. IDBLUE, CathexisInnovationsInc., idblue.com, 6 June 2012.
5. IDBlue^{TM}, Home page, http://www.baracoda.com/baracoda/products/p_21.html
6. EPC Global Ratified Standard, Reader protocol standard, Version 1.0, June 2006.
7. EPC Global Ratified Standard, The application level events (ALE) Specification, Version 1.0, September 2005.
8. EPC Global Working Draft, EPC information services (EPCIS), Version 1.0, September 2005.
9. EPC Global Ratified Standard, Object naming service (ONS), Version 1.0, October 2005.
10. D. Kim, T. Sanchez, S. Yoo, J. Sung, J. Kim, Y. Kim, and Y. Doh, ANTS: An evolvable network of tiny sensors, *EUC'05*, Nagasaki, Japan, 2005.
11. K. Kumar, S. Hariri, and N.V. Chalfoun, Autonomous middleware framework for sensor networks, *ICPS 2005*, Santorini, Greece, July 11-14, 2005.
12. S. Liaquat K., M. Riaz, S. Lee, and Y.-K. Lee, Context awareness scale ubiquitous environments with a service oriented distributed middleware, *ICIS 2005*, Jeju, South, Korea, 2005.
13. A. Ranganathan and R.H. Campbell, A middleware for context-aware agents in ubiquitous computing environments, *Middleware 2003*, LNCS 2672, p. 143161, Urbana, IL, 2003.

14. H.L. Chen, An intelligent broker architecture for pervasive context-aware systems, Thesis dissertation, University of Maryland, Baltimore, MD, 2004.

15. H.-S. Park, D.-H. Choi, Y.-H. Jeong, and T.-U. Cho, A CMQ middleware architecture for multimedia application in ubiquitous environment, *ICACT 2006*, Phoenix Park, South Korea, February 20–22, 2006.

16. S.S. Yau and F. Karim, Context-sensitive middleware for real-time software in ubiquitous computing environments, *ISORC 2001*, Magdeburg, Germany, 2001.

17. H. Takahashi, T. Suganuma and N. Shiratori, AMUSE: An agent-based middleware for context-aware ubiquitous services, *ICPADS'05*, Fukuoka, Japan, July 20–22 2005.

18. F. Kawsar, K. Fujinami, and T. Nakajima, Experiences with developing context-aware applications with augmented artifacts, *Ubicomp 2005*, Tokyo, Japan, 2005.

19. H.W. Gellersen, A. Schmidt, and M. Beigl, Adding some smartness to devices and everyday things, *WMCSA'00*, Monterrey, Mexico, 2000.

20. I. Siio, J. Rowan, N. Mima, and E. Mynatt, Digital decor: Augmented everyday things, *Graphics Interface*, Halifax, NS, Canada, pp. 159-166, 2003.

21. Y. Isoda, S. Kurakake, and H. Nakano, Ubiquitous sensors based human behavior modeling and recognition using a spatio-temporal representation of user states, *AINA'04*, Vol. 1, p. 512, Fukuoka, Japan, 2004.

22. S. Kandula, J. Hou, and L. Sha, A case for resource heterogeneity in large sensor networks, *MilCom 2004*, Monterey, CA, 2004.

23. Zigbee Network Specification, Version 1.0, December 2004.

24. K. Taub et al., The EPC Global architecture framework, White Paper, EPC Global 2005, 1 July 2005.

HUMAN–COMPUTER INTERACTION (HCI) AND MULTIMEDIA COMPUTING

Chapter 9

Designing Multi-Sensory Displays for Mobile Devices

Keith V. Nesbitt and Dempsey Chang

Contents

9.1 Introduction

It is probably fair to say that small mobile computing devices, such as personal digital assistants (PDAs), have been slow to claim the large market that was predicted in the late 1980s. The reasons behind the slow consumer uptake are not clear. What is clear is that the latest generation of consumers have been more than willing to embrace mobile phones, portable music players, and hand-held gaming devices. It is also becoming clear that these stand-alone applications of mobile technology have been converging. For example, we now might expect our mobile phones to play music and provide games, video, and picture-taking facilities. We might also expect seamless access to wireless Internet and the required functionality for both e-mail and web browsing. In an unexpected way, it seems the convergence of these mobile technologies is finally signaling that the time of the PDA has arrived.

Mobile devices are increasingly being used to perform a broad range of professional activities. For example, pharmaceutical industry uses PDA for ordering pharmaceuticals and managing workflow (Karampelas et al., 2002). Real-time stock-taking tasks can further be enhanced by integrating bar-code technology to reduce human errors during data entry. Maintenance engineers can access working checklists, review histories, and order replacement parts on location. Visitors to some museums can also expect to have the option of using a PDA to enhance their experience by accessing context-relevant information about exhibits (Micha and Economou, 2005). In other realms, global positioning devices have become indispensable for activities that require accurate navigation.

Despite the many uses for mobile devices, a key trade-off when designing applications is the need to keep devices small and compact and yet also provide as much visual screen space as possible. This presents a challenge for designers who would like to present large amounts of information on small screens. For example, people can make more mistakes when using a small screen compared to a full-size display when searching for information on the web (Kim and Albers, 2001). Other issues of applied usability (Masoodian and Budd, 2004; Ringbauer, 2005) and information visualization (Chittaro, 2006) on mobile devices have also been addressed.

One approach that has yet to be fully leveraged is the use of multisensory displays to provide greater levels of feedback. The motivation is simple enough; by designing displays that present information to the different senses, we might increase the available bandwidth for displaying information. Apart from visual display, we usually see auditory feedback used to provide alarms in mobile devices. However, more sophisticated displays such as auditory monitoring of the stock market are also possible. In terms of tactile and haptic displays, suggestions for using the sense of touch to interact with the mobile applications have previously been made (O'Modhrain, 2004). Indeed vibration as a form of haptic alarm is sometimes used for mobile phone alarms. Banatre et al. (2004) have also demonstrated a mobile application designed to assist visually impaired people use public transport.

Unfortunately the design of multisensory displays is complex, as it is necessary to carefully consider the perceptual capabilities of humans. Indeed the understanding of how we perceive and process multisensory perceptions is still not well understood (Calvert et al., 2004). Because of the embryonic nature of the field, we believe it is important to gather as much assistance as possible for designers of multisensory displays. While some PDA studies have discussed the system architecture and visualizing information research for small mobile devices, none of them have focused on design guidelines that consider multisensory display (visual, auditory, and haptic).

In our work, we tend to distinguish between the idea of *design principles* and *design guidelines*. A *design principle* is considered to provide a higher level of advice for general design decisions. A design principle provides a "general recommendation on the process of design" (ISO 14915-1, 2002). A *design guideline* is a lower-level instruction that provides designers with an exact design rule (Faulkner, 1998; Shneiderman and Plaisant, 2005). Therefore, guidelines provide designers with detailed instructions during the design process.

There are many general principles that can be applied for designing any human–computer interaction (AppleComputerInc, 1996; Faulkner, 1998; Dix et al., 2004; Shneiderman and Plaisant, 2005). There are also some design principles for designers to use in specific area, for example, there are detailed principles for multimedia user interfaces (ISO 14915-1, 2002; ISO 14915-2, 2003), general user interface design principles for educational applications (Najjar, 1998), detailed user interface principles for web-based context, and detailed design principles for multisensory displays (Nesbitt, 2003).

Consistency is an example of a commonly used design principle. This principle provides some general advice for designers that the user interface should always be designed to seem consistent. This can be interpreted in many ways. For example, the designer should ensure that the user enters commands in a consistent way and receives output in a consistent manner (Faulkner, 1998; Shneiderman and Plaisant, 2005). It can also be interpreted more precisely to imply that any text used in the interface maintains a consistency with regard to character size, spacing, punctuation, and colors (ISO 14915-2, 2003; Ozok and Slvendy, 2004).

Flexibility is another common design principle that has been suggested for user interface design. Once again, flexibility is a general suggestion that reminds designers to provide various methods for the user to operate the system (Dix et al., 2004). For example, the system should be able to maximize the user's performance by allowing the user to modify the way commands are entered. This might allow the user to change the font size or create personal shortcut keys (Faulkner, 1998).

The described principles of consistency and flexibility are general and can be interpreted in many ways. Such principles contrast with guidelines, where the design advice is more precise. For example, a design guideline to help with the screen design of a user interface might stipulate that the date entry format must be DD-MM-YY in Australia (Preece, 1994). Another example might be to always have a "confirm" message box appear, to the user, in the center of the window before quitting the system (Brown, 1988; ETSI EG 202 048, 2002).

Although the distinction between "what is a principle" and "what is a guideline" is somewhat blurred, the two levels of advice are useful for categorizing design support. Principles tend to be few in number and also widely applicable, which makes them more accessible to designers. Unfortunately their generality can also lead to problems of misinterpretation. Guidelines are more precise in nature, but this can make them less broadly applicable. This lack of generality means that collections of guidelines tend to be large in an effort to cover all scenarios. Having a large body of guidelines can lead to problems with locating the correct one when it is needed. The pragmatic approach we take is to develop frameworks that group precise, low-level guidelines by more general, high-level principles. The principles then act as both a useful general advice and an index for designers into the large collection of more precise guidelines.

The intention of this chapter is not to present an exhaustive list of detailed guidelines for multisensory display but rather to introduce some key concepts, or high-level principles that are relevant to designers of multisensory interfaces for mobile devices. These more general concepts are useful tools for designers who are faced with the challenge of displaying large amounts of information, especially where there is limited room for visual display. It is also assumed that many readers may not be familiar with the possibilities and difficulties of using multisensory display techniques, and so this chapter is also designed to be an introduction to the science of such displays.

We begin by providing some background to the use of multisensory displays. This is followed by an introduction to the important concepts of spatial, temporal, and direct metaphors. These concepts are applicable across the senses and so provide a useful framework for comparing and contrasting design strategies that make use of the different senses. We will then introduce the use of Gestalt principles as further key concepts that are again applicable across the senses and so allow designers to more easily move designs between the sensory modalities. In both discussions, we will focus on the high-level principles while presenting some relevant guidelines that relate to the design of mobile devices.

9.2 Multisensory Display of Information

A problem facing many applications is the large number of data and how to display and interact with these data. This is a problem not just in mobile technology but also in many applications from data mining of business and medical data to educational applications and computer games. In our more digital world, there are increasing access to larger amounts of information and also a desire to present and understand as much of these data as possible. We broadly describe such applications as *human perceptual tools*, as their intention is to display information to the user and allow the user to search for useful patterns. These systems take advantage of the human capability to perform subtle pattern-matching tasks. It is the design of these perceptually based tools for finding patterns that motivates much of our work behind multisensory display.

The general field that studies the visual presentation of abstract data is called *information visualization* (Card et al., 1999). *Information sonification* is a newer field of study that uses sound rather than vision to represent abstract data (Kramer, 1994a). By contrast with the visual domain, less work has been done in the auditory domain. However, a variety of applications have been described (Kramer, 1994a), and detailed work has occurred in specific areas, for example, the design of auditory alarms for aircrafts (Patterson, 1982). The use of haptic (touch) displays for displaying abstract data is uncommon, although they have been employed in novel ways to investigate force fields (Brooks et al., 1990) and fluid flows (Nesbitt et al., 2001) and ISO guidelines are currently being developed to support haptic display (Carter et al., 2005).

There have also been a number of investigations into using multisensory displays for abstract data; indeed, many applications attempt to enhance interaction by the addition of sound or haptic feedback. A simple motivation for this style of display is that multisensory interfaces might allow the user to perceive and assimilate multiattributed information more effectively. For example, by mapping different attributes of the data to different senses, it may be possible to find patterns in larger data sets. Unfortunately issues of memory, attention, and association become even more complicated when more than one sense is involved. Indeed at present, the fundamental mechanisms of multisensory perception are not completely understood (Calvert et al., 2005).

The pragmatic approach for designers is to categorize performance with multisensory displays in one of three ways: *complementary, conflicting*, or *redundant* (McGee et al., 2001) (Figure 9.1).

Complementary displays map additional, and different, information to each sense. Because the user is receiving extra information, it is expected that task performance with a complementary display should be superior to the performance with a single-modality display. A *conflicting* display maps contradictory information to each sense. Because a user must deal with the conflicting information, we should expect a user's performance to be worse with the multisensory display compared to their performance with the single-modality display. With *redundant* displays, the same information is displayed to each sense. In a redundant display, the user performs equally well with the single-sensory and multisensory displays. However, users may report a reduction in workload or an increase in confidence with multisensory displays that show redundancy.

A question that naturally arises with multisensory display is whether one sense is superior to another for displaying information? Many people seem to suggest that vision is our dominant sense. While it is true that vision is highly detailed and well suited to comparing objects arranged in space, it is equally true that hearing is effective for monitoring sounds from all directions and alerting our visual attention. In fact, all senses are well suited for different kinds of tasks. This is supported by what is known as the *modal specific theory* (Friedes, 1974), which states that each sensory modality has distinct patterns of transduction. This implies that each sense detects unique sensory and perceptual qualities that are adept with certain kinds of complex information.

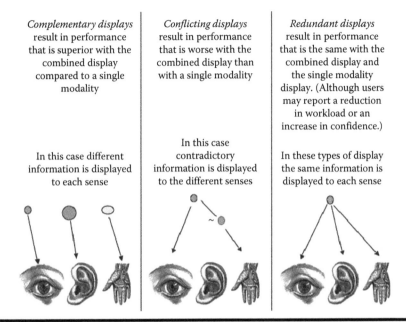

Figure 9.1 **Performance with multisensory displays can be categorized in one of three ways:** *complementary,* *conflicting,* **or** *redundant.* **(From McGee, M.R. et al., The effective combination of haptic and auditory textural information,** *Haptic Human-Computer Interaction, First International Workshop,* **Glasgow, U.K., August 31–September 1, 2000,** *Proceedings,* **Lecture Notes in Computer Science 2058 Springer, 2001.)**

A display designer must become familiar with different types of complex information. Another critical area of knowledge is the range of perceptual channels that each sense has available for receiving information. This can be thought of as the multisensory design space. Despite more rigorous attempts to categorize the visual display space (Bertin, 1981; Card and Mackinlay, 1997) and the emergence of standard auditory techniques, such as earcons (Blattner et al., 1989) and auditory icons (Gaver, 1986), it is still not clear when designing a display what mapping should be used between what types of data and what sensory channels.

One immediate difficulty in comparing, contrasting, and reusing design concepts across sensory boundaries is that the languages to describe visual, auditory, and haptic display have arisen in isolation. To overcome this problem, the MS-Taxonomy was proposed as an alternative classification of the multisensory design space (Nesbitt, 2003). The MS-Taxonomy is hierarchical and describes the multisensory design space at multiple levels of abstraction. At the higher levels of abstraction, the same terminology can be used for describing visualizations, sonifications, and haptic displays. In software engineering terms, the MS-Taxonomy allows a designer to consider reuse of designs at abstract general level and also the more detailed level of individual perceptual units. Reusable design patterns can be discussed independent of sensory modes. This allows for the same design pattern to be implemented for any sense and performance to be directly compared. The structure of the MS-Taxonomy has also been used to group guidelines and to provide a process for designing and evaluating multisensory displays (Nesbitt, 2003).

The next section focuses at the most abstract level of the MS-Taxonomy, describing the concepts of spatial, temporal, and direct metaphors. These concepts are discussed in relation to mobile interface design. Some example guidelines are also presented. We have found that the MS-Taxonomy is one useful conceptual framework to help designers of multisensory displays understand and explore the design space. Another useful framework based on Gestalt principles will be introduced later, and at that stage the concepts of spatial, temporal, and direct metaphors will be used to examine Gestalt principles in a more general way. This examination will suggest further guidelines for multisensory display design on mobile devices.

9.3 Spatial, Temporal, and Direct Metaphors

The MS-Taxonomy is a framework that describes multiple levels of abstraction and covers the range of perceptual qualities that can be used for multisensory display. Rather than claiming to be a definitive conceptual model of the multisensory design space, the MS-Taxonomy is intended to provide pragmatic assistance to designers of multisensory display. A key feature of the taxonomy is that the high-level concepts transcend traditional approaches that tend to focus on individual senses. The usefulness of the taxonomy rests on its ability to provide a simple viewpoint of the multisensory design space that allows knowledge to be transferred across sensory modalities.

The MS-Taxonomy provides a detailed breakdown of perceptual qualities (Figures 9.2 through 9.4). We will discuss some of these detailed concepts but focus on the high-level concepts of spatial, temporal, and direct metaphors that provide the key descriptions of the multisensory design space.

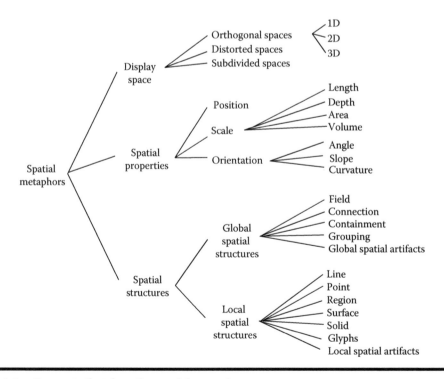

Figure 9.2   Concepts that describe spatial metaphors.

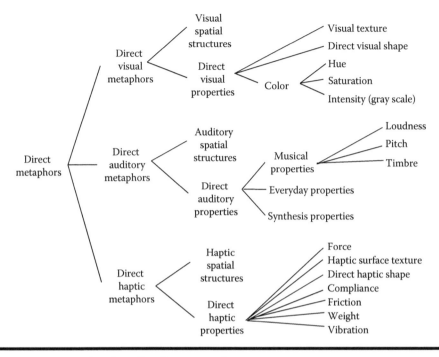

Figure 9.3   Concepts that describe direct metaphors.

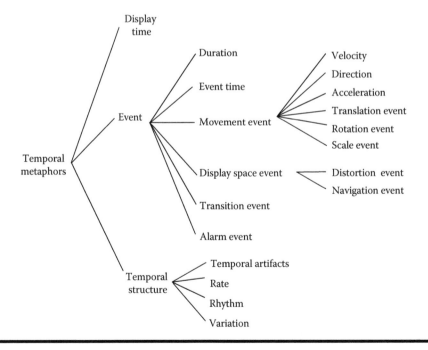

Figure 9.4   Concepts that describe temporal metaphors.

9.3.1 Spatial Metaphors

In the real world, a great deal of useful information is dependent on the perception of space. For example, driving a car requires an understanding of the relative location of other vehicles. Parking the car requires a comparison of the size of the car with the size of the parking space. Navigating the car requires an understanding of the interconnections and layout of roadways. Real-world information is often interpreted in terms of spatial concepts like position, size, and structure. Information designs can also be interpreted in terms of these spatial concepts.

Because the concepts apply across the senses, it is possible for spatial metaphors to be directly compared between senses. For example, the ability of the visual sense to judge the position of objects in space can be compared with the ability of hearing to locate a sound in space. This sensory independence also enables concepts to be reused between senses. This is useful for designers of mobile devices who may be most familiar with designing visual displays. Visual displays often focus on presenting information in a spatial way. As noted with mobile devices, the available display space is often limited, and this can restrict the use of common spatial metaphors such as position, scale, or orientation for displaying information. Another important spatial metaphor is the display of structure. Grouping objects together, connecting objects (e.g., networks, graphs), and enclosure (e.g., tables) are all important structural devices for display. Again the limited display space impacts on the designer's ability to use these traditional approaches often used in the design of visual displays.

One possible solution is for designers to reconsider how the display space is designed. Subdividing the design space is frequently used. By dividing the design space into a number of disconnected parts, each part can be displayed in different parts of the space. With a large screen, it is normal to subdivide the screen and display each separate part in different locations. With a mobile device, these parts can be spread across screens so only a single part is seen at one time. While this is a simple strategy, it introduces the problem that the parts become more disconnected. While they are being displayed on a single screen, there is some inherent structure or relationship between the parts. If this higher-level structure is important, then it may be necessary to introduce additional structuring or grouping mechanisms so that the separate parts can be displayed on different screens and yet the user can still be made aware of the higher-level structure. A common way to do this is to provide an overview map. Unfortunately this is also a spatial metaphor and so requires more screen space. Spatial metaphors are displayed most effectively using the visual sense. While the sense of touch is an adequate replacement for spatial metaphors, our auditory sense is particularly poor at interpreting many spatial metaphors. Color might also be used effectively to provide some grouping. We will categorize color as a direct metaphor. Fortunately direct metaphors are more readily interchanged between senses, and so we might replace color with a sound property such as pitch to display this information.

Apart from subdividing the design, another common strategy is to use a single large and homogenous space. This has the advantage of being very natural as we typically expect the real-world space to be somehow consistent. That is, space is the same everywhere, and so we can make reliable judgments about spatial qualities such as distance, position, or size of objects in space. You can think of the display space as the context on which we interpret all spatial properties and most spatial structures. One approach that is not used very frequently in mobile interfaces is the use of distorted spaces. In a distorted space, properties such as size and position will be different depending on where they are in the space. For example, in a "fish-eye" display (Furnas, 1995) with objects in the center of the display, the space will be more spread out and allow for more detail to be presented. At the periphery, the space is squashed together and detail is not visible. Such displays are often called "focus and context" as they allow detailed information to be displayed at a focal point and yet also maintain the overall structure. Some spatial metaphors, such as connections and grouping, may still be very effective in distorted spaces as long as the user can interactively change the focus point. Some useful examples of this type of display have been developed for drawing graphs (Lamping and Rao, 1996), and while such display spaces may seem unusual, they are in fact well matched to the way our visual system is organized around highly detailed foveal vision and more contextual, low-detail peripheral vision (Goldstein, 2009).

There is nothing inherent about spatial metaphors that limit them to visual display. Indeed even some examples of using distorted spatial space for auditory display have been described (Shinn-Cunningham and Durlach 1994; Nesbitt and Barrass, 2002; Nesbitt and Barross, 2004). However, visual display design is frequently associated with the use of spatial metaphors, and because the visual sense is organized in a spatial way, it is very difficult to transfer these metaphors to other sense. The haptic sense is also adept at interpreting spatial qualities and can be considered, although often haptic interpretations take significantly longer (Chang et al., 2007). The use of auditory display space should not altogether be ruled out, for example, with a pair of headphones it is quite simple to display information in stereo by panning sounds between left and right channels. This may be a useful way to provide peripheral spatial context that can easily be integrated with small-sized visual displays, by using headphones.

One mistake that designers often make is to transfer visual spatial metaphors into haptic and auditory domains and unwittingly substitute time for space. Vision allows for very quick comparisons of objects in space, where as many haptic and auditory investigations of space often rely on scanning the space over time. That is, the information about the objects in space must be integrated over time, and this puts a significant load on short-term memory. With these senses, it may be better to consider alternative mechanisms for displaying the information.

We have discussed some key concepts for designing spatial metaphors, and some relevant guidelines from this discussion are summarized in Table 9.1.

9.3.2 Direct Metaphors

In the real world, a great deal of useful information is perceived directly from the properties of sights and sounds and by touching. For example, a sound may have a certain loudness or pitch. Objects in the real world may be recognized on the basis of visual properties such as color or lighting. The sense of touch can also provide information about surface texture, shape, and hardness. Real-world information is often interpreted in terms of properties like pitch, color, and hardness. Abstract information can also be interpreted in terms of these direct properties.

An important distinction between spatial metaphors and direct metaphors is that direct metaphors are interpreted independently from the perception of space. While the concepts of spatial metaphors apply generally for each sense, this is not true for direct metaphors. There is very little intersection, for example, between the low-level concepts of direct visual metaphors and the low-level concepts of direct auditory metaphors. This is not surprising as direct metaphors relate to the properties that the individual sensory organs can detect.

However, direct metaphors tend to be simple, for example, you can map information directly to a color or to a sound's pitch or a particular haptic property such as vibration. For the designer, it is relatively simple to exchange direct metaphor between the senses. Instead of using color, you may substitute a sound property such as timbre. For a designer of mobile interfaces, where the opportunities for using visual properties are restricted, using sound or haptic properties is a relatively simple way to add more information without major changes to the overall design.

Table 9.1   Some Examples of Useful Guidelines for Spatial Metaphors

Visual space dominates our perception of space
The spatial visual metaphor is the critical design component as it provides the base for integrating the auditory and haptic displays. As Card et al. note, "Space is perceptually dominant" (Card et al., 1999)
Use each sensory modality to do what it does best
The "modal specific" theory states that each modality has distinct patterns of transduction. So, each sense has unique sensory and perceptual qualities that are adept with certain kinds of complex information (Friedes, 1974). Because of this, the various sensing modalities are differentially well suited for different kinds of perceptual tasks (Welch and Warren, 1980). With complex information, specific modalities are adept with certain kinds of spatial or temporal patterns (Friedes, 1974). There are three primary categories of patterns in sensory information: one involves space, one involves time, and the other involves movement (space and time). For each of these categories (space, time, movement), one of the senses (visual, auditory, haptic) is the most adept modality (Friedes, 1974)
Vision emphasizes spatial qualities
Vision is well suited to finding spatial patterns. The visual sense is most accurate at judging spatial relationships
Hearing emphasizes temporal qualities
Hearing is a primarily temporal sense as sound evolves over time. Hearing is well suited to judging temporal relationships
Haptics emphasizes movement
Haptic perception involves the integration of spatial and temporal patterns. Haptic sense can act on the environment, unlike vision and audition (Durlach and Mavor, 1995)
Consider sensory substitution
With complex information, the senses are adept in their preferred modality. However, if the information is simple, then different modalities are equivalent and the senses can be substituted for each other. In this case, there may be opportunities to adapt the spatial mappings of vision and the temporal mappings of audition to alternative senses
Adapt spatial visual metaphors to spatial auditory metaphors
Spatial visual metaphors have been well explored and can be transferred to the auditory domain provided the reduced spatial location ability and temporal nature of hearing are considered
Adapt spatial visual metaphors to spatial haptic metaphors
Spatial visual metaphors can be transferred to the haptic domain provided the reduced spatial location ability of haptics is considered. Touch has previously been used as a substitute for vision. For example, the Optacon was developed to use a number of small vibrators to provide a display of photographs or text (Goldstein, 2009)

Note: A complete list is available elsewhere (Nesbitt, 2003).

It is of course important to remember that each property has unique characteristics that need to be considered. Two important factors are whether the properties have an ordering and how well we can distinguish between different categories of a property.

Sensory properties such as hue and timbre do not have a natural ordering (low to high), while other properties such as saturation (less to more), surface texture (smooth to rough), and pitch (low to high) are perceived in an ordered way. If you want to display unordered information, then an unordered property such as timbre or hue is probably a better choice. Similarly if your information has some ordering, then an ordered sensory property would be better for conveying the underlying order in your information (Table 9.2).

While direct metaphors provide a useful way to display categories of information, they are less appropriate for conveying exact information. Very few people can exactly recognize a sound's pitch, for example, or identify a particular shade of red. While it is possible to convey continuous information, such as a stock price, by a property like pitch or color, it is unlikely that the user will be able to interpret its exact value. Therefore displaying categories of information is perhaps a better approach. You could instead map low, medium, and high stock prices to low, medium, and highly pitched sounds. Such distinct categories are more likely to be identified. What is important for the designer is that they identify appropriate categories of a sensory property that will not be misinterpreted. (You don't want to sell your stocks because you thought the price was high, when in fact the price was low.)

9.3.3 Temporal Metaphors

In the real world, a great deal of useful information is dependent on the perception of time. For example, a pedestrian crossing a busy road is required to interpret the amount of time between vehicles. The rate and frequency of traffic may also impact on the pedestrian's decision of when to cross. Temporal concepts like duration, rate, and frequency can also be used to encode information.

Temporal metaphors relate to the way we perceive changes to pictures, sounds, and haptic sensations over time. The emphasis is on interpreting information from the changes in the display. Temporal metaphors are also closely related to both spatial and direct metaphors. For example, it is changes that occur to a particular spatial metaphor or direct metaphor that may display the information. We might, for example, interpret a point changing position in space as meaning something. We could also interpret an object's changing color over time as having some particular meaning.

Of course, all the senses require some amount of time to interpret a stimulus. This is very fast for vision, while with hearing most sounds are more prolonged events with some temporal structure. A sound stimulus is perceived by interpreting changes that occur in air pressure over time. Even a single sound event, such as a bottle breaking, contains a complex temporal pattern that is perceived over a short period of time. With temporal auditory metaphors, the designer's focus is on how changes that occur in sound events can be used to represent information. A key aspect of temporal metaphors is how we perceive events in time with different sensory modalities.

Temporal metaphors are very complex to design and yet probably offer the most scope for increasing the amount of information displayed in a mobile interface. This is because there is limited space for displaying information and the designer has little choice but to display the information over time. This is a common approach that sonification designers use to transfer visual displays to the auditory domain. For example, in the auditory histogram (Scaletti, 1994), a user can listen to the pitch of each bar in the graph over time. So for a histogram with five bars, you would hear five sounds in succession, each with a pitch corresponding to the value of the bar.

Table 9.2   Some Examples of Useful Guidelines for Direct Metaphors

Direct visual metaphors are the first choice for displaying categorical data
While direct visual properties are not suitable for accurate display of quantitative information, they are very useful for displaying ordinal and nominal categories. Objects with direct visual properties can be easily compared when arranged in space (Tufte, 1990)
Color aids in object searching (speed, accuracy, memorability)
In visual search and recognition tasks, colors add speed, accuracy, and memorability (Hardin, 1990)
Users can distinguish about 24 steps of hue
The number of JNDs of hue that an observer can detect is about 24 (Trumbo, 1981)
Visual texture can be used to represent ordinal categories
There are neurons in the striate nucleus that respond directly to different spatial frequencies (Goldstein, 2009). Visual texture is ordered from smooth to rough and so is best used to represent ordinal categories
Direct visual shape can be used to represent nominal categories
Neurons in the cortex respond to both simple forms and more complex stimuli such as complex forms of different sizes (Goldstein, 2009). Shapes are readily recognized and can be used to represent categories. Because shapes have no ordering, they are best used for nominal categories
Pitch and loudness can represent ordinal categories
Loudness and pitch can be used to display continuous quantitative data. However, these properties cannot be interpreted accurately and so are better used to represent distinct ordinal categories
Timbre can represent nominal categories
Timbre is unordered and so is useful for nominal categories. There is a wide range of available timbres that are discriminable. These include musical instruments and sampled natural sounds such as voices. Use timbres with multiple harmonics as this helps perception and avoids one sound masking another (Brewster et al., 1994)
Pitch is an ordered property
The perceived property of pitch is closely related to the frequency of the sound stimulus. Different frequencies are signaled by activity at different places in the auditory system. In what is known as place coding, the inner hair cells located at different places along the cochlea fire to specific frequencies. The organization of the cortex maintains this tonotopic map (Goldstein, 2009). Pitch is ordered from low to high, and pitch differences naturally represent up and down (high and low pitch)
Individuals have very different haptic perceptions
The individual differences in many measures of haptic perception are large (Stuart, 1996). Because of the large differences between individuals, it is safer to use large categorical differences between haptic properties

Table 9.2   (continued) Some Examples of Useful Guidelines for Direct Metaphors

Haptic surface texture is an ordinal property
Surface texture can be experienced as slip on a smooth surface like glass through to the roughness of more abrasive surfaces such as sandpaper. This property is ordered from smooth to rough but it is not judged precisely. This makes it useful for displaying ordinal categories
Vibration is an ordinal property
Vibration is ordered but it is not judged precisely; this makes it useful for displaying ordinal categories

Note: A complete list is available elsewhere (Nesbitt, 2003).

The sounds might also substitute timbre for color to help the listener identify each bar. The two main problems with using this approach are that short-term memory is required to compare the sounds and hence it may be difficult to compare between the first and last sounds. With visual spatial metaphors, it is simple to quickly compare between different values by moving the eyes. To do this with temporal displays requires replaying the events, and this tends to be time consuming and more demanding on the user's level of concentration. Of course, identifying this problem is the first step in designing a solution, and we will discuss some strategies dealing with better temporal metaphors when we discuss Gestalt principles.

Despite some difficulties, temporal metaphors are less commonly used and yet can be very powerful ways to display structure. For example, you might think of music as an example of a complex and rich way to display information over time. We would argue that for designers of mobile interfaces have yet to fully explore the possibilities for displaying visual, auditory, and haptic information over time. Most auditory and haptic information is in the form of alarm events. Using a stream of events to monitor information is also a natural extension of how we use these senses in the real world. While this is a natural way to use these senses, there is also room for developing richer and more complex temporal displays akin to the way information is displayed using music.

One important principle to bear in mind when designing these displays is the law of constancy. With all our perceptions, we tend not to perceive small changes to a signal over time. Therefore, a slowly continuous changing signal may not register as different until it reaches some significant threshold. This is important consideration when designing monitoring applications. This principle, along with other guidelines related to temporal metaphors, is provided in Table 9.3.

9.3.4 Discussion of Spatial, Direct, and Temporal Metaphors

The previous sections have introduced the concepts of spatial, direct, and temporal metaphors. These abstract concepts are useful high-level tools that allow designers to think about multisensory display. More detailed examples of how these principles have been applied and a full list of guidelines are available elsewhere (Nesbitt, 2003). The intention of this discussion is to highlight alternative design approaches that might be used for displaying information. These are of particular relevance to mobile interface designers as there are restrictions on the size of screen displays and hence how much information can be displayed visually and particularly using spatial metaphors.

The next section will introduce another useful framework for designers working in multisensory display. This framework is based on gestalt principles. Although these principles were originally

Table 9.3 Some Useful Guidelines for Temporal Metaphors

Adapt temporal auditory metaphors to temporal visual metaphors   Temporal auditory metaphors have been well established in the domain of music, and these principles can be transferred to the visual domain for encoding temporal patterns
Adapt temporal auditory metaphors to temporal haptic metaphors   Temporal haptic metaphors have not been well explored; however, concepts from temporal auditory metaphors may be transferred to the haptic domain for encoding temporal patterns
Use temporal visual metaphors to understand change   Consider temporal visual metaphors for temporal data, such as data that change over time or occur as events in time. Transitions or steps of change in the data can be illustrated by changing position or properties of objects in space. Animation also helps the user maintain their understanding of spatial relationships
Perceptual constancy resists change   Properties of objects that change slowly over time may not be noticed, as we tend to perceive object properties as constant (GP-3)
Use temporal auditory metaphors to display time series data   Hearing is mainly a temporal sense and so is good for detecting changes over time (Kramer, 1994a). Ears are good for certain types of data analysis, for example, picking up correlations and repetitions of various sorts. Hearing detects sounds repeated at regular rhythms. It is especially useful where a sudden change to constant information needs to be detected
Auditory events can be compressed in time   Time compression of audio signals reduces the time to replay a series of data. Because of the high temporal resolution, auditory events can still be distinguished. The temporal resolution ranges from milliseconds to several thousand milliseconds (Kramer, 1994a; Stuart, 1996)
Sound is good for monitoring data   We hear things we do not see, and so it is good for monitoring what is happening in the background. Sound is useful for monitoring a situation when eyes are busy. People can process audio information while simultaneously engaged in an unrelated task (Cohen, 1994)
Sound is good for alarms   Sound is insistent and is very suitable for alarms and signaling an event. Sound is good at getting attention, and different temporal and spectral patterns can be used to encode different levels of urgency. "Acoustic signals tend to produce an alerting and orienting response and can be detected more quickly than visual signals" (Wenzel, 1994)
Sound can convey complex messages with temporal structures   A number of ways have been devised to provide structure in complex sound messages. For example, in music, the concepts of rhythm, meter, and inflection can impose structure. Other hierarchically structured sound messages have been designed and shown to be effective in some instances. These include earcons (Blattner et al., 1989) and parameter nesting (Kramer, 1994b)

Table 9.3   (continued) Some Useful Guidelines for Temporal Metaphors

Haptic feedback can detect a wide range of frequencies
A wide range of force frequencies can be perceived, from fine vibrations at 5,000–10,000 Hz up to coarse vibrations of 300–400 Hz (Stuart, 1996). This allows haptics to be used for detecting a wide range of temporal patterns

applied in the realm of visual display, more recently they have also been investigated for auditory display (Williams, 1994). Although our investigations have just begun, we are also finding that the concepts are all very applicable to the design of haptic displays (Chang et al., 2007). We will introduce some of these principles and provide some example design guidelines that are relevant to mobile interface designers. At the conclusion of this chapter, we discuss the concepts of spatial, direct, and temporal metaphors in relation to these gestalt principles and offer some suggestions about how to use the gestalt principles to improve the design of temporal metaphors for mobile devices.

9.4 Gestalt Principles

Gestalt theory is one of the well-known perceptual theories for perceptual organization. Gestalt theory tries to explain how humans organize individual elements into groups and how humans perceive and recognize patterns and forms. In many applications, it is important to assist users to find patterns or recognize relationships and structures. As a result, Gestalt principles suggest themselves as another useful high-level framework for understanding multisensory display and for structuring guidelines (Chang and Nesbitt, 2005). In this section, we shall provide some background to the Gestalt principles and also discuss the relevance of gestalt principles to visual, auditory, and haptic display and, more particularly, their relevance for designers of mobile information displays.

Gestalt psychology developed in the 1900s to help counter the associationist view that stimuli are perceived as parts and then built into complete images. German researchers Max Wertheimer (Wertheimer, 1924), Wolfgang Köhler (Köhler, 1920), and Kurt Koffka (Koffa, 1935) rejected the prevailing models of scientific analysis in psychology and used the principles of field theory to explain cognitive processes, which could not previously be explained without a holistic viewpoint.

Perception of visual structure and form, even in stationary images, is difficult. A complicated issue is that the overall picture we perceive often seems unrelated to its component parts. This phenomenon is often described by the saying "the whole is more than the sum of the parts" (Köhler, 1920, p. 17). Every individual perceptual element has its own nature and characteristics, but the nature of individual elements alone cannot account for how a group of elements will be perceived. The essential point of Gestalt theory is that the perception of the whole pattern (or gestalt) cannot be explained from the sum of its parts.

Gestalt theory developed principles that try to explain how we organize individual elements into groups. One of the most appealing things about these principles is that they are themselves simple to state, understand, and apply. The main criticisms for the Gestalt approach are a lack of a computational theory to explain how it works and the shortage of an identified physiology to explain the theory. Despite these problems, the Gestalt principles are useful tools for designers of information displays.

A key part of Gestalt theory is a number of principles first introduced by Max Wertheimer (Wertheimer, 1924) and later elaborated on by Kurt Koffka (Koffa, 1935). Wertheimer

(Wertheimer, 1924) identified a number of foundational Gestalt principles for perceptual grouping, such as the principles of closure, common fate, continuation, proximity, and similarity. Koffka (Koffa, 1935) studied three more principles for perceptual organization: the principles of figure–ground, pragnanz, and simplicity. Over time, a number of other principles have been suggested.

Most work on the Gestalt principles has focused on visual perception (Chang and Nesbitt, 2006), and the principles are typically used to explain how users perceive the form of visual objects (Table 9.4) (Goldstein, 2009). In the following sections, we will focus our discussion on the four principles of similarity, proximity, common fate, and good continuation. These principles have been explicitly applied to animated networks (Nesbitt and Friedrich, 2002) and sonification (Williams, 1994). In sound display, the emphasis is not on how visual elements group together but on how sounds segment themselves into streams (Table 9.5) (Bregman, 1990). At this stage, little work has been done to apply Gestalt principles to the haptic display of data (Chang and Nesbitt, 2006). The use of these principles for haptic display (Table 9.6) is part of our ongoing work, although results from a series of experiments already indicate that both visual and haptic elements are grouped in the same way (Chang et al., 2007).

Table 9.4 Principles of Visual Grouping

Gestalt Principle	Impact on Visual Grouping
Similarity	Similar things appear to be grouped together
Proximity	Things that are near to each other appear to be grouped together
Common fate	Things that are moving in the same direction appear to be grouped together
Good continuation	Points that, when connected, result in straight or smoothly curving lines are seen as belonging together, and the lines tend to be seen in such a way as to follow the smoothest path

Source: Goldstein, E.B., *Sensation and Perception*, Brooks/Cole, Pacific Grove, CA, 2009, pp. 104–109.

Table 9.5 Principles of Sound Streaming

Gestalt Principle	Impact on Sound Streaming
Similarity	Sounds that share attributes are perceived as related. An element in a stream may be captured by another stream with elements that are similar
Proximity	Sounds that are close to each other are more likely to be grouped together
Common fate	Sounds that undergo the same type of changes at the same time are perceived as related
Good continuation	When a sound undergoes a smooth transition to another sound, those sounds are perceived as related

Source: Williams, S., Perceptual principles in sound grouping, auditory display, *Proceedings of SFI Studies in the Sciences of Complexity*, Vol. XVIII, Addison-Wesley, Reading, MA, pp. 95–125.

Table 9.6   Proposed Principles of Haptic Grouping

Gestalt Principle	Impact on Haptic Grouping
Similarity	Objects that share the same haptic such as surface texture are related. This principle applies in a similar way for both visual and haptic grouping
Proximity	Objects that are close to each other are more likely to be grouped together. This principle applies in a similar way for both visual and haptic grouping
Common fate	Haptic objects that are moving in the same direction are perceived as related. Haptic stimuli that change in the same way over time are perceived as related
Good continuation	Haptic points that, when connected, result in straight or smoothly curving lines are seen as belonging together, and the lines tend to be interpreted in a way as to follow the smoothest path. When a haptic stimulus undergoes a smooth transition to another state, those states are perceived as related

9.4.1 Similarity

Elements will tend to be grouped together if their attributes are perceived as related (Goldstein, 2009; Moore, 2003). For example, with visual displays, elements will be grouped together if the intensity (Figure 9.5), hue, size, orientations, and shape are closely related with each other (Goldstein, 2009; Palmer et al., 2003). People also group similar sounds together if the timbre, pitch, subjective location, and loudness are closely related to each other (Bregman, 1990). For example, people may group string instruments together within an orchestra, because they have similar timbres. With haptic perception, it is also possible to group similar shapes, forces, surface textures, weights, and vibrations. For example, visually disabled people are able to separate cutlery by similar shapes, for example, grouping forks and spoons into two different groups.

Some examples of visual and haptic guidelines for this principle are shown in Table 9.7.

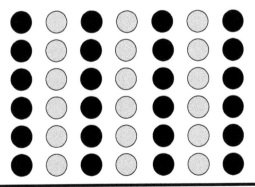

Figure 9.5   Elements in this figure are visually grouped by the similar intensity into four groups of dark circles and three groups of lighter circles.

Table 9.7   Some Example Guidelines Organized by the Similarity Principle

Similarity Principle
Visual Guideline for Similarity
"Provide consistent visual appearance within icon sets" (ETSI EG 2002 048, 2002) In screen design, a designer should keep the graphical style consistent for items in the same icon set. This is because users will naturally group icons if the icons are similar. For example, elements with the same color and the same size can be used to keep icons consistently within the same icon group (Palmer et al., 2003)  *Two distinct groups of icons are shown. Similar color icons form a group*
Auditory Guideline for Similarity
Rhythms can represent data categories
Message categories to be identified by the sounds rhythm. If the rhythm is used to represent categories they should be made as different as possible. Putting a different number of notes in each rhythm is very effective for distinguishing different rhythms, but very short notes should be avoided (Brewster et al., 1994)
Haptic Guideline for Similarity
Use same properties for the element to indicate the same purpose
When people perceive the same surface texture, they will tend to identify them as the same group (Chang et al., 2007). A designer may use the same surface texture to indicate interface elements that serve the same purpose

9.4.2 Proximity

The principle of proximity states that elements that are close to each other will be grouped together (Figure 9.6) (Fisher and Smith-Gratto, 1998–1999; Goldstein, 2009). Sound events are also grouped together if the sounds are related to one another in time (Bergman, 1990). For example, three flute sounds playing the same melody at the same time that are close together temporally will be grouped together. If they were to play the melody in an unsynchronized way, then the three flutes would form separate groups. The example of haptic grouping can be found in Braille; each character in Braille is made of a group of characters that are grouped together because they are close to one another (Figure 9.7). The example of visual, auditory, and haptic guidelines for the principle of proximity is in Table 9.8.

9.4.3 Common Fate

The principle of common fate suggests that display elements that change at the same time or move in a similar way will be grouped together (Goldstein, 2009). For example, animated visual elements that move in the same direction with the same speed will be seen as related (Figure 9.8). In auditory displays, people tend to group sounds together if they change in pitch in a similar way. Sounds that

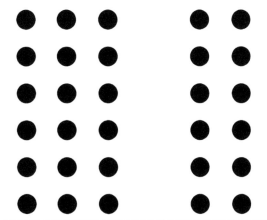

Figure 9.6 In this image, we typically perceive two distinct groups based on their visual proximity.

Figure 9.7 Raised dots that form Braille letters are grouped by proximity.

begin and finish at the same time are also likely to be perceived as related (Moore, 2003). We also experience common fate with the haptic sense. For example, if you turn a door handle and feel the door handle on the other side of the door also turn, then you can perceive their common fate. An example of an existing guideline for the principle of common fate is shown in Table 9.9.

9.4.4 Good Continuation

People tend to perceive a smooth and continuous connection between points rather than lines with sudden or irregular changes in direction. Thus elements will be grouped together if a continuous pattern can be interpreted and this pattern will be assumed to continue even if some parts are hidden (Moore and Fitz, 1993).

For example, in a painting, if smoke covers some part of a curving road on the mountain, we still assume the curving road continues and will reach the top of the mountain (Figures 9.9 and 9.10) (Goldstein, 2009). If a sound slowly changes in pitch, loudness, or timbre in a very smooth manner, then the sound will still be perceived as the one sound (Moore, 2003). By contrast, people will perceive different sounds if the timbre, pitch, or loudness changes abruptly. When we use the sense of touch, we tend to perceive a smooth and continuous outline even though some parts are hidden with unfamiliar patterns. For example, people can exit a completed dark room by touching the walls, perceiving the wall as continuous and ignoring irrelevant objects such as light switches. Some examples of relevant guidelines for this principle are shown in Table 9.10.

9.4.5 Discussion of Gestalt Principles

We have presented a second framework designed for categorizing guidelines to assist designers of multisensory displays. The framework is based on Gestalt principles. These well-known principles are used to create high-level categories that act as general advice for designers but also provide

Table 9.8   Some Example Guidelines Organized by the Proximity Principle

Proximity Principle
Visual guidelines for proximity
"Arrange data to make relationships clear" (Brown, 1988) Designers need to take care when arranging elements in space. For example, put related displayed data next to each other as people tend to group elements based on their location to each other (Fisher and Smith-Gratto, 1998–1999) *Provide the appropriate space between visual elements* It is important to organize the space presentation to any of the visualization, especially for mobile device due to the small screen and space limitation. People will perceive the information easily if the obvious spaces have been provided (Chang et al., 2007)
Auditory Guidelines for Proximity
Provide an appropriate gap between auditory elements
It is important to organize the appropriate gap when presenting auditory information in order to prevent the unnecessary information delivery. People will tend to perceive all the auditory information as a whole if there is no obvious gap between them. The appropriate gap can group the different auditory information into groups
Haptic Guidelines for Proximity
Provide the appropriate space between haptic elements
People tend to group the spatially close elements into the same groups. The palpable gap can help our haptic sense to group the different haptic information into groups (Chang et al., 2007). People will not recognize the structures as different if there is not enough space between haptic elements

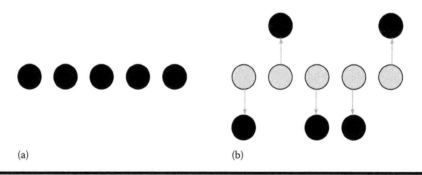

(a)   (b)

Figure 9.8   In (a), the elements are grouped together. If you can imagine these visual elements moving as indicated by the arrows (b), then you will perceive two groups based on their common fate.

a novel way of structuring existing guidelines. These principles can provide important design advice for designers of mobile devices. Typically in large screen displays, we would expect the principle of proximity to be used more frequently than is possible with smaller screen displays. A simple strategy is to use the principle of similarity to provide the same type of structure in small-screen devices. For example, instead of grouping icons by spatial location, they can be related

Table 9.9   Some Example Guidelines Organized by the Common Fate Principle

Common Fate Principle
Visual Guidelines for Common Fate
"Use only two levels of blink coding (flashing)" (Brown, 1988) Blink coding is often used for indicating warning displays. Designers need to be aware that they should not use different blink levels for displaying categories of information as this will confuse people (Brown, 1988)
Auditory Guideline for Common Fate
"Sounds which experience the same kinds of changes at the same time are perceived as related" When designing displays for monitoring two streams of sound, the designer needs to be aware that sounds that change in the same way may be grouped into a single stream (Williams, 1994)
Haptic Guideline for Common Fate
Haptic properties that change at the same time are perceived as related When designing displays for monitoring two haptic properties, the designer needs to be aware that haptic properties that change at the same time may be grouped together

Figure 9.9   Continuous edge of the path is broken by shadow, but the path is perceived to continue in a smooth and regular pattern.

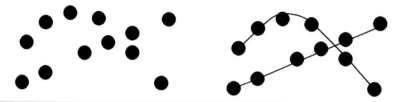

Figure 9.10 Elements tend to be grouped into continuous lines (either curved or straight).

Table 9.10 Some Example Guidelines Organized by the Continuation Principle

Continuation Principle
Visual Guideline for Continuation
Provide a mapping to group visual information
It is important to consider how to lead the user through the information being presented and how to ravigate the contents (Chittaro, 2006), especially when presenting large amounts of information for the small-screen device. Jones et al. (1999) suggest the content should be well structured for PDAs
Auditory Guideline for Continuation
To prevent users connecting unrelated auditory icons, put a gap between consecutive auditory icons (ETSI EG 2002 048, 2002)
People will tend to perceive a continuous auditory signal and ignore small gaps between consecutive auditory icons. In some cases, the perceived continuation is not desirable and designers should use an appropriate gap (0.1 s) to ensure people perceive separate sound events (Brewster et al., 1992)
Haptic Guideline for Continuation
Haptic maps can be used to connect elements and provide structure
People perceive haptic structures as continuous in the same way they perceive visual continuous structures. A property such as surface texture will enhance this grouping, and therefore a designer may use this haptic property to connected structures in the interface (Chang et al., 2007)

by color. Even though the icons may be displayed on different screens, they will still be perceived as connected or related due totheir color.

9.5 Relating Gestalt Principles to Spatial, Direct, and Temporal Metaphors

Because the principle of similarity deals with attributes or perceived qualities, it described groups based on direct properties. For display designers, it suggests that direct properties such as color, pitch, or hardness might be interchanged to promote grouping by similarity. For example, for blind users, can we replace a direct visual property, such as color, with a direct haptic property, such as hardness. For multisensory display designers, an interesting question that arises is how strongly the different direct properties enforce grouping. For example, where conflict exists would a user group elements based on color, rather than hardness.

Because the principle of proximity deals with the relative position of objects in space, it is strongly associated with the MS-Taxonomy concept of spatial metaphors. It would seem that this principle would apply in a similar way to each sense. However, a designer would need to consider the different resolution or accuracy of each sense when applying this principle. For example, the perception of when objects are close or apart is likely to be different for visual, auditory, and haptic display. For multisensory display, conflicts might occur in proximity and users may ignore a mode as it occurs with the ventriloquist effect.

The principle of common fate describes groupings related to common changes over time and hence relates to the category of temporal metaphors. Temporal metaphors are not independent of the other types of metaphor as they involve changes to both spatial and direct properties. Where information in a display is encoded in changes to direct properties, it may be possible to simply transfer the design to another modality. For example, information displayed by changing color may be displayed by changing timbre in a sound display. When information is encoded as a change in spatial properties, then each sense's ability to detect position in space needs to be considered. For example, a small change in position may be detected visually but not with an auditory display. However, the auditory sense is attributed with higher temporal resolution (Kramer, 1994c), and so it may be possible to use finer changes in direct properties to convey information. Overall, the perception of time is still not well understood. However, it seems sensible that limitations in short-term memory must also be considered when designing temporal metaphors. If a conflict occurs between the timing of events on the different senses, for example, if a video and an audio track for a film are not synchronized, then the effect is likely to be disconcerting. We might also expect that a mismatch in the timing of sensory events would quickly break common fate groupings. Such disjoint timings also occur naturally, for example, in a storm the thunder and lightning may be perceived as disjoint. For multisensory display, it also seems reasonable to assume that conflicts in temporal events will quickly break grouping (Table 9.11).

The principle of good continuation can apply to both spatial and temporal metaphors. Goldstein describes this principle in terms of the way visual elements are arranged in space (Goldstein, 2009), while Williams describes the principle in terms of temporal continuation (Williams, 1994), that is, if a signal varies a direct property slowly, it will remain in the same stream. Conversely a sudden change to a direct property will cause the sound to be interpreted as a different stream. That the good

Table 9.11 Relevance of Gestalt Principles to the Metaphor Types of the MS-Taxonomy

Gestalt Principle	*MS-Taxonomy*		
	Spatial	*Temporal*	*Direct*
Similarity			✓
Proximity	✓		
Common fate		✓	
Good continuation	✓	✓	

continuation principle is defined differently for each sense highlights a problem for multisensory designers. Both good spatial continuation and good temporal continuation are both important issues for display designers, but it may be better to treat them as separate or even competing grouping principles. For example, we might expect groupings based on spatial continuation to break more quickly with vision compared to audition. We might also expect auditory groups based on temporal continuation to break more quickly compared with visual groups related by changes that occur over time.

The principle of familiarity is the most generic and nonspecific principle, being relevant to all three categories of spatial, temporal, and direct metaphors. This is perhaps because familiarity is arguably more cognitive than perceptual in nature. We may expect to "find" certain patterns at familiar locations in space and time, and we might also expect to see familiar relationships between direct properties. All three of these factors may conflict or complement each other. This interdependence suggests that it may be more difficult to control familiarity within a display design. Indeed, because familiarity crosses over the entire design space, it may be best for the designer to adopt it as a more general principle. For example, with perceptual data mining tools, it is important to check that detected patterns have not been biased by familiarity.

9.6 Conclusion

The user interfaces for mobile devices such as PDA provide new opportunities to develop more complex multisensory displays. Multisensory displays attempt to take advantage of a range of human senses. For example, such displays would include not only visual information but also auditory and haptic (tactile) feedback. One broad application of such displays is to allow more channels on which information can be displayed, especially in mobile devices where one visual channel may be restricted due to lack of screen space. This is a new and emerging computer discipline, and we have been working to provide both low-level guidelines and higher-level conceptual principles that can assist designers understand and take advantage of the multisensory design space.

In this chapter, we have provided some low-level guidelines that are useful for designers. However, the emphasis has been on the higher-level categorizations of Gestalt principles and the MS-Taxonomy, which can be used to structure these guidelines. We believe that well-structured guidelines are more useful for designers because they provide both context and detail. This matches well to the top-down (high-level) and bottom-up (low-level) approaches used in software design (Pfleeger, 1998).

Of course, much further work needs to be done in gathering, collating, and evaluating design guidelines. It would also be desirable to develop more quantitative models of the perceptual and cognitive processes involved in interpreting information displays. These two activities form the basis of our ongoing work. As a final comment, we would like to remind readers that design is a process that involves many trade-offs to be made. This was summarized nicely by Tufte (1983) in his first book on designing visual presentations of statistical information: "Design is choice. The theory of the visual display of quantitative information consists of principles that generate design options and that guide choices among options. The principles should not be applied rigidly or in a peevish spirit; they are not logically or mathematically certain; and it is better to violate any principle than to place graceless or inelegant marks on paper. Most principles of design should be greeted with some skepticism, for word authority can dominate our vision, and we may come to see only through the lenses of word authority rather than with our own eyes" (Tufte, 1983).

References

AppleComputerInc. (1995): *Macintosh Human Interface Guidelines*. Cupertino, CA, Addison-Wesley.

Banatre, M., Couderc, P., Pauty, J., and Becus, M. (2004): Ubibus: Ubiquitous computing to help blind people in public transport. *Mobile HCI2004*. Berlin, Germany, Springer-Verlag, pp. 310–314.

Bertin, J. (1981): Graphics and graphic information processing. *Readings in Information Visualization: Using Vision to Think*. S. K. Card, J. D. Mackinlay, and B. Shneiderman (Eds.), San Francisco, CA, Morgan Kaufmann, pp. 62–65.

Blattner, M., Sumikawa, D., and Greenberg, R. (1989): Earcons and icons: Their structure and common design principles. *Human Computer Interaction* 4(1): 11–14.

Boring, E. G. (1942): *Sensation and Perception in the History of Experimental Psychology*. New York, Appleton Century Crofts.

Bregman, A. (1990): *Auditory Scene Analysis: The Perceptual Organization of Sound*. Cambridge, MA, The MIT Press.

Brewster, S., Wright, P. C., and Edwards, A. D. N. (1992): A detailed investigation into the effectiveness of earcons. *Proceedings of the First International Conference on Auditory Display*, Santa Fe, NM, Addison-Wesley.

Brewster, S., Wright, P., and Edwards, A. (1994): A Detailed Investigation into the Effectiveness of Earcons. *Auditory Display: Sonification, Audification and Auditory Interfaces*. G. Kramer (Ed.), Reading, MA, Addison-Wesley.

Brooks, F. P., Ouh-Young, M., Batter, J. M., and Kilpatrick, P. J. (1990): Project GROPE—Haptic displays for scientific visualization. *Proceedings ACM SIGGRAPH*, Dallas, TX, pp. 177–185.

Brown, C. (1988): *Human-Computer Interface Design Guidelines*. Norwood, NJ, Alex.

Calvert, G. A., Spence, C., and Stein, B. E. (2004): Introduction. *The Handbook of Multisensory Processes*. Cambridge, MA, MIT Press Cambridge, pp. xi–xvii.

Calvert, G., Spence, C., and Stein, B. (2005): *The handbook of multisensory processes*. MIT Press, Cambridge MA, USA.

Card, S. K. and Mackinlay, J. D. (1997): The structure of the information visualisation design space. *Proceedings of IEEE Symposium on Information Visualization*, Phoenix, AZ, IEEE Computer Society.

Card, S. K., Mackinlay, J. D., and Shneiderman, B. Eds. (1999): *Information Visualization. Readings in Information Visualization*. San Francisco, CA, Morgan Kaufmann Publishers, Inc.

Carter, J., Fourney, D., Fukuzumi, S., Gardner, J. A., Yasuo Horiuchi, Y., Gunnar, J., Jürgensen et al. (2005): The GOTHI model of tactile and haptic interaction. *Proceedings of GOTHI'05 Guidelines on Tactile and Haptic Interaction*, October 24–26, 2005, Saskatoon, Saskatchewan, Canada.

Chang, D. and Nesbitt, K. (2005): Developing Gestalt-based design guidelines for multi-sensory displays. *The Proceedings of NICTA-HCSNet Multimodal User Interaction Workshop, the ACS Conferences in Research and Practice in Information Technology, CRPIT*, Sydney, New South Wales, Vol. 57, pp. 9–16.

Chang, D. and Nesbitt, K. (2006): Identifying commonly-used Gestalt principles as a design framework for multi-sensory displays. *IEEE International Conference on Systems, Man, and Cybernetics*, Taipei, Taiwan, October 8–11, 2006.

Chang, D., Nesbitt, K., and Wilkins, K. (2007): The Gestalt principles of similarity and proximity apply to both the haptic and visual grouping of elements. *ACSW 2007*, Ballarat, Victoria, Australia, January 31–February 2, 2007.

Chang, D., Wilson, C., and Dooley, L. (2003–2004): Toward criteria for visual layout of instructional multimedia interfaces. *Journal of Educational Technology Systems*, **32**(1): 3–29.

Chittaro, L. (2006): Visualizing information on mobile devices. *IEEE Computer*, **39**: 40–45.

Cohen, J. (1994): Monitoring background activities. *Auditory Display: Sonification, Audification and Auditory Interfaces*. G. Kramer (Ed.), Vol. XVIII, Reading, MA, Addison-Wesley Publishing Company, pp. 499–534.

Dix, A., Finlay, J., Abowd, G., and Beale, R. (2004): *Human-Computer Interaction*. London, U.K., Prentice Hall.

Durlach, N. I. and Mavor, A. S. Eds. (1995): *Virtual Reality: Scientific and Technological Challenges*. Washington, DC, National Academy Press.

ETSI EG 2002 048 (2002): Human factors: Guidelines on the multimodality of icons, symbols and pictograms. European Telecommunications Standards Institute. Report No. ETSI EG202 048, ETSI, Sophia Antipolis, France.

Faulkner, C. (1998): *Human-Computer Interaction*. London, U.K., Prentice Hall.

Fisher, M. and Smith-Gratto, K. (1998–1999): Gestalt theory: A foundation for instructional screen design. *Journal of Education Technology Systems*, 27(4): 361–371.

Friedes, D. (1974): Human information processing and sensory modality: Cross-modal functions, information complexity, memory and deficit. *Psychological Bulletin*, **81**(5): 284–310.

Furnas, G. W. (1995): Generalized Fisheye views. *Human Factors in Computing Systems CHI '95*, Denver, CO.

Gaver, W. W. (1986): Auditory icons: Using sound in computer interface. *Human Computer Interaction*, **2**: 167–177.

Goldstein, E. B. (2009): *Sensation and Perception*. Cengage Learning. California, U.S.A., Wadsworth Publishing.

Hardin, C. L. (1990). Why color? *Perceiving, Measuring and Using Color*. Santa Clara, CA, SPIE-The International Society for Optical Engineering.

ISO 14915-1 (2002): *Software ergonomics for multimedia user interfaces—Part 1: Design principles and framework*. 15-1. ISO International Standard.

ISO 14915-2 (2003): *Software ergonomics for multimedia user interfaces—Part 2: Multimedia navigation and control*. 15-2. ISO International Standard.

Jones, M., Marsden, G., Mohd-Nasir, N., Boone, K., and Buchanan, G. (1999): Improving web interaction on small displays. *Proceedings WWW8*, Toronto, Ontario, Canada.

Karampelas, P., Akoumianakis, D., and Stephanidis, C. (2002): User interface design for PDAs: Lessons and experience with the WARD-IN-HAND prototype, theoretical perspectives, practice, and experience. *7th ERCIM International Workshop on User Interfaces for All*, Paris, France, October 24–25, 2002, pp. 474–485.

Kim, L. and Albers, M. J. (2001): Web design issues when searching for information in small screen display. *SIGDOC'01*, Santa Fe, NM, October 21–24, 2001, pp. 193–200.

Koffa, K. (1935): *Principles of Gestalt Psychology*. London, U.K., Routledge & Kegan Paul Ltd.

Köhler, W. (1920): Physical Gestalten (Die Physischen Gestalten in Ruhe und im stationaren Zustand). Erlangen: Eine naturphilosophische Untersuchung: ix-259. [Ellis, W. D. (Ed.) (1938): *A Sourcebook of Gestalt Psychology*. London, U.K., Routledge & Kegan Paul, pp. 17–54].

Kramer, G. (1994a): An introduction to auditory display. *Auditory Display: Sonification, Audification and Auditory Interfaces*. G. Kramer (Ed.), Vol. XVIII, Reading, MA, Addison-Wesley Publishing Company, pp. 1–78.

Kramer, G. (1994b): Some organizing principles for representing data with sound. *Auditory Display: Sonification, Audification and Auditory Interfaces*. G. Kramer (Ed.), Vol. XVIII, Reading, MA, Addison-Wesley Publishing Company, pp. 185–222.

Kramer, G. (1994c): An introduction to auditory display. *Auditory Display: Sonification, Audification and Auditory Interfaces*. G. Kramer (Ed.), Reading, MA, Addison-Wesley Publishing Company.

Lamping, J. and Rao, R. (1996): The hyperbolic browser: A focus + context technique for visualizing large hierarchies. *Journal of Visual Languages and Computing*, 7(1): 33–55.

Masoodian, M. and Budd, D. (2004): Visualization of travel itinerary information on PDAs. *Conferences in Research and Practice in Information Technology, The 5th Australasian User Interface Conference (AUIC 2004)*, Dunedin, New Zealand, 28: 65–71.

Mcgee, M. R., Gary, P., and Brewster, S. (2001): The effective combination of haptic and auditory textural information. *Haptic Human-Computer Interaction, First International Workshop*, Glasgow, U.K., August 31–September 1, 2000, *Proceedings*. Lecture Notes in Computer Science 2058 Springer 2001.

Micha, K. and Economou, D. (2005): Using personal digital assistants (PDAs) to enhance the museum visit experience *Proceedings of Advances in Informatics: 10th Panhellenic Conference on Informatics, PCI 2005*, Volas, Greece, November 11–13, 2005, pp. 188–198.

Moore, B. (2003): *An Introduction to the Psychology of Hearing*. 5th edn., London, U.K., Academic Press.

Moore, P. and Fitz, C. (1993): Gestalt theory and instructional design. *Journal of Technical Writing and Communication*, **23**(2): 137–157.

Najjar, L. (1998): Principles of educational multimedia user interface design. *Human Factors*, **40**(2): 311–323.

Nesbitt, K. (2003): Designing multi-sensory displays for abstract data. PhD thesis, School of Information Technology, University of Sydney, Sydney, New South Wales, Australia.

Nesbitt, K. and Barrass, S. (2002): Evaluation of a multimodal sonification and visualisation of depth of market stock data. *International Conference on Auditory Display—ICAD 2002*, Kyoto, Japan.

Nesbitt, K. and Barrass, S. (2004): Finding trading patterns in stock market data. *IEEE Computer Graphics and Applications*, **24**(5): 45–55.

Nesbitt, K. and Friedrich, C. (2002): Applying Gestalt principles to animated visualizations of network data. *Proceedings of the Sixth International Conference on Information Visualisation*, London, U.K., IEEE Computer Society, pp. 737–743.

Nesbitt, K. V., Gallimore, R., and Orenstein, B. J. (2001): Using force feedback for multi-sensory display. *2nd Australasian User Interface Conference AUIC 2001*, Gold Coast, Queensland, Australia, IEEE Computer Society.

O'Modhrain, S. (2004): Touch and go—Designing haptic feedback for a hand-held mobile device. *BT Technology Journal*, **22**(4): 13–145.

Ozok, A. and Slvendy, G. (2004): Twenty guidelines for the design of web-based interfaces with consistent language. *Computers in Human Behavior*, **20**: 149–161.

Palmer, S. E., Brooks, J. L., and Nelson, R. (2003): When does grouping happen? *Acta Psychologica*, **114**: 311–330.

Patterson, R. D. (1982): *Guidelines for Auditory Warning Systems on Civil Aircraft*. London, U.K., Civil Aviation Authority.

Pfleeger, S. L. (1998): *Software Engineering: Theory and Practice*. Englewood Cliffs, NJ, Prentice Hall.

Preece, J. (1994): *Human-Computer Interaction*. London, U.K., Addison-Wesley.

Ringbauer, B. (2005): Smart home control via PDA: An example of multi-device user interface design. *IFIP International Federation for Information Processing*, **178**: 101–120.

Scaletti, C. (1994): Sound synthesis algorithms for auditory data representations. *Auditory Display: Sonification, Audification and Auditory Interfaces*. G. Kramer (Ed.), Vol. XVIII, Reading, MA, Addison-Wesley Publishing Company, pp. 223–252.

Shinn-Cunningham, B. and Durlach, N. I. (1994). Defining and redefining limits on human performance in auditory spatial displays. *International Conference on Auditory Display*, Santa Fe, NM.

Shneiderman, B. and Plaisant, C. (2005): *Designing the User Interface*. College Park, MD, Addison Wesley.

Stuart, R. (1996). *The Design of Virtual Environments*. New York, McGraw-Hill.

Trumbo, B. E. (1981): A theory for coloring bivariate statistical maps. *The American Statistician* **35**(4): 220–226.

Tufte, E. (1983). *The Visual Display of Quantitative Information*. Cheshire, CT, Graphics Press.

Tufte, E. (1990): *Envisioning Information*. Cheshire, CT, Graphics Press.

Welch, R. B. and Warren, D. H. (1980). Immediate perceptual response to intersensory discrepancy. *Psychological Bulletin* **88**(3): 638–667.

Wenzel, E. M. (1994). Spatial sound and sonification. *Auditory Display: Sonification, Audification and Auditory Interfaces.* G. Kramer (Ed.), Vol. XVIII, Reading, MA, Addison-Wesley Publishing Company, pp. 127–150.

Wertheimer, M. (1924): Gestalt theory (Uber Gestalttheorie): 39–59, [Ellis, W. D. (Ed.) (1938): *A Sourcebook of Gestalt Psychology.* London, U.K., Routledge & Kegan Paul, pp. 1–11].

Williams, S. (1994): Perceptual principles in sound grouping, auditory display. *Proceedings of SFI Studies in the Sciences of Complexity.* Vol. XVIII, Reading, MA, Addison-Wesley, pp. 95–125.

Chapter 10

User's Perception of Web Multimedia Interfaces

An Exploratory Experiment

Eric Y. Cheng

Contents

10.1 Introduction

The conventional wisdom is that a multimedia program makes the interface easier to use, more enjoyable, and more understandable. However, empirical results indicate that there is not enough theoretically based evidence to guide how we shall design the interface to improve human performance. Not much is known about the design and effective use of interactive features in user interfaces. In this regard, we believe instead of using traditional usability evaluation methods, using a well-defined factorial design experiment could lead to more discoveries and understanding of the relationships between user interface design and usability.

Perception is a fundamental aspect of human and computer interaction. It is important that users are able to visually perceive information in a meaningful way. Every day we are bombarded with a mass of information, which needs to be filtered out, perceived, and understood unambiguously. This has important implications for the design of effective user interface. An effective computer interface needs to facilitate the following functions:

- Present information in a logical and meaningful way
- Keep users' attention focused on the subject within the presentation windows
- Provide easy to control, consistent, predictable operations
- Offer easy to follow, direct access organization
- Give clear navigation aids

Usability evaluation is often used to evaluate a computer interface. Usability is defined by the following five attributes: learnability, efficiency, memorability, errors, and satisfaction. This term is aimed to replace the concept widely known as "user friendliness." Clarifying the measurable aspects of usability is much better than aiming at a warm, fuzzy feeling of "user friendliness." The general characteristics of a poor usability website include difficulties in finding the information sought, disorganized, confusing, or contradicting information, lost hyperlinks, fuzzy graphics, long download time, and lack of navigation support. When users encounter a poor web page, normally they will feel frustrated, discouraged, like they are wasting time and other resources. Especially for web based multimedia programs with real time feeding and receiving, it is still unclear how those real-time interface features on the Web would differ from previous HCI studies that mainly focus on standalone personal computers.

Our research builds upon knowledge accumulated on the human computer interfaces. Much of the previous work in the development of the human computer interfaces was focus on the personal computer or electronic devices, though user interface researchers have started to move their focus on mobile computing devices such as PDAs and mobile phones. Historically, some previous remarkable work was done to study what makes things fun to play computer games, and some focused on the use of personal computer programs. Not until the late 1990s have we seen web usability become popular. Many books explain the best way to display information and point out what should be avoided, how it does not work, and what would work better. They provide a handy guide for web publishers, but a solid scientific proof is still missing. Although the 3W consortium has regularly updated web design guidelines such as cascading style sheets (CSS), they are not based on user reaction experiments. Despite these previous works, there is not much closely related study regarding the use of multimedia on the web. How to better design a web program is important to address the user's needs and provide better information for the web designers to consider during the web development stage.

From a cost-benefit perspective, interactive technology is expensive and requires more human resources, software, and hardware support. It is important to understand the effect of any intervention and to consider the utility of interventions. In this study, our objective is to investigate how to design a web multimedia interface appropriately and consequently to evaluate user interaction satisfaction. This research tries to find answers to the following questions: What interface features would users prefer to have in a web program? How would users feel about the differences among various designs and in what aspects?

10.2 Related Work and Evaluation Methods

Experience has shown that multimedia techniques can improve users' interaction with the application, especially within the areas of information acquisition, learning, presentation, and process control [1,2]. An important contribution multimedia can make is in the formation of simulated environments, also referred to as telepresence, "the experience of presence in an environment by means of a communication medium" [3]. This evokes involuntary sensory experiences of being there, or of "immersion" in a mediated environment. These simulated environments can encourage role-playing and vicarious performances in various health contingency situations, such as a hypolycemid attack.

Another important implication of multimedia capabilities of interactive technologies is demonstrated by research on how individuals with different levels of prior knowledge differ in how they process information presented in a multimedia format. Those who already possess a sufficient knowledge base tend to look at pictures only initially, whereas those without an adequate knowledge base tend to refer to pictures throughout the reading process. It could be hypothesized that the former group refers to pictures initially to invoke relevant mental models. Once these models are activated, new information is simply deposited onto the same framework and the learning process proceeds fairly efficiently. For the latter group, lacking the requisite mental models, the learning process involves two processes: creation of new models and assimilation of new information into these models. This implies that presentation of new information should be supplemented by external aids that facilitate the formation of mental models, as has been suggested elsewhere.

For learning to occur, a balance needs to be struck between redundancy levels and novelty of information [4–6]. Multimedia can facilitate this process. Pictures and sound can present material in novel ways through dramatization, role-playing, or simulation of sensory-rich environments to elicit the user's attention initially. Once attention has been elicited, opportunities open up for presenting cognitively demanding material. The challenges and goals setting often engage users in deep cognitive processing, learning-for-its-own-sake, and motivation increasing, which can be pleasurable experiences.

There are two general features of media, vividness, and interactivity [7]. The vividness of the multimedia feature represents the richness of the mediated environment in terms of the number of media components used (e.g., video, text, music). An information environment that utilizes more components can create a more vivid presentation in terms of the number of senses engaged in message processing (e.g., sight, sound, and touch), the sensory quality of the stimulus (e.g., color, movement, resolution, and loudness), and the extent to which message content is accentuated (e.g., using narration and video to demonstrate a behavior).

Perhaps the most powerful aspect of computer technology is the ability to combine text, graphics, sounds, and video in meaningful way [8]. Multimedia content enhances many web presentations. Due to bandwidth limitations multimedia have been slow to reach the web, though nowadays cable modem and ADSL are quite popular in North America. Advances in technology bring the new design capability of multimedia solutions to computer systems. Multimedia is, by definition, an integration of a variety of natural human communication channels. A users of multimedia systems can thus read visually presented text, view graphics and pictures, watch videos, and listen to speech, music, and other sounds. User's input to a multimedia system may consist not only of typed text, but also of selecting and manipulating 2D and 3D objects, using body gestures, and speaking in natural language. Furthermore, it is essential that the multimedia systems support a variety of methods for navigating through its information space. If the juxtaposition of media is carefully designed, multimedia can improve the quality of man–machine communication in many application areas. There is not much research addressing the different design issues within computer programs for web multimedia.

According to human information processing theory in cognitive psychology [9], the maximum number of chunks of information most people can hold in short-term memory is seven words. Thus, past research suggests: "The goal of most organizational schemes is to keep the number of local variables the reader must keep in short-term memory to a minimum, using a combination of graphic design and layout conventions along with editorial division of information into discrete units." The way people seek out and use information also suggests that smaller, discrete units of information are more functional and easier to navigate through than long, undifferentiated units. In the domain of educational psychology, Moreno and Mayer [10] studied a coherence effect in multimedia learning with an attempt to minimize irrelevant sounds in the design of multimedia instructional messages. They [11] also proposed nine ways to reduce cognitive load in multimedia learning. One problematic aspect of evaluation is the uncertainty that remains even after exhaustive testing by multiple methods. It is impossible to implement a perfect evaluation for complex human endeavors. As for other products, a web project has a deadline for marketing. Although problems may be found during the evaluation, at some point the product has to serve in the field. Thus, evaluation should be a continuous process during this program's lifecycle. Information overload is an important issue in human information processing. Very often the environment is full of noise or too much information that is usually more than the brain can handle. Broadbent's information flow model explained how the human perception system processes information input. Based on the model demonstrated in Figure 10.1, psychologists have done much research to compare different stimuli such as words, auditory, and pictorial memory.

Many previous researchers in human information processing have shown that visual information is easier to memorize than words [12–14]. Potter and Levy [15] showed that subjects have a good memory for pictorial information even when this information is presented at relatively fast rates of presentation. Nickerson [16] and Shepard [17] showed extremely good recognition memory for lists of hundreds of pictures. Haber [18] carried Shepard's study to an extreme by asking his subjects to look at 2560 photographic slides over the course of several days. Haber's subjects averaged about 90% correct in a forced choice recognition task. These experiments show that visual memory for complex concepts or scenes is extremely good when tested with a recognition procedure. Paivio [19–21] concluded that this visual memory also seems to improve memory performance substantially when subjects form images of words rather than trying to remember the words in a purely linguistic form.

In the search of how human memory functions, it has been long and widely accepted that researchers use computer memory as research basis. However, Glenberg [22] suggested psychologists

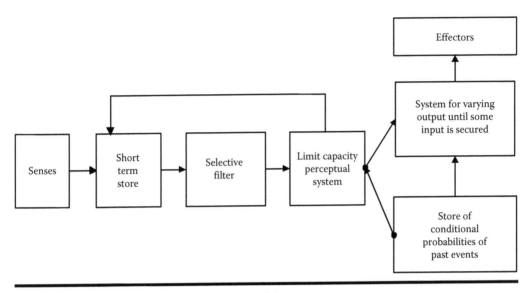

Figure 10.1   Broadbent's information flow model.

drop the widely accepted view that human memory works like computer memory, which stores abstract symbols designed to be reproduced with verbatim accuracy. Glenberg argued that human memory is a direct result of action. Memory exists to help us walk, talk, run, drive a car, answer the phone, and all of the tasks of getting along in the environment. In one study, Glenberg taught volunteer participants how to identify landmarks using a compass. One student group received only written and oral descriptions of the components, whereas another group watched video clips of a person interacting with a compass to identify landmarks. The video group had no problem later when asked to read and use instructions for how to use the compass. Based on Glenberg's theory, in our study we anticipate that the subjects, who are in the video group, should have better understanding of medical decision making than the subjects who are assigned to text or audio group.

In the following sections, we will first briefly review five major evaluation methods that have being developed and used to evaluate a web program, then a comparative critical analysis of those evaluation methods, as well as the method used in this study.

10.2.1 Usability Laboratory Testing

In the early 1980s, usability laboratory testing was an innovative approach and a truly user-needs focused assessment method. The method has the subject and observer sit at a workstation. While the subject is operating the system, running video recorders capture the user's actions and the contents of the computer screens. The subject and observer can address their concerns about any relevant issue during the test. With the help of video capturing, it is easy to identify what causes the subject to stop, pause, or ponder which action he/she should take. This process can be repeated after the problem that causes confusion or failure is fixed. Human subjects chosen should represent the intended future user group. The key point to a successful test is having the subject speak out loud about what he/she feels when operating the system. This method is promising. However, there is no guarantee that the performance generated from a first time, typically 2 h, monitored test could be applied to the performance in repetitive, long term, and random uses of the system.

The test also requires more resources than other methods. Moreover, it is not convenient and may be expensive to get subjects to come to the testing laboratory especially if the subject is at a remote location.

With its evident pros and cons, nevertheless, usability is still playing a steadily more important role in software development. This can be proved by the growing budgets for usability engineering. According to Nielsen [23], in 1971 it was estimated that a reasonable share for usability budgets for nonmilitary systems was only about 3%. However, in 1993, Nielsen surveyed 31 development projects and found that the median share of their budgets allocated to usability engineering was 6%. Nevertheless, in terms of percentage growth rate, it looks like 3% is not much. But if you try to image how much money per year has been spent for software development in business, the amount is actually huge. There are also new companies solely working on the usability testing and provide services for companies who do not have their own lab so as to use as an outsourcing service.

10.2.2 Expert Reviews

For some institutions with more web design experts available, expert reviews are a good alternative to informal and random user feedback. They have proved to be effective. Expert reviews can be conducted in the markup, designing, or refinement phase. It is also feasible to setup a meeting to examine the web design on a regular basis. The goal could be problem identification such as missing links, bugs, layout, style, wording, typos, inconsistent control operations, inappropriate graphics or multimedia use, slow downloading, browser plug-ins, expired database connection, or recommendations for future improvement. One or two outsiders should also be included in the expert panel to increase the diversity. Web design is more art than science. The design team should be open to any question the expert panel may have. There is no "silly" question, because sometimes it could lead to an unexpected solution. Shneiderman and Plaisant [24] categorized a variety of expert-review methods as: heuristic evaluation, guidelines review, consistency inspection, cognitive walkthrough, and formal usability inspection.

10.2.3 Heuristic Evaluation

Indeed some Web design principles have their theoretical standpoints and could be useful in the practice. However, design does not provide a scientific means for evaluating the interface in terms of usability. Shneiderman [23] commented that "He [Lynch] [25] has sorted out the issues better than most but still leaves designers very much in the dark about what to do." Thus, it is obvious that a more scientific way of evaluating a web page is demanded. Nielsen in SUN Microsystems [26] proposed the most well-known website evaluation method, which is the heuristic evaluation method. Although other researchers have attempted to generate universal guidelines, they still apply heuristic evaluation methods as their fundamental approach.

Nielsen and Sano [27] conducted two studies proposing the heuristic evaluation method. Heuristic evaluation is used to examine an interface that tries to generate an opinion about what is good and bad about the interface. It requires a set of evaluators, who are used to examine the interface and judge its compliance with recognized usability principles. Nielsen and Mack [28] suggested what a typical set of heuristics should include: use simple and natural dialogue, speak the user's language, minimize user memory load, be consistent, provide feedback, have clearly marked exits and shortcuts, have good error messages, prevent errors, and help documentation. The number of evaluators depends on a cost-benefit analysis. Nielsen [26] implemented a study trying to depict the relationship between the number of evaluators and the proportion of usability problems found. According to his research, five is most cost beneficial and three is the minimum. Evaluation participants, normally users, will inspect the interface using each individual evaluator, and finally the results will be aggregated.

10.2.4 Empirical Guidelines

The development of empirical guideline started in the late 1990s. Borges, Morales, and Rodriguez [29] criticized the usability laboratory testing method for its requirement of running a lab usability test, which is time consuming and not practical for most web page designers. Instead, they adopt heuristic evaluation as their approach, implement user testing, and try to generate universal applicable guidelines for designing web pages. From an experiment conducted with two versions of home pages of three websites, randomly selected from 10 colleges and universities, they compiled a list of guidelines for designing web pages. On the other hand, W3C has actively promoted the use of style sheets on the web since the Consortium was founded in 1994. Right now, W3C even provide CSS Validation Service online, a free service that checks CSS in (X)HTML documents or standalone for conformance to W3C recommendations. Unfortunately, what CSS cannot do is to help designers develop user-friendly web pages or detect critical layout flaws. It is not used to investigate how and why an individual design component affects a user's perception as well as the way each design component interacts with each other. Table 10.1 demonstrates the empirical guidelines proposed by Borges, Morales, and Rodriguez in 1996.

10.2.5 Questionnaire Surveys

Questionnaire surveys are widely used. They are convenient, inexpensive, generally acceptable, and less biased because of the typically large number of participants. A survey questionnaire should be prepared, refined, reviewed among the development team, and pilot tested with a small sample of users before a large-scale survey is conducted. Statistical analyses should also be developed before the final survey is distributed. Because this experiment was conducted over the web, an on-line survey is a good approach. Designers can provide a hyperlink and bring users directly to the survey form right after they have been through a particular program. Many people prefer to answer a brief survey displayed on a screen, rather than to write on a printed form and mail it back to the surveyor. Users can express their opinions by selecting checkboxes using scales such as: (1) strongly agree, (2) moderately agree, (3) slightly agree (4) neutral, (5) slightly disagree (6) moderately disagree, and (7) strongly disagree. Moreover, by carefully designing a step by step error-checking procedure for participants to fill out the questionnaire, the problem of missing data, which is usually a time-consuming job for the designer to clean up, could be solved.

10.2.6 Factorial Design Experiment

Factorial experiments investigate the effects of two or more factors or input parameters on the output response of a process [30]. Factorial experiment design, or simply factorial design, is a systematic method for formulating the steps needed to successfully implement a factorial experiment. Estimating the effects of various factors on the output of a process with a minimal number of observations is crucial to being able to optimize the output of the process. In a factorial design, the effects of varying the levels of the various factors affecting the process output is investigated. Each complete trial or replication of the experiment takes into account all the possible combinations of the varying levels of these factors. Effective factorial design ensures that the least number of

Table 10.1   Example of Empirical Guidelines

	For Any Page	
1.	Headers should not take more than 25% of a letter-size page	
2.	Headers and footers should be clearly separated from the body of the page	
3.	Names of links should be concise and provide a hint of the content of the page they link to	
4.	Avoid adding explanatory comments to textual links	
5.	Avoid linking mania (making a link every time a keyword of a page is mentioned in a text	
6.	Verify that links connect to existing pages	
7.	Linking icons should have a distinctive feature of the page they link to	
8.	Maintain consistency when using icons. The same icon should be used for the same purpose	
9.	Colors should be selected so that the page can be clearly displayed and reproduced on black and white displays and printers	
10.	It is desirable to include the date of when the page was last modified, the mail address of the person who maintains the page, and the Uniform Resource Locator (URL) address of the page on a footer	
	For the Home Page of the Repository	
11.	Descriptive information about the institution should be placed just below the header and kept to a minimum: a link to a secondary page is preferable	
12.	Pages should not be overcrowded with links	
13.	Pages should be short (about a letter size page).	
14.	Links should be to primary aspects or characteristics of the institution. Textual information should be left for secondary pages	
15.	Organize links as primary and secondary topics	
16.	Links to resources or other repositories on the Internet should be placed on a secondary page. This page should be reached with a link on the primary page	
17.	A more extensive index of links, properly grouped, can be provided on a secondary page for fast access to a wide range of the institution's repositories	

experiment runs are conducted to generate the maximum amount of information about how input variables affect the output of a process.

Factorial design experiment has been further developed and successfully adopted in the domain of quality engineering. The application on quality engineering is called Robust Design method. This method is also called the Taguchi Method [31], which is pioneered by Dr. Genichi Taguchi. The major purpose of quality engineering is to improve engineering productivity by statistically controlling the manufacturing processes. By consciously considering the noise factors (environmental variation during the product's usage, manufacturing variation, and component deterioration) and the cost of failure in the field, the Robust Design method helps ensure customer satisfaction. Robust Design focuses on improving the fundamental function of the product or process, thus facilitating flexible designs and concurrent engineering. Therefore, it is recognized as the most powerful method available to reduce product cost, improve quality, and simultaneously reduce development interval [32].

The method originally developed after World War II has evolved over the last five decades. Many companies around the world have saved hundreds of millions of dollars by using the method in diverse industries: automobiles, xerography, telecommunications, electronics, software, etc. According to Brenda Reichelderfer of ITT Industries, it was reported on their benchmarking survey of many leading companies, "design directly influences more than 70 perent of the product life cycle cost [32]; companies with high product development effectiveness have earnings three times the average earnings; and companies with high product development effectiveness have revenue growth two times the average revenue growth." She also observed, "40% of product development costs are wasted!" Though, we do not have specific statistics for software development, it is obvious that a huge amount of effort has been wasted in user interface design process.

If we were to treat the designing of any computer user interface as an industrial manufacturing process, then it is feasible to adopt factorial design experiment to reduce interface design cost, improve quality, reduce software development interval, and most importantly increase user satisfaction. Ideally, any computer user interface could be decomposed into a number of independent components. In order to apply factorial design method, statistically these independent components have to be mutually exclusive to each other. These independent components are the factors in a factorial design. The most difficult part to apply factorial design as a method to investigate or optimize a specific user interface is the defining of levels. In a factorial design, each factor has two distinct levels: low and high. The investigator has to subjectively decide for each specific factor what level of settings are categorized as low and high. The difficulty exists because one may argue that the design of user interface is a form of both arts and science, and that the overall performance of a specific user interface could not be judged by such decomposing. Indeed, the preceding argument is definitely true that one should not take the experiment result of this approach as the final verdict for any well-designed user interface. It is absolutely not our intention to kill creativity by applying factorial design experiment. However, we do believe this approach could help a lot in preventing poor design and providing clues for finding the optimized design just like most of the evaluation methods developed. This method has the advantages of being comprehensive, quantitative, repeatable, systematic, and more cost effective than using a usability testing laboratory, especially for a small number of projects.

10.2.7 Comparison of Evaluation Methods

Each evaluation method has its pros and cons. Although empirical guidelines provide clear instructions that give designers a direction to follow, those guidelines are not universally applicable or exhaustive. For example, the guidelines say nothing about web projects that contain multimedia components such as animation or streaming video. However, they do include many considerations that should be included in a general assessment. Thus, we would suggest the empirical guidelines should be used as a preliminary assessment method, but need to be customized or expanded for examining pages with different features.

Second, usability testing laboratories are a scientifically good method, but they require more time and resources to implement. Other than large funded institutes that are well equipped and regularly doing usability laboratory tests, most web designers do not have an access to the equipment or devices that would be appropriate to implement a reliable usability laboratory test. They would either need a huge investment or seek outsourcing. From the perspective of cost effectiveness, we will not adopt this method as our approach.

Expert review seems a good trade-off between a usability lab test and a relatively narrow approach of empirical guidelines. However, we also run into a problem of finding experts. For an institute with many web developers, it is possible to get a panel discussion. However, except in some well-established organizations, designers normally cannot find experts.

Besides the convenience and low cost, the web survey has more advantages over other methods. The web survey is dynamic. Researchers can use an expert system and design online questionnaires tailored to respondents' particular situations. The program will automatically skip and redirect the respondent to the next question. One researcher found that there is less missing data using the web survey compared with mail and e-mail surveys.

Though there are six evaluation methods being discussed separately, research could adopt a combination of evaluation methods together depending on budget, time, availability of facility, and resources. Based on the preceding comparison and under the constraint of available resources, we decided to use a combination of factorial design method and online survey as our assessment method in this study. A comparative critical analysis of evaluation methods is summarized in Table 10.2.

10.3 Research Hypothesis

Based on human information processing theories and other empirical studies in the field of cognitive psychology, our fundamental hypothesis is that the higher the levels of interaction the more effective the interface, where effectiveness is defined as the level of recognition procedure it can reach. The more interactions the more likely the information would be stored in the longterm memory instead of temporarily held in the memory buffer. We are also assuming that a high level of interaction increases the ease of use. However, we cannot make the assumption that satisfaction will increase when more features were added in this program. Moreover, three independent variables, video transcription, control panel, and hyperlink, contribute in different aspects and levels on user satisfaction. Thus, the results should be examined separately.

We should also keep in mind that the differences in performance might be due to the use of different presentations, not to the dynamic nature of the presentation of video. Moreover, Williams [33] suggests that although combining visual and auditory information can lead to enhanced comprehension, it is also possible that having both visual and audio modes may result in no performance improvements, and may or may not increase user satisfaction. The following are the experimental hypotheses. User interface satisfaction is defined by the following four dimensions: user reactions (UR), layout and content (LC), video use (VID), and quality of multimedia (QAV).

- User interface satisfaction will be higher in the experimental designs with hyperlinks and their associated information
- User interface satisfaction will be higher in the experimental designs with multimedia control panels
- User interface satisfaction will be higher in the experimental designs with video transcriptions

Table 10.2   Pros and Cons of Evaluation Methods

Evaluation Method	Advantage	Disadvantage
Usability laboratory	• Observer can inspect users' movement in specific details • Able to review and investigate further	• Requires more equipment and resources • Not suitable for a diverse base of users • Normally a small size of sample users, may not cover all the details • First time test may not apply to the result of using the program in a regular manner • Monitored lab behavior may not reflect home usage pattern
Heuristic evaluation	• Does not require experiment hardware • Easy to perform and generate reports	• Requires human resources • May not cover all aspects of the program • Expert users cannot represent the behavior of average users, the result may be biased
Expert review	• Professional feedback • Qualitative analysis oriented	• Qualification of experts is ambiguous and arguable • Availability of experts • Cost might be expensive or not affordable • Not comprehensive • Subjective
Empirical guidelines	• Clear direction • Convenient • Low cost	• Subjective • Not universally applicable • Lack of multimedia evaluation guidelines • Needs to be updated frequently
Web-based questionnaire survey	• Convenient • Objective and inexpensive • Represents general users' behavior • Able to implement remotely • May have missing data but relatively larger sample size than other methods, less biased • Advantages over paper format of survey	• Does not have specific details of how the user interacts with the program • Cannot provide prompt help to users not familiar with how to use the program

(continued)

Table 10.2   (continued) Pros and Cons of Evaluation Methods

Evaluation Method	Advantage	Disadvantage
Factorial experiment	• Comprehensive • Objective and inexpensive • Represents general users' behavior • Able to implement remotely • Repeatable • Provides clues for optimized performance	• May not be able to decompose a user interface into independent components • May have difficulties in defining two distinct levels • Not cost effective if simultaneously investigated by many interface components

10.4 Methods

10.4.1 Experimental Design

To achieve our goal, a factorial design method was employed in this experiment [30,34]. A factorial design is used to evaluate two or more factors simultaneously. The treatments are combinations of levels of the factors. With factorial designs, we do not have to compromise when answering these questions. We can have it both ways if we cross each of our two times in instruction conditions with each of our two settings. The advantages of factorial designs over one-factor-at-a-time experiments is that they are more efficient and they allow interactions to be detected. Factorial design has several important features. First, it has great flexibility for exploring or enhancing the signal (treatment) in our studies. Whenever we are interested in examining treatment variations, factorial designs should be strong candidates as the designs of choice. Second, factorial designs are efficient. Instead of conducting a series of independent studies, we are effectively able to combine these studies into one. Finally, factorial designs are the only effective way to examine interaction effects.

The factors and their levels are given in Table 10.3. Factor A represents video transcription; B represents video control panel; and C represents hyperlink correspondently. These factors were chosen because they represent important features of interactive technologies discussed earlier. The definition of a hyperlink is as follows:

A reference (link) from some point in one hypertext document to (some point in) another document or another place in the same document. A browser usually displays a hyperlink in some

Table 10.3   Definition of Factors and Levels

Factor	Low Level (−)	High Level (+)
Text (A)	Present video only	Present video and transcription
Control (B)	No control panel, video will start automatically	Fully functional control panel, provides stop, pause, play buttons, plus progress bar, and volume adjustment, requires user control to start video
Link (C)	No hyperlink and their associated hypertext	With hyperlink and their associated hypertext

Table 10.4 Full 2 by 3 Factorial Design and Effects

Order	Factor A	Factor B	Factor C	Effects
1	−	−	−	(1)
2	+	−	−	A
3	−	+	−	B
4	+	+	−	AB
5	−	−	+	C
6	+	−	+	AC
7	−	+	+	BC
8	+	+	+	ABC

distinguishing way, e.g. in a different color, font or style. When the user activates the link (e.g. by clicking on it with the mouse) the browser will display the target of the link.

In order to test lack of hyperlink effect, in this experiment, a hyperlink includes not only the reference link but also the associated hypertext. Thus, web pages with hyperlinks in this experiment will provide both a higher level of interactivity and more information.

Hyperlinks provide higher interactivity and also represent a critical feature of web navigation. The video control panel provides users the control they need for using web multimedia. The control features such as stop, play, and pause also provide higher temporal flexibility and allow self-pacing. Based on human information process theory [13,22], the more senses we use to acquire information the more effective it will be. The addition of video transcription is supposed to create a higher level of multimodality. A complete 2 by 3 factorial design was employed to implement this experiment. Every design includes a streaming video plus a different combination of other three factors we are interested in. Experimental design for each design is given in Table 10.4.

10.4.2 Three Variant Interface Design Factors

The without video transcription version provides nothing but the video presentation. For the with video transcription version, the text content of the video presentation was provided under the video window. Users could read the text transcription while they were watching the video.

For the version without the video control panel, there is no media control panel on the presentation window. Users could not control the streaming video in any way either to start, pause, stop the video presentation, or to change the audio volume. The design example is illustrated in Figure 10.2. But in the full functional control panel version, we provided the following control functions: play, pause, stop, a meter bar for fast moving, and a voice volume control button. Therefore, users had a full control of the pace they wanted. The popular streaming video software "Real Player" was used, but the control panel was customized by the experiment designer. The design example is illustrated in Figure 10.3.

For the with hyperlink version, each advanced level word has a hyperlink in the short article on the web page. For example, the word "ontology" is considered as an advanced level word for

Figure 10.2   Interface design without video control panel.

Figure 10.3   Interface design with video control panel.

English readers as it is not normally used in daily life. Therefore, such words were embedded with hyperlinks. If a user clicks on the hyperlinked word, another window will pop up to show the explanation of it. On the other hand, in the without hyperlink version, experiment participants were given a short article to read that included those advanced level words but without any built-in hyperlink. As previously discussed, since we cannot break the compound effect, this factor is to test both the impacts of lack of hyperlinks as well as insufficient information associated with them.

10.4.3 Experiment Implementation

10.4.3.1 Subjects and Laboratory Procedure

The subjects were recruited from the University of Wisconsin at Madison (UW-Madison). The experiment was implemented individually in the research laboratory of research center at the UW-Madison. The lab has been divided as a number of cubicles with a standard PC on the desktop. Before implementing the experiment, each subject was given a study number, randomly assigned to one of those eight designs with different combinations of factors. There are equal numbers of subjects in each design. A total of 80 UW-Madison students participated and completed the experiment, which means there are 10 identical runs for each of the 8 different interface designs.

There is no missing data due to the error-checking design of a web-based online survey. If there was a missing data, the program would ask the subject to fill it out again.

The subjects were recruited through on-campus e-mail lists managed and maintained by the UW-Madison information division. The e-mail lists allowed us to have access to the subgroups within the university. Respondent recruiting e-mails were sent to the students in the following schools: engineering, letters and science, business, medical sciences, agriculture, and life sciences. To motivate students to participate in this study, we offered $10 reward to participants who had successfully completed the online experiment and survey questionnaires. All recruited subjects first filled out the demographics questionnaires. Then they were directed to experience the program. Finally, they filled out the post-test questionnaires.

10.4.3.2 Survey Instrument and Data Collection

Table 10.5 presents the construct of variables in this experiment. Each question has a preference measure from one to nine. We modified the user interaction satisfaction (UIS) questionnaire, which was originally developed by the University of Maryland, as our measuring instrument. Following are the dimensions we covered in the post-test questionnaire:

- User reaction (interactivity, general impressions, perceptions)
- Content (easy to understand, learning, error), layout (page sequence and arrangement)
- Use of the multimedia (helpful, satisfying, stimulating, easy to use, pleasant to use)
- Quality of the multimedia (quality of pictures, audio or video, sound output)

User reaction questions were written to measure the subjective feelings after users had been through the program. Some of questions are related to interactivity, as we wanted to find out if users found this program interesting and stimulating. Content and layout questions were written to measure if users found the content, layout, and the amount of information was appropriate. Because this is a web-based multimedia program, the use of the multimedia question was written to measure if users found it easy to use the multimedia. Due to bandwidth limitations, we were not able to provide the best quality of pictures and sounds we could have provided. Qualities of the multimedia questions were written to measure if users were satisfied with the multimedia quality. We also examined user previous experiences on using computers. One of the reasons that people feel frustrated owes to lack of experiences using technology or computers. In the pre-test phase, we emphasized on the following areas: past experience of using computers (using operation system and Internet, installing and using computer software) and demographics (education, age, ethnic group, familiarity with Internet).

10.4.3.3 Evidence of Validity and Reliability

Factor analysis attempts to identify underlying variables, or factors, that explain the pattern of correlations within a set of observed variables. The factor analysis model specifies that variables be determined by common factors and unique factors. The computed estimates are based on the assumption that all unique factors are uncorrelated with each other and with the common factors.

There are four dimensions we covered in the post-test questionnaire of user satisfaction survey. They are user reaction, content and layout, use of the multimedia, and quality of the multimedia. The extraction method we used to implement the factor analysis is principal component analysis.

Table 10.5 Construct of Variables

Variable	Construct (Measure: Scale 1–9)
UR_1	User reactions: Terrible/wonderful
UR_2	User reactions: Frustrating/satisfying
UR_3	User reactions: Dull/stimulating
UR_4	User reactions: Difficult/easy
UR_5	User reactions: Rigid/flexible
LC_1	Web page layouts were appropriate: Never/always
LC_2	Amount of information that can be displayed in this program : Inadequate/adequate
LC_3	Arrangement of information can be displayed in this program: Illogical/logical
LC_4	Next screen in a sequence: Unpredictable/predictable
LC_5	Do you consider the design of this program user friendly? Never/ always
LC_6	Information in this program is easily understood: Never/always
VID_1	The use of web streaming video is: Helpless/helpful
VID_2	The use of web streaming video is: Frustrating/satisfying
VID_3	The use of web streaming video is: Dull/stimulating
VID_4	The use of web streaming video is: Difficult/easy
VID_5	The use of web streaming video is: Annoying/pleasant
QAV_1	Quality of movies: Fuzzy/clear
QAV_2	Quality of movies: Dim/bright
QAV_3	Sound output: Choppy/smooth
QAV_4	Sound output: Garbled/clear

The varimax with Kaiser Normalization rotation method was also employed. A rotation converged in six iterations. The number of extraction is four. It was decided by the proportion of variance explained by the factors, subject matter knowledge, and reasonableness of the solution. As shown in Table 10.6, the factor analysis results confirmed that the questions grouped into four dimensions. These four dimensions are exactly what was expected.

Reliability was assessed for the user satisfaction questionnaires in this study. For a total of 22 survey questions, the overall reliability coefficient Alpha is 0.898. Every scale seemed to be reliable. The estimated use of the multimedia was quite reliable ($r \geq 0.90$). The estimated user subjective reaction ($r = 0.872$), content and layout ($r = 0.841$), and quality of the multimedia ($r = 0.836$) was also reliable. The scale for each question is from 1 to 9. There is no question that needs to be reversed.

There are four dimensions we covered in the post-test questionnaire of user satisfaction survey. They are user reaction, content and layout, use of the multimedia, and quality of the multimedia.

Table 10.6 Factor Analysis

Variables	Video Use	Content Layout	User Reaction	Multimedia Quality
UR_1	.327	.382	.574	−.215
UR_2	.378	.145	.792	.051
UR_3	−.018	.293	.707	−.018
UR_4	.056	.261	.853	.105
UR_5	.057	.204	.829	−.090
LC_1	.221	.627	.342	.136
LC_2	.009	.591	.330	.062
LC_3	.202	.791	.143	−.036
LC_4	.090	.732	.177	−.032
LC_5	.308	.797	.239	.178
LC_6	.222	.583	.124	.094
VID_1	.795	.375	.016	.198
VID_2	.868	.157	.082	.182
VID_3	.850	.150	.249	−.017
VID_4	.787	.130	.062	.299
VID_5	.898	.201	.122	.097
QAV_1	.199	.025	.021	.831
QAV_2	.099	−.033	.027	.810
QAV_3	.131	.149	−.061	.755
QAV_4	.081	.077	−.023	.820

The extraction method we used to implement the factor analysis is principal component analysis. We also implemented the Varimax with Kaiser Normalization rotation method. A rotation converged in six iterations. The number of extraction is four. It was decided by the proportion of variance explained by the factors, subject matter knowledge, and reasonableness of the solution. Though not demonstrated, an eigenvalue plot was used in visually assessing the importance of factors. The optimal number of factors is where the eigenvalue equals to one. It indicated that the extraction of four principal components is appropriate.

10.4.3.4 Demographics

Among the total of 80 subjects, 75% of subjects are Caucasian, 21.25% Asian, and 3.75% Hispanic. Eighty eight percent of subjects expressed they are either extremely or very familiar with the Internet. Nine percent of subjects are familiar with the Internet, and only 3% of subjects felt a little familiar with the Internet. Ninety four percent of subjects expressed they are either in excellent or good health, and only 6% of subjects expressed their health status is fair. Students coming from the engineering school dominated the subject population of 65%. Letters and science students account for 24%. Only 3% of the subjects are from business school and the rest are 8%. Moreover, it is relatively safe to assume that engineering students have better understanding of using computer technologies.

10.5 Result and Analysis

10.5.1 Descriptive Statistics

Table 10.7 shows the details of descriptive statistics for interface survey results. The total sample size N is 80 that satisfies the power analysis requirement. The sample average is 7.158. Since the scale is from 1 to 9, in general users have positive reactions to this program. It is slightly higher than the pilot test result, which is 6.683. Based on the written feedback we received, many users agree that this is a creative way to present breast cancer information and the program is very educational. The standard deviation is 0.921, which is smaller than the pilot standard deviation 1.42 due to the increasing sample size.

10.5.2 Main Effect Analysis

A main effect is an outcome that is a consistent difference between levels of a factor. Table 10.8 shows the results of main effect analysis. It is interesting to find out that there is a negative significant effect on the UIS in the layout and content dimension when a design has a video transcription added with the video presentation ($LC_5 : t = -2.38, p = .02$). This means if video transcription was added below the video presentation window, it is considered not "user friendly" by the experiment participants. There are two possible explanations for such unexpected outcomes. First, many websites now have online videos, but most of them do not have video transcription. It may contradict users' mental model that with video transcription is an odd design. From the demographics statistics, it also indicates that every respondent participating in this experiment has previous experiences using the Internet. A total of 88% of subjects feel they are either extremely or very familiar with the Internet. The use of interface itself is a learning process, and the learned behaviors of using a specific format of interface could become a habit, which naturally and automatically leads users to be in favor of a design once users get used to it. The other explanation is video transcription does become a visual obstacle when users tried to focus on the video presentation. Moreover, the content of video transcription is exactly the same as the video; it might be considered redundant by the users when they were watching the video.

Table 10.7 Descriptive Statistics

Sample Size	Mean	Median	Std Dev
80	7.158	7.264	0.921
Minimum	Maximum	Quantile 1	Quantile 3
3.136	8.409	6.818	7.773

Table 10.8 Significant Main Effects

Response	Factor	P-Val[a]
LC_5: Do you consider the design of this program user-friendly?	A (−)	.02
LC_6: Information in this program is easily understood?	C (+)	.033
VID_2: The use of web streaming video is frustrating or satisfying?	B (+)	.036
VID_4: The use of web streaming video is difficult or easy?	A (−)	.035
QAV_3: Sound output is choppy or smooth?	A (−)	.01

[a] Critical value $\alpha \equiv 0.05$.

The result from LC_6 ($t = 2.17, p = .033$) shows that hyperlinks and their associated hypertexts are critical and helpful for users to understand the web page content. Again, it reconfirms the importance of associating related information with appropriate hyperlink, though it is regarded as a poor design if using an overwhelming number of hyperlinks in a single web page. Moreover, lack of sufficient information has led to negative results as expected. It also indicates that the audience really cares about the information provided and wants to learn more. Nowadays large corporations normally use outsourcing services for web interface design, but for individuals it is recommended to better know the targeting audience before designing web sites to avoid complaints.

When a design includes video control panel, it is significantly satisfactory for users (VID_2: $t = 2.14, p = .036$). A video control panel not only lets users control the pace of watching but also provides a choice to review the video when they hope so. This is an interesting finding that explains why control is important when people are browsing. People want to have control over the content they receive rather than passively given the information. The ability of controlling is a key part of interactivities. Such behavior is similar to people who like to use a remote control to switch channels when they are watching TV even there is nothing particularly of their interest.

On the contrary, results from VID_4 ($t = -2.15, p = .035$) and QAV_3 ($t = -3.56, p = .001$) show us a totally different story. When an interface has video transcription, it has a negative effect on the use of streaming video. It is considered difficult to use the streaming video if we provide video transcription under the video window. Moreover, the sound quality degrades when video transcription is added. As we can see from previous discussions, video transcription has a negative effect on user friendliness. This once again confirms that using video transcription is not a good idea when designing a web multimedia program.

10.5.3 *Interaction Effect Analysis*

An interaction effect exists when differences on one factor depend on the level you are on another factor. It is important to recognize that an interaction is between factors, not levels. Table 10.9 summarizes the results of significant interaction effects for interface preference survey. An interaction between factors occurs when the change in response from the low level to the high level of one factor is not the same as the change in response at the same two levels of a second factor. That is, the effect of one factor is dependent upon a second factor. Notice the result of UR_1, hyperlink alone has a positive effect ($t = 2.24, p = .029$) on overall user reactions (terrible / wonderful), but when a control panel was added in the design, it created a negative effect ($t = -2.49, p = .015$). This means if a control panel and hyperlink were added together, the interface would be considered more

Table 10.9   Significant Interaction Effects

Response	Factor[a]	P-Val[b]
UR_1: User reaction: terrible or wonderful?	C (+)	.029
	B × C (−)	.015
UR_5: User reaction: rigid or flexible?	B × C (−)	.041
LC_2: Amount of information that can be displayed in this program is inadequate or adequate?	A (−)	.018
	C (+)	.001
	A × B (+)	.001
VID_1: The use of web streaming video is helpless or helpful?	A (−)	.014
	A × B (+)	.047
QAV_1: Quality of movies: fuzzy/clear?	B × C (+)	.038

[a] Sign × means interaction.
[b] Critical value $\alpha \equiv 0.05$.

terrible. A possible explanation is that both control panel and hyperlink are related to user actions to control what they want to see. Thus, if we put them together it might make the interface too complicated and users might get confused among too many actions they need to take to accomplish the browsing task.

The response UR_5 is about if a user feels the design of this program is rigid or flexible, which is related to interactivity. Similar to the preceding result, the result of UR_5 ($t = -2.08, p = .041$) also indicated if control panel and hyperlink were added together, the interface would be considered more rigid. Though in this case hyperlink does not show a significant effect, the interaction effect between hyperlink and control panel indicates a similar pattern as UR_1. This once again confirms that if we add hyperlink and control panel together within an interface, it has a significant negative effect on the interactivity.

From LC_2 results, the design with only hyperlink factor and the other design with both video transcription and control panel demonstrates positive effects on the appropriate amount of information that can be displayed in this program ($t = 3.47, p = .001$)($t = 3.55, p = .001$). This indicates how users feel strongly that without having hyperlinks the amount of information provided in this program is inadequate. However, the design with only video transcription has a negative effect ($t = -2.42, p = .018$) on users, perception of adequate amount of information that can be displayed in this program. Respondents not only dislike the interface with video transcription but also consider video transcription demos as inadequate amount of information.

For using web streaming video VID_1, video transcription alone has a negative effect ($t = -2.52$, $p = .014$). But if we add a control panel with it, it significantly improves the result ($t = 2.03$, $p = .047$). This suggests that it is better not to add a video transcription when using web video but if we must have video transcription we probably should also add a control panel with it. Respondents also felt the quality of movies increases when both video transcription and control were added to the design ($t = 2.11, p = .038$).

10.5.4 Three Dimensional Analysis

So far we have demonstrated both the individualized main and interaction effect analysis. However, it seems there is no single design factor that could dominate the user satisfaction experiment. Indeed, the above analysis has indicated that interface design is both science and art. Consequently, one might ask which one among those eight designs is the best? Is there an optimized design? In order to present the overall effects with regard to those three design factors, 3D graphic technique was used in Figure 10.4.

According to the 3D cube plot, the design with only hyperlink included (−1, −1, +1), data means = 7.45, is the best overall performer. The worse performance comes from the design combination (+1, −1, −1), data means = 6.514. This seems a little surprising in the first place, however it makes sense as we look more deeply. First, the result is consistent for both the best and worst cases because the best is just on the opposite corner of the worst. This means participants are most satisfied with the interface than with hyperlinks embedded but without the other two components being available. On the other hand, participants feel most frustrated with the interface than without hyperlinks but with video transcription and advanced control panel.

When we were designing a web page, usually we were aware that we should put a hyperlink wherever or whenever we "feel" we should. However, we did not realize how important it is to the users and how much difference it would make. Moreover, we often have doubts if it is really necessary to have a hyperlink or even argue about where to put it. The design of a web page is mostly based on intuition. Even within a web design team, it is not unusual everyone has different preferences. We also have no idea that if hypertext is missing will users really care about it. What happens if we do not provide a hyperlink when users need it? There exists a gap between web developers and users. From this research, it is clear that we found some answers to these questions.

Moreover, the second best performance comes from the design combination (+1, −1, −1), means = 7.40. It indicates that even without hyperlinks, participants felt the interface good enough if there is a video control panel. From TV study, we know that people enjoy using controllers to switch channels. We also cannot deny that the function of a web browser has gone beyond what it

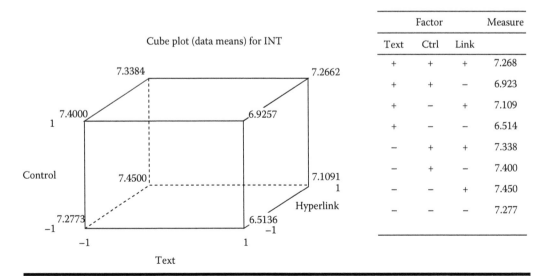

	Factor		Measure
Text	Ctrl	Link	
+	+	+	7.268
+	+	−	6.923
+	−	+	7.109
+	−	−	6.514
−	+	+	7.338
−	+	−	7.400
−	−	+	7.450
−	−	−	7.277

Figure 10.4   3D cube plot of overall effects for three factors.

was designed for. With Internet bandwidth obstacles being overcome, people now use web browsers not just to acquire information, online shopping, but for entertainment as well. A web browser provides more freedom to control than a TV. A higher degree of freedom to control contributes to the success and popularity of web multimedia. In this experiment, we found that providing more control options could significantly ease the frustration from lacking sufficient hyperlinks.

10.6 Discussion and Lesson Learned

Hyperlink is a critical element for web browsing. People will feel frustrated if you do not add it when they need it. In this experiment, we found that our subjects feel bad if the need to explore advanced level words is not satisfied. The results indicate that those subjects do care about hyperlinks provided in this experiment. The use of control panel does play an important role when people try to use streaming video on the web, but in general it does not have significant impact. Experiment results indicate it is not a good idea to add video transcription with web video. Though there is no proof, such behavior is similar to people's reactions to TV or movie video transcription; people found it interfering when watching except for people who have difficulty watching the audiovisual format of content and need to read the video transcription to understand the content. When we go to a movie theater, there is no video transcription but we never hear people complain about it. However, if the average health status of the target audience is below the average of the general public, it might be helpful if we add video transcription upon request.

History does repeat, TV was a great invention and new technology in the 1960s, as it can present images rather than just an audio that radios can provide. It was also a very popular research subject in the field of mass communication. The 1960s witnessed a marked change as a positivistic-empirical approach overtook the dominant position of mass communication studies for a decade. As for the research orientation, there were no significant regional differences. For instance, the researchers from the departments of sociology, psychology, education, TV and mass communication of four different universities were summoned to join the first large-scale audience study, and shared a similar view of mass communication research. Now, when a web browser is capable of providing a variety of media formats and absolute freedom to control, it seems predictable that the usage of web browser is going to win over TV. Consequently, the study of web interface is supposed to have a similar position a TV had. Optimistically, though a totally different subject matter, this reminds us that a lot of research ideas could be inspired from TV study.

Lastly, obviously we did not include all web components in this experiment. Ideally, we should exhaust every interaction pattern and explore its effect so as to have a complete understanding of web interactivity and different effects due to different designs. Due to the time and cost restrictions, the sample size is limited. Thus, a larger sample size is required in order to enhance the power of statistical analysis. We hope that in the future similar researches could be done to further investigate this issue in more detail such as the best number of hyperlinks to put on a web page and in what aspects they will have impacts. Other interaction patterns such as animation, voice recognition, and linear or nonlinear browsing should be also investigated in detail in the following studies. Moreover, the definition of hyperlink includes not only the interaction but also the information associated with hypertext. We will be interested to see the effect generated by interaction.

Acknowledgments

This experiment was part of a research project funded and supported by the Center for Health Systems Research and Analysis in the UW-Madison. The original purpose of this study was to find an optimized design to better help people with physical disabilities or living remotely acquiring information via multimedia on the web for making medical treatment decisions. I also would like to thank Professor David H. Gustafson and Robert Hawkins in UW-Madison for their insights, kind assistance, and helpful comments.

References

1. R. L. Street, W. Gold, and T. Manning, *Health Promotion and Interactive Technology: Theoretical Applications and Future Directions* (Lawrence Erlbaum Associates, Mahwah, NJ, 1997).
2. S. Palmiter, *Journal of Visual Languages and Computing* **4**, 71 (1993).
3. J. Steuer, *Journal of Communication*, **24**(4), 73–93 (1992).
4. T. W. Malone, What makes things fun to learn? A study of intrinsically motivating Computer games. Technical report, Xerox Palo Alto Research Center, Palo Alto, CA, (1980).
5. S. Tindall-Ford, P. Chandler, and J. Sweller, *Journal of Experimental Psychology: Applied* **3**, 257 (1997).
6. J. T. Stasko, Animation in user interfaces: Principles and techniques, in *Trends in Software*, Special issue on User Interface Software (John Wiley P. Dewan & Sons, New York, 1993), pp. 81–101.
7. W. Langston, *Memory and Cognition* **26**, 247 (1998).
8. J. Preece, *Human Computer Interaction* (Addison-Wesley, New York, 2002).
9. D. E. Broadbent, Stimulus set and response set: Two kinds of selective attention, *Attention: Contemporary Theory and Analysis* (Wiley, New York, 1970).
10. R. Moreno and R. E. Mayer, *Journal of Educational Psychology* **92**, 117 (2000).
11. R. E. Mayer and R. Moreno, *Educational Psychologist* **38**, 43 (2003).
12. D. E. Broadbent, *Journal of Experimental Psychology* **47**, 191 (1954).
13. D. E. Broadbent, *Acta Psychological* **50**, 153 (1982).
14. D. E. Broadbent and M. Gregory, *Proceedings of the Royal Society of London (Biology)*, **68**, 81–93 (1967).
15. M. C. Potter and E. I. Levy, *Journal of Experimental Psychology* **81**, 10 (1969).
16. R. S. Nickerson, *Canadian Journal of Psychology* **19**, 155 (1965).
17. R. N. Shepard, *Journal of Verbal Learning and Verbal Behavior* **6**, 156 (1967).
18. R. N. Haber, *Scientific American* **222**, 104 (1970).
19. A. Paivio, *Imagery and Verbal Processes* (Holt, Rinehart, and Winston, New York, 1971).
20. A. Paivio, *Mental Representations: A Dual Coding Approach* (Oxford University Press, Oxford, England, 1986).
21. J. R. Anderson, *Cognitive Psychology and Its Implications* (Worth Publishers, New York, 2005).
22. M. A. Glenberg, *The American Journal of Psychology* **111**, 466 (1998).
23. J. Nielsen, *Designing Web Usability: The Practice of Simplicity* (New Riders, Indianapolis, IN, 2000).
24. B. Shneiderman and C. Plaisant, *Designing the User Interface: Strategies for Effective Human-Computer Interaction* (Addison-Wesley, New York, 2005).
25. P. J. Lynch and S. Horton, *Web Style Guide: Basic Design Principles for Creating Web Sites* (Yale University Press, London, U.K., 1999).
26. J. Nielsen, *Usability Engineering* (Academic Press, San Francisco, CA, 1993).
27. J. Nielsen and J. Sano, in *Proceeding of 2nd World Wide Web Conference: Mosaic and the Web* (1994), Chicago, 1: pp. 547–557.
28. J. Nielsen and R. L. Mack, *Usability Inspection Methods* (John Wiley & Sons, New York, 1994).
29. J. A. Borges, I. Morales, and N. J. Rodriguez, in *Conference Companion of the Computer-Human Interaction Conference,* Vancouver, BC, Canada (1996), pp. 277–278.

30. G. E. P. Box, J. S. Hunter, and W. G. Hunter, *Statistics for Experimenters: An Introduction to Design, Data Analysis, and Model Building* (John Wiley & Sons, New York, 1978).

31. G. Taguchi, S. Chowdhury, and Y. Wu, *The Mahalanobis-Taguchi System: Case Studies from Fuji Photo Film, Mitsubishi Software, Nissan, Seiko Epson, Sharp, Xerox and More* (The American Supplier Institute, New York, September 2000).

32. M. S. Phadke, *Quality Engineering Using Robust Design* (Prentice Hall PTR, Upper Saddle River, NJ, 1989).

33. J. R. Williams, Guidelines for the use of multimedia in instruction, in *Proceedings of the Human Factors and Ergonomics Society 42nd Annual Meeting* (1998), pp. 1447–1451. Santa Monica, CA.

34. R. A. Johnson and D. W. Wichern, *Applied Multivariate Statistical Analysis* (Prentice Hall, Upper Saddle River, NJ, 2002).

Chapter 11

Integrated Multimedia Understanding for Ubiquitous Intelligence Based on Mental Image Directed Semantic Theory

Masao Yokota

Contents

11.1 Introduction

At present, the realization of wireless sensor and actor network (WSAN) is one of the challenging topics in the concerned research fields, and a considerable number of important issues have been proposed, especially from the viewpoint of networking [1–3]. From the viewpoint of artificial intelligence and cognitive robotics [4], a WSAN can be considered as an intelligent robot system with distributed sensors and actuators that can gather information of high density and perform appropriate actions upon its environment over wide areas. Furthermore, as an ideal ubiquitous computing environment, a WSAN must be intelligent enough to work autonomously and to interact with ordinary people when their aids are needed. In order to realize such an environment, we have proposed the concept of distributed intelligent robot network (DIRN [5]).

As shown in Figure 11.1, a DIRN is one kind of WSAN, consisting of one brain node and numerous sensor and actor nodes with human-friendly interfaces. It is assumed, for example, that sensors and actuators can collaborate autonomously to perform appropriate actions just like reflexive actions in humans and that the brain node works exclusively for complicated computation

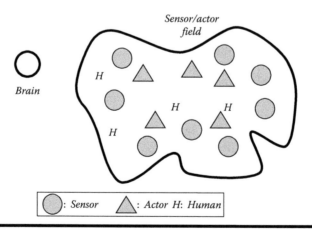

Figure 11.1 Physical architecture of DIRN.

based on profound knowledge in order to control the other kinds of nodes, to communicate with people, etc.

In order to realize well-coordinated DIRNs, it is very important to develop a systematically computable knowledge representation language (KRL) [5–7] as well as efficient networking technologies [3]. This type of language is indispensable to *knowledge-based* processing such as *understanding* sensory events, *planning* appropriate actions and *knowledgeable* communication even with humans, and therefore it needs to have at least a good capability of representing spatiotemporal events that correspond to human/robotic sensations and actions in the real world.

Conventionally, such quasi-natural language expressions as "move(10 meters)," "find(box, red)" and so on, uniquely related to computer programs, were employed for deploying sensors/motors in robotic systems [e.g., [8,9]]. These kinds of expressions, however, were very specific to devices and apt to have miscellaneous syntactic variants among them, such as "move(*Distance, Speed*)," "move(*Speed, Distance, Direction*)," etc., for motors and "find(*Object, Color*)" "find(*Object, Shape, Color*)," etc., for sensors. This is very inconvenient for communications especially between devices unknown to each other, and therefore it is very important to develop such a language as is universal among all kinds of equipments.

Yokota, M., has proposed a semantic theory for natural languages, so-called Mental Image Directed Semantic Theory (MIDST) [10,11]. In the MIDST, word concepts are associated with omnisensory mental images of the external or physical world and are formalized in an intermediate language L_{md} [10] This language is employed for many-sorted predicate logic with five types of terms. The most remarkable feature of L_{md} is its capability of formalizing both temporal and spatial event concepts on the level of human sensations, while the other similar knowledge representation languages are designed to describe the logical relations among conceptual primitives represented by natural-language words [12–14] or formally defined tokens [15–17].

The language L_{md} was originally proposed for formalizing the natural semantics, that is, the semantics specific to humans, but it is general enough for the artificial semantics, that is, the semantics specific to each artificial device such as a robot. This language has already been implemented on several types of computerized intelligent systems [4,5,18,19] and there is a feedback loop between them for their mutual refinement, unlike other similar ones [20,21].

This chapter presents the concept of DIRN and a sketch of the formal language L_{md} and focuses on the semantic processing of multimedia information represented in L_{md}, simulating the interactions between robots and their environments including humans.

11.2 Mental Image Description Language L_{md}

11.2.1 Omnisensory Image Model

In the MIDST, word meanings are treated in association with mental images, not limited to visual but omnisensory, modeled as "Loci in Attribute Spaces." An attribute space corresponds with a certain measuring instrument just like a barometer, thermometer, or so, and the loci represent the movements of its indicator.

For example, the moving gray triangular object shown in Figure 11.2a is assumed to be perceived as the loci in the three attribute spaces, namely, those of "Location," "Color," and "Shape" in the observer's brain. A general locus is to be articulated by "Atomic Locus" with the duration $[t_i\ t_f]$ as depicted in Figure 11.2b and formulated as (11.1):

$$L(x, y, p, q, a, g, k) \qquad (11.1)$$

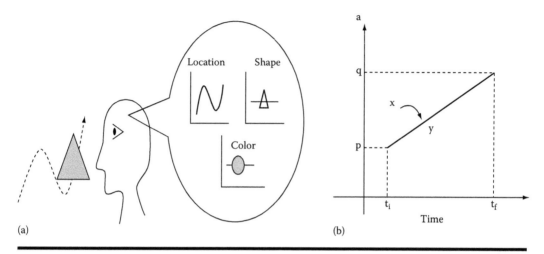

Figure 11.2   Mental image model (a) and Atomic Locus in Attribute Space (b).

This is a formula in many-sorted predicate logic, where "L" is a predicate constant with five types of terms: "Matter" (at "x" and "y"), "Attribute Value" (at "p" and "q"), "Attribute" (at "a"), "Event Type" (at "g"), and "Standard" (at "k"). Conventionally, matter variables are headed by "x," "y," and "z." This formula is called "Atomic Locus Formula" whose first two arguments are sometimes referred to as "Event Causer" (EC) and "Attribute Carrier" (AC), respectively, while ECs are often optional in natural concepts such as intransitive verbs. For simplicity, the syntax of L_{md} allows matter terms [e.g., "Tokyo" and "Osaka" in (11.2) and (11.3)] to appear at attribute values or standard in order to represent their values at the time. Moreover, when it is not so significant to discern ECs or standards, anonymous variables, usually symbolized as "_," can be employed in their places [See (11.39a)].

The intuitive interpretation of (11.1) is given as follows:

Matter 'x' causes attribute 'a' of matter 'y' to keep (p = q) *or change* (p ≠ q) *its values temporally* (g = Gt) *or spatially* (g = Gs) *over a time interval, where the values 'p' and 'q' are relative to the standard 'k'.*

When g = Gt and g = Gs, the locus indicates monotonic change or constancy of the attribute in time domain and that in space domain, respectively. The former is called "temporal event," and the latter, "spatial event." For example, the motion of the "bus" represented by S1 is a temporal event, and the ranging or extension of the "road" by S2 is a spatial event whose meanings or concepts are formulated as (11.2) and (11.3), respectively, where "A12" denotes the attribute "Physical Location." These two formulas are different only at the term "Event Type":

(S1) The bus runs from Tokyo to Osaka.

$$(x, y, k)L(x, y, \text{Tokyo}, \text{Osaka}, \text{A12}, \text{Gt}, k) \wedge \text{bus}(y) \tag{11.2}$$

(S2) The road runs from Tokyo to Osaka.

$$(\exists x, y, k)L(x, y, \text{Tokyo}, \text{Osaka}, \text{A12}, \text{Gs}, k) \wedge \text{road}(y) \tag{11.3}$$

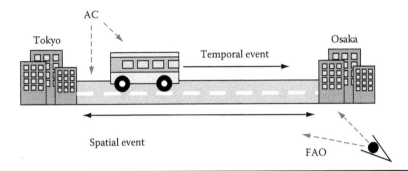

Figure 11.3   FAO movements and event types.

A considerable number of works [e.g., 22–25] have shown that human active sensing processes may affect perception and in turn conceptualization and recognition of the physical world. The author has hypothesized that the difference between temporal and spatial event concepts can be attributed to the relationship between the Attribute Carrier (AC) and the Focus of the Attention of the Observer (FAO). To be brief, the FAO is fixed on the whole AC in a temporal event but *runs* about on the AC in a spatial event. Consequently, as shown in Figure 11.3 the *bus* and the FAO move together in the case of S1 while the FAO solely moves along the *road* in the case of S2. That is, this hypothesis can be rephrased that *all loci in attribute spaces correspond one to one with movements or, more generally, temporal events of the FAO.*

11.2.2 Tempo-Logical Connectives

The duration of an atomic locus, suppressed in the atomic locus formula, corresponds to the time interval over which the FAO is put on the corresponding phenomenon outside. The MIDST has employed "tempo-logical connectives" (TLCs) representing both logical and temporal relations between loci at a time.

A tempo-logical connective K_i is defined by (11.4), where τ_i, χ, and K refer to one of the temporal relations indexed by an integer "i," a locus, and an ordinary binary logical connective such as the conjunction "$\wedge$," respectively. This is more natural and economical than explicit indication of time intervals, considering that people do not consult chronometers all the time in their daily lives. The definition of $\tau_i(-6 \leq i \leq 6)$ is given in Table 11.1 from which the theorem (11.5) can be deduced. This table shows 13 types of temporal relations between two events χ_1 and χ_2 whose durations are $[t_{11}, t_{12}]$ and $[t_{21}, t_{22}]$, respectively. This is in accordance with Allen's notation [15], which, to be strict, is exclusively for "temporal conjunctions" ($= \wedge_i$) such as introduced in the following:

$$\chi_1 \, K_i\chi_2 \leftrightarrow (\chi_1 K\chi_2) \wedge \tau_i(\chi_1, \chi_2) \tag{11.4}$$

$$\tau_{-i}(\chi_2, \chi_1) \equiv \tau_i(\chi_1, \chi_2)(\forall i \in \{0, \pm1, \pm2, \pm3, \pm4, \pm5, \pm6\}) \tag{11.5}$$

The TLCs used most frequently are "SAND" ($\wedge_0$) and "CAND" ($\wedge_1$) standing for "Simultaneous AND" and "Consecutive AND" and conventionally symbolized as "Π" and "$\cdot$," respectively.

Table 11.1   List of Temporal Relations

$\tau_i(\chi_1, \chi_2)$	Allen's Notation	Definition
$\tau_0(\chi_1, \chi_2)$	Equals(χ_1, χ_2)	$t_{11} = t_{21} \wedge t_{12} = t_{22}$
$\tau_0(\chi_1, \chi_2)$	Equals(χ_2, χ_1)	
$\tau_1(\chi_1, \chi_2)$	Meets(χ_1, χ_2)	$t_{12} = t_{21}$
$\tau_{-1}(\chi_2, \chi_1)$	Met-by(χ_2, χ_1)	
$\tau_2(\chi_1, \chi_2)$	Starts(χ_1, χ_2)	$t_{11} = t_{21} \wedge t_{12} < t_{22}$
$\tau_{-2}(\chi_2, \chi_1)$	Started-by(χ_2, χ_1)	
$\tau_3(\chi_1, \chi_2)$	During(χ_1, χ_2)	$t_{11} > t_{21} \wedge t_{12} < t_{22}$
$\tau_{-3}(\chi_2, \chi_1)$	Contains(χ_2, χ_1)	
$\tau_4(\chi_1, \chi_2)$	Finishes(χ_1, χ_2)	$t_{11} > t_{21} \wedge t_{12} = t_{22}$
$\tau_{-4}(\chi_2, \chi_1)$	Finished-by(χ_2, χ_1)	
$\tau_5(\chi_1, \chi_2)$	Before(χ_1, χ_2)	$t_{12} < t_{21}$
$\tau_{-5}(\chi_2, \chi_1)$	After(χ_2, χ_1)	
$\tau_6(\chi_1, \chi_2)$	Overlaps(χ_1, χ_2)	$t_{11} < t_{21} \wedge t_{21} < t_{12} \wedge t_{12} < t_{22}$
$\tau_{-6}(\chi_2, \chi_1)$	Overlapped-by(χ_2, χ_1)	

For example, the concepts of the English verbs "carry" and "return" are to be defined as (11.6) and (11.7), respectively. These formulas can be depicted as Figure 11.4a and b, where the optional ECs can be omitted as shown in Figure 11.5a and b, respectively:

$$(\lambda x, y)\text{carry}(x, y) \leftrightarrow (\lambda x, y)(\exists p, q, k)L(x, x, p, q, A12, Gt, k)\Pi$$

$$L(x, y, p, q, A12, Gt, k) \wedge x \neq y \wedge p \neq q \tag{11.6}$$

$$(\lambda x)\text{return}(x) \leftrightarrow (\lambda x)(\exists p, q, k)L(x, x, p, q, A12, Gt, k).$$

$$L(x, x, p, q, A12, Gt, k) \wedge x \neq y \wedge p \neq q \tag{11.7}$$

The expression (11.8) is the definition of the English verb concept "fetch" depicted as Figure 11.6a This implies such a temporal event that "x" goes for "y" and then comes back with it:

$$(\lambda x, y)\text{fetch}(x, y) \leftrightarrow (\lambda x, y)(\exists p_1, p_2, k)L(x, x, p_1, p_2, A12, Gt, k).$$

$$\left((\underline{L(x, x, \underline{p}_2, \underline{p}_1, A12, Gt, k)}\Pi\underline{L(x, y, \underline{p}_2, \underline{p}_1, A12, Gt, k)} \right)$$

$$\wedge x \neq y \wedge p_1 \neq p_2 \tag{11.8}$$

In the same way, the English verb concept "hand" or "receive" depicted as Figure 11.6b is defined uniformly as (11.9a) or its abbreviation (11.9b):

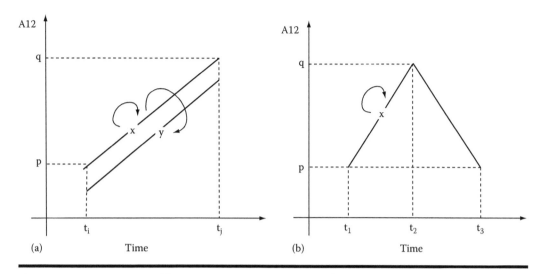

Figure 11.4   Depictions of loci: "carry" (a) and "return" (b).

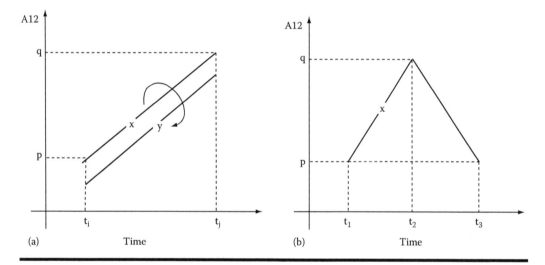

Figure 11.5   Simplified depictions of loci: "carry" (a) and "return" (b).

$$(\lambda x, y, z)\text{hand}(x, y, z)$$

$$\leftrightarrow (\lambda x, y, z)\text{receive}(z, y, x)$$

$$\leftrightarrow (\lambda x, y, z)(\exists k)L(x, y, x, z, A12, Gt, k) \Pi L(z, y, x, z, A12, Gt, k) \wedge$$

$$x \neq y \wedge y \neq z \wedge z \neq x \tag{11.9a}$$

$$\equiv (\lambda x, y, z)(\exists k)L(\{x, z\}, y, x, z, A12, Gt, k)$$

$$\wedge x \neq y \wedge y \neq z \wedge z \neq x \tag{11.9b}$$

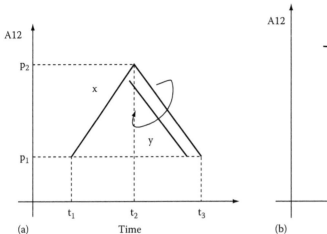
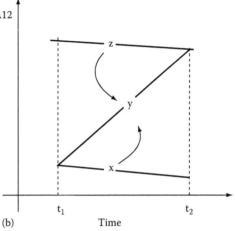

(a) Time

(b) Time

Figure 11.6 Loci of "fetch" (a) and "hand/receive" (b).

Such locus formulas as correspond with natural event concepts are called "Event Patterns," and about 40 kinds of event patterns have been found concerning the attribute "Physical Location" (A12), for example, *start, stop, meet, separate, carry,* and *return* [11].

Furthermore, a very important concept called "Empty Event" (EE) and denoted by "ε" is introduced. An EE stands for nothing but for time elapsing and is explicitly defined as (11.10) with the attribute "Time Point" (A34). According to this scheme, the duration $[t_a, t_b]$ of an arbitrary locus χ can be expressed as (11.11):

$$\varepsilon([t_1, t_2]) \leftrightarrow (\exists x, y, g, k)L(x, y, t_1, t_2, A34, g, k) \tag{11.10}$$

$$\chi \, \Pi \, \varepsilon([t_a, t_b]) \tag{11.11}$$

Any pair of loci temporally related in certain attribute spaces can be formulated as (11.12) through (11.16) in exclusive use of SANDs, CANDs and EEs. For example, the loci shown in Figure 11.7a and b correspond to the formulas (11.13) and (11.16), respectively:

$$\chi_1 \wedge_2 \chi_2 \equiv (\chi_1 \cdot \varepsilon)\Pi\chi_2 \tag{11.12}$$

$$\chi_1 \wedge_3 \chi_2 \equiv (\varepsilon_1 \cdot \chi_1 \cdot \varepsilon_2)\Pi\chi_2 \tag{11.13}$$

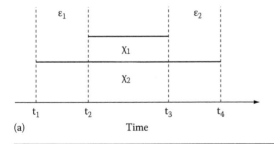
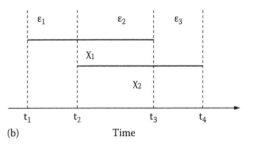

(a) Time

(b) Time

Figure 11.7 Tempo-logical relations: during(χ_1, χ_2) (a) and overlaps (χ_1, χ_2) (b).

$$\chi_1 \wedge_4 \chi_2 \equiv (\varepsilon \cdot \chi_1)\Pi\chi_2 \tag{11.14}$$

$$\chi_1 \wedge_5 \chi_2 \equiv \chi_1 \cdot \varepsilon \cdot \chi_2 \tag{11.15}$$

$$\chi_1 \wedge_6 \chi_2 \equiv (\chi_1 \cdot \varepsilon_3)\Pi(\varepsilon_1 \cdot \chi_2)\Pi(\varepsilon_1 \cdot \varepsilon_2 \cdot \varepsilon_3) \tag{11.16}$$

Employing these TLCs, tempo-logical relationships between miscellaneous event concepts can be formulated without explicit indication of time intervals. For example, an event "fetch(x,y)" is necessarily *finished by* an event "carry(x,y)" as indicated by the underline at (11.8). This fact can be formulated as (11.17), where "$\supset_{-4}$" is the "implication ($\supset$)" furnished with the temporal relation "finished-by (τ_{-4})." This kind of formula is not an axiom but a theorem deducible from the definitions of event concepts in our formal system:

$$\text{fetch}(x, y) \supset_{-4} \text{carry}(x, y) \tag{11.17}$$

11.2.3 Attributes and Standards

The attribute spaces for humans correspond to the sensory receptive fields in their brains. At present, about 50 attributes concerning the physical world have been extracted exclusively from English and Japanese words as shown in Table 11.2. They are associated with all of the five senses (i.e., sight, hearing, smell, taste, and feeling) in our everyday life, while those for information media other than languages correspond to limited senses. For example, those for pictorial media, marked with "†" in Table 11.2, associate limitedly with the sense "sight" as a matter of course. The attributes of this sense occupy the greater part of all, which implies that the sight is essential for humans to conceptualize the external world by. And this kind of classification of attributes plays a very important role in our cross-media operating system [4].

Correspondingly, six categories of standards shown in Table 11.3 have been extracted that are assumed necessary for representing values of each attribute in Table 11.2. In general, the attribute values represented by words are relative to certain standards as explained briefly in Table 11.3. For example, (11.18) and (11.19) are different formulations of a locus due to the different standards "k_1" and "k_2" for scaling as shown in Figure 11.8a and b, respectively. That is, whether the point (t_2, q) is significant or not, more generally, how to articulate a locus depends on the precisions or the granularities of these standards, which can be formulated as (11.20) and (11.21), so-called *Postulate of Arbitrariness in Locus Articulation*. This postulate affects the process of conceptualization on a word based on its referents in the world [26] and is applied in a DIRN as "Data Interpretation Function" (F_d) that translates a set of sensory data into a locus formula at the precision of a standard (see 3.2):

$$(L(y, x, p, q, a, g, k_1) \, \Pi \, \varepsilon([t_1, t_2])) \cdot (L(y, x, q, r, a, g, k_1)$$

$$\Pi \varepsilon([t_2, t_3])) \tag{11.18}$$

$$L(y, x, p, r, a, g, k_2)\Pi \, \varepsilon([t_1, t_3]) \tag{11.19}$$

$$(\forall p, q, r, k)(L(y, x, p, q, a, g, k) \cdot L(y, x, q, r, a, g, k) \supset$$

$$(\exists k')L(y, x, p, r, a, g, k') \wedge k' \neq k \tag{11.20}$$

$$(\forall p, r, k)(L(y, x, p, r, a, g, k) \supset$$

$$(\exists q, k')L(y, x, p, q, a, g, k') \cdot L(y, x, q, r, a, g, k') \wedge k' \neq k \tag{11.21}$$

Table 11.2   Examples of Attributes

Code	Attribute [Property*]
A01†	Place of existence [N]
A02†	Length [S]
A03†	Height [S]
A04†	Width [S]
A05†	Thickness [S]
A06†	Depth1 [S]
A07†	Depth2 [S]
A08†	Diameter [S]
A09†	Area [S]
A10†	Volume [S]
A11†	Shape [N]
A12†	Physical location [N]
A13†	Direction [N]
A14†	Orientation [N]
A15†	Trajectory [N]
A16†	Velocity [S]
A17†	Mileage [S]
A18	Strength of effect [S]
A19	Direction of effect [N]
A20	Density [S]
A21	Hardness [S]
A22	Elasticity [S]
A23	Toughness [S]
A24	Feeling [S]
A25	Humidity [S]
A26	Viscosity [S]
A27	Weight [S]
A28	Temperature [S]
A29	Taste [N]

Table 11.2 (continued)
Examples of Attributes

A30	Odor [N]
A31	Sound [N]
A32[†]	Color [N]
A33	Internal sensation [N]
A34	Time point [S]
A35	Duration [S]
A36	Number [S]
A37	Order [S]
A38	Frequency [S]
A39	Vitality [S]
A40	Sex [S]
A41	Quality [N]
A42	Name [V]
A43	Conceptual category [V]
A44[†]	Topology [V]
A45[†]	Angularity [S]

* S, scalar value; N, nonscalar value.
[†] Attributes concerning the sense of sight.

Table 11.3   List of Standards

Categories	*Remarks*
Rigid standard	Objective standards such as denoted by measuring *units* (meter, gram, etc.)
Species standard	The attribute value ordinary for a species. A short train is ordinarily longer than a long pencil
Proportional standard	"Oblong" means that the width is greater than the height at a physical object
Individual standard	Much money for one person can be too little for another
Purposive standard	One room large enough for a person's sleeping must be too small for his jogging
Declarative standard	The origin of an order such as "next" must be declared explicitly just as "next to him"

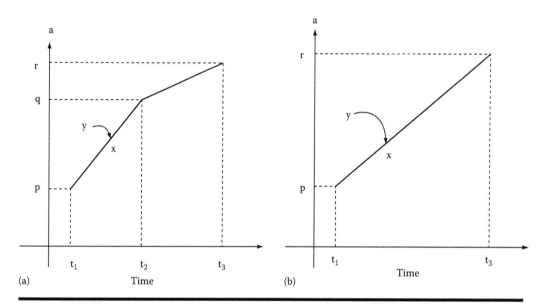

Figure 11.8 Arbitrariness in locus articulation due to standards: standard *k1* (a) is finer than *k2* (b).

11.2.4 Attribute Values and Atomic Loci

In our formal system, a constant term of attribute value is to be assigned a point set, possibly, with a fuzzy boundary due to its semantic vagueness. This is the case especially when such a term is associated with a certain word concept and the boundary of its point set assigned is to be controlled by the standard specific to the term. For example, a word of color such as *red* is semantically vague possibly depending on a certain individual standard and can be assigned a certain region in the attribute space, conventionally called "Color Solid," with the three dimensions of "Chrome," "Hue," and "Value."

In general, the relation between a word of attribute value "v_i" and its corresponding point set "$S(p_i, a_i, k_i)$" can be formalized as (11.22) by the function "Assign", where "p_i", "a_i", and "k_i" are "Attribute Value", "Attribute", and "Standard" specific to "v_i", respectively. For example, the word "*long* ($= v_i$)" can be assigned a point set "$\{p_i | p_i > k_i\}$", where k_i is some standard for being long:

$$Assign(v_i) = S(p_i, a_i, k_i) \tag{11.22}$$

According to this assumption of attribute value, the formal interpretation of an atomic locus formula such as underlined in (11.23) can be given as follows:

$$L(x, y, p_1, p_2, a, g, k) \Pi \varepsilon([t_1, t_2]) \tag{11.23}$$

Firstly, the real sensations and the terms for them are related as (11.24), where "$Vsense(y, a, t_j)$" is the region in the attribute space "a" *onto* which the real sensation of "y" is projected at the time t_j:

$$Vsense(y, a, t_1) \subseteq S(p_1, a, k),$$
$$Vsense(y, a, t_2) \subseteq S(p_2, a, k) \tag{11.24}$$

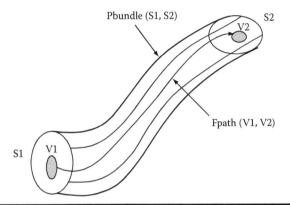

Figure 11.9   Fpath as a member of Pbundle.

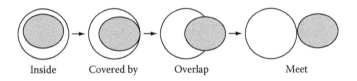

Inside   Covered by   Overlap   Meet

Figure 11.10   Monotonic change from "Inside" to "Meet" by translation.

Secondly, the scan path of the FAO from V_1 to V_2, "Fpath(V_1, V_2)", is related to the bundle of the shortest paths from S_1 to S_2, "Pbundle(S_1, S_2)", as shown in Figure 11.9 and formulated by (11.25) and (11.26), where "Spath(q_1, q_2)" is the shortest path from q_1 to q_2:

$$\text{Fpath}(\text{Vsense}(y, a, t_1), \text{Vsense}(y, a, t_2) \subseteq$$
$$\text{Pbundle}(S(p_1, a, k),\ S(p_2, a, k)) \tag{11.25}$$
$$\text{Pbundle}(S_1, S_2) = \{\text{Spath}(q_1, q_2) | \forall q_1 \in S_1\ \forall q_2 \in S_2\} \tag{11.26}$$

As easily imagined, "Fpath" corresponds with "Atomic Locus" without the causation (i.e., "x → y"), and the paths are to follow the structure specific to the attribute space. For example, the topology between two regions can be represented as the Closest-Topological-Relation-Graph [27], where the Fpath from the node "Inside" to the node "Meet" must pass the nodes "CoveredBy" and "Overlap" on the way of such monotonic translation as shown in Figure 11.10 without any other deformation.

11.3 Specification of DIRN's World

"The world for a DIRN" (**W**) refers to "the set of matters observable for the DIRN" and is defined by (11.27) as the union of the set of its nodes (**D**) and the set of the objects in its environment. The set **D** is the union of the sets of a brain node ({B}), sensor nodes (**Se**), and actor nodes (**Ac**) as represented by (11.28), while the set *O* includes possibly humans and the other DIRNs:

$$\mathbf{W} = \mathbf{D} \cup \mathbf{O} \tag{11.27}$$
$$\mathbf{D} = \{B\} \cup \mathbf{Se} \cup \mathbf{Ac} \tag{11.28}$$

"A constituent C_k of the world for a DIRN" (i.e., $C_k \in \mathbf{W}$) can be specified by the loci in the attribute spaces distinguishable by the sets of attributes and standards unique to the DIRN.

11.3.1 Specification of Objects

An object in the environment of a DIRN (i.e., $C_k \in \mathbf{O}$) can be characterized by the loci of its structure and so on. For example, the characteristics of a tree "C1" in the environment can be represented by such a locus formula as (11.29), reading its height (A03) is between 4 and 5 m, its location (A12) is in the park "C2",.... For another example, a road "C3" that runs from a town "C4" to a town "C5" via a town "C6" can be defined by (11.30):

$$tree(C1) \leftrightarrow (\exists x, p, k, \ldots)L(x, C1, p, p, A03, Gt, Me)$$

$$\wedge (4m \le p \le 5\,m) \wedge L(x, C1, C2, C2, A12, Gt, k)$$

$$\wedge\, park(C2) \wedge \ldots \tag{11.29}$$

$$road(C3) \leftrightarrow (\exists x, k, \ldots)L(x, C3, C4, C6, A12, Gs, k).$$

$$L(x, C3, C6, C5, A12, Gs, k) \wedge town(C4) \wedge town(C5) \wedge town(C6) \tag{11.30}$$

11.3.2 Specification of a Sensor Node

A sensor node (i.e., $C_k \in \mathbf{Se}$) can be specified by the loci of its structure and its collectable sensory data. In general, a sensor can be distinguished by the definition (11.31) from another kind of constituent, reading that a sensor "x" is what takes in some data "y" from some constituent. A data set is to be translated into a locus formula by "Data Interpretation Function" (F_d) (see 8.2) as defined by (11.32):

$$(\lambda x)sensor(x) \leftrightarrow (\lambda x)(\exists y, z, g_1, k_1)L(x, y, z, x, A12, g_1, k_1)$$

$$\wedge\, data(y) \tag{11.31}$$

$$F_d(y) = (\exists z, z_1, \ldots, z_n, a, g, k, p_0, \ldots, p_n)L(z_1, z, p, p_1, a, g, k) \cdot \ldots \cdot$$

$$L(z_n, z, p_{n-1}, p_n, a, g, k) \tag{11.32}$$

The left hand of (11.32) is given as such a locus formula as characterized by the attribute "a" and the standard "k" unique to the sensor. For example, a thermometer with the measurable range $[-10°C, +100°C]$ can be characterized by (11.33) with the attribute "temperature" (A28) and the rigid standard of "Celsius" (Ce):

$$(\exists z, z_1, \ldots, z_n, p_0, \ldots, p_n)L(z_1, z, p_0, p_1, A28, Gt, Ce) \cdot \ldots \cdot$$

$$L(z_n, z, p_{n-1}, p_n, A28, Gt, Ce) \wedge (-10°C \le p_i \le +100°C$$

$$0 \le i \le n) \tag{11.33}$$

11.3.3 Specification of an Actor Node

An actor (i.e., $C_k \in \mathbf{Ac}$) can be specified by the loci of its structure, performable actions and, if any sensors with it, collectable sensory data. For example, a tanker "C8" with the coverage [0,

100 km] can be characterized by (11.34) with the attribute "Mileage" (A17) at the rigid standard of "Meter" (Me):

$$(\exists x, p)L(C8, x, 0, p, A17, Gt, Me) \wedge (0\,km \leq p \leq 100\,km)$$
$$\wedge \; liquid(x) \tag{11.34}$$

11.3.4 Specification of the Brain Node

The brain node (i.e., *B*) can be specified by its commonsense knowledge and world knowledge including such specifications of the other constituents as mentioned previously. For example, (11.35) is an example of commonsense knowledge piece, reading that *a matter has never different values of an attribute at a time*:

$$L(x, y, p_1, q_1, a, g, k)\Pi L(z, y, p_2, q_2, a, g, k) \supset p_1 = p_2 \wedge q_1 = q_2 \tag{11.35}$$

The intelligence of the brain node must be conscious of all about the other constituents but can be unconscious of its own structure (e.g., hardware configuration) and computational performance specification (e.g., CPU speed) because they are what only metasystems such as OS and metabrain node have to concern. In our case, the brain node is a personal computer with the intelligent system IMAGES-M [28] installed under the OS WINDOWS/XP.

11.4 Interaction between DIRN and Its World

11.4.1 Intelligent System IMAGES-M

As shown in Figure 11.11, a DIRN is to gather information pieces from its world, interpret them into locus formulas, find/solve problems, and act appropriately upon its field. All these intelligent performances are executed by IMAGES-M possibly with aids of humans. The intelligent system IMAGES-M [28], still under development, is intended to facilitate integrated multimedia information understanding, including miscellaneous cross-media operations. This system has employed

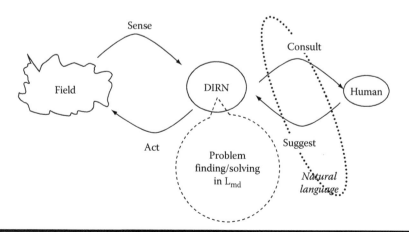

Figure 11.11   Interaction between DIRN and its world.

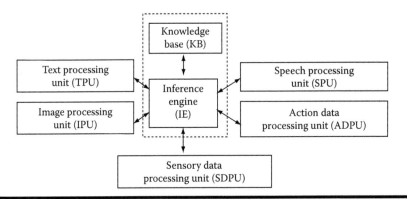

Figure 11.12   Configuration of IMAGES-M.

locus formula as intermediate knowledge representation, through which it can integrally understand and generate sensor data, speech, visual image, text, and action data. IMAGES-M is to work as the main intelligence of the brain node of a DIRN while the intelligence of each sensor or actuator is a small-scaled IMAGES-M adapted for its specialized function.

IMAGES-M as shown in Figure 11.12, is one kind of expert system equipped with five kinds of user interfaces for multimedia communication, that is, Sensory Data Processing Unit (SDPU), Speech Processing Unit (SPU), Image Processing Unit (IPU), Text Processing Unit (TPU), and Action Data Processing Unit (ADPU) besides Inference Engine (IE) and Knowledge Base (KB). Each processing unit in collaboration with IE performs mutual conversion between each type of information medium and locus formulas.

11.4.2 *Fundamental Computations on* L_{md}

The fundamental computations on L_{md} by IMAGES-M are to detect semantic anomalies, ambiguities and identities in data or expressions. These are performed as inferential operations on locus formulas at IE.

Detection of semantic anomalies is very important to avoid succession of meaningless computations or actions. For an extreme example, consider such a report from certain sensors as (11.36) represented in L_{md}, where "..." and "A29" stand for descriptive omission and the attribute "Taste." This locus formula can be translated into the English sentence S3 by TPU, but it is semantically anomalous because a "desk" has ordinarily no taste:

$$(\exists x, y, k)L(y, x, Sweet, Sweet, A29, Gt, k) \wedge desk(x) \tag{11.36}$$

(S3) The desk is sweet.

These kinds of semantic anomalies can be detected in the following processes.

Firstly, assume the postulate (11.37) as the commonsense or default knowledge of "desk", stored in KB, where "A39" is the attribute "Vitality." The special symbol "*" represents "always" as defined by (11.38), where "$\varepsilon([t_1, t_2])$" is a simplified atomic locus formula standing for time elapsing with an interval $[t_1, t_2]$. Furthermore, "_" and "/" are anonymous variables employed for descriptive simplicity and defined by (11.39a) and (11.39b), respectively:

$$(\lambda x)\text{desk}(x) \leftrightarrow (\lambda x)(\ldots L^*(\_, x, /, /, A29, Gt, \_) \wedge \ldots \wedge$$

$$L^*(\_, x, /, /, A39, Gt, \_) \wedge \ldots) \tag{11.37}$$

$$X^* \leftrightarrow (\forall[t_1, t_2])X\Pi\varepsilon([t_1, t_2]) \tag{11.38}$$

$$X(\_) \leftrightarrow (\exists u)X(u) \tag{11.39a}$$

$$X(/) \leftrightarrow \neg(\exists u)X(u) \tag{11.39b}$$

Secondly the postulates expressed by (11.40) and (11.41) in KB are utilized. The formula (11.40) means that *if one of two loci exists every time interval, then they can coexist.* The formula (11.41) states that *a matter has never different values of an attribute at a time*:

$$X \wedge Y^* \supset X\Pi Y \tag{11.40}$$

$$L(x, y, p, q, a, g, k)\Pi L(z, y, r, s, a, g, k) \supset p = r \wedge q = s \tag{11.41}$$

Lastly, IE detects the semantic anomaly of "sweet desk" by using (11.37) through (11.41). That is, the following formula (11.42) is finally deduced from (11.37) through (11.41), which violates the postulate (11.37), that is, *Sweet* $\neq$ /:

$$L(\_, x, \textit{Sweet}, \textit{Sweet}, A29, Gt, \_)\Pi L(z, x, /, /, A29, Gt, \_) \tag{11.42}$$

The processes mentioned earlier are also employed for dissolving syntactic ambiguities in people's utterances such as S4. IE rejects "sweet desk" and eventually adopts "sweet coffee" as a plausible interpretation:

(S4) Bring me the coffee on the desk, which is very sweet.

If multiple plausible interpretations of a text or another type of information are represented in different locus formulas, it is semantically ambiguous. In such a case, IMAGES-M will ask for further information in order for disambiguation.

Furthermore, if two different representations are interpreted into the same locus formula, they are paraphrases of each other. Such detection of semantic identities is very useful for deleting redundant information, for cross-media translation, etc. [28].

11.5 Problem Finding and Solving by DIRN

11.5.1 Definition of Problem and Task for DIRN

The problems for a DIRN can be classified roughly into two categories as follows:

(CP) Creation problem, e.g., house building, food cooking

(MP) Maintenance problem, e.g., fire extinguishing, room cleaning

In general, an MP is relatively simple one that the DIRN can find and solve autonomously, while a CP is relatively difficult one that is given to the DIRN, possibly, by humans and to be solved in cooperation with them. A DIRN must determine its task to solve a problem in the world.

The conventional AI defines a problem as the difference or gap between a "Current State" and a "Goal State" and a task as its cancellation. Here, the term "Event" is preferred to the term "State", and "State" is defined as static "Event", which corresponds to a level locus. On this line, the DIRN needs to interpolate some transit event X_T between the two events, namely, "Current Event" (X_C) and "Goal Event" (X_G) as (11.43):

$$X_C \cdot X_T \cdot X_G \tag{11.43}$$

According to this formalization, a problem X_P can be defined as $X_T \cdot X_G$ and a task for the DIRN can be defined as its realization.

The events in the world are described as loci in certain attribute spaces, and a problem is to be detected by the unit of atomic locus. For example, employing such a postulate as (11.44) implying "Continuity in attribute values", the event X in (11.45) is to be inferred as (11.46):

$$L(x, y, p_1, p_2, a, g, k) \cdot L(z, y, p_3, p_4, a, g, k) \supset p_3 = p_2 \tag{11.44}$$

$$L(x, y, p_1, p_2, a, g, k) \cdot X \cdot L(z, y, p_3, p_4, a, g, k) \tag{11.45}$$

$$L(z', y, p_2, p_3, a, g, k) \tag{11.46}$$

11.5.2 CP Finding and Solving

Consider a verbal command such as S5 uttered by a human. Its interpretation is given by (11.47) as the goal event X_G. If the current event X_C is given by (11.48), then (11.49) with the transit event X_T underlined can be inferred as the problem corresponding to S5:

(S5) Keep the temperature of "room C9" at 20.

$$L(z, C9, 20, 20, A28, Gt, k) \wedge room(C9) \wedge (z \in \mathbf{O}) \tag{11.47}$$

$$L(x, C9, p, p, A28, Gt, k) \wedge room(C9) \tag{11.48}$$

$$\underline{L(z1, C9, p, 20, A28, Gt, k)} \cdot L(z, C9, 20, 20, A28, Gt, k) \wedge$$

$$room(C9) \wedge (z_1 \in \mathbf{O}) \tag{11.49}$$

For this problem, the DIRN is to execute a job deploying a certain thermometer and actors "z_1" and "z." The selection of the actor "z_1" is performed as follows:

> *If 20-p < 0 then z_1 is a cooler, otherwise*
>
> *if 20-p > 0 then z_1 is a heater, otherwise*
>
> *20-p = 0 and no actor is deployed as z_1.*

The selection of "z" is a job in case of MP described in the next section.

11.5.3 MP Finding and Solving

In general, the goal event X_G for an MP is that for another CP such as S3 given possibly by humans and solved by the DIRN in advance. That is, the job in this case is to autonomously restore the goal event X_G created in advance to the current event X_C as shown in (11.50), where the transit event X_T is the reversal of such X_{-T} that has been already detected as "abnormal" by the DIRN.

For example, if X_G is given by (11.47) in advance, then X_T is also represented as the underlined part of (11.49) while X_{-T} as (11.51). Therefore the job here is quite the same that was described in the previous section:

$$X_G \cdot X_{-T} \cdot X_C \cdot X_T \cdot X_G \tag{11.50}$$

$$L(z_1, C9, 20, p, A28, Gt, k) \wedge room(C9) \wedge (z_1 \in \mathbf{O}) \tag{11.51}$$

11.6 Natural Language Understanding

11.6.1 Word Meaning Description

Natural language is the most important information medium because it can convey the exact intention of the sender to the receiver due to its syntax and semantics common to its users, which is not necessarily the case for another medium such as picture. IMAGES-M can translate systematically natural language, either spoken or written, and L_{md} expression into each other by utilizing syntactic rules and word meaning descriptions of natural language [10].

A word meaning description M_w is given by (11.52) as a pair of "Concept Part" (C_p) and "Unification Part" (U_p):

$$M_w \leftrightarrow [C_p : U_p] \tag{11.52}$$

The C_p of a word W is a locus formula about properties and relations of the matters involved such as shapes, colors, functions, potentialities, etc., while its U_p is a set of operations for unifying the C_ps of W's syntactic governors or dependents. For example, the meaning of the English verb "carry" can be given by (11.53):

$$[(\exists x, y, p_1, p_2, k)L(x, x, p_1, p_2, A12, Gt, k)\Pi$$

$$L(x, y, p_1, p_2, A12, Gt, k) \wedge x \neq y \wedge p_1 \neq p_2 : ARG(Dep.1, x); ARG(Dep.2, y);] \tag{11.53}$$

The U_p mentioned earlier consists of two operations to unify the first dependent (Dep.1) and the second dependent (Dep.2) of the current word with the variables x and y, respectively. Here, Dep.1 and Dep.2 are the "subject" and the "object" of "carry", respectively. Therefore, the surface structure *Mary carries a book* is translated into the conceptual structure (11.54) via the surface dependency structure shown in Figure 11.13:

$$(\exists y, p_1, p_2, k)L(Mary, Mary, p_1, p_2, A12, Gt, k)\Pi$$

$$L(Mary, y, p_1, p_2, A12, Gt, k) \wedge Mary \neq y \wedge p_1 \neq p_2 \wedge book(y) \tag{11.54}$$

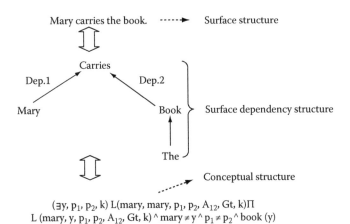

Figure 11.13   Mutual conversion between natural language and L_{md}.

For another example, the meaning description of the English preposition "through" is also given by (11.55):

$$[(\exists x, y, p_1, z, p_3, g, k, p_4, k_0)(L(x, y, p_1, z, A12, g, k) \cdot L(x, y, z, p_3, A12, g, k))\Pi$$

$$L(x, y, p_4, p_4, A13, g, k_0) \wedge p_1 \neq z \wedge z \neq p_3 : ARG(Dep.1, z);$$

$$IF(Gov = Verb) \rightarrow PAT(Gov, (1, 1));$$

$$IF(Gov = Noun) \rightarrow ARG(Gov, y);] \tag{11.55}$$

11.6.2 Mutual Conversion between Text and Locus Formula

The U_p mentioned earlier is for unifying the C_ps of the very word, its governor (Gov, a verb or a noun) and its dependent (Dep.1, a noun). The second argument (1,1) of the command PAT indicates the underlined part of (11.55), and in general (i,j) refers to the partial formula covering from the ith to the jth atomic formula of the current C_p. This part is the pattern common to both the C_ps to be unified. This is called "Unification Handle" (U_h), and when missing, the C_ps are to be combined simply with "$\wedge$."

Therefore, the sentences S6, S7, and S8 are interpreted as (11.56), (11.57), and (11.58), respectively. The underlined parts of these formulas are the results of PAT operations. The expression (11.59a) is the C_p of the adjective "Long" implying "there is some value greater than some standard of 'Length (A02)'", which is often simplified as (11.59b):

(S6) The train runs through the tunnel.

$$(\exists x, y, p_1, z, p_3, k, p_4, k_0)(\underline{L(x, y, p_1, z, A12, Gt, k)}.$$

$$L(x, y, z, p_3, A12, Gt, k))\Pi L(x, y, p_4, p_4, A13, Gt, k_0)$$

$$\wedge p_1 \neq z \wedge z \neq p_3 \wedge train(y) \wedge tunnel(z) \tag{11.56}$$

(S7) The path runs through the forest

$$(\exists x, y, p_1, z, p_3, k, p_4, k_0)(\underline{L(x, y, p_1, z, A12, Gs, k)}$$

$$L(x, y, z, p_3, A12, Gs, k))\Pi L(x, y, p_4, p_4, A13, Gs, k_0)$$

$$\wedge p_1 \neq z \wedge z \neq p_3 \wedge path(y) \wedge forest(z) \tag{11.57}$$

(S8) The path through the forest is long.

$$(\exists x, y, p_1, z, p_3, x_1 k, q, k_1, p_4, k_0)$$

$$(L(x, y, p_1, z, A12, Gs, k) \cdot L(x, y, z, p_3, A12, Gs, k))$$

$$\Pi L(x, y, p_4, p_4, A13, Gs, k_0) \wedge L(x_1, y, q, q, A02, Gt, k_1)$$

$$\wedge p_1 \neq z \wedge z \neq p_3 \wedge q > k_1 \wedge path(y) \wedge forest(z) \tag{11.58}$$

$$(\exists x_1, y_1, q, k_1)L(x_1, y_1, q, q, A02, Gt, k_1) \wedge q > k_1 \tag{11.59a}$$

$$(\exists x_1, y_1, k_1)L(x_1, y_1, Long, Long, A02, Gt, k_1) \tag{11.59b}$$

The process mentioned previously is completely reversible except that multiple natural expressions as paraphrases can be generated as shown in Figure 11.14 because event patterns

(Input)
    With the long red stick Tom precedes Jim.
(Output)
    Tom with the long red stick goes before Jim goes.
    Jim goes after Tom goes with the long red stick.
    Jim follows Tom with the long red stick.
    Tom carries the long red stick before Jim goes.
    ... ...

Figure 11.14 Language to language translation.

are shareable among multiple word concepts. This is one of the most remarkable features of IMAGES-M and is also possible between different languages as understanding-based translation [29].

11.7 Cross-Media Translation

11.7.1 Functional Requirements

The core technology for integrated multimedia information understanding in IMAGES-M is that for cross-media translation via intermediate representation in L_{md}. The author has considered that systematic cross-media translation must have such functions as follows:

(F1) To translate source representations into target ones as for contents describable by both source and target media. For example, positional relations between/among physical objects such as "in", "around", etc., are describable by both linguistic and pictorial media.

(F2) To filter out such contents that are describable by source medium but not by target one. For example, linguistic representations of "taste" and "smell" such as "sweet candy" and "pungent gas" are not describable by usual pictorial media although they would be seemingly describable by cartoons, etc.

(F3) To supplement default contents, that is, such contents that need to be described in target representations but not explicitly described in source representations. For example, the shape of a physical object is necessarily described in pictorial representations but not in linguistic ones.

(F4) To replace default contents by definite ones given in the following contexts. For example, in such a context as "There is a box to the left of the pot. The box is red. . . ", the color of the box in a pictorial representation must be changed from default one to red.

For example, the text consisting such two sentences as "There is a hard cubic object" and "The object is large and gray" can be translated into a still picture in such a way as shown in Figure 11.15

11.7.2 Formalization

The MIDST assumes that any content conveyed by an information medium is to be associated with the loci in certain attribute spaces, and in turn that the world describable by each medium can be characterized by the maximal set of such attributes. This relation is conceptually formalized by the expression (11.60), where Wm, Am_i, and F mean "the world describable by the information medium m", "an attribute of the world", and "a certain function for determining the maximal set of attributes of Wm", respectively:

$$F(Wm) = \{Am_1, Am_2, \ldots, Am_n\} \tag{11.60}$$

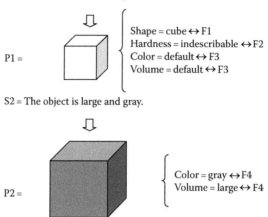

S1=There is a hard cubic object.

P1 =
Shape = cube ↔ F1
Hardness = indescribable ↔ F2
Color = default ↔ F3
Volume = default ↔ F3

S2 = The object is large and gray.

P2 =
Color = gray ↔ F4
Volume = large ↔ F4

Figure 11.15 Systematic cross-media translation.

Considering this relation, cross-media translation is one kind of mapping from the world describable by the source medium (m_s) to that by the target medium (m_t) and can be defined by the expression (11.61):

$$Y(Sm_t) = \psi(X(Sm_s)) \tag{11.61}$$

where

Sm_s is the maximal set of attributes of the world describable by the source medium m_s

Sm_t is the maximal set of attributes of the world describable by the target medium m_t

$X(Sm_s)$ is a locus formula about the attributes belonging to Sm_s

$Y(Sm_t)$ is a locus formula about the attributes belonging to Sm_t

ψ is the function for transforming X into Y so-called Locus formula paraphrasing function

The function ψ is designed to realize all the functions F1–F4 by inference processing at the level of locus formula representation.

11.7.3 Locus Formula Paraphrasing Function ψ

In order to realize the function F1, a certain set of *Attribute Paraphrasing Rules (APRs)*, so called, are defined *at every pair of source and target media* (See Table 11.4).

The function F2 is realized by detecting locus formulas about *the attributes without any corresponding APRs* from the content of each input representation and replacing them by *empty events*.

For F3, *default reasoning* is employed. That is, such an inference rule as defined by the expression (11.62) is introduced, which states *if X is deducible and it is consistent to assume Y, then conclude Z.*

This rule is applied typically to such instantiations of X, Y, and Z as specified by the expression (11.63), which means that the indefinite attribute value "p" with the indefinite standard "k" of the indefinite matter "y" is substitutable by the constant attribute value "P" with the constant standard "K" of the definite matter "$O_\#$" of the same kind of "M":

Table 11.4   APRs for Text–Picture Translation

APRs	Correspondences of Attributes (Text: Picture)	Value Conversion Schema (Text ⇔ Picture)
APR-01	A12: A12	$p \Leftrightarrow p'$
APR-02	{A12, A13, A17}: A12	$\{p, d, l\} \Leftrightarrow p' + l'd'$
APR-03	{A11, A10}: A11	$\{s, v\} \Leftrightarrow v's'$
APR-04	A32: A32	$c \Leftrightarrow c'$
APR-05	{A12, A44}: A12	$\{p_a, m\} \Leftrightarrow \{p'_a, p'_b\}$

$$X^{\circ} Y \to Z \tag{11.62}$$

$$\{X/(L(x, y, p, p, A, G, k) \wedge M(y)) \wedge (L(z, O_\#, P, P, A, G, K) \wedge M(O_\#)),$$

$$Y/p = P \wedge k = K, Z/L(x, y, P, P, A, G, K) \wedge M(y)\} \tag{11.63}$$

Lastly, the function F4 is realized quite easily by memorizing the history of applications of default reasoning.

11.8 Miscellaneous Cross-Media Operations

11.8.1 Mixed-Media Dialogue by Text and Picture

As easily imagined, IMAGES-M and humans can perform mixed-media dialogue employing text and picture as shown in Figure 11.16 where a drawing tool is utilized as graphical interface.

It is one of the most essential tasks for the system to determine how many pictures a locus formula should be interpreted into. Consider such somewhat complicated sentences as S9 and S10. The underlined parts are considered to refer to some events neglected in time and in space, respectively. These events are called "Temporal Empty Event" and "Spatial Empty Event", denoted by "ε_t" and "ε_s" as EEs with g = Gt and g = Gs at (11.10), respectively. The concepts of S9 and S10 are given by (11.64) and (11.65), where "A15" and "_" represent the attribute "Trajectory" and abbreviation of the variables bound by existential quantifiers, respectively.

In general, an atomic locus formula with g = Gt is to be depicted as a pair of pictures of the formula and that with g = Gs, as one still picture. Therefore, (11.65) is depicted as the still picture in Figure 11.17 while (11.64) as a series of still pictures, namely, a motion picture. Figure 11.18 is an example of map generation from text via locus formula representation:

(S9) The *bus* runs 10 km straight east from A to B, and *after a while*, at C it meets the street with the sidewalk.

$$(\exists x, y, z, p, q)(L(\_, x, A, B, A12, Gt, \_)\Pi L(\_, x, 0, 10\,km, A17, Gt, \_)$$

$$\Pi L(\_, x, Point, Line, A15, Gt, \_)\Pi L(\_, x, East, East, A13, Gt, \_))$$

$$\cdot \varepsilon_t \cdot (L(\_, x, p, C, A12, Gt, \_)\Pi L(\_, y, q, C, A12, Gs, \_)\Pi$$

$$L(\_, z, y, y, A12, Gs, \_)) \wedge bus(x) \wedge street(y) \wedge sidewalk(z) \wedge p \neq q \tag{11.64}$$

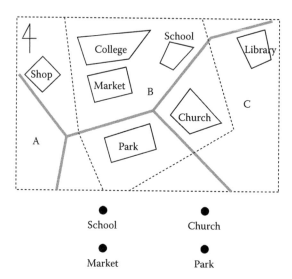

H: Where are the school, church, market, and park
   located as shown above?
S: District b.
H: How is the market located in district b?
S: To the north of park
   between college and park.

Figure 11.16   Q–A by mixture of text and picture between humans (H) and IMAGES-M (S).

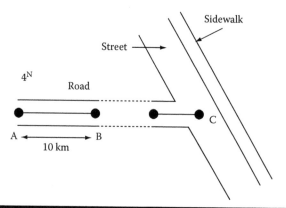

Figure 11.17   Pictorial interpretation of (65).

(S10) The *road* runs 10 km straight east from A to B, and *after a while*, at C it meets the street with the sidewalk.

$$(\exists x, y, z, p, q)(L(\_, x, A, B, A12, Gs, \_)\Pi L(\_, x, 0, 10\,km, A17, Gs, \_)$$

$$\Pi L(\_, x, Point, Line, A15, Gs, \_)\Pi L(\_, x, East, East, A13, Gs, \_))$$

$$\cdot \varepsilon_s \cdot (L(\_, x, p, C, A12, Gs, \_)\Pi L(\_, y, q, C, A12, Gs, \_)\Pi$$

$$L(\_, z, y, y, A12, Gs, \_)) \wedge road(x) \wedge street(y) \wedge sidewalk(z) \wedge p \neq q \qquad (11.65)$$

Figure 11.18 **Map generated from a text via a locus formula.**

The APRs must be set up as relationships between the attributes concerning the paired media to be translated by each other. For our experiment on text–picture translation [4], there were employed five kinds of APRs such as shown in Table 11.4, where p, s, c,... and p′, s′, c′,... are linguistic expressions and their corresponding pictorial expressions of attribute values, respectively. Further details are as follows:

1. APR-02 is used especially for a sentence such as "The box is 3 meters to the left of the chair." The symbols p, d and l correspond to "the position of the chair", "left", and "3 meters", respectively, yielding, the pictorial expression of "the position of the box", namely, "p′+l′d′".
2. APR-03 is used especially for a sentence such as "The pot is big." The symbols s and v correspond to "the shape of the pot" (default value) and "the volume of the pot" (big), respectively. In pictorial expression, the shape and the volume of an object are inseparable and therefore they are represented only by the value of the attribute "shape", namely, "v′s′".
3. APR-05 is used especially for a sentence such as "The cat is under the desk." The symbols p_a, p_b, and m correspond to "the position of the desk", "the position of the cat", and "under", respectively, yielding a pair of pictorial expressions of the positions of the two objects.

11.8.2 *Linguistic Interpretation of Human Motion Data*

The human body can be described in a computable form using locus formulas. That is, the structure of the human body is one of spatial event where the body parts such as head, trunk, and limbs extend spatially and connect with each other. The expressions (11.66) and (11.67) are examples of these descriptions using locus formulas, which read roughly that an arm extends from the hand to the shoulder and that a wrist connects the hand and the forearm, respectively:

$$(\lambda x)arm(x) \leftrightarrow (\lambda x)(\exists y_1, y_2, k)$$
$$L(x, x, y_1, y_2, A12, Gs, k) \wedge shoulder(y_1) \wedge hand(y_2) \qquad (11.66)$$

$$(\lambda x)\text{wrist}(x) \leftrightarrow (\lambda x)(\exists y_1, y_2, y_3, y_4, k)$$
$$(L(y_1, y_1, y_2, x, A12, Gs, k)\Pi L(y_1, y_1, x, y_3, A12, Gs, k))$$
$$\wedge \text{ body} - \text{part}(y_1) \wedge \text{forearm}(y_2) \wedge \text{hand}(y_3) \tag{11.67}$$

The structural description in the computable form is indispensable to mutual translation between human motion data and linguistic expressions. For example, it enables the system to recognize the anomaly of such a sentence as S11 in such a process described in Section 11.4.2:

(S11) The left arm moved away from the left shoulder and the left hand.

Various kinds of human motions have been conceptualized as specific verbs in natural languages, such as "nod" and "crouch." For example, the conceptual description of "nodding" is given by (11.68), which reads roughly that a person lets the head fall forward. The conceptual description of a verb gives the framework of the meaning representation of the sentence where the very verb appears. This kind of meaning representation is called "Text Meaning Representation" (TMR) as mentioned in the following:

$$(\lambda x)\text{nodding}(x) \leftrightarrow (\lambda x)(\exists y_1, y_2, k_1, k_2, k_3)$$
$$L(y_1, \{y_1, y_2\}, x, x, A01, Gt, k_1) \Pi L(y_1, y_2, \text{Down}, \text{Down}, A13, Gt, k_2)$$
$$\Pi L(y_1, y_2, \text{Forward}, \text{Forward}, A13, Gt, k_3) \wedge \text{person}(y_1)$$
$$\wedge \text{head}(y_2) \wedge \text{motion}(x) \tag{11.68}$$

As for our experiment, colored markers were put on the upper half part of the human body, namely, head, neck, shoulders, elbows, hands, and navel and their position data (i.e., 3D coordinates) were taken in through a motion capturing system at a sampling rate. Figure 11.19 shows the structure of the wire frame model of the upper half of the human body. This model was implemented by using locus formula representation just like (11.66) and (11.67). Real motion data were graphically interpreted according to the model as shown in Figure 11.20

The datum unit can be formalized by a quadruple (S, B, P, T), where S, B, P and T mean "name of the subject", "name of the body part", "position of the body part", and "time point of data sampling", respectively.

The Data Interpretation Function (F_d) digests a large number of motion data of the subject's head over a time interval into a locus formula such as (11.69), where *Tom* is the default name of the subject and P_is are characteristic points of the movement of the head such as turning points.

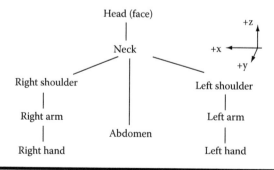

Figure 11.19 Wire frame model of upper half of human body.

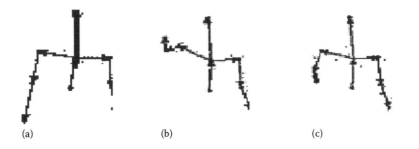

(a)   (b)   (c)

Figure 11.20 Graphical interpretations of real motion data: (a) data at t_1, (b) data at t_2, and (c) data at t_3.

This type of expression is called "Motion Meaning Representation" (MMR), where the standard constant Mc means one of certain rigid standards specific to the motion capturing system:

$$L(Tom, Head, P_1, P_2, A12, Gt, Mc)$$

$$\cdot L(Tom, Head, P_2, P_3, A12, Gt, Mc)$$

$$\cdot \ldots \cdot L(Tom, Head, P_{n-2}, P_{n-1}, A12, Gt, Mc)$$

$$\cdot L(Tom, Head, P_{n-1}, P_n, A12, Gt, Mc) \tag{11.69}$$

Human motion data gained through a motion capturing system associate limitedly with the sense "sight" and its related attributes are A12 (physical location) and A34 (time point).

In translation between motion data and texts, these two attributes and the others are to be paraphrased with each other according to "Attribute paraphrasing rules" (APRs) such as (11.70) through (11.72), where the left and right hands of the symbol "$\Leftrightarrow$" refer to the attributes concerning to MMRs and TMRs, respectively, and the attributes "A01" and "A13" refer to "Place of existence" and "Direction", respectively:

$$(\exists \mathbf{p}, \mathbf{q})L(y_1, y_2, \mathbf{p}, \mathbf{q}, A12, Gt, Mc) \wedge \mathbf{q} \neq \mathbf{p} \wedge \mathbf{p} = (p_x, p_y, p_z)$$

$$\wedge \mathbf{q} = (q_x, q_y, q_z) \Leftrightarrow (\exists x, k)L(y_1, \{y_1, y_2\}, x, x, A01, Gt, k)$$

$$\wedge motion(x) \tag{11.70}$$

$$(q_z - p_z < 0, A12) \Leftrightarrow (Down, A13) \tag{11.71}$$

$$(q_y - p_y > 0, A12) \Leftrightarrow (Forward, A13) \tag{11.72}$$

Based on APRs (11.70) through (11.72), the MMR (11.69) is unified with (11.68), namely, the conceptual description of the verb "nod", which yields the TMR (11.73):

$$(\exists x, k_1, k_2, k_3) \, L(Tom, \{Tom, Head\}, x, x, A01, Gt, k_1)$$

$$\Pi L(Tom, Head, Down, Down, A13, Gt, k_2)$$

$$\Pi L(Tom, Head, Forward, Forward, A13, Gt, k_3)$$

$$\wedge person(Tom) \wedge head(Head) \wedge motion(x) \tag{11.73}$$

The sentence "Tom nodded." is to be generated from this TMR using the sentence pattern of "nod", which is generalized as "$\mathbf{y_1}$ **nod**" indicating the correspondence between the subject of the verb and the term "y_1" in its conceptual description (11.68).

Tom moved the right hand.
Tom moved the right arm.
Tom moved the right elbow.
....
Tom put the right hand up.
Tom raised the right arm.
Tom bent the right arm.
Tom put the right hand up and simultaneously bent the right arm.
....

(a) Text for motion data from t_1 to t_2.

....
Tom put the right hand down.
Tom lowered the right arm.
Tom stretched the right arm and simultaneously lowered the right hand.
....

(b) Text for motion data from t_2 to t_3.

Figure 11.21 Texts generated from real motion data.

Figure 11.20a through c are graphical interpretations of the real motion data at the time points, t_1, t_2, and t_3 respectively. The sets of real motion data over time intervals $[t_1, t_2]$ and $[t_2, t_3]$ were translated into the texts in Figure 11.21a and b, respectively.

11.8.3 Robot Manipulation by Natural Language

The intelligent system IMAGES-M can deploy Sony AIBOs, dog-shaped robots, as actors and gather information about the physical world through their microphones, cameras, and tactile sensors. Communications between IMAGES-M and humans are performed through the keyboard, mouse, microphone, and multicolor TV monitor of the personal computer.

Consider such a verbal command as S12 uttered to the robot, Sony AIBO, named "John":

(S12) John, walk forward and wave your left hand.

Firstly, late in the process of cross-media translation from text to AIBO's action, this command is to be interpreted into (11.74) with the attribute "shape" (A11) and the values "Walkf$_1$", and so on, at the standard of "AIBO", reading that John makes himself walk forward and wave his left hand. Each action in AIBOs is defined as an ordered set of shapes (i.e., time-sequenced snapshots of the action) corresponding uniquely with the positions of their actuators determined by the rotations of the joints. For example, the actions "walking forward" (**W**alkf) and "waving left hand" (**W**avelh) are defined as (11.75) and (11.76), respectively:

$$L(John, John, Walkf_1, Walkf_m, A11, Gt, AIBO)$$

$$\wedge\, L(John, John, Wavelh_1, Wavelh_n, A11, Gt, AIBO) \tag{11.74}$$

$$\mathbf{W}alkf = \{Walkf_1, Walkf_2, \ldots, Walkf_m\} \tag{11.75}$$

$$\mathbf{W}avelh = \{Wavelh_1, Wavelh_2, \ldots, Wavelh_n\} \tag{11.76}$$

Secondly, an AIBO cannot perform the two events (i.e., actions) simultaneously, and therefore the transit event X_T between them is to be inferred as the underlined part of (11.77), which is the goal event X_G here:

$$L(John, John, Walkf_1, Walkf_m, A11, Gt, AIBO)$$

$$\cdot \underline{L(John, John, Walkf_m, Wavelh_1, A11, Gt, AIBO)}$$

$$\cdot L(John, John, Wavelh_1, Wavelh_n, A11, Gt, AIBO) \tag{11.77}$$

Figure 11.22   AIBO (Sony) behaving in accordance to the command, "Walk forward and wave your left hand."

Thirdly, (11.78) is to be inferred, where the transit event, underlined, is interpolated between the current event X_C and the goal event X_G [=(11.77)]:

$$L(John, John, p_1, p_2, A11, Gt, AIBO)$$
$$\underline{\cdot L(John, John, p_2, Walkf_1, A11, Gt, AIBO)} \cdot X_G \tag{11.78}$$

Finally, (11.78) is interpreted into a series of joint rotations in the AIBO as shown in Figure 11.22.

11.9 Discussion and Conclusion

"Action planning" (AI planning) deals with the development of representation languages for planning problems and with the development of algorithms for plan construction [6,7,30]. The author formalized the performances of a DIRN as predicate logic in the formal language L_{md} and applied it to robot manipulation by text and so on as simulation of DIRN–world interaction.

This is one kind of cross-media operation via locus formulas as already reported. At my best knowledge [e.g., 31,32], there is no other theory or system that can perform cross-media operations in such a seamless way as ours, which leads to the conclusion that employment of locus formulas has made both spatial and temporal event concepts remarkably computable in an integrated way and has proved to be very adequate to systematize cross-media operations. This is due to their medium-freeness and good correspondence with the performances of miscellaneous devices, which in turn implies that locus formula representation may make it easier for the devices to share a task than any other representation, even if, based on some precise ontology or mathematical definition [33,34].

In this simulation, a problem for a DIRN to solve is defined as a goal event (X_G) and a transit event (X_T) between the current event (X_C) and the goal event. The task sharing and assignment among the nodes or agents are executed based on the information of a problem described as locus formulas in L_{md}. The most useful keys to task assignment are the attributes involved and about 50 kinds of attributes have been found in association with natural languages and human sensory organs as shown in Table 11.2.

Furthermore, most of the computations on L_{md} are simply for unifying (or identifying) atomic locus formulas and for evaluating arithmetic expressions such as "$p = q$", and therefore we believe that our formalism can reduce the computational complexities of the others [15–17] when applied to the same kinds of problems described here.

The simulation results lead to the conclusion that L_{md} can be a universal language appropriate for WSANs including DIRNs. Our future work will include establishment of learning facilities for automatic acquisition of word concepts from sensory data and human–robot interaction by natural language under real environments.

Acknowledgments

This work was partially funded by the grants from Computer Science Laboratory, Fukuoka Institute of Technology, and Ministry of Education, Culture, Sports, Science and Technology, Japanese Government, numbered 14580436, 17500132 and 23500195.

References

1. I.F. Akyildiz, W. Su, Y. Sankarasubramaniam, and E. Cayirci. Wireless sensor networks: A survey. *Computer Networks*, 38(4), 393–422 (2002).
2. M. Haenggi. Mobile sensor-actuator networks: Opportunities and challenges. *Proceedings of 7th IEEE International Workshop,* Frankfurt, Germany, pp. 283–290 (2002).
3. I.F. Akyildiz and I.H. Kasimoglu. Wireless sensor and actor networks: Research challenges. *Ad Hoc Networks*, 2, 351–367 (2004).
4. M. Yokota and G. Capi. Integrated multimedia understanding for ubiquitous intelligence based on mental image directed semantic theory. *IFIP EUC'05 UISW2005*, Nagasaki, Japan, pp. 538–546 (2005).
5. M. Yokota. Towards a universal knowledge representation language for ubiquitous intelligence based on mental image directed semantic theory. J. Ma et al. (Eds.), *Ubiquitous Intelligence and Computing 2006 (UIC 2006)*, LNCS 4159, Wuhan, China, pp. 1124–1133 (2006).
6. D.E. Wilkins and K.L. Myers. A common knowledge representation for plan generation and reactive execution. *Journal of Logic and Computation*, 5(6), 731–761 (1995).
7. F. Kabanza. Synchronizing multiagent plans using temporal logic specifications. *Proceedings of the First International Conference on Multi-Agent Systems (ICMAS-95)*, San Francisco, CA, pp. 217–224 (1995).
8. J. Bos and T. Oka. A spoken language interface to a mobile robot. *Proceedings of the Eleventh International Symposium on Artificial Life and Robotics (AROB-11)*, Oita, Japan (2006).
9. S. Coradeschi and A. Saffiotti. An introduction to the anchoring problem, *Robotics and Autonomous Systems*, 43, 85–96 (2003).
10. M. Yokota. An approach to natural language understanding based on a mental image model. *Proceedings of the Second International Workshop on Natural Language Understanding and Cognitive Science*, Miami, FL, pp. 22–31 (2005).
11. M. Yokota et al. Mental-image directed semantic theory and its application to natural language understanding systems. *Proceedings of Natural Language Processing Pacific Rim Symposium (NLPRS'91)*, National Computer Board (NCB), 71 Science Park Drive, NCB Building, Singapore, pp. 280–287 (1991).
12. J.F. Sowa. *Knowledge Representation: Logical, Philosophical, and Computational Foundations* (Brooks Cole Publishing Co., Pacific Grove, CA, 2000).
13. G.P. Zarri. NKRL, a knowledge representation tool for encoding the 'Meaning' of complex narrative texts. *Natural Language Engineering—Special Issue on Knowledge Representation for Natural Language Processing in Implemented Systems*, 3, 231–253 (1997).
14. B. Dorr and J. Bonnie. Large-scale dictionary construction for foreign language tutoring and interlingual machine translation. *Machine Translation*, 12(4), 271–322 (1997).
15. J.F. Allen. Towards a general theory of action and time. *Artificial Intelligence*, 23(2), 123–154 (1984).

16. Y. Shoham. Time for actions: On the relationship between time, knowledge, and action. *Proceedings of the Eleventh International Joint Conference on Artificial Intelligence (IJCAI89)*, Detroit, MI, pp. 954–959 (1989).

17. P. Haddawy. A logic of time, chance, and action for representing plans. *Artificial Intelligence,* 80(2), 243–308 (1996).

18. S. Oda, M. Oda, and M. Yokota. Conceptual analysis description of words for color and lightness for grounding them on sensory data. *Transactions of JSAI,* 16(5-E), 436–444 (2001).

19. M. Amano et al. Linguistic interpretation of human motion based on mental image directed semantic theory. *Proceedings of IEEE AINA-2005,* Taipei City, Taiwan, pp. 139–144 (2005).

20. R. Langacker. *Concept, Image and Symbol* (Mouton de Gruyter, Berlin, Germany/New York, 1991).

21. G.A. Miller and P.N. Johnson-Laird. *Language and Perception* (Harvard University Press, Cambridge, MA, 1976).

22. E. Leisi, *Der Woltinhalt—Seine Struktur in Deutchen und Englischen* (Quelle & Meyer, Heidelberg, Germany, 1961).

23. E. De Bono. *The Mechanism of Mind* (C.Tuffle Co. Inc., Tokyo, Japan, 1969).

24. D. Noton and L. Stark. Scanpaths in eye movements during pattern perception. *Science,* 171(3968), 308–311 (1971).

25. I.A. Rybak, V.I. Gusakova, A.V. Golovan, L.N. Podladchikova, and N.A. Shevtsova. A model of attention-guided visual perception and recognition, *Vision Research,* 38, 2387–2400 (1998).

26. M. Yokota. An approach to integrated spatial language understanding based on mental image directed semantic theory. *Proceedings of the 5th Workshop on Language and Space,* Bremen, Germany (October 2005).

27. M. Egenhofer and K. Khaled. Reasoning about gradual changes of topological relations, *Proceedings of International Conference GIS—From Space to Territory: Theories and Methods of Spatio-Temporal Reasoning,* Pisa, Italy, pp. 196–219 (1992).

28. M. Yokota and G. Capi. Cross-media operations between text and picture based on mental image directed semantic theory, *WSEAS Transactions on Information Science and Applications,* 10(2), 1541–1550 (2005).

29. M. Yokota, H. Yoshitake, and T. Tamati. Japanese-English translation of weather reports by ISOBAR. *Transactions of IECE Japan,* E67(6), 315–322 (1984).

30. R. Lundh, L. Karlson, and A. Saffiotti. Plan-based configuration of a group of robots. *Proceedings of the Seventeenth European Conference on Artificial Intelligence (ECAI),* Riva del Garda, Italy (2006).

31. G. Adorni, M. Di Manzo, and F. Giunchiglia. Natural language driven image generation. *Proceedings of COLING'84,* Stanford, CA, pp. 495–500 (1984).

32. J.P. Eakins and M.E. Graham. Content-based image retrieval. A report to the JISC Technology Applications Programme (Institute for Image Data Research, University of Northumbria, Newcastle upon Tyne, U.K., 1999).

33. N. Guarino and C.A. Welty. An overview of onto clean. *Handbook on Ontologies,* pp. 151–172 (Springer-Verlag, Berlin, Germany, 2004).

34. P. Gardenfors. Representing actions and functional properties in conceptual spaces. T. Ziemke, and J. Zlatev (Eds.), *Body, Language and Mind* (Nature Publishing Group, Basingstoke, U.K., 2004).

SECURITY, PRIVACY, AND TRUST MANAGEMENT

Chapter 12

Agent-Based Automated Trust Negotiation and Management for Pervasive Computing

Dichao Peng, Daoxi Xiu, and Zhaoyu Liu

Contents

12.1 Introduction

Trust is a relation between two entities such that one entity (i.e., the host) believes, expects, and accepts that a trusted entity (i.e., a client) will act or intend to act beneficially [14]. In other words, trusted clients will perform according to the host's expectation. Based on this trust concept, the general application of trust to information systems and computer networks is in the following areas: *information services*, which provide relevant and valid information to the other systems and human beings, and *security assurance*, which ensures and enhances the security of the computing systems through trust management.

Valid information services including providing data and executing actions are what people expect from the computing systems and networks. In this sense, a computing system can be trusted only if it meets the expectations. In order to provide valid information services, the computer systems and networks not only rely on correct and effective algorithms and computing architectures, which are the topics for algorithm designers and computer system architects, but also need to ensure the security of the systems to prevent themselves from external interferences or compromises on the data and processes across the networks, which is the topic that computer security is concerned.

Trust and security are closely related in following aspects [6]:

■ Security is viewed as one objective of trust in which only the trusted clients are allowed to access the restricted resources or get the information services from a system.

■ Security measures (e.g., cryptography) keep the trust information or evidences from being compromised during information distribution processes among entities.

■ Security can be realized more effectively using dynamic trust management, by which a client with a certain trust level will be allowed to execute certain actions that match its trust level in the host environments, and the host is confident that the client will not compromise host environments.

Therefore, trust is the prerequisite for a computing system to interact securely with any external entities and provide reliable services. As a result, trust management is identified as one of the important components in computer security assurance, which comprises of the activities for establishing and maintaining the trust relation between entities based on the trust evidences available, and controlling the execution of the trusted actions requested by the clients.

Before a client wants to access to the resources or get the services in a system, trust must be established between the client and the system so that the security of the system and the client can be ensured. Trust establishment is mainly achieved through the trust formation process, in which the system collects the trust evidences of its clients, defines the trust polices for the resources in the system, and builds up the trust levels of the clients based on the trust evidences and policies [6]. In traditional distributed systems, trust establishment is based on the mechanisms of authentication and authorization that identify clients' identities to determine their access rights. The practical implementations for these mechanisms include access control and cryptography, which are further extended in the forms of role-based access control, capability-based access control, and public key infrastructure. In general, the trust formation processes in these systems are static considering the following facts:

- The systems and the clients may be all known to each other based on the sharing information in the networks.
- The clients may preregister in the systems through off-line credential verification and system setup procedures.
- Trust levels of the clients cannot be modified online, that is, the trust level of the clients are determined by the trust control policies and the credentials available at the setup time.

Although these trust mechanisms provide the effective ways for securing the systems, they are obviously inadequate for the increased flexibility and scalability that are required in the coming pervasive computing environments.

The pervasive computing environments, including the mobile computing and dynamic distributed computing systems, have the following properties [6]:

- Mobility—the entities may move anywhere and request to access to services and information from other entities anytime.
- No central control—the mobile entities are likely disconnected from their home network.
- Partial available information—the entities may be unknown to each other.
- Heterogeneity—different system structures may be implemented in different domains.
- Autonomy—autonomous and dynamic actions are executed among massive mobile entities.

All these properties make it much more difficult for dynamic trust establishment. Therefore, more effective trust establishment mechanisms for security assurance are needed [6,22]. The objective of our current research is to explore feasible and effective solutions to the problems in this area.

In this chapter, an agent-based automation trust negotiation framework is proposed that can build up the trust between the entities dynamically. By introducing an active trust negotiation scheme and using the mobile agents called security capsules to encode the trust negotiation strategies, the agent-based automated trust negotiation framework not only provides a solution to adaptive and dynamic trust negotiation between the entities in pervasive computing environments, but also introduces a uniform infrastructure that can be applied to both distributed systems and Internet applications.

The remainder of this chapter is organized as follows. Section 12.2 discusses the concepts regarding trust negotiation and its challenge in pervasive computing environments. Section 12.3 proposes the concept for the active trust negotiation scheme. In Section 12.4, the infrastructure for agent-based trust negotiation framework is designed based on the proposed active trust negotiation scheme. In Section 12.5 we describe more on the maintenance of trust relations. Section 12.6 discusses the challenging issues in this research and our future work in the implementation of the agent–based trust management infrastructure. We conclude the discussion in Section 12.7.

12.2 Background and Motivation

In general, trust establishment is the process by which the system determines the trustworthiness of a client after collecting and evaluating the client's trust evidences. Trust establishment between two entities is grounded in one's evaluation on another's ability, benevolence, and integrity, and trust evaluation is based on trust evidences that describes the client's intention or behavior pattern [14]. Three types of client trust evidences are identified to be used for trust establishment [22]: credentials, environment context, and behavior records.

The concept of trust has been used in distributed computing systems for a while [3,8,14,16,19]. To ensure the system security through trust management, many researches are proposed to help unknown parties establish trust between each other [1,2,4,9,10,15–18,20,21].

KeyNote [1] is a well-known trust management language designed to work for a variety of Internet-based applications. It provides a unified language for both local policies and credentials. KeyNote policies and credentials, called "assertions," contain predicates describing delegations in terms of actions that are relevant to a given application. KeyNote applies a policy to a set of credentials and outputs a decision.

Stajano and Anderson's resurrecting duckling [18] model represents a peer-to-peer trust establishment framework. Based on the natural fact that a duckling emerging from its egg will recognize as its mother the first moving object it sees that makes a sound, their model advises exchanging keying material via an out-of-band physical contact. Assuming the security of this physical contact channel, this model provides a simple but effective trust establishment scheme, although it is usually not applicable in many pervasive computing scenarios.

Yu [23] has investigated issues of automated trust negotiation with privacy control. In this approach, a large set of negotiation strategies is defined, each of which chooses the set of disclosures in a different way. If two parties independently pick any two strategies from the set, then their negotiation is guaranteed to establish trust under some condition. Then the problem becomes that the two parties have to agree on which set of strategies they will use. In this research work, it is assumed that there exist trust negotiation strategies and protocols that both the clients and the system know and agree on beforehand. But in the pervasive computing environments, this assumption does not always hold considering the following factors:

- Clients may be strangers and they may not know the system or the resources in the system before they meet. Therefore, they may not have the shared trust negotiation strategies and protocols related to the resources either.
- Different resources in the system may implement different trust control policies, so trust negotiation strategies and protocols for the resources may be different too. Further, it is not possible for the clients to keep all the trust negotiation strategies and protocols beforehand since there are unlimited unknown systems.
- It is not reasonable for the clients to take up a lot of space to keep all the trust negotiation strategies and protocols for all the systems or resources due to the economic consideration and space limitation.
- The trust policies for the systems or the resources in the systems may change dynamically, so do trust negotiation strategies.

These listed factors raise the following questions on the automated trust negotiation in pervasive environments:

- How to define and maintain the trust negotiation strategies and protocols so that the trust strategies can dynamically adapt to the change of the trust control polices and the system environments?

- How to get the trust negotiation strategies and protocols dynamically so that both the client and the system can have the shared strategies for the negotiation?
- How to ensure the interoperability for the trust negotiation strategies between the clients and the systems?

The agent-based automated trust negotiation framework proposed in this chapter will provide an answer to these questions as described in the following sections.

12.3 Active Automated Trust Negotiation

The main challenges for automated trust negotiation are how to define, maintain, and acquire the interoperable trust negotiation strategies and protocols. In this section, we first look at the general considerations for defining , maintaining, and acquiring the trust negotiation strategies, then present an effective approach, active trust negotiation scheme, for automated trust negotiation.

12.3.1 Sources of Trust Negotiation Strategies

As mentioned in Section 12.2, the trust negotiation process requires both the client and the system to have the interoperable negotiation strategies and protocols in place before the negotiation starts. Due to the computing space or domain limitations, the client usually does not have all the trust negotiation strategies when the client intends to negotiate for accessing the resources in the system. In this case, the client needs to acquire them from the system or other resources when trust negotiation is on demand. Otherwise, trust negotiation cannot proceed.

The interoperable trust negotiation strategies are in a family, and any two of them can be used by the client and the system to ensure the trust negotiation process to proceed. One special case is that the client and system are using one same trust negotiation strategy so that the interoperability is definitely ensured. This property is what we expect and provide an option for use to choose trust negotiation strategies.

The trust negotiation strategy and protocol for a resource in the system is mainly determined by the trust control polices of the resource. In other words, each trust policy in the system with negotiation ability should specify a trust negotiation strategy to be used when trust negotiation is on demand. If the client intends to negotiate to access a resource in a system, both the client and the system can use the same trust negotiation strategy specified by the trust control policy for the resource in the system.

For performance and security reason, negotiation strategies and protocols for the resources are usually defined and maintained in the same system as the policies. Because trust negotiation strategies are used locally to help maintain and control the release sequence of local credentials, they must be under the control of local environments for efficiency and security. One effective and dynamic way for the client to get the trust negotiation strategy from the system is to download the trust negotiation strategy online when the client cannot find an interoperable trust negotiation strategy and protocol locally. This leads to our active automated trust negotiation scheme, which operates in two stages: setup stage—trust negotiation strategy download phase; and execution stage—trust negotiation execution phase.

12.3.2 Active Automated Trust Negotiation Scheme

The active trust negotiation scheme includes two stages: trust negotiation strategy download and trust negotiation execution. The process in each stage poses challenges in the implementation.

In the following sections, we will discuss the general procedures in each stage and the main issues concerned.

12.3.2.1 Setup Stage: Trust Negotiation Strategy Download

In this stage, the client and system will go through the following procedures in sequence:

1. The client requests to access a resource in the system, and both the client and the system realize that trust negotiation is needed.
2. The system notifies the client what trust strategy and protocol is used for the trust negotiation, and the client is searching for an interoperable trust strategy and protocol locally.
3. If the client does not find an interoperable trust negotiation strategy and protocol, it requests to download the trust strategy from the system so that the client and system share the same trust negotiation strategy and protocol. Else, the client will use its local interoperable trust negotiation strategy.
4. Both the client and the system notify each other after loading the trust negotiation strategies, respectively.

During this setup stage, the following issues should be put into consideration for security and compatibility:

■ Download process should be secure enough so that it ensures that the downloaded codes from a system do not include the malicious codes. This needs a secure and reliable download protocol.
■ The trust strategy and protocol being downloaded should be applicable and compatible with the client's system, that is, the client's system can understand what the strategy and protocol denote and use them to execute the trust negotiation.

12.3.2.2 Execution Stage: Negotiation for Access to the Resources

After the client has found a local interoperable trust negotiation strategy or downloaded the trust negotiation strategy from the system, both the client and the system will load the trust negotiation strategy and protocol and start the trust negotiation. The client and the system will go through the following procedures:

1. Both the client and the system exchange credentials as requested by the trust control polices, which is guided by the trust negotiation strategy and uses the trust negotiation protocol for messages transfer.
2. If the trust negotiation code executer (normally the client) is not confident of the code security, it may use sandbox technologies to help protect itself from malicious code.
3. If the client's credentials meet the requirement of the trust polices for accessing the requested resource, the system grants the access rights for the client and the client performs the allowable actions on the resource. Else, no trust can be established.

Trust negotiation strategies and protocols are related to the trust control policies in the system. In our case, trust negotiation strategies will be downloaded to the clients, so one important thing to be considered is to make the trust negotiation strategies compatible with the local trust control policies the client.

Although there is still more research work to do in providing secure and efficient trust negotiation strategies, it is not our focus in this chapter. We assume that there are valid trust negotiation strategies being stored in the system and specified by trust control policies.

12.4 Agent-Based Infrastructure for Trust Negotiation

The concept for the active automated trust negotiation provides a way for trust negotiation for the mobile entities in pervasive computing environments theoretically. As mentioned in Section 12.2, there are several issues concerned such as download protocols, trust policy language, and trust negotiation strategy encoding. In this section, an agent-based infrastructure for trust management is proposed to implement the active trust negotiation scheme and provide the corresponding solutions to the related issues.

We first present the overview of the framework and the main components in this framework, then concentrate on the trust negotiation agent design and the detailed mechanisms to resolve the implementation issues for the active trust negotiation scheme in the infrastructure.

12.4.1 Architecture Overview

Figure 12.1 illustrates our agent-based infrastructure for trust management in pervasive computing environments. As described in the following, this infrastructure is aimed to realize trust management with different trust management strategies for different application scenarios such as trust negotiation, trust recommendation, etc. In this chapter, we will focus on trust negotiation and discuss the design of trust negotiation agent and the implementation of the active trust negotiation scheme.

The main components in this trust management infrastructure include service control, communication control, security agent, various repositories, and different trust agents such as trust negotiation agent, trust recommendation agent, and so on.

12.4.1.1 Service Control

Service control works as a general controller and is the main interface to communicate with the external entities. All the service requests from clients should go through the service control.

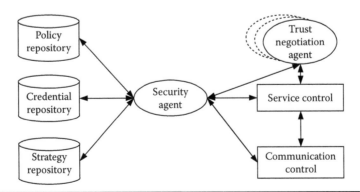

Figure 12.1 Overview of agent-based infrastructure for trust management.

The service control will consult with the security agent for the security enforcement. It is service control that initiates the corresponding trust management processes based on the consulting result from the security agent. The service control requests the communication control to allocate the communication protocols and channels for specific tasks or specific components in the systems (e.g., trust negotiation is done through the negotiation communication channels). In this way, some components (e.g., trust negotiation agents) will communicate with the external entities through the specified channels. The service control also keeps all the current clients and manages communication session information.

12.4.1.2 Security Agent

A security agent enforces the security of the local system. It maintains the policy repository, the credential repository, and the strategy repository and uses the trust control polices and digital signatures of credentials to perform trust management. All the regular authentication and authorization are supported by the security agent. The security agent also delegates the trust evaluation tasks to the specific agents based on different trust evidence types and application scenarios. By synthesizing the evaluation results from all the available agents, it concludes the trustworthiness of the clients and makes the decision for granting access rights to the client so that the client can perform on the resources.

12.4.1.3 Communication Control

Communication control is a component for allocating the communication channels and protocols. It usually receives the requests from the service control or the security agent for allocating the communication channels and protocols and monitoring data traffic. The commonly used channels and protocols are specified by the service control when the system starts. Specific agents such as trust negotiation agents are allocated specific channels, respectively.

12.4.1.4 Policy Repository, Credential Repository, and Negotiation Strategy Repository

The policy repository stores the rules and policies that are used by the security agent to enforce the system security. The credential repository keeps the credential information and the account information for the system itself and clients. The negotiation strategy repository keeps all the trust negotiation strategies and protocols, which are specified by the trust control policies and used for trust negotiation. All these repositories and the information stored in them are controlled by the security agents and can be queried by other agents.

12.4.1.5 Trust Negotiation Agent, Trust Recommendation Agent, and Other Agents

These are specific agents who handle trust evaluation tasks for different application scenarios and trust evidence types. Since different application scenarios need different methods for collecting trust evidences and use different criteria for evaluating trust evidences, different agents are used to implement these different trust establishment mechanisms, respectively. In this chapter, our focus is on the trust negotiation process, so we will describe its detailed implementation and how it realizes our two-stage active trust negotiation strategies.

12.4.2 *Trust Negotiation Agent and Trust Negotiation Process*

Trust negotiation agent in the trust management infrastructure is responsible for trust negotiation. It receives the delegated task from the security agent and service control for negotiating to establish a trust relationship with the other party. Following our active trust negotiation scheme, the trust negotiation agent will do the following:

- Downloading the trust negotiation strategies
- Loading the trust strategies locally
- Executing the trust negotiation based on the trust strategies

The detail implementations are related to trust control policies and languages, trust negotiation protocols, trust negotiation strategies, secure download protocols, and negotiation agent designs. In the following sections, detailed implementation considerations are presented for each component and protocol.

12.4.2.1 *Basic Considerations and Assumptions*

The trust negotiation process is to exchange the credentials (i.e., trust evidences) for trust establishment. It gets involved in the request for credentials and the release of credentials, which in turn needs the trust policies to specify what the requested credentials should be, and relies on the trust negotiation strategies and protocols to determine how to guide and control the request and release of the credentials.

Trust negotiation must start with some common noncritical credential exchanges, then proceed on critical credential exchanges guided by trust control policies and trust negotiation strategies. In other words, trust negotiation is not a zero knowledge–based process. Trust control policies, trust negotiation strategies, and protocols are related to how to describe the credentials or handle the credentials, so compatible languages are needed for semantic consistency when describing the credentials and action types. Therefore, based on the discussion in Section 12.3, the following assumptions are made for simplicity.

- The trust negotiation discussed in this chapter is not a zero knowledge–based process. The client and the system must have some common base or noncritical credentials (e.g., known identifications) to start with.
- The same trust policy language is used to define trust or access control policies so that the concept or vocabulary in the trust control policies can be understood by the trust negotiation strategies.
- The same protocol language is used to define the trust negotiation protocol and encoding the messages, whose concept and vocabulary are the same as (or compatible with) the ones in trust control policies.
- These trust negotiation strategies and protocols are defined and stored in the systems and downloaded to the client.

12.4.2.2 *Trust Control Policies, Disclosure Policies, and Policy Language*

Trust control policy of a resource in the system specifies the context and the required credentials for the client to present for trust evaluation. The system can grant the client to access the resource if the client's credentials meet the requirements specified by the trust control policy. In the trust negotiation process, in order not to release the content of trust control policies of the system totally, the trust control policies should be designed to request the client's credentials gradually from the noncritical ones, less critical ones, to the more critical ones. Meanwhile, the trust policies in the client may also specify the gradual request sequence for some system credentials before the client's credentials are released to the system. The trust control policies are also called *disclosure polices (or negotiation policies)*, which define the gradual disclosure sequences of the contents of policies and credentials for trust negotiation. We still call it *trust control policies* in this chapter.

Trust control policies are defined by policy languages. For simplicity, we assume that trust policies are defined using one same policy language in our implementation. An XML-based trust policy language similar to the trust Language TRN-X defined in [2] or TPL in [7] can be an appropriate option due to its applicability and flexibility.

12.4.2.3 Trust Negotiation Protocol

Trust negotiation protocol defines the format for trust negotiation messages, which includes the requests for credentials, replies for requests, and command to start or halt trust negotiations as well as the contents of messages such as credentials and policies. We assume that protocols are written in XML to encode all the commands, credentials, and polices in the messages, which match the trust policy language in both vocabulary and semantics. The negotiation protocols are stored in the system together with the trust negotiation strategy and will be downloaded to the clients.

12.4.2.4 Trust Negotiation Strategy and Security Capsule

Trust negotiation strategies are the algorithms for controlling the credential disclosure based on the trust control policies. So this requires the trust negotiation strategies to interpret the trust control polices, maintain the acquired credentials, and arrange the credentials disclosure sequence during trust negotiation. Since a trust negotiation strategy is an algorithm, it should be implemented as an integral program that can be executed by the calling environment. A client will download the program, load it into the local running environment, and execute the program. Also, the trust negotiation strategies should be under the control of trust negotiation agents and security agent.

For this purpose, we extend the concept of active capability given in [12,13] and define security capsule to encode the trust negotiation strategies. A security capsule is an executable code that encodes the negotiation strategy and protocol. It includes the data types and algorithm of the trust negotiation strategy and can be loaded and executed by the negotiation agent. For the compatibility and integrity of the trust negotiation strategy, security capsules implement a common interface with the negotiation agent. The negotiation agent queries or receives the information about the credential request or disclosure sequence to form the negotiation messages according to the negotiation protocols. In this way, the security capsule is transparent to negotiation agents and can be loaded to any trust negotiation system if the negotiation agents import the common interface.

The relationship between the negotiation agent and security capsule is illustrated using the UML component diagram in Figure 12.2, in which *Istrategy* represents the common interface that is implemented by all the Security Capsule components, *StrategyMaintainer* object is responsible for interpreting the trust control policies and controlling the credential release based on the control polices, and *CredentialMaintainer* is responsible to maintain the external credentials provided by the counter party and the local credentials that have been released to the counter party.

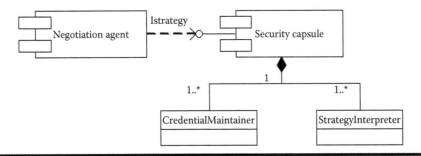

Figure 12.2   The relationship between the trust negotiation agent and security capsule.

In general, the security capsules have the following advantages by extending the concept of active capability:

- Mobility—they can be downloaded from the system online dynamically.
- Ready for use—they are programs that can be executed when being loaded into the run-time and can be reused by moving to any systems or clients.
- Configurability—a security capsule interacts with the negotiation agent through the common interface. Any updating in the internal objects is transparent to the negotiation agent. So a security capsule can be used for different applications by reconfiguring it if it is designed so.

12.4.2.5 Trust Negotiation Strategy Download Protocol

When the client downloads the security capsules, it should ensure that security capsules do not contain any malicious codes. This involves the following two-step operations:

- The download protocol should ensure the integrity of the downloaded security capsule, which requires the protocol to have a checking mechanism for verification. The security hash method of MD5 can provide this functionality.
- The security capsule can only communicate with the negotiation agent locally through the common interface and it cannot leak any information and compromise local environments. This can be realized by the local security policies and security domain such as the sandboxes.
- The security capsule will be verified using some bench trust control policies for its validity.

12.4.2.6 Trust Negotiation Agent

The trust negotiation agent is responsible for the trust negotiation task in our agent-based trust management infrastructure. During the two-stage trust negotiation process, when the service control receives a request for access to resources, it consults with the security agent and concludes that trust negotiation is needed, then delegates the task to the trust negotiation agent. Then the trust negotiation agents from both sides will check whether the trust negotiation strategies required by the trust policy can be found locally. If the client cannot find a compatible one locally, its trust negotiation agent will request to download the security capsule for the trust strategies from the host. When the trust strategies are present with the security capsules and ready for use, the trust negotiation agents will start to execute the negotiation process and begin exchanging credentials, which is guided by the trust negotiation strategies. The general trust negotiation process is illustrated in Figure 12.3.

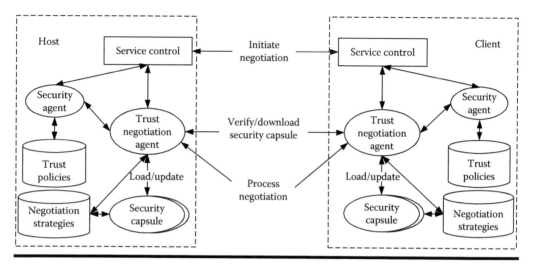

Figure 12.3   Negotiation process.

As illustrated in Figure 12.3, the client most times needs to download the trust negotiation strategies. Therefore, the trust negotiation agent needs to maintain a reliable download protocol in addition to the trust negotiation protocols. All these protocols should be compatible with the vocabulary and semantics of the trust control policies and trust negotiation strategies, which are usually defined using policy languages.

12.5 Trust Maintenance and Management

Besides an effective trust negotiation scheme, the next important research is to ensure trust maintenance after trust relations are established. In pervasive computing systems, because the nodes are moving in and out the computing space constantly, this dynamic characteristic requires mighty trust maintenance schemes to guarantee that the established trust is not abused.

The proposed trust maintenance and management module is composed of several collaborative components (Figure 12.4), which resides on different computing devices and works in a on-demand-assembling fashion to fulfil the security demand of certain services. Some of the components indentify a user, monitor its behavior to detect potential threats, and intervene the service execution; while others manage the established trust in a distributed manner.

The trust synchronization component keeps trust repositories on different virtual servers up-to-date. To ensure the system stableness and reliableness, one pervasive computing space may adopt multiple virtual servers running at the same time, each virtually providing all available services. These proxies may host trust establishment with several clients at the same time. Since there is high probability that a newly enrolled client may immediately access other services residing in the same space, the client profile on each virtual server needs to be synchronized with the other to avoid a client repeatedly negotiating trust with these services.

The trust rating component uses incentives that reward good behavior while punishing bad ones. The trust rating system provides a feedback interface for each service to evaluate the client's behavior in the transaction when the service session is completed. Besides the service's evaluation, a client's rating score is also adjusted by the virtual server's monitoring component. The rating score is distributed in the system repositories. The client may also request a copy of its rating

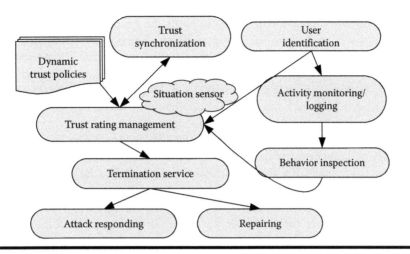

Figure 12.4   Trust maintenance and management.

certificate signed by the pervasive computing space and keep it as a future credential for other services. Correspondingly, a low trust rating of a client can have negative effects on its future resource accessing or even result in the shutdown of the ongoing service.

To cut the realtime-detected malicious users from further endangering the system, *termination service* is designed to shut down an ongoing service session, under the circumstance that one's trustworthiness remarkably declines to an unacceptable level during the service session. The trust level drop may result from an abnormal behavior, detected by the behavior inspection component, or from the user identification component suggesting that the client tries to hide its dishonest history. Because the termination service shuts down ongoing services, which can seriously affect service quality if the user is innocent, it should only be called with high enough confidence.

Complete prevention of bad behavior is infeasible. When the system is compromised or damaged, the *attack responding and repairing* components are activated to minimize the loss. When the attack comes from a client, the responding and repairing components immediately shut down its service and build up a special profile for the attacker, including its conventional behavior and preferences, for future identification even further lawsuit purposes. If the attack is from an ill-behaved service, the responding and repairing components notify all distributed virtual servers to block it from clients and perform further inspections.

Because pervasive computing environments involves very dynamic population, it is difficult to identify a user or their devices. Attackers can change their identity after their reputation (in terms of trust rating scores) has fallen lower than initial value [5]. We propose a *user identification* component to address this issue to recognize a user. The user identification component is designed based on the assumption that every user behaves uniquely during a connection session. The hours of day one usually comes, the displaying devices one prefers, the networking traffic compositions, the preferred access point (may suggest preferred position/location), the service set one uses, the temperature one prefers, and so on; all the information together can uniquely identify a user. Our user identification component creates profiles for each detected ill-behaved device and its user, with a matrix of user preferences. When a self-claimed newcomer is trying to enter the pervasive computing space, a signature-based inspection will be applied to the matrix of every attacker profile. A dishonest veteran that claims itself as a newcomer will be denied further services.

To recognize dishonest behavior that tries to harm the system and to support trust rating component's evaluation, the *activity monitoring and logging* component is dedicated to track user behavior and log this information for further inspection and evaluation. The designing principle of the activity monitoring component is to make the logging information thin and efficient. To keep the logging database thin, only critical activities such as login failures should be stored for a reasonably long time (e.g., 1 year). Other logging information will only stay in the repository for a short time and then be deleted.

Based on the logging information collected by the activity monitoring component, *behavior inspection* applies both abnormality-based and signature-based inspection methods to recognize security breaches. For example, if a supermarket customer is exploring the RFID product querying system by scanning hundreds of items per minute, the abnormality analysis engine on the activity monitoring component will generate an alarm, indicating the customer is behaving abnormally. This alarm may be applied to reevaluate client trustworthiness as a feedback to the trust rating management component.

Trust maintenance and management components may be physically residing on distributed computing devices, collaborating to provide trust support. Though the trust maintenance as a whole is an on-demand-assembled component in the architecture, its subcomponents are themselves optionally recombinable. According to the requirements of a service session, these subcomponents will aggregate to form a trust maintenance component to fulfil the security demand.

12.6 Discussion and Future Work

In the previous sections, an automated trust negotiation framework using the concept of active trust negotiation scheme is proposed for the dynamic trust establishment between two entities in pervasive computing environments. This agent-based infrastructure presents one effective and novel solution for automated trust negotiation, by which the client can share the same trust negotiation strategy defined in the host so that the inter-operability of trust negotiation strategies is ensured.

The trust negotiation is a core part of the trust management framework. In [11], we have proposed a security middleware architecture for pervasive computing system. The trust management component in this middleware includes two subsystems: trust establishment and trust maintenance. The trust establishment subsystem will adopt the strategies discussed in this chapter to help build up initial trust between two parties that are unfamiliar with each other. On the other hand, the trust maintenance system will adopt schemes such as user behavior identification, trust rating management, trust value synchronization, etc., to handle the trust relationship management after the initial establishment.

The implementation of the trust negotiation infrastructure depends on practical policy languages and valid trust negotiation strategies. So, the following are key topics for our future research:

■ Adopting an appropriate trust policy language. A versatile and applicable trust policy language can provide a uniform description of the things and actions including credentials and policies for trust negotiation purpose. This can eliminate the confusion and mismatch in vocabulary and semantics defined in different trust policy languages.

■ Designing and implementing the security capsule to encode various trust negotiation strategies. Also, the privacy of the trust negotiation policy and the corresponding credentials during the trust negotiation process will be considered in this implementation.

■ Providing effective validators to validate the trust negotiation strategies in the following part: (1) they do not contain the malicious codes that can compromise the local environments

and negotiation logic; and (2) they provide valid and efficient logic for trust negotiation. This requirement is not only for the cases downloading trust negotiation strategies, but also for the cases incorporating the trust negotiation strategy products from third-party vendors into the local trust negotiation environments.

Our research work will be carried out in these areas and provide a testbed for trust negotiation based on our agent-based trust management infrastructure. Therefore, different trust negotiation strategies can be proposed, encoded as security capsules, and tested in the testbed.

12.7 Conclusions

In this chapter, an agent-based automated trust negotiation framework is proposed to dynamically build up trust between the mobile entities in pervasive computing environments. It provides an effective and novel solution to a challenging problem in dynamic trust formation in pervasive computing environments: how to get interoperable trust negotiation strategies and protocols to negotiate for dynamic trust establishment if the two unknown entities meet at anytime and anywhere.

In this agent-based automated framework, the proposed concept of active automated trust negotiation scheme addresses when and how to get the trust negotiation strategies. This active automated trust negotiation scheme enables the clients and the system to start the negotiation with the shared trust negotiation strategies at anytime and at their wish. By using the mobile agents called *security capsules* to encode the definition and interpretation of the trust negotiation strategies, the negotiation strategies can be distributed and loaded dynamically and transparently. The mobility, configurability, and transparency of the security capsules make this trust negotiation framework flexible, adaptive, and dynamic for trust negotiation between different clients and systems.

This framework also introduces a uniform infrastructure for dynamic trust negotiation in pervasive computing environments as well as distributed systems and Internet environments. From the practical point of view, the mobility and ubiquity in pervasive computing environments allow devices or applications to move to new network domains frequently. Therefore, automated trust negotiation could be the first step to build the trust relationship between the mobile devices or applications and the network domains. The agent-based automated trust negotiation framework provides a practical, flexible and uniform solution to fulfill this first step.

12.8 Acknowledgments

The authors would like to thank the anonymous reviewers for their valuable comments. The research is supported by the NSF Grant 0406325.

References

1. AT&T Labs and University of Pennsylvania. The KeyNote trust-management system, 1999. http://crypto.com/trustmgt/
2. E. Bertino, E. Ferrari, and A.C. Squicciarinii, Trust-X: A peer-to-peer framework for trust establishment, *IEEE Trans. Knowledge and Data Engineering*, 16(7), 827–842, 2004.
3. M. Blaze, J. Feigenbaum, and J. Lacy. Decentralized Trust Management, *IEEE Conference on Privacy and Security*, Oakland, CA, 1996.

4. M. Burrows, M. Abadi, and R. Needhamm. A logic of authentication, *Proceedings of the Royal Society, Series A*, 426, 233–271, 1989.

5. D. Denning. *Information Warfare and Security*, Addison Wesley / ACM Press Books, New York, 1999.

6. C. English, P. Nixon, S. Terzis, A. McGettrick, and H. Lowe. Dynamic trust models for ubiquitous computing environments, *UBICOMP2002: Workshop on Security in Ubiquitous*, Goteborg, Sweden, September 2002.

7. A. Herzberg, J. Mihaeli, Y. Mass, D. Naor, and Y. Ravid. Access control meets public infrastructure, or: Assigning roles to strangers, *Proceedings of the IEEE Symposium on Security and Privacy*, Berkeley, CA, May 2000.

8. P. Lamsal. Understanding trust and security, 2001. http://citeseerx.ist.psu.edu/viewdoc/download?doi=10.1.1.17.7843&rep=rep1&type=pdf

9. N. Li, W.H. Winsborough, and J.C. Mitchell. Distributed credential chain discovery in trust management, *Journal of Computer Security*, 11(1), 35–86, February 2003.

10. J. Linn. Trust models and management in public-key infrastructures, Technical report, RSA Data Security, Inc., Redwood City, CA, November, 2000.

11. Z. Liu and D. Peng. A security middleware architecture for pervasive computing systems, *Proceedings of the 2nd IEEE International Symposium on Dependable, Autonomic and Secure Computing*, Indianapolis, IN, 2006.

12. Z. Liu, R.H. Campbell, and M.D. Micunas. Active security support for active works, *IEEE Transactions on Systems, Man, and Cybernetics - Part C: Applications and Reviews*, 33(4), 432–445, November 2003.

13. Z. Liu, T. Joy, and R. Thomson. A dynamic trust model for mobile ad hoc networks, *10th IEEE International Workshop on Future Trends in Distributed Computing Systems (FTDCS 2004)*, Suzhou, China, May 26-28, 2004.

14. D.H. McKnight and N.L. Chervany. The meanings of trust, *Trust in Cyber-Societies*, Vol. LNAI 2246, New York, 2001.

15. A. Pirzada and C. McDonald. Establishing trust in pure ad-hoc networks, *Proceedings of the 27th Australasian conference on Computer science*, Berlin, Germany, Darlinghurst, NSW, Australia, 2004.

16. V. Shmatikov and C. Talcott. Reputation-based trust management, *Workshop on Issues in the Theory of Security (WITS)*, Warsaw, Poland, 2003.

17. H. Skogsrud, B. Benatallah, and F. Casati. Trust-serv: Model-driven lifecycle management of trust negotiation policies for web services, *Proceedings of the 13th international Conference on World Wide Web, SESSION: Security and Privacy*, New York, 2004.

18. F. Stajano and R. Anderson. The resurrecting duckling: Security issues for ad-hoc wireless networks, *Proceedings of the 7th International Workshop on Security Protocols*, Cambridge, U.K., 1999.

19. J.A. Stinson, S.V. Pellissier, and A.D. Andrews. Trust model—defining and applying generic trust relationships in a networked computing environment, ATI special report pp. 00–06, North Charleston, SC, 2000, http://ips.aticorp.org/Reports/TrustModel.pdf

20. W.H. Winsborough, K.E. Seamons, and V.E. Jones. Automated trust negotiation, *DARPA Information Survivability Conference and Exposition*, Hilton Head, SC, January 2000.

21. W.H. Winsborough and N. Li. Towards practical automated trust negotiation, *Proceedings of the 3rd International Workshop on Policies for Distributed Systems and Networks (POLICY 2002)*, Monterey, CA, June 2002. IEEE Computer Society Press, Los Alamitos, CA, pp. 92–103.

22. D. Xiu and Z. Liu. A dynamic trust model for pervasive computing environments, *Fourth Annual Security Conference*, Las Vegas, NV, March 30-31, 2005.

23. T. Yu. Automated trust establishment in open systems, PhD dissertation, University of Illinois, Urbana-Champaign, IL, 2003.

Chapter 13

ÆTHER

Securing Ubiquitous Computing Environments

Patroklos Argyroudis and Donal O'Mahony

Contents

13.1 Introduction

The ubiquitous computing paradigm envisioned by Mark Weiser approaches computing from a human-centric, nonintrusive viewpoint [1]. Computing devices are being embedded into everyday appliances and become part of our environment. The same vision is shared among many academic and industrial projects that aim to realize it [2,3]. These technologies facilitate the construction of smart environments that are able to interact with users in a natural way and aid them in their everyday activities. However, the open nature of such environments raises security and privacy concerns that need to be addressed in a coherent manner along with the development of the required underlying infrastructure. Although the traditional security requirements remain the same, this new approach to computing has introduced additional challenges. The main problem in addressing the security requirements of smart environments is the large number of ad hoc interactions among previously unknown entities, hindering the reliance on predefined associations. Another equally important problem is that the employed security solution should follow the ubiquitous computing vision and be naturally integrated with the actions the users perform in order to complete their objectives. A user that carries a multitude of devices must be able to establish spontaneous secure communication channels with the devices embedded into the environment or carried by other users without extensive reconfiguration tasks. The problem is aggravated since data transmissions in smart environments use wireless media, such as Bluetooth and IEEE 802.11, whose integrity and confidentiality can easily be undermined by malicious entities.

In order to specify and build an authorization management framework that is able to support pervasive computing, a number of requirements must be taken into consideration:

- Decentralized management: The dynamic nature and the great number of participating devices in pervasive computing imply the need for frequent establishments of communication channels between entities that belong to different administrative domains. Centralized architectures are not able to support the autonomous nonhierarchical building of a secure pervasive computing infrastructure. Furthermore, external centralized entities constitute a single point of attack/failure and it is not realistic to assume that they can be universally trusted.

- Disconnected operation: This requirement is closely related to the previous one. An authorization architecture for pervasive computing must be able to support the establishment of secure relationships between devices that are not able, due to physical location, network partitions, or other reasons, to connect to a third party for mutually authenticating and authorizing each other.

- Unobtrusiveness: Human intervention for the required administrative tasks should be naturally and gracefully integrated with the physical workflow of users in order to facilitate the unobtrusive nature of pervasive computing.

■ Context-awareness: The use of contextual information, such as physical location, activity, and others, should be utilized to allow the dynamic adaptation of the management framework to a variety of situations and applications. Moreover, context-awareness minimizes human intervention and user distractions for configuration purposes alleviating unobtrusiveness.

Our proposal, named ÆTHER, is an authorization management framework designed specifically to address trust establishment and access control in pervasive computing environments where *a priori* knowledge of the complete set of participating entities and global centralized trust registers cannot be assumed. The basis of our work is the Role-Based Access Control (RBAC) model [4], according to which entities are assigned to roles and roles are associated with permissions. ÆTHER extends RBAC in order to support decentralized administration, disconnected operation, and context-awareness. Furthermore, we use the well-defined concept of location-limited channels to specify an unobtrusive usage model for the required administrative tasks.

Based on this general framework we have instantiated two different systems:

1. $ÆTHER_0$ has been designed to address the authorization needs of small pervasive environments whose management requirements are simple. It utilizes only symmetric key cryptography in order to provide security services. Consequently, it is appropriate for devices that have particularly limited processing capabilities (such as simple sensors).
2. $ÆTHER_1$ addresses the authorization requirements of large pervasive computing domains that have multiple owners with complicated security relationships. It relies on asymmetric cryptography and therefore is more fitting to domains that consist of devices that have sufficient information processing capabilities.

In the following sections, we start by exploring solutions that have been proposed in the literature for securing ubiquitous computing systems and discuss their advantages and disadvantages in Section 13.2. In Section 13.3 we describe in detail our proposed framework and its extensions to traditional RBAC. Section 13.4 analyzes our two instantiations of the framework, namely, $ÆTHER_0$ and $ÆTHER_1$. A thorough evaluation of our work is presented in Section 13.5 that focuses on both implementation aspects and not strictly technical issues. We conclude in Section 13.6 by summarizing our contributions and present possible directions for future work.

13.2 Related Work

13.2.1 Traditional Computer Security

The X.509 authentication infrastructure [5,6] offers little help in supporting complex access control decisions. Although it can be used to establish the identity of an entity and provide authentication services by employing a centralized architecture, most systems are not interested in the name or the identity of an entity. The real requirement is to know what operations on which resources an entity is authorized to perform. To address this problem with the standard X.509 architecture we need to enumerate all the identifiers of the participating entities in a system, create Access Control Lists (ACLs) for every resource in the system and for every supported action, and bind identities to permissions. Obviously when we have to manage authorization information in large systems with great numbers of users this approach fails to scale. X.509 Attribute Certificates (ACs) were introduced to address this problem [7]. They bind a user's identifier with one or more *privilege attributes*, which are defined as attribute type/value pairs. ACs are issued by Attribute Authorities (AAs) that play they same role as CAs, but focus on authorization instead of authentication certifications. The X.509 AC infrastructure is called a Privilege Management Infrastructure (PMI) and it can also follow a hierarchical trust model, as a normal PKI. The root element of a PMI is called the Source of Authority (SOA). However, PMIs still rely on the existence of a PKI.

X.509 PKI and PMI models are not able to address the problem of authorization in pervasive computing environments, mainly because they were not designed for such a purpose. Their main disadvantage is the requirement of a globally trusted central server that is assumed to always be online and readily accessible. In pervasive computing we have usage scenarios that take place in all possible locations that may or may not have access to an Internet gateway, hence the ability to support disconnected operation is crucial. Another limitation of this approach lies in the use of identities as a basis for building trust and authorizing service requests. PKI and PMI credentials bind the owner's public key to a name. As we move to pervasive computing authority domains, naming becomes increasingly locally scoped. Therefore, a name may not have meaning to a different domain than the one that it was certified in. Scalability problems also exist in PKI when authorization decisions are based on identities in dynamic environments since enumeration of all the participating entities is nearly impossible.

The Role-Based Access Control (RBAC) model [4] provides the ability to specify and enforce domain-specific access control policies and simplifies the process of authorization management in computing environments with a great number of users. This is achieved by associating roles with access permissions and users with roles. In essence, RBAC defines all access rights through roles. A role is an entity that acts as a collection of permissions. Users are assigned to roles, usually by an administrator on a central system, and receive access permissions only based on the roles they have been assigned to. Of course, before users are able to access services and applications the security administrator must configure the required permissions, create associations between roles and these permissions, and finally assign roles to users. Although RBAC is far more flexible in management and in detail of control when compared to other access control management mechanisms, such as Mandatory Access Control (MAC) and Discretionary Access Control (DAC), it still requires centralized management via an online server and direct human intervention, as proposed by the National Institute of Standards and Technology (NIST) RBAC specification [8]. According to MAC, access to resources is defined based on the specific resource's *sensitivity* as it is defined by a central authority. Therefore, RBAC relates to MAC in the sense that end users have no control over the permissions defined by the security administrator and enforced by the central authorization server. On the other hand, DAC permits the granting and the revocation of permissions to be left to the discretion of the individual end users of a system. A user that controls certain resources is able to assign and revoke rights for them without the need to contact a central server, or be authorized to do so by an administrator. Consequently, the NIST-proposed RBAC standard is a *non-discretionary* access control model. Role assignment remains a centralized operation and requires the manual intervention from an administrative authority. This requirement as well as the requirement of having an always accessible online central server prohibit RBAC to be directly used in securing pervasive computing environments.

Trust management, first defined in [9], addresses the problem of decentralized authorization by avoiding the use of identities and instead binds access rights directly to public keys. In such *key-based* access control systems, like PolicyMaker [9], KeyNote [10] and SPKI/SDSI 2.0 [11,12], authorization certificates are used to delegate permissions directly from the key of the issuer to the key of the subject enabling access control decisions between strangers without the need for a universally trusted third party. The syntax of the signed statements we use in ÆTHER is similar to the syntax of the signed assertions in KeyNote, but we bind attributes instead of access rights to keys in order to achieve flexibility and scalability in defining authorizations for dynamic sets of entities. Furthermore, we adopt the notion of identifying principals by their public keys instead of identities since it decouples the problem of authorization from the problem of secure naming. However, we argue that the delegation of a specific set of access permissions to a key is not appropriate for the ubiquitous computing paradigm in which the initiation of new services is frequent and the number of users particularly large. According to our view, a user that wishes to provide a new service should not be burdened with the additional administration overhead of initiating delegation chains and distributing the credentials that define the access rights for the new service.

13.2.2 Pervasive Computing Security

In the area of pervasive computing security the Resurrecting Duckling model was initially proposed to cover master-slave type of relationships between two devices [13]. The relationship is established when the master device exchanges over an out-of-band secure channel a secret piece of information with the slave device. Stajano and Anderson call this procedure *imprinting*. The imprinted device can authenticate the master through the common secret and accept service requests from it. The idea of bootstrapping trust relationships in adversarial environments over secure out-of-band channels was examined further by Balfanz et al. [14] and their concept of location-limited channels, which have the property that human operators, can precisely control which devices are communicating with each other thus ensuring the authenticity requirement. They propose the use of contact and infrared as a way of exchanging pre-authentication information, such as the public keys of the peers, before the actual key exchange protocol. The extended Resurrecting Duckling security model proposed the imprinting of devices with policies that define the type of relationships the slave device is allowed by its master to have with others in order to address peer-to-peer interactions [15]. However, the author simply proposed the use of trust management systems as a way to define the imprinted policies without offering any specific implementation details. Furthermore, even the extended Resurrecting Duckling system defines a static association model between the master and the imprinted devices, limiting its direct application in situations where associations are established in an ad hoc manner. In ÆTHER we have extended these concepts and designed a complete authorization management framework for dynamic pervasive computing environments.

The Gaia project aims to define a generic computational environment that integrates physical spaces and the pervasive computing devices that exist in them into a programmable system [16]. Cerberus is a core service in the Gaia platform that provides identification, authentication, context-awareness, and reasoning [17]. It utilizes an infrastructure of sensors that are installed throughout the environment in order to capture and store as much context information as possible. This information is then used to identify entities and reach decisions regarding the actions they request to perform. The main disadvantage of Cerberus lies in its monolithic architecture and its reliance on a central server to collect context information, evaluate policies, and make authorization decisions on behalf of resource providers. This implies the requirement that all the resource providers of the environment trust the server to give them correct authorization decisions and not to reveal their personal data, which may be private in nature. However, pervasive computing entities (both resource providers and consumers) establish short-lived associations and frequently roam between administrative domains and hence do not necessarily trust each other. Additionally, the central component that reaches authorization decisions needs to be constantly accessible and therefore forbids disconnected operation.

Covington et al. proposed the concept of *environment roles* in order to develop an access control system for pervasive computing that can utilize environmental and contextual information [18]. Environment roles are in essence one component of the Generalized Role-Based Access Control (GRBAC) model [19]. GRBAC is an extension of the traditional RBAC model where object and environment roles are defined in addition to subject roles. The resources provided in an environment can be assigned to object roles and environmental (contextual) conditions are used to define environment roles. GRBAC uses a logic language similar to Prolog to express access control constraints on context variables. Authorizations are based both on traditional subject roles and environment roles. This allows the system to be sensitive to contextual information, which is captured by sensors and translated to specific environment conditions. These conditions are then used to activate environment roles for the subjects in the smart space based on logic predicates.

GRBAC constitutes a flexible set of extensions to RBAC; the concept of object roles greatly simplifies the administration of resources and the definition of access control policies. Based on captured context, environment roles can be activated automatically for subjects in the smart space giving them certain access rights without the need of manual configuration tasks. However, GRBAC relies on a central management service. This service has two responsibilities. The first is to collect context data and maintain environment conditions. The second is to store the logic predicates that implement environment role activation decisions and authorizations. All resource consumers and providers that participate in the system need to have persistent access to the management service, otherwise they cannot activate environment roles or reach authorization decisions.

13.3 ÆTHER Architecture

The ÆTHER architecture has been designed to address the security concerns associated with Pervasive Authority Domains (PADs) and specifically to support authorization management and trust establishment in the presence of a multitude of constrained devices and frequent interactions with unknown principals. A PAD is an abstraction that organizes policy and privilege within a domain and is defined differently in our two instantiations (see Section 13.4).

Threat model: Our security model aims to address the threats associated with authorization management both from a technical and a usability perspective. Consequently, we consider several classes of threats to be outside the scope of our target area. For example, denial-of-service threats assume that the attacker is a properly authorized user of the system, or of the communication medium when the attack focuses on the network or physical layers, while we concentrate on authorization itself. Moreover, we do not consider the threats, although we discuss the repercussions, of stolen devices since we believe that when an attacker has physical access to a device she can eventually bypass all security mechanisms and recover confidential data and key material [20,21]. Another assumption we make is that the sensors that are embedded into the environment collect and report trustworthy contextual information, that is, we are not concerned with context authentication. However, we protect against unauthorized context consumption.

13.3.1 User-Based Authorization Management

In pervasive computing systems users and their requirements are placed at the center of every design choice. The ultimate goal is to develop systems that are first of all usable, meaning that they fulfill the expectations of their users, and unobtrusive in the sense of not distracting users from their primary physical tasks. We believe that a pervasive computing system's security needs are included in the usability requirement. For example, when a user decides to share a document stored on his PDA with a trusted colleague, he expects that the system will ensure the confidentiality of the exchange so that a nearby malicious entity cannot access it. Traditional security protocols rely on digital data, like credentials and policies, to establish secure associations and reach access control decisions. However, in pervasive computing, human users are directly involved in the performed tasks as these occur in the physical space. Each participating user has a mental model formulated in her mind that represents the particular circumstance of the task at hand. The system must allow the user to express this perceptive model of the situation to the pervasive computing devices in order to define security relationships and this must be accomplished in an unobtrusive manner. We believe that this expression of the users' mental security model can be integrated gracefully with the performance of the primary task at hand. Continuing the example with the shared file, the owner can state his intent that he trusts the person next to him by pointing his PDA to his colleague's handheld device. The authorization data, like credentials, necessary for securing the exchange can be established through this physical action that states the intent of the user. Figure 13.1 illustrates the differences between the traditional authorization management approach and our own *user-based* authorization management model for pervasive computing. While the traditional model is concerned only with the protocols between the devices and the static security policies, the ÆTHER management model also includes dynamic policies inferred from the physical actions of the users, each playing a specific role.

We believe that the concept of location-limited channels as first defined by Stajano and Anderson [13] and later extended by Balfanz et al. [14] can be used to establish policy data, like for example the issuing of credentials from one user to another or the exchange of a secret key, and be integrated gracefully to the users' physical actions. These actions are part of the effort a user is already undertaking in order to accomplish the primary task and they state the intent of the user regarding authorization. Therefore, we accomplish unobtrusive user-based authorization management that does not distract users with complicated security-related dialogs, decisions, or creation of new accounts.

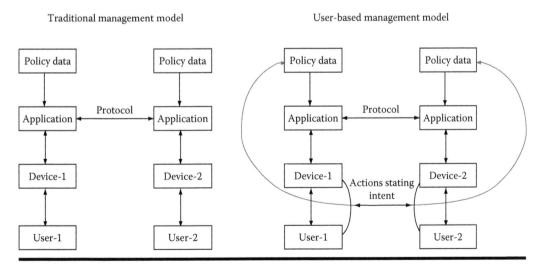

Figure 13.1 Traditional vs. user-based authorization management.

13.3.2 ÆTHER Model

The general ÆTHER model extends the core Ferraiolo–Kuhn RBAC model to provide authorization management for pervasive computing environments. While RBAC depends on a fully centralized management model in which a system administrator is responsible for user and permission assignment, ÆTHER utilizes the user-based model described in the previous section. This high-level model is instantiated into two distinct architectures, namely, ÆTHER_0 and ÆTHER_1. We believe that the decentralized and highly dynamic nature of pervasive computing is compatible with authorization relationships based on roles, given that role authority does not follow a centralized management model. This requirement for decentralization relates both to the ability to make assertions concerning user assignments to roles and the delegation of authority over a specific role.

The ÆTHER model makes the following semantic distinction between subjects and users. A user may have multiple subjects in operation. Each subject is uniquely referenced by an identifier (a message digest of the hosting device's hardware addresses concatenated with a randomly generated number in ÆTHER_0 or a public key in ÆTHER_1). There is no way for the system to map subjects to the identities of physical users, as is the case in RBAC, thus providing a layer of privacy. Moreover, ÆTHER uses *attributes* instead of roles. An attribute is defined as a property of an entity, while a role is a type of attribute used to define the position an entity has in an organization. The ÆTHER model is composed of the following components and we use the (‡) mark to denote our extensions to the core RBAC model (a graphical representation of the components and their relationships is shown in Figure. 13.2):

- A set of subjects: S
- A set of resources: R
- A set of operations: O

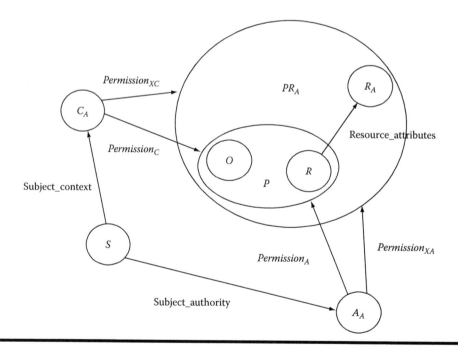

Figure 13.2 The ÆTHER model.

- A set of authority attributes: A_A
- A set of context attributes: C_A (‡)
- A set of resource attributes: R_A (‡)
- A set of permissions: $P = \mathcal{P}(O \times R)$
- Authority attribute to subject assignment: $SA_A \subseteq A_A \times S$
- Context attribute to subject assignment: $SC_A \subseteq C_A \times S$ (‡)
- Resource attribute to resource assignment: $RA \subseteq R_A \times R$ (‡)
- Basic permission to authority attribute assignment: $PA_A \subseteq P \times A_A$
- Basic permission to context attribute assignment: $PC_A \subseteq P \times C_A$ (‡)
- Basic permission to resource attribute assignment: $PR_A \subseteq P \times R_A$ (‡)
- Composite permission to authority attribute assignment: $XPA_A \subseteq PR_A \times A_A$ (‡)
- Composite permission to context attribute assignment: $XPC_A \subseteq PR_A \times C_A$ (‡)
- A function mapping an authority attribute to a set of permissions: $permission_A = A_A \rightarrow \mathcal{P}(P)$, or more formally: $permission_A(a : A_A) = \{p : P \mid (p, a) \in PA_A\}$
- A function mapping a context attribute to a set of permissions: $permission_C = C_A \rightarrow \mathcal{P}(P)$, or more formally: $permission_C(c : C_A) = \{p : P \mid (p, c) \in PC_A\}$ (‡)
- $permission_{XA} = A_A \rightarrow \mathcal{P}(PR_A)$, or more formally: $permission_{XA}(a : A_A) = \{p : PR_A \mid (p, a) \in XPA_A\}$ (‡)
- $permission_{XC} = C_A \rightarrow \mathcal{P}(PR_A)$, or more formally: $permission_{XC}(c : C_A) = \{p : PR_A \mid (p, c) \in XPC_A\}$ (‡)
- The mapping of a subject to a set of authority attributes: $subject\_authority(s : S) \rightarrow \mathcal{P}(A_A)$, or more formally: $subject\_authority(s : S) \subseteq \{a \in A_A \mid (a, s) \in SA_A\}$
- The mapping of a subject to a set of context attributes: $subject\_context(s : S) \rightarrow \mathcal{P}(C_A)$, or more formally: $subject\_context(s : S) \subseteq \{c \in C_A \mid (c, s) \in SC_A\}$ (‡)
- The mapping of a resource to a set of resource attributes: $resource\_attributes(r : R) \rightarrow \mathcal{P}(R_A)$, or more formally: $resource\_attributes(r : R) \subseteq \{ra \in R_A \mid (ra, r) \in RA\}$ (‡)
- Access authorization: A subject s can perform an operation o on a resource r for either of the following:
  1. There exists an authority attribute a that has been assigned to subject s and there exists a permission p that is assigned to a such that the permission authorizes the performance of o on r
  2. There exists a context attribute c that subject s possesses and there exists a permission p that is assigned to c such that the permission authorizes the performance of o on r
  3. There exists an authority attribute a that has been assigned to subject s and there exists a resource attribute ra that has been assigned to resource r and there exists a composite permission ap that is assigned to a such that the permission authorizes the performance of o on ra
  4. There exists a context attribute c that has been assigned to subject s and there exists a resource attribute ra that has been assigned to resource r and there exists a composite permission ap that is assigned to c such that the permission authorizes the performance of o on all resources that have the resource attribute ra.
  5. Formally: $access : S \times O \times R \rightarrow BOOLEAN$, $s : S, o : O, r : R$, $access(s, o, r) \Rightarrow \exists a : A_A, c : C_A, p : P, ra : R_A, ap : PR_A, ((a \in subject\_authority(s) \wedge p \in permission_A(a) \wedge (o, r) \in p) \vee (c \in subject\_context(s) \wedge p \in permission_C(c) \wedge (o, r) \in p) \vee (a \in subject\_authority(s) \wedge ap \in permission_{XA}(a) \wedge ra \in resource\_attributes(r) \wedge (p, ra) \in ap) \vee (c \in subject\_context(s) \wedge ap \in permission_{XC}(c) \wedge ra \in resource\_attributes(r) \wedge (p, ra) \in ap))$

We have four different kinds of attributes: *authority* attributes, *context* attributes, *aggregated context* attributes, and *resource* attributes. An attribute is defined as an ordered 2-tuple of the form *(name, value)* where *name* is the identifier of the attribute and *value* is its value. Authority attributes are the traditional security attributes used to describe the properties of subjects. These properties are relatively static, meaning that their values are obtained during the establishment of a session. Authority attributes are similar to the concept of roles in traditional RBAC. A subject is a member of a role if and only if it has been assigned the authority attribute that represents the corresponding role. The values of authority attributes must be configured through manual human intervention and are then obtained from a subject's credentials. In that sense, authority attributes are static security metadata whose values can be used in order to reach authorization decisions. On the other hand, context attributes describe the dynamic properties of subjects associated with their current context. Although manual intervention is still required to designate the use of context attributes in authorization decisions, their values are obtained dynamically every time an access request is made. In contrast to authority attributes, the values of context attributes may change during the lifetime of an active session since they reflect a subject's environmental context. An aggregated context attribute is a context attribute defined as an aggregation of one or more context attributes. The aggregation is performed by a Context Aggregator (CA) that is defined as the following function:

$$CA : C_{A_1} \times C_{A_2} \times \cdots \times C_{A_v} \to C_A, \, v \geq 1$$

Resource attributes are assigned to resources to simplify their administration and the definition of authorization policies. Basic permissions are assigned to resource attributes and define composite permissions. Authority or context attributes are then associated with composite permissions in order to allow the creation of authorizations addressing a whole set of resources. For example, the context attribute *(location, office_008)* can be associated with all output devices allowing anyone in office 008 to access all devices that have been defined as output devices.

The concept of context attributes provides a high-level interpretation of the low-level context information collected by sensors embedded into the environment. The assignment of context attributes to subjects is maintained dynamically based on the raw data of context dissemination sensor service infrastructures, like for example the ones presented in [3] and [22]. Thus, whenever an access control decision is required that has been defined using a context attribute, the corresponding service is queried regarding the status of the requesting subject, *pulling* for context data.

13.3.3 Context-Awareness

The ÆTHER framework, according to the categories defined by Chen and Kotz in [23], follows the *active* context-awareness category. Therefore, it is able to dynamically adapt its behavior based on sensed contextual information from the environment. This ability relates to the reached authorization decisions, and the context utilized is both that of the requesting entity and that of the service provider. For example, ÆTHER allows the specification of context-aware policies that allow the execution of a specific action to entities that are in the same physical location as the service provider. If the service provider moves to a new location the policy does not need to be modified; it adapts and operates in the new location as well.

In ÆTHER permissions are modeled as the rights of authority or context attributes. We associate rights with actions, so possession of an authority or context attribute permits the corresponding principal to perform a certain action. Authorization policies define context attribute

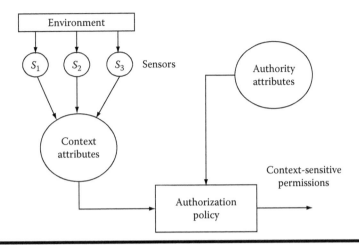

Figure 13.3   Context-sensitive permissions.

requirements or restrictions in addition to the ones using authorization attributes (see Figure. 13.3). This allows the specification of context-adaptive policies that control access to protected resources.

For example, the owner of a printer may allow all people physically present in the same room as the printer to use it. However, the information regarding associations between subjects and context attributes may be considered sensitive or private. In these cases we view the context-dissemination service as a normal resource to be protected and accessed only by properly authorized principals. The authorization can be handled by authorization attributes or even context attributes which are not considered sensitive in certain situations. Furthermore, each device maintains its own current context in what we call a Local Context Profile (LCP). When the value of a context attribute is defined as a *dynamic local context value* then this value is retrieved from the corresponding context attribute stored in the device's LCP.

The principals that are responsible of maintaining the associations between context attributes and principals are specified in policy statements called Context Attribute Sets (CASs). Such principals are sensor devices and through CAS statements are given authority over specific context attributes. Sensor devices maintain the assignments of context attributes to principals using *dynamic sets* defined for specific context attributes. A normal nondynamic set can be described according to set theory as a static assortment of elements. On the other hand, a dynamic set can be described as an assortment of elements that varies with respect to time. Hence, a nondynamic set can be viewed as an instance of a dynamic set in a specific time point. We formally define a dynamic set as follows:

> Let V be a set of principals and $\mathcal{P}(V)$ the power set of V. Let T be a time period, i.e., a set consisting of specific points in time. We define the function $f: T \rightarrow \mathcal{P}(V)$ as a dynamic set that consists of elements that belong to V in the time period T.

For example, if we want to allow access to a provided resource only to entities that are in the same location as the provider we could create a context attribute (defined as f) for this location, e.g., the context attribute *(location, office_008)*. The elements of this dynamic set at time $t_1 \in T$

would be given by $f(t_1)$. In other words, a principal $v_1 \in V$ would be allowed access if and only if the statement $\exists t_1 \in T \cdot v_1 \in f(t_1)$ holds.

13.3.4 Authority Revocation

In certain occasions the authority given with an issued credential needs to be revoked. Such occasions may include loss or compromise of the key used to issue a credential, loss or compromise of the key to which a credential has been issued to, or realization on the part of the issuer that the information contained in a previously issued credential is no longer valid. There are four mechanisms to achieve the revocation of a credential:

1. By using the validity period of the credential itself and setting it to a short time period. This mechanism effectively limits the window of revocation to the validity period specified as part of the credential.
2. Another mechanism is the use of Certificate Revocation Lists (CRLs), which are basically lists that include all credentials that have been revoked. The issuer must create a CRL and distribute it to all participants of the system.
3. Current Identity Lists (CILs) follow the opposite approach of the CRLs mechanism. CILs are signed lists that include all the valid credentials that have been issued by a given credential authority.
4. The final revocation mechanism is the one advocated by OCSP [24] and requires the credential verifier to check the revocation status of a credential at every verification.

The advantages and disadvantages of these mechanisms have been discussed extensively in the literature; for a summary see [25]. In ÆTHER we have chosen to adopt the first mechanism of short expiration time periods for the issued credentials. We have rejected the use of CRLs since their correct implementation and management is particularly complex, especially in environments with a lot of participating entities. Online credential status mechanisms require the issuer to always be accessible by every verifier, a requirement that is not compatible with the disconnected nature of pervasive computing. In addition, the involvement of the credential issuer in every access control decision makes its use very expensive from a computation point of view. ÆTHER attribute credentials are issued with short expiration time periods over a location-limited channel established between the issuer and the subject. The issuer is able to specify in the *renewable* field if the credential can be renewed automatically before it expires or not. If the field is set, then the subject is free to contact the issuer any time before the expiration of the credential and request a renewal. This process takes place over a wireless transmission and requires the subject to be within the communication range of the issuer. If the *renewable* field is not set, or if the issuer cannot be contacted the credential expires. If the subject needs another certificate of the same type then a location-limited channel with the issuer must be established again.

The main problem with the refreshing mechanism is the choice of the expiration time period. If it is too short then the issuer becomes frequently involved in renewing procedures; if it is too long the possibility of having credentials that have to be revoked but cannot increases. The ideal design choice would be to let the end user decide the validity period of the credentials she issues. However, this is an unrealistic approach as it assumes that every user is able to analyze all the risks involved in such a decision. In our current design the typical validity period we suggest is 1 h. We believe that this choice satisfies the above requirements, however, we must note that this is just a suggestion.

The optimal validity period depends on the exact application scenario, requires extensive research and analysis, and is therefore outside the scope of our study.

13.4 Instantiations

13.4.1 ÆTHER$_0$

In ÆTHER$_0$ we have two classes of devices. The first one is comprised of normal pervasive computing devices that participate in the environment and offer services that have to be protected from unauthorized access. These devices can be simple without extensive processing power or large amounts of available memory. Examples of such devices include printers, switches that control various functionalities like doors and lamps, television sets, different types of sensors, and audio speakers. The second class consists of devices that directly represent a user in the management model. These are equipped with adequate physical interaction interfaces, such as keypads and screens, as well as more advanced processing capabilities and memory capacities than the devices of the previous class. Suitable devices for this purpose can be PDAs, mobile phones, or more traditional mobile computing devices such as laptops and tablet PCs. Users use these devices to store cryptographic data that can be used for authentication purposes as well as policy data that are used for authorization. In the ÆTHER$_0$ instantiation we call these devices *alpha-pads*. Each and every device that participates in an ÆTHER$_0$ PAD is uniquely identified by a hexadecimal string. This is constructed by concatenating the hardware addresses of the location-limited channel interface and the hardware address of the normal networking interface with a randomly generated number. The resulting value is given as input to a hash function and the output is encoded to hexadecimal format. This string is stored in the device itself, is public, and is used to identify the device that generated it in the utilized policy statements.

13.4.1.1 Management Model

Although ÆTHER$_0$ does not depend on any external centralized infrastructure for its operation, it follows a locally centralized management model. Before a pervasive device of the first class can participate in a domain and offer its services it must be bound by the owner's alpha-pad. The binding takes place over a location-limited channel established between the two devices with the help of the owner and is enforced through a shared secret. The alpha-pad device generates a symmetric secret key (included in a binding policy statement) and sends it to the new device over the location-limited channel. Now the new device is bound by the owner's alpha-pad and can authenticate it through this shared secret. A user is free to have more than one alpha-pad and a slave device can be bound to more than one of them as well. However, a binding procedure can only take place over a location-limited channel and only if a user has expressed her intent to do so via some physical method (like pressing a button) on the device to be bound. The owner must repeat the same process for all the devices of the first class she owns that participate in the domain and have services to offer. Hence, at the end of this process all devices are bound to one or more alpha-pads and share secret keys with the alpha-pad that was used to bind them.

Using the alpha-pad the owner can now create further policy statements for the bound devices. These policy statements are sent to the bound devices over a wireless transmission medium and are authenticated, as well as encrypted, using the previously established shared secrets. The issuing alpha-pad as well as the subject device are identified in the policies with the unique hexadecimal strings. Each alpha-pad in ÆTHER$_0$ has its own localized namespace and is responsible to bind and define authorization policies for normal devices that offer services. Each bound device follows the namespace and participates in the domain defined by its alpha-pad. Therefore, a Pervasive Authority Domain (PAD) in ÆTHER$_0$ is represented by a specific alpha-pad that has been used to bind and configure a set of ordinary devices.

When a user wishes to access a service provided by a device she owns, she can simply use the alpha-pad that has been used to bind the target device (it is also possible to use a non-alpha-pad device she owns, the following paragraph discusses the relevant protocols). The target identifies the alpha-pad through the binding shared secret key and performs the requested operation. If a user wants to access a service of a device that belongs to another user, then both users need to be at the same physical space. When the two users meet we assume that each one carries his alpha-pad device on him. Following our general user-based trust model, for a user to allow another user to use one of his devices that offers some services, he has to state his intent that the other user is trusted to do so through some physical action with his alpha-pad device. To express this trust the two users establish a location-limited channel with their alpha-pads and agree on a symmetric secret key over it. We call this procedure *mutual binding*. Furthermore, each user assigns authority attributes to the other user. After this procedure both alpha-pads are authenticated through the established shared secret and authorized with the assigned authority attributes.

A user can now use her alpha-pad (or another device she owns) to access a service provided by a device of the user she performed the mutual binding with. The requesting user's device securely contacts its alpha-pad (this step is obviously not performed if the requesting user directly uses his alpha-pad) and states the request to access the target device. The source device also generates and sends a new secret key. The alpha-pad finds the mutual binding policy statement with the target's alpha-pad and forwards the access request and the new key to it protected with their mutual binding key. The target alpha-pad transmits everything to the target device encrypted using their shared binding key. Additionally, the target alpha-pad transmits to the target device the authority attributes it has assigned to the requesting alpha-pad during their mutual binding process. The target device based on the authorization policies it locally has and the authority attribute assignments it has been forwarded can now reach an access control decision for the requested action. Figure 13.4

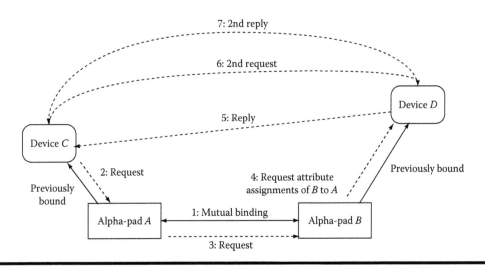

Figure 13.4   Overview of the ÆTHER$_0$ management model.

summarizes this process (dashed lines represent wireless transmissions, while continuous lines represent location-limited channels).

Example: Alice visits Bob's home. She carries her mobile phone which plays the role of her alpha-pad and a pervasive computing wrist watch (which is a normal device that has been previously bound to her mobile phone). Bob has a PDA as his alpha-pad and a remotely controlled lamp. When the two users meet they mutually bind their alpha-pads, establishing a shared secret key, and they assign authority attributes to each other. We assume that Bob assigns to Alice the authority attribute *(group, visitor)*. These assignments stay with the issuer (Bob in this case); they are not quantities that are given to the subject (Alice). Moreover, Bob has sometime ago bound his lamp to his PDA and issued a positive authorization statement to it allowing principals with the authority attribute *(group, visitor)* to control it. At some point Alice wants to switch the lamp on using her wrist watch. The wrist watch forwards the access request and a newly generated secret key to Alice's mobile phone encrypted with their shared key. The mobile phone decrypts the request, finds the mutual binding with Bob's PDA and the key they have established, encrypts the request and the new key with it, and in turn forwards them to Bob's PDA. The PDA decrypts the request and identifies that the target of the request is the lamp. It then sends the *(group, visitor)* attribute assignment that Bob has given to Alice to the lamp along with the access request and the new secret key. The subject identifier in the attribute credential that is sent to the lamp is the one of the device that initiated the request, in this case Alice's wrist watch. This message is protected with the secret key that the lamp and the PDA share. All the messages are encrypted and authenticated using the respective previously established secret keys. The lamp can now authorize the request based on the authority attribute assignments it has been forwarded and its local policy. The lamp transmits its authorization decision to the wrist watch. The communication with Alice's wrist watch is protected with the new key. Further access requests from the wrist watch can be issued directly to the lamp (and vice versa), until the new shared key expires and the lamp deletes it from its policy database.

13.4.1.2 Service Discovery and Access Protocols

When a user visits a new pervasive computing environment she initially needs to find what services are offered so that can consequently issue access requests. In $ÆTHER_0$ alpha-pad devices are used to securely bind all service offering devices that belong to a particular owner. Furthermore, owners use their alpha-pads to issue authorization policies that control access to the services that they want their other devices to offer. Hence, alpha-pads are aware, and store in their local databases, what services are provided and by which devices. Two owners that establish a mutual binding between their alpha-pads state with this action that they trust each other, not completely but up to the level implied by the authority attributes they assign to each other. The services that an owner has to offer can be securely advertised to the other party, using the secret key the two alpha-pads share after the performance of the mutual binding procedure. Service advertisements include the identifier of the provider, a service identifier ($Service_{ID}$) which is meaningful only to the provider and is used to locally distinguish the different provided services, and a textual description of the service. Similarly, when an alpha-pad is used to bind a normal device service advertisements and the identifiers of the devices that offer them can be released from the alpha-pad to the newly bound device. Each device maintains what we call a Current PAD View (CPV). The CPV is a list of the services and their associated details (the service's identifier, the service's textual description, the identifier of the provider, and the service's access interface if the device managed to get access to the service in question) that gets populated with entries when a device receives service advertisements.

Services can also be provided by the alpha-pads themselves. Therefore, there are five different access protocols. The first one considers the case in which a normal device wants to access a service provided by the alpha-pad that bound it (and the opposite; an access request from an alpha-pad to a normal device it has bound before). The second one when both the requester and the provider are normal devices bound by the same alpha-pad. The third one when an alpha-pad is used to access a service provided by another alpha-pad with which it shares a mutual binding. The fourth case involves an access request from a normal device to an alpha-pad with which the requesting device does not share a binding, but the alpha-pad of the requester shares a mutual binding with the target alpha-pad (and the opposite). The final protocol addresses an access request from a normal device to another normal device that have been bound by different alpha-pads that share a mutual binding. Due to space constraints we will only present the final protocol, as the most complicated one. For a complete discussion we refer the interested reader to [25].

The protocol we present below covers the situation where an access request is made from a normal device (C) to another normal device (D). However, the two devices have been bound by different alpha-pads. Specifically, device C has been bound by alpha-pad A and device D by alpha-pad B. The assumption here is that these two alpha-pads share a mutual binding and that the owner of B has assigned authority attributes to the owner of A. Figure 13.5 illustrates the protocol.

Device C sends the request to its alpha-pad protected with their shared key S_{AC} (we use the *encrypt-then-authenticate* mechanism of Krawczyk [26]). The request includes a new symmetric secret key that C generated, namely, S_{CD}, and a timestamp T_1 to protect against replay attacks. Alpha-pad A forwards the request and the new key to alpha-pad B. This message is protected with the mutual binding key A and B share. B identifies that the target of the request is D and forwards to it the new key along with the authority attributes (AA_{BA}) it has previously assigned to A. The credentials that are sent to D by B have as their *subject* field value the identifier of the requester, in this case ID_C. D decrypts this message with the key it shares with its alpha-pad B and passes to its inference engine the request, the set of authority assignments it has been forwarded, and its local policies. The result is sent directly to C protected with the new secret key (S_{CD}). If the result is positive the *REP* message also contains the access interface of the requested service. The new key S_{CD} can now be used for further communications between C and D (and vice versa). However, after S_{CD} expires the whole process has to be performed again.

13.4.1.3 Inference Engine

The inference engine of ÆTHER$_0$ is invoked by the provider of a service after the requester has made an access request according to the access protocols. The algorithm used is the following (illustrated by the flowchart presented in Figure 13.6):

1. When a request is received from a principal the default value of the access decision is set to *false*.
2. All the authorization policies relevant to the request are collected from the local policy database of the service provider.
3. If context attributes are used in the authorization policies, the engine checks to see if any of these attributes have dynamic local context values. If there are context attributes with dynamic values used to define the authorization policies, their current values are retrieved from the provider's LCP. If no context attributes are required the execution jumps to step 5.
4. Based on the CAS statements the provider locally has it identifies the principals that are responsible for maintaining the membership of principals in the dynamic sets that represent

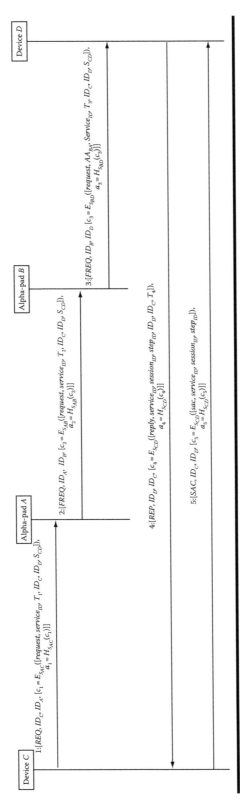

Figure 13.5 Normal device to normal device (different alpha-pad) access protocol.

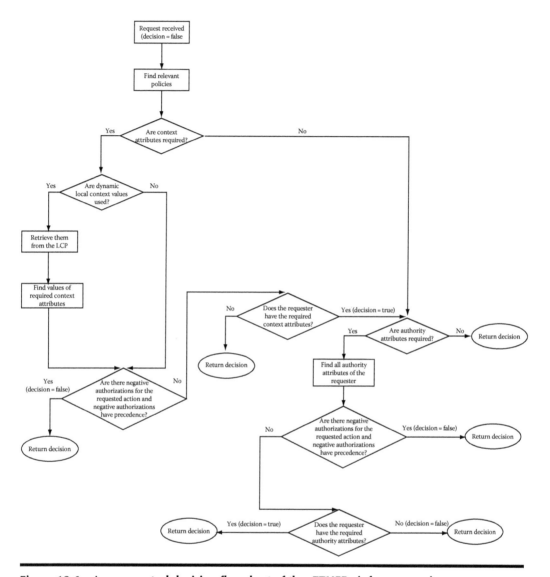

Figure 13.6   Access control decision flowchart of the ÆTHER₀ inference engine.

the context attributes used in the identified policies. Then the membership status of the requester for the specific context attributes is requested from the relevant principal (e.g., a sensor device). This is a normal access request that is accomplished using the access protocols we have already presented. If there are negative authorizations for the requested operation and negative authorizations have precedence over positive authorizations the value of the access decision is set to *false* and returned. If the requester has the necessary context attributes and no negative authorizations exist (or they do not have precedence over positive ones) the decision value is set to *true*.

5. Then the engine checks to see if there are authority attributes required. If none are required the decision value is returned.

6. If there are authority attributes required, the engine searches the local policy database to find all the authority attributes that have been assigned to the requesting principal. If there are negative authorizations for the requested operation and negative authorizations have precedence over positive authorizations the value of the access decision is set to *false* and returned. Otherwise the execution continues with the next step.
7. If the authority attributes of the requester satisfy the authorization policies, the decision value is set to *true* and returned. Otherwise it is set to *false* and returned.

13.4.1.4 Security Considerations

The access protocol of ÆTHER$_0$ we have presented in the previous paragraph is based on the Wide-Mouth Frog protocol [27]. Its basic assumption is that it requires loosely synchronized clocks between the participating entities in order to protect against replay attacks by utilizing timestamps. Other similar solutions, like for example the Otway–Rees [28] or the Needham–Schroeder [29] protocols, avoid the use of timestamps by using nonces and therefore more message exchanges to achieve the same goals. The only requirement of ÆTHER$_0$ regarding access protocols is the previous establishment of symmetric keys over location-limited channels according to the management model we have already presented. All access request protocols can be modified to be based on other similar symmetric authentication and key exchange protocols. Furthermore, the general ÆTHER model already makes the assumption of loosely synchronized clocks among the participants by using expiration time periods for authority revocation purposes. The main reason behind the selection of the Wide-Mouth Frog protocol is its simplicity and the small number of exchanged messages that it requires.

13.4.2 ÆTHER₁

In ÆTHER$_1$ all participating devices have a built-in asymmetric cryptography key pair. According to the design approach that is advocated by SPKI/SDSI [12], we also assume that an entity is identified, and therefore named, by the public key of this key pair (the same association between entities' names and public keys can easily be satisfied with the use of an Identity-Based Encryption (IBE) system, like the one presented in [30]). The corresponding private key is kept protected within the device. ÆTHER$_1$ allows every device to issue bindings and assign authority attributes to other devices by generating authority attribute certificates signed with their private key. Therefore, every device has its own local namespace and is allowed to define its own names for attributes. Public keys are encoded in hexadecimal format and are used to identify principals in the policy statements used in ÆTHER$_1$.

13.4.2.1 Management Model

A Pervasive Authority Domain (PAD) in ÆTHER$_1$ is defined as the initial set of relationships between attributes and principals specified in a security policy and is a logical representation of a ubiquitous computing environment. The owner of several devices creates a PAD by specifying in policies which principals are trusted to certify which authorization and context attributes. Moreover, the owner creates policy entries for controlling what authorization and context attributes a principal must possess in order to get specific access rights to a resource provided by a device. These policies are distributed to subject principals by issuers over wireless transmissions. In order to bootstrap security relationships and authenticate the distribution of these policies we again rely on location-limited channels for creating bindings between principals. Binding policies in ÆTHER$_1$ serve the purpose of binding the subject principal to the issuing principal. This is similar to the imprinting functionality of the Resurrecting Duckling system, but instead of sharing a common secret (as we do in ÆTHER$_0$) we utilize a signed statement to achieve the same goal. The embedding of the binding policy to the subject device must take place over a location-limited channel in order to ensure its authenticity. After a device has been bound, it can accept new policy entries remotely by the principal that bound it. These new policies are fully authenticated since the subject has already been bound by the issuer and knows its public key. As in ÆTHER$_0$, an ÆTHER$_1$ device can be bound to more than one devices.

When a principal wishes to issue an attribute certificate for another principal, the two devices establish a location-limited channel with the help of their human owners. Although any principal is allowed to issue a certificate for any authority attribute, only the ones that are issued by members of the AAS responsible for the specific authority attribute are considered valid. An Authority Attribute Set (AAS) is a policy statement that identifies the principals that are the sources of authority for a specific attribute. Membership in an AAS means that the corresponding principal is trusted to issue attribute credentials of the specific type. A member of an AAS is defined as a principal that has been explicitly denoted in the signed AAS statement as such, or an unknown principal that has been given the required attribute certificates (ACs) by at least a threshold number of members of the set. This threshold number is called Membership Threshold Value (MTV) and the accepted values are integers greater or equal to 2. This is in essence a delegation of the authority specified in the policy statement. The allowed depth of delegation is controlled by a *delegation depth* entry in the AAS definition statement that in essence implements integer delegation control. Although there have been arguments in the literature against integer control regarding the inability to predict the proper depth of delegation [11], the application domain of ÆTHER$_1$ makes its use particularly attractive. The owner of a PAD can use representative values for integer delegation control when she defines the AASs of the domain according to the importance of the attributes they authorize. Important attributes that convey a high value of trust can have small delegation depth values while more general attributes that authorize less important actions can have bigger values.

Example: In a domestic ubiquitous computing scenario the AAS policy statement for the attribute *(group, visitor)* that authorizes visitors to the house could specify a delegation depth and a membership threshold value of 2. This would allow a trusted visitor to introduce another visitor in the house, and in order for a visitor to be able to do so she must be trusted by two existing members of the AAS. This AAS with principals Key$_0$ and Key$_1$ as the initial sources of authority is illustrated in Figure 13.7. At some point the user of the device identified by key Key$_0$ establishes a location-limited channel with the user of the device identified by key Key$_2$. Key$_0$ issues an attribute certificate for the attribute *(group, visitor)* to Key$_2$ over this channel. Since the membership threshold value of the corresponding AAS statement has been defined as 2, ACs for the attribute *(group, visitor)* issued by Key$_2$ at this point are not valid. Later on Key$_1$ also establishes a location-limited channel with Key$_2$ and issues a *(group, visitor)* AC to it. Now Key$_2$ has the required number of certificates and dynamically joins the *(group, visitor)* AAS. It is now able to issue valid certificates for authority attributes of this type. Key$_2$ does so by establishing a location-limited channel with Key$_3$. This AC is valid and Key$_3$ can use it to support access requests.

The ÆTHER$_1$ management model does not have semantic categories of devices based on their rights to issue attribute assignments or to create bindings as is the case in ÆTHER$_0$. As we have already explained, every device in ÆTHER$_1$ is free to issue authority attribute assignments and also to bind other devices to itself. However, some devices may not have the required physical interaction interfaces to do so. As an example, a PDA can easily be used to bind other devices

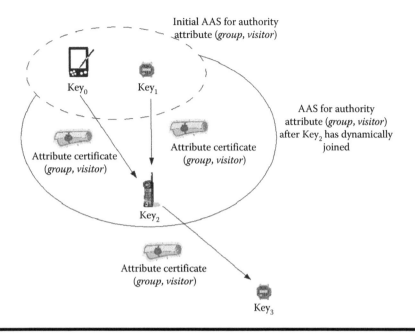

Initial AAS for authority attribute (*group, visitor*)

AAS for authority attribute (*group, visitor*) after Key_2 has dynamically joined

Attribute certificate (*group, visitor*)

Attribute certificate (*group, visitor*)

Attribute certificate (*group, visitor*)

Key_0 Key_1

Key_2

Key_3

Figure 13.7   A dynamic Authority Attribute Set (AAS).

since it has a touch screen and/or a keypad to interact with the user who can use these facilities to express this intent. On the other hand, a small sensor with a diameter of a few centimeters lacks the physical interface capabilities to be used for this purpose. Such devices are bound by other devices that have the required interfaces over location-limited channels and are then issued policies over wireless transmissions. We assume that they have some physical way (like a small switch or a button) to allow the owner to indicate her wish to initiate a location-limited channel and bind them to some other device.

13.4.2.2 Service Discovery and Access Protocols

In ÆTHER_1 the discovery of available services is performed according to two different methods, namely, the *eager* and the *lazy* service discovery and access protocols. As in ÆTHER_0, every ÆTHER_1 device maintains a list of all the available provided services it knows about. We call this list the Current PAD View (CPV) and its maintenance strategy depends on the utilized service discovery method. The following paragraphs present these methods and analyze the access protocols.

A concept that we use in both strategies is that of the Reduced Access Control List (RACL). The RACL is constructed by the service provider and is sent to a requester in order to aid the latter in identifying the authority attribute credentials that must be released to support a request. Since the general ÆTHER model supports negative authorizations, a provider cannot simply release the normal ACL of a service to a requester. If this was the case, the requester could simply avoid releasing to the provider its credentials that subtract privileges. Therefore, the provider does not release the ACL itself, but a *reduced* ACL that is constructed from the ACL. In the RACL the provider includes just the names of the authority attributes that appear in the ACL without their values or the predicates that apply on them. Context attributes are not included in the RACL.

The concept of the RACL both helps a requester to identify the credentials it must release to support a request and it protects the provider from making public the exact ACL of a service. Of course, given a sufficiently simple ACL and a committed attacker the exact predicates can be discovered through successive trials. To protect against this a provider can blacklist a requester that makes a certain number of successive failed attempts within a certain amount of time.

Eager: The main goal of the eager service discovery and access strategy is to protect the service providers from disclosing their advertisements to unauthorized parties. One of the main threats to service discovery is the enumeration of all available services by attackers. This can directly lead to the selection of vulnerable services to attack and constitutes an invasion of privacy. In order to secure the service discovery process, the eager strategy utilizes Public Resource Advertisements (PRAs). A PRA is a message which simply states that a service is provided. Specifically, a PRA contains the public key of the service provider (K_P) and the service identifier ($Service_{ID}$) which is a number used only by the service provider to distinguish between the different services it provides. PRAs are periodically broadcasted over a wireless communication channel every t seconds by providers for each service they have to offer (the exact value of t depends on the application scenario). When a device (the requester in our example with public key K_R) receives a PRA it checks to see if it already managed to get access to the service with the corresponding $Service_{ID}$. If this is the case then the PRA is ignored. Otherwise, it tries to access the provided service and enter it into its CPV. This is performed for every PRA a device receives.

The requester sends an access request to the service provider. The initial request message contains a packet identifier (REQ), the received service identifier ($Service_{ID}$), a session identifier ($Session_{ID}$), a protocol step identifier ($Step_{ID}$), and the service request. Furthermore, the public key of the requester (K_R), a digitally signed tuple containing the public keys of the two entities, a generated symmetric key (S_{RP}) and a timestamp (T) are also included. The entire message (except the packet identifier) is encrypted with the public key of the provider (K_P). Our key agreement protocol is based on [31]. The provider replies with the RACL for the requested service (assuming that such a service is indeed provided), the same $Session_{ID}$ and an incremented $Step_{ID}$. The message is encrypted using S_{RP}, the negotiated session key. Furthermore, message authentication is performed by computing an HMAC using c_1 (the ciphertext of the message) and S_{RP}. Based on the received RACL, the requester builds a reply with a set of authority attribute credentials it holds that can support its request and sends it to the provider. When the complete set of credentials is sent, the requester also sends a *CERTFIN* message to denote that it has sent all the relevant credentials it holds. The service provider passes the set of credentials and the relevant policy statements it locally has to its ÆTHER₁ inference engine. The result is a boolean value that is transmitted to the requester in step 6 (for the sake of the example we assume that the inference engine reached a positive decision) along with the Detailed Resource Advertisement (DRA) and the access interface of the requested service. A DRA contains a textual description of the related service. At the end of step 6 we consider the ÆTHER₁ handshake to be over. When the requester receives the DRA it enters the information it contains into its CPV. Therefore, in the eager strategy when an owner lists the entries of her CPV she sees all the services in the current PAD that she is allowed to access based on her authority and current context attributes. When the owner decides to use a particular service we have the final step of the protocol where the requester constructs and sends a *SAC* message based on the received access interface and the session ends. At this point the provider deletes the symmetric key S_{RP}. An established session also expires and the corresponding session key is deleted after a certain amount of time has elapsed when no *SAC* message is sent from the requester to the provider. In our current design we have set the session expiration time period to 120 s. The protocol is illustrated in Figure 13.8.

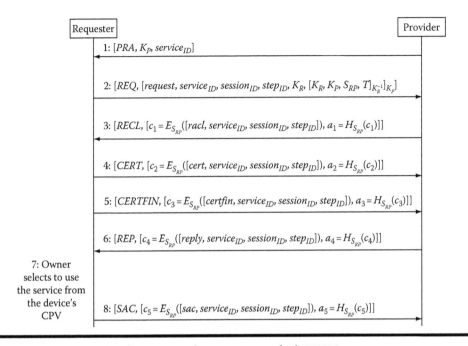

Figure 13.8 Eager service discovery and access protocol of ÆTHER₁.

In order to enhance the performance of the eager access protocol we use an *authorization cache*. At the end of step 5 when the provider reaches an authorization decision for a request from a specific principal with a specific set of credentials it caches the result. When the same requester wishes to access a service that already is in its CPV it simply sends a new request for it. This request is the same as step 2 of the above protocol. The requester generates a new secret key (S'_{RP}), a new session identifier ($Session'_{ID}$), and a new timestamp (T'). The provider checks its cache to see if the specific requester has been successfully authorized in the past for the specific request. If it has then it replies with a *REP* message (like the one in step 6), in essence confirming the new session key. Now the requester can access the provided service with a *SAC* message (as in step 8) and after it does the session ends and the provider again deletes the symmetric key associated with this session (S'_{RP}). The entries of the authorization cache expire after a certain amount of time. This time period is directly related to the nature of the corresponding provided service. A typical value we suggest is 10 min, a compromise between performance and the risk of granting access based on a previously valid AC that has expired during this time period. Another important issue has to do with context-sensitive permissions. When a permission requires the possession of context attributes, the service provider asks (this is a normal access request performed according to the previously presented protocol) the principal that is the source of authority for the specific context attribute (identified through the CAS statements it locally has in its policy database) about the membership status of the requester. Authorization decisions that have been reached using context attributes are never cached. We have taken this design choice since the context of an entity may change at any time and therefore previously valid authorization decisions may no longer be valid if the relevant context has changed.

Lazy: In the lazy access strategy pervasive devices remain passive regarding the service discovery process. Service providers broadcast their services and the pervasive device only responds when necessary. Hence, PRAs are not used. Instead service providers periodically broadcast DRAs for the services they want to offer. As we have described in the previous paragraph, DRAs contain a textual description of the corresponding service. Devices that receive DRAs enter them into their CPVs. When the owner of a device wants to access a particular service she lists the contents of the CPV maintained by the device she carries. Based on the descriptions she selects the appropriate service and the access protocol begins. This protocol is the same as the previous one with one difference. After the provider returns the authorization decision and the access interface of the service (assuming that the decision is positive) the requester directly sends a *SAC* message to use the service and the session ends. As in the eager strategy, we again rely on an authorization cache in order to improve performance.

The lazy strategy sacrifices the protection of service advertisements in order to avoid having devices in a PAD constantly trying to get access to all provided services. This helps devices to conserve battery energy, however, it allows attackers to enumerate all services offered in a PAD. Whether a device operates in eager or lazy mode is an option that can be directly controlled by its owner.

13.4.2.3 Inference Engine

The inference engine of ÆTHER_1 takes as input the set of attribute credentials that a requester has sent to a provider in order to support an access request, the request itself and the set of local policies that a provider possesses. If there are authorizations that use context attributes the principals responsible for maintaining the corresponding dynamic sets are queried regarding the membership status of the requester. It provides as output a boolean value that represents whether the request is allowed to be performed or denied. Since the algorithm of the ÆTHER_1 inference engine has a lot of similar steps with the ÆTHER_0 inference engine algorithm we only present here the steps that differ (however, Figure 13.9 presents the complete decision flowchart of the ÆTHER_1 inference engine).

1. After a request is received the engine checks the authorization cache to find whether the requester has made the same access request previously. If it has it returns the decision from the cache.
2. The engine also checks the validity of the requester's ACs, i.e., whether they have been issued by members of the corresponding AASs or not. If the ACs are valid the decision value is set to *true* and returned. Otherwise it is set to *false* and returned.

13.4.2.4 Security Considerations

As in ÆTHER_0, the key agreement protocol we use in ÆTHER_1 as the basis for our access protocol relies on loosely synchronized clocks to protect against replay attacks. However, loosely synchronized clocks are required between the participants in any case since we use short expiration time periods for the issued ACs. Any other key agreement protocol that also provides secure transport services can be used instead of the one we have presented. For example, the SSL/TLS protocol can be used. In this case the requester and the provider generate self-signed X.509 certificates for their public keys. These certificates ascertain verifiers that the issuer indeed controls the private key that corresponds to the presented public key. After the end of the SSL/TLS key agreement and the establishment of the secure channel, the requester presents its ÆTHER_1 ACs to the provider normally. Before the provider passes these to its inference engine it makes sure that the *subject* field of the requester's ACs is the same as the public key contained in the requester's self-signed X.509 certificate used to establish the SSL/TLS channel. This is done in order to detect and avoid

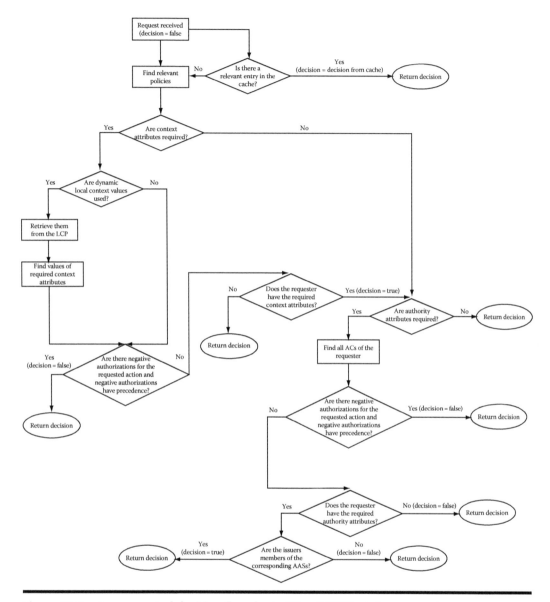

Figure 13.9 Access control decision flowchart of the ÆTHER₁ inference engine.

man-in-the-middle attacks. Afterward, the authorization process proceeds as we have previously described. In fact, the application of ÆTHER₁ in the domain of web services uses this approach. For more details see [32].

Another security issue we have to address is that of checking whether a service provider is authorized to provide a particular service or not. For example, a requester may want to be sure that a file to be stored in a file server is going to reach an authorized server and not a rogue one. The ÆTHER₁ access control protocol only provides to the service provider the guarantee that the public key of the requester is allowed or not to access the service it offers. However, the requester is not able to know if the service provider is a legitimate or a malicious entity. In case the requester has to release information of sensitive nature while accessing a service (such as storing a personal file or printing an important document) it is important for the utilized protocol to be able to support *server-side* authorization.

This problem can addressed in ÆTHER₁ by extending the previously presented access protocols so that the requester demands the possession of certain authority or context attributes before it releases any kind of sensitive information to the service provider. The requester must construct authorizations for the resources it wants to protect. These authorizations are regular ÆTHER₁ authorization statements. After the requester finishes sending its ACs to the provider and receives a positive reply for its access request, it sends its own RACL to the provider. Then the provider sends its own ACs to satisfy this RACL to the requester. The requester passes all these to its inference engine and if the output is positive then it sends the sensitive information to the provider who is now fully authorized.

13.5 Implementation and Evaluation

13.5.1 Platform

We have implemented prototypes of the ÆTHER₀ and ÆTHER₁ instantiations as *layers* of the Networks and Telecommunications Research Group (NTRG) ad hoc networking stack developed at the University of Dublin, Trinity College [33]. In the NTRG stack components are assembled using a layered architecture achieving abstraction borders between the different implementation elements. The main goal of the NTRG stack is "to produce a general-purpose mobile node capable of running a large range of network applications that can adapt its mode of operation to the prevailing wireless network architecture" [33]. It has been implemented using the C++ programming language and it currently supports all Windows-based operating systems (including Windows CE) and Linux. The reasons we selected it for implementing our prototypes are:

- The NTRG stack includes layers that implement several building blocks of mobile ad hoc networks, such as routing protocols (like the Ad hoc On-demand Distance Vector (AODV) [34] and Dynamic Source Routing (DSR) [35]), autoconfiguration addressing schemes, and decentralized location services, that can be directly reused for setting up the communications infrastructure of pervasive computing environments. This fact allows us to focus solely on the implementation of ÆTHER.
- The NTRG stack is simple and extensible allowing us to quickly build and evaluate the service discovery and access protocols of ÆTHER.
- It has been designed to work on modern handheld devices giving us the opportunity to experiment with real-world pervasive computing scenarios and applications.

Another facility provided by the stack is the *upmux* layer, a layer designed for upward multiplexing. It allows multiple independent layers to sit below a single stack layer. This finds a lot of applications, for example when we want to create network entities that bridge two distinct domains built on different wireless networking technologies, or to support location-limited channels. Figure 13.10 presents the ÆTHER layer as part of the NTRG stack.

13.5.2 Policy Language

The policy language we use is similar to the assertion language of the KeyNote trust management system. However, we have extended the syntax in order to allow the use of ÆTHER-specific policy

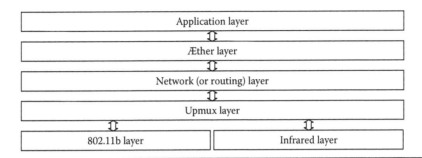

Figure 13.10   The ÆTHER layer as part of the NTRG stack.

data structures. Space constraints do not allow the presentation of our extensions to KeyNote; we refer the interested reader to [25].

Another important note regards the utilized namespace. We believe that given the distributed and decentralized nature of pervasive computing it would be difficult to enforce a global namespace. Instead ÆTHER follows the design approach that is advocated by SPKI/SDSI and relies on local namespaces. The vocabulary for defining policies (either for the data, the operations, or the conditions) is outside the scope of the work presented in this thesis. Existing research efforts to create ontologies for authorization management and pervasive computing can be used for this purpose [36,37].

13.5.3 *Performance Analysis*

In order to evaluate the two instantiations of our architecture in a quantitative manner we have completed a detailed performance analysis based on modern handheld devices. Specifically, the hardware platform we use is the HP iPAQ H6340 [38] with a Texas Instruments OMAP 1510 processor at 168 MHz and 64 MB RAM (64 MB ROM), running the Windows CE Mobile 2003 [39] operating system. We use the infrared interface of the handhelds to implement the required location-limited channel and the IEEE 802.11b WLAN interface for normal communication links.

13.5.3.1 *ÆTHER$_0$*

We have evaluated ÆTHER$_0$ in respect to the time required for the completion of the different access protocols we have previously presented. In all the experiments we use AES with a 256-bit key (in CBC mode) as the symmetric encryption algorithm and HMAC-SHA-1 as the HMAC algorithm. Furthermore, we assume that the authorization policy for the requested operation specifies only one required authority attribute.

We also assume that the alpha-pad (or the alpha-pad that bound the device that offers the requested service in the last two experiments) has assigned to the owner of the requesting device the required authority attribute and therefore reaches a positive decision. In the normal device to alpha-pad protocol we have measured the time required for a requester (the normal device) to construct a request, send it over the wireless interface to the alpha-pad that provides a service, and receive a reply. The average time required for the protocol to complete is 0.175 s (175 ms). The second case examines the normal device to normal device (same alpha-pad) access protocol. The required average time for the completion of this protocol is 0.189 s (189 ms). In the final access

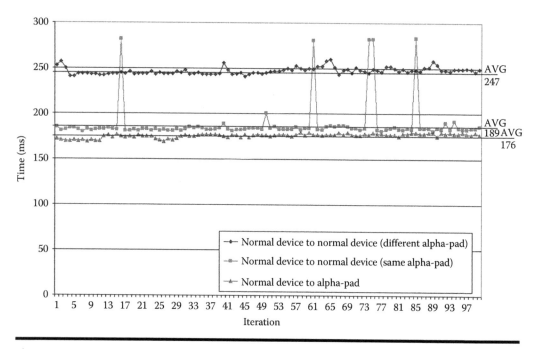

Figure 13.11 Timing measurements for the ÆTHER$_0$ access protocols.

protocol experiment for ÆTHER$_0$ we have examined the situation where a request is initiated from a normal device to another normal device. The assumption is that the alpha-pads that bound the devices share a mutual binding. The average time is 0.247 s (247 ms) in this case. All three access protocol measurements for ÆTHER$_0$ are presented in Figure 13.11.

The timing measurements of the ÆTHER$_0$ access protocols demonstrate the feasibility of using them on mobile-constrained devices. Even the normal device to normal device (different alpha-pad) protocol which is the most complicated of the three and involves message exchanges between four different devices requires approximately only 0.25 s in average. Based on previous work done on benchmarking cryptographic algorithms [40], like the AES cipher we use, we believe that the ÆTHER$_0$ instantiation can easily be deployed even in low-end sensor networks consisting of devices with 8-bit processors.

In order to evaluate the inference engine of the ÆTHER$_0$ instantiation we have performed successive experiments with an increasing number of required authority attributes in the authorization policy that applies to the requested action. In this experiment we have only measured the time required for a service provider to reach an authorization decision, without taking into consideration message exchanges over the wireless communication medium. The results are shown in Figure 13.12 and are the averages of 100 iterations for each experiment.

When only one authority attribute is required the average time for the inference engine to reach an authorization decision is 0.148 s (148 ms). As the number of required authority attributes increases, so does the average time needed for the inference engine to reach a decision, by an average of 5 ms. For five authority attributes the required time is 0.168 s (168 ms). Therefore, the overhead introduced by the ÆTHER$_0$ inference engine is not prohibitive for pervasive computing applications.

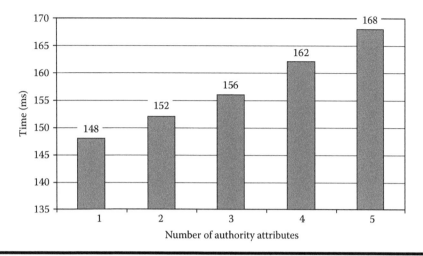

Figure 13.12   Timing measurements for the ÆTHER$_0$ inference engine.

13.5.3.2 ÆTHER$_1$

Our experiments for the ÆTHER$_1$ access protocol are based on the eager service discovery and access strategy, since it does not require human intervention for its completion as does the lazy strategy. We measured the time required from the point a device receives a public resource advertisement from a service provider and sends a request message for it, until the point it gets back a *REP* message that contains an authorization decision plus the provided service's interface, if the decision is positive, and enters this information into its CPV. We assume that there is an authorization policy for the provided service that specifies that a requester needs a single authority attribute to access the service. Moreover, that the requester has been issued an AC of the required type from a principal that is directly trusted by the provider to do so; i.e., the AC's issuer is listed in the *sources of authority* field of the AAS statement for the specific authority attribute that the provider has. The experiment was performed with the service provider's authorization cache both disabled and enabled in order to have demonstrative results (see Figure 13.13).

When keys of 512 bits are used the average required time is 0.3 s (300 ms). The utilization of the authorization cache decreases this to 0.132 s (132 ms), a performance gain of 56%. With 1024-bit keys the average required time for the completion of the protocol is 0.536 s (536 ms), and when we enable the authorization cache of the service provider this becomes 0.366 s (366 ms). The decrease of the average required time is in the order of 31.7%. We believe that the timing results of both key sizes are realistic for a pervasive computing security system which relies on public key cryptography. Even in the eager service discovery and access strategy in which a device tries to access all services provided in an environment irrespectively of whether its user is interested in them, the time required for a successful completion of the protocol is half a second when 1024-bit keys are used. This time is small enough not to be prohibitive.

The inference engine of the ÆTHER$_1$ instantiation has been evaluated according to two metrics: the number of authority attributes that are required to reach an authorization decision, and the delegation depth that applies on principals that dynamically joined the AAS responsible for a required authority attribute. Both experiments have been performed with RSA keys of 512 and 1024 bits. In the first one we have measured the time required for the inference engine to reach

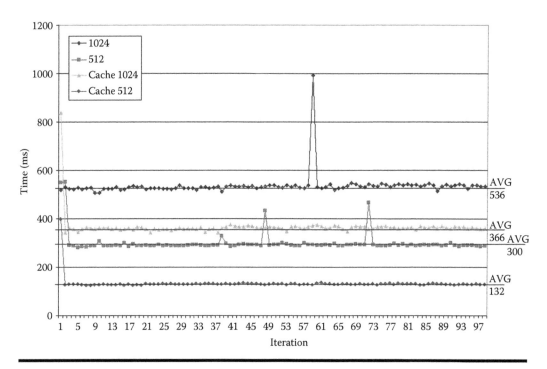

Figure 13.13 Timing measurements for the ÆTHER₁ access protocol.

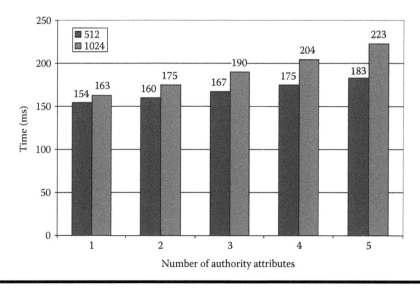

Figure 13.14 Impact of number of authority attributes on the ÆTHER₁ inference engine.

a positive decision when the number of the necessary authority attributes increases. Figure 13.14 shows the results as averages of 100 iterations for each number of authority attributes.

As expected the impact is more considerable when 1024-bit keys are utilized. Specifically, with 512-bit keys the average increase of the required time when the number of authority attributes increases is 7.25 ms, while with 1024-bits the average increase is 15 ms.

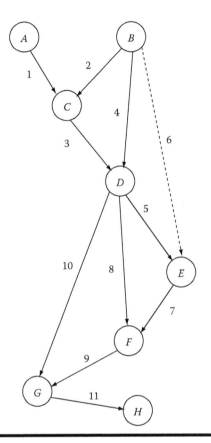

Figure 13.15   Delegation depth example.

We have also evaluated the overhead introduced by the AAS's delegation depth parameter. In order to make this more clear consider the example presented in Figure 13.15.

We have an AAS for the authority attribute *(group, family_member)* with a delegation depth value of 0 (denoting no restrictions on the depth of delegation) and a membership threshold value of 2. This AAS has as initial sources of authority the principals identified by the public keys A and B. In step 1 principal A issues an AC for the authority attribute *(group, family_member)* to principal C. This AC can be used by C to support the access requests it makes. When B issues an AC of the same type to C (step 2), C dynamically joins the *(group, family_member)* AAS (since we have an MTV of 2) and is now able to issue valid ACs of this type. It does so by issuing an AC to D (step 3) which makes an access request to a service provider. This is delegation depth 1 and the service provider must now verify three ACs, from steps 1, 2, and 3. The depth of delegation increases up to 5 where a service provider needs to perform 11 verification operations.

We have performed the experiment 100 times for each delegation depth step with RSA keys of 512 and 1024 bits. In the first case we have found that the average increase in the required time to reach a positive authorization decision is 16.25 ms, and in the second case 56 ms. The specific results for each delegation depth step are shown in Figure 13.16.

We believe that the observed performance results of the ÆTHER₁ inference engine illustrate the feasibility of using it on mobile constrained devices. Even when the delegation depth is 5, with

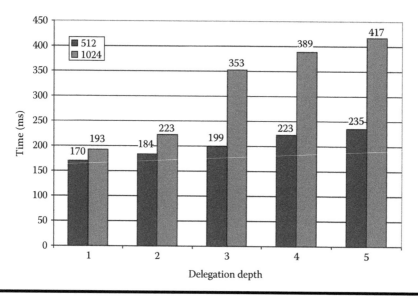

Figure 13.16   Impact of delegation depth on the ÆTHER₁ inference engine.

an MTV of 2 and keys of 1024 bits, the average time required to reach a decision is less than half a second (417 ms) and therefore has minimal impact on the applications employed at a higher layer.

13.5.4 Usability and Maintainability

One of the most difficult problems that every security solution for pervasive computing has to address is that of usability. Since pervasive computing aims to make the information processing capabilities of the environment integrate naturally with the workflow of the human users, it must not require from them to constantly answer complicated dialogs or fill input forms. Flaws in the human–computer interface design may result in the complete failure of otherwise perfectly secure systems. Users prefer to ignore, or even completely disable, security settings they do not fully understand [41]. For example, in early versions of ÆTHER we were querying the user whether a given previously established policy decision (like a binding or an authority attribute assignment) should be allowed to be renewed or not. However, we found out that this was particularly distracting and eventually we decided to fully integrate renewal mechanisms in our model.

Policy composition is another area of security usability that requires careful attention. We realize that only a very small percentage of users will actually fully explore all the functionalities of their devices and compose authorization policies for them. Users do not want to be bothered with managing their devices, yet they also do not want to leave their devices unprotected or let an outside, global and centralized entity to handle their management. To address this we suggest that manufacturers should provide default sets of policies with their devices according to the services they provide and the context in which they are going to be used. Users can then just create simple authorization policies (like "allow anyone in the room to use the printer," or "all my friends can access my mobile phone's files") probably with a graphical utility. Of course the ÆTHER policy language gives the option to power users to modify the default policies and create their own much more flexible and appropriate to specific situations policies.

Maintainability refers to the ease with which any maintenance task can be carried out on an information processing device. Therefore, its role is essential in pervasive computing where human users have countless devices that have to be maintained in the most unobtrusive manner possible.

One of the maintainability problems that we have to consider is that of lost or stolen devices. In $ÆTHER_0$ when a user device is stolen it can either be an alpha-pad or a normal device. In the case it is a normal device then the problem can easily be addressed by the user. The secret keys the device has established with other devices during initiating or replying to service access requests will eventually expire and therefore do not present a direct system compromise. In $ÆTHER_1$ the problem of stolen devices is similar to the problem of stolen alpha-pads in $ÆTHER_0$. Since every device in $ÆTHER_1$ is allowed to bind other devices, a stolen device means that the owner must revoke all the bindings that the stolen device has been used to create. This can be accomplished either with the hardware switch that wipes all policy state from a bound device, or with another device that has been used to create an additional binding with the target device. We do not believe that multiple bindings are an inconvenience to end users. When a user has a device that is of particular importance and has a significant chance to be stolen (due to its size and the fact that the user always carries it with him) it makes sense not to bind all other devices in a PAD just with this particular device.

13.5.5 Applications

In this paragraph we will briefly present several application examples of the two instantiations of the general ÆTHER model. For more details, please see [25].

- Location-based access control: The context-awareness features of ÆTHER enable the specification of policies that allow access to a printer only by entities that are in the same physical location as the printer itself. If the printer is moved to another location the policy does not need to changed as the location context variable takes values in a dynamic manner.
- Home automation security: Using $ÆTHER_0$ a home owner can set up a front-door lock that only family members can open using their handheld, or other pervasive, devices. He can specify that visitor to the house can turn on the television set, but not change the settings of the house heating installation.
- Building access control: Current building access control systems rely on particularly static, inflexible, and centrally managed technologies. The typical approach is to connect all door locks to a central server where access policy is regulated and register the identification numbers of specific swipe cards as authorized to open specific locks. Using $ÆTHER_1$ an organization or institution will be able to control access to physical spaces such as rooms and buildings in a dynamic manner, allowing disconnected operation while still maintaining sophisticated access control policies.
- Pervasive automobiles: Current automobiles have hundreds of embedded computers in order to provide services such as traction control, air conditioning, seat positioning, navigation, and entertainment, among others. As technology progresses with a particularly fast pace in this area, future automobiles will be full-fledged pervasive computing environments. By employing $ÆTHER_1$ the owner can specify, for example, that a parking driver is allowed to drive her car up to 20 miles per hour but not allowed to open the trunk.

13.6 Conclusion and Future Work

A detailed analysis of the existing traditional and pervasive security management systems focused on the identified requirements revealed that there is a need for an authorization framework for pervasive computing which is able to provide fine-grained access control capabilities, without depending on external centralized infrastructure. Moreover, such a framework must be able to support disconnected operation and gracefully integrate the human intervention required for the administrative tasks with the active physical task at hand in order to minimize distractions.

ÆTHER is a framework for pervasive computing authorization management that directly extends the traditional RBAC model for supporting decentralized administration, disconnected operation, context-awareness and follows an unobtrusive usage model. Based on this general framework we have instantiated two different systems. ÆTHER_0 has been designed to address the authorization requirements of small pervasive environments which consist of particularly constrained devices. It relies only on symmetric key cryptography by sacrificing the local decentralization requirement. Our second instantiation, ÆTHER_1, is both globally and locally decentralized. However, this greater flexibility comes at greater computational requirements from the participating devices since we use public key cryptography and certificates for the required attribute assignments. In ÆTHER_1 the sets of principals that act as sources of authority for specific attributes are allowed to grow dynamically supporting the establishment of trust with unknown principals and authorizing their actions in the local domain. Our implementation and evaluation of the two ÆTHER instantiations demonstrated the feasibility of deploying and using them in pervasive computing environments.

During the design, implementation, and evaluation of the ÆTHER framework, we have identified several directions that future research on the subject can follow. These are briefly summarized here:

- Although we have performed a quantitative evaluation of the ÆTHER framework, we have not evaluated it from a qualitative perspective. Future work must focus on ensuring that our user-based authorization model and the integration of the establishment of location-limited channels with the primary physical actions of users conform to recognized usability guidelines.
- The ÆTHER framework uses short expiration time periods for handling the problem of authority revocation. The fundamental problem of this approach is the selection of the expiration time period. Further research on this area can lead to a detailed risk management analysis of particular pervasive computing applications and ultimately to the recommendation of specific validity periods. These should be validated with both formal justification and practical experimentation.

Furthermore, we have identified two general pervasive computing areas that future research should focus on. Pervasive computing is a relatively new, multidisciplinary paradigm for defining the future of the role that information processing devices will play in our lives. As such its goals are broad and ambitious. Human users and the interactions they have based on their social relationships must always be at the center of any pervasive computing system; indeed this goal effectively summarizes the vision of pervasiveness. However, all our current system design methodologies and guidelines fail to capture this essential requirement. The main reason is that although humans and their interactions are included in the design process, their behavior remains unpredictable and outside the control of system designers. Therefore, all efforts to design management systems, including security management systems such as ÆTHER, are problematic.

It is our belief that considerable research is required on two fields of pervasive computing: the reexamination of traditional design approaches regarding the unpredictability introduced by human users as system components, and the definition of clear and concise metrics that will allow us to compare the different proposed solutions. This work will allow us to both build and evaluate dependable pervasive computing security systems that fully integrate their intended users as system components.

References

1. M. Weiser. The computer for the twenty-first century. *Scientific American*, 265(3):94–104, 1991.
2. B. Brumitt, B. Meyers, J. Krumm, A. Kern, and S. Shafer. EasyLiving: Technologies for intelligent environments. In *Proceedings of 2nd International Symposium on Handheld and Ubiquitous Computing (HUC2K)*, Bristol, U.K., pp. 12–29, 2000.
3. S.R. Ponnekanti, B. Lee, A. Fox, P. Hanrahan, and T. Winograd. ICrafter: A service framework for ubiquitous computing environments. In *Proceedings of 3rd International Conference on Ubiquitous Computing (UBICOMP'01)*, London, U.K., pp. 56–75, 2001.
4. D.F. Ferraiolo and D.R. Kuhn. Role-based access controls. In *Proceedings of NIST-NSA National Computer Security Conference*, Baltimore, MD. pp. 554–563, 1992.
5. ISO/ITU-T Recommendation X.509. The Directory: Authentication Framework, 2001.
6. R. Housley, W. Ford, W. Polk, and D. Solo. Internet X.509 public key infrastructure certificate and CRL profile. RFC 2459, 1999.
7. S. Farrell and R. Housley. An Internet attribute certificate profile for authorization. RFC 3281, 2002.
8. D. Ferraiolo, R. Sandhu, S. Gavrila, D.R. Kuhn, and R. Chandramouli. Proposed NIST standard for role-based access control. *ACM Transactions on Information and System Security*, 4(3):224–274, 2001.
9. M. Blaze, J. Feigenbaum, and J. Lacy. Decentralized trust management. In *Proceedings of 1996 IEEE Symposium on Security and Privacy*, Oakland, CA. pp. 164–173, 1996.
10. M. Blaze, J. Feigenbaum, and A.D. Keromytis. The KeyNote trust management system (version 2). RFC 2704, 1999.
11. C. Ellison, B. Frantz, B. Lampson, R.L. Rivest, B. Thomas, and T. Ylonen. SPKI certificate theory. RFC 2693, 1999.
12. D. Clarke, J.-E. Elien, C. Ellison, M. Fredette, A. Morcos, and R.L. Rivest. Certificate chain discovery in SPKI/SDSI. *Journal of Computer Security*, 9(4):285–322, 2001.
13. F. Stajano and R. Anderson. The resurrecting duckling: Security issues for ad hoc wireless networks. In *Proceedings of 7th International Workshop on Security Protocols*, Cambridge, U.K., pp. 172–182, 2000.
14. D. Balfanz and D.K. Smetters. Talking to strangers: Authentication in ad hoc wireless networks. In *Proceedings of 9th Network and Distributed System Security Symposium (NDSS'02)*, San Diego, CA, 2002.
15. F. Stajano. The resurrecting duckling—What next? In *Proceedings of 8th International Workshop on Security Protocols*, Cambridge, U.K., pp. 204–214, 2001.
16. M. Roman, C.K. Hess, R. Cerqueira, A. Ranganathan, R.H. Campbell, and K. Nahrstedt. Gaia: A middleware infrastructure to enable active spaces. *IEEE Pervasive Computing*, 1(4):74–83, 2002.
17. J. Al-Muhtadi, A. Ranganathan, R. Campbell, and D. Mickunas. Cerberus: A context-aware security scheme for smart spaces. In *Proceedings of 1st IEEE International Conference on Pervasive Computing and Communications (PERCOM'03)*, Fort Worth, TX. pp. 489–496, 2003.
18. M.J. Covington, W. Long, S. Srinivasan, A. Dey, M. Ahamad, and G. Abowd. Securing context-aware applications using environment roles. In *Proceedings of 6th ACM Symposium on Access Control Models and Technologies (SACMAT'01)*, Chantilly, VA. pp. 10–20, 2001.
19. M.J. Covington, M.J. Moyer, and M. Ahamad. Generalized role-based access control for securing future applications. In *Proceedings of 23rd National Information Systems Security Conference (NISSC'00)*, Baltimore, MD. 2000.

20. R. Anderson and M. Kuhn. Tamper resistance—A cautionary note. In *Proceedings of 2nd USENIX Workshop on Electronic Commerce*, Oakland, CA. 1996.

21. R. Anderson and M. Kuhn. Low cost attacks on tamper resistant devices. In *Proceedings of 5th International Workshop on Security Protocols*, Paris, France. pp. 125–136, 1997.

22. D. Salber, A.K. Dey, and G.D. Abowd. The context toolkit: Aiding the development of context-enabled applications. In *Proceedings of SIGCHI Conference on Human Factors in Computing Systems*, Pittsburgh, Pennsylvania, pp. 434–441, 1999.

23. G. Chen and D. Kotz. A survey of context-aware computing research. Technical Report TR2000-381, Dartmouth College, Department of Computer Science, Hanover, Germany, 2000.

24. M. Myers, R. Ankney, A. Malpani, S. Galperin, and C. Adams. X.509 Internet public key infrastructure online certificate status protocol - OCSP. RFC 2560, 1999.

25. P. Argyroudis. Authorization management for pervasive computing. PhD thesis, School of Computer Science and Statistics, University of Dublin, Trinity College, Dublin, Ireland, 2006.

26. H. Krawczyk. The order of encryption and authentication for protecting communications (or: How secure is SSL?). In *Proceedings of 21st Annual International Cryptology Conference (CRYPTO'01)*, London, U.K., pp. 310–331, 2001.

27. M. Burrows, M. Abadi, and R. Needham. A logic of authentication. *ACM Transactions on Computer Systems*, 8(1):18–36, 1990.

28. D. Otway and O. Rees. Efficient and timely mutual authentication. *Operating Systems Review*, 21(1):8–10, 1987.

29. R.M. Needham and M.D. Schroeder. Using encryption for authentication in large networks of computers. *Communications of the ACM*, 21(12):993–999, 1978.

30. D. Boneh and M. Franklin. Identity-based encryption from the weil pairing. In *Proceedings of 21st Annual International Cryptology Conference (CRYPTO'01)*, London, U.K., pp. 213–229, 2001.

31. M. Abadi. Private authentication. In *Proceedings of 2002 Workshop on Privacy Enhancing Technologies (PET'02)*, San Fransisco, CA. pp. 27–40, 2002.

32. P. Stefas. Decentralized authorization for web services. MSc thesis, School of Computer Science and Statistics, University of Dublin, Trinity College, Dublin, Ireland, 2005.

33. D. O'Mahony and L. Doyle. An adaptable node architecture for future wireless networks. *Mobile Computing: Implementing Pervasive Information and Communication Technologies*, Kluwer Publishing, 2001.

34. C.E. Perkins and E.M. Royer. Ad hoc on-demand distance vector routing. In *Proceedings of 2nd IEEE Workshop on Mobile Computing Systems and Applications*, New Orleans, LA, pp. 90–100, 1999.

35. D.B. Johnson, D.A. Maltz, and J. Broch. DSR: The dynamic source routing protocol for multi-hop wireless ad hoc networks, *Ad hoc Networking*, Addison-Wesley, pp. 139–172. 2001.

36. M. Donner. Toward a security ontology. *IEEE Security & Privacy*, 1(3):6–7, 2003.

37. H. Chen, T. Finin, and A. Joshi. An ontology for context-aware pervasive computing environments. *Knowledge Engineering Review*, 18(3):197–208, 2003.

38. HP iPAQ H6340. http://www.ipaqchoice.com/, 2005.

39. Microsoft Windows CE. http://www.microsoft.com/windowsce/, 2005.

40. Y.W. Law, J. Doumen, and P. Hartel. Benchmarking block ciphers for wireless sensor networks. In *Proceedings of 2004 IEEE International Conference on Mobile Ad hoc and Sensor Systems*, Fort Lauderdale, FL. pp. 447–456, 2004.

41. A. Whitten and J.D. Tygar. Why Johnny can't encrypt: A usability evaluation of PGP 5.0. In *Proceedings of 8th USENIX Security Symposium*, Washington, DC. pp. 169–183, 1999.

Chapter 14

Enhancing Privacy of Universal Re-Encryption Scheme for RFID Tags

Junichiro Saito, Sang-Soo Yeo, Kouichi Sakurai,
Jin Kwak, and Jae-Cheol Ryou

Contents

14.1 Introduction

A RadioFrequency Identification (RFID) tag is a small and inexpensive device that consists of an IC chip and an antenna that communicate by radio frequency. A radio communication device called as a reader emits a query to RFID tags and reads their ID. Some readers also transmit power to RFID tags when they emit a query. In this case, RFID tags do not have power supply. Therefore, RFID tags are expected to be used as a substitute for a bar code in the future [1–6]. In order to use as the replacement of the bar code, the cost of RFID tags is $0.05/unit, and tags are as small as $0.4\,mm \times 0.4\,mm$ and thin enough to be embedded in paper [2,6–9]. For this reason, the processing capacity of a RFID tag is limited.

The RFID system using this tag and a reader is used for the automobile object identification. Since the goods attached to the RFID tags in a cardboard box can be checked even if the box is not open, it is used for the management of goods [2,8,10]. A RFID tag is attached to to goods, and it is expected that its function like a bar code is achieved and it is useful to theft detection. Moreover, after goods are purchased, a RFID system gives a useful function for a consumer. For example, a refrigerator with the reader will be able to recognize expired foodstuffs, and a closet will be able to offer a few of the enticing possibilities of its contents [1,2]. Moreover, the European Central Bank (ECB) has proposed to embedded RFID tags in Euro banknotes [3]. By using identification combined ID on RFID tags and serial number printed on banknotes, it is expected to prevent forgery or money laundering. Therefore, the RFID system becomes popular for automated identification in EPC network application. However, this technology creates new problems such as the invasion of user's privacy. Thus, several methods for protecting the user's location privacy have been proposed [5,11–17].

In this chapter, we discuss the RFID system using universal re-encryption based on ElGamal proposed by Golle et al. [18]. By using universal re-encryption, we can re-encrypt ciphertext without the knowledge of a public key. This property is suitable for the RFID system.

However, since the system cannot protect a modification of the information on RFID tags, it can be exploited by an attacker. According to [18], an attacker may alter the information on RFID tags to the contents that cancel re-encryption. Moreover, we point out that an attacker may alter the information to the ciphertext encrypted by the attacker's public key.

To avoid these attacks, we propose two schemes. One is the re-encryption protocol with a check. This scheme checks the information written in a RFID tag. Our scheme prevents infringement of the location privacy by the modification of the information on a RFID tag. The other is the re-encryption protocol using a one-time pad for anonymity in RFID tags. We introduce the simple access control using hash function to prevent modification of the information on RFID tag. Moreover, the protection capability from infringement of location privacy can be effective because

a RFID tag re-encrypts its ID information by using a one-time pad. A reader updates a one-time pad on a RFID tag to prevent from spoiling its effect. Furthermore, we discuss the security of our methods from a viewpoint of privacy protection.

14.2 Related Works

14.2.1 General RFID System

RFID systems are basically composed of tags and readers. An RFID tag, also known as a transponder, is an RFID device consisting of a microchip and antenna attached to a substrate. The microchip in the tag is used for data storage and logical operations, whereas the coiled antenna is used for communication with the reader. When the RFID reader queries, the RFID tag transmits identification information, such as an ID, to the RFID reader by means of a Radio Frequency signal. An RFID reader is known as a transceiver or interrogator. The reader generally consists of an RF module, control unit, and coupling element to interrogate electronic tags via RF communication. An RFID reader receives identification information from an RFID tag and subsequently delivers this information to the RFID middleware. The RFID reader can read and write data on the RFID tags [19–24].

The RFID tag is either an active or a passive tag. The *active tag* possesses a battery and actively transmits information to the reader for communication. The *passive tag* must be inductively powered from the RF signal of the reader since RFID tags usually do not possess their own battery power supply. The EPC is stored on this tag. Tags communicate their EPCs to readers using a RF signal [19–23].

Figure 14.1 shows a general RFID system.

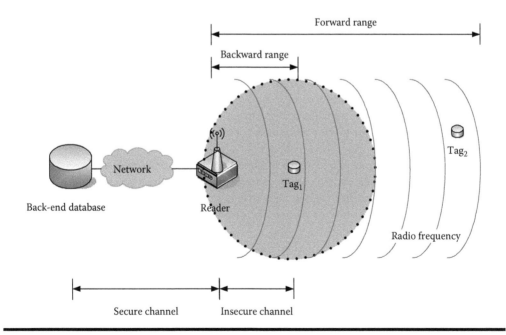

Figure 14.1 General RFID system.

14.2.2 Cryptography in RFID System

The wireless communication between a reader and RFID tag is performed by radio. So it is easily tapped by an attacker. The reader can simply derive information from the RFID tag and it can be used to infringe privacy. Moreover, the location of the owner can be traced by tracing the information on the specific RFID tag even if the attacker cannot understand the contents of ID. For this reason, there are some problems such as a retail store pursues a consumer and the circulation information on goods is revealed [2–8,11–17].

There are many researches related to the RFID privacy problems. The simplest scheme is the kill-command method that destructs the tag's hardware after purchasing an item. The kill-command scheme is the basic protection scheme of EPCglobal's tag [2,4,5]. Kill-command is operated by inputting a specific PIN number for each tag. After a tag is destructed by kill-command, it loses all of its RFID functionalities. This strategy seems to be a perfect solution for privacy protection, but it makes useless all of the potential valuable services of RFID. In other words, it is not a good solution, because this method makes tags to be isolated from networking environments.

Weis et al. suggested a hash locking scheme, randomized hash locking scheme, and silent tree walking scheme [6]. In the hash locking scheme, a tag can be either in locked or unlocked mode, and only the authenticated reader can unlock a tag. When a tag is in the locked mode, all the functions except transmission of the metaID of the tag are disabled. However, it does not solve the problem of location tracking, since a tag in locked mode always emits the same metaID value. In the randomized hash locking scheme, the tag has a hash function and a pseudo-random number generator. This scheme partly prevents location tracking, but has some security holes in reader-tag communication and very heavy computational load in the back-end server [27,28].

Ohkubo et al. proposed a self-refreshment scheme [13] that the tag refreshes its secret data by itself, using two one-way hash functions. However, this scheme is impractical and not scalable, because it has a very high computational complexity of back-end server to identify tags. There are two schemes that use a precomputation approach and/or a grouping approach for supporting Ohkubo's scheme [15,16].

Juels et al. proposed the blocker-tag scheme [8] and suggested a privacy protection scheme for RFID-enabled banknotes [3]. The blocker-tag scheme uses anticollision protocols reversely, which are originally intended for tag singulation. The blocker-tag always responds to the reader ping query in both 0 and 1 with the previous prefix. Therefore, it forces the reader to give up trying to recognize the tag and then the privacy is protected. However, the blocker tag presents the risk of being misused.

There are agent schemes whose approach is arbitrating the communication between the reader and tag using the mobile device [17,29–31]. Because the mobile devices have more powerful processors than the tag does, they can use high level cryptographic modules that have been used in the general networking systems. However, agent schemes have some problems related to tag forgery and ownership transferring.

In addition to encrypt the information on the RFID tag, there is an approach of re-encrypting the encrypted information on the RFID tag periodically [3]. Re encryption means encrypting a ciphertext again. It is performed by using public key cryptography. Even if a ciphertext is re-encrypted repeatedly, we can obtain the plaintext by decrypting only once with using a private key. By using symmetric key cryptography, we must decrypt the re-encrypted ciphertext many times or the reader has to synchronize with the RFID tag. Since the information on a RFID tag is changed by re-encryption, it can prevent from tracing the information on the specific RFID tag.

Moreover, if the reader processes re-encryption, a RFID tag does not need to carry out complicated computation.

However, if a reader processes re-encryption with a public key, the owner has to deliver information about the public key for the reader in the case of re-encryption. In that case, the attacker will be able to trace the RFID tag relevant to the public key [18].

14.3 RFID Systems Using Universal Re-Encryption

Universal re-encryption does not need knowledge of a public key in the case of re-encryption, but re-encryption is performed by determining a random number. Moreover, re-encryption does not need the information about plain text unlike encryption. The ciphertext that was processed re-encryption repeatedly can be once decrypted to the original plain text using the private key. Furthermore, universal re-encryption has semantic secrecy required for the privacy protection. These properties are effective in privacy protection of RFID tags [18].

14.3.1 Model and Operating Schemes

We define a model in the RFID system using universal re-encryption based on ElGamal. The model consists of a RFID tag, a database, a reader for reading ID information, a reader for re-encryption, and an attacker. The owner always possesses the reader who can process re-encryption. Moreover, if the property of universal re-encryption is used in the case of re-encryption, then a trusted third party, such as a bank and a public institution, can process the re-encryption procedure as a service [18].

RFID tag: It sends its ID information (ciphertext C) in response to query from a reader. The ID data is encrypted by universal re-encryption.

Database: It has private key x for ID information on a RFID tag, and the information on the item relevant to the RFID tag. Private key x is saved securely by an existing access control scheme.

Reader for reading for ID data: Interrogates a RFID tag and receives the tag's ID data. Then, it asks a database by an existing authentication scheme, and acquires the information about the item in which the RFID tag was attached. Then it offers service, a function, etc., based on the ID information.

Reader for re-encryption: Interrogates a RFID tag and receives the tag's ID data. It is asked to a database by an existing authentication scheme, and receives new ciphertext C'. Then, it updates the ID information of the RFID tag. Using universal re-encryption, if a reader for re-encryption saved ID information, it becomes difficult for tracing a RFID tag by semantic security when the next re-encryption is performed by another reader [18].

The protocol of universal re-encryption based on ElGamal process follow operations.

14.3.1.1 Key Generation

Generates secret key x and public key $(y = g^x)$.

14.3.1.2 Encryption

Ciphertext $C = [(\alpha_0, \beta_0); (\alpha_1, \beta_1)]$ is generated from the following formulas using message m, public key y, and random number $r = (k_0, k_1)$. Ciphertext C is written in a RFID tag:

$$C = [(\alpha_0, \beta_0); (\alpha_1, \beta_1)],$$
$$\alpha_0 = my^{k_0}, \beta_0 = g^{k_0}, \alpha_1 = y^{k_1}, \beta_1 = g^{k_1}$$

14.3.1.3 Decryption

The reader for reading ID information receives ciphertext C from a RFID tag, and sends it to a database. A database calculates the decryption algorithm described as follows. Compute $m_0 = \alpha_0/\beta_0^x$ and $m_1 = \alpha_1/\beta_1^x$ using ciphertext $C = [(\alpha_0, \beta_0); (\alpha_1, \beta_1)]$ under public key y from a RFID tag and secret key x. If $m_1 = 1$, then output message $m = m_0$. Otherwise the decryption fails, and a special symbol $\perp$ is output. A given key can be decrypted only under one given key. It will get a message m_0 as ID of the RFID tag. Even if ciphertext C is re-encrypted many times, it can return to plaintext by decryption once. And a database searches the information on a RFID tag using its ID, and transmits it to the reader for reading ID information.

14.3.1.4 Re-Encryption

The reader for re-encryption derives ciphertext $C = [(\alpha_0, \beta_0); (\alpha_1, \beta_1)]$ from a RFID tag, and sends it to a database. A database selects random number $r' = (k_0', k_1')$. And a database generates new ciphertext C' by calculating the formula described as follows. $C' = [(\alpha_0', \beta_0'); (\alpha_1', \beta_1')]$. re-encrypted ciphertext C' is written in a RFID tag by reader for re-encryption.

14.4 Security Analysis

In the RFID system using universal re-encryption, since the RFID tag can be written by an attacker, they can exploit it using a reader who can rewrite the contents of a RFID tag. Golle et al. pointed out that an attacker may alter the information on RFID tags to the contents that cancel re-encryption. Moreover, we introduce that an attacker may alter the information on RFID tags to the ciphertext encrypted by the attacker's public key. We show two attacks exploiting these problems.

14.4.1 Attack Model 1

The contents of a RFID tag are rewritten to the ciphertext C_A encrypted by the attacker's public key $y_A = g_A^{x_A}$:

$$C_A = [(\alpha_0, \beta_0); (\alpha_1, \beta_1)] = [(m_A y_A, g_A); (y_A, g_A)]$$

A trusted third party re-encrypts the ciphertext C_A to the new ciphertext C_A' using random number $r = (k_0, k_1)$.

$$C'_A = [(\alpha_0\alpha_1^{k_0}, \beta_0\beta_1^{k_0}); (\alpha_1^{k_1}, \beta_1^{k_1})]$$
$$= [(m_A y_A^{k_0}, g_A^{k_0}); (y_A^{k_1}, g_A^{k_1})]$$

Even if re-encryption is performed, the attacker can decrypt the new ciphertext C'_A using the attacker's private key x_A.

$$m = \alpha_0\alpha_1^{k_0}/(\beta_0\beta_1^{k_0})^{x_A} = m_A y_A^{k_0}/(g_A^{k_0})^{x_A} = m_A$$

14.4.2 Attack Model 2

When rewritten by the contents that cancel re-encryption, an attacker may rewrite the ciphertext C to the new ciphertext like $C_A = [(\alpha_0, \beta_0); (\alpha_1, \beta_1)] = [(\alpha'_0, \beta'_0); (1, 1)]$ described in [7]. Even if it re-encrypts the ciphertext C_A to the new ciphertext C'_A using a random number $r = (k'_0, k'_1)$, the new ciphertext C'_A does not change as follows:

$$C'_A = [(\alpha_0\alpha_1^{k'_0}, \beta_0\beta_1^{k'_0}); (\alpha_1^{k'_1}, \beta_1^{k'_1})]$$
$$= [(\alpha_0, \beta_0); (\alpha_1, \beta_1)]$$
$$= C_A$$

Such attacks pass through the key check of a cryptosystem. In addition, the key check is performed as $m_1 = \alpha_1/\beta_1^x$.

When m_1 is 1, it passes the key check. There is the following in the value that passes along this check.

1. If $\alpha_1 = y^r$ and $\beta_1 = g^r$, then $y = g^x$
2. $\alpha_1 = \beta_1 = 1$
3. If x is even, then $(\alpha_1, \beta_1) = (1, -1)$
4. If x is odd, then $(\alpha_1, \beta_1) = (-1, -1)$

In the case of 2–4, a ciphertext does not change, or since a ciphertext changes regularly, it is used for an attack.

By *attack model 1*, since an attacker can decrypt the information on a RFID tag, they can trace a specific RFID tag by pursuing the information that they can decrypt. Moreover, by *attack model 2*, since it does not change even if the information on a RFID tag is re-encrypted, an attacker can trace it.

14.5 Proposed Schemes

14.5.1 Re-Encryption Protocol with a Check

In order to solve the modification problem, we propose the re-encryption protocol with a check. The scheme checks the contents by a RFID tag when re-encryption is performed. In our scheme, authentication of a reader is not performed. But a part of decryption procedure is performed by a RFID tag for checking the contents of the information sent from a reader in the case of re-encryption. Therefore, a RFID tag possesses the private key x. Our system uses the same model and algorithms in Section 14.3. After the re-encryption of Section 14.3, a RFID tag performs the following procedures.

14.5.1.1 Check of Re-Encryption

A RFID tag receives new ID information (re-encrypted ciphertext) $C = [(\alpha_0', \beta_0'); (\alpha_1', \beta_1')]$. If α_1' or β_1' is 1 or -1, the re-encryption has failed, and the contents of a RFID tag does not change. Compute $m_1 = \alpha_1'/\beta_1'^{x}$, if $m_1 = 1$, then the contents of a RFID tag is changed to C.

The protocol of the re-encryption with this scheme is shown in Figure. 14.2. When the check of re-encryption has gone wrong, information on a RFID tag does not change. In our scheme, *attack model 2* is prevented by checking whether α_1' or β_1' is 1. Next, the contents of α' and β', that is, the information of the key of encryption, are checked using a private key. In other words, it is checked whether the ciphertext is encrypted by the public key $y = g^x$. Thereby, *attack model 1* can be prevented.

14.5.1.2 Attack Against Our Scheme

An attacker derives the ciphertext $C = [(\alpha_0, \beta_0); (\alpha_1, \beta_1)]$. The following ciphertext C_A is written in the RFID tag using message m_A that the attacker set up:

$$C_A = [(\alpha_0', \beta_0'); (\alpha_1', \beta_1')] = [(m_A, m_A); (\alpha_A, \beta_A)]$$

Since (α_A, β_A) of this ciphertext C_A is (α_1, β_1) itself received from the RFID tag or the contents processed the suitable re-encryption procedure, this writing cannot be prevented by our scheme.

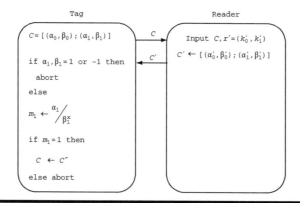

Figure 14.2 Re-encryption with a check.

14.5.1.3 Re-encryption by a Trusted Third Party

A trusted third party gets a ciphertext $C = [(\alpha_0, \beta_0); (\alpha_1, \beta_1)] = [(m_A, m_A); (\alpha_A, \beta_A)]$. Compute C' using random number $r' = (k'_0, k'_1)$ and ciphertext C, and write it to a RFID tag:

$$C' = [(\alpha'_0, \beta'_0); (\alpha'_1, \beta'_1)]$$

$$= [(\alpha_0 \alpha_1^{k'_0}, \beta_0 \beta_1^{k'_0}); (\alpha_1^{k'_1}, \beta_1^{k'_1})]$$

$$= [(m_A \alpha_A^{k'_0}, m_A \beta_A^{k'_0}); (\alpha_A^{k'_1}, \beta_A^{k'_1})]$$

The ciphertext C' cannot be decrypted without the private key x. Since the information on a RFID tag differs from the original contents, the identification ability of a RFID system will be spoiled.

However, it is impossible for an attacker to decrypt re-encrypted ciphertext C', so they cannot trace the specific RFID tag by the semantic security of universal re-encryption. Therefore, location privacy is protected.

14.5.2 Re-Encryption Protocol Using a One-Time Pad

We introduce the re-encryption protocol using a one-time pad. This scheme prevents revealing ID information of a RFID tag by encryption, and the information on a RFID tag changed by using universal re-encryption, so location privacy is protected. Moreover, it is possible to change the ID frequently by using the memory called one-time pad in a RFID tag. The one-time pad is renewed by a reader for updation. Therefore, a RFID tag does not pay the calculation cost of generating the one-time pad. In our system, it is necessary to classify a reader. The components for our system are shown in the following text.

14.5.2.1 RFID Tag

A tag saves a one-time pad Δ for changing ID information, which is encrypted (ciphertext C), secret information S, and session number i. A RFID tag transmits ID information to a query from a reader. Re-encryption of ID information is performed by universal re-encryption. Moreover, it authenticates a reader by access key X using the Hash Function at the time of updating a one-time pad. A RFID tag saves secret information S and session number i, and access key X generated by hash value $X = h(S, i, \Delta)$.

14.5.2.2 Database

A database has private key x for ID information C on a RFID tag, the information on the item relevant to the RFID tag, ID information, access key $X = h(S, i, \Delta)$ for updating one-time pad and Hash Function h. Secret information S of the access key that is used to update a one-time pad and session number i are saved securely by an existing access control scheme. Moreover, it is necessary to use an existing authentication scheme for accessing them.

14.5.2.3 Reader for Reading ID Information

This emits a query to a RFID tag and receives ID information C. Then, it asks to a database by an existing authentication scheme, and obtains the information about the item to which the RFID tag is attached. And it offers a service, a function, etc., based on the ID information.

14.5.2.4 Reader for Updating a One-Time Pad

This emits a query to a RFID tag and receives ID information C. It asks a database by an existing authentication scheme, and receives access key X and new one-time pad Δ'. Then, it updates the one-time pad of the RFID tag. Using universal re-encryption, if a reader for updating a one-time pad saves ID information and the value of a one-time pad, it becomes difficult to trace a RFID tag by semantic security when the next re-encryption is performed by another reader [18]. However, the reader for updating a one-time pad has to process it correctly.

Unlike the existing system, our scheme has two readers because of the difference in the degree of secrecy of the information to treat. The procedure of key generation, encryption, and decryption are the same as Section 14.3. Then a database saves the secret key x generated by key generation. Our scheme follows the following procedures.

14.5.2.5 Generate One-Time Pad

A database generates a one-time pad Δ from ciphertext C of a RFID tag and random number $r = (l_1, \ldots, l_{2n})$:

$$\Delta = [(\alpha_1^{l_1}, \beta_1^{l_1}), \ldots, (\alpha_1^{l_{2n}}, \beta_1^{l_{2n}})]$$

In the state of the first stage, the one-time pad Δ and ciphertext C are written in a RFID tag by a reader. After that, a one-time pad is updated by the reader for updating a one-time pad.

14.5.2.6 Re-Encryption

Whenever a RFID tag emits ID information to the query from a reader, it re-encrypts ciphertext C. In that case, a RFID tag chooses two sets of elements $[(\alpha_1^{l_k}, \beta_1^{l_k}), (\alpha_1^{l_{k+1}}, \beta_1^{l_{k+1}})](k = 1, 2, \ldots, n)$ from a one-time pad Δ. And the RFID tag obtains new ciphertext C' re-encrypted by the following calculation:

$$C' = [(\alpha_0', \beta_0'); (\alpha_1', \beta_1')]$$
$$= [(\alpha_0 \alpha_1^{l_k}, \beta_0 \beta_1^{l_k}); (\alpha_1 \alpha_1^{l_{k+1}}, \beta_1 \beta_1^{l_{k+1}})]$$

Since the element of a one-time pad is a $2 * n$ set, re-encryption is possible n times. After carrying out n times re-encryption, the element of the one-time pad will be reused. Thereby, although the security by re-encryption will be spoiled, this is avoided by updating the one-time pad shown in the following text.

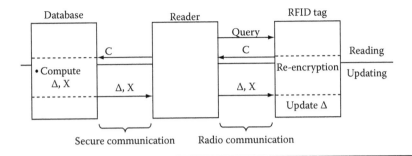

Figure 14.3   Communication with a reader.

14.5.2.7 Update One-Time Pad

The reader for updating a one-time pad interrogates a RFID tag, and gets ciphertext C. It sends ciphertext C to a database and requires the new one-time pad to update.

By decrypting ciphertext C, a database searches secret information S. And according to the generation procedure of a one-time pad, a database generates new one-time pad Δ' using ciphertext C. Access key $X = h(S, i, \Delta')$ is generated from new one-time pad Δ', secret information S, and session number i.

The reader for updating a one-time pad receives a new one-time pad Δ' and access key X, and transmits them to a RFID tag. The RFID tag calculates access key $X = h(S, i, \Delta')$ by the same calculation as a database. If the access key X accords with the received access key, the RFID tag updates the one-time pad and saves next session number $i' = i + 1$. About the case where ID information is drawn out by the reader for reading ID information and the reader for updating a one-time pad, the situation of each communication is shown in Figure 14.3.

14.5.2.8 Analysis of the Proposed Scheme

In our scheme, the RFID tag performs multiplication on ElGamal and calculation of a hash value. Moreover, it is possible to apply flexibly by changing the size of a one-time pad according to the capacity of a RFID tag. The contents written with a reader are authenticated using a Hash Function. Therefore, the modification of the information on the RFID tag by the attacker can be prevented.

Moreover, since the scheme has high anonymity, it has high protection capability to infringement of the location privacy by long-term tracing of a RFID tag and short-term tracing.

However, since secret information is centralized on the database, its employment and management become important. Furthermore, in order to request updating a one-time pad for a third party, they have to fulfil the character stated by the model of the system. Moreover, detecting modification of a RFID tag is possible although it cannot protect the alteration of the one-time pad in the tampering of a RFID tag.

14.6 Conclusions

We proposed two schemes for addressing the problems of RFID systems using universal re-encryption. Re-encryption protocol with a check can check a modification of the information on a RFID tag. However, since it checks a part of contents, an attacker can alter the other contents. Moreover, the calculation cost of RFID tags is too much.

On the other hand, in the re-encryption protocol using a one-time pad, the calculation cost of RFID tags is reduced by performing some calculations by a database. But restrictions of its model are more severe. That is, since a database authenticates a reader in the case of updating a one-time pad, a reader must update it in the right procedure.

Our schemes are well adapt to management of goods for consumers. If a consumer has a closet that can read RFID tags, it will be able to offer a few of the enticing possibilities of its contents. Moreover, if the closet can update RFID tags, whenever clothes are stored in the closet, the content of the RFID tag attached to the clothes is changed. So they need not worry about infringement of location privacy.

You may consider a private key is shared and symmetric key cryptosystem is used. In this case, although a ciphertext is made to change using a random number, etc., it is necessary to take synchronization for the value between databases. Since our proposal uses universal re-encryption, it can decrypt ciphertext to plaintext at once without taking synchronization.

Therefore, it is possible to reduce the cost of processing on database. Moreover, it is necessary to manage the private key of each RFID tag safely in a database by the method using the symmetric key cryptosystem. However, by our proposal system using universal re-encryption, a database should manage only one private key safely. Therefore, we believe the management cost can also be reduced.

Acknowledgments

This work was done during the Core University Program on Next Generation Internet between Kyushu University and Chungnam National University supported by Japan Society for the Promotion of Science and by Korea Science and Engineering Foundation. The fourth author, Jae-Cheol Ryou, was supported in part by the ITRC program of the Ministry of Information and Communication, Korea. We would like to thank Kenji Imamoto for many useful discussions.

References

1. K. Finkerseller. *RFID Handbook*. Carl Hanser Verlag, Munchen, Germany, September 2002.
2. A. Juels. *Privacy and Authentication in Low-Cost RFID Tags*. Technical Report, RSA Lab., 2003.
3. A. Juels and R. Pappu. Squealing Euros: Privacy protection in RFID-enabled banknotes. *Financial Cryptography 2003, FC'03*, Vol. 2742 of LNCS, pp. 103–121, Springer-Verlag, Berlin, Germany, 2003.
4. S. E. Sarma, S. A. Weis, and D. W. Engels. RFID systems, security and privacy implications. Technical report MIT-AUTO-WH-014, AutoID Center, MIT, Cambridge, MA, 2002.
5. S. E. Sarma, S. A. Weis, and D. W. Engels. Radio-frequency-identification security risks and challenges. *CryptoBytes*, Vol. 6, RSA Laboratories, Bedford, MA, 2003.
6. S. Weis, S. Sarma, R. Rivest, and D. Engels. Security and privacy aspects of low-cost radio frequency identification Systems, *Security in Pervasive Computing - SPC 2003*, Vol. 2802 of LNCS, pp. 454–469, Springer-Verlag, Berlin, Germany, March 2003.
7. D. M. Ewatt and M. Hayes. Gillette razors get new edge: RFID tags. *Information Week*, January 13 2003. Available from http://www.informationweek.com

8. A. Juels, R. L. Rivest, and M. Szydlo. The blocker tag: Selective blocking of RFID tags for consumer privacy. *Tenth ACM Conference on Computer and Communications Security, CCS 2003*, pp. 103–111, Washington, DC, 2003.

9. S. E. Sarma. Towards the five-cent tag. Technical Report MIT-AUTOID-WH-006, MIT Auto ID Center, Cambridge, MA, 2001. Available from http://www.autoidcenter.org

10. Security Technology. Where's the smart money? *The Economist*, pp. 69–70, February 9, 2002.

11. H. Knospe and H. Pobl. RFID security. Information security technical report, Vol. 9, Issue 4, pp. 39–50, Elsevier, Philadelphia, PA, 2004.

12. J. Kwak, K. Rhee, S. Oh, S. Kim, and D. Won. RFID system with fairness within the framework of security and privacy. *Second European Workshop on Security and Privacy in Ad hoc and Sensor Networks, ESAS 2005*, Vol. 3813 of LNCS, Springer-Verlag, Berlin, Germany, 2005.

13. M. Ohkubo, K. Suzuki, and S. Kinoshita. A cryptographic approach to "Privacy-Friendly" tag. *RFID Privacy Workshop*, Cambridge, MA, November 2003. http://www.rfidprivacy.org/

14. K. Rhee, J. Kwak, S. Kim, and D. Won. Challenge-response based RFID authentication protocol for distributed database environment. *Second International Conference on Security in Pervasive Computing, SPC 2005*, Vol. 3450 of LNCS, pp. 70–84, Springer-Verlag, Berlin, Germany, 2005.

15. G. Avoine and P. Oechslin. A scalable and provably secure hash-based RFID protocol, *IEEE PerSec 2005*, Kauai Island, HI, March 8, 2005.

16. S.-S. Yeo, and K. Kim, Scalable and flexible privacy protection scheme for RFID systems, *Second European Workshop on Security and Privacy in Ad Hoc and Sensor Networks (ESAS 2005)*, Vol. 3813 of LNCS, pp. 153–163, Springer-Verlag, Berlin, Germany, July 2005.

17. S.C. Kim, S.S. Yeo, and S.K. Kim, MARP: Mobile agent for RFID privacy protection, *International Conference on Smart Card Research and Advanced Applications - CARDIS 06*, Vol. 3928 of LNCS, pp. 300–312, Springer-Verlag, Berlin, Germany, April 2006.

18. P. Golle, M. Jakobsson, A. Juels, and P. Syverson. Universal re-encryption for mixnets, To be presented at RSA 2004, *Cryptographer's Track*. http://crypto.stanford.edu/~pgolle/, RSA Laboratory, Bedford, MA, 2004.

19. D. L. Brock. The electronic product code (EPC): A naming scheme for objects. Technical report MIT-AUTOID-WH-002, MIT Auto ID Center, Cambridge, MA, 2001. Available from http://www.autoidcenter.org

20. D. L. Brock. EPC tag data specification. Technical report MIT-AUTOID-WH-025, MIT Auto ID Center, Cambridge, MA, 2003. Available from http://www.autoidcenter.org

21. D. Engels. The reader collision problem. Technical report. MIT-AUTOID-WH-007, MIT Auto ID Center, Cambridge, MA, 2001. Available from http://www.autoidcenter.org

22. D. Engels. EPC-256 : The 256-bit electronic product code representation. Technical report MIT-AUTOID-WH-010, MIT Auto ID Center, Cambridge, MA, 2003. Available from http://www.autoidcenter.org

23. EPCglobal. The EPCglobal network: Overview of design, benefits, and security. September 24 Troy, MA, 2004. Available from http://www.epcglobalinc.org

24. K. S. Leong and M. L. Ng. A simple EPC enterprise model. *Auto-ID Labs Workshop Zurich*, Zurich, Switzerland, 2004. Available from http://www.m-lab.ch

25. D. McCullagh. RFID tags: Big brother in small packages. CNet, January 13, Chicago, IC, 2003. Available at http://news.com.com/2010-1069-980325.html

26. T. Scharfeld. An analysis of the fundamental constraints on low cost passive radio-frequency identification system design. MS thesis, Department of Mechanical Engineering, Massachusetts Institute of Technology, Cambridge, MA, 2001.

27. J. Saito, J.C. Ryou, and K. Sakurai, Enhancing privacy of universal re-encryption scheme for RFID tags. *Embedded and Ubiquitous Computing - EUC '04*, Vol. 3207 of LNCS, pp. 879–890, Springer-Verlag, Berlin, Germany, August 2004.

28. G. Avoine, Adversarial model for radio frequency identification, Cryptology ePrint archive, Report 2005/049, http://eprint.iacr.org, Lausanne, Switzerland, 2005.

29. M. Rieback, B. Crispo, and A. Tanenbaum, RFID guardian: A battery-powered mobile device for RFID privacy management. *Australasian Conference on information Security and Privacy - ACISP 2005*, Vol. 3574 of LNCS, pp. 184–194, Springer-Verlag, Berlin, Germany, July 2005.

30. S. Konomi, Personal privacy assistants for RFID users, *International Workshop Series on RFID 2004*, Tokyo, Japan, November 2004.

31. A. Juels, P. Syverson, and D. Bailey, High-power proxies for enhancing RFID privacy and utility, Center for High Assurance Computer Systems—CHACS 2005, Washington, DC, August 2005.

EMBEDDED
REAL-TIME SYSTEMS

Chapter 15

Reconfigurable Microarchitecture-Based Low Power Motion Compensation for H.264/AVC

Cheong-Ghil Kim, Dae-Young Jeong,
Byung-Gil Kim, and Shin-Dug Kim

Contents

15.1 Introduction

The demand of low-power and high-speed computing architecture for mobile systems has been increased dramatically due to the dominant popularity of multimedia and video processing. At the same time, with a fast growing VLSI technology, the architecture of digital integrated systems has evolved at a very fast pace. Thus, today's chips can integrate several complex functions, memories, and logic on the same die, called system-on-chip (SoC), which allows designing multimedia systems within a short period. Furthermore, this technical evolution grows into SoC platform technology targeting a class of applications [1]. Mobile multimedia systems are mostly targeted for real-time applications based on computationally intensive video coding algorithms on limited storage capacity and bandwidth-constrained wireless networks. H.264/AVC [2], the latest international video coding standard, is the most remarkable codec at the present time since it can make high-quality motion pictures transmitted at low bit rates, and defines three profiles: Baseline, Main, and Extended. The Baseline profile is the simplest profile; it targets mobile applications with limited processing resources such as digital multimedia broadcasting (DMB) [3] in Korea. The Main profile is intended for digital television broadcasting and next-generation DVD applications by adding several features to improve video quality at the expense of increasing computational complexity greatly. The Extended profile targets streaming video possessing features to improve error resilience and to facilitate switching between different bitstreams.

H.264/AVC is based on a hybrid video coding scheme that combines inter picture prediction to exploit temporal redundancy with transform-based coding of the prediction errors to exploit the spatial redundancy within this signal. Therefore, it is similar in spirit to previous ones such as H.263 and MPEG-4 but with considerably higher coding efficiency. This is mainly due to the enhanced motion estimation (ME) comprised of variable block size ME ($16 \times 16, 16 \times 8, 8 \times 16, 8 \times 8, 8 \times 4, 4 \times 8$, and 4×4), sub-pixel ME (1/2, 1/4, and 1/8-pel accuracy), and multiple frame referencing. However, they inherently accompany additional complexity in encoding and decoding. The primitive operation of H.264/AVC decoder is motion compensation (MC) that requires extensive computations with heavy memory accesses the ratio of which is up to 75% among four major modules: reference picture store, deblocking, display feeder, and MC [4,5]. Here, we can notice that the motion compensation module dominates memory accessing in decoding.

In general, multimedia applications involve data-dominant algorithms that require large amounts of data and complex arithmetic processing. It means that the data transfer and memory organization will have a dominant impact on the power and area cost of the realization [6]. In multimedia systems, this is why architectural and algorithmic level approaches for low power could get the biggest result compared with other levels of abstractions such as technology, layout, circuit, and gate [7]. Therefore, an efficient implementation method of the algorithms to reduce power consumption is required at the system level by designing application specific memory organization and control flow. This chapter [8] recalls the importance of this idea by showing that I/O energy can be as high as 80% of the total energy consumption of the chip. To tackle this problem, there have been several methods on low power video and signal processing applications [9–13].

This research proposes a new computing architecture for motion compensation with low power in H.264/AVC decoder. It can achieve large savings in the system power in MC module by decreasing memory accesses without sacrificing the performance or the system latency. For this purpose, we devise two techniques: merging two different stages, half-pel interpolation and averaging, into one and is reordering execution sequences of the interpolation. These are mapped onto SIMD (single instruction multiple data)-style architecture equipping with small intermediate memory, which result in much fewer memory accesses while executing interpolation stages.

In the next section, we describe H.264/AVC decoder and the motion compensation algorithm. Section 15.3 introduces the basic architecture of SoC platform. In Section 15.4, the proposed method for quarter-pel interpolation is described. In Section 15.5, the power model and the simulation results are introduced. Finally, this chapter ends with conclusions in Section 15.6.

15.2 Background

15.2.1 Overview of H.264/AVC Decoder

As other hybrid video coding standards, H.264/AVC decoder generates the predicted video blocks and decodes the coded residual blocks. At the end, the results are summed and clipped to the valid range for pixel data from the reconstructed blocks. Figure 15.1 shows a block diagram of H.264/AVC decoder. It receives incoming bitstream from network abstraction layer (NAL). And then the data pass through entropy decoding and reordering stage to produce a set of quantized coefficients X. After that, they are rescaled and inverse transformed and added to the predicted data from the previous frames using the header information. Finally, the original block is obtained after the deblocking filter.

15.2.2 Motion Compensation

Video coding is achieved by removing redundant information from the raw video sequence. In general, pixel values are correlated with their neighbors both within the same frame and between consecutive frames, which is known as spatial and temporal redundancy, respectively. Those redundancies can be reduced by motion estimation and compensation that are often based on rectangular blocks (M × N or N × N). A 16 × 16 pixel area is the basic data unit for motion compensation in current video coding standards, which is called a macro block. Consequently, motion compensation reconstructs the current macro block using a motion vector, which is the difference between the current macro block and the macro block of the previous frame. The difference value is quantized inversely and performed using an inverse integer transform. Motion compensation performs interpolation operations for chrominance components.

H.264/AVC is based on the block-based motion estimation and compensation and similar with previous standards; however, it can achieve significant gains in coding efficiency over them. This may come from the enhanced key features to motion estimation and compensation: (a) variable block-size motion compensation with small block sizes, (b) quarter-pel motion estimation

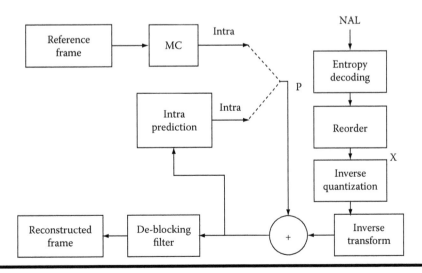

Figure 15.1   Block diagram of H.264/AVC decoder.

```
for (j=0; j < block_size; j++)          // block_size=4
    for (i=0; i < block_size; /++)
        for (result = 0, x = -2; x < 4; x++)
        //6_tab FIR filter
        result+=imgY[max(0,min(y,y_pos+j))][max(0,max(x,x_pos+i+x))]*COEF[x+2];
```

Figure 15.2   Fractional pixel interpolation in inter motion compensation.

accuracy, and (c) multiple reference picture selection. Unfortunately, it is inherently accompanying with increased complexities. Figure 15.2 shows an MC kernel that contains nested loops using six-tab FIR filter. There are several conditional branches at the last loop.

15.2.3 Half-Pel/Quarter-Pel Interpolation

In H.264/AVC, quarter-pel interpolation is presented for motion estimation accuracy. The quarter-pel interpolation that makes 1/4 pixels to offer more detailed images is made up with conventional half-pel interpolations that make 1/2 pixels and averaging operation that calculates the average of half pixel and integer pixel. As shown in Figure 15.3, the quarter-pel and half-pel interpolations support for a range of block sizes (from 16×16 down to 4×4) and quarter resolution motion vectors. The motion vector consists of two coordinates, x and y. If the motion vector of current macro block has two half values, half-pel interpolation is needed alone. But, if motion vector has one or two quarter values, both half-pel interpolations and averaging are needed. The half-pel interpolation needs six adjacent pixel values that lie on a straight line to decide the middle value of two adjacent pixels, as shown in Figure 15.4. Gray rectangles are integer-pels and white rectangles with strips are half-pels. The equation of half-pel interpolation is presented as

$$b = [(E - 5F + 20G + 20H - 5I + J)/32] \tag{15.1}$$

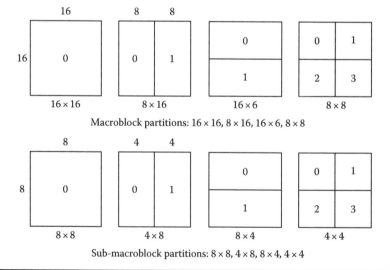

Macroblock partitions: 16×16, 8×16, 16×6, 8×8

Sub-macroblock partitions: 8×8, 4×8, 8×4, 4×4

Figure 15.3   Macro block partitions.

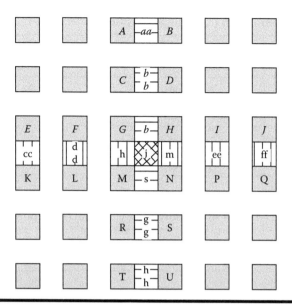

Figure 15.4   Interpolation of half-pel positions (gray: integer-pels; lattice: half-pels).

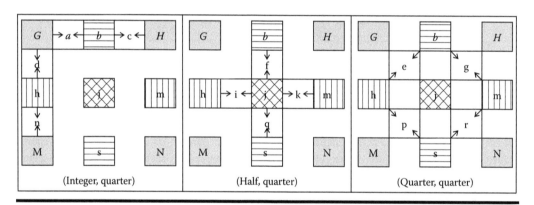

Figure 15.5   Interpolation of quarter-pel positions (gray: integer-pels; lattice: half-pels, white: quarter-pels).

Several instructions for addition, multiply, and shift are required to compute *b*. Although the equation looks very simple, it has to be repeated many times and requires a lot of data transfers. As shown in Figure 15.5, in averaging stage, the value is calculated with the average of integer-pel value and half-pel value that are already computed at the half-pel interpolation stage. The equation of averaging is presented as

$$a = (G + b)/2 \qquad (15.2)$$

From the memory access point of view, the system has to bring the referenced macro block from off-chip memory to local memory while executing each interpolation. Moreover, local memory access is also required since the interim values have to be stored in half the quarter-pel interpolation execution. Table 15.1 shows the number of repeated interpolation executions on each video

Table 15.1 Interpolation Count by Resolution

Resolution	Max. Half-Pel Interpolation Count Per Frame	Max. Quarter-Pel Interpolation Count Per Frame
QCIF(176*144)	50,688	25,344
CIF(352*288)	202,752	101,376
VGA(640*480)	614,400	307,200
10801(1920*1080)	4,147,200	2,073,600

Table 15.2 The Number of Memory Access by Macro block Size

Macroblock Size	Necessary Memory Accesses for Quarter-Pel Interpolation	Max. Count of Memory Accesses Per Frame VGA Resolution
16×16	4921	5,905,200
16×8, 8×16	2456	5,894,400
8×8	1368	6,566,400
8×4, 4×8	684	6,566,400
4×4	412	7,910,400

resolution. The result shows that the number of half-pel interpolations is twice as many as that of quarter-pel interpolations. However, each quarter-pel interpolation consists of one or two half-pel interpolation stages and averaging stages. Therefore, quarter-pel interpolations need more performance than half-pel interpolations, and quarter-pel interpolations are the most complex parts in motion compensation. Table 15.2 shows how many accesses are required from the memory according to macro block sizes. Memory access is generated frequently and there is no specific locality among memory accesses. Therefore, memory accesses are occurring on each interpolation. This is a significant problem in H.264/AVC decoding scheme, since memory access not only extends the execution time but also causes the oligopoly of the internal bus that other memory requests from other system parts cannot be accepted.

15.3 Basic System Architecture

As depicted in Figure 15.6, the basic architecture is composed of several IP cores dedicated for target applications and components at system level. There is a 32B RISC-based main processor that executes all decoding operations, performs interface, and transfers data with other modules including DSP, memory subsystem, and on-chip system bus. In addition, various I/O peripherals can be configured through system bus. All memories are independent of each other and can be accessed concurrently. The target application discussed here is motion compensation of H.264/AVC; the shaded module shown Figure 15.6 is MC IP consisting of several processing elements (PEs) operating on SIMD mode and small intermediate memory that is application specific and shared by all PEs. Therefore, each PE can read the used data from the local shared memory instead of accessing

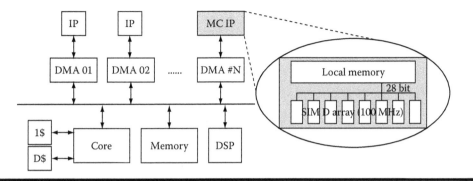

Figure 15.6 Overall system architecture.

the large external memory, which consumes energy heavily. Each PE can operate in parallel without data waiting cycles since operations are independent and the data is fed from a shared local memory through an internal 128-bit wide bus at every cycle.

Memory hierarchy is configured with off-chip memory and local memory. We assume that off-chip memory has 1MB capacity and local memory is 1KB. This configuration is sufficient to store six frames on off-chip memory and interim values of motion compensation on local memory. Result values of entropy coding, that is, the first decoding stage of H.264/AVC, are saved in off-chip memory. The motion compensation stage gets these values from off-chip memory and uses local memory as interim reservoir. If all interpolation execution is over, then the results are saved to the off-chip memory again. So it is impossible to reduce off-chip memory access unless compensation accuracy is diminished. Decreased memory access is possible on local memory with reduced interim load/store.

To simulate the proposed method, a new instruction set architecture (ISA) is devised running in SIMD mode. This ISA is minimized to reduce instruction fetch overhead. And instructions receive 8, 16, 32B immediate values, registers, and global registers for memory access. As shown in Table 15.3, this consists of the instruction set, memory access instruction set, and control instruction set. The numeral instruction set manages simple numeral calculations. This contains *ADD, SUB, MUL, SHL,* and *SHR* instructions. Because instructions are optimum, all numeral instructions take only 8B and 32B arguments. The memory access instruction set manages the control unit (CU) to guarantee that PEs in SIMD could get correct data. To achieve this, the instruction set contains instructions about global register control to supply adequate memory addresses. These instructions are *GMOV, GADD, GSUB, STORE,* and *LOAD* instructions. The control instruction set manages the control flow of code for loop. This contains *JMP* and *JN* instructions.

15.4 Proposed Methodology of Quarter-Pel Interpolation

The proposed quarter-pel interpolation consists of two techniques resulting in the reduction of memory accesses. One is merging two interpolation stages into one for the (integer, quarter) interpolation as shown in Figure 15.7. This figure presents the flow diagrams of the conventional and proposed quarter-pel interpolation in conjunction with the pixel representation shown in Figure 15.3. The other is reordering the execution sequences for the (half, quarter) or (quarter, quarter) interpolation as shown in Figure 15.8. The two proposed methods are internally taking advantage of the temporal locality to remove the redundant memory accesses.

Table 15.3   Instruction Set Architecture

Instruction Group	instruction	Supported Operand	Address Mode
Numeral	ADD	8, 32 bit register and immediate	r, r, r r, mm, r, r
	SUB	8, 32 bit register and immediate	r, r, r r, mm, r, r
	MUL	8, 32 bit register and immediate	r, r, r r, mm, r, r
	SHL	8, 32 bit register and immediate	r, r, r r, mm, r, r
	SHR	8, 32 bit register and immediate	r, r, r r, mm, r, r
Memory access	GMOV	32 bit global register and immediate	gr, mm
	GADO	32 bit global register and immediate	gr, mm
	GSLB	32 bit global register and immediate	gr, mm
	LOAD	8, 32 bit global register and 32 bit global register immediate	r, gr, gr, r, mm, r, gr
	STORE	8, 32 bit global register and 32 bit global register immediate	r, gr, gr, r, mm, r, gr
Control	JMP	N/A	N/A
	JN	8, 32 bit register	r

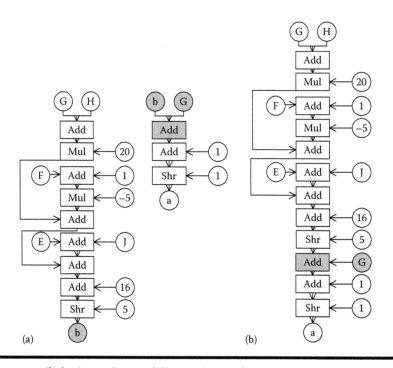

Figure 15.7   Detailed view of two different interpolations. (a) Conventional quarter-pel interpolation. (b) Proposed quarter-pel interpolation.

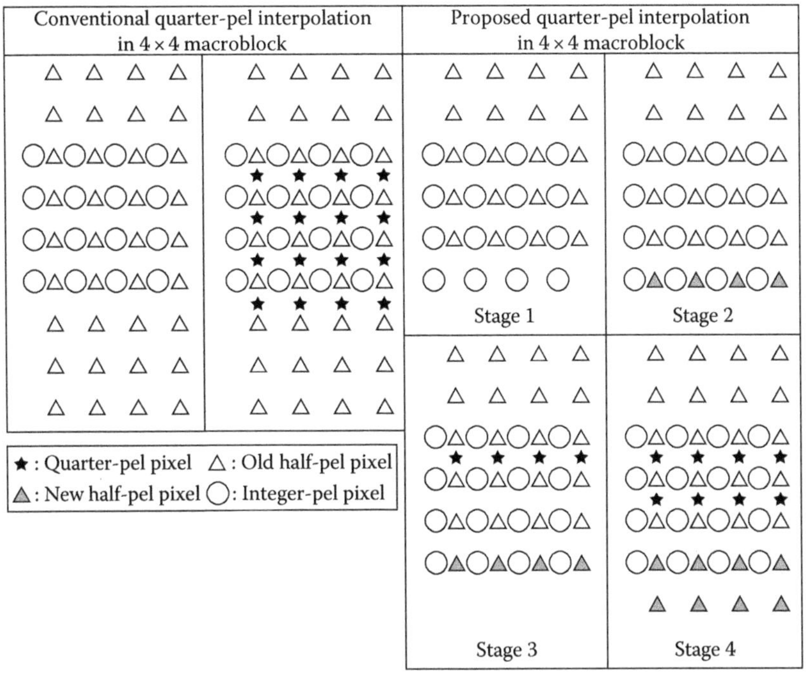

Figure 15.8 **Rearrangement of interpolation executions.**

15.4.1 The (Integer, Quarter) Case

The quarter-pel interpolation requires half-pel interpolations to be executed for an entire macro block. Then it can go further either for the averaging stage directly or averaging stage following one more execution of half-pel interpolations in another direction. The decision is made by motion vector value types. In the case of the motion vector of (integer, quarter), (half, half) values are not necessary (*J* in Figure 15.3), so only one half-pel interpolation is needed. Otherwise, another half-pel interpolation is required for *J* as the case of (half, quarter) or (quarter, quarter).

The conventional quarter-pel interpolation is composed of two stages executed separately. However, in the case of motion vector (integer, quarter), two stages are not necessarily executed separately. Therefore, we propose a method of merging two interpolation processes into one. The modified equation is obtained as follows:

$$a = [([(E - 5F + 20G + 20H - 5I + J)/32] + G)/2] \tag{15.3}$$

The shaded blocks shown in Figure 15.7 represent the result of the proposed modification mapped on the SIMD module. In this way, memory accesses can be reduced by removing interim loads/stores with simple controls. The effect will be covered in Section 15.5.

15.4.2 The (Half, Quarter)/(Quarter, Quarter) Case

For this stage, we devise a new execution flow by reordering conventional 17 steps as shown in Figure 15.9b, and then it is mapped onto the array effectively to reduce local memory accesses. Figure 15.9a shows the conventional flow. In this figure, the arrows signify the reordering positions;

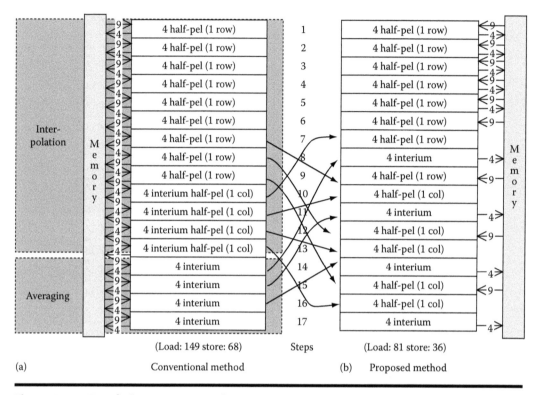

Figure 15.9 Reordering sequence and memory access counts. (a) Conventional method. (b) Proposed method.

at the bottom the counts show the number of reads/writes of each flow. At this point, the reducing method of folded memory access is not contained because a mix of two methods is very complex and hard to understand. The reordered executions are summarized as follows:

1. Execute five rows to make old half pixels in a horizontal direction, and store the results.
2. Execute one row interpolation to make a new half pixel.
3. Load the interim results and necessary pixel values, and execute proposed interpolations (Equations 15.3) in vertical direction with loaded pixels and in pixel executed stage 2.
4. Repeat steps 2 and 3.

The conventional quarter-pel interpolation method is shown in the left of Figure 15.8. First, all horizontal interpolations are executed to make interim values for vertical interpolations, and then all vertical interpolations that need the interim values are executed. Consequently, there are unnecessary memory accesses that can be reduced. The proposed interpolation method shown in the right of Figure 15.8 can minimize unnecessary interim memory accesses by changing the order of interpolation sequence. Stage 1 in Figure 15.8 is the part that cannot be modified, so this stage is the same as the conventional interpolation. Stage 2 makes one row half pixels as the interpolation results that are going to be needed for vertical half-pel interpolations requiring six variables in column. These six variables in column are interpolated in stage 3. Stage 3 utilizes the merging method generated in Section 15.4.1, so it can make quarter pixels though there is no interim memory access. Because the five variables in column are not required to be loaded in vertical half-pel interpolation by utilizing temporal locality mentioned in Section 15.4.3, six variables for horizontal half-pel interpolation to make black triangle and one variable for quarter-pel interpolation to make black star are loaded for each interpolation in stage 4.

15.4.3 Temporal Locality in Interpolation Stages

The temporal data locality of interpolation stages is utilized to achieve reductions of interim loads/stores by keeping variables read in the previous interpolation stages. As shown in Figure 15.10, the first interpolation needs pixels starting from A to F; the second interpolation from B to G, and so on. In this case, there are unnecessary folded memory accesses that may increase power consumption. Those accesses could be removed at this stage, which can be realized not by hardware configuration but by code optimization. Figure 15.11 shows the result of code regeneration, in which all conventional interpolation codes are merged and folded data loads are removed. This method is utilized in overall interpolation stages.

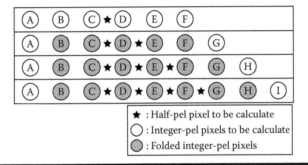

Figure 15.10 Folded memory accesses between interpolations.

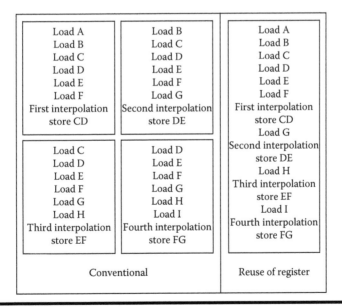

Figure 15.11 Difference between conventional method and reuse of register method.

15.5 Power Model Performance Evaluation

To obtain power estimation from high level system description, we used the methodology similar with [6], such that we conduct a relative comparison that selects the most promising candidates. The application discussed here is H.264/AVC, a data-intensive application in which the power required by memory transfer is dominant, and both conventional and proposed algorithms are mapped onto the same target architecture. Therefore, we can neglect the power consumption in operators, controls, and so on. In this way, our emphasis is on the achieved power reduction that comes from decreasing data transfer, in another words, memory accesses.

The power of data transfer is a function of the size of the memory, the frequency of access, and the technology [6]. Here, the technology can be a memory type that will be excluded on the assumption of on-chip, and then the simple power model function can be expressed as follows:

$$P_{Transfer} = E_{Tr} \times \frac{\#Transfer}{Second} \qquad (15.4)$$

$$E_{Tr} = f(\#words, \#bits) \qquad (15.5)$$

In [14], a function f is proposed to estimate the energy per transfer E_{Tr} in terms of the number of words and with width in bits. Total system power can be considered linearly proportional to the numbers of data transfer. Supposing we compare the power reduction ratio between two different operational models on the same system, we have to know the required energy for one data transfer and the total number of data transfers of input video streams for two operation models. Based on the power model of [14], the energy for one read to a memory of 256×256 words of 8B is estimated to be 1.17 µJ. We selected four input video streams: salesman, carphone, container, and mthr_dotr. Their frame size is QCIF and 178 frames of each video stream are used. For the simulation, an algorithm mapping method was used and programmed the two operational models with C++ based on the hardware architecture shown in Figure 15.6.

The result shows that the proposed method can reduce memory accesses greatly by changing control flows of quarter-pel interpolation. It did not bring out the performance degradation and system latency because there is no deleting operation. In all input video streams, the proposed method attained up to 87% of reduced local memory accesses, which resulted in decreasing power consumption at the same ratio.

As referenced earlier, these proposed methods are only for local memory access reduction. In Figure 15.12, because both off-chip and local memory accesses are shown, only about 46% of memory accesses are reduced. But, our result shows local memory access reduction only.

15.6 Conclusions

We have presented a new architecture for motion estimation of H.264/AVC with power reduction by decreasing data transfers. In data intensive applications such as H.264/AVC, data transfers dominate the power consumption. For this objective, we used a microarchitecture level configurability according to the value of motion vectors, which allow the system to operate on different control flows while executing quarter-pel interpolation. As a result, we can achieve power reduction at system-level significantly. The simulation result shows that the proposed interpolation method could reduce up to 87% of power consumption compared with a conventional method on the target architecture without scarifying performance.

Video image	Interpolation method	Off-chip memory read count	Off-chip memory write count	Local memory READ count	Local memory WRITE count	Consumed energy to READ from local memory (µJ)	READ power saving ratio (1-proposed /conventional)	Consumed energy to WRITE from local memory (µJ)	WRITE power saving ratio of (1-proposed /conventional)
Salesman	Conventional	9972288	4432128	1466940	900780	26.19	87.93%	16.08	86.94%
	Proposed	9972288	4432128	177216	117760	3.16		2.10	
Carphone	Conventional	10008000	4448000	5951480	3795928	106.25	86.85%	67.77	88.15%
	Proposed	10008000	4448000	782231	449780	13.97		8.03	
Container	Conventional	9505152	4224512	930084	483892	16.60	97.65%	8.64	98.03%
	Proposed	9505152	4224512	21951	9720	0.39		0.17	
Mthr_dotr	Conventional	9828288	4368128	3447044	2070564	61.54	88.61%	36.97	90.70%
	Proposed	9828288	4368128	392697	192820	7.01		3.44	

Figure 15.12   Simulation results.

References

1. Vahid, F., Givargis, T., Platform tuning for embedded systems design. *Computer*, 34(3), 112–114, March(2001).
2. ISO/IEC 14496-10:2003: Coding of audiovisual objects-Part 10: Advanced video coding, 2003, also ITU-T Recommendation H.264: Advanced Video Coding for Generic Audiovisual Services ISO/IEC and ITU-T (2003).
3. TTA Standard, Radio broadcasting systems; Specification of the video services for VHF digital multimedia broadcasting (DMB) to mobile, portable and fixed receivers. Telecommunications Technology Association (TTA) of Korea, TTAS.KO-07.0026 seongnam-si, (2004).
4. Sato, K., Yagasaki, Y., Adaptive MC interpolation for memory access reduction in JVT video coding. *Proceedings of Seventh International Symposium on Signal Processing and Its Applications*, Paris, France, Vol. 1, July (2003), pp. 77–80.
5. Wang, R., Li, M., Li, J., Zhang, Y., High throughput and low memory access sub-pixel interpolation architecture for H.264/AVC HDTV decoder. *IEEE Transactions on Consumer Electronics*, 51(3), 1006–1013, August (2005).
6. Smith, R., Fant, K., Parker, D., Stephani, R., Ching-Yi, W., System-level power optimization of video codecs on embedded cores: A systematic approach. *Processing of Journal of VLSI Signal*, 18, 89–109, (1998).
7. Kakerow, R., Low power design methodologies for mobile communication computer design. *Proceedings of 2002 IEEE International Conference on VLSI in Computers and Processors*, Freiburg, Germany, September (2002), pp. 8–13.
8. Musoll, E., Lang, T., Cortadella, J., Exploiting the locality of memory references to reduce the address bus energy. *Proceeding of 1997 International Symposium on Low Power Electronics and Design*, August (1997), pp. 202–207.
9. Kim, H., Park, I. C., High-performance and low-power memory-interface architecture for video processing applications. *IEEE Transactions on Circuits and Systems for Video Technology*, 11(1), 1160–1170, (2001).
10. Wang, R., Li., J.T., Huang, C., Motion compensation memory access optimization strategies for H.264/AVC decoder. *Proceedings of 2005 IEEE International Conference on Acoustics, Speech, and Signal Processing*, Atlanta, GA, Vol. 5, 18–23 March (2005) pp. 97–100.
11. Lee, T. M., Lee, W. P., Lin, T. A., Lee, C. Y., A memory-efficient deblocking filter for H.264/AVC video coding, *IEEE International Symposium on Circuits and Systems, ISCAS 2005*, Kobe, Japan, Vol. 3, 23–26 May (2005), pp. 2140–2143.
12. Mo, L., Ronggang, W., Wuchen, W., The high throughput and low memory access design of sub-pixel interpolation for H.264/AVC HDTV decoder. *IEEE Workshop on Signal Processing Systems Design and Implementation*, Austin, TX, 2-4 November (2005) pp. 296–301.
13. Lingfeng, L., Goto, S., Ikenaga, T., An efficient deblocking filter architecture with 2-dimensional parallel memory for H.264/AVC. *Proceedings of the ASP-DAC 2005 Asia and South Pacific Design Automation Conference*, Shanghai, China, Vol. 1, January (2005), pp. 623–626.
14. Landman, P., Low-power architectural design methodologies. PhD thesis, University of California, Berkeley, CA, August (1994).

Chapter 16

Automatic Synthesis and Verification of Real-Time Embedded Software for Mobile and Ubiquitous Systems

Pao-Ann Hsiung and Shang-Wei Lin

Contents

16.1 Introduction

With the proliferation of embedded mobile and ubiquitous systems in all aspects of human life, we are making greater demands on these systems, including more complex functionalities such as pervasive computing, mobile computing, embedded computing, and real-time computing.

Currently, the design of real-time embedded software is supported partially by modelers, code generators, analyzers, schedulers, and frameworks [1–21]. Nevertheless, the technology for a completely integrated design and verification environment is still relatively immature. Furthermore, the methodologies for design and for verification are also poorly integrated relying mainly on the experiences of embedded software engineers. Motivated by the above status quo, this work demonstrates how the integration of software engineering techniques such as software component reuse, formal software synthesis techniques such as scheduling and code generation, and formal verification technique such as model checking can be realized in the form of an integrated design environment targeted at the acceleration of real-time embedded software construction.

Mobile and ubiquitous systems involve the dynamic reconfiguration of applications in response to changes in their environments. Middlewares such as network layer mobility support, transport layer mobility support, traditional distributed systems applied to mobility, middleware for wireless sensor networks, context awareness-based middleware, and publish-subscribe middleware are required for efficient development of mobile and ubiquitous applications. A user can develop an application using such middlewares, however it can sometimes be too tedious and complex to consider all the different possible environments and application features. Examples of environments include office and domestic spaces, educational and healthcare institutions, and in general urban and rural environments. Examples of applications include domestic and industrial security applications, education and learning-type applications, healthcare applications, traffic management, commercial advertising, games and arts, and more extreme applications, such as applications for rescue operations and the military. Given such complex combinations of environments and applications, one would desire a higher level of reuse than that allowed by object-oriented design and middlewares. We are thus proposing an integrated design framework that allows such higher level of reuse.

Several issues are encountered in the development of an integrated design framework. First and foremost, we need to decide upon an architecture for the framework. Since our goal is to integrate reuse, synthesis, and verification, we need to have greater control on how the final generated application will be structured, thus we have chosen to implement it as an object-oriented application framework [22], which is a "semi-complete" application, where users fill in application-specific objects and functionalities. A major feature is "inversion of control," that is the framework decides on the control flow of the generated application, rather than the designer. Other issues encountered in architecting an application framework for real-time embedded software are as follows.

1. To allow software component reuse, how do we define the syntax and semantics of a reusable component? How can a designer uniformly and guidedly specify the requirements of a system to be designed? How can the existing reusable components with the user-specified components be integrated into a feasible working system?

2. What is the control-data flow of the automatic design and verification process? When do we verify and when do we schedule?

3. What kinds of model can be used for each design phase, such as scheduling and verification?

4. What methods are to be used for scheduling and for verification? How do we automate the process? What kinds of abstraction are to be employed when system complexity is beyond our handling capabilities?

5. How do we generate portable code that not only crosses real-time operating systems (RTOS) but also hardware platforms. What is the structure of the generated code?

Briefly, our solutions to the above issues can be summarized as follows.

1. *Software component reuse and integration*: A subset of the Unified Modeling Language (UML) [23] is used with minimal restrictions for automatic design and analysis. Precise syntax and formal semantics are associated with each kind of UML diagram. Guidelines are provided so that requirement specifications are more error-free and synthesizable.
2. *Control flow*: A specific control flow is embedded within the framework, where scheduling is first performed and then verification because the complexity of verification can be greatly reduced after scheduling [4].
3. *System models*: For scheduling, we use variants of Petri Nets (PN) [6,7] and for verification, we use Extended Timed Automata (ETA) [7,24], both of which are automatically generated from user-specified UML models that follow our restrictions and guidelines.
4. *Design automation*: For synthesis, we employ quasi-static and quasi-dynamic scheduling methods [6,7] that generate program schedules for a single processor. For verification, we employ symbolic model checking [25–27] that generates a counterexample in the original user-specified UML models whenever verification fails for a system under design. The whole design process is automated through the automatic generation of respective input models, invocation of appropriate scheduling and verification kernels, and generating reports or useful diagnostics. For handling complexity, abstraction is inevitable, thus we apply model-based, architecture-based, and function-based abstractions during verification.
5. *Portable efficient multilayered code*: For portability, a multilayered approach is adopted in code generation. To account for performance degradation due to multiple layers, system-specific optimization and flattening are then applied to the portable code. System dependent and independent parts of the code are distinctly segregated for this purpose.

In summary, this work illustrates how an application framework may integrate all the above proposed design and verification solutions. Our implementation has resulted in a Verifiable Embedded Real-Time Application Framework (VERTAF) whose features include formal modeling of real-time embedded systems through well-defined UML semantics, formal synthesis that guarantees satisfaction of temporal as well as spatial constraints, formal verification that checks if a system satisfies user-given properties or system-defined generic properties, and code generation that produces efficient portable code.

This chapter is organized as follows. Section 16.2 described previous related work. Section 16.3 describes the design and verification flow in VERTAF along with an illustration example. Section 16.4 presents the experimental results of an application example. Section 16.5 gives the final conclusions with some future work.

16.2 Previous Work

The software in mobile and ubiquitous systems has both traditional features of real-time embedded systems and also contemporary features such as adaptive resource management, proactive service discovery, context-aware coordination, multiagents, and models for heterogeneous platforms [28]. This is mainly due to the unique requirements of such systems including interoperability, heterogeneity, mobility, survivability, security, adaptability, ability of self-organization, augmented reality, and scalable content. Though there are numerous work on the software in mobile and ubiquitous systems, besides VERTAF, there is practically no design environment that can encompass the whole design and verification flow of such systems. In the following, we briefly survey two main areas of research in this domain, including middleware and frameworks.

Middleware design is important for ubiquitous systems because it is through this software that an application connects to the network and exchanges data with other applications. Typical examples include the OSA+ middleware architecture [29], the Reconfigurable Context-Sensitive Middleware (RCSM) [30], and the T-Engine architecture [31]. The OSA+ middleware facilitates the development of distributed real-time applications in a heterogeneous environment. Some essential features of OSA+ include quality of service information requirement for each service, explicit support for asynchronous communication, real-time memory services, and small memory footprint. OSA+ has been applied to e-health management services including patient identification, location monitoring, remote checking, and continuous accurate monitoring of patient's vital signs. The RCSM architecture facilitates the development of real-time context-aware software in ubiquitous computing environments. This architecture mainly combines CORBA and FPGA such that CORBA allows mobility and FPGA allows dynamic reconfiguration (ubiquity). RCSM has been applied to sensor networks such that object interactions are context-triggered. The T-Engine architecture is an open, real-time embedded systems platform aimed at improving software productivity. The T-Engine consortium includes computer hardware and software vendors, telecommunication carriers, and computer-using companies. T-Engine adopts a layered architecture including application, middleware, kernel, monitor, and hardware layers.

Software architectures can be specified by architecture description languages (ADLs); however as pointed out in [32], most existing ADLs provide scant support for transferring desired architectural decisions into source code. Our proposed VERTAF framework incidentally bridges this gap between ADL and source code, which is important because many implementation-level issues have direct implications on a system's success. There are several either fixed architectures or variable frameworks that have been proposed in the literature for mobile and ubiquitous systems. Typical examples include the Connected Multimedia Services (CMS) framework [33], the Earl Gray JVM-based Component (EGC) framework [34], and the Static Composition Framework (SCF) of service-based real-time applications [35]. The CMS framework is based on SIP and X.10 protocols and allows multimedia sessions to be preserved when a user moves from one computing environment to another. The EGC framework analyzes component dependencies using a component-based JVM called Earl Gray. The SCF framework allows to announce services, to discover services, and to select services for an application.

From the above survey, one can picture the current state of the art in software for mobile and ubiquitous systems. There is no tool or environment that can encompass all the design and verification phases of the design of such software; hence VERTAF is unique, as described in the rest of this chapter.

16.3 Design and Verification Flow in VERTAF

Before going into the component-based architecture of VERTAF, we first introduce the design and verification flow. As shown in Figure 16.1, VERTAF provides solutions to the various issues introduced in Section 16.1.

In Figure 16.1, the control and data flows of VERTAF are represented by solid and dotted arrows, respectively. Software synthesis is defined as a two-phase process: a machine-independent software construction phase and a machine-dependent software implementation phase. This separation helps us to plug-in different target languages, middleware, real-time operating systems, and hardware

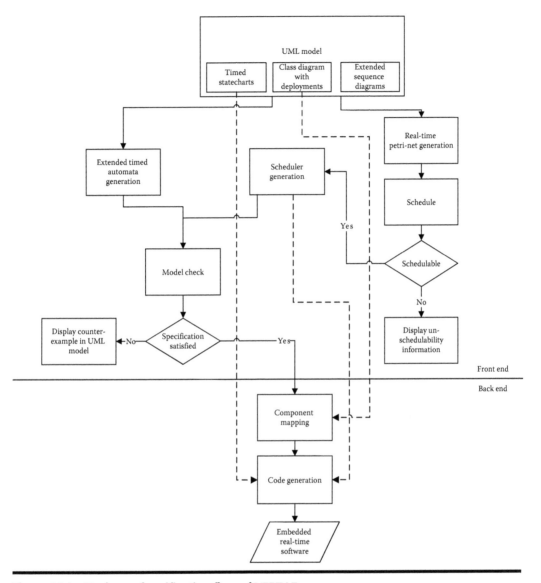

Figure 16.1 Design and verification flow of VERTAF.

device configurations. We call the two phases as front-end and back-end phases. The front-end phase is further divided into three subphases, namely, UML modeling phase, real-time embedded software scheduling phase, and formal verification phase. There are two subphases in the back-end phase, namely, component mapping phase and code generation phase. We will now present the details of each phase in the rest of this section illustrated by a running example called Entrance Guard System with Mobile and Ubiquitous Control (EGSMUC). EGSMUC is a real-time embedded system that controls any entrance with a programmable electronic lock installed. Two ways of control accesses are allowed: (1) registered users can be authenticated locally at the entrance itself, and (2) guest users may obtain a remote authentication through master acknowledgment. Here, a master could be the owner of the building to which the entrance system is protecting and he or she can have mobile and ubiquitous control access to EGSMUC. The master can grant entry access to the guest user irrespective of how he or she is connected to EGSMUC (mobile access) and also irrespective of where he or she is located (ubiquitous access). We will model EGSMUC and VERTAF will automatically synthesize and verify the code for the system.

16.3.1 UML Modeling

UML [23] is one of the most popular modeling and design languages in the industry. It standardizes the diagrams and symbols used to build a system model. After scrutiny of all diagrams in UML, we have chosen three diagrams for a user to input as system specification models, namely, class diagram, sequence diagram, and statechart. These diagrams were chosen such that information redundancy in user specifications is minimized and at the same time adequate expressiveness in user specifications is preserved. UML is a generic language and its specializations are always required for targeting at any specific application domain. In VERTAF, the three UML diagrams are both restricted as well as enhanced along with guidelines for designers to follow in specifying synthesizable and verifiable system models (just as synthesizable HDL code for hardware designs).

The three UML diagrams extended for real-time embedded software specification are as follows.

- *Class diagrams with deployment*: A deployment relation is used for specifying a hardware object on which a software object is deployed. There are two types of methods, namely, event-triggered and time-triggered that are used to model real-time behavior.
- *Timed statecharts*: UML statecharts are extended with real-time clocks that can be reset and values checked as state transition triggers.
- *Extended sequence diagrams*: UML sequence diagrams are extended with control structures such as concurrency, conflict, and composition, which aid in formalizing their semantics and in mapping them to formal Petri net models that are used for scheduling.

For our running EGSMUC example, the system class diagram with deployment is shown in Figure 16.2, a timed statechart for the system controller class is shown in Figure 16.3, and an extended sequence diagram for one of the use cases dealing with guest entry and master acknowledgment is shown in Figure 16.4.

UML is well known for its informal and general-purpose semantics. The enhancements described above are an effort at formalizing semantics preciseness such that there is little ambiguity in user-specified models that are input to VERTAF. Furthermore, design guidelines are provided to a user such that the goal of correct-by-construction can be achieved. Typical guidelines are given here.

- Hardware deployments are desirable as they reflect the system architecture in which the generated real-time embedded software will execute and thus generated code will adhere to designer intent more precisely.
- If the behavior of an object cannot be represented by a simple statechart that has no more than four levels of hierarchy, then decompose the object.
- To maximize flexibility, a sequence diagram can represent one or more use-case scenarios. Overlapping behavior among scenarios often results in significant redundancy in sequence diagrams, hence either control structures may be used in a sequence diagram or a set of nonoverlapping sequence diagrams may be interrelated with precedence constraints.

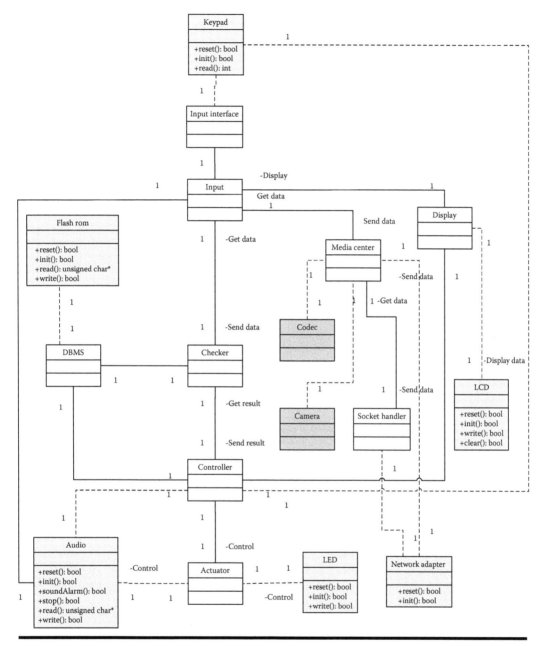

Figure 16.2   Class diagram with deployment for EGSMUC.

- Ensure the logical correctness of the relationships between class diagram and statecharts and between statecharts and sequence diagrams. The former relationship is represented by actions and events in statecharts that correspond to object methods in class diagram. The latter relationship is represented by state markers in sequence diagrams that correspond to statechart states.

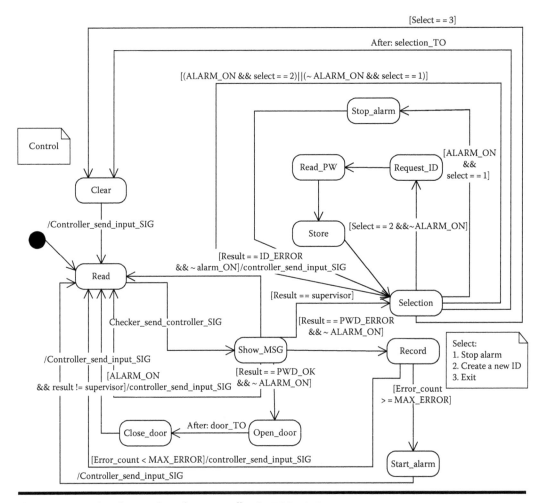

Figure 16.3   Timed statechart for controller in EGSMUC.

The set of UML diagrams input by a user, including a class diagram with deployments, a timed statechart corresponding to each class, and a set of extended sequence diagrams, constitutes the requirements for the real-time embedded software to be designed and verified by VERTAF. The formal definition of a system model is as follows.

Definition 16.1 Real-time embedded software system model   Given a class diagram $D_{class} = \langle C, \delta \rangle$, a statechart $D_{schart}(c) = \langle Q, q_0, \tau \rangle$ for each class c in C, and a set of sequence diagrams $\{D_{seq}|D_{seq} = \langle C', M \rangle, C' \subseteq C\}$, where C is a set of classes, δ is the mapping for inter-class relationships and deployments, Q is a set of states, q_0 is an initial state, τ is a transition relation between states, and M is a set of messages, a real-time embedded software system S is defined as a set of objects as specified in D_{class}, the behavior of which is represented by the individual statecharts $D_{schart}(c)$, and which interact with each other by sending/receiving messages $m \in M$ as specified in the set of sequence diagrams $\{D_{seq}\}$. A formal behavior model of the system S is defined as the parallel composition of the set of statecharts along with the behavior represented by the sequence

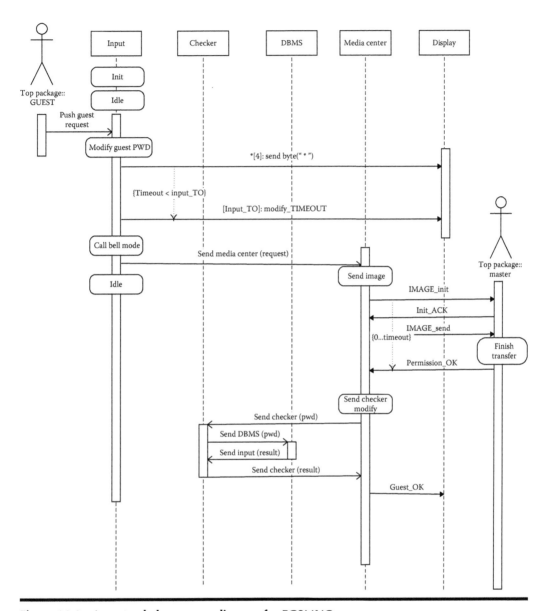

Figure 16.4   An extended sequence diagram for EGSMUC.

diagrams. Notationally, $D_{schart}(c_0) \times \cdots \times D_{schart}(c_{|C|}) \times B(D_{seq}^1, \ldots, D_{seq}^k)$ denotes the system behavior semantics, where B is the scheduler ETA as formalized in Section 16.3.2.

16.3.2 Real-Time Embedded Software Scheduling

There are two issues in real-time embedded software scheduling, namely, how are memory constraints satisfied and how are temporal specifications such as deadlines satisfied. Based on whether the system under design has an RTOS specified or not, two different scheduling algorithms are applied to solve the above two issues.

■ *Without RTOS*: *Quasi-dynamic scheduling* (QDS) [6,7] is applied, which requires *Real-Time Petri Nets* (RTPN) as system specification models. QDS prepares the system to be generated as a single real-time executive kernel with a scheduler.

■ *With RTOS*: *Extended quasi-static scheduling* (EQSS) [36] with real-time scheduling [37] is applied, which requires *Complex Choice Petri Nets* (CCPN) and set of independent real-time tasks as system specification models, respectively. EQSS prepares the system to be generated as a set of multiple threads that can be scheduled and dispatched by a supported RTOS such as MicroC/OS II or ARM Linux.

In order to apply the above scheduling algorithms, we need to map the user-specified UML models into Petri nets, RTPN or CCPN. RTPN enhances the standard Petri net with code execution characteristics associated with transitions. Given a standard Petri net $N = \langle P, T, \phi \rangle$, where P is a set of places, T is a set of transitions, and ϕ is a weighted flow relation between places and transitions, $N_R = \langle N, \chi, \pi \rangle$ is an RTPN, where χ maps a transition t to its worst-case execution time α_t and deadline β_t and π is the period for N_R. CCPN allows non-free choices at transitions [36], but does not allow the computations from a branch place to synchronize at some later place. Further, CCPN only allows a loop that has at least a single token in some place along the loop. These restrictions imposed by CCPN also apply to RTPN and are set mainly for synthesizability. Here, we briefly describe how RTPN and CCPN models are generated automatically from user-specified UML sequence diagrams, through a case-by-case construction.

1. A message in a sequence diagram is mapped to a set of Petri net nodes, including an incoming arc, a transition, an outgoing arc, and a place. If it is an initial message, no incoming arc is generated. If a message has a guard, the guard is associated to the incoming arc.
2. For each set of concurrent messages in a sequence diagram, a fork transition is first generated, which is then connected to a set of places that lead to a set of message mappings as described in Step 1 above.
3. If messages are sent in a loop, the Petri nets corresponding to the messages in the loop are generated as described in Step 1 and connected in the given sequential order of the messages. The place in the mapping of the last message is identified with the place in the mapping of a message that precedes the loop, if any. This is called a branch place. The loop iteration guard is associated with the incoming arc of the first message in the loop, which is also an outgoing arc of the branch place. Another outgoing arc of the branch place points to a transition outside the loop, which corresponds to the message that succeeds the loop.
4. Different sequence diagrams are translated to different Petri nets. If a Petri net has an ending transition which is the same as the initial transition of another Petri net, they are concatenated by merging the common transition.
5. Sequence diagrams that are interrelated by precedence constraints are first translated individually into independent Petri nets, which are then combined with a connecting place, that may act as a branch place when several sequence diagrams have a similar precedent.
6. An ending transition is appended to each generated Petri net because otherwise there will be tokens that are never consumed resulting in infeasible scheduling.

By applying the above mapping procedure, all user-specified sequence diagrams are translated and combined into a compact set of Petri nets. All kinds of temporal constraints that appear in the sequence diagrams such as time-out, time interval between two events (sending and receiving of messages), periods and deadlines associated with a message, and timing guards on messages are translated into guard constraints on arcs in the generated Petri nets. This set of RTPN or CCPN is then input to QDS or EQSS, respectively, for scheduling. Details on the scheduling procedures can be found in [6,7], and [36]. The basic strategy is to decompose each Petri net into conflict-free components that are scheduled individually for satisfaction of memory constraints. A conflict-free component is a reduction of a Petri net into one without any branch place. This is EQSS. QDS applies EQSS first and then because the resulting memory satisfying schedules may have some sequencing flexibilities, they are taken advantage of for satisfaction of temporal constraints. Finally, we have a set of feasible schedules, each of which corresponds to a particular behavior configuration of the system. A behavior configuration of a system is a feasible computation that results from the concurrent behaviors of the conflict-free components of its constituent Petri nets. For example, a system with two Petri nets, N_1 and N_2, which have two conflict-free components each, namely, N_{11}, N_{12}, and N_{21}, N_{22}, can have totally at most four different behavior configurations: $N_{11}||N_{21}, N_{12}||N_{21}, N_{11}||N_{22}$, and $N_{12}||N_{22}$.

For systems without RTOS, we need to automatically generate a scheduler that controls the system according to the set of transition sequences generated by QDS. In VERTAF, a scheduler is constructed as a separate class that observes and controls the status of each object in the system. Temporal constraints are monitored by the scheduler class using a global clock. Further, for verification purposes, an extended timed automaton is also generated by following the set of transition sequences. For uniformity, this scheduler automaton can be viewed as a timed statechart for the generated scheduler class and thus the scheduler is just another object in the system. Code generation becomes a lot easier with this uniformity.

For our running EGSMUC example, as shown in Figure 16.5, a single Petri net is generated from the user-specified set of statecharts, which is then scheduled using QDS. In this example, scheduling is required only for the timers associated with the actuator, the controller, and the input object. After QDS, we found that EGSMUC is schedulable.

16.3.3 Formal Verification

VERTAF employs the popular model checking paradigm for formal verification of real-time embedded software. In VERTAF, formal ETA models are generated automatically from user-specified UML models by a flattening scheme that transforms each statechart into a set of one or more ETA, which are merged, along with the scheduler ETA generated in the scheduling phase, into a state-graph. The verification kernel used in VERTAF is adapted from *State Graph Manipulators* (SGM) [20], which is a high-level model checker for real-time systems that operate on state-graph representations of system behavior through manipulators, including a state-graph merger, several state-space reduction techniques, a dead state checker, and a TCTL model checker. There are two classes of system properties that can be verified in VERTAF: (1) system-defined properties including dead states, deadlocks, livelocks, and syntactical errors, and (2) user-defined properties specified in the *Object Constraint Language* (OCL) as defined by OMG in its UML specifications. All of these properties are automatically translated into TCTL specifications for verification by SGM.

Automation in formal verification of user-specified UML models of real-time embedded software is achieved in VERTAF by the following implementation mechanisms.

1. User-specified timed statecharts are automatically mapped to a set of ETA.
2. User-specified extended sequence diagrams are automatically mapped to a set of Petri nets that are scheduled and then a scheduler ETA is automatically generated.

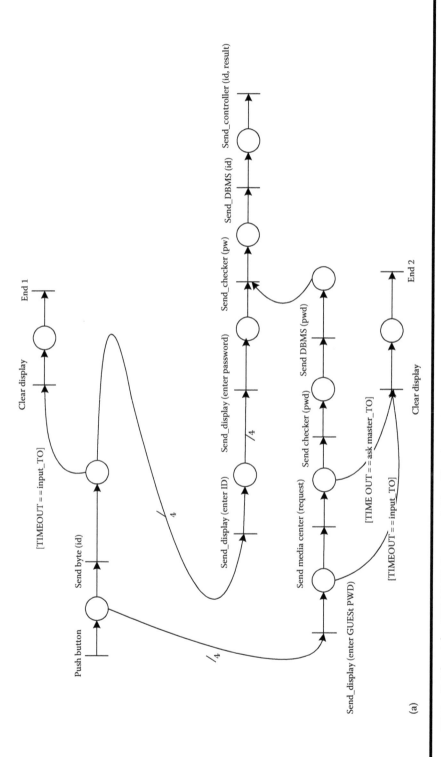

Figure 16.5 Petri net for EGSMUC: Password Checking (part 1).

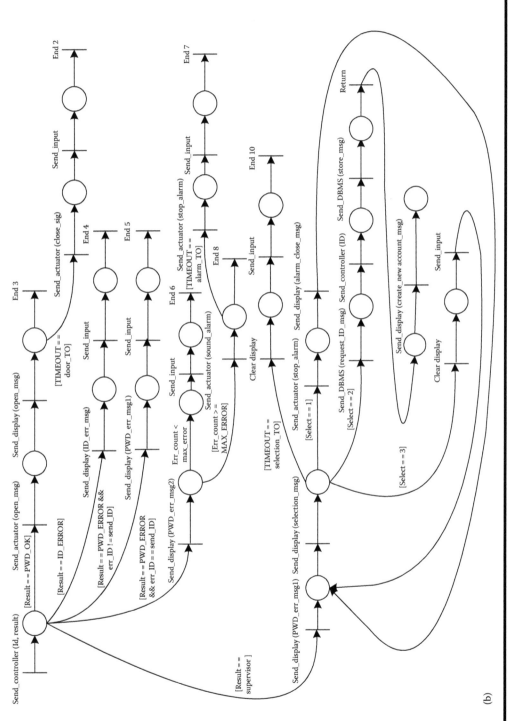

Figure 16.5 **Petri Net for EGSMUC: Controller (part 2).**

(b)

3. Using the state-graph merge manipulator in SGM, all the ETA resulting from the above two steps are merged into a single state-graph representing the global system behavior.
4. User-specified OCL properties and system-defined properties are automatically translated into TCTL specification formulas.
5. The system state-graph and the TCTL formulas obtained in the previous two steps are then input to SGM for model checking.
6. When a property is not satisfied, SGM generates a counterexample, which is then automatically translated into a UML sequence diagram representing an erratic trace behavior of the system. This approach provides a seamless interface to VERTAF users such that the formal models are all hidden and the users need to interact only with what they have specified in UML models.

Design complexity is a major issue in formal verification, which leads to unmanageable and exponentially large state-spaces. Both engineering paradigms and scientific techniques are applied in VERTAF to handle the state-space size explosion issue. The applied techniques include the following.

1. *Model construction guidelines*: The kind of specification models that a designer inputs to any tool always has a great effect on how the tool performs, thus guidelines aid designers to get the most out of a tool. Some typical guidelines that a VERTAF user is suggested to follow are:
   a. Reuse existing components as much as possible.
   b. Maximize the explicit definition of all hardware deployments in the class diagram.
   c. A class should have only one statechart representing its behavior.
   d. A statechart should have no more than four levels of hierarchy.
   e. Make explicit the relations among all sequence diagrams.
   f. Both event-triggered and time-triggered methods in each class should appear somewhere in its statechart or sequence diagram.
2. *Architectural abstractions*: An assume-guarantee reasoning (AGR)-based approach is adopted, whereby a complex verification task of a system is broken down into several smaller verification tasks of constituent subsystems. The theory of AGR is beyond scope here, but details can be found in [38,39]. For the purpose of automation, we have proposed and implemented the automatic generation of assumptions and guarantees for each ETA based on their interface traces, which are then verified individually [5]. This divide and conquer approach overcomes the exponential state-space size issue to a significant extent. The benefit of AGR becomes limited when we are trying to verify properties that crosscut the entire system. Thus, VERTAF users are suggested to decompose their properties into several smaller parts. The formal verification of component-based software is made feasible through the hierarchical decomposition of system properties into sub-properties for each software component [19]. Related issues such as memory reference, object reference, and reentrance [40] are handled using a Call-Graph which records all component invocations.
3. *Functional abstractions*: The smaller tasks of verifying each module obtained in the architectural abstraction step is further simplified through a series of user-guided functional abstractions, including communication abstraction (communication methods such as protocols are individually verified), bit-width abstraction (instead of a 32-bit wide bus, a 1-bit or 2-bit abstract model is used), transactor models (an abstract model of other components in

the system is used to verify a specific functionally detailed component), transaction-level verification (both hardware and software signals are abstracted), and assertion-based verification (only interface assertions are verified).

4. *State-space reductions*: Several of the state-space reduction manipulators provided by SGM have been either directly applied to the ETA models generated in VERTAF or modified for adaptation to embedded systems. Since the scope here does not allow us to go into details of the reduction techniques, we merely list the techniques available and refer designers to related work [20]. The techniques applicable are: read-write reduction, discrete variable hiding reduction, clock shielding reduction, internal transition bypassing, and timed symmetry reduction.

The above abstraction techniques are applied to a user-specified UML model as follows. While constructing the UML models, users not following the guidelines are warned of the possible intricacies. Upon completion of model construction, first Petri net models are generated, which are then scheduled to produce feasible system schedules that are represented by a scheduler ETA. Then, for each ETA generated from the statecharts, its assumptions and guarantees are generated. The guarantees of an ETA are verified by first merging the ETA with functional abstractions of the other ETA in the system and then reducing the state-spaces of the merged state-graph using SGM reduction manipulators. We can see that not only is verification automated but abstraction techniques such as AGR and state-space reductions are also automatically performed, which makes VERTAF scalable to large applications.

For our running EGSMUC example, the ETA for each statechart were generated and then merged with the scheduler ETA. For illustration, we show in Figure 16.6 the ETA that is generated by VERTAF corresponding to the controller statechart of Figure 16.3. There are seven other ETA in this system example. All ETA were input to SGM and AGR was applied. Reduction techniques were then applied to each state-graph obtained from AGR. OCL constraints were then translated into TCTL and verified by the SGM model checker kernel.

16.3.4 Component Mapping

This is the first phase in the back-end design of VERTAF and starts to be more hardware dependent. All hardware classes specified in the deployments of the class diagram are those supported by VERTAF and thus belong to some existing class libraries. The component mapping phase then becomes simply the configuration of the hardware system and operating system through the automatic generation of configuration files, make files, header files, and dependency files. The corresponding hardware class API will be linked in during compilation.

The main issue in this phase occurs when a software class is not deployed on any hardware component or not deployed on any specific hardware device type, for example, the type of microcontroller to be used is not specified. Currently, VERTAF adopts an interactive approach whereby the designer is warned of this lack of information and he/she is requested to choose from a list of available compatible device types for the deployment. An automatic solution to this issue is not feasible because estimates are not easy without further information about the non-deployed software classes.

Another issue in this phase is the possible conflicts among hardware devices specified in a class diagram such as interrupts, memory address ranges, I/O ports, and bus-related characteristics such as device priorities. Users are also warned in case of such conflicts.

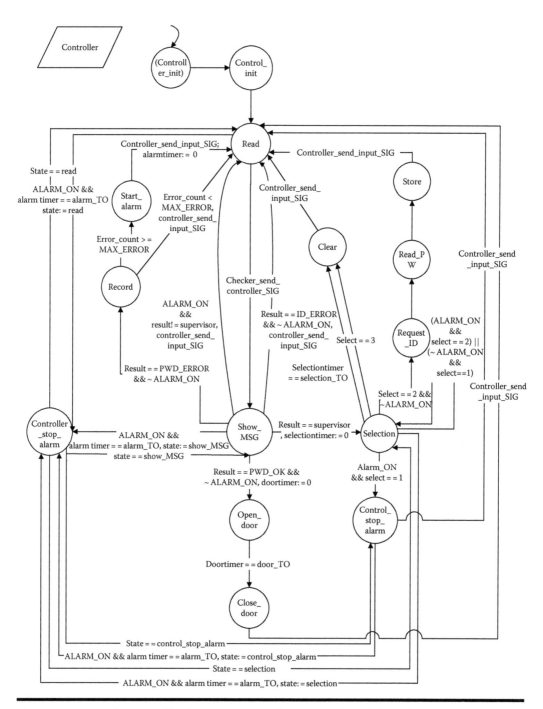

Figure 16.6 ETA for controller in entrance guard system with mobile and ubiquitous control

For our running EGSMUC example, all software classes in the class diagram given in Figure 16.2 are deployed on one or more hardware or software classes supported by VERTAF.

16.3.5 Code Generation

There are basically three issues in this phase including hardware portability, software portability, and temporal correctness. We adopt a multitier approach for code generation: an operating system layer, a middleware layer, and an application with scheduler layer, which solves the above three issues, respectively. Currently supported underlying hardware platforms include dual core ARM-DSP based, single core ARM, StrongARM, or 8051 based, and Lego RCX-based Mindstorm systems. For hardware abstraction, VERTAF supports MicroHAL and the embedded version of POSIX. For operating systems, VERTAF supports MontaVista Linux, MicroC/OS, Embedded Linux, and eCOS. For middleware, VERTAF is currently based on the Quantum Framework [13]. For scheduler, VERTAF creates a custom ActiveObject according to the Quantum API. Included in the scheduler is a temporal monitor that checks if any temporal constraints are violated. A sample configuration is shown in Figure 16.7, where the multitier approach decouples application code from the operating system through the middleware and from the hardware platform through the operating system layer.

Each ETA that is generated either from UML statecharts or from the scheduled Petri nets (sequence diagrams) is implemented as an ActiveObject in the Quantum Framework. The user-defined classes along with data and methods are incorporated into the corresponding ActiveObject. The final program is a set of concurrent threads, one of which is a scheduler that can control the other objects by sending messages to them after observing their states. For systems without an OS, the scheduler also takes the role of a real-time executive kernel.

For our running example, the final application code consisted of nine activeobjects derived from the statecharts and one activeobject representing the scheduler. Makefiles were generated for linking in the API of the eight hardware classes and configuration files were generated for the ARM-DSP dual microprocessor platform called DaVinci from Texas Instruments with MontaVista Linux as its operating system on the ARM processor and DSP/BIOS real-time kernel as the operating system on the DSP TMS6646DSP processor. There were totally 2340 lines of C code for the full EGSMUC system, out of which the system designers had to write only around 263 lines of C code, which is only 11.2% of the full system code.

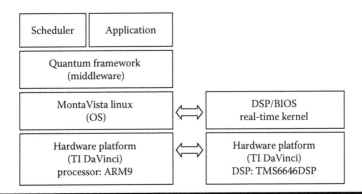

Figure 16.7   Multitier code architecture.

16.4 Analysis and Evaluation

For the running example EGSMUC, we now analyze why VERTAF is capable of generating a significant part of the system implementation code, thus alleviating the designer from the tedious and error-prone task of manual coding. Due to its application framework architecture, VERTAF supports software components that are commonly found in mobile, ubiquitous, real-time, and embedded application domains. We classify the components supported by VERTAF into the following.

- *Storage and I/O devices:* This class includes all the storage and I/O devices that are supported by VERTAF and required for implementing a real-time embedded system. Examples from the EGSMUC system include FlashRom, Keypad, LCD, Audio, LED, and Camera.
- *Communication interfaces:* This class includes all the interface components that allow connection with the external world, for example, wired and wireless network connection, Bluetooth, and GSM/GPRS. Network adapter is an example from EGSMUC system.
- *Multimedia processing:* This class includes all the components providing API for multimedia encoding and decoding through codecs specific to hardware platforms such as the codecs provided by TI for DaVinci multimedia platform. The DSP class in the EGSMUC system is an example.
- *Control and management interfaces:* This class includes all the components for controlling and managing system components, such as the socket handler in the EGSMUC example.

To implement mobile and ubiquitous control access in a real-time embedded system, a user normally, without VERTAF, would have to install a web server, write multimedia processing code, write network code, and integrate everything together, along with application-specific context awareness or publish-subscribe middlewares. With VERTAF, most of these tedious work are not required as long as the user configures the correct components from the framework for use in his or her application.

For illustration purposes, we show how the Media Center class in the EGSMUC example was implemented using VERTAF. The Media Center class is responsible for getting acknowledgment from a mobile master ubiquitously, which means whenever a guest wants to enter the building that the EGSMUC system is guarding, the Media Center notifies the DSP class to use the Camera to capture an image of the guest and then send the guest image to a master (the owner of the building or house). The master can send an acknowledgment through the web after which the guest can enter the building. A password is setup by a guest so that the guest can enter the building within the span of time set by the master beforehand.

Figure 16.4 gave the sequence diagram that a user needs to specify in order for VERTAF to generate corresponding code. The architecture of the code generated by VERTAF is shown in Figure 16.8, where QF ActiveObject is an active object from the Quantum Framework. The code consists of three parts, namely, a web server, a QF activeobject, and an image processing interface. The web server allows a master to connect to EGSMUC using a web browser that can run Java applets. The applet opens a socket connection between the media center and the client machine of the master. The image of the guest requesting entrance is captured and processed through the image processing interface. When a master acknowledges, the guest is notified through the input class. The control and data flows of the media center are automatically generated by VERTAF and the user has to merely specify the sequence diagrams as shown in Figure 16.4 and deploy the related

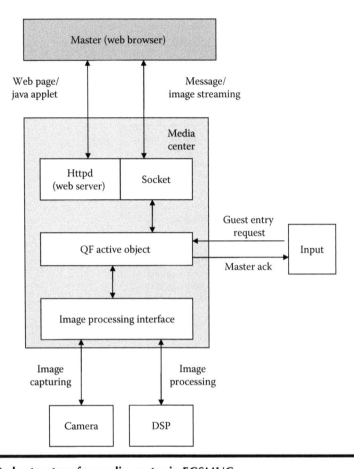

Figure 16.8   Code structure for media center in EGSMUC.

classes to hardware or software components in the class diagram as shown in Figure 16.2. This is exactly the reason why VERTAF can save a lot of coding and design efforts.

Figure 16.9 shows the detailed flow of image capture and processing as implemented in the TI DaVinci platform, which has several image and video codecs. After a capture device is initialized, a required codec engine is opened and an algorithm created to run on the DSP. The camera is initialized and if the camera times out for three times, an error message is shown and the program exits. Otherwise, the raw data from the camera buffer is read, input to encode engine, and finally the image is written to disk. The DSP engine and the camera device are then closed.

There were totally 18 objects in the final application generated by VERTAF, out of which the user or designer had to only model seven classes. The remaining 11 classes included components from all the four categories as described at the start of Section 16.4. Empirical results obtained from comparing two different implementations of the EGSMUC system, one using VERTAF, and one without using VERTAF, showed that not only the user written code reduced to 11.2% and the number of objects reduced to 41%, but the total time required to develop the application also reduced by more than 60%. The average learning time for each designer using VERTAF was approximately 0.1 day. The experimental and empirical results all show that VERTAF is beneficial to designers of real-time embedded software with mobile and ubiquitous control access.

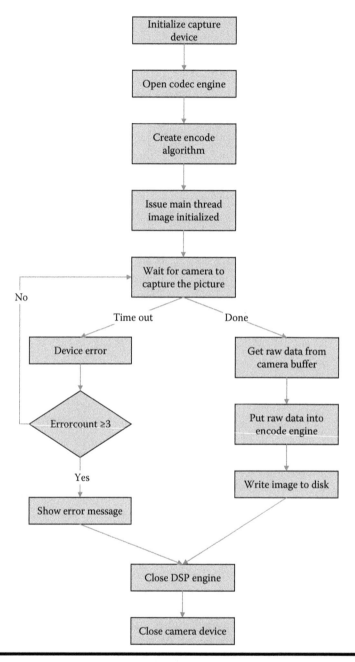

Figure 16.9   Flow chart for media center Code in EGSMUC.

By employing various construction guidelines for design and several reduction techniques for verification as described in Section 16.3.3, VERTAF is scalable to large and complex applications. Since VERTAF was constructed in a component-oriented way, one can also easily extend its features by modifying existing components or introducing new components into the framework. The flow of VERTAF can also be easily modified to incorporate the changes.

16.5 Conclusions and Future Work

An object-oriented component-based application framework, called VERTAF, was proposed for the development of real-time embedded system applications with mobile and ubiquitous control access. It was a result of the integration of three different technologies: software component reuse, formal synthesis, and formal verification. Starting from user-specified UML models, automation was provided in model transformations, scheduling, verification, and code generation. VERTAF can be easily extended since new specification languages, scheduling algorithms, etc., can easily be integrated into it.

Future extensions will include support for share-driven scheduling algorithms. More applications will also be developed using VERTAF. VERTAF will be enhanced in the future by considering more advanced features of real-time applications, such as network delay, network protocols, and online task scheduling. Performance related features such as context switch time and rate, external events handling, I/O timing, mode changes, transient overloading, and setup time will also be incorporated into VERTAF in the future.

References

1. T. Amnell, E. Fersman, L. Mokrushin, P. Petterson, and W. Yi, in *Proceedings of the 1st International Workshop on Formal Modeling and Analysis of Timed Systems (FORMATS)*, Marseille, France, 2003.

2. B. Douglass, *Doing Hard Time: Developing Real-Time Systems with UML, Objects, Frameworks, and Patterns*, Addison Wesley Longman, Inc., Reading, MA, 1999.

3. P. Hsiung, in *Proceedings of the 27th International Conference on Technology of Object-Oriented Languages and Systems (TOOLS'98)*, Beijing, China, IEEE Computer Society Press, 1998, pp. 138–147.

4. P. Hsiung, *Journal of Systems Architecture—The Euromicro Journal* **46**, 1435 (2000).

5. P. Hsiung and S. Cheng, in *Proceedings of the 16th International Conference on VLSI Design (VLSI' 03)*, New Delhi, India, IEEE CS Press, 2003, pp. 249–254.

6. P. Hsiung and C. Lin, in *Proceedings of the 1st ACM/IEEE/IFIP International Conference on Hardware-Software Codesign and System Synthesis (CODES+ISSS'03)*, Newport Beach, CA, ACM Press, 2003, pp. 114–119.

7. P. Hsiung, C. Lin, and T. Lee, in *Proceedings of the 9th International Conference on Real-Time and Embedded Computing Systems and Applications (RTCSA'03)*, Tainan, Taiwan, 2003.

8. A. Knapp, S. Merz, and C. Rauh, in *Proceedings of the 7th International Symposium on Formal Techniques in Real-Time and Fault-Tolerant Systems*, Oldenburg, Germany, Springer Verlag, 2002, Vol. 2469 of LNCS, pp. 395–414.

9. T. Kuan, W. See, and S. Chen, in *Proceedings of OOPSLA'95 Workshop #18*, Austin, TX, 1995.

10. S. Kodase, S. Wang, and K. Shin, in *Proceedings of Design, Automation and Test in Europe Conference*, Munich, Germany, 2003, pp. 170–175.

11. L. Lavazza, A methodology for formalizing concepts underlying the DESS notation, EUREKA-ITEA project (http://www.dess-itea.org), D1.7.4, 2001.

12. D. de Niz and R. Rajkumar, in *Proceedings of the International Workshop on Languages, Compilers, and Tools for Embedded Systems*, San Diego, CA, 2003, pp. 133–143.

13. M. Samek, *Practical Statecharts in C/C++ Quantum Programming for Embedded Systems*, CMP Books, Lawrence, KS, 2002.

14. D. Schmidt, in *Handbook of Programming Languages*, ed. P. Salus, vol. I, MacMillan Computer Publishing, Indianapolis, IN, 1997.

15. W. See and S. Chen, *Object-Oriented Real-Time System Framework*, John Wiley, New York, 2000, pp. 327–338.

16. B. Selic, in *Proceedings of the 4th International Workshop on Parallel and Distributed Real-Time Systems*, Honolulu, HI, 1996, pp. 11–18.

17. B. Selic, in *Proceedings of the IFIP Conference on Hardware Description Languages and Their Applications*, Ottawa, Ontario, Canada, 1993.

18. B. Selic, G. Gullekan, and P. Ward, *Real-Time Object Oriented Modeling*, John Wiley & Sons, Inc., New York, 1994.

19. T. Shen, Master's thesis, Dept. of CSIE, National Chung Cheng University, Minhsiung, Taiwan, 2003.

20. F. Wang and P. Hsiung, *IEEE Transactions on Computers* **51**, 61 (2002).

21. S. Wang, S. Kodase, and K. Shin, in *Proceedings of International Conference of Euro-uRapid*, Frankfurt, Germany, 2002.

22. F. M. and D. Schmidt, *Communications of the ACM, Special Issue on Object-Oriented Application Frameworks* **40**, (1997).

23. J. Rumbaugh, G. Booch, and I. Jacobson, *The UML Reference Guide*, Addison Wesley Longman, Reading, MA, 1999.

24. R. Alur and D. Dill, *Theoretical Computer Science* **126**, 183 (1994).

25. E. Clarke and E. Emerson, in *Proceedings of the Logics of Programs Workshop*, Yorktown Height, NY, Springer Verlag, 1981, Vol. 131 of LNCS, pp. 52–71.

26. E. Clarke, O. Grumberg, and D. Peled, *Model Checking*, MIT Press, Cambridge, MA, 1999.

27. J. Queille and J. Sifakis, in *Proceedings of the International Symposium on Programming*, Torino, Italy, Springer Verlag, 1982, Vol. 137 of LNCS, pp. 337–351.

28. E. Niemelä and J. Latvakoski, in *Proceedings of the 3rd International Conference on Mobile and Ubiquitous Multimedia*, College Park, MD, ACM Press, 2004, pp. 71–78.

29. U. Brinkschulte, A. Bechina, B. Keith, F. Picioroaga, and E. Schneider, in *Proceedings of the IAR Workshop*, Institute for Automation and Robotic Research, Grenoble, France, 2002.

30. S. S. Yau and F. Karim, in *Proceedings of the 4th International Symposium on Object-Oriented Real-Time Distributed Computing (ISORC)*, Magdeburg, Germany, IEEE CS Press, 2001, pp. 163–170.

31. K. Sakamura and N. Koshizuka, *IEEE Micro* **22**, 48 (2002).

32. N. Medvidovic, *IEEE Software* **22**, 83 (2005).

33. J.-Y. Kwak, D.-M. Sul, S.-H. Ahn, and D.-H. Kim, in *Proceedings of the IEEE Workshop on Software Technologies for Future Embedded Systems*, Hakodate, Japan, IEEE CS Press, 2003, pp. 61–64.

34. H. Ishikawa, Y. Ogata, K. Adachi, and T. Nakajima, in *Proceedings of the IEEE Workshop on Software Technologies for Future Embedded Systems*, Hakodate, Japan, IEEE CS Press, 2003, pp. 9–12.

35. I. Estevez-Ayres, M. Garcia-Vails, and P. Basanta-Val, in *Proceedings of the 3rd IEEE Workshop on Software Technologies for Future Embedded and Ubiquitous Systems*, Seattle, WA, IEEE CS Press, 2005, pp. 11–15.

36. F. Su and P. Hsiung, in *Proceedings of the 10th IEEE/ACM International Symposium on Hardware/Software Codesign (CODES'02)*, Estes Park, Co, ACM Press, 2002, pp. 211–216.

37. C. Liu and J. Layland, *Journal of the Association for Computing Machinery* **20**, 46 (1973).

38. T. Henzinger, S. Qadeer, and S. Rajamani, in *Proceedings of the IEEE/ACM International Conference on Computer-Aided Design (ICCAD'00)*, San Jose, CA, 2000, pp. 245–252.

39. M. Zulkernine and R. Seviora, in *Proceedings of the 15th International Parallel and Distributed Processing Symposium*, San Francisco, CA, 2001, pp. 1552–1560.

40. C. Szyperski, *Component Software: Beyond Object-Oriented Programming*, Addison-Wesley, Boston, MA, 2002.

Chapter 17

Agent-Based Architecture on Reconfigurable System-on-Chip for Real-Time Systems

Yan Meng

Contents

17.1 Introduction

The definition of real-time system is any system that is both logically and temporally correct. Logically correctness means the system satisfies all functional specifications. Temporal correctness means the system is guaranteed to perform these functions within explicit time frames. Real-time systems are classified as hard or soft. Hard systems have catastrophic consequences if the temporal requirements are not met—up to and including complete system destruction. Conversely, soft systems only have degraded performance if the temporal requirements are not met.

Conventionally, all of the tasks of a complex real-time system are implemented on a general purpose processor in software due to its great flexibility, where a powerful processor is required to guarantee that hard deadlines can always be met, even for very rare conjunctions of events. Sometimes the sensors of a real-time system are supposed to work in parallel and independently; for example, in a mobile robot system, it would be impossible to implement the parallel sensing in software due to its inherent sequential processing. Furthermore, with multitasks, signals, and events, it would be a challenging job for a software engineer to implement all of the tasks under real-time constraints, especially with some computation-intensive information, say video images.

On the other hand, the specific hardware can respond to external inputs more efficiently since the multiple hardware units can operate in parallel, concurrently, and asynchronously, so that individual response times are much less variable and easier to guarantee, even as the number of tasks increases. Parallel hardware is not so affected by issues such as task swapping, scheduling, interrupt service, and critical sections, which complicate real-time software solutions. The expensive ASIC can fulfill the speed criteria. However, it is a complicated and expensive procedure if a slight change occurs.

Reconfigurable computing is intended to fill the gap between hardware and software, achieving potentially much higher performance than software, while maintaining a higher level of flexibility than hardware. Field-programmable gate arrays (FPGAs) is one of the most popular reconfigurable devices with various configurable elements and embedded blocks providing new solutions for high-density and high-performance embedded system design. These platforms not only enable system architects to design and develop complex custom systems using embedded processors and interoperable IP cores but also provide technologies such as dynamic reconfiguration. The advances of FPGA technology make the hybrid hardware–software architecture of reconfigurable system-on-chip (rSoC) feasible in implementation, which can provide both runtime flexibility and real-time performance for complex applications. These two features are not simultaneously achievable by traditional pure software or hardware implementations.

The newest FPGA technology allows a designer to use a single reconfigurable platform to instantiate both the processors and the required logic units, which directly leads to feasibility of the rSoC architecture. This rSoC architecture raises the possibility of different hardware/software (HW/SW) codesign approaches. To speed up the system integration and HW/SW codesign process on an rSoC platform, a unified HW/SW interface is necessary. Based on this rSoC platform, a belief-desire-intention (BDI) agent model is proposed in this project as a unified structure for both hardware and software, which is intended to design high-level aspects of systems and to simplify the HW/SW partitioning and communication. The system is first decomposed into agents based on system specifications, where these agents can accomplish some specific tasks independently and can communicate with each other. These agents are then partitioned into software agents and hardware agents based on the heuristic approaches. Compared to the regular module-based architecture, the BDI agent model can provide more flexibility, intelligence, autonomy, and scalability.

Since all of the agents share the same mechanism, on top of this unified agent-based representation, it is possible to provide a unified communication mechanism between the hardware and software parts. A novel *on-demand message passing (ODMP)* communication protocol for this multi-agent system is then proposed. This communication mechanism provides transport independence, and, therefore, it is portable to different agent-based real-time systems without significant modification to both the software and hardware. The only parts need to be customized are developing different transport-dependent adapters to new systems.

This rSoC platform is desirable for those real-time systems that must be highly responsive to the dynamic environments. Since hardware component is more efficient to handle external events compared to software component, it can be effectively managed by a dynamic reconfigurable hardware logic units. *Dynamic reconfiguration* is based on the idea that parts of the system remain available during the reconfiguration. Although disruption is unavoidable, the impact of the disruption should be minimized, as well as the duration of the effects of this disruption. The reconfiguration module should introduce minimal overhead during normal operation. Deploying dynamic runtime reconfiguration in systems results in reduced chip area and power consumption. Therefore, several hardware reconfiguration modules are proposed in this project.

To evaluate the real-time performance of the proposed agent-based architecture on an rSoC platform, two case studies are presented: one is a mobile robot system, and the other is a mobile vision system for people detection and tracking. Vision sensors are the powerful sensors for a mobile robot for situation awareness and target detection. However, most computer vision approaches have underlying assumptions, such as large memory, high computation power, static vision system, or off-line processing.

When a vision system is on a mobile platform, there are some constraints that are specific compared to the static vision system. First, the system has to be highly robust to adapt to the changing illumination and environmental structures. Second, the real-time processing is very important to realize a natural reactive behavior and prevent any serious damages. Furthermore, a mobile robot system consists of not only a vision system but also the navigation and control systems.

Most developed mobile robot systems are designed to work efficiently under some certain environments, such as structured indoor buildings, or outdoor open environments. However, in some emergency response situations, such as in an urban search and rescue (USAR) environment or some unknown hazardous environments, the working environment of the mobile robot is unpredictable and dynamically changed. All of these tasks have to be processed by the onboard processor of a robot under the real-time constraints. A software-only solution will push the limit of the processing capability, which may lead to a very conservative computation solution and a very challenging real-time scheduler. Furthermore, with one or multiple microprocessor(s) in the system, it is difficult to handle the high frequency of the external I/O. The overhead for the interrupts reduces the microprocessor performance. To tackle these issues, an rSoC platform is applied in this project. This rSoC platform is expected to respond with greater flexibility and processing capacity, and be suitable for most real-time systems, especially for those with dynamic environments and may encounter various emergencies.

17.2 Related Work

Most previous work on reconfigurable embedded platform for real-time systems has been implemented on software only. Gafni et al. [1] proposed an architectural style for real-time systems, in which the dynamic reconfiguration was implemented by a synchronous control task in response to the sensed conditions. To handle runtime dependencies between components, Almeida et al. [2] employed runtime analysis of interactions to selectively determine which interactions should be allowed to proceed, in order to reach a safe state for change, and which to block until reconfiguration is completed. Cobleigh et al. [3] presented a hierarchically self-adaptation model for a robot system to provide fault tolerance. MacDonald et al. [4] proposed some design options for dynamic reconfiguration with their service-oriented software framework. Stewart et al. [5] described a programming paradigm that uses domain-specific elemental units to provide specific guidelines to control engineers for creating and integrating software components by using port-based object. A self-reconfigurable robot design with a distributed software architecture that followed the physical reconfiguration of hardware in response to the needs of task or environment was proposed in [6,7].

The recent advance of dynamic reconfigurable FPGA platform attracts more attentions in large applications. Most of the real-time systems use FPGAs to speed up the portions of an application that fail to meet required real-time constraints. The approach of applying reconfigurable logic for data processing has been demonstrated in some areas such as video transmission, image recognition, and various pattern-matching operations [8,9]. In [10], the system used FPGAs that were dynamically reconfigured on dynamically programmable gate arrays at runtime to implement different functions. The fastest solution for reconfiguration was through context switching, which was able to store a set of different configuration bitstreams and make the context switching in a single clock cycle. Contest switching FPGA architectures were proposed in [11,12], which emphasized on both proving fast context switching as well as fast random access to the configuration memory. This is important because you may want to change one or more of your configuration streams inside the FPGA at runtime.

To fully take advantage of the reconfigurable characteristics of FPGAs, some self-reconfiguration mechanisms were proposed. Blodget et al. [13] proposed a self-reconfiguration platform for embedded system using ICAP and an embedded processor core within the FPGA. Fong et al. [14] developed a system that used the RS-232 port to transfer configuration bitstreams to the FPGA, which may easily lead to a bottleneck for the transfer speed and an increase in the reconfiguration latency.

As the soft processor cores for FPGAs become available in the rSoC platform, the HW/SW codesign issues become critical in the overall system design procedure. Some of the more recent HW/SW codesign research used reconfigurable processors as the platform for implementation [9,15], where their architectures combined a reconfigurable functional unit with the microprocessor. Partitioning is a well-known NP-complete problem, and many heuristics have been proposed to guide the partitioning of a system into hardware and software components [16,17]. In [17], the HW/SW partitioning problem was modeled as a constraint satisfaction problem (CSP), and a genetic algorithm-based approach was proposed to solve the CSP to obtain the partitioning solution.

The major innovative architecture design proposed in this project is applying the multiagent paradigm to the reconfigurable embedded system on both software and hardware design and establishing a framework by which hardware agents can be dynamically reconfigured to meet the system's real-time constraints. The proposed reconfigurable architectures can provide the consistency preservation, low reconfiguration latency, and fast random access to the configuration memory. Due to its generality, the proposed architecture can be extended to large real-time applications.

17.3 BDI-Based Multiagent Architecture

An *agent* is an independent processing entity that interacts with the external environment and the other agents to pursue its particular set of goals. The agent pursues its given goals adopting the appropriate plans, or intention, according to its current beliefs about the state of the world, so as to perform the role it has been assigned. Such an intelligent agent is generally referred to as a BDI agent. In short, belief represents the agent's knowledge, desire represents the agent's goal, and intention lends deliberation to the agent.

In the agent model, as shown in Figure 17.1, beliefs and desires influence each other reciprocally. Furthermore, beliefs and desires both influence intentions. The model includes three external ports: interagent communication, control, and input/output. The control port is for the agent to synchronize with the system. The input/output port is used to send and receive information to and from the host environment. The interagent communication port allows the agents to send/receive information to/from other agents and cooperate with others.

The agent control loop is as follows: first, determine current beliefs, goals, and intensions; find applicable plans; then decide which plan to apply; and finally start executing the plan. Both the software agents and hardware agents in our system adopted the BDI architecture model.

The BDI agent can be expressed in a structure as follows:

```
<Agent>{<Beliefs>
Constraints; Data Structures;
<Desires>
Values; Condition; Functions;
<Intentions and Commitment>
Methods; Procedures;}
```

This structure can be implemented by high-level software languages on microprocessors or Verilog hardware description language (VHDL) on FPGA. The BDI agent is capable of providing the following features that are impossible for the regular module-based architecture:

■ Autonomy: Once launched with the information describing the bounds and limitations of their tasks, BDI agents are able to operate independently of and unaided by their use.

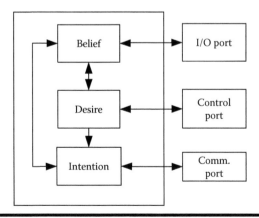

Figure 17.1   BDI agent model.

- Social ability: To effectively change or interrogate their environment, BDI agents possess the ability to communicate with the outside world through their input/output ports and communicate with other agents through intercommunication ports.
- Reactivity: BDI agents are able to perceive their environment and respond to changes in a timely fashion.
- Proactivity: To help BDI agents to be adaptive to new situations, they are able to exhibit proactivity, that is, the ability to effect actions that achieve their goals by taking the initiative.

Some agent-oriented implementations have been developed, such as JACK [18] and distributed multiagent reasoning system (DMARS) [19]. In our systems, the software agent follows the similar idea of JACK. The BDI agents with beliefs, desires, intentions, and plan library are created. Events alert the agent to changes in the world or its internal state. When an event is raised, the agent looks through its plan library and finds a plan that is relevant to the event and its current beliefs. It then creates an instance of this plan and executes it. However, if it fails, another plan is tried until all relevant plans are exhausted.

The hardware agent implementation is similar to the software agent except that, in software, all the parameters of the agents can be transmitted by function calls in high-level software languages, while, in hardware, we have to specifically define all of the hardware agent entities, their associate ports, and their parameters in VHDL. The hardware agent interface should be simple for customization to any specific applications without significant rework.

17.4 Hardware/Software Codesign

HW/SW codesign provides a way of customizing the hardware and software architectures to complement one another in ways that improve system functionality, performance, reliability, survivability, and cost effectiveness. Figure 17.2 shows the codesign flow performed in order to transform a system-level specification of a mixed HW/SW implementation. The codesign process is composed of four steps: system-level specification, system-level partitioning, communication synthesis, and software and hardware synthesis. The goal of HW/SW codesign is to produce an efficient implementation that satisfies the performance and minimizes the cost, starting from the initial specification. The mixed HW/SW system will be mapped from system behavior description.

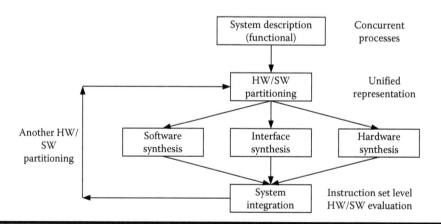

Figure 17.2 Typical HW/SW codesign process.

The proposed agent-based architecture provides a unified model for HW/SW cosimulation and cosynthesis. To achieve high performance of overall system, an effective HW/SW partitioning methodology and a competent communication protocol among the agents are required.

17.4.1 HW/SW Partitioning

It is desirable to propose a method that can automatically partition the agents into hardware and software on the fly based on the task parameters of each agent, such as execution time, release time, and deadline. However, automatically partitioning is a challenge problem not only because it is a well-known NP-complete problem but also due to the lack of efficient and accurate performance profiling tools. Furthermore, the dynamic partitioning becomes more complex if it has to handle some sporadic tasks due to environmental changes.

In this project, since our research focus is on the architecture design, an off-line heuristic partitioning method is applied to simplify the HW/SW codesign procedure. In general, the following partitioning policies are required to follow. First, more regularly structured agents that have highly repetitive and extensive time-consuming operations are suitable for implementation in reconfigurable hardware, whereas the more complex and irregularly structured agents should be programmed in software. Second, due to the different implementation platforms of hardware and software, the agents who have heavy communication or dependencies should be all partitioned either to hardware or software to minimize the communication costs. Third, the agents with hard deadlines are distributed to hardware, while the agents with soft deadlines may be implemented in software.

17.4.2 Multiagent Communication Mechanism

Agents are generally designed with a specific purpose in mind. They can do one or perhaps several tasks very well but are not often designed as a jack-of-all-trades. If agents must perform more tasks, we can either increase their complexity (which may increase the development effort), or we can make them work cooperatively. Usually a complex real-time system can be constructed as a set of independent and cooperating agents, where each agent owns its own intension and pursuit its own sets of goals. For cooperation between agents to succeed, effective communication is required. We can view a collection of agents that work together cooperatively for a global goal. For a global goal to function coherently, we need a common language and communication medium.

Various agent communication languages have been developed, such as Knowledge Query Manipulation Language (KQML) [18] and Foundation for Intelligent Physical Agent (FIPA) [19]. But they have high overheads in terms of memory usage and computation time that are not always acceptable for real-time systems. Numerous proprietary methods have also been developed, but more and more often they encounter maintenance, porting, and enhancement issues.

Since all of the agents work in an asynchronous mode, the messages will only be issued on demand, which leads to a new ODMP protocol proposed in this project. A typical example of two agents connected by a communication path is shown in Figure 17.3. In this mechanism, each agent creates a message queue in FIFO order with priorities, where the messages marked as urgent are attached to the head of the queue, and the ones marked as normal are lined in an FIFO order.

Each agent may have multiple current tasks need to be implemented. To reduce the routing complexity of hardware agents and memory requirement, the database only maintains the list of all other related agents in the system along with their current status. When we say "related," we mean the agents who need to receive/send the message from/to the current agent. The current status of

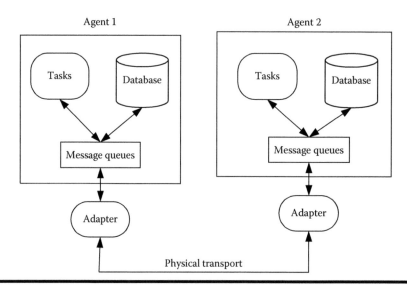

Figure 17.3 Multiagent communication structure.

the agents would help to improve the fault tolerance. For example, before a task in agent 1 sends a message to agent 2, agent 1 needs to check the status of agent 2 in its database. If the agent 2 is not on service due to some failures, agent 1 stops all its messages to agent 2. On the other hand, when agent 1 receives a message from agent 2, it makes its decision based on the new message and updates its own database as well.

To provide a media-independent agent communication mechanism, an *adapter* mechanism is proposed. One adapter is designed to be associated with each different physical transport, which can be a bus driver for software agent, or other high-level communication protocols, such as UDP or Ethernet. The adapter mechanism provides a uniform interface for the agents, making the agent to be independent from the specific transport being used. One adapter may be connected to multiple agents if necessary. When an agent needs to send a message to other agents, it passes the message buffer to the adapter first, and then the adapter sends data out through the physical transport. Similarly, when an agent receives a message from other agents, the adapter reconstructs a message buffer from the incoming data and sends buffer to the agent.

The minimum message transmitting/receiving unit is called *message frame*. A message may contain multiple message frames. The maximum number of message frames for a single message is platform dependent, which is restricted by the available internal and external memory. A message frame includes protocol header, service header, and data to be transmitted. The protocol header is defined by the physical transport, which was built from values provided by the adapters. The contents of this header may vary from protocol to protocol but should include fields such as source ID, destination ID, sender ID, receiver ID, data size, and priority of messages if the transport supports priority. Service header usually identifies the internal service sending the message and the message type, which are application dependent.

When an agent sends a message to a remote agent, this data can be a fragment of the whole message if the total message size exceeds the size of a message frame. A message buffer is needed to reconstruct the whole message if it has been segmented before the transmission. The agent does not acknowledge individual message unit; instead, only the whole messages are acknowledged. If the event of a transmission error or if a telegram is lost, the whole message must be retransmitted.

17.5 Hardware Agent Reconfiguration Modules

Performing reconfiguration on a running system is an intrusive process. Reconfiguration may imply, for example, interference with ongoing interactions between entities. One of the main issues of dynamic reconfiguration is consistency preservation [20]. Changing management functionality must assure that system parts that interact with entities under reconfiguration do not fail because of reconfiguration, that is, system consistency needs to be preserved. Furthermore, the runtime reconfiguration latency can be significant. To maximum application execution performance, such loading overhead must be minimized. In this section, various techniques are proposed to reduce/tolerate the configuration latency.

The Virtex-II Pro device is used as the FPGA platform in our project, which is capable of implementing flexible, high performance, and low-cost system-on-chip designs by combining a variety of features embedded in the FPGA fabric with specially developed HW/SW IP cores [21]. Particularly, it cooperates fully embedded IBM PowerPC 405 processor core [22]. The embedded PPC 405 core is a 32-bit Harvard architecture processor with functional units such as cache unit and memory management unit (MMU). It is capable of more than 300 MHz clock frequency and 420 Dhrystone MIPS and is contained in a processor block. The processor block also contains on-chip memory controllers and integration circuitry compatible with IBM CoreConnect bus architecture that enables the compliant IP cores to integrate with this block.

17.5.1 Virtual Agent Reconfiguration Module

The consistency preservation could become a very complicated problem. In order to preserve the consistency while keep the dynamic reconfiguration procedure simple, we make the following assumptions for the reconfiguration models:

1. Agents work independently.
2. Agents can communicate with other agents.
3. Each agent can have a finite number of schemes that is a description of possible events. Each scheme of the agent can be created a priori before the application is implemented and can be stored in the external memory.

Based on the previous assumptions, a *virtual agent reconfiguration (VAR)* module is proposed, which is based on the work of Sckanina and Ruzicka [12] by implementing a user-defined FPGA inside an ordinary FPGA. As shown in Figure 17.4, this method consists of a number of *virtual* states that each agent could possibly go through at runtime. These virtual states are defined as configuration bitstreams that are loaded into the reconfigurable hardware memory when the system starts up. Thus, in this scheme, the FPGA configuration registers are fixed at runtime. The device is reconfigured only by simply selecting an active state by a multiplexer based on its local point of view. There is no reconfiguration latency due to the static reconfiguration. If the device contains plenty number of agents, the cost of this method is extremely high, and the efficiency is very poor because all possible agent states must be implemented, which requires more reconfiguration resources. However, the number of logic blocks in the Vertex blocks is large (e.g., XC2VP30 has 30k logic cells), and each block can be configured in a set of ways. Therefore, if the number of agents in a real-time system is relatively small, and the computation of data processing is not extremely intensive, this reconfiguration method would be suitable due to its low overhead and low reconfiguration latency.

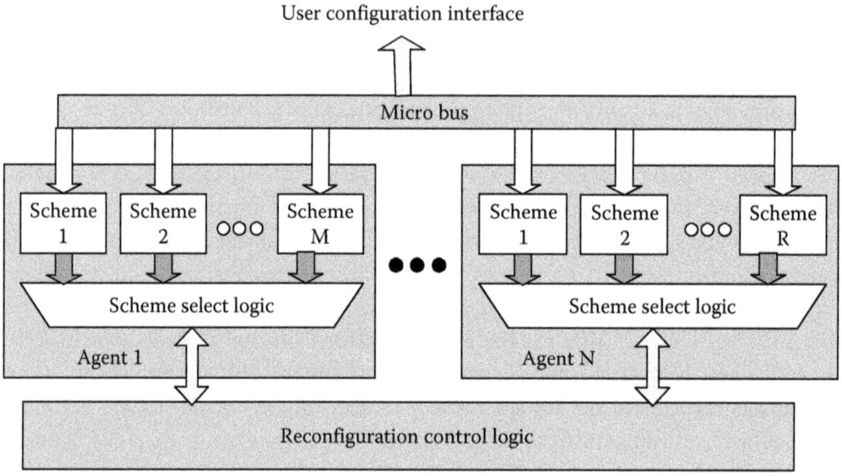

Figure 17.4   Implementing a virtual agent reconfiguration.

17.5.2 Pipeline Agent Reconfiguration Module

In order to use the reconfigurable resources more efficiently when the number of reconfigurable agents is large, one option is that the replacement agent reconfiguration overwrites the previous agent state while suspending the functionality of the reconfigurable logic. This mode supports several temporal modes of operation. However, there is high reconfiguration latency because the system must spend lots of time to load the selected agent states from high-capacity/low-cost hardware configuration storage to the partially reconfigurable logic. This method is only suitable for those applications that do not have critical real-time constraints.

Instead of waiting for the long latency time of loading configurations from low-cost hardware to the reconfiguration memory, we propose the *pipelined agent reconfiguration (PAR) module*. First, the FPGA is partitioned into a set of static and dynamic sub-FPGAs, where each sub-FPGA can be configured separately using partial reconfiguration. The static sub-FPGA is fixed and always active at runtime and is used to manage the dynamic sub-FPGAs that are configured at runtime and are not always in operation. That is, only one of the dynamic sub-FPGAs is active at any given time. Thus, the one that is not active can load its new configuration through reconfiguration control logic by partial reconfigurations, while the other one is executing. In this way, the FPGA becomes a pipelined unit where the configuration is pipelined with execution, as shown in Figure 17.5.

The number of dynamic sub-FPGAs could be increased if the execution time is shorter than the configuration time. By applying this pipelined method, the reconfiguration latency is greatly diminished, which is critical for the real-time systems. However, only part of the reconfigurable resources can be used at a given time. This method can obtain the optimized balance between the reconfiguration latency and resource occupancy.

17.5.3 Self-Reconfiguration Module Using Virtex-II Pro FPGA

The previous VAR and PAR modules can be applied to any partially reconfiguration platform. Since Virtex-II Pro FPGA is adopted in our project, a self-reconfiguration method is proposed based on Virtex-II Pro FPGA. *Self-reconfiguration* is a special case of dynamic reconfiguration where the configuration control is hosted within the logic array that is being dynamically reconfigured.

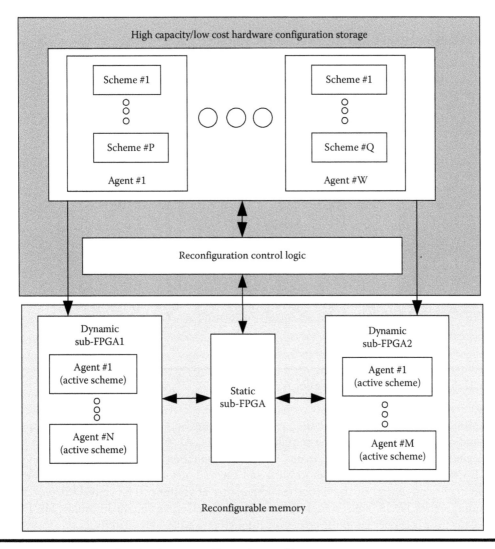

Figure 17.5 Pipelined redundant reconfiguration architecture.

The part of the logic array containing the configuration control remains unmodified throughout execution. Since the control logic is located as close to the logic array as possible, the latencies associated with accessing the configuration port are minimized.

The Virtex-II Pro configuration architecture features an internal configuration access port (ICAP) that provides the user logic with access to FPGA configuration interface and therefore access to memory bits of configuration memory [23]. In active partial reconfiguration, new data can be loaded through ICAP to dynamically reconfigure a particular area of FPGA while the rest of FPGA is still optional. Some self-reconfiguration platforms proposed in [13,14] used a limited amount of on-chip memory to store the reconfiguration data, which may easily lead to a performance bottleneck for a complex real-time system.

To overcome this performance bottleneck, a new self-reconfiguration framework using DDR external memory is proposed here. The goal of this framework is to provide a flexible method

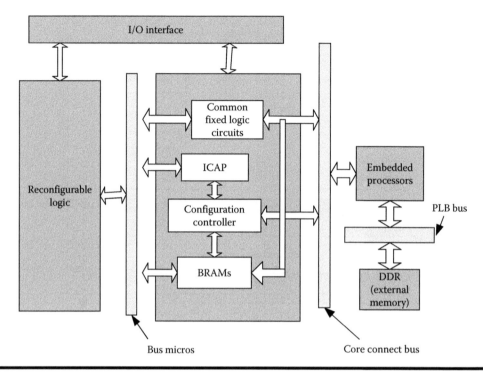

Figure 17.6 Self-reconfiguration platform.

for implementing field self-reconfiguration while minimizing both system hardware and required configuration data, and reducing the reconfiguration time. As shown in Figure 17.6, the fixed logic includes common fixed logic circuits, ICAP, configuration controller, multiple dual-port block RAMs (BRAMs), embedded processors, and external memory. The embedded processors provide intelligent control of the device reconfiguration at runtime. The embedded processors communicate with FPGA logic through IBM-designed CoreConnect bus architecture. The DDR external memory is connected to the processors through process local bus (PLB).

The fixed logic part is located at the rightmost columns because ICAP in Virtex-II Pro FPGA is located at the lower-right corner while the reconfigurable logic part is located on the left-side columns. The configuration controller uses the ICAP to reconfigure the reconfigurable logic parts. The dual-port characteristics allow the BRAM to be accessed from different clock domains on each side. One port is accessed from the configuration controller, while the other port is accessed from the reconfiguration logic part.

To improve the reconfiguration performance, we can take advantage of the BRAMs in two ways. First, they can be set up as registers with individual address and easily to read from and write to by both configuration controller and reconfiguration logic part. This feature would allow the reconfiguration logic part get setup values from embedded processor via the configuration controller and also report back the current status to the processors, acting like a shared memory between the processors and reconfiguration logic parts. Second, the BRAMs can be applied as buffers between the configuration controller and reconfigurable part. This buffer feature can reduce the reconfiguration latency due to its pipeline mechanism.

The common fixed logic circuits can implement some specific hardware agents as well as interrupt generation logic. The BRAMs can also be accessed by common fixed logic directly. These

BRAMs should be different memory locations than those accessed by configurable logic parts. This feature provides the shared memory mechanism between the common fixed logic circuits and the embedded processors.

To improve the system robustness, two trigger conditions for the dynamic reconfiguration are applied here, that is, bottom-up and top-down. The bottom-up method is through event interrupts invoked by some specific agents. The top-down method is that the processors start the reconfiguration due to decisions based on the input information.

Another benefit from this self-reconfiguration platform is that the multiagent communication mechanism can easily take advantage of the shared memory feature provided in this platform. Due to the dual-port characteristics of the BRAM, the software agents can send messages to BRAM through configuration controller, while the hardware agents can access their own message queues directly from the same BRAM. On the other hand, if a hardware agent sends a message to a software agent either through the configuration controller if it is a configurable hardware agent, or through CoreConnect bus directly if it is fixed hardware agent. The message can be stored on the DDR SDRAM via CorConnect PLB.

17.6 Case Study

17.6.1 Prototype Platform

A Pioneer 3DX mobile robot system from ActivMedia Robotics is adopted as a mobile vehicle in our project, as shown in Figure 17.7a, which is 44 cm × 38 cm × 22 cm aluminum body with 16.5 cm drive wheels. The two motors use 38.3:1 gear ratios and contain 500-tick encoders.

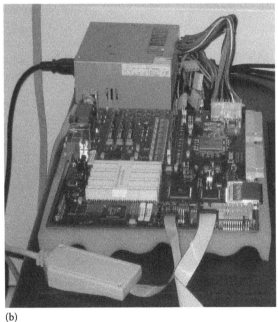

(a)   (b)

Figure 17.7 Experimental platform: (a) Pioneer 3DX robot, (b) ML310 embedded platform.

The onboard sensors of this robot system include one SONY CCD camera, one laser rangefinder and eight SONARs for range measurement, and two motor encoders for odometry measurement.

The Xilinx ML310 embedded development platform is a Virtex-II Pro–based platform suitable for rapid prototyping and system verification and is installed to the robot as the onboard processor and our prototype platform. The ML310 offers designers a Virtex-II Pro XC2VP30-based embedded platform. It provides 30k logic cells, over 2400 kb BRAM, and dual PPC405 processors, as well as onboard Ethernet MAC/PHY, 256 M DDR memory, multiple PCI slots, FPGA UART, standard PC I/O ports, and high-speed I/O through RocketIO multi-gigabit transceivers (MGTs). By using ML310 embedded platform, hardware agents are configured as fixed or reconfigurable FPGA logic circuits in VHDL, where bus macros are used as fixed data paths for signals between reconfigurable logic parts and other logic parts (fixed or reconfigurable). Software agents are implemented on the embedded soft cores PowerPC 405 in C/C++ language.

The IBM-developed hierarchical CoreConnect bus architecture is used to connect the external memory, I/Os, and other peripherals. Although a unified design representation is applied to hardware and software agents, the physical communication mechanism between these two has to go through the CoreConnect bus. Based on the proposed ODMP communication mechanism, the message queues are created individually for each agent (hardware or software). The message queues are saved on internal BRAM for hardware agents and external DDR SDRM for software agents. The self-reconfigurable module is applied for dynamic reconfiguration of hardware agents in our experiment.

17.6.2 Mobile Robot System

To evaluate the proposed architecture, we develop the agent-based architecture for a mobile robot navigation system [24]. The block diagram is shown in Figure 17.8. Usually one mobile robot system has multiple sensor systems, where each sensor agent is designed to acquire some specific information from the external environment. All of the sensor agents work parallel and independently. Sensor fusion agent analyzes the sensor information acquired by all of the sensor agents, fuses the sensor data if necessary, and sends the fused sensor data to the higher-level

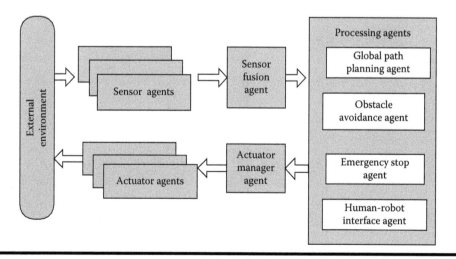

Figure 17.8 Multiagent architecture for a mobile robot navigation system.

processing agents. Global path planning agent plans an optimized path for the robot to navigate across the global environment given a starting point and a destination point. Obstacle avoidance agent detours the robot when there are obstacles in the way to prevent the robot hitting the obstacle.

Actuator management agent receives requests from processing agents and distributes the control commands to individual actuator agents. Human–robot interface agent receives the events or processed environmental information from sensor fusion agents and then sends corresponding commands to the robot. Emergency stop agent usually takes commands from human and sends request to actuator manager agent to stop the robot.

According to the HW/SW codesign technology, the hardware and software agents are partitioned as follows: the sensor agents, sensor fusion agent, actuator management agent, and actuator agents are configured in hardware on FPGA, while all of the processing agents, including path planning, obstacle avoidance, emergency stop, and human–robot interface agents, are implemented in software on embedded processor core.

Based on the previous multiagent architecture, when a mobile robot navigates to different environments or sensors encounter some malfunctions, only affected sensor agents need to be reconfigured, while the other agents will continue their own tasks without any influence. The reconfiguration can be either triggered by the sensor agents by invoking an interrupt to processor or by the processor itself based on its data checking algorithm.

The real-world experimental scenario is shown in Figure 17.9. The mobile robot (circle) needs to navigate itself from one fixed starting point at room B to the fixed destination point at room A while searching a specific feature target (a green cube) and avoiding all of the obstacles on its way. A simple color detection algorithm is applied to the vision system to detect the target. The map of two office rooms is installed to the robot before the navigation, and the obstacles (triangle and oval) are randomly scattered.

By implementing both the HW/SW codesign and software-only approaches to the mobile robot, the robot has successfully navigated itself to the destination point and found the target without hitting any obstacles on its way. Five configurations with different target locations in room A and different obstacles distributed in room B are designed in the previous environment. Ten runs have been conducted on each configuration for both HW/SW and software-only approaches. The average navigation time of experimental results is shown in Figure 17.10. The resource usage of hardware agents is listed in Table 17.1.

It can be seen from Figure 17.10 that there is a dramatic decrease in average navigation time in a dynamic environment with HW/SW codesign comparing with SW-only approach. The HW/SW

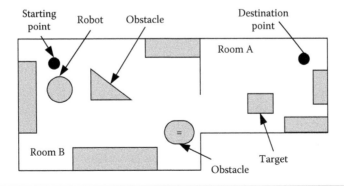

Figure 17.9   Navigation environment.

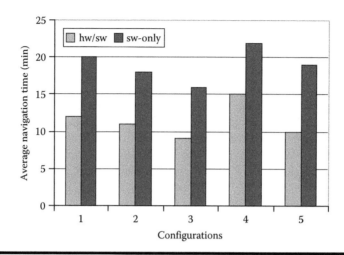

Figure 17.10 Average navigation time comparison for real-world experiments.

Table 17.1 Resource Usage of Hardware Agents

Resources	Self-Reconfiguration Model (%)
IOs	56.34
Function generators	73.34
CLB slides	65.32
DEEs or latches	5.23

approach increases the speed of action and reaction to dynamic environment because it makes decisions much faster than the software competitors. Another reason for the speedup is because hardware responds favorably to dynamic changes during runtime due to its parallel structures.

Since only the simple color-based target detection is applied in our experiment, we did not take full advantage of the proposed HW/SW architecture. For the more complex vision-based robots, such as service robots, the overall system responsiveness should be improved significantly using the proposed platform because the vision-related algorithms are extensive time consuming with software-only approach. Therefore, a mobile vision system for people detection and tracking is discussed in the next section.

17.6.3 Mobile Vision System for People Detection and Tracking

Some researches have been conducted for people detection systems using one single cue at the detection, such as motion [25], color [26], contour [27], or face's features (such as eyes, nose, and mouth, etc.) [28]. Due to the natural, complex, and dynamic working environment of a mobile robot, the robot vision system has to be highly robust and independent with the application scenario. Therefore, some researches proposed the combinations of multiple cues for people segmentation [29–31]. In this project, we propose a detection method combining the color and motion cues.

First, skin-color classification is applied to find faces in images since it is independent from egomotion of the camera systems. To represent skin color, the dichromatic r-g-color space

$[r = R/(R + G + B), g = G/(R + G + B)]$ is normalized in brightness and thus is widely independent from variations in luminance. To improve the system real-time performance, we have built up a look-up table with manually classified skin color pixels in the r-g-color space for 30 people under different illumination conditions.

Then, motion detection is applied to distinguish the people moving in the scene from the stationary background. For a stationary camera system, motion detection can be conducted by subtraction of two successive image frames. However, if the camera system is moving, the egomotion has to be taken into account. The algorithm proposed in [32] is applied to estimate the compensation for the egomotion of the mobile robot. Assume the robot motion parameters can be obtained from the navigation and control systems. The next successive image can be predicted by using the epipolar transformation. The predicted image is matched with the actual image from the scene, and all nonmatching areas represent the moving objects.

It may occur frequently that multiple Regions of Interests (ROIs) are detected in one image, where a fusion algorithm is necessary to merge those ROIs. To simplify the merging procedure, we assume that there are relatively small changes in size and position of people from the successive image frames; in other words, the people are moving in a relatively slow speed. Then a probabilistic approach based on the color and motion cues as well as the geometric parameters of the object in the images is applied to select the most likely ROI.

The condensation tracking algorithm in [33] is adopted here for face tracking. The task of calculating the probability of the presence of a face for every pixel and tracking the resulting density function over time is solved by an approximation of the density function. Then, a recursive filter is applied to update the density function on each sample. Once a person is tracked, the samples of the condensation algorithm are concentrated on his or her face.

For a mobile vision system, a large amount of images from vision system need to be processed under real-time constraints. Image matching is critical for navigation system since 3D environment knowledge can only be obtained through multiple views. With the partial reconfiguration features of FPGA, it is possible to run the image loading and image matching simultaneously. This parallel scheme can reduce the implementation time dramatically since the image loading from external memory to reconfigurable hardware is usually extremely computational extensive. When the image is loaded onto the buffer from I/O, the other buffer is applied for high-speed matching algorithm. When the image loading is finished, the control logic will check the status of another buffer. If the matching is done, the two buffers exchange operations. Otherwise, the control logic will inform the loading buffer to wait until the other buffer finishes the matching and then exchange the operations.

Furthermore, traversing the image is necessary to identify the Region of Interests (ROIs). To speed up this traversing procedure, each loaded image can be divided into several regions, where each region can be assigned to different FPGA logic block. A separate control logic is responsible for whole-image information integration. By paralleling the searching procedure, the overall speed of the image processing can be significantly reduced. Since the FPGA blocks are limited for each rSoC platform, the dynamic reconfiguration is conducted so that the new image regions will be continuously loaded into each block after the previous block has been processed. If some of the regions finish their traversing faster than others, it is possible that new image regions can be loaded in the finished regions to start processing. Therefore, a pipeline scheme can be formulated to improve the processing efficiency by using the FPGA logic blocks greedily.

To evaluate the proposed architecture, we develop the agent-based architecture for a mobile vision [34], and the block diagram is shown in Figure 17.11, where rectangles represent hardware agents and ovals represent software agents. Since robot path planning agent and robot/human

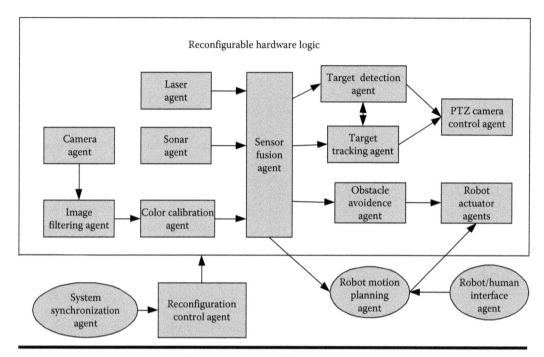

Figure 17.11   Agent-based architecture for a mobile vision system.

interface agents are complex and not critical in real-time reaction, they are partitioning into the software agents, where all of other time-critical and extensive time-consuming agents, such as image processing, target detection and tracking, camera control, laser and sonar senor, sensor fusion, and robot actuator control, are assigned to hardware agents. Unlike the other hardware agents that are located on the reconfigurable logic blocks, the reconfiguration control agent locates on the fixed logic block, which needs to control the reconfiguration procedure of the other hardware agents.

First, the camera captures the image from the environment and sends the images to the image filtering agent. After the color calibration agent finishes its processing to remove the illumination affects, the sensor fusion agent combines the image information with the laser agent and sonar agent. The output of the sensor fusion agent can serve for target detection agent, target tracking agent, obstacle avoidance agent, and robot motion planning agent.

The target detection and tracking agent will control the movement of the PTZ camera to keep the target inside its field of view. It is worth noting that at any given time, the obstacle avoidance agent and motion planning agent can work in a complementary mode, which means that only one agent can work at a time. If obstacles are detected, obstacle avoidance agent works, otherwise, motion planning agent works.

The motion control signal will be sent to robot actuator agents. Since the target detection and tracking agents are all configured in the hardware FPGA, they can be implemented concurrently with the obstacle avoidance agent, which means that the vision system can detect and track the target while avoiding the obstacles on its way. In this project, we use people as our target for detection and tracking by the mobile vision system.

To evaluate how much performance to be gained by the HW/SW codesign platform compared to the software only platform, both software-only and HW/SW codesign approaches have been implemented for face detection and tracking using the vision system installed on the mobile robot.

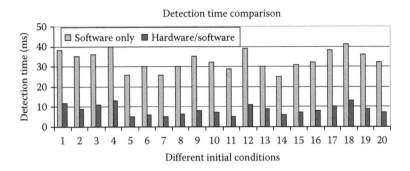

Figure 17.12   Detection time comparison.

The method in [31] was adopted in the experiments for face detection due to its high detection rate and very low false-positive rate. The detection rate is 61.83%, and false-positive rate is 0.0004%.

The experiments for both approaches have been conducted for 20 times in terms of different initial conditions of the vision system. To make the experimental results to be more comparable, for the same initial condition, both the software-only and HW/SW approaches are implemented, respectively. One person is sitting on a chair at a fixed location facing to the direction where the robot locates, where the robot initially starts from different locations with different facing direction of the camera system. There are some obstacles dynamically sitting on the way between the vision system and the person. Before the robot detects the face and tracks the face, since the robot has no idea where the person is, therefore, random movements are initiated until the vision system detects the face, then the robot moves toward the face while controlling the PTZ camera to track the face afterward.

The detection time is defined as the time difference from the initial time to the time when the face is detected. The experimental results are shown in Figure 17.12. The average detection time for software-only is 33.05 ms, while the average time for HW/SW codesign approach is 8.375 ms. It is obvious that the codesign approach can improve the system performance significantly in its responsiveness and fault tolerance for real-time systems, especially in the situations where the environments are dynamically distributed by some obstacles.

17.7 Conclusion and Future Work

This chapter proposes an agent-based architecture on an rSoC platform for real-time systems. The unified BDI agent structure significantly simplifies the HW/SW partitioning and communication. Some heuristic HW/SW portioning methods are developed as the first step for HW/SW codesign. Then a novel ODMP protocol is proposed for the agent-to-agent communication. By developing different adapters for physical transports, the agent-based architecture design can be transport independent. To take full advantage of the partial reconfiguration provided by FPGA platforms, some reconfiguration schemes are proposed to provide the consistency preservation, low reconfiguration latency, and fast random access to the configuration memory.

To evaluate the proposed agent-based architecture on rSoC platform, two real-time systems have been conducted using P3-DX robot with vision systems. First system is a mobile robot system with simple detection capability, and the second one is a mobile vision system for people detection and tracking. The experimental results show that the proposed architecture is feasible and much more efficient compared to the software-only approach. Due to its generality and flexibility of the proposed agent-based architecture, it can be easily extended to large various real-time applications.

The proposed BDI agent-based architecture provides a very flexible platform for the future extension, such as learning capability and proactivity to adapt to the dynamic environments. Our future research will focus on dynamic HW/SW partitioning and scheduling instead of just using an off-line heuristic method. In addition, we will also investigate the learning capability of the BDI agent architecture to make the overall system be more robust and adaptive to the environmental changes.

References

1. V. Gafni. Robots: A real-time systems architectural style. *Proceedings of 7th European Software Engineering Conference*, pp. 57–74, Toulouse, France, 1999.
2. J. Almeida, M. Wegdam, M. Sinderen, and L. Nieuwenhuis. Transparent dynamic reconfigurable for CORBA. *Proceeding of the Third International Symposium on Distributed Objects and Applications (DOA 2001)*, Rome, Italy, September 17–20, 2001.
3. J. Cobleigh et al. Containment units: A hierarchically composable architecture for adaptive systems. *ACM SIGSOFT Software Engineering Notes*, 27(6):159–165, November 2002.
4. B. MacDonald, B. Hsieh, and I. Warren. Design for dynamic reconfiguration for robot software. *Second International Conference on Autonomous Robots and Agents*, Palmerston North, New Zealand, December 13–15, 2004.
5. D. Stewart, R. Volpe, and P. Khosla. Design of dynamically reconfigurable real-time software using port-based objects. *IEEE Transactions on Software Engineering*, 23(12):759–776, December 1997.
6. E. Yoshida, S. Murata, A. Kamimura, K. Tomita, H. Kurokawa, and S. Kokaji. Self-reconfiguration modular robots—Hardware and software development in AIST. *Proceedings, IEEE International Conference on Robotics, Intelligent Systems and Signal Processing*, Vol. 1, pp. 339–346, Ghangsha, China, October 8–13, 2003.
7. S. Patterson, K. Knowles, and B. E. Bishop. Toward magnetically-coupled reconfigurable modular robots. *Processing of IEEE International Conference on Robotics and Automation*, New Orleans, LA, April 2004.
8. Y. Meng. A dynamic self-reconfiguration mobile robot navigation system. *IEEE/ASME International Conference on Advanced Intelligent Mechatronics (AIM 2005)*, KaoHsiung City, Taiwan, 2005.
9. Y. Li, T. Callahan, E. Darnell, R. Harr, U. Kurkure, and J. Stockwood. Hardware-software co-design of embedded reconfigurable architectures. *Design Automation Conference*, Los Angeles, CA, June 2000.
10. J. Fleischmann, K. Buchenrieder, and R. Kress. Codesign of embedded systems based on java and reconfigurable hardware components. *Design Automation and Test in Europe*, Munich, Germany, March 1999.
11. S. M. Scalcra and J. R. Vazques. The design and implementation of a context switching FPGA. *IEEE Symposium on FPGAs for Custom Computing Machines*, pp. 78–85, Napa, CA, 1998.
12. L. Sckanina and R. Ruzicka. Design of the special fast reconfigurable chip using common FPGA. *Proceedings of Design and Diagnostics of Electronic Circuits and Systems—IEEE DDECS'2000*, pp. 161–168, Smolenice, Slovakia, 2000.
13. B. Blodget, P. James-Roxby, E. Keller, S. McMillan, and P. Sundararajan. A self-reconfiguring platform. *Proceeding of the 13th International Conference on Field Programmable Logic and Applications (FPL'03)*, pp. 565–574, Lisbon, Portugal, 2003.
14. R. Fong, S. Harper, and P. Athanas. A versatile framework for FPGA field updates: An application of partial self-reconfiguration. *Proceedings of the 14th IEEE International Workshop on Rapid Systems Prototyping (RSP'03)*, San Diego, CA, 2003.

15. M. Baleani, F. Gennari, Y. Jiang, Y. Patel, R. K. Brayton, and A. Sangiovanni-Vincentelli. HW/SW partitioning and code generation of embedded control applications on a reconfigurable architecture platform. *Proceedings of the Tenth International Symposium on HW/SW Codesign*, Estes Park, CO, May 2002.

16. R. Niemann and P. Marwedel. An algorithm for HW/SW partitioning using mixed integer linear programming. *Design Automation of Embedded Systems*, 2(2):165–193, 1997.

17. D. Saha, R. S. Mitra, and A. Basu. HW/SW partitioning using genetic algorithm. *Proceeding of 10th International Conference on VLSI Design*, pp. 155–160, Hyderabad, India, 1997.

18. Agent-Oriented Software Pty Ltd, JACK Manual (v4.1), http://www.agent-oriented.com, 2003.

19. M. d'Inverno, D. Kinny, M. Luck, and M. Wooldridge. A formal specification of dMARS. *Proceedings of the 4th International Workshop on Agent Theories, Architecture, and Language*, Vol. 1365 of LNAI, pp. 155–176, Berlin, Germany, 1998.

20. J. Kramer and J. Magee. The evolving philosophers problem: Dynamic change management. *IEEE Transactions on Software Engineering*, 16(11):1293–1306, 1990.

21. Virtex-II Platform FPGA Handbook, version 2.0, Xilinx, Inc., 2004.

22. PowerPC Processor Reference Guide, version 2.0, Xilinx, Inc., 2003.

23. Virtex-II Platform FPGA User Guide, version 4.0, Xilinx, Inc., 2005.

24. Y. Meng. An agent-based reconfigurable system-on-chip architecture for real-time systems. *The 2nd International Conference on Embedded Software and Systems (ICESS 2005)*, Xian, China, December 16–18, 2005.

25. K. Rohr. Towards model-based recognition of human movements in image sequences. *CVGIP: Image Understanding*, 59(1):94–115, 1994.

26. Y. Raja, S. J. McKenna, and S. Gong. Tracking and segmenting people in varying lighting conditions. *Proceedings of the 3rd International Conference on Automatic Face and Gesture Recognition*, pp. 228–233, Nara, Japan, 1998.

27. J. Davis and V. Sharma. Robust detection of people in thermal imaging. *Seventeenth International Conference on Pattern Recognition (ICPR'04)*, pp. 713–716, Cambridge, U.K., 2004.

28. H. A. Rowley, S. Baluja, and T. Kanade. Neural network-based face detection. *IEEE Transactions on Pattern Analysis and Machine Intelligence*, 20(1):23–38, 1998.

29. S. Feyrer and A. Zell. Detection, tracking, and pursuit of humans with an autonomous mobile robot. *Proceedings of the International Conference on Intelligent Robots and Systems (IROS'99)*, pp. 864–869, Gyeongju-si South Korea, 1999.

30. B. Froba and C. Kublbeck. Face detection and tracking using edge orientation information. *SPIE Visual Communications and Image Processing*, pp. 583–594, San Jose, CA, 2001.

31. P. Viola and M. Jones. Robust real-time object detection. *Proceedings of the Second International Workshop on Statistical and Computational Theories of Vision*, Vancouver, British Columbia, Canada, 2001.

32. D. Murray and A. Basu. Motion tracking with an active camera. *IEEE Transactions on Pattern Analysis and Machine Intelligence*, 16(5):449–459, 1994.

33. M. Isard and A. Blake. Condensation—Conditional density propagation for visual tracking. *International Journal on Computer Vision*, 29(1):5–28, 1998.

34. Y. Meng. A mobile vision system with reconfigurable intelligent agents, *IEEE World Congress on Computational Intelligence (WCCI 2006)*, Vancouver, British Columbia, Canada, July 16–21, 2006.

NETWORKING SENSING AND COMMUNICATIONS

Chapter 18

Hardware Task Scheduling for an FPGA-Based Mobile Robot in Wireless Sensor Networks

Tyrone Tai-On Kwok and Yu-Kwong Kwok

Contents

18.1 Introduction

The regular structure of field programmable gate arrays (FPGAs) enables the utilization of massively parallel circuit design for hardware acceleration of software tasks. Specifically, we can offload some computation-intensive tasks in software applications, which are originally executed in the instruction set processor, to FPGAs so as to increase the overall system efficiency. Specifically, an FPGA-based system offers the following advantages over a traditional processor-based system:

Computation efficiency: FPGAs, a kind of (re)programmable chips, have traditionally been used for hardware prototyping or as glue logic for connecting different components in a system. With today's advanced chip technology, FPGAs' gates are clocked at high rate and of high density and have abundant dedicated hardware resources, which brings about an amazing computer architecture for high performance computing—reconfigurable computing. FPGAs can typically achieve 20–100 speedup factors over processors.

Energy efficiency: Figure 18.1 shows the analysis of energy efficiency versus flexibility trade-off for a number of common architectural styles [19]. As we can see in the graph, dedicated hardware is the most energy efficient but the least flexible. Microprocessors are the reverse. FPGAs sit between the two extremes. Because the architecture of CPUs is instruction based, the circuit of a CPU is designed to support different instructions (in fact, this accounts for the flexibility of CPUs). However, when the CPU is executing some instructions at a particular

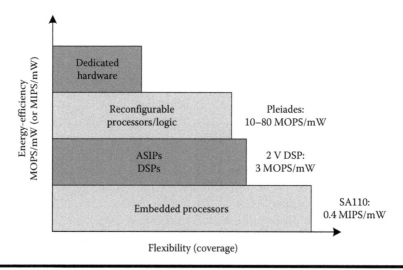

Figure 18.1 Energy efficiency versus flexibility. (From J.M. Rabaey, Silicon platforms for the next generation wireless systems—what role does reconfigurable hardware play? in *Proceedings of Programmable Logic and Applications (FPL'00)*, Villach, Austria, pp. 277–285, Sep. 2000.)

instant, the circuitry for other instructions becomes idle but is still consuming power, which is the source of low energy efficiency of instruction-based microprocessors. FPGAs avoid this kind of inefficiency by implementing circuits that are only dedicated for executing the tasks a user wants at a particular instant. Specifically, critical software loops to be executed by the instruction-based processor are identified and then re-implemented in the reconfigurable logic [13,19,25].

Flexibility: The flexibility of FPGAs is that they can be reconfigured to implement different circuits. Just like a processor switches different executables at different times, different bitstreams can be downloaded on an FPGA to implement different hardware functions. The reconfigurability of FPGAs, as in the Xilinx Virtex series FPGAs [35], emerges from the fact that FPGAs can be dynamically reconfigured at run-time. With this ability, a portion of the FPGA can be partially reconfigured without stopping the functionality of the unchanged sections, enabling the FPGA to fully and rapidly adapt to user needs. This also makes it easy for upgrading or changing the functionality of a part of the FPGA. Simply put, it is reconfiguration-on-demand. With FPGAs, a promising design paradigm for embedded systems is that a complete system can be built on an FPGA, known as System-on-Programmable-Chip (SoPC). Due to the programmability of FPGAs, such a SoPC-based system can be tailor-made to meet the requirements of different users.

The aforementioned features of FPGAs make it beneficial to incorporate FPGAs in systems that need high-performance demands and need to adapt to the changing workload requirements. For example, it suits ideally for a pervasive computing system where it needs to execute numerous context-aware tasks. Moreover, it is also suitable for systems where energy efficiency is a key concern, since executing computation-intensive tasks in hardware can be more energy efficient than executing in the processor [19,25]. Specifically, such a platform is good for a mobile sensor system that is battery operated (i.e., having limited energy) and needs to adapt to the changing environment and user needs, so as to carry out some on-demand tasks (e.g., surveillance by taking some pictures with feature extraction).

As an example, inside NASA's Mars rovers Opportunity and Spirit there are radiation-tolerant FPGA chips for controlling the rovers' motors, cameras, antennas, and other instruments [22]. The use of FPGAs, on top of the performance advantage, is to reduce the size and weight of the whole system. Specifically, the FPGA can be reconfigured to implement different hardware logic to carry out different tasks, instead of shipping all hardware logic of all tasks. Moreover, the reprogrammability of FPGAs allows for design changes and updates after the rovers are launched.

In order to execute different tasks on an FPGA, similar to the scheduler in an operating system (OS), we need a judicious mechanism to handle the FPGA reconfiguration process and resolve some system constraints. In this chapter, we focus on discussing the design of a hardware task scheduler, which is opposed to the software task scheduler in an OS kernel, in an FPGA-based embedded reconfigurable computing system with its applications in wireless sensor networks.

In a wireless sensor network, a considerable amount of sensor nodes work collaboratively in a distributed manner so as to accomplish a global task (e.g., environmental monitoring for tracking intruders). However, each node is typically power-constrained and has limited processing power and memory, and energy consumed by the communication part dominates that by the computation part [20,31]. In addition, since such a network is deployed in a massive scale of sensor nodes, a layered communication tree structure is formed. Thus, it is advantageous to introduce a more capable sensor node to process data locally at a lower layer of the sensor node structure and then extract interested information to send to the upper layer. In this chapter, we demonstrate how such a capable sensor node can be realized on an embedded reconfigurable computing platform, so as to increase the efficiency of a wireless sensor network.

18.2 FPGA Hardware Task Scheduling

18.2.1 Dynamic and Partial Reconfiguration

As Figure 18.2 illustrates, the structure of an FPGA is basically a two-dimensional (2D) array of Configurable Logic Blocks (CLBs) and routing elements. Logic circuits in FPGAs are implemented in the CLBs. In Xilinx Virtex-II series devices, each CLB has four identical slices. Moreover, each slice contains two Look-Up Tables (LUTs), which are used to implement logic functions, two flip-flops and other components. On the other hand, input/output blocks (IOBs) are used for the FPGA to interface to its pins or pads.

To implement a circuit, specifically, a bitstream is downloaded to the FPGA to configure the CLBs and routing elements. The configuration of CLBs and routing elements are stored in the memory of the FPGA. Thus, by downloading a bitstream to the FPGA, in effect, the configuration memory is overwritten to reflect a new configuration. Figure 18.2 depicts the configuration process.

The configuration memory of the FPGA is arranged in a rectangular array of bits. The array is segmented into frames, which are 1 bit wide and extend from the top of the array to the bottom. A frame is the smallest unit of reconfiguration. Frames are grouped together into larger units called columns. Figure 18.3 shows the CLB and IOB columns.

Dynamic and partial reconfiguration is the ability of an FPGA to reconfigure only a portion of the FPGA to realize a new function in the portion, without affecting the functionality of the unchanged section of the FPGA. The term "partial" means that we only produce a part of the whole

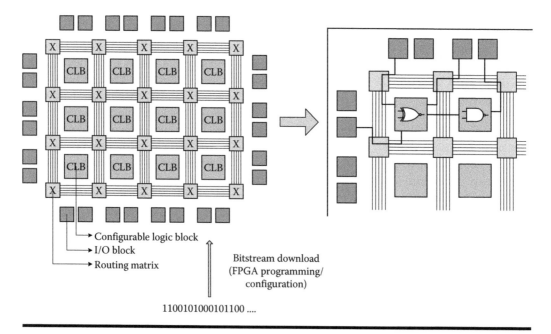

Figure 18.2   FPGA configuration using a bitstream.

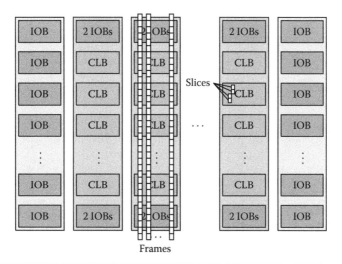

Figure 18.3   Configuration columns.

configuration memory of the FPGA, while "dynamic" means that we can reconfigure the intended portion on-demand. That is, whenever we need to realize a new logic circuit, we just reconfigure (portion of) the FPGA on-the-fly. Specifically, the affected frame columns are reconfigured on a column-by-column basis.

18.2.2 Multitasking

Figure 18.4 illustrates the idea of hardware task multitasking through dynamically partial reconfiguration. In the figure, a complex software task is partitioned into five subtasks. Any three (but not all) of the five subtasks can be implemented and then executed on the FPGA chip. In order to execute the complex task, first, the partial configurations (for configuring the corresponding portions of the FPGA) of the corresponding hardware tasks of the five subtasks are stored in some memory device. Then, according to the execution order (assumed that it is defined beforehand), different partial configurations are downloaded onto the FPGA for executing the subtasks.

18.2.3 The Need for a Scheduler

A reconfigurable computing system with multitasking support provides users with high-performance computing capability, while at the same time provides adaptability to changing workload requirements. However, we need a judicious mechanism to handle the FPGA reconfiguration process and resolve some system constraints. With reference to the illustrative example in Figure 18.4 again, we would require the system to finish executing the five subtasks as early as possible, yet without consuming more power.

In order to build a reconfigurable computing system with multitasking support, we need to address a number of issues [26]. Firstly, a scheduling model is needed, which defines the hardware tasks and the underlying reconfigurable architecture for executing the hardware tasks. Secondly, a runtime system is needed, which efficiently operates the system and resolves conflicts (e.g., resource

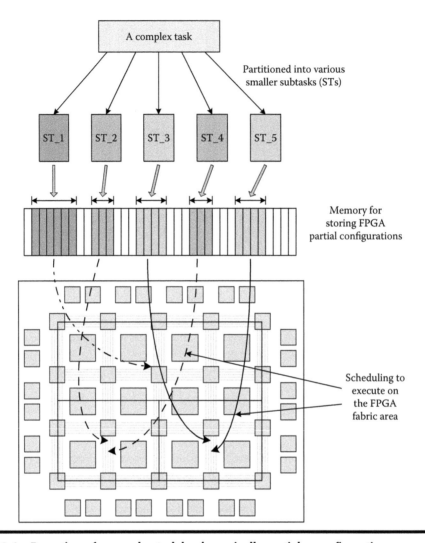

Figure 18.4 Execution of a complex task by dynamically partial reconfiguration.

conflicts) between hardware tasks. Specifically, a hardware task scheduler is needed. Hence, it is of high interest to build some system services supporting hardware task and resource management into the operating system, so as to increase the efficiency of the system. In this regard, Brebner has done some pioneering work on operating system support for reconfigurable hardware [3,4], in which swappable logic units (SLUs) are defined for executing tasks of application programs in the FPGA fabric.

18.2.4 Resource Placement Models

Figure 18.5 [30] shows the four commonly-used FPGA resource placement models, which define how the 2D CLB array of an FPGA (i.e., the FPGA fabric area) can be partitioned for multitasking of different hardware tasks. Based on the four models, a number of hardware task scheduling algorithms have been designed [1,9,11,12,26,30].

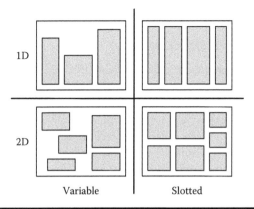

Figure 18.5   Different FPGA resource placement models. (From H. Walder, Operating system design for partially reconfigurable logic devices, PhD thesis, Swiss Federal Institute of Technology (ETH), Zurich, Switzerland, April 2005.)

In the four resource placement models, 2D-variable provides the greatest flexibility and therefore can result in maximum scheduling efficiency. On the other hand, 1D-slotted gives the least flexibility, but it is the most viable resource model that we can adopt in practice. The reason is that current development tools provide sufficient support only to the 1D-slotted resource model (primarily due to the fact the current FPGAs are configured on a column-by-column basis). We are advised to divide the FPGA area into partitions having the same height, as detailed in [34]. Thus, scheduling of algorithms using 2D resource models (e.g., the 2D scheduler in [26] chooses the smallest area among all rectangles that are large enough to accommodate a task) and merging of partitioned blocks [12] (i.e., two consecutive areas can be merged together) are unable or difficult to be implemented using the design flows in current development platform. For example, one of the noticeable challenges is to reroute input/output signals when a new area is created for executing hardware tasks. This is analogous to the case that when a house is inevitably to be built on a busy road, we need to build new roads to replace the interrupted segment of the road to maintain the normal traffic flow.

To apply the 2D-variable resource model, we might require future FPGA devices to support block-based reconfiguration (i.e., instead of reconfiguring a whole column spanning the full height of the FPGA device, we can just reconfigure a section of the column). This can reduce the size of the bitstream that is indeed required for a partial reconfiguration, and hence speed up the reconfiguration process. Moreover, we might need a built-in communication network such that when a block is reconfigured, the communication between other blocks is not affected [26]. As we would like to discuss the design of a hardware task scheduler with practical implementation in mind, in subsequent discussion we would focus on the 1D-slotted resource model.

On designing hardware task scheduling algorithms, besides practicability, we also need to consider scheduling efficiency. We need to consider the execution time and energy consumption of a set of hardware tasks, which is formed as a result of the scheduling process. For example, Khan and Vemuri [11] designed a battery-aware task scheduling algorithm for a battery-operated system. In their scheme, the FPGA area is divided into a number of fixed reconfigurable slots called configurable tiles. Then battery usage can be adjusted by varying the number of tiles used for scheduling tasks. However, they partitioned the FPGA area into a set of fixed tiles, which cannot adapt efficiently to the tasks of having different FPGA area requirements.

18.3 The Design of a Hardware Task Scheduler

18.3.1 Reconfigurable Architecture

We consider reconfigurable architecture that is feasible for implementation using current technology. Figure 18.6 shows the target reconfigurable architecture. As can be seen from the figure, a host CPU is connected to a reconfigurable device. The host CPU is used for configuring the reconfigurable device and transferring data to/from the device. The reconfigurable device is divided into two areas, static area and dynamic area. The dynamic area implements some reconfigurable modules (RMs) for executing some hardware tasks, while the static area implements a static module that servers as a bridge between the host CPU and the RMs. The function of the static area is fixed after the whole reconfigurable device is configured. On the other hand, each RM can be partially reconfigured, without affecting the rest of the device.

Logic circuits in an FPGA device are implemented in the CLBs, which are arranged as a 2D array inside the FPGA. We suppose that the dynamic area comprises $W \times H$ CLBs, where H is the height of the reconfigurable device and W, the width, is variant to the size of the dynamic area. On the other hand, we assume that the size of the smallest RM is of $W_{MIN} \times H$. In practical implementation, the height of each RM spans the full height of the device [34], and W_{MIN} depends on the hardware task that requires the smallest number of CLBs. Then, the maximum number of RMs is:

$$N_{MAX} = \frac{W}{W_{MIN}} \tag{18.1}$$

Here, we assume that W is a multiple of W_{MIN}. Figure 18.7 shows some of the possible partitioning strategies of the dynamic area for the case where $N_{MAX} = 8$. As mentioned earlier, because we consider practical implementation of the reconfigurable architecture, we do not allow merging/splitting of RMs, which is the reason why the partitioning of the dynamic area needs to be explicitly

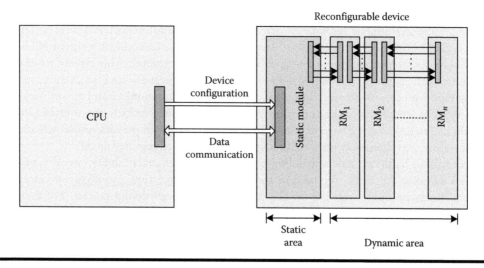

Figure 18.6 Target reconfigurable architecture.

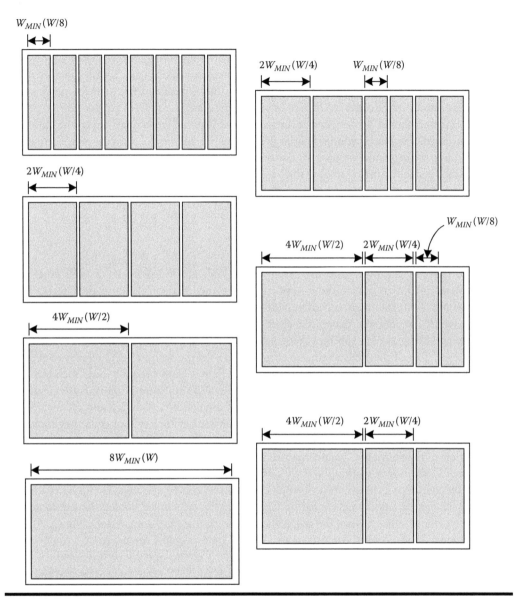

Figure 18.7 Some possible partitioning strategies of the dynamic area for the case where $N_{MAX} = 8$.

defined. This also follows that, to change from one partitioning strategy to another, the whole dynamic area needs to be reconfigured.

Our view on allowing different partitioning strategies at run-time, instead of a particular one after the system starts, is that the system needs to adapt to the changing workloads, which are characterized by different computational complexities of tasks. In general, a more complex task needs more CLBs to implement, and hence needs an RM of wider width. Analogous to a cluster computing environment where a number of parallel programs ask for different number of machines for execution, in the target reconfigurable architecture, a number of tasks ask for execution in the RMs of different widths.

On the other hand, it should be emphasized that Figure 18.7 only shows a particular subset of all the possible partitioning strategies for the case where $N_{MAX} = 8$. For the sake of practical implementation in an embedded system, only a set of partitioning strategies as in Figure 18.7 would be implemented. The reason is that it needs space in a memory device (e.g., a flash memory chip) to store the FPGA device configuration aforementioned files for each of the partitioning strategy, whose size is typically in the order of 100 kB. In view of the reasons, the set of partitioning strategies to be implemented should be designed to be as generic as possible, so as to cope with different workload patterns. For instance, one design criterion is that, for every hardware task, there must be a partitioning strategy implemented for the execution of the hardware task. Another criterion is that, if during a particular period of time the hardware tasks in the system only require one kind of RM to execute on, then there must be a partitioning strategy implemented such that the dynamic area can accommodate the maximum number of such RM.

18.3.2 Task Definition

We assume that the hardware tasks executing in the RMs are independent to each other and cannot be preempted, that is, cannot be stopped and resumed later on the same or different RM. A hardware task T_i in the target reconfigurable architecture is characterized by two parameters: (1) f_i^{MAX}, which specifies the maximum clock frequency, in MHz, the hardware task can be executed at; and (2) w_i, which specifies the width of the reconfigurable module needed (it should be remembered that all the RMs share the same height). In the website of Opencores.org [16], there are a number of open source cores implementing different hardware functions. Such cores are in fact the hardware tasks mentioned in this paper. In the web site, there is information about the logic resources required when a core is implemented on a target FPGA device and the maximum frequency the core can be clocked at.* From this kind of information, we can get a rough idea on the values of the two parameters of a hardware task.

In addition to the above two parameters, each hardware task has execution time t_i and energy consumption E_i when it is working at f_i^{MAX} MHz for a predefined amount of work. t_i can be found out by considering the amount of work to be carried out by the task. E_i can be estimated by adopting the energy model developed by Choi et al. [5]. Moreover, when a hardware task is scheduled to execute in the system, the task has an arrival time A_i and completion time C_i. All the above-mentioned attributes and parameters of a task T_i are stored in a vector $\mathbf{v_i}$.

In the system, a special type of hardware task called NOP (no-operation) is defined for each RM of different widths. When an RM is configured with an NOP task, we assume that the RM will consume negligible energy. The use of NOP tasks will be elaborated in Section 18.3.4.

Formally, the following two functions are defined to extract the estimated execution time and energy consumption information of each task T_i executing at different frequencies:

1. $g^{time}(\mathbf{v_i}, f)$: It returns the estimated time needed by task T_i to finish execution. f specifies the frequency the hardware task is clocked at. It is defined as:

$$g^{time}(\mathbf{v_i}, f) = \frac{f_i^{MAX}}{f} \cdot t_i \qquad (18.2)$$

---

* For example, an excerpt of the description of a DES IP core in CBC mode (available at http://www.opencores. org/projects.cgi/web/des/overview):

```
Xilinx Spartan IIe-50: >100 MHz 1339 LUTs (87% device utilization)
```

Table 18.1   Comparison of Software and Hardware Tasks

Feature	Software Task	Hardware Task
Development language	Assembly or high-level language like C/C++	Hardware description language like VHDL
Building block	A number of instructions	A number of CLBs
Execution mode	Sequential execution of instructions	Parallel execution of CLBs
Activation	Activated by setting the program counter of the CPU to address of the program entry point	Activated by resetting the circuit

2. $g^{energy}(\mathbf{v_i}, f)$: It returns the estimated energy consumption by task T_i during its execution. When applying the energy model developed by Choi et al. [5], it is shown that the energy consumption is directly proportional to the working frequency. Hence, the function is defined as:

$$g^{energy}(\mathbf{v_i}, f) = \frac{f}{f_i^{MAX}} \cdot E_i \tag{18.3}$$

Now, we have discussed the definition of a hardware task. We can see that a hardware task has different characteristics than a software task. Table 18.1 summarizes the differences between the two.

18.3.3 Runtime System

Figure 18.8 depicts the block diagram of the target runtime system. Specifically, a user's application program is partitioned into two types of tasks, namely, software task to be executed by the instruction set processor and hardware task to be executed by the reconfigurable device. After the application program is submitted for execution, the corresponding software and hardware tasks are diverted to the respective queues of the software task and hardware task schedulers. In subsequent discussion, we focus on the design of a scheduling algorithm for the hardware task scheduler. Moreover, the hardware/software partitioning of an application is not the concern in this chapter. The long-term goals of the runtime system are:

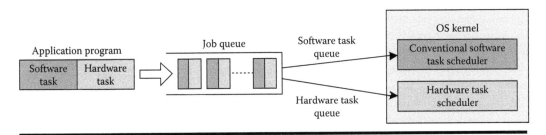

Figure 18.8   Target runtime system.

1. To minimize the energy consumption of the set of hardware tasks:

$$Minimize \left(\sum g^{energy}(\mathbf{v_i}, f_i) \right), \qquad (18.4)$$

where f_i is the actual working frequency of task t_i when it is scheduled on the dynamic area.

2. To minimize the total execution time of the set of hardware tasks:

$$Minimize \left(\max_i(C_i) - \min_i(A_i) \right) \qquad (18.5)$$

When scheduling tasks to execute in the RMs, there is one cost that needs to be considered— the cost of (re)configuring the RMs, which depends on the width of individual RM. It also needs to consider the cost of changing the partitioning strategy of the dynamic area, where the whole reconfigurable device needs to be reconfigured. However, for our proposed scheduling algorithm, we consider scheduling tasks having execution time, which is longer than the reconfiguration time of the RMs and/or the reconfigurable device such that we can neglect the cost of reconfiguration.

18.3.4 Frequency Adaptive Hardware Task Scheduler

18.3.4.1 Overview

Figure 18.9 describes the flowchart of our proposed hardware task scheduling algorithm, namely, *ECfEE* (*E*nergy-*E*fficient and *C*omputation-*E*fficient algorithm with *f*requency adaptation). The algorithm is divided into two parts. In the first part, it starts by choosing a partitioning strategy with the goal of executing a maximum number of hardware tasks. Then, it finds a working frequency for all the tasks selected to be executed in the chosen partitioning strategy, with the goal of minimizing energy consumption. Choosing a suitable working frequency is the key step of *ECfEE* for reducing energy consumption, because the tasks will run at a slower speed. Moreover, as we will show later, it will also shorten the total completion time of all tasks.

In the second part, it mainly deals with the case when a task finishes execution. In this case, if not all the tasks have finished execution, the scheduler needs to either find a new task to execute on the unoccupied RM or schedule an NOP task on the RM; otherwise, the reconfigurable device will be reconfigured to execute another set of new tasks.

18.3.4.2 Details of the Scheduler

Algorithms 18.1 through 18.3 describe the pseudo-code of *ECfEE*. The algorithm works on a hardware task queue, T_{queue}, whose size is changed dynamically, that is, hardware tasks are continuously being injected into the queue. We assume that the queue is implemented using some data structure such as a linked list, and hence the algorithm takes as input the pointer pointing to the head of the queue, T_{queue_head}. The algorithm considers the first q_{range} hardware tasks when it carries out the scheduling procedure. To prevent some tasks from starving, we introduce another parameter, s_i, to each task T_i. s_i is set to zero initially, and it is incremented by one, if $T_i \in T_{queue}[1..q_{range}]$ was not scheduled during a round of choosing candidate tasks for scheduling, that is, Steps 1–6 of Algorithm 18.2.

In Step 3 of Algorithm 18.1, F denotes the set of frequencies that *all* the RMs in the reconfigurable device may be clocked at. All RMs sharing the same working frequency is a consideration

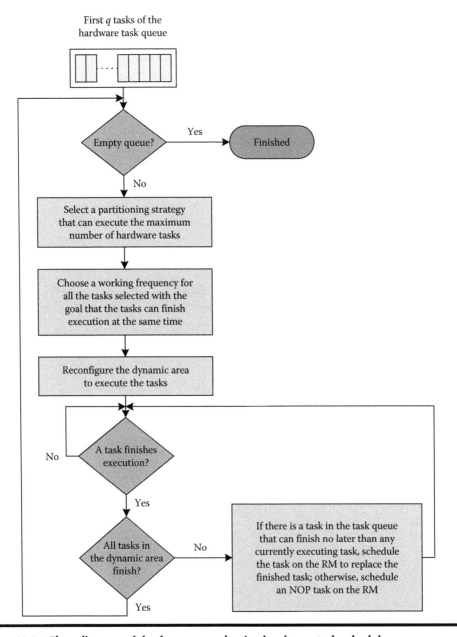

Figure 18.9   Flow diagram of the frequency adaptive hardware task scheduler.

for practical implementation, as recommended by [34]. Moreover, the fact that the available frequencies are stepped by five is also a practical consideration. The primary reason is to reduce the memory space required for storing the configuration files of the reconfigurable device. In *SelectWorkingFrequency*() of Algorithm 18.3, we choose a working frequency for the scheduled tasks such that all the tasks can finish execution all together, with the minimal average difference of execution time. The reason for doing so is that it can reduce the amount of executing NOP tasks in Steps 13–24 of Algorithm 18.1. Since the execution of NOP tasks (instead of normal hardware

---

Algorithm 18.1 ECfEE

Schedule(T_{queue_head})

1:    $q_{range} \leftarrow \alpha \cdot N_{MAX}$ /* α is a constant for setting the value of q_{range} */
2:    $s_{threshold} \leftarrow \beta \cdot N_{MAX}$ /* β is a constant for setting the value of $s_{threshold}$ */
3:    $F \leftarrow \{f_{MIN}, f_{MIN} + 5, f_{MIN} + 10, ..., f_{MAX}\}$
4:    $P \leftarrow GeneratePartitioningStrategies(N_{MAX})$
5:    $p_{current} \leftarrow null$ /* current partitioning strategy used */
6:    $T_{scheduled} \leftarrow null$ /* set of scheduled tasks */
7:    $T_{not\_scheduled} \leftarrow null$ /* set of tasks which are considered but not scheduled */
8:    $f_{working} \leftarrow null$ /* selected working frequency */
9:    **while** (TRUE) **do**
10:      $(p_{current}, T_{scheduled}, T_{not\_scheduled}) \leftarrow SelectPStrategyTasks(T_{queue\_head}, q_{range}, s_{threshold},$
       $P, T_{not\_scheduled})$
11:      $f_{working} \leftarrow SelectWorkingFrequency(T_{scheduled}, F)$
12:      **while** (not all the tasks in the RMs have finished execution) **do**
13:        **if** a task has finished execution in RM_j **then**
14:          **if** there is some other task still executing **then**
15:            **if** $\exists$ task $T_i \in T_{queue}[1..q_{range}]$ and T_i will finish execution no later than any current executing task **then**
16:            Schedule T_i to execute in RM_j. If more than one such T_i exists, select the one which is the nearest from T_{queue_head}.
17:            **if** $T_i \in T_{not\_scheduled}$ **then**
18:              $T_{not\_scheduled} \leftarrow T_{not\_scheduled} - T_i$
19:            **end if**
20:          **else**
21:            Schedule NOP task to execute in RM_j.
22:          **end if**
23:        **end if**
24:        **end if**
25:      **end while**
26: **end while**

---

tasks), is considered a waste of resources, by reducing the amount of executing NOP tasks, it can better use the resources of RMs, and hence better use the energy of the system.

In Steps 13–24 of Algorithm 18.1, after a task has finished execution, we do not choose any task that can fill the RM. There are two reasons for that. Firstly, we want to avoid the loop of scheduling NOP task when there is no task suitable for execution under the current partitioning strategy but we need to wait for the executing tasks to finish. Secondly, after all the tasks under the current partitioning strategy finish their execution, the system can consider another partitioning strategy so as to schedule the maximum possible number of tasks for execution.

The dominating computation of *ECfEE* is in the *SelectPStrategyTasks*() function, where for each partitioning strategy, q_{range} tasks have to be looked up so as to choose a suitable partitioning strategy. Thus, the complexity of *ECfEE* is $O(p \cdot q_{range})$, where p is the maximum number of partitions in the dynamic area.

Algorithm 18.2 ECfEE: Selection of Partitioning Strategy and Tasks for Scheduling

SelectPStrategyTasks(T_{queue_head}, q_{range}, $s_{threshold}$, P, $T_{not\_scheduled}$)

1: **if** $\exists\, s_i \geq s_{threshold}$ and $T_i \in T_{not\_scheduled}$ **then**
2:    $s_j \leftarrow \max(s_i | T_i \in T_{not\_scheduled})$
3:    Select $p_{current} \in P$ which allows the task having s_j and the maximum number of other tasks $\in T_{queue}[1..q_{range}]$ to execute in the RMs.
4: **else**
5:    Select $p_{current} \in P$ which can accommodate the maximum number of tasks within $T_{queue}[1..q_{range}]$.
6: **end if**
7: Denote the set of tasks which contributes to the selection of $p_{current}$ as $T_{scheduled}$.
8: $T_{not\_scheduled} \leftarrow T_{not\_scheduled} \bigcup (T_{queue}[1..q_{range}] - T_{scheduled})$
9: $T_{queue} \leftarrow T_{queue} - T_{scheduled}$
10: $s_j \leftarrow s_j + 1$, where $T_j \in T_{not\_scheduled}$
11: **Outputs:**

  - $p_{current}$: Selected partitioning strategy.
  - $T_{scheduled}$: Set of scheduled tasks.
  - $T_{not\_scheduled}$: Set of tasks which are considered but not scheduled.

Algorithm 18.3 ECfEE: Selection of Working Frequency for the Tasks Selected for Scheduling

SelectWorkingFrequency($T_{scheduled}$, F)

1: $f_{scheduled}^{MAX} \leftarrow \min(f_i^{MAX} | T_i \in T_{scheduled})$
2: For each $f \in F$ and $f_{MIN} \leq f \leq f_{scheduled}^{MAX}$, calculate:

  - $G_{average}^{time} = \frac{1}{|T_{scheduled}|} \cdot \sum g^{time}(\mathbf{v_i}, f)$, where $T_i \in T_{scheduled}$
  - $\sum (G_{average}^{time} - g^{time}(\mathbf{v_i}, f))^2$, where $T_i \in T_{scheduled}$

3: Denote the f which contributes to the smallest value of $\sum (G_{average}^{time} - g^{time}(\mathbf{v_i}, f))^2$ in the previous step as $f_{working}$.
4: **Output:**

  - $f_{working}$: Selected working frequency.

18.3.5 Simulation Results

18.3.5.1 Simulation Setting

Because currently there is no benchmark package for a reconfigurable system, similar to [12,26], the performance of the proposed scheduling algorithm is studied through simulation. The simulation program is written in C. We construct T_{queue} by randomly generating a set of tasks of the following parameters:

1. *Task width*: We set $N_{MAX} = 8$ and $W_{MIN} = 1$, meaning that the task width $w_i \in [1, 2, 4, 8]$.
2. *Executing frequency*: We set $F_{MIN} = 20$ and $F_{MAX} = 50$ such that $f_i^{MAX} \in [20..50]$ and $f_{working} \in [20, 25, 30, 35, 40, 45, 50]$.
3. *Execution time*: $t_i \in [500, 5000]$ ms.
4. *Energy consumption*: $E_i \in [100, 1200]$ mJ and is proportional to w_i.

18.3.5.2 Trade Off between Execution Time and Energy Consumption

In the simulations, we set $\beta = 2$, which means that $s_{threshold} = 16$. Figure 18.10a and c show, respectively, the total execution time and total energy consumed in executing 1000 randomly generated tasks. As can be seen from the figures, for q_{range} smaller than 15, the total execution time and total energy consumed decrease with q_{range}. The reason for this is that by using a larger q_{range}, the scheduler can choose a better partitioning strategy to schedule more tasks for better utilization of the RMs. This also means that the tasks are likely to execute at a frequency lower than f_i^{MAX} and hence energy is saved. On the other hand, for q_{range} larger than 15, an increase in q_{range} causes the total execution time to increase and the total energy consumed to decrease. The cause of this is due to the fact that by further increasing q_{range}, the scheduler can further schedule more tasks in a partitioning strategy. However, by considering more tasks, the $f_{scheduled}^{MAX}$ chosen

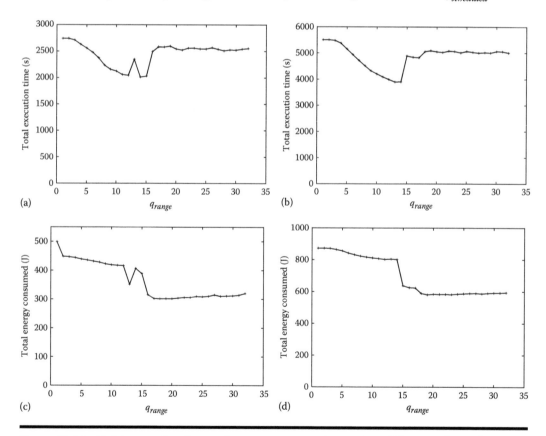

Figure 18.10   Total execution time and energy consumption against different values of q_{range} in ECfEE. (a) Execution time—1000 tasks. (b) Execution time—2000 tasks. (c) Energy consumption—1000 tasks. (d) Energy consumption—2000 tasks.

(*SelectWorkingFrequency*(), Algorithm 18.3) will be of a lower value. As a result, the $f_{working}$ chosen will also be of a lower value, causing the tasks to take more time to finish, but save more energy. Thus, from Figure 18.10a and c, we can see that $q_{range} = 15$ is a good trade-off between the total execution time and total energy consumed. The same value of $q_{range} = 15$ is observed in the execution of 2000 tasks, as shown in Figure 18.10b and d.

18.3.5.3 Comparison with Existing Algorithms

Figure 18.11 shows an illustrative example of *ECfEE*. The configurations of the seven hardware tasks (Figure 18.11a) are adapted from [26]. The parameter a_i is the time when the hardware task is

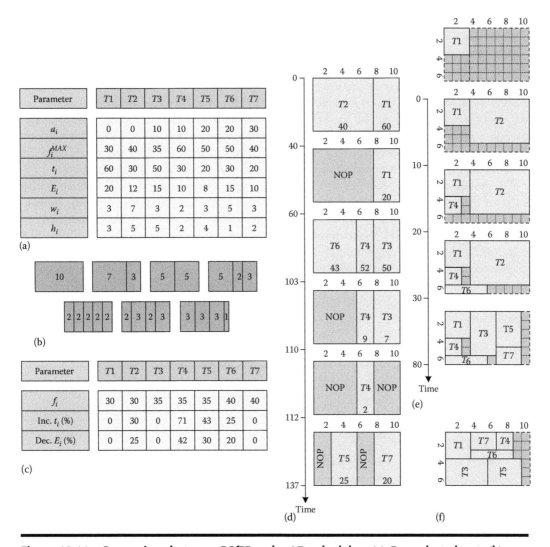

Figure 18.11 Comparison between ECfEE and a 2D scheduler. (a) Example task set, (b) partitioning strategies, (c) change of execution time and energy consumption, (d) ECfEE, (e) 2D scheduling, and (f) 2D defragmentation. (From Steiger C. et al., *IEEE Trans. Comput.*, 53(11), 1393, Nov. 2004.)

sent to the hardware task queue. Figure 18.11b illustrates the partitioning strategies of the dynamic area of the FPGA, according to which the scheduler will perform the allocation of hardware tasks to the RMs.

In the example, we set q equal 4, meaning that we only consider the first four tasks in the hardware task queue during each scheduling round. At the beginning, that is, time step 0, only tasks $T1$ and $T2$ are in the hardware task queue, and therefore they are scheduled on the dynamic area. Originally, the execution times of $T1$ and $T2$ are 60 and 30, respectively, when they are executed at frequencies 30 and 40 MHz, respectively. Thus, the difference of execution times is 30, which means that we need to schedule an NOP task on the RM occupied by $T2$ before for a time period of 30. To reduce the amount of executing the NOP task, the scheduler chooses a common working frequency of 30 MHz for both tasks. Figure 18.11c shows the percentages of increased execution time and decreased energy consumption. Similarly, Figure 18.11c and d show the detail of scheduling the rest of the hardware tasks.

As a comparison, Figure 18.11e illustrates the scheduling steps of a 2D scheduler proposed in [26]. To schedule a hardware task, the scheduler will choose the smallest area among all rectangles that are large enough to accommodate the task. We can see from the scheduling steps that one issue arises in 2D scheduling is the fragmentation problem. However, there are some defragmentation techniques [10,23,29] that we can use to make the hardware tasks more compactly packed in the dynamic area, as shown in Figure 18.11f. Because of having more flexibility, a 2D scheduler can produce a more efficient scheduling result than a 1D scheduler. Despite this, in the practical aspect, 1D schedulers are the viable approach due to the limitations of current development tools.

The problem of placing reconfigurable modules onto the dynamic area of the 1D-slotted model as shown in Figure 18.5 is similar to the 1D bin-packing problem. Best-fit and first-fit are two well-known online algorithms for the 1D bin-packing problem, and they have been considered for hardware task placement [1]. A variant of best-fit has been adapted to compare the performance with *ECfEE*. We name this algorithm *BF* and the algorithm is outlined in Algorithm 18.4. On the other hand, it should be noted that *BF* is effectively *ECfEE* with $q_{range} = 1$. To choose a suitable partitioning strategy, *BF* only considers the head task of the task queue for each partitioning strategy. Thus, the complexity of *BF* is $O(p)$ where p is the maximum number of partitions in the dynamic area.

Algorithm 18.4 BF

Schedule(T_{queue_head})

1: $F \leftarrow \{f_{MIN}, f_{MIN} + 5, f_{MIN} + 10, ..., f_{MAX}\}$

2: $P = GeneratePartitioningStrategies(N_{MAX})$

3: **while** (TRUE) **do**

4:   Select $p_{current} \in P$ where the partitioning strategy $p_{current}$ allows the head task of T_{queue} to execute and contains the maximum number of RMs.

5:   With reference to the f_i^{MAX} of the head task, select the largest possible $f_{working} \in F$.

6:   Schedule the head task to execute and remove it from T_{queue}.

7:   For any free RM in current partitioning strategy, if the head task of T_{queue} can execute on it, schedule the head task to execute and remove the task from T_{queue}. This step is repeated until no suitable task is found.

8:   Run Steps 4–7 when a task finishes execution.

9: **end while**

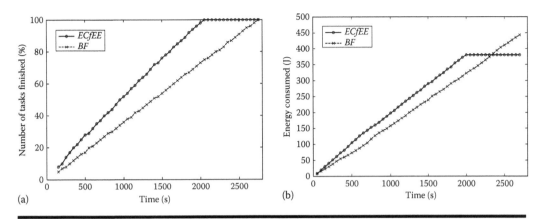

Figure 18.12 **Performance comparison between *ECfEE* and *BF*. (a) Task completion and (b) energy consumption.**

In Figure 18.12, the performance of *ECfEE* is compared with *BF*. For the comparison, 1000 tasks are executed. From the figures, we can see that, when compared to *BF*, *ECfEE* can significantly reduce the energy consumption and execution time. Specifically, *ECfEE* can reduce execution time and energy consumption by 26% and 14%, respectively. From these results, it is interesting to see that by exploiting adaptive working frequency of hardware tasks, we can not only reduce the energy consumption, but also reduce the execution time of the tasks.

18.4 Practical Illustration: Mobile Robots in a Wireless Sensor Network

18.4.1 SoPC-Based Mobile Robots

18.4.1.1 Use of Mobile Robots in a Wireless Sensor Network

While wireless sensor networks are found useful in many applications, there are still a number of challenges that remain to be should in order to make such networks more viable in practice. The challenges, which are mainly due to the inherent properties of sensor nodes and the deployment nature of wireless sensor networks, include issues related to security, power management, failure detection, calibration, etc. [14] In order to meet the challenges or mitigate the situation, researchers have proposed the incorporation of mobile robots into wireless sensor networks [7,14,17,33].

As illustrated in Figure 18.13, there are two kinds of sensor nodes: (1) low-power sensor nodes which are deployed in a large amount and are used to take up some sensing jobs and to detect some events; and (2) mobile robots which can move around a region of the wireless sensor network locally. When a mobile robot wanders in the wireless sensor network, it can detect the health of the specific region of interest of the network, by visual inspection of sensor nodes or monitoring the traffic within the network. Specifically, it can detect whether some nodes have become failure and need calibration. Additionally, the mobile robot can also retrieve information from other sensor nodes and then perform post-processing. Using such localized information processing, the robot can react to some events quickly. This is because the robot can operate in an autonomous way, which reduces the communication overhead incurred in human intervention.

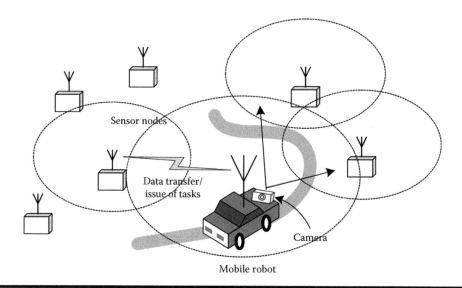

Figure 18.13   Mobile robot in a wireless sensor network.

LaMarca et al. [14] demonstrated that mobile robots have the potential to greatly increase the feasibility of practical wireless sensor networks, including sustaining energy resources of sensor nodes, automating sensor calibration, and detecting of senor failure and inappropriate deployment. On the other hand, Dantu et al. [7] designed a robot platform, called Robomote, which functions as a single mobile node in a mobile sensor network.

From the aforementioned discussion, we see that mobile robots can play an important role to increase the efficiency of a wireless sensor network. Because of the nice features of FPGAs (i.e., computation efficiency, energy efficiency and flexibility), we believe that applying reconfigurable computing technique to mobile robots using FPGAs can further increase the efficiency of the wireless sensor network, by offloading computation-intensive tasks from the processor to the FPGA fabric.

18.4.1.2 Prototype Design of a SoPC-Based Mobile Robot

Figure 18.14 shows the block diagram of our prototype of a mobile robot. The prototype consists of two FPGAs: the master FPGA is for implementing a MicroBlaze-based system, while the slave FPGA is for carrying out some hardware tasks.

MicroBlaze is a 32 bit soft processor core optimized for FPGA implementation. Figure 18.15 shows the block diagram of a typical MicroBlaze system. There is μClinux [28] port for MicroBlaze systems [15]. The entire system is a system-on-programmable-chip (SoPC) system. In the master FPGA, there are two servo motors attached to it to control the movement of the robot. In addition, a Cricket senor node [18] is also attached via a UART interface. The sensor node is installed with TinyOS [27], which is a lightweight operating system for sensor nodes and provides a programming framework for developing applications in sensor nodes. With the Cricket sensor node, the mobile robot can communicate with other sensor nodes in the wireless sensor network. Another application of the Cricket sensor node is the on-demand remote reconfiguration of the slave FPGA via the wireless sensor network. Specifically, a user can transfer the bitstream file of a hardware task via the wireless sensor network so as to reconfigure the slave FPGA for realizing new functions.

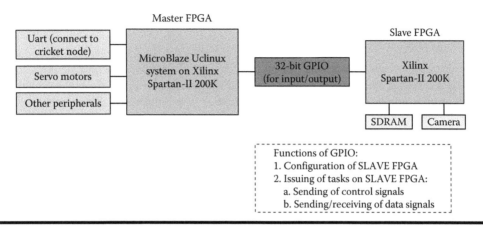

Figure 18.14 Block diagram of the mobile robot prototype.

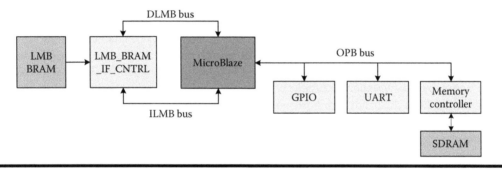

Figure 18.15 Block diagram of a typical MicroBlaze embedded system. (From xilinx, Inc., EDK MicroBlaze Tutorial, V2.1 edn., March 2004.)

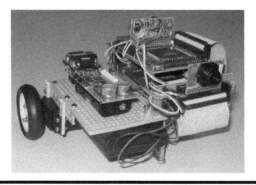

Figure 18.16 Mobile robot prototype.

A camera is attached to the slave FPGA so that the robot can capture some images of the monitoring environment. The camera used is a COMedia C3038 camera module [6], which can capture a color image of up to 356 × 292 pixels. Moreover, some SDRAM memory is attached to the slave FPGA for storing image data so that the image can be processed quickly.

Figure 18.16 shows the picture of the prototype of the mobile robot.

18.4.2 Image Processing Example

18.4.2.1 Application Scenario

Suppose in a wireless sensor network as shown in Figure 18.17, the leaf nodes continuously capture some images and send them to upper layers, and finally to the base station to the user side. Suppose the application is simply a pattern recognition application where the user wants to track the shape of objects in the monitoring area. If all the images captured are sent to the base station, then a large traffic volume and hence a large transmission delay will be induced, because the size of a typical image is in the order of 100 kB. However, in a wireless sensor node, energy consumed by the communication part dominates that by the computation part, and more importantly a sensor node is energy constrained. For example, in a wireless sensor node produced by Rockwell, Inc., the energy expended in transmitting 1 bit is around 2000 times of that for executing one instruction [20,31].

Therefore, it is beneficial to reduce the size of data sent from the leaf nodes to the base station, so as to reduce bandwidth and hence energy consumption. Consequently, it is advantageous to process data locally to extract interested information and then send to the base station.

To process image data locally, we can apply reconfigurable computing using FPGAs at a layer higher than the leaf nodes, that is, the square nodes in Figure 18.17. These square nodes are in fact mobile robots we mentioned before. Then, the images captured from the leaf nodes can be processed locally using FPGA hardware acceleration at the square nodes. In this way, from the square nodes to the base station, the traffic volume for an image is only in the order of 1 or 10 kB. This greatly reduces the bandwidth burden on the sensor network, and reduces the computation burden at the user side. In this application scenario, one major work that a square node needs to handle is the square node receives a batch of jobs (e.g., edge detection and AES/DES encryption) from the leaf nodes and it needs to decide when and how to process the jobs, since different jobs

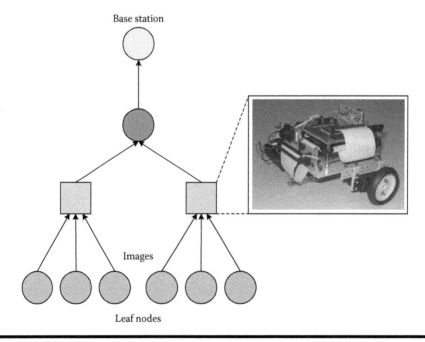

Figure 18.17 Image processing example in a wireless sensor network.

have different execution times and need different resources. In essence, a hardware task scheduler is needed to handle the resource management problem.

The target application in the wireless sensor network is edge detection, a well-known image processing technique. The edge detection process is a bottleneck in the application scenario, and therefore the process is worth being carried out in an FPGA so as to increase the system performance. FPGAs have long been used in implementing various image processing algorithms not only because of the computation-intensive nature of such algorithms, but also because of the well-matched between data type of image data and computational structure of FPGAs [2,21,32]. Specifically, image data usually consist of a 2D array of pixels, and FPGAs consist of a 2D array of CLBs. If a set of CLBs (e.g., 4 CLBs) is responsible for handling a pixel, and each set of CLBs directly communicates with its neighbor sets, then there is a high parallelism in processing the image data, where the operations are repetitive in nature.

The square sensor node itself will constantly capture some images to perform edge detection. In addition, it will also receive images from leaf nodes to perform edge detection. The resulted images after edge detection will then be sent to upper layers and finally to the base station. To enhance data security, the resulted images after edge detection can be encrypted using the DES encryption algorithm before they are sent out. Thus, the following hardware tasks are defined for the square sensor node:

1. Capture an image and save it to the SDRAM.
2. Perform edge detection.
3. Perform DES encryption.

The partitioning of the slave FPGA for executing the aforementioned three hardware tasks is shown in Figure 18.18. Despite that the chosen FPGA device can only allow us to execute a quite limited number of hardware tasks at the same time, which does not allow the full utilization of the proposed scheduler, we would like to demonstrate the potential of using such a scheduler.

18.4.2.2 Implementation Results

Using the TIFF format and a resolution of 356×292 pixels, a captured image is of size 310 kB while the resulted edge detection image is of size 57 kB. There is about 82% reduction in file

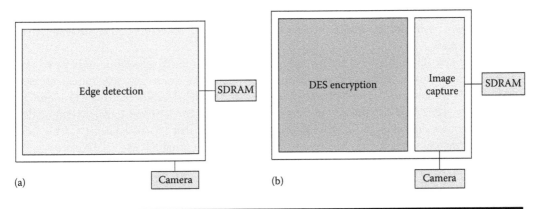

Figure 18.18 Partitioning of the slave FPGA for executing different hardware tasks. (a) One partition and (b) two partitions.

Table 18.2 Speedup Factors When Executing Tasks in the Slave FPGA

Task	Speedup Factor
Image capture	60
DES encryption	40
Edge detection	22

transfer size when we transmit the resulted edge detection image instead of the original image, which can greatly reduce the bandwidth and energy consumption.

On the other hand, Table 18.2 gives the speedup factors when executing the aforementioned tasks using the slave FPGA, on which the DES engine runs at 33 MHz while the edge detection algorithm runs at 20 MHz, instead of using the 48 MHz MicroBlaze microprocessor. The speedup factor is defined as follows:

$$Speedup = \frac{execution\_time(MicroBlaze)}{execution\_time(FPGA)} \tag{18.6}$$

We have also compared the execution of the aforementioned three hardware tasks in an execution environment with and without multitasking support. Without multitasking support, the three tasks are executed alternately in the FPGA. With multitasking support, DES encryption and image capture are performed concurrently in the FPGA. It is found that, with multitasking support, the total execution time of the three tasks is reduced by about 15%.

18.5 Conclusions and Future Work

Reconfigurable computing using FPGAs is becoming more and more popular nowadays, from the small-scale embedded systems to the large-scale high-performance computing systems. Because of the software-like flexibility and high-performance capability of FPGAs, we might expect that, in such high-performance computing systems, more and more hardware tasks would coexist with software tasks. Inevitably, we would incorporate hardware task scheduling into the traditional scheduling process.

Hardware task scheduling is radically different from software task scheduling because of the different nature between hardware and software tasks. As for hardware task scheduling, we can model the FPGA area as a 1D or 2D resource placement model. Despite that the 1D model is the practically viable approach due to the constraints of current development tools, the 2D model provides the most flexibility and hence efficiency. We believe that 2D scheduling will become feasible in forthcoming years, and that reconfigurable computing using FPGAs will be a promising approach for high-performance computing, especially in mobile and ubiquitous computing.

Currently, in our target runtime system, the software task scheduler and the hardware task scheduler are independent of each other. Specifically, two separate job queues are created for executing software and hardware tasks of application programs. In our future work, we would like to combine the two job queues for more efficient execution of application programs. Like the work in [8] and the BORPH (Berkeley Os for ReProgrammable Hardware) project [24], we are targeting on developing a hardware/software co-scheduling scheme. However, we further want to strike a proper balance between computation time and energy consumption when developing the co-scheduling scheme. In addition, we would take the reconfiguration time of reconfigurable modules into consideration. This allows for developing a more realistic hardware/software co-scheduler, which can decide whether executing a task on the FPGA fabric is more efficient than executing on the processor, when considering also the reconfiguration overhead of using the FPGA fabric.

References

1. K. Bazargan, R. Kastner, and M. Sarrafzadeh, Fast template placement for reconfigurable computing systems, *IEEE Design and Test of Computers*, 17(1), 68–83, Jan 2000.
2. R. A. C. Bianchi and A. H. R. Costa, Implementing computer vision algorithms in hardware: An FPGA/VHDL-based vision system for a mobile robot, *Proceedings of RoboCup 2001*, Seattle, WA, LNAI 2337, pp. 281–286, 2002.
3. G. Brebner, A virtual hardware operating system for the Xilinx XC6200, *Proceedings of Field Programmable Logic and Applications (FPL'96)*, Darmstadt, Germany, pp. 327–336, Sep 1996.
4. G. Brebner, The swappable logic unit: A paradigm for virtual hardware, *Proceedings of 5th IEEE Symposium on FPGA-Based Custom Computing Machines (FCCM'97)*, Napa Valley, CA, pp. 77–86, Apr 1997.
5. S. Choi, J.-W. Jang, S. Mohanty, and V. K. Prasanna, Domain-specific modeling for rapid system-wide energy estimation of reconfigurable architectures, *Proceedings. Engineering of Reconfigurable Systems and Algorithms (ERSA'02)*, Las Vegas, NV, Jun 2002.
6. COMedia C3038 Camera Module Datasheet, http://home.pacific.net.hk/~comedia/c3038.pdf, Oct 2006.
7. K. Dantu, M. Rahimi, H. Shah, S. Babel, A. Dhariwal, and G. S. Sukhatme, Robomote: Enabling mobility in sensor networks, *Proceedings. Information Processing in Sensor Networks*, Los Angeles, CA, pp. 404–409, Apr 2005.
8. Q. Deng, S. Wei, H. Xu, Y. Han, and G. Yu, A reconfigurable RTOS with HW/SW co-scheduling for SOPC, *Proceedings of Second International Conference on Embedded Software and Systems (ICESS'05)*, Xian, China, Dec 2005.
9. O. Diessel, H. ElGindy, M. Middendorf, H. Schmeck, and B. Schmidt, Dynamic scheduling of tasks on partially reconfigurable FPGAs, *Proceedings of IEE Computers and Digital Techniques*, 147, 181–188, May 2000.
10. M. G. Gericota, G. R. Alves, M. L. Silva, and J. M. Ferreira, On-line defragmentation for run-time partially reconfigurable FPGAs, *Proceedings of Field Programmable Logic and Applications (FPL'02)*, Montpellier, France, pp. 302–311, Sep 2002.
11. J. Khan and R. Vemuri, An efficient battery-aware task scheduling methodology for portable RC platforms, *Proceedings of Field Programmable Logic and Applications (FPL'04)*, Antwerp, Belgium, pp. 669–678, Sep 2004.
12. B. Krishnamoorthy, J. G. Wu, and T. Srikanthan, Hardware partitioning algorithm for reconfigurable operating system in embedded systems, *Proceedings of Sixth Real-Time Linux Workshop*, Singapore, pp. 117–123, Nov 2004.
13. J. Lach, D. Evans, J. McCune, J. Brandon, and L. Hu, Power-efficient adaptable wireless sensor networks, *Proceedings of 6th Annual Military and Aerospace Programmable Logic Devices International Conference*, Washington, DC, Sep 2003.
14. A. LaMarca, W. Brunnete, D. Koizumi, and M. Lease, Making sensor network practical with robots, *Proceedings of International Conference on Pervasive Computing*, Zurich, Switzerland, pp. 152–166, Apr 2002.

15. MicroBlaze μClinux Project, http://www.itee.uq.edu.au/~jwilliams/mblaze-uclinux/, Oct 2006.

16. Opencores.org, http://www.opencores.org, 2006.

17. P. N. Pathirana, N. Bulusu, A. V. Savkin, and S. Jha, Node localization using mobile robots in delay-tolerant sensor networks, *IEEE Transactions on Mobile Computing*, 4(3), 285–296, May 2005.

18. N. Priyantha, A. Chakraborty, and H. Balakrishnan, The cricket location-support system, *Proceedings of ACM MOBICOM*, Boston, MA, pp. 32–43, Aug 2000.

19. J. M. Rabaey, Silicon platforms for the next generation wireless systems—What role does reconfigurable hardware play? *Proceedings of Field-Programmable Logic and Applications (FPL'00)*, Villach, Austria, pp. 277–285, Sep 2000.

20. V. Raghunathan, S. Ganeriwal, and M. Srivastava, Energy efficient wireless packet scheduling and fair queuing, *ACM Transactions on Embedded Computing Systems*, 3(1), 3–23, Feb 2004.

21. N. K. Ratha and A. K. Jain, Computer vision algorithms on reconfigurable logic arrays, *IEEE Transactions on Parallel and Distributed Systems*, 10(1), 29–43, Jan 1999.

22. D. Ratter, FPGAs on mars, *Xcell Journal*, Available from: http://www.xilinx.com/publications/xcellonline/xcell_50/xc_pdf/xc_mars50.pdf, Jul 2005.

23. J. Septién, H. Mecha, D. Mozos, and J. Tabero, 2D Defragmentation heuristics for hardware multi-tasking on reconfigurable devices, *Proceedings of 13th Reconfigurable Architectures Workshop (RAW'06)*, Rhodes, Greece, Apr 2006.

24. H. K. H. So, A. Tkachenko, and R. Brodersen, A unified hardware/software runtime environment for FPGA-based reconfigurable computers using BORPH, *Proceedings of 4th International Conference on Hardware/Software Codesign and System Synthesis (CODES+ISSS'06)*, Seoul, Korea, pp. 259–264, Oct 2006.

25. G. Stitt, B. Grattan, J. Villarreal, and F. Vahid, Using on-chip configurable logic to reduce embedded system software energy, *Proceedings of 10th IEEE Symposium on Field-Programmable Custom Computing Machines (FCCM'02)*, NaPa Valley, CA, pp. 143–151, Apr 2002.

26. C. Steiger, H. Walder, and M. Platzner, Operating systems for reconfigurable embedded platforms: Online scheduling of real-time tasks, *IEEE Transactions on Computers*, 53(11), 1393–1407, Nov 2004.

27. TinyOS, http://www.tinyos.net/, Oct 2006.

28. μClinux—Embedded Linux/Microcontroller Project, http://www.uclinux.org/, Oct 2006.

29. J. C. van der Veen, S. P. Fekete, M. Majer, A. Ahmadinia, C. Bobda, F. Hannig, and J. Teich, Defragmenting the module layout of a partially reconfigurable device, Available from: http://arxiv.org/pdf/cs.AR/0505005, May 2005.

30. H. Walder, Operating system design for partially reconfigurable logic devices, PhD thesis, Swiss Federal Institute of Technology (ETH), Zurich, Switzerland, Apr 2005.

31. A. Woo and D. E. Culler, A transmission control scheme for media access in sensor networks, *Proceedings of MOBICOM 2001*, Rome, Italy, pp. 221–235, Jul 2001.

32. J. Woodfill and B. Von Herzen, Real-time stereo vision on the PARTS reconfigurable computer, *Proceedings of 5th IEEE Symposium on FPGA-Based Custom Computing Machines (FCCM'97)*, NaPa Valley, CA, pp. 201–209, 1997.

33. J. Wu (Ed.), *Handbook on Theoretical and Algorithmic Aspects of Sensor, Ad Hoc Networks, and Peer-to-Peer Networks*, Auerbach Publications, Boca Roton, FL, pp. 317–346, 2006.

34. Xilinx, Inc., Xilinx application note XAPP290: two flows for partial reconfiguration: Module-based or difference-based, V1.2 edn., Sep 2004.

35. Xilinx, Inc., Xilinx application note XAPP151: Virtex series configuration architecture user guide, V1.7 edn., Oct 2004.

36. Xilinx, Inc., EDK MicroBlaze tutorial, V2.1 edn., Mar 2004.

Chapter 19

Logical Location-Based Multicast Routing in Large-Scale Mobile Ad Hoc Networks

Guojun Wang, Minyi Guo, and Lifan Zhang

Contents

19.1 Introduction

Ad hoc networks are self-organizing, rapidly deployable, and dynamically reconfigurable networks, which require no fixed infrastructure. Ad hoc networks in which the nodes are connected by wireless links and can be mobile are referred to as Mobile Ad hoc NETworks (MANETs), where all the

MNs function as *hosts* and *routers* at the same time. Two MNs communicate directly if they are within the radio transmission range of each other; otherwise, they reach each other via a multi-hop route.

Many existing and forthcoming applications in MANETs require the collaboration of groups of mobile users. Communication in battlefield and disaster relief scenarios, video conferencing and multiparty gaming in conference room or classroom settings, and emergency warnings in vehicular networks are example applications. As a consequence, multicast routing in MANETs becomes a hot research topic in recent years. Multicast routing is a communication scheme for sending the same messages from a source to a group of destinations in an efficient way. MANETs are inherently ready for multicast communication due to their broadcast nature. However, limited bandwidth between MNs and highly dynamic topology due to unpredictable node mobility make the design of scalable and QoS-aware multicast routing protocols much more complicated than that in the traditional networks.

Most research works focus on small-or medium-scale MANETS, which consist of up to several hundreds of MNs, as proposed by the IETF MANET Working Group [17]. In recent years, some research works focus on large-scale MANETs with thousands, even hundreds of thousands of MNs, for example, the landmark routing with mobile backbones for digitized battlefield [35], the CarNet system [20], the Terminodes system [2], and the Ad Hoc City [10] for metropolitan environment.

As MANETs are infrastructure-less, many virtual backbone-based routing schemes have been proposed to seek for similar capabilities of the high speed and broadband backbone in the Internet in supporting efficient data transportation. In the literature, two major techniques are used to construct a virtual backbone, namely, connected dominating set [29,34] and clustering [26,35].

Because the search space for route discovery in MANETs is reduced to the nodes in the virtual backbone consisting of the dominating set or the Cluster Heads (CHs), routing based on the virtual backbone scales better than that based on flat structure. However, the virtual backbone-based routing protocols still cannot scale well in large-scale MANETs when the number of nodes in the backbone becomes large.

In theory, a multi-tier hierarchy can potentially solve the scalability problem in the two-tier hierarchy. Therefore, a natural way is to further organize the backbone nodes into multiple tiers in large-scale MANETs. However, this scalability is not automatically guaranteed if too many tiers exist in the hierarchy: (1) Due to the mobility and failure of nodes, all the backbone nodes may join or leave the hierarchy at any time, which makes the maintenance of multi-tier routing tables quite challenging; (2) Most traffic loads go through the nodes in the higher tiers of the hierarchy, and these nodes become the bottlenecks; and (3) There are some hardware limitations, for example, different types of radio capabilities are required at different tiers. Although multiple radios in some backbone nodes are common practice in military applications, this may not be practical in many commercial applications if too many tiers of radios are required. Due to these reasons, one generally uses a backbone with only a few tiers (say, two) [35].

Ad hoc routing protocols can be generally categorized into two kinds: topology-based and location-based. Topology-based routing protocols use the network topology information to compute routes, but it is hard to extend such protocols to large-scale MANETs due to the highly dynamic property of the network topology. Many location-based routing protocols have been developed in order to solve the scalability problem. Recent surveys on these protocols can be found in [19,30]. In location-based routing, each node determines its own location through the use of Global Positioning System (GPS) or some other type of positioning service. A location service is used by the sender of a packet to determine the location of the destination and to include it in the header of the packet. The routing decision at each forwarding node is then based on the locations of the forwarding node's neighbors and the destination node. In this way, the location-based routing does not need to maintain routing tables. Therefore, location-based routing can scale quite well in large-scale MANETs.

In this chapter, we propose a novel logical Hypercube-based Virtual Dynamic Backbone (HVDB) model using location information to support QoS-aware multicast routing in MANETs. A preliminary version of this model appeared in [33]. The proposed model is derived from n-dimensional hypercubes, which have many desirable properties, such as high fault tolerance, small diameter, regularity, and symmetry. Due to these properties, the proposed model meets the new QoS requirements of high availability and good load balancing.

This model uses some clustering techniques such as the mobility prediction and location-based clustering technique in [26] to form stable clusters, which elects an MN as a CH when it satisfies the following criteria: (1) it has the highest probability, in comparison with other MNs within the same cluster, to stay for longer time within the cluster; and (2) it has the minimum distance from the center of the cluster. Based on this technique, this model further abstracts a flat structure into three tiers: the Mobile Node Tier (MNT), the Hypercube Tier (HT), and the mesh tier, where each CH elected by the clustering algorithm can be simply mapped to a hypercube node at the HT.

Based on the HVDB model, we propose a QoS-aware multicast routing protocol, which consists of three parts: proactive local logical route maintenance, summary-based membership update, and logical location-based multicast routing. Simulation studies show good performance of the proposed algorithms even in the presence of a large number of faulty nodes in the network.

The remainder of this chapter is organized as follows. We present some related works in Section 19.2. The proposed model is introduced in Section 19.3. Section 19.4 describes the design of our HVDB-based multicast routing protocol. We evaluate the proposed protocol in Section 19.5 through simulations. Section 19.6 concludes this chapter.

19.2 Related Works

19.2.1 Preliminaries of Hypercubes

An n-dimensional hypercube has 2^n nodes. Each node is labeled by a bit string $k_1 \cdots k_n$ ($k_i \in \{0, 1\}, 1 \leq i \leq n$). Two nodes are connected by a link if and only if their labels differ by exactly 1 bit. The Hamming distance between two nodes u and v, denoted by $H(u, v)$, is the number of bits in which u and v differ.

An n-dimensional hypercube has many desirable properties: (1) High fault tolerance: The hypercube offers n node disjoint paths between each pair of nodes; therefore, it can sustain up to $n - 1$ node failures; (2) Small diameter: The diameter of the hypercube is defined as the maximal Hamming distance between any pair of nodes in the hypercube, which is n; (3) Regularity: The hypercube has a very regular structure, in which every node plays exactly the same role, and no node is more loaded than any others to achieve load balancing; and (4) Symmetry: The hypercube is symmetrical in graph terminology. In particular, any $(k+1)$-dimensional subcube in the hypercube consists of two k-dimensional subcubes for all $1 \leq k < n$, each of which is also symmetrical.

The hypercube used to be a very hot research topic. It was originally proposed as an efficient interconnection network topology for Massively Parallel Processors (MPPs). In recent years, much research has been done to apply the hypercube to other network environments, such as multicast communication on the Internet [7,16], hypercube-like prefix routing in P2P networks [21,27,36], and hypercube-based overlay formation for P2P computing [23].

In [12], the authors propose the *incomplete hypercube*, which may contain any number of nodes. We generalize the incomplete hypercube by allowing that any number of nodes/links can be absent due to many reasons such as mobility, transmission range, and failure of nodes.

19.2.2 Multicast Routing in MANETs

As is mentioned above, topology-based routing protocols do not scale well. So, topology-based multicast routing protocols, for example, flooding-based, tree-based, and mesh-based, cannot scale well in large-scale MANETs either. In [32], an adaptive flooding protocol is developed, where nodes can dynamically switch their routing mechanisms to one of the three modes, namely, scoped flooding, plain flooding, and hyper flooding modes, based on their perception of network conditions, by using relative velocity as the switching criterion. The advantage of flooding is that even if there are faults in transmission, as long as there is at least one path toward any of the receivers, all the receivers can potentially receive multicast data packets. However, it consumes too much bandwidth and it does not scale well.

MAODV (Multicast Ad hoc On-Demand Distance Vector routing protocol) [22] builds multicast trees as needed to connect multicast group members. Each node maintains two routing tables: Route Table (RT) and Multicast Route Table (MRT). The RT is used for recording the next hop for routes to other nodes in the network. The MRT contains entries for multicast groups of which the node is a router, that is, a node in the multicast tree. A source node that is not a part of the multicast tree can join the multicast group by broadcasting a request message. Any node in the multicast tree can respond to a join request by unicasting a reply back to the source node. Multicast tree maintenance includes three parts: selecting and activating the link to be added to the tree when a new node joins the group, pruning the tree when a node leaves the group, and repairing a broken link. The protocol is able to obtain a high packet delivery success rate and to offer a minimum control packet overhead for both unicast and multicast communication. But the multicast tree cannot respond to the changing topology quickly, and it is hard to maintain the multicast tree.

Typical mesh-based protocols are ODMRP (On-Demand Multicast Routing Protocol) [14], which delivers packets to destinations on a mesh topology using scoped flooding of data. So, it provides better packet delivery than tree-based protocols but pays an extra cost for mesh maintenance. ODMRP uses the concept of a forwarding group to establish a mesh for each multicast group. While a multicast source has packets to send, it periodically broadcasts a join request to the entire network. When a node receives a non-duplicate join request, it stores the upstream node ID and rebroadcasts the packet. When the join request packet reaches a multicast receiver, the receiver creates or updates the source entry in its member table. While valid entries exist in the member table, join tables are broadcast periodically to the neighbors. A node that receives a join table will check if it is on the path to the source and thus is part of the forwarding group. The join table is thus propagated by each forwarding group member until it reaches the multicast source via the shortest path. However, packet loss is a problem during mesh reconfiguration, and scalability is also a problem.

A comparative study of some typical multicast routing protocols in MANETs has been carried out in [15]. This study also gives a general conclusion that, in a mobile scenario, mesh-based protocols outperform tree-based protocols. The reason is that the availability of alternate routes provides robustness to mobility.

In recent years, location-based unicast routing has attracted much attention because it scales quite well in large-scale MANETs. Accordingly, researchers have proposed to use location information in multicast routing protocols. In the Dynamic Source Multicast (DSM) protocol [1], when a packet is to be multicast, the sender first locally computes a snapshot of the global network topology according to the location and transmission radius information collected from all the nodes in the network. A multicast tree for the addressed multicast group is then computed locally based on the snapshot. The resulting multicast tree is then optimally encoded and is included in the packet header. This protocol improves the scalability because it eliminates the maintenance of the multicast session state in each router, which has to be done in traditional multicast tree or multicast mesh-based protocols. However, its scalability is still limited because the location and transmission radius information has to be periodically broadcast from each node to all the other nodes in the network.

In [6], the Small Group Multicast (SGM) protocol based on packet encapsulation is proposed. This protocol builds an overlay multicast packet distribution tree on top of the underlying unicast routing protocol. Different from the DSM protocol that computes the multicast tree at each sender, this protocol constructs the tree in a distributed way: each node only constructs its outgoing branches to the next-level subtrees and forwards the packet to the roots of the subtrees. This process repeats until all the destinations have been reached. This protocol is more scalable than the DSM protocol because the nodes in a group need not to know the global network topology. Instead, they are only aware of each other in terms of the group membership and the location information of the group nodes. However, this protocol does not specify a method for dynamic joins and leaves in terms of location update among the group nodes. Therefore, this protocol is more suitable for the groups in which the group membership is static.

In [18], the Position-Based Multicast (PBM) protocol is proposed using only locally available location information about the destination nodes. This protocol provides a solution in order to approximate the optima for two potentially conflicting properties of the multicast distribution tree: (1) the length of the paths to the individual destinations should be minimal, and (2) the total number of hops needed to forward the packet to all the destinations should be as small as possible. If not properly handled, a greedy multicast forwarding may lead to a problem when a packet arrives at a node that does not have any neighbor providing progress for one or more destinations. This problem is solved in location-based unicast routing, such as using the right hand rule-based recovery strategy in [11]. This protocol extends the strategy to support the packet with multiple destinations. This protocol can deal with group members distributed in large-scale MANETs. However, it cannot scale well in terms of the number of group nodes due to the fact that the location and group membership information is required at each sender of the multicast group.

In [31], the Scalable Position-Based Multicast (SPBM) protocol is proposed to extend PBM. SPBM uses a hierarchical aggregation of membership information: The further away a region is from an intermediate node, the higher the level of aggregation should be for this region. This hierarchical scheme improves scalability. However, because all the nodes in the network are involved in the membership update, it still cannot scale well in large-scale MANETs. In this chapter, we solve this problem by summarizing the group membership information in a novel way and disseminating this information to only a portion of nodes in the network. Therefore, our scheme can potentially scale well in terms of both the number of groups and the number of group nodes in each group in large-scale MANETs.

19.2.3 QoS-Aware Routing Issues

Generally speaking, QoS is a loosely defined term. There are some metrics that affect QoS, such as delay, bandwidth, packet loss, and energy consumption. QoS-aware routing has been studied extensively in the wired networks such as the Internet. Due to the node mobility and the scarcity of resources such as energy of nodes and bandwidth of wireless links, it is much more difficult to provide QoS guarantee in MANETs than on the Internet. In fact, guaranteeing QoS in such a network may be impossible if the nodes are *too mobile* [3]. In the literature, there are only a few works tackling this problem in MANETs.

In [5], a hard-QoS protocol based on the well-known IntServ model is proposed in MANETs, which searches multiple paths in parallel in order to find the most qualified one. In [28], the authors propose to use location information in QoS routing decisions, and consider connection time (estimated lifetime of a link) as a QoS constraint. In [9], the authors present a protocol for TDMA-based bandwidth reservation for QoS routing in MANETs. It solves the race condition and parallel reservation problems by maintaining three-state information (free/allocated/reserved) at each MN.

In [13], a soft-QoS protocol based on the well-known Differentiated Services model is proposed in MANETs. It extends the Dynamic Source Routing (DSR) protocol to embed the QoS constraints in route discovery, route maintenance, and traffic management. In highly dynamic MANETs, soft-QoS protocols may have better overall performance than hard-QoS protocols due to the highly unpredictable topological change of MANETs.

In MANETs, network nodes/links may be broken sometimes, disrupting the continuity of an ongoing session and potentially terminating the session, thus inducing the QoS problem. Many papers view the QoS as a scheme in providing fault tolerance [4,8]. In particular, in [25], the authors propose to precompute some routes before existing routes break and thus avoid route recomputation delay. In this sense, the HVDB model proposed in this chapter helps to provide fault tolerance due to the high fault tolerance of hypercubes.

The QoS problem is hard to tackle even in the wired network. In [24], the authors point out that high availability and even distribution of traffic over the network are a prerequisite for the economical provisioning of QoS. We complement that it is especially true in MANETs due to limited bandwidth and energy of MNs. Here high availability indicates that a network has the capability of hiding or quickly responding to faults, giving users no sense of faults in the network. Load balancing indicates that traffic loads can be distributed evenly in the network to the greatest extent in order to eliminate hot spots in the network. Based on these, traditional QoS models, such as IntServ and DiffServ models, can perform much more effectively in MANETs.

19.3 QoS Multicast Model

In this section, we introduce our logical HVDB model shown in Figure 19.1, which has high availability and good load balancing properties in large-scale MANETs.

The MNT consists of MNs that move in and out of reach to each other with regard to the radio propagation range. The MNs are grouped into clusters according to some criteria. Every cluster has one CH and multiple cluster members. These CHs are responsible for forwarding packets and communicating between clusters, and managing their cluster members. In this tier, we use the same way to divide a geographical area (or even the whole earth) into equal regions of circular shape as in [26]. Each MN can determine the circle where it resides if location information is available. We call the circle a Virtual Circle (VC), accordingly the center of the VC is called Virtual Circle Center (VCC). If there is a CH in a VC, then we view the VCC as the CH; if not, it is only a placeholder. We also use the mobility prediction and location-based clustering technique in [26], which has been shown to be able to form clusters much more stably than other schemes.

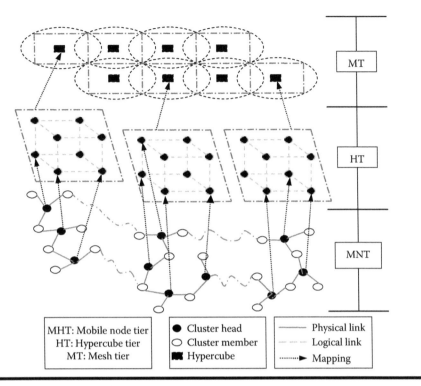

| MHT: Mobile node tier
HT: Hypercube tier
MT: Mesh tier | ● Cluster head
○ Cluster member
▰ Hypercube | ——— Physical link
– – – Logical link
·······▶ Mapping |

Figure 19.1   The HVDB model.

The HT comprises multiple logical n-dimensional hypercubes (n is relatively small in our consideration, e.g., 3, 4, 5, or 6), whose nodes are actually the CHs. A logical hypercube node becomes an actual one only when a CH exists in the VC. There is a one-to-one mapping relation between a CH and a hypercube node. The CHs located within a predefined region build up a logical n-dimensional hypercube, which is probably an incomplete hypercube. The hypercube is logical in the sense that the logical link between two adjacent logical hypercube nodes possibly consists of multi-hop physical links.

The Mesh Tier (MT) is a logical 2-dimensional mesh network obtained by viewing each n-dimensional hypercube as one mesh node. In the same way, the 2-dimensional mesh is possibly an incomplete mesh, and the link between two adjacent mesh nodes is logical and physically multi-hop. A mesh node becomes an actual mesh node only when a logical hypercube exists in it.

In Figure 19.1, the mesh tier is drawn in circle regions, and the HT is not drawn with circles for clarity. The higher two tiers comprise the HVDB. In particular, the HVDB has the *non-virtual* and *non-dynamic* properties, which are similar to *reality* and *stability* properties of the backbone of the Internet, respectively.

In order to realize the non-virtual property of the HVDB, we assume each MN can acquire its location information, such as geographical position, moving velocity, and moving direction, using some devices such as a GPS. Then each MN can determine the VC where it resides. Each VC represents a node in the HVDB. If at least one MN capable of functioning as a CH exists in a VC, then there is an actual node in the HVDB. These VCs are overlapped with each other and an MN within the overlapped regions can be a cluster member of only one cluster for easy management of QoS-aware routing.

In order to realize the non-dynamic property of the HVDB, we mitigate the dynamic behavior of MANETs by making two assumptions: (1) We assume to use the clustering technique in [26]; and (2) We assume MNs have different computation and communication capabilities, with the CHs having stronger capability than others, especially the multilevel radios. The former guarantees to form stable clusters in large scale MANETs. The latter guarantees to form a stable HVDB. We argue that the latter is reasonable in practice and easy to realize, for example, in a battlefield, a mobile device equipped on a tank can have stronger capability than the one equipped for a foot soldier.

19.4 QoS-Aware Multicast Protocol

Based on the HVDB model, we assume the CHs have two-level radios. The first level is short, which is used for communication between normal MHs within each cluster. The second level is long, which is used for direct communication among CHs.

19.4.1 Proactive Local Logical Route Maintenance

Many traditional routing protocols cannot scale well in large-scale MANETs because global topology has to be known either by all the MNs for proactive protocols or by the senders for reactive protocols. In order to scale well in large-scale MANETs, our protocol requires only the CHs to maintain local topology in a distributed way. In this subsection, we show how to proactively maintain local logical routes at each CH in our HVDB model.

In our model, the whole network is divided into many VCs of equal size. The VCs are grouped to form logical hypercubes. We consider logical hypercubes with small dimension, which is set as a system parameter. A simple function is used to map each CH to a hypercube node, using system parameters such as central coordinate, length and width of the whole network, diameter of VCs, and dimension of logical hypercubes. We define four kinds of logical identifiers: Cluster Head ID (CHID), Hypercube Node ID (HNID), Hypercube ID (HID), and Mesh Node ID (MNID). The relation between CHID and HNID is one-to-one mapping, the relation between HNID and HID is many-to-one mapping, and the relation between HID and MNID is one-to-one mapping. In this chapter, the logical identifier of each logical node is also called *logical location*.

Figure 19.2 shows an example MANET with 8×8 VCs, which is further divided into four 4-dimensional logical hypercubes. One logical hypercube is shown in Figure 19.3, with additional logical links connecting some VCs according to the logical relationship among these VCs.

The CHs are classified into Border Cluster Heads (BCHs) and Inner Cluster Heads (ICHs). A BCH is a CH that may have logical link between two adjacent mesh nodes, that is, two adjacent logical hypercubes. A BCH forwards traffic among logical hypercubes, while an ICH does that within a logical hypercube.

We define the *number of logical hops* of a logical link as follows. If a logical link satisfies two conditions—(1) it connects two CHs, and (2) it does not rely on any other CH to route packets along the link—then the number of logical hops of the logical link is 1, and the logical link is called a *1-logical hop route*. Accordingly, the number of logical hops of a logical link connecting any two CHs is the total number of concatenated 1-logical hop routes that comprise the logical link between the two CHs. For example, the number of logical hops that comprise 1-logical hop routes of $1000 \rightarrow 1100 \rightarrow 1101$ is 2.

Our proactive local logical route maintenance algorithm is shown in Figure 19.4. Each CH periodically exchanges its local logical route information with those CHs that are at most $k \geq 1$

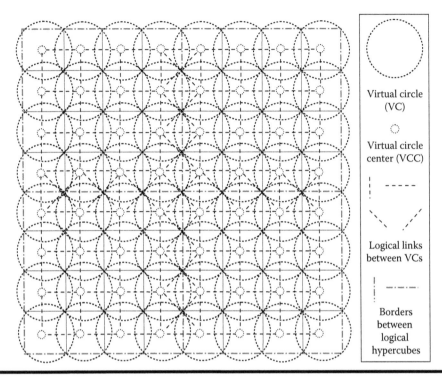

Figure 19.2   An example MANET with 8 × 8 VCs.

logical hops away. In particular, the information such as delay and bandwidth is maintained in each specific local logical route, which is used for QoS routing. Here k is a system parameter; as for how large is proper for the value of k, we will discuss it in the next section.

Finally, we show some local logical routes maintained by the algorithm at the hypercube node labeled with 1000. The 1-logical hop routes include: $1000 \rightarrow 1001, 1000 \rightarrow 1010, 1000 \rightarrow 0010, 1000 \rightarrow 1100, 1000 \rightarrow 0000$, and some route(s) to its adjacent logical hypercube(s). The 2-logical hop routes include: $1000 \rightarrow 1001 \rightarrow 1100, 1000 \rightarrow 1100 \rightarrow 1101, 1000 \rightarrow 0010 \rightarrow 0011, 1000 \rightarrow 0010 \rightarrow 0110$, and many others.

19.4.2 Summary-Based Membership Update

Based on the HVDB model, we propose to summarize the group membership information at three tiers. At the MNT, each node in each cluster knows which multicast groups it has currently joined, which is called *Local-Membership*. Each MN periodically sends Local-Membership to its CH. Each CH then summarizes the group membership according to all the information got from all its cluster members, which is called *MNT-Summary*.

At the HT, each hypercube node, that is, each CH in the hypercube, periodically sends MNT-Summary to all the CHs within the hypercube. Each CH then summarizes the group membership information according to all the information got from all the CHs within the hypercube, which is called *HT-Summary*.

If we neglect the delay for transmitting the Local-Membership and MNT-Summary messages, then each CH in a logical hypercube has the same HT-Summary information. Therefore, any one

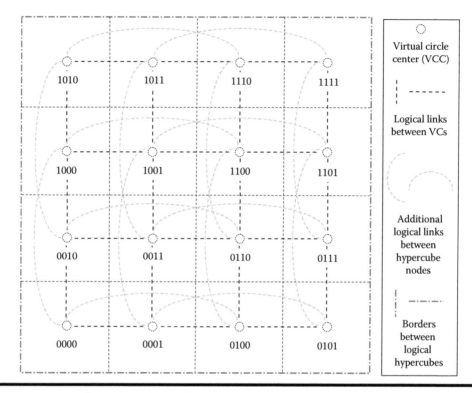

Figure 19.3   A 4-dimensional logical hypercube.

Proactive local logical route maintenance:
1. Each CH periodically sends beacon messages with its local logical route information such as delay and bandwidth to its *k*-logical hop neighboring CHs;
2. Each CH updates its local logical routes when receiving a beacon message.

Figure 19.4   The proactive local logical route maintenance algorithm.

of the CHs can be designated to periodically broadcast the HT-Summary information to all the CHs in the whole network. There are many ways to choose one to do such a task. The simplest way is to always designate the same CH to do the task. However, it may be not reliable, and the CH may become the bottleneck. Another way is to request each CH to be responsible for this task one by one. It eliminates the above two problems, but it may incur too much overhead in order to coordinate among the CHs.

We propose a solution to reduce the overhead. Each CH uses its up-to-date information about its own MNT-Summary and the collected MNT-Summary messages to decide whether it will itself be responsible for doing the task or not according to some criteria. One criterion is to choose the CH that contains the largest number of multicast groups, or the largest number of group members. Another criterion is to choose the CH such that the total number of multicast groups, or the total number of group members, contained by itself and all its 1-logical hop neighboring CHs, is the largest one. When considering the delay for transmitting membership messages, we argue that the latter criterion can work well because the probability for only one CH to satisfy the same criterion will be very high in most of the time.

Summary-based membership update:

1. Each MN updates its Local-Membership when it joins or leaves a multicast group.
2. Each MN periodically sends its Local-Membership to its CH in its cluster.
3. Each CH summarizes Local-Membership messages into MNT-Summary message and periodically sends it to all the CHs in the hypercube where it resides.
4. Each CH summarizes MNT-Summary messages into a HT-Summary message and decides whether to broadcast it to all the CHs in the whole network.
5. Each CH summarizes HT-Summary messages into MT-Summary message, which is used by the logical location-based multicast routing algorithm.

Figure 19.5 The summary-based membership update algorithm.

At the mesh tier, each mesh node, in fact, each of the CHs in the network, summarizes the HT-Summary messages into its MT-Summary membership information. Since each CH in the network only needs to know which logical hypercubes contain which groups of members, the timeout interval for broadcasting HT-Summary messages can be set much larger than that for sending MNT-Summary or Local-Membership messages. Therefore, we argue that the proposed algorithm does not incur too much overhead.

Finally, we present the summary-based membership update algorithm in Figure 19.5.

19.4.3 Logical Location-Based Multicast Routing

In our membership scheme, each CH maintains highly summarized membership information about all the groups in the network. In particular, the MT-Summary information maintained by all the CHs becomes identical if the group membership does not change too drastically.

Based on this membership scheme, our multicast routing scheme is as follows. If an MN needs to send multicast messages to a specific group, it sends them to its CH. Then the CH checks its MT-Summary to determine which logical hypercubes contain members of this group, and the logical identifiers, that is, the logical locations of these logical hypercubes, are used to compute a multicast tree for the messages. The multicast tree is then cached for future use. Then the information about the multicast tree is encapsulated into the messages. Finally, the messages are forwarded along the multicast tree.

The multicast tree is built at the mesh tier, and each node in the tree is a mesh node, that is, a logical hypercube. In the multicast tree, we assume to use some location-based unicast routing algorithm to send a packet from one logical hypercube to its next hop logical hypercube. When a packet enters a logical hypercube at a certain logical node at its first time, two tasks are executed at the hypercube node: (1) The packet is re-encapsulated (possibly duplicated), then it is unicast to its next hop logical hypercube(s); and (2) within the logical hypercube, the packet is forwarded to those hypercube nodes that contain group members.

In order to forward the packet within the logical hypercube, the hypercube node computes a multicast tree using its HT-Summary. The multicast tree is cached at the hypercube node for future use. The multicast tree is then encapsulated into the packet header in order to forward the packet within the logical hypercube. The packet forwarding using the multicast tree at the HT is different from that using the multicast tree at the mesh tier. In each logical hypercube, each hypercube node has already maintained all the local logical routes in advance that are at most k logical hops away. If all the local logical routes at the multicast tree are at most k logical hops away, then the packet forwarding can directly use them. In some extreme cases, for example, if k is too small but the logical hypercube is too large, then it is possible for some of the local logical routes at the multicast

Logical location-based multicast routing:

1. Any MN can act as a multicast source to send multicast messages to a multicast group through its CH;
2. The CH computes a mesh-tier multicast tree using its MT-Summary or uses its cached multicast tree; The CH encapsulates the multicast tree into the header of the packet;
3. The packet is forwarded along the multicast tree, each branch of which is routed by some location-based unicast routing algorithm;
4. When the packet enters a logical hypercube first time, the packet is re-encapsulated at the CH for next hop logical hypercube(s); The CH computes a hypercube-tier multicast tree using its HT-Summary or uses its cached tree, and encapsulates it into the packet header;
5. The packet is forwarded along the hypercube-tier multicast tree, each branch of which is routed using the local logical routes in each CH;
6. When the packet enters a CH at its first time, and MNT-Summary shows group members exist, it sends the packet to these group members.

Figure 19.6   The logical location-based multicast routing algorithm.

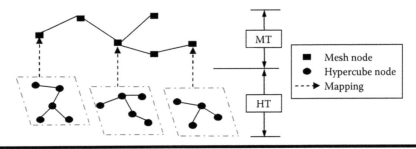

Figure 19.7   The two-tier multicast tree in the HVDB model.

tree to go beyond *k* logical hops away. In this chapter, we suppose *k* is sufficiently large and the hypercube is relatively small in order to avoid such extreme cases.

Besides the multicast routing at the mesh tier and the HT, it is also needed to route a packet to the group members (if exist) at the MNT when a CH, that is, a hypercube node, receives the packet. Many methods, such as local broadcast, can be used to route a packet to the group members within a cluster.

Finally, we show our logical location-based multicast routing algorithm in Figure 19.6, and show the multicast tree constructed by the algorithm in Figure 19.7, which consists of the mesh-tier multicast tree and the HT multicast tree.

19.5 Performance Evaluation

In order to evaluate the performance of the proposed logical location-based multicast routing algorithm, we conduct simulations to test some properties of the multicast tree formed by the proposed algorithm. Here, we emphasize on the evaluation of building the multicast tree in each logical hypercube. Notice that the logical hop discussed in this section is indeed the physical hop, and we use them interchangeably.

We simulate the 6-dimensional hypercube, the nodes of which are initially placed uniformly at random in VCs on a 2-dimensional rectangular region. Then, we map those nodes into the 6-dimensional hypercube by a mapping function (see Figure 19.3). We suppose the hypercube nodes (i.e., CHs) have two kinds of transmission ranges in terms of the long radio. The two kinds of transmission ranges are shown in Figure 19.8, with a 6-dimensional hypercube consisting of 8 × 8

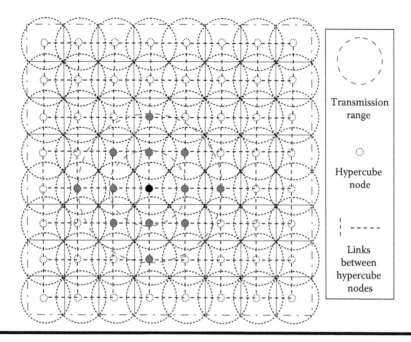

Figure 19.8   The two kinds of transmission ranges in terms of the long radio.

hypercube nodes. The short transmission range is drawn in a small circle, which enables the direct communication between a CH and its four adjacent neighboring CHs. The long transmission range is drawn in a big circle, which enables the direct communication between a CH and its 12 adjacent neighboring CHs.

Since the hypercube node can be viewed as not changed if only a CH exists in the corresponding VC, we evaluate two metrics related to the multicast tree built at the HT, namely, *average hop count* and *maximum hop count*. The average hop count is defined as the average value of all the minimum hop counts of routing paths between all pairs of hypercube nodes. The maximum hop count is defined as the maximum value among all those minimum hop counts. Here, the average hop count stands for the average delay when routing on the multicast tree, and the maximum hop count stands for the maximum depth of the multicast tree.

Based on the proactive local route maintenance algorithm, each hypercube node maintains a routing table of hypercube nodes in its k-hop vicinity by the long radio. Here, we assume a small k, such as 1, 2, or 3. Greedy forwarding routing is performed based on the local k-hop routing table: each node chooses a next hop node among its neighbors, which has the shortest geographical distance to the destination. In order to decrease the probability of routing loops caused by faulty nodes, each node does not choose the node that sends packets to itself as the next hop node. That is, for each intermediate node on routing, the proximate upstream node and the next hop node are not the same. Since the hypercube is 6-dimensional and each hypercube node maintains a k-hop routing table, we observed in the simulations that the probability of routing loops is very small. This is because the size of each hypercube network is not large, and the k-hop routing table maintained by each hypercube node facilitates the avoidance of the routing loops.

We run simulations with different k and different node fault probability p in different transmission ranges of long radio. Here, the node fault probability is expressed as the number of faulty nodes in the 6-dimensional hypercube. For each pair of k and p, we run the simulations with 1000 times, and then get the average of *the average hop count* and the average of *the maximum hop count*. k is simulated with the value of 1 and 2, respectively. The value of p is simulated from $0/64 = 0$ to $10/64 \approx 16.0\%$. Here, $p = 0$ denotes an ideal network situation without any faults, and we always get the best simulation results in this situation. Simulation results are given from Figures 19.9 through 19.12.

Figure 19.9 shows the needed average hop count when routing in the 6-dimensional hypercube in the case that each hypercube node can communicate with its four adjacent neighbors. When $k = 1$, that is, each hypercube node only maintains the local routes to its direct neighbors, the average hop count is equal to 5.333 in case that no faulty node exists. Along with the increase of the number of faulty nodes, the average hop count also increases, especially after the number of faulty nodes reaches to 6. When $k = 2$, that is, each hypercube node maintains the local routes to the nodes in its 2-hop vicinity, along with the increase of the number of faulty nodes, the average hop count almost does not change, which approaches to the ideal value of 5.333.

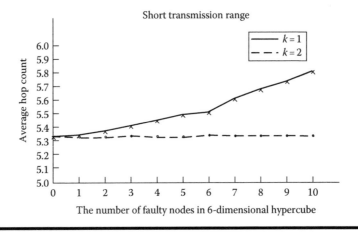

Figure 19.9   The average hop count using short range.

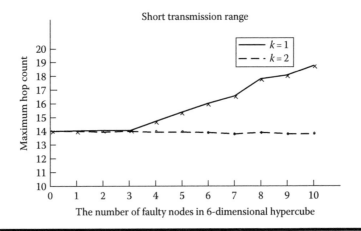

Figure 19.10   The maximum hop count using short range.

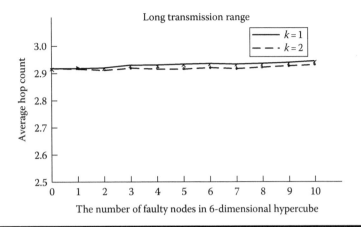

Figure 19.11   The average hop count using long range.

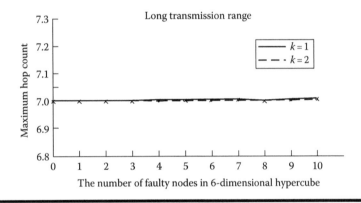

Figure 19.12   The maximum hop count using long range.

Figure 19.10 shows the needed maximum hop count when routing in the 6-dimensional hypercube in the case that each hypercube node can communicate with its four adjacent neighbors. When $k = 1$, the maximum hop count increases quickly after the number of faulty nodes reaches to 3. When $k = 2$, the maximum hop count almost does not change with the increase of the number of faulty nodes, it approaches the ideal value of 14.

From Figures 19.9 and 19.10, we observed that if using the defined short transmission range of the long radio in 6-dimensional hypercubes, 2-hop is enough for the proactive local route maintenance, and then a good performance can be achieved. In particular, both the average hop count and the maximum hop count (i.e., the average routing delay and the maximum depth of multicast tree) approximate the ideal value generated in corresponding hypercube network without any faults.

Figure 19.11 shows the needed average hop count when routing in the 6-dimensional hypercube in the case that each hypercube node can communicate with its 12 adjacent neighbors. The average hop count is equal to 2.921 when the hypercube network has no faults. Along with the increase of the number of faulty nodes, we observed that the two curves of $k = 1$ and $k = 2$ are close to each other, but the curve of $k = 1$ is slightly above to the curve of $k = 2$, and the later can be considered as keeping constant.

Figure 19.12 shows the needed maximum hop count when routing in the 6-dimensional hypercube in the case that each hypercube node can communicate with its 12 adjacent neighbors. The two curves of $k = 1$ and $k = 2$ are almost the same with each other. Both of them keep nearly constant, with the ideal value of 7.

Thus, according to Figures 19.11 and 19.12, we observed that if using the defined long transmission range of the long radio in 6-dimensional hypercubes, only 1-hop is enough for the proactive local route maintenance. In particular, both the average hop count and the maximum hop count (i.e., the average routing delay and the maximum depth of multicast tree) approximate the ideal value generated in corresponding hypercube network without any faults. It can be seen that the longer the transmission range of the long radio, the smaller the k needed.

From these four figures, we can see that better properties can be achieved in the case of $k = 2$ than in the case of $k = 1$. It is easy to explain that if the node has more routing information, then routing can easily be done and better properties can be achieved.

The simulations show good performance of the proposed logical-location-based multicast routing algorithm in terms of the average routing delay and maximum depth of the multicast tree generated at the HT. Furthermore, the simulations also show a strong fault tolerance of the HVDB model. The dimension of the hypercube can even be larger, such as 8-dimensional hypercube with 256 hypercube nodes. Since each hypercube node (i.e., CH) has several cluster members, an 8-dimensional hypercube can have thousands of MNs. In a sense, the hypercube can be viewed as an autonomous system in the Internet routing. Thus, the whole MANETs can be divided into several domains, and multicast routing in MANETs can be considered as inter-domain multicast routing and intra-domain multicast routing. To state another way, the former is multicast routing at the mesh tier and the latter is that at the HT.

19.6 Conclusions

We have proposed a novel HVDB model to support QoS-aware multicast routing in large-scale MANETs. The proposed model is derived from n-dimensional hypercubes, which have many desirable properties, such as high fault tolerance, small diameter, regularity, and symmetry. The proposed model uses the location information of MNs and meets the new QoS requirements: high availability and good load balancing. Firstly, in an incomplete logical hypercube, there are multiple disjoint local logical routes between each pair of CHs, and the high fault tolerance property provides multiple choices for QoS routing. That is, if the current logical route is broken, multiple candidate logical routes become available immediately to sustain the service without QoS being degraded. Secondly, small diameter facilitates a small number of logical hops on the logical routes. Thirdly, due to the regularity and symmetry properties of hypercubes, no leader is needed in a logical hypercube, and every node plays almost the same role except for the slightly different roles of BCHs and ICHs. Thus, no single node is more loaded than any other node, and no problem of bottlenecks exist, which is likely to occur in tree-based architectures.

Acknowledgments

This research is supported by the National High-Tech Research and Development Plan of China (863 Plan) under Grant Nos. *2006AA01Z202* and *2006AA01Z199*, and the National Natural Science Foundation of China under Grant Nos. *60503007* and *60533040*.

References

1. S. Basagni, I. Chlamtac, and V.R. Syrotiuk, Location aware, dependable multicast for mobile ad hoc networks, *Computer Networks (Elsevier)*, 36, 659–670, 2001.
2. L. Blazevic, S. Giordano, and J.-Y. Le Boudec, Self organized terminode routing, *Cluster Computing (Kluwer Academics)*, 5, 205–218, 2002.
3. S. Chakrabarti and A. Mishra, QoS issues in ad hoc wireless networks, *IEEE Communications Magazine*, 39, 2, 142–148, Feb 2001.
4. T.-W. Chen, P. Krzyzanowski, M.R. Lyu, C. Sreenan, and J.A. Trotter, Renegotiable quality of service—A new scheme for fault tolerance in wireless networks, *Proceedings of 27th Annual International Symposium on Fault-Tolerant Computing (FTCS-27)*, Seattle, WA, pp. 21–30, Jun 1997.
5. S. Chen and K. Nahrstedt, Distributed quality-of-service routing in ad hoc networks, *IEEE Journal on Selected Areas in Communications*, 17 (8), 1488–1505, Aug 1999.
6. K. Chen and K. Nahrstedt, Effective location-guided tree construction algorithms for small group multicast in MANET, *Proceedings of IEEE 21th Annual Joint Conference of the IEEE Computer and Communications Societies (INFOCOM 2002)*, New York, Vol. 3, pp. 1180–1189, Jun 2002.
7. R. Friedman, S. Manor, and K. Guo, Scalable stability detection using logical hypercube, *IEEE Transactions on Parallel and Distributed Systems*, 13 (9), 972–984, Sep 2002.
8. S.S. Gokhale and S.K. Tripathi, Effect of unreliable nodes on QoS routing, *Proceedings of 7th International Conference on Network Protocols (ICNP 1999)*, Toronto, Ontario, Canada, pp. 173–181, Oct–Nov 1999.
9. I. Jawhar and J. Wu, A race-free bandwidth reservation protocol for QoS routing in mobile ad hoc networks, *Proceedings of 37th Annual Hawaii International Conference on System Sciences*, Big Island, HI, pp. 293–302, Jan 2004.
10. J.G. Jetcheva, Y.-C. Hu, S. PalChaudhuri, A.K. Saha, and D.B. Johnson, Design and evaluation of a metropolitan area multitier wireless ad hoc network architecture, *Proceedings of 5th IEEE Workshop on Mobile Computing Systems and Applications (WMCSA 2003)*, Monterey, CA, pp. 32–43, Oct 2003.
11. B. Karp and H.T. Kung, GPSR: Greedy perimeter stateless routing for wireless networks, *Proceedings of 6th Annual International Conference on Mobile Computing and Networking (MobiCom 2000)*, Boston, MA, pp. 243–254, Aug 2000.
12. H.P. Katseff, Incomplete hypercubes, *IEEE Transactions on Computers*, 37 (5), 604–607, May 1988.
13. H. Labiod and A. Quidelleur, QoS-ASR: An adaptive source routing protocol with QoS support in multihop mobile wireless networks, *Proceedings of IEEE 56th Vehicular Technology Conference (VTC 2002-Fall)*, Vancouver, British Columbia, Canada, 4, 1978–1982, Sep 2002.
14. S.J. Lee, W. Su, and M. Gerla, On-demand multicast routing protocol in multihop wireless mobile networks, *Mobile Networks and Applications*, 7, 441–453, Dec 2002.
15. S.J. Lee, W. Su, J. Hsu, M. Gerla, and R. Bagrodia, A performance comparison study of ad hoc wireless multicast protocols, *Proceedings of IEEE Infocom 2000*, Tel Aviv, Israel, pp. 566–574, 2000.
16. J. Liebeherr and B.S. Sethi, A scalable control topology for multicast communications, *Proceedings of 17th Annual Joint Conference of the IEEE Computer and Communications Societies (INFOCOM 1998)*, San Francisco, CA, Vol. 3, pp. 1197–1204, Mar–Apr 1998.
17. Mobile Ad-hoc Networks (MANETs) Working Group, http://www.ietf.org/html.charters/manet-charter.html
18. M. Mauve, H. Fubler, J. Widmer, and T. Lang, MobiHoc Poster: Position-based multicast routing for mobile ad-hoc networks, *ACM SIGMOBILE Mobile Computing and Communications Review*, 7 (3), 53–55, Jul 2003.
19. M. Mauve, A. Widmer, and H. Hartenstein, A survey on position-based routing in mobile ad hoc networks, *IEEE Network*, 15 (6), 30–39, Nov–Dec 2001.

20. R. Morris, J. Jannotti, F. Kaashoek, J. Li, and D. Decouto, CarNet: A scalable ad hoc wireless network system, *Proceedings of 9th Workshop on ACM SIGOPS European Workshop: Beyond the PC: New Challenges for the Operating System*, pp. 61–65, 2000.

21. A. Rowstron and P. Druschel, Pastry: Scalable, distributed object location and routing for large-scale peer-to-peer systems, *Proceedings of IFIP/ACM International Conference on Distributed Systems Platforms (Middleware)*, Heidelberg, Germany, pp. 329–350, Nov 2001.

22. E.M Royer and C.E. Perkins, Multicast operation of the ad-hoc on-demand distance vector routing protocol, *Proceedings of 5th Annual ACM/IEEE International Conference on Mobile Computing and Networking (MobiCom 1999)*, pp. 207–218 , Seattle, WA, Aug 1999.

23. M. Schlosser, M. Sintek, S. Decker, and W. Nejdl, A scalable and ontology-based P2P infrastructure for semantic web services, *Proceedings of Second International Conference on Peer-to-Peer Computing (P2P 2002)*, Linkoping, Sweden, pp. 104–111, Sep 2002.

24. G. Schollmier and C. Winkler, Providing sustainable QoS in next-generation networks, *IEEE Communications Magazine*, 42 (6), 102–107, Jun 2004.

25. S.H. Shah and K. Nahrstedt, Predictive location-based QoS routing in mobile ad hoc networks, *Proceedings of IEEE International Conference on Communications (ICC 2002)*, New York, Vol. 2, pp. 1022–1027, Apr–May 2002.

26. S. Sivavakeesar, G. Pavlou, and A. Liotta, Stable clustering through mobility prediction for large-scale multihop intelligent ad hoc networks, *Proceedings of IEEE 2004 Wireless Communications and Networking Conference (WCNC 2004)*, Atlanta, GA, Vol. 3, pp. 1488–1493, Mar 2004.

27. I. Stoica, R. Morris, D. Liben-Nowell, D.R. Karger, M.F. Kaashoek, F. Dabek, and H. Balakrishnan, Chord: A scalable peer-to-peer lookup protocol for Internet applications, *IEEE/ACM Transactions on Networking*, 11 (1), 17–32, Feb 2003.

28. I. Stojmenovic, M. Russell, and B. Vukojevic, Depth first search and location based localized routing and QoS routing in wireless networks, *Proceedings of 2000 International Conference on Parallel Processing (ICPP 2000)*, Toronto, Ontario, Canada, pp. 173–180, Aug 2000.

29. I. Stojmenovic, M. Seddigh, and J. Zunic, Dominating sets and neighbor elimination-based broadcasting algorithms in wireless networks, *IEEE Transactions on Parallel and Distributed Systems*, 13 (1), 14–25, Jan 2002.

30. I. Stojmenovic, Position-based routing in ad hoc networks, *IEEE Communications Magazine*, 40 (7), 128–134, Jul 2002.

31. M. Transier, H. Fubler, J. Widmer, M. Mauve, and W. Effelsberg, Scalable position-based multicast for mobile ad-hoc networks, *Proceedings of First International Workshop on Broadband Wireless Multimedia: Algorithms, Architectures and Applications (BroadWim 2004)*, San Jose, CA, Oct 2004.

32. K. Viswanath and K. Obraczka, An adaptive approach to group communications in multi hop ad hoc networks, *Proceedings of 7th International Symposium on Computers and Communications (ISCC 2002)*, Taormina, Italy, pp. 559–566, Jul 2002.

33. G. Wang, J. Cao, L. Zhang, K.C.C. Chan, and J. Wu, A novel QoS multicast model in mobile ad hoc networks, *Proceedings of 19th IEEE International Parallel and Distributed Processing Symposium (IPDPS 2005)*, Denver, CO, pp. 206–213, Apr 2005.

34. J. Wu, Extended dominating-set-based routing in ad hoc wireless networks with unidirectional links, *IEEE Transactions on Parallel and Distributed Systems*, 13, (9), 866–881, Sep 2002.

35. K. Xu, X. Hong, and M. Gerla, Landmark routing in ad hoc networks with mobile backbones, *Journal of Parallel and Distributed Computing (Elsevier)*, 63, 110–122, 2003.

36. B.Y. Zhao, L. Huang, J. Stribling, S.C. Rhea, A.D. Joseph, and J.D. Kubiatowicz, Tapestry: A resilient global-scale overlay for service deployment, *IEEE Journal on Selected Areas in Communications*, 22 (1), 41–53, Jan 2004.

Chapter 20

Performance Analysis of Dynamic Routing Protocols for Mobile Ad Hoc Networks

Cláudia J.B. Abbas and L.J. García Villalba

Contents

20.1 Introduction

For more than a century, researches have been carried out in order to make possible the communication among devices in motion. Those studies are being developed with success ever since because new methods of communication and services are being enthusiastically adopted by people all over the world.

Actually, the development level of those systems allows almost the complete independence of moving terminal location at the time the calls are made.

The arrival of full mobility in the telephony area increases the wishes for the mobile computation. Nowadays, the establishment of a standard by IEEE (Institute of Electrical and Electronic Engineers) for the wireless communication networks (WLAN—wireless local area network) is making the extensive use of those technologies quite popular. Consequently, new applications, necessities, and technical challenges are arising.

The IEEE 802.11 standard for WLANs includes the establishment of infrastructured or independent networks (ad hoc) [25]. The infrastructured network is based on the use of a fixed network, with one or more access points. The ad hoc is a kind of network without an appropriate infrastructure, where a node communicates directly with another one in the network. This kind of network that does not need a fixed infrastructure is able to give more flexibility, for instance, in inhospitable areas. Another occasion on which the ad hoc network is particularly important arises when an immediate infrastructure for information flows is required, such as disasters, battlefields, deliverances, or explorations of uninhabited areas. In those situations, the ad hoc networks are extremely helpful since they are merely based on wireless communication, allowing in that way a greater mobility of hosts. Those networks might be also used in meetings or events, which demand a communication structure but their duration does not really justify installing equipments and cable infrastructure in order to meet such needs.

One of the main problems in ad hoc networks is the routing issue [26]. Scalable and efficient routing scheme plays important role in ad hoc networks. In this chapter, we will show a simulation-based analysis over three mobile network routing protocols, focusing on their behavior under different traffic volumes and node mobility.

We chose to analyze and compare the DSR (*Dynamic Source Routing*) [1], AODV (*Ad Hoc on-Demand Vector Routing*) [2], and OLSR (*Optimized Link State Routing Protocol*) [3] dynamic routing protocols because they were submitted by the Internet Engineering Task Force (IETF) for publication as experimental RFC (Request For Comments—Official Internet Protocol Standards) and are implemented for the Network Simulator version 2 (NS-2), which we decided to use in our research. All of them had been approved as RFC [2,4,5].

20.2 Motivation and Related Works

This work was developed this work because related works about performance of routing protocols in ad hoc networks, some shown later, have not considered different movement speeds versus number of nodes in doing a comparative study in IEEE 802.11 networks [27].

In this work we had a motivation to know which routing protocol works better in case of low (5 m/s) and high speed (10 m/s) in standard IEEE 802.11 and also when it increases the number of wireless users.

One related work is [6], where they used the simulator tool GloMoSim [7]. The simulation models the network of 30 mobile hosts migrating within a 20 m × 20 m space with a transmission range of 5 m. Every node moves in a random fashion with a static time of 5 s. The channel capacity is 2 Mbps, which means the old standard IEEE 802.11. They used the DCF algorithm as the medium access mode. Besides using the free-space propagation models, they also developed SIRCIMs (simulation of indoor radio channel impulse response models), which consider fading, barriers, multipath interference, foliages, etc., so are more accurate than the free-space propagation model.

Another work is [8], where they do not consider the routing protocol OLSR that became a standard in MANET IETF working group. The wireless standard simulated was the old IEEE 802.11. They also do not consider routing packet overhead during performance analysis. This work presents experimental evidence from two wireless testbeds, which show that there are usually multiple minimum hop-count paths, many of which have poor throughput. As a result, minimum-hop-count routing often chooses routes that have significantly less capacity than the best paths that exist in the network.

In the work [9] they did not analyze OLSR protocol, and they consider the old standard IEEE 802.11. Although they have analyzed various movement speeds, from 1 to 20 m/s, and considered until 50 wireless nodes, in the random waypoint mobility model, they did not consider the problem that, after a certain time, the nodes become concentrated in the center of the scenario [10].

Authors of [11] did performance analysis of DSDV, DSR, and AODV protocols, not including the OLSR protocol, in the old standard IEEE 802.11. The only metrics studied were delay and throughput. Besides simulation study, they did a real study of a conference scenario with speed movement of 1 Mbps, 50 wireless nodes, and the CBR (*constant bitrate*) data traffic.

In [11], they use a discrete event, packet-level, routing simulator called *MaRS* (Maryland Routing Simulator) [12]. They assume a channel bandwidth of 1.5 Mbps, which is not compatible with none of the IEEE 802.11 standards and also no multiple-access contention or interference is modeled, an assumption that can lead to unreal performance results.

In [13], we have a study of performance just about on-demand routing protocols, with principal focus on DSR routing protocol. Their interest was to investigate about the chosen best routing path. They conclude that while multipath routing is significantly better than single-path routing, the performance advantage is small beyond a few paths and for long-path lengths. It also shows that providing all intermediate nodes in the primary (shortest) route with alternative paths has a significantly better performance than providing only the source with alternate paths.

The novelty of doing a comparative study of AODV, DSR, and OLSR protocols in IEEE 802.11 network, in different scenarios related to movement speeds and data traffic rates is an analysis of different performance metrics: end-to-end delay, throughput, percentage of packet loss, and routing packet overhead.

20.3 Wireless LANs and the IEEE 802.11

It is known that wireless LANs are being widely deployed, thanks to its ease of installation, flexibility, and decreasing price. The IEEE 802.11 [14] standard is one of the responsible for this technology popularization because it led to the development of new products, guaranteeing the interoperability between distinct manufacturers.

There are several different technologies for wireless medium access control and physical data transmission. Bluetooth, HomeRF, and HiperLAN are examples of such technologies, but we are going to explain the 802.11 standard a little further because WLAN equipment vendors are largely adopting it. All the performance comparison we make is based on this protocol that released its first version in 1997 and is constantly evolving since then.

This first version defined transmission rates of 1 and 2 Mbps utilizing diffused infrared technology or radio frequency. It defined FHSS (frequency-hopping spread spectrum) and DSSS (direct-sequence spread spectrum) for operation in the unlicensed ISM 2.4–2.4835 GHz radio-frequency band. It also defined a standard for medium access control (MAC) techniques, as well as the system architectures infrastructured mode and ad hoc mode.

The infrared technology is reliable because it does not suffer from electromagnetic interference from other communications. It is also physically more secure because the transmission range is confined to a room, but it is rarely used because there are only a few vendor products available and they are usually based on proprietary technology.

The radio-frequency spread spectrum technologies standardized in 802.11 are broadband modulation techniques that use low-power transmissions. Its data appears like noise to someone analyzing the frequency spectrum.

The FHSS is a modulation technique where a carrier that hops in frequency in a pseudo-random order modulates the information signal. This carrier is called the pseudo-noise code. It makes the resulting bandwidth many times that of the baseband signal. This modulation scheme reduces interference from other communications because the interfering signal is sitting in only one of the wide range of possible carrier frequencies of receiving stations. The DSSS works in a similar way, but the frequency does not hop in this case. It varies in a direct sequence, obtaining an even wider bandwidth. It reduces even more the interference from other systems.

The 802.11 standard also specifies a MAC layer that manages communication between wireless stations, coordinating the access to a shared radio channel, and provides protocols that optimize this access. Sometimes said to be the wireless network brain, the MAC layer relies on the PHY layer for carrier detection, transmission, and reception of 802.11 frames.

Research and development of wireless networks did not stop after the first 802.11 standard was published and new transmission technologies came up. The IEEE created new working groups, and new PHY layer specifications were published. Each IEEE working group is represented by a different alphabet letter and is responsible for fixing and/or enhancing some standard section.

The 802.11b/g [15] working group enhanced the DSSS PHY layer for transmissions at 11 Mbps rates in the 2.4 GHz ISM band and 54 Mbps also in 2.4 GHz ISM, respectively. The 802.11a [16] group published a standard for transmission at up to 54 Mbps rates at the 5 GHz unlicensed frequency band utilizing the OFDM modulation scheme. Also approved by the FCC for operation in the United States, the 802.11g [17] standard defines communication in the 2.4 GHz ISM band at rates of up to 54 Mbps while maintaining interoperability with older 802.11b and 802.11 wireless products.

There are a lot of current working groups at the IEEE publishing about several issues like 802.11c (*Bridge Operation Procedures*), 802.11d (*Global Harmonization*), 802.11e (*Quality of Service*), 802.11f (*Inter Access Point Protocol*), 802.11h (*Coexistence Management*), and 802.11i (*Security*), but they are not going to be further discussed in this chapter.

20.4 Routing in Mobile Ad Hoc Networks

The role of the routing protocol is to define where to send a packet to make it get to a determined address. The network entities that participate in the best route decision process are the routers. To achieve the goal of being an ad hoc network, a multihop network, some (generally all) nodes work as routers, receiving and processing packets. The routers make decisions based on the information got from other nodes while forwarding packets to its destinations.

To carry through this function correctly, the routers change information with each other in order to remain aware of the available routes and of part (or totality) of the network topology. Initially, the routers only possess local knowledge, or either, they know their own addresses and the situation of the links to which they are on. Through the exchange of messages, each router constructs its knowledge of the net. This process continues until all the excellent information has been gotten and the network enters in balance, when we say that the network has converged.

Although this state of balance is not a constant situation, links or nodes may stop working or moving, forcing the routers to make new decisions and to establish new routes to forward the packets. The routing mechanisms must be capable of reestablishing the network balance as fast as possible. The way it happens depends on the routing protocol implemented in the nodes, but it is desirable that the routing algorithm used be simple and does not cause overhead.

Because of the mentioned characteristics of the MANETs, these algorithms must be distributed, since the network does not depend on any infrastructure. They must be capable of dealing with fast changes in the network topology. According to the route discovering process, the main algorithms are grouped in two categories:

1. *Proactive:* The algorithm tries continuously to evaluate the routes so that when a packet needs guiding, the route is already known and can be used immediately. In this case, the nodes keep one or more tables with information about the network and react to topology changes propagating updates in order to maintain the network consistency. These updates are started by temporization mechanisms, which cause a constant number of transmissions even when the network is in balance.
2. *Reactive:* The algorithm defines the routes on demand. When a route is required, it starts the route discovery procedure. This way, a *route request* packet starts the process. Once this route is discovered, some management procedure maintains the route active. As the arrival of a packet requesting a route happens randomly in time, these protocols do not trade messages in a regular basis, what saves bandwidth and battery. However, these algorithms cause higher delay times.

Knowing these concepts, we classify DSR [18] as a reactive protocol, AODV [2] also as a reactive protocol, and OLSR [5] as a proactive protocol.

In the following, we have a brief explanation of each of these routing protocols.

20.4.1 DSR Protocol

The DSR protocol is a simple routing protocol designed specifically for use in multihop wireless ad hoc networks of mobile nodes. DSR allows the network to be completely self-organizing and self-configuring, without the need for any existing network infrastructure or administration. The protocol is composed of the two main mechanisms of "route discovery" and "route maintenance," which work together to allow nodes to discover and maintain routes to arbitrary destinations in the ad hoc network [1].

All aspects of the protocol operate entirely on demand, allowing the routing packet overhead of DSR to scale automatically to only that needed to react to changes in the routes currently in use. The protocol allows multiple routes to any destination and allows each sender to select and control the routes used in routing its packets, for example, for use in load balancing or for increased robustness. Other advantages of the DSR protocol include easily guaranteed loop-free routing, operation in networks containing unidirectional links, use of only "soft state" in routing, and very rapid recovery, when routes in the network change.

The principal disadvantage of the DSR routing protocol, as others source routing protocols, is the scalability problem because of the flooding method used to discover the path route until destination.

20.4.2 AODV Routing Protocol

The AODV routing protocol is based on another routing protocol called DSDV (Destination-Sequenced Distance-Vector Routing Algorithm) [18]. The AODV protocol changed the DSDV protocol in respect to eliminate global routing broadcast dissemination and minimized end-to-end latency when new path routes are requested.

The AODV routing is based on vector distance algorithm. The source node, before requesting a route path, looks for its own routing table. In the case of not finding it, it asks for the path route to its neighbor nodes. When the path routes have been disseminated along the ad hoc network, all nodes use this information to update its own routing tables. If the source node does not receive an answer during a certain time, it does another solicitation or assumes that the destination is unreachable.

The AODV algorithm enables dynamic, self-starting, multihop routing between participating mobile nodes wishing to establish and maintain an ad hoc network. AODV allows mobile nodes to obtain routes quickly for new destinations and does not require nodes to maintain routes to destinations that are not in active communication. AODV allows mobile nodes to respond to link breakages and changes in network topology in a timely manner. The operation of AODV is loop-free, and by avoiding the Bellman–Ford "counting to infinity" problem offers quick convergence when the ad hoc network topology changes (typically, when a node moves in the network). When links break, AODV causes the affected nodes to be notified so that they are able to invalidate the routes using the lost link [2].

20.4.3 OLSR Protocol

The OLSR protocol was developed for mobile ad hoc networks. It operates as a table-driven, proactive protocol, that is, exchanges topology information with other nodes of the network regularly. Each node selects a set of its neighbor nodes as "multipoint relays" (MPRs). In OLSR, only nodes, selected as such MPRs, are responsible for forwarding control traffic, intended for diffusion into the entire network. MPRs provide an efficient mechanism for flooding control traffic by reducing the number of transmissions required.

Nodes, selected as MPRs, also have a special responsibility when declaring link-state information in the network. Indeed, the only requirement for OLSR to provide shortest path routes to all destinations is that MPR nodes declare link-state information for their MPR selectors. Additional available link-state information may be utilized, for example, for redundancy.

Nodes have been selected as MPRs by some neighbor node(s) to announce this information periodically in their control messages. Thereby a node announces to the network that it has reach ability to the nodes that have selected it as an MPR. In route calculation, the MPRs are used to form the route from a given node to any destination in the network. Furthermore, the protocol uses the MPRs to facilitate efficient flooding of control messages in the network [4].

20.5 Simulation Environment

In this section will be shown the methodology used to make the comparative performance analysis of the studied routing protocols. As said before, it is a novel comparative study, and includes AODV, DSR, and OLSR protocols altogether.

20.5.1 Network Simulator 2

The NS-2 [19] is a simulator of discrete events of computer networks. In our study, the simulated physical layer is radio transmission; the NS simulates the propagation of radio controls. The used values of transmission power, thresholds and used frequency band had been chosen in accordance with the wireless interface of the Lucent manufacturer, WaveLAN model, giving to the nodes a reach of approximately 250 m in the two-ray ground propagation model, outdoor environment with omnidirectional antennas lifted 1.5 m from the ground.

The medium access schema is DCF (distributed coordinated point) with CSMA/CA (carrier sense multiple access/collision avoidance) as the algorithm to MAC, also from the IEEE 802.11 standard.

The output file of a simulation in the NS-2 [19] is a trace file having a line for each event that has occurred during the simulation. The possible events are transmission, reception, discarding (dropping) and forwarding of a packet. The events are recorded for all the layers in the stack of each node. Each generated packet for a given node carries a unique identification, which allows the packet to be followed through the trace files. It allows us to measure, for example, the delay of the packet as being the time it takes to reach its destination.

The environment simulated in the NS-2 is simplified. It does not have physical obstacles for nodes and/or the radio controls nor any type of external interference to the radio controls. With this, it is presumed that the transmission medium is more reliable in the simulation than in the real world. All links are bidirectional.

20.5.1.1 Setup

Each simulation is executed from a script in OTcl language describing the simulation parameters (propagation model, simulation area, number of nodes, traffic generated, etc.). The initial location of each node, the simulation and warm time, and the movement model are also specified.

20.5.1.2 Output

The NS-2 output file is a trace file that has a line with each event that occurred along the simulation. These events can be transmission, reception, dropping, and forwarding of packets. The events can be from any TCP/IP protocol level. For example, a transmission of a UDP packet from one node to another generates events also in network and medium access levels.

For each packet generated from a specific node exits an identifier that allow the packet to be traced along the trace file. It allows us to calculate, for example, the packet delay as the time from leaving a source node and arriving in a destination node.

20.5.1.3 Limitations

The simulations are deterministic that means that to the same scenario, the file trace is always the same. In a real MANET network, various factors can contribute to make the events to occur in a random manner as processors' capacity and memory latency. Overcoming this weakness is a common practice to simulate the scenario various times. In our study, we simulate each scenario for 10 times.

20.5.2 Mobility Model: Random Waypoint

In summary, this model works in the following form [20]: before the simulation, the following parameters are defined by the user: number of nodes, area of simulation, maximum speed, and maximum pause time. When simulation starts, the model defines, on a random mode, the destination and the speed of each node. The chosen speed is necessarily between the previously defined values of maximum and minimum speed. When the node reaches its destination, it stays motionless during a random time, chosen between zero and the maximum defined pause time, and then it follows in another random direction, with speed chosen randomly too.

The problem with the random waypoint model is that it does not get a steady state and that the velocity of movement tends to diminish along the simulation, unless the minimum speed is configured not null [10]. In this work, we used a scenario generator tool called BonnMotion 1.1 and take care about the minimum speed values [21].

Besides this problem, 1600 s of warm-up time was set to avoid the nodes getting too concentrated in the center of the scenario, which is another problem of this model [10].

20.5.3 Scenarios and Simulations

The objective of this work is to do a performance study of routing protocols in IEEE 802.11 networks in situations that we have movements of wireless nodes. As mentioned earlier, this work is a novelty because we have a comparative study of AODV, DSR, and OLSR protocols and in IEEE 802.11 network scenarios.

The main parameters that affect mobility are the maximum speed of the hosts and the pause time between each moving. In our simulations, the maximum speed chosen was 2 m/s, simulating a person walking; 10 m/s, simulating a car at a very low speed; and 20 m/s, also a car but in a greater speed.

The pause times were zero (nodes in a constant movement during simulation time) and 300 s at each final movement; nodes stay stopped during a random time between 0 and 300 s, before going to another direction.

The simulation area was a 1500 m × 300 m, without any kind of physical obstacle that could impede the movement of the nodes. We have chosen a rectangular area to have a greater average number of hops from source to destination.

The transmission rate was 3, 6, 9, and 12 packets/s, each packet with a length of 512 bytes (including UDP and IP headers), so we have transmissions rates from 12.288 bps to 49.152 bps, rates of typical IP applications.

We have 20 nodes with the maximum of 10 nodes generating traffic simultaneously. We have 24 simulations of each routing protocol, with a total of 72 different simulations. The simulation time was 300 s with a warm-up time of 1600 s.

The parameters used in our simulations are resumed in Table 20.1.

20.6 Performance Data Analysis

Simulation results are huge trace files containing registration of all events occurred during simulation time. We wrote then some AWK [22] scripts to read these trace files and plot the relevant metrics

Table 20.1   Simulation Parameters

Routing Protocols	OLSR, AODV, and DSR
Number of nodes	20 nodes
Maximum speed	2, 10, and 20 m/s
Maximum pause times	0 and 300 s
Transmission rates	3, 6, 9, and 12 packets/s
Packet size	512 bytes
"Warm-up" time	1600 s
Simulation time	1900 s

for comparing the dynamic routing protocols: end-to-end delay, throughput, dropped packets, and routing overhead. The following paragraphs describe these metrics and how they were estimated.

20.6.1 Metrics

From the trace file, we can calculate the fundamental metrics [23] used to compare the routing protocols. Next, we describe how we calculate them.

End-to-end delay: It is the delay calculated from the time a packet leaves a source node until it reaches the destination node. In this time, waiting time for receiving the route (reactive routing protocol) and the propagation time in the medium are included. The processing time is not included, as NS-2 does not have this timer.

Throughput: It is the bit rate of data packet that was received in a specific node. Packets that do not reach their destination are not included in this calculation, as routing packets and control packets. This metric reflects the protocol efficiency; that is, how well the protocol is doing the task of routing data packets.

Packet loss: Packets can be lost in their paths for a variety of reasons as there is no path to the destination node (in case of proactive protocol) because of a time-out in getting the route information. This metric reflects the number of lost packets in comparison to the number of packets sent.

Overhead: This is a metric of the efficiency of the routing protocol. When the routing protocol has done its tasks of delivering packets, we have to measure its efficiency in doing this. A great quantity of control packets can interfere in data traffic. In this metric, we measure the absolute number of routing packets.

Next, we are going to provide some plots of our results followed by brief explanations. We are going to contrast a low-average mobility scenario where we simulate walking people carrying mobile wireless devices at 1 m/s statistical average speed with a higher mobility scenario where we simulate faster vehicles equipped with similar devices in motion communicating with those people in the same rectangular area, giving a statistical average speed of 10 m/s. There are no physical obstacles in none of our scenarios.

We will show the plots divided per metric.

20.6.1.1 Delay

Delay is shown in Figures 20.1 to 20.3.

20.6.1.2 Throughput

Throughput is shown in Figures 20.4 to 20.6.

20.6.1.3 Packet loss

Packet loss is shown in Figures 20.7 to 20.9.

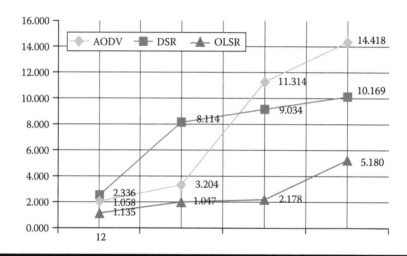

Figure 20.1   Source to destination average packet delay measured in milliseconds according to the volume of data traffic (packets/s) that nodes transmitted. Packets of 512 bytes with 10 m/s of mobility.

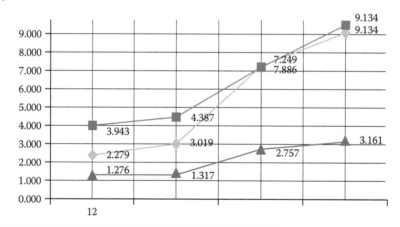

Figure 20.2   Source to destination average packet delay measured in milliseconds according to the volume of data traffic (packets/s) that nodes transmitted. Packets of 512 bytes with 2 m/s of mobility.

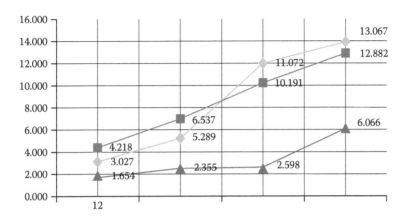

Figure 20.3 **Source to destination average packet delay measured in milliseconds according to the volume of data traffic (packets/s) that nodes transmitted. Packets of 512 bytes with 20 m/s of mobility.**

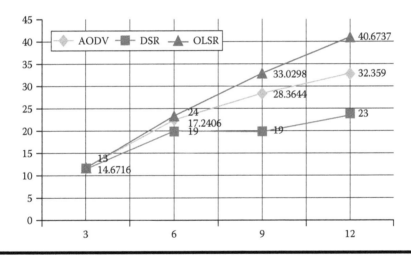

Figure 20.4 **Data throughput according to the volume of data traffic (packets/s) that nodes transmitted. Packets of 512 bytes with 10 m/s of mobility.**

20.6.1.4 Overhead

Overhead is shown in Figures 20.9 to 20.12.

In the following, we have a discussion that we can infer from the plots showed earlier, according to the chosen metric.

The results show that for the specific simulated scenarios, the DSR is usually responsible for the highest packet loss rate (Figures 20.7 through 20.9), and it gets worse as it increases the average mobility and the traffic volume. Consequently, it is usually the protocol with the lowest achieved throughput. For example, in a scenario with 2 m/s and pause time of 300 s (so a low mobility), the packet loss percentage was more than 53% with a data rate of 12 packets/s (Figure 20.7). This packet loss percentage was greater, more than 75%, when the speed was 20 m/s (Figure 20.9).

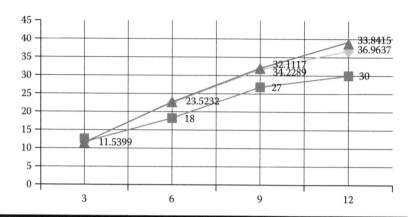

Figure 20.5 Data throughput according to the volume of data traffic (packets/s) that nodes transmitted. Packets of 512 bytes with 2 m/s of mobility.

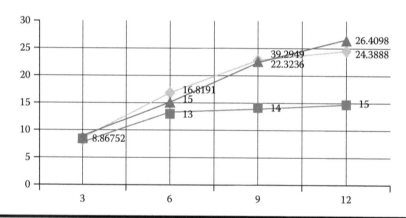

Figure 20.6 Data throughput according to the volume of data traffic (packets/s) that nodes transmitted. Packets of 512 bytes with 20 m/s of mobility.

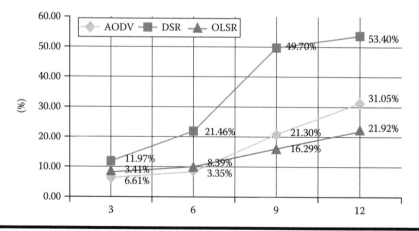

Figure 20.7 Packet loss percentage relative to the amount of packets sent (packets/s) in the 10 m/s mobility scenario.

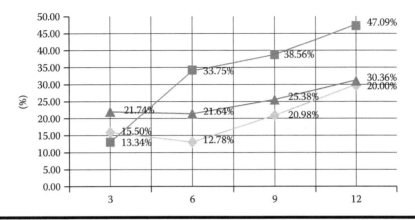

Figure 20.8 Packet loss percentage relative to the amount of packets sent (packets/s) in the 2 m/s mobility scenario.

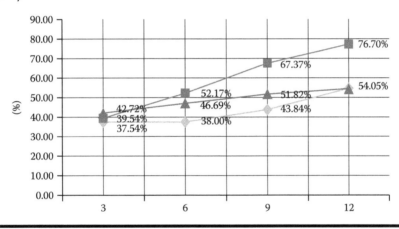

Figure 20.9 Packet loss percentage relative to the amount of packets sent (packets/s) in the 20 m/s mobility scenario.

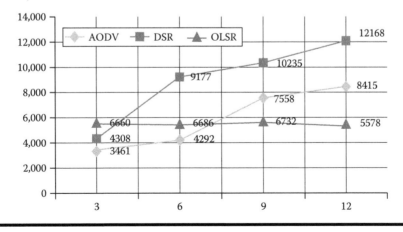

Figure 20.10 Quantity of routing control packets relative to the amount of packets sent (packets/s) in the 10 m/s mobility scenario.

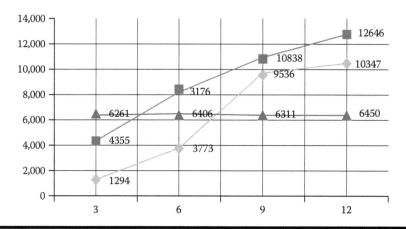

Figure 20.11   Quantity of routing control packets relative to the amount of packets sent (packets/s) in the 2 m/s mobility scenario.

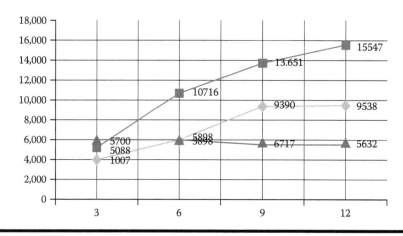

Figure 20.12   Quantity of routing control packets relative to the amount of packets sent (packets/s) in the 20 m/s mobility scenario.

We can say that DSR protocol was created for small networks, from 5 to 10 nodes. This study shows that with 20 nodes, we do not have good performance. This is probably due to the high volume of routing information that DSR nodes exchange, what can also be seen in the plots (Figures 20.10 through 20.12), and is caused by a lack of route maintenance mechanism, present in AODV routing protocol.

In general, the OLSR protocol achieved the best performance. It had the low end-to-end delay in all scenarios (Figures 20.1 through 20.3). In comparison to AODV routing protocol, OLSR is better when increasing data rate (Figures 20.3 and 20.9).

In both mobility scenarios (low and high mobility), the AODV routing protocol has a similar performance (sometimes, better, for example, loss packet in 20 m/s scenario—Figure 20.9) to OLSR protocol at low data rate.

But in high data rate scenarios, OLSR protocol looks more stable (Figures 20.3, 20.9, and 20.12). OLSR protocol is proactive, the volume of its traffic control packets does not vary in relation to the volume of data traffic packets (Figures 20.10 through 20.12). The increasing traffic in AODV routing protocol that is reactive causes more routing overhead, so AODV routing protocol is less efficient when increasing data traffic volume.

So the OLSR protocol is better than AODV and DSR, except in a low-mobility scenario with low-data-traffic situation where the OLSR protocol sends more routing control packets and would consume more power, that is less battery lifetime. In a high mobility and high-data-traffic situation (Figure 20.3), the recurrent AODV route discovery process would cause a high-average end-to-end packet delay time causing some applications to time out, so being worse than OLSR protocol.

20.7 Conclusion

The main aspect of wireless ad hoc networks is the complete lack of infrastructure. The missing fixed access points make communication more difficult because network topology is constantly changing, so there is no way to predefine packet routes. In this chapter, we compared three of the proposed protocols for these mobile networks.

Comparative analysis of MANET routing protocols was already done by various researches, as in [6,9,11,24,28,29] where simulation environments were used for performance evaluation and comparison of older protocols with some of the protocols analyzed in this chapter, but there is no previous paper comparing directly the three protocols discussed here. Our results are not directly comparable to the results of any previous paper because we used different mobility, traffic, and network size parameters.

There is no answer about which one is the best protocol. The protocol performance depends on network conditions and environment. If the MANETs were going to be set up to a meeting or conference, for example, in a scenario where each attendant will follow presentations and workshops using a wireless PDA or laptop receiving a relatively large traffic volume and low mobility, the best choice would be the OLSR protocol.

However, if the MANET implementation is for a scenario with high mobility and low traffic, such as a battlefield where control information of weapons and diverse equipment will traverse the wireless medium between all kinds of military vehicles (tanks, helicopters, and so on) and soldiers, the best choice would be the AODV routing protocol that only finds new routes on demand.

Although the protocols studied did not reach large-scale industry deployment, since they are not able to scale for network composed of a few hundred nodes, a network that demands multicast support, networks sensible to high packet loss, or networks that demand high levels of security and availability, a lot of advance was already obtained.

We showed that it is possible to have a throughput greater than 40 kbps for each transmitting node in a network composed by 20 mobile nodes when 10 of them were transmitting simultaneously, totalizing 400 kbps of network traffic. This represents 20% of utilization from the 2 Mbps nominal transmission rate.

These values are acceptable for implementation in the real world, since we can use technologies that provide standardized transmission rates of up to 54 Mbps, and usually much less than 50% of network nodes are transmitting data at the same time.

Acknowledgments

This work has been supported by the Ministerio de Ciencia e Innovación (MICINN) through Projects TEC2007-67129 and TEC2010-18894 and the Ministerio de Industria, Turismo y Comercio (MITyC) through the Projects AVANZA I+D TSI-020100-2008-365 and TSI-020100-2009-374. This work has also been supported by the Arab Academy for E-Business, Aleppo, Syria.

References

1. J. Broch, D. B. Johnson, and D. A. Maltz, The dynamic source routing protocol for mobile ad hoc networks. Internet-Draft, draft-ietf-manet-dsr-10.txt, July 2004.
2. C. E. Perkins, E. M. Belding-Royer, and S. R. Das, Ad hoc on-demand distance vector (AODV) routing, IETF, RFC 3561, July 2003.
3. A. Laouiti, P. Muhlethaler, A. Qayyum, L. Viennot, T. Clausen, and P. Jacquet, Optimized link state routing protocol, in *IEEE INMIC*, Lahore, Pakistan, 2001.
4. T. Clausen and P. Jacquet, Optimized link state routing protocol, IETF, San Francisco, CA, RFC3626, October 2003.
5. T. Clausen and P. Jacquet, Optimized link state routing protocol (OLSR), RFC 3626, Project Hipercom (INRIA), France, October 2003.
6. S.-J. Lee, M. Gerla, and C.-K. Toh, A simulation study of table-driven and on-demand routing protocols for mobile ad-hoc networks, 13(4): 48–54, July 1999.
7. UCLA Parallel Computing Laboratory and Wireless Adaptive Mobility Laboratory, GlomoSim, a scalable simulation environment for wireless and wired network systems, April 2012, http://pcl.cs.ucla.edu/projects/glomosim/
8. D. S. J. De Couto, D. Aguayo, B. A. Chambers, and R. Morris, Performance of multihop wireless networks: Shortest path is not enough, MobiCom'03, pp. 66–80, October 2002.
9. J. Broch, D. A. Maltz, D. B. Johnson, Y. C. Hu, and J. Jetcheva, A performance comparison of multi-hop wireless ad hoc network routing protocols, Mobicom'98, October 1998.
10. J. Yoon, M. Liu, and B. Noble, Random waypoint considered harmful, in *Proceedings of INFOCOM*, San Francisco, CA, IEEE, 2003.
11. P. Johansson, T. Larsson, N. Hedman, B. Mielczarek, and M. Degermark, Scenario-based performance analysis of routing protocols for mobile ad-hoc networks, Mobicom'99, pp. 195–206, August 1999.
12. C. Alaettinoglu, A. U. Shankar, K. Dussa-Zieger, and I. Matta. Design and implementation of MaRS: A routing testbed. *Journal of Internetworking: Research and Experience*, 5(1):17–41, 1994.
13. A. Nasipuri and R. Castañeda, Performance of multipath routing for on-demand protocols in mobile ad hoc networks. *ACM/Baltzer Mobile Networks and Applications (MONET) Journal*, 6(4), 2001.
14. IEEE Computer Society LAN MAN Standards Committee, *Wireless LAN Medium Access Protocol (MAC) and Physical Layer (PHY) Specification*, IEEE Std 802.11-1997, IEEE, New York, 1997.
15. IEEE Std 802.11b-1999, April 2012, http://standards.ieee.org/
16. IEEE Std 802.11a-1999, April 2012, http://standards.ieee.org/
17. IEEE Std 802.11g-2003, April 2012, http://standards.ieee.org/
18. C. E. Perkins and P. Bhagwat. Highly dynamic destination-sequenced distance-vector routing (DSDV) for mobile computers, August 1994.
19. Information Sciences Institute (ISI), The network simulator–ns-2, University of Southern California (USC), Los Angeles, CA, April 2012, http://www.isi.edu/nsnam/ns/
20. D. Johnson and D. Maltz, Dynamic source routing in ad hoc wireless networks, *Mobile Computing*, Kluwer Academic Publishers, Dordrecht, the Netherlands, pp. 153–181, 1996.

21. BonnMotion, University of Bonn, A mobility scenario generation and analysis tool, April 2012, http://web.informatik.uni-bonn.de/IV/Mitarbeiter/dewaal/BonnMotion/

22. The AWK Manual, April 2012, http://www.cs.uu.nl/docs/vakken/st/nawk/nawk_toc.html

23. S. Corson and J. Macker, Mobile ad hoc networking (MANET): Routing protocol performance issues and evaluation considerations, RFC 2501, January 1999.

24. S. R. Das, R. Castaneda, J. Yan, and R. Sengupta, Comparative performance evaluation of routing protocols for mobile, ad hoc networks, Mobile Networks and Applications, October 1998.

25. J. Geier, *Wireless LANs, Implementing High Performance IEEE 802.11 Networks*, 2nd edn., SAMS, Indianapolis, IN, 2001.

26. S. Haykin, *Communication Systems*, 3rd edn., Wiley, New York, 1994.

27. B. P. Lathi, *Modern Digital and Analog Communication Systems*, Oxford University Press, London, U.K., 1998.

28. D. B. Johnson, Routing in ad hoc networks of mobile hosts, IEEE Workshop, December 1994.

29. V. D. Park and M. S. Corson, A highly adaptive distributed routing algorithm for mobile wireless networks, 1997.

Chapter 21

Analysis of Connectivity for Sensor Networks Using Geometrical Probability

Jianxin Wang, Weijia Jia, and Yang Zhou

Contents

21.1 Introduction and Motivation

Wireless sensor networks have recently emerged as a premier research topic. Sensor networks pose a number of new conceptual and optimization problems such as location, deployment, and tracking. Wireless sensors are self-creating, self-organizing, and self-administering [1,2]. The absence of fixed infrastructure means that nodes of a sensor network can communicate directly with one another in a peer-to-peer fashion. Sensors can be mobile in the area concerned or deployed in a certain location. A typical application for this "larger" kind of network concerns distributed micro-sensors: Each sensor node of the network is capable of monitoring a given surrounding area (sensing), and coordinating with the other nodes to achieve a large sensing task [3]. In such an application, each node in the small area will connect with other nodes in the same area more frequently than in the different areas. A node may just roam in a given small area in a large network, such as sensor networks in which the sensor nodes may be deployed in a specific zone which are particularly useful in the hazardous areas for measuring critical and real-time data. In such situation, the nodes are distributed with local uniformity, which is discussed in detail in this chapter. Recent advances in processors, memory, and radio technology have made it possible to enable sensor networks with potentially very large number of small, lightweight, and low-cost nodes.

In sensor networks, all nodes cooperate in routing each other's packets, so that packets are transported in a multi-hop fashion from source to a destination. In general, the transmission range of the nodes is significantly smaller than the routers in conventional networks. Multi-hop routing is normally used to achieve the high degree of network connectivity and nodes form the links by choosing the power levels at which they are transmitting. Hence, it is essential to control the transmitter's power such that the information signals reach their intended receivers, while causing minimal interference for other receivers sharing the same channel [4]. Therefore, it is critical for nodes equipped with limited energy supply to reduce the energy consumption while maintaining maximum possible connections.

One of the focused research issues in sensor network is to maximize the probability of connectivity with minimum possible node energy. In particular, it has been shown that determining an optimal range assignment is solvable in polynomial time in the one-dimensional case, while it is NP-hard in the two-dimensional case. Ref. [5] showed that, if n nodes are displaced at random in $[0, 1]$ in one-dimension, then the r-homogeneous range assignment connects with probability at least $1 - (l - r)(1 - r/l)^n$. But they did not give the probability analysis for two-dimension area.

There are several protocols for controlling the transmission range of the nodes in sensor networks that have been presented recently in the literature [3,4]. Ref. [3] proposed a method to compute the critical transmission range for a wireless network, given that the locations of the mobile nodes are known. The protocol presented in [4] aimed at minimizing the energy required to communicate with a given master node. They assume that nodes are equipped with low-power GPS receivers, which provide position information to each node. These protocols are used for controlling the energy in a concreted wireless network without theoretically analyzing related transmission problem.

A question arises naturally as how many nodes must be directly connected and what is the minimum energy to maintain desirable connection. Ref. [6] shows that for every finite number, the probability of network disconnectivity converges to 1 as the number of nodes the network increases to infinity. It is also shown that the number of neighbors of each node needs to grow proportional to $O(\log n)$ if the network is to be connected, where n is the number of nodes in the network. In [7], for the case of n points uniformly distributed in an area, it was shown that a range chosen as $r(n) = \sqrt{A(\log n + r_n)/\pi \cdot n}$ will lead to the probability of connectedness converging to 1 as $n \to \infty$ if and only if $r_n \to \infty$.

Theoretical works above are asymptotic in nature. In contrast to the related works, this chapter presents a novel approach for computing the lower bound connectivity probability (CP) in a sensor network under uniform distribution of sensors. At the same time, we do not require that the number of sensors must be infinity. Even with the limit number of nodes, we can still derive the CP effectively. We use a divide-and-conquer approach to partition the whole network into mesh-block grid. Then the CP in each two adjacent mesh grids are calculated by geometrical probability analysis methods. Then we progressively combine the blocks into a large block for deriving the CP for the entire network is achieved. More specially, this chapter made the following contributions: with desired CP requirement, we design scheme that can effectively determines (1) the minimum transmission energy required for each node when n nodes are deployed in area connected; (2) the minimum number of nodes required to maintain the expected CP with fixed transmission energy.

21.2 Related Work

Coverage and connectivity are two important metrics for sensor networks. The goal of coverage is to have each location in physical space of interest within the sensing range of at least one sensor node, which is a measure of quality of service (QoS) of the sensing functions. Due to limited communication capabilities, each node has to act as router to help forward data packets to remote base station. Network connectivity then is of paramount importance to ensure successful data forwarding. Refs. [8,9] both proved that "communication range is at least twice of sensing range" is the sufficient condition to ensure that a full coverage of a convex area implies connectivity among active nodes inside the convex area, which is often the case for physical sensor nodes. Thus, coverage and connectivity problem for sensor networks can be discussed together to some extent.

Ref. [10] surveyed recent contributions addressing energy-efficient coverage problems in the context of static wireless sensor networks, including various coverage formulation, their assumptions, as well as an overview of the solutions proposed.

In [11], the problem of finding maximal number of covers in a sensor network, where a cover is defined as a set of nodes that can completely cover the target area is addressed. They proved NP-completeness of such problem and provided a centralized solution. They showed that the proposed algorithm approaches the upper bound of the solution under most cases. Ref. [12] presented how to use ILP (Integer Linear Programming) technique to solve several optimal 0/1 node scheduling problems in wireless sensor networks, for example, what is the minimum active nodes that can completely cover a sensor field. A heuristic algorithm is designed to find k-sets of node so that each node set can cover the whole monitored area and the cardinality of node sets is maximal [11]. In the context of providing network coverage, ref. [13] examined the relationship reduction in sensor duty cycle and required level of redundancy for a fixed performance measure. In particular, two types of mechanisms, Random Sleep and Coordinated Sleep, are taken into account. The approach presented in the above papers is both centralized, therefore requiring collection of global information.

Refs. [14–16] provided alternative, suboptimal, and distributed algorithms to solve same or similar problems. In [14], a subset of nodes is selected out initially and operates in active nodes until they run out of their energy or are destroyed. Other nodes fall asleep and wake up occasionally to probe their local neighbors. If there is no working node with its probing range, a sleeping node starts to work. Probing range can be adjusted to control sensing redundancy. However, original sensing coverage and connectivity may be reduced after node scheduling. Ref. [15] presented a distributed node-scheduling algorithm that can ensure full coverage preservation. In the work, each node arithmetically calculates the union of all sectors covered by its neighbors and determines its working status according to calculation results. In [16], the authors designed another distributed node-scheduling algorithm, wherein each node divides its sensing range into small grids. For a grid, each node sorts their neighbor that can cover the grid in the ascending order of their associated reference points. Based on the sequence of the reference points, the nodes in the sequence can coordinate their on-duty time to continuously monitor that concerned grid. By merging the results of all covered grids, each node can finally determine its working time.

Their algorithm can support k-degree coverage and connectivity preservation by extending each node's workspace proportionally along the sequence of reference points. Also, refs. [28–30] also extensively explored the connected coverage problem.

All work mentioned above focus on full coverage. In full coverage, each point in the network is covered by at least one sensor node. In some applications, only a subset of points in the network need to be covered and hence the number of nodes that need to be awake is reduced. Ref. [17] proposed a protocol for partial (but high) coverage in sensor networks, and demonstrated that it is feasible to maintain high coverage ($\geq 90\%$) while significantly increasing coverage duration when compared with protocols for full coverage. Ref. [18] studied the connected coverage problem with a given coverage guarantee. The work firstly introduced partial coverage concept and analyzed its property for the first time to prolong network lifetime. Then a heuristic algorithm was presented which takes into account the partial coverage and sensor connectivity simultaneously.

Idealized coverage is often adopted in most of the current work, where the sensing coverage of a node is always assumed uniform in all directions (represented by unit disc) following the binary detection model. An event that occurs within sensing radius of a node is always assumed detected with probability 1, while an event outside the circle is assumed not detected. However, sensing capabilities of networked nodes are affected by environment factors in real deployment. Thus, references [19–22] explored probabilistic coverage for sensor networks from different aspects. Assuming probabilistic coverage for sensors, an error model targeting location estimation application was proposed [19]. A signal strength based approach is used that model a probabilistic function that depends on distance between sensor and object. In [20], the authors gave an analytic model based on probabilistic coverage to track a moving object in the sensor field. This approach assumed that node deployment is dense enough to support duty cycling to save energy at the cost of providing probabilistic coverage. Ref. [21] presented a grid-based clustered approach to evaluate the detection probability. The cluster head is responsible for calculating the probability of detection at grid point. This approach assumes all nodes are mobility capable and cluster head can direct nodes to adjust their positions in the topology for the gain of detection probability. And ref. [22] designed a probabilistic coverage algorithm to evaluate area coverage in randomly deployed wireless sensor networks. The algorithm takes into account the variation in sensing behavior of deployed nodes and adopts a probabilistic approach in contrast to widely used idealized unit disk model.

Apart from area coverage mentioned earlier, another kind of coverage-barrier coverage is also of great importance in some scenarios, where sensors form a barrier rather than regular region (disc or square) for intrusion detection. The concept of barrier coverage first appeared in [23]. Simulations were performed in [24] to find the optimal number of sensors to be deployed to achieve barrier coverage. Ref. [25] proposed efficient algorithms which can quickly determine whether a region is k-barrier covered after deploying the sensors, and established the optimal deployment pattern to achieve k-barrier coverage when deploying sensors deterministically.

In many working environment, it is necessary to make use of mobile nodes, which can move to appropriate places to provide required coverage. Based on Voronoi diagrams, ref. [26] designed two sets of distributed protocols for controlling movement of nodes to meet required coverage. To maximize coverage, ref. [27] described an algorithm for deploying nodes one-at-a-time into an unknown environment with each node make use of information gathered by previous deployed nodes to determine proper deployment location. All such work above concern much more on protocol and algorithm design than analysis. The fundamental problems facing the deployment of sensor networks are: how many sensors need to be deployed so that each point is covered by at least one node or k nodes and network connectivity is ensured; what's the relationship between sensing or communication radius and network coverage as well as connectivity.

All such work above concern much more on protocol and algorithm design than analysis. The fundamental problems facing the deployment of sensor networks are: how many sensors need to be deployed so that each point is covered by at least one node or k nodes and network connectivity is ensured; what's the relationship between sensing or communication radius and network coverage as well as connectivity. In [28], author derived necessary and sufficient conditions for 1-coverage and 1-connectivity when n nodes are deployed in a $\sqrt{n} \times \sqrt{n}$ grid and each node is allowed to fail independently with probability $(1 - p)$. Ref. [29] first obtained the sufficient and necessary condition for random grid network to cover the unit square area as well as ensure that the active nodes are connected. Ref. [30] extended its work and considered three kinds of deployment for sensor networks: $\sqrt{n} \times \sqrt{n}$ grid, random uniform (for all n nodes) and Poisson (with density n). Furthermore, in [31], author derived sufficient condition for k-coverage under three deployments above. Ref. [32] concluded that there exists a critical threshold of neighbors with which each sensor should connect to achieve network connectivity. Most of existing approaches are based on asymptotic analysis, which requires number of nodes goes to infinity to achieve corresponding properties. The upper and lower bound results derived from these approaches do not reveal the exact performance of an network with given number of nodes, in particular when the number is small. In [33], author studied the coarse capacity of wireless mesh networks, when number of nodes are finite. In this chapter, geometrical probability is first introduced into connectivity probability analysis and even under the circumstance that the number of nodes is finite, connectivity probability (CP) is still able to be derived effectively.

21.3 Network Model and Problems

Denote a dense sensor network on a given two-dimensional area as G with total N nodes. For simplicity, we model the entire area G as a $(A \times A)$ square where A is the length of the square. Thus A is considered as the size of the network. The network G is further divided into m^2 blocks thus each block has side-length $L = \lfloor A/m \rfloor$. We called m as the block order. In each of the blocks we assume that there exist at least n nodes and the nodes are uniformly distributed in the block with the same transmission radius (energy or power) R. Thus, the network is modeled as a 4-tuple $G(A, m, n, R)$, and $N = m^2 n$ and n is the number of nodes that may reside in each block. For the given network model, we are going to tackle the following problems for the sensor network:

Problem 21.1 Given the network $G(A, m, n, R)$, what is the lower bound of the connection probability P?

Problem 21.2 Given desired connection probability P for the network $G(A, m, n, R)$, what is the minimal transmission range R required for each node in $G(A, m, n, R)$ that P can be maintained?

Problem 21.3 Given desired connection probability P for the network $G(A, m, n, R)$, the network size A, what is the minimum number of nodes n that the connection probability can be held?

In the following section, we intend to provide our solutions to tackle the problems.

21.4 Analysis of Connectivity for Two Adjacent Blocks

Let us consider a simple case with only two square blocks. The CP between the two adjacent blocks (called level-0 block) is taken as the base toward the computation of CP for the entire sensor network. In this section, we first use the geometrical probability approach to compute the CP of two adjacent blocks with one node residing in each block. Based on the result, we approximate the lower bound CP when n nodes reside in each level-0 blocks.

21.4.1 Geometrical Probability Approach

We consider two adjacent blocks with one node as shown in Figure 21.1. To ensure the acceptable transmission quality or reliability, a link between two mobile nodes should remain a sufficient signal-to-noise ratio. Therefore, the link distance d between the two nodes is set less than the transmission range R.

To derive CP of the nodes $p(r)$, we first introduce Lemma 21.1. It can be seen that when $R = 1.414L$, then on sensor's transmission range cover the whole level-0 block.

Lemma 21.1 As shown in Figure 21.1. Let P and Q be two independent nodes, when P and Q are distributed uniformly in the left and right block, respectively. Let d be the distance between P and Q and the connectivity density function be $g(u)$ where $u = d(2)$. Thus $g(u)$ is given by

$$g(u) = \begin{cases} \sqrt{u} - \frac{u}{2}, & 0 \le u \le 1, \\[4pt] u - 2\sqrt{u} - 2\sqrt{u-1} - 2\arcsin(1/\sqrt{u}) \\ \quad +\pi + \frac{3}{2}, & 1 \le u \le 2, \\[4pt] 2\sqrt{u-1} - 2\sqrt{u} - 2\arcsin\left(\sqrt{\frac{u-1}{u}}\right) \\ \quad +\pi - \frac{1}{2}, & 2 \le u \le 4, \\[4pt] 2\sqrt{u-1} + \sqrt{u-4} + \arcsin(2/\sqrt{u}) \\ \quad -2\arccos(1/\sqrt{u}) - \frac{u}{2} - \frac{5}{2}, & 4 \le u \le 5, \\[4pt] 0, & elsewhere. \end{cases}$$

Proof Taking $a = b = 1$ in special case of theory about random points and random distance in Mathai's work [13]. We have $g_1(w_1)$, $g_2(w_2)$, and $g(u)$ as following

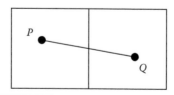

Figure 21.1 **1-Nodes case.**

$$g_1(w_1) = \begin{cases} \frac{1}{2} & 0 \le w_1 \le 1 \\ 2w_1^{-\frac{1}{2}} - 1 & 1 \le w_1 \le 4 \\ 0 & otherwise \end{cases}$$

$$g_2(w_2) = \begin{cases} w_2^{-\frac{1}{2}} - 1 & 0 \le w_2 \le 1 \\ 0 & elsewhere \end{cases}$$

$$g(u) = \int_v g_1(v)g_2(u-v)dv$$

Therefore, we can achieve an expression of $g(u)$ with respect to the duration of u as follows

Case 1. $0 \le u \le 1$

$$g(u) = \int_0^u \frac{1}{2}[(u-v)^{-\frac{1}{2}} - 1]dv = \sqrt{u} - \frac{u}{2}$$

Case 2. $1 \le u \le 2$

$$\begin{aligned} g(u) &= \int_{u-1}^1 \frac{1}{2}[(u-v)^{-\frac{1}{2}} - 1]dv + \\ &\quad \int_1^u [2v^{-\frac{1}{2}} - 1][(u-v)^{-\frac{1}{2}} - 1]dv \\ &= u - 2\sqrt{u} - 2\sqrt{u-1} - 2\arcsin\frac{1}{\sqrt{u}} + \\ &\quad \pi + \frac{3}{2} \end{aligned}$$

Case 3. $2 \le u \le 4$

$$\begin{aligned} g(u) &= \int_{u-1}^u [2v^{-\frac{1}{2}} - 1][(u-v)^{-\frac{1}{2}} - 1]dv \\ &= 2\sqrt{u-1} - 2\sqrt{u} - 2\arcsin\left(\sqrt{\frac{u-1}{u}}\right) + \pi - \frac{1}{2} \end{aligned}$$

Case 4. $4 \le u \le 5$

$$\begin{aligned} g(u) &= \int_{u-1}^u [2v^{-\frac{1}{2}} - 1][(u-v)^{-\frac{1}{2}} - 1]dv \\ &= 2\sqrt{u-1} + \sqrt{u-4} + \arcsin\left(\frac{2}{\sqrt{u}}\right) \\ &\quad - 2\arccos\left(\frac{1}{\sqrt{u}}\right) - \frac{u}{2} - \frac{5}{2} \end{aligned}$$

Case 5. $u < 0$ or $u > 5$

$$g(u) = 0$$

Based on Lemma 21.1, we have the following theorem for the derivation of $p(r)$. □

Theorem 21.1 Let P and Q be two independent nodes distributed uniformly in the left and right block, respectively. Define $p(r) = P(d \leq \sqrt{r} \cdot L)$ as the probability of the node distance within power transmission range, then $p(r)$ can be represented as:

$$p(r) = \begin{cases} 0, & r \leq 0 \\[2mm] \frac{(8r^{\frac{3}{2}} - 3r^2)}{12}, & 0 \leq r \leq 1, \\[2mm] (\frac{3}{2} + \pi)r + \frac{r^2}{2} - \frac{4r^{\frac{3}{2}}}{3} - 2r\arcsin(\frac{1}{\sqrt{r}}) \\[1mm] \quad -2\sqrt{r-1} - \frac{4(r-1)^{\frac{3}{2}}}{3} - \frac{1}{4}, & 1 \leq r \leq 2, \\[2mm] (\pi - 0.5)r + 2\sqrt{r-1} - 2r\arcsin\left(\sqrt{\frac{r-1}{r}}\right) \\[1mm] \quad -\frac{4r^{\frac{3}{2}}}{3} + \frac{4(r-1)^{\frac{3}{2}}}{3} - \frac{11}{12}, & 2 \leq r \leq 4, \\[2mm] 4\sqrt{r-4} + 2\sqrt{r-1} - 2r\arccos\left(\frac{1}{\sqrt{r}}\right) \\[1mm] \quad +2r\arcsin\left(\frac{2}{\sqrt{r}}\right) + \frac{2(r-4)^{\frac{3}{2}}}{3} + \frac{4(r-1)^{\frac{3}{2}}}{3} \\[1mm] \quad -\frac{5r}{2} - \frac{r^2}{4} + \frac{5}{12}, & 4 \leq r \leq 5, \\[2mm] 1, & 5 \leq r. \end{cases}$$

Based on Lemma 21.1, applying an integral on $g(u)$ from 0 to r results in $p(r)$. It is easy to see when the transmission of the two nodes is R, then the disconnectivity probability is derived as: $q(r) = 1 - p(r)$.

21.4.2 CP of Two Adjacent Blocks with More Nodes

Consider two adjacent blocks as shown in Figure 21.2 and $n > 1$ nodes in each of the blocks with uniform distribution. To calculate CP of the case for the two adjacent blocks, we have Lemma 21.2 below.

Lemma 21.2 Let $P_1, P_2, P_3, \ldots, P_n$ be independently random nodes in the left block and $Q_1, Q_2, Q_3, \ldots, Q_n$ be independently distributed random nodes in the right block, as shown in Figure 21.2. Denote the event A as that there exists at least one pair of nodes (P_i, Q_j) $(i, j = 1, 2, 3, \ldots, n)$ with $d \leq R$. Then the CP of the two adjacent blocks is $P(A) \geq 1 - q^n$.

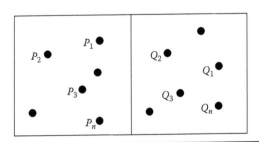

Figure 21.2 *n*-Nodes case.

The proof is straightforward and we omit the detail here. For simplicity, in the following section, we denote the $q(r)$ as q.

21.4.3 Connectivity Probability Derivation for Large Area

For a given system $G(A, m, n, R)$, we take the progressive approach to calculate the connection probability (CP) of any two nodes in the two-dimensional area G. Firstly, the whole area G is divided into $m \times m$ small blocks as depicted in Figure 21.3. We categorize the blocks into levels and denote as B_s where $s = 0, 1, 2, \ldots$. Thus, B_0 is called level-0 block. As illustrated before, the disconnection probability between the two adjacent level-0 B_0 in the same row or on the same column is q^n.

As shown in Figure 21.4, a B_1 block is composed of four level 0 blocks as B_0^1, B_0^2, B_0^3, and B_0^4 respectively. We consider two blocks as neighbors if two blocks are adjacent to each other.

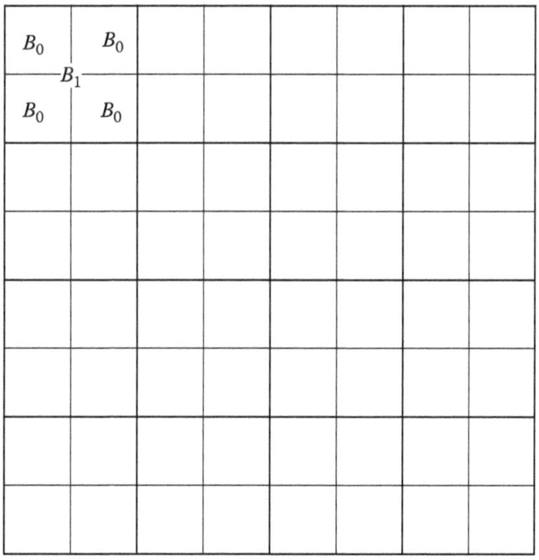

Figure 21.3   **The whole network is divided into $m \times m$ B_0 blocks.**

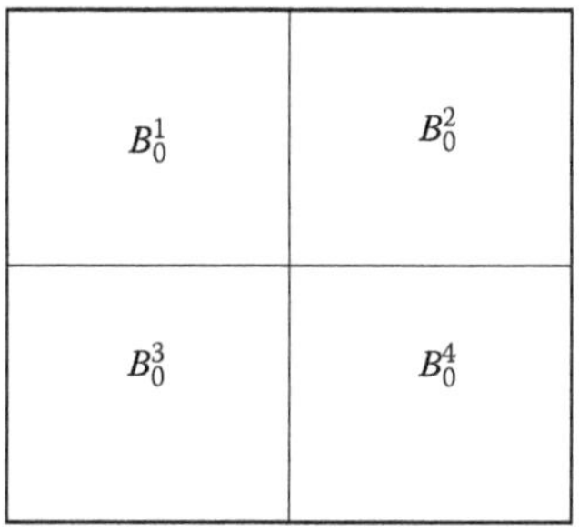

Figure 21.4   **A B_1 block is composed of four B_0 blocks.**

Therefore, B_0^1 and B_0^2, B_0^1 and B_0^3, B_0^3 and B_0^4 are neighbors, but B_0^1 and B_0^4, B_0^2 and B_0^3 are not. To derive CP for large blocks, the following events are defined:

Definition 21.1 $E(B_s^i, B_s^j)$: The event occurs when the nodes in block B_s^i and B_s^j are able to connect with each other by single or multi-hop routing, where B_s^i and B_s^j are level-s blocks and $i \neq j$.

Definition 21.2 $E(B_s^i)$: Any node in B_s^i block can connect/communicate with any other nodes in B_s^i block with single or multi-hop routing.

The progressive derivation for CP of G follows the following steps:

Step 1: As depicted in Figure 21.4, it can be seen that $E(B_0^i, B_0^j)=E(B_0^j, B_0^i)$. Let $P(E(B_0^i, B_0^j))$ be the probability that event $E(B_0^i, B_0^j)$ is true. Consider the following events: $E(B_0^1, B_0^2)$, $E(B_0^1, B_0^3)$, $E(B_0^2, B_0^4)$, and $E(B_0^3, B_0^4)$. If the four events are true, then B_1 is considered to be connected. Thus the connectivity of B_1 is defined based on the connectivity of the four B_0 blocks. Let the number nodes $n > 2$ denote CP in B_1 block as $P(E(B_1))$. Based on lemma, the following inequality holds

$$P(E(B_1)) \geq (1 - q^{\frac{n}{2}})^4 \qquad (21.1)$$

The rationale of the inequality can be explained as follows. Because each B_1 block is composed of four level 0 blocks and each level 0 block has two adjacent neighbors. For simplicity, we assume the connections of B_0 blocks are independent, thus, $n/2$ nodes may have the possibility to connect to a node in its neighbor approximately. Since there are total $m^2/4$ level 1 blocks that cover the network G, assume that q_1 is the disconnectedness probability in the sense that there exists at least one disconnected B_1 block in the network. It can be seen that

$$q_1 = 1 - P(E(B_1))^{\frac{m^2}{4}} \leq 1 - (1 - q^{\frac{n}{2}})^{m^2} \qquad (21.2)$$

Step 2: Similarly, as shown in Figure 21.5, a level 2 block B_2 is composed of four level 1 blocks: B_1^1, B_1^2, B_1^3, and B_1^4. Thus, G can be further partitioned into $m^2/16 B_2$ blocks. Let $CP = P(E(B_1^i))$ for block B_1^i, where $i = 1, 2, 3, 4$. To derive CP for B_2, we assume that all level 1 blocks are fully connected. Thus, $P(E(B_1^i)) = P(E(B_1^j)) = 1$ where $i, j = 1, 2, 3, 4, i \neq j$. Let event $E(B_1^i, B_1^j)$ be that all the nodes within B_1^i and B_1^j are connected with each other. Then we may calculate CP of level 2 blocks based on CPs of the lower level blocks, i.e., levels 0 and 1. Denote the conditional probability $P(E(B_1^i, B_1^j)|E(B_1^i)E(B_1^j))$ as the event that $E(B_1^i, B_1^j)$ occurs under the conditions that $E(B_1^i)$ and $E(B_1^j)$ are true. Consider the independency of $E(B_1^i)$, by Lemma 21.2, we have

$$P(E(B_1^i, B_1^j)|E(B_1^1)E(B_1^2)E(B_1^3)E(B_1^4)) = P(E(B_1^i, B_1^j)|E(B_1^i)E(B_1^j)0) \geq 1 - q^n$$

where $(i, j) = (1, 2), (1, 3), (2, 4), (3, 4)$.

Step 3: Let $E(A_1)$ be the event that all B_1 blocks are connected. It is easy to see that all $E(B_1)$ events are independent to each other. Based on Step 2, CP of B_2 block can be derived as:

$$P(E(B_2)|E(A_1)) = P(E(B_2)|E(B_1^1)E(B_1^2)E(B_1^3)E(B_1^4)) \geq (1 - q^n)^4$$

Similarly, there are $m^2/16 B_2$ blocks in G, let $q2$ be the disconnection probability in the sense that there exists at least one B_2 block which is not connected. Then we can infer that

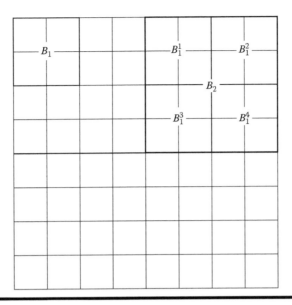

Figure 21.5   G is composed of $m^2/16$ B_2 blocks, which are formed by four B_1 blocks.

$$q_2 = P\left(\bigcup_j \overline{E(B_2^j)}\right) = P\left(\bigcup_j \overline{E(B_2^j)}|E(A_1)\right) P(E(A_1))$$

$$= P\left(\overline{\bigcap_j E(B_2^j)}|E(A_1)\right) P(E(A_1))$$

$$= \left[1 - P(E(B_2)|E(A_1))^{\frac{m^2}{16}}\right] P(E(A_1))$$

$$\leq \left[1 - (1 - q^{\frac{n}{2}})^{\frac{m^2}{4}}\right] P(E(A_1)) \tag{21.3}$$

Following the same argument above, we can get q_i.

Step 4: Consider B_i, $1 \leq i \leq k$ which is composed of four B_{i-1} blocks: B_{i-1}^1, B_{i-1}^2, B_{i-1}^3, and B_{i-1}^4. And $P(E(B_{i-1}^j))$ is the CP of block B_{i-1}^j, where $j = 1, 2, 3, 4$. Let event $E(A_{i-1})$ represent all block B_{i-1} are connected within themselves. Again it is assumed that all $E(B_{i-1})$ events are independent with each. Therefore, CP in B_i block under the condition $E(A_{i-1})$ can be calculated as

$$P(E(B_i)|E(A_{i-1})) = P(E(B_i)|E(B_{i-1}^1)E(B_{i-1}^2)E(B_{i-1}^3)E(B_{i-1}^4))$$

$$\geq [1 - (q^{\frac{n}{2}})^{2^{i-1}}]^4 = [1 - (q^n)^{2^{i-2}}]^4 \tag{21.4}$$

Consequently, we have

$$q_i = P\left(\bigcup_j \overline{E(B_i^j)}\right) = P\left(\overline{\bigcap_j E(B_i^j)}|E(A_{i-1})\right) P(E(A_{i-1}))$$

$$\leq \left[1 - (1 - (q^{\frac{n}{2}})^{2^{i-1}})^{\frac{m^2}{4^{i-1}}}\right] P(E(A_{i-1})) \tag{21.5}$$

Final Step: Let $G(A, m, n, R) = B_k$ block with $m = 2^k$ and it consists of four B_{k-1} blocks. Let events $E(C)$ and $E(\overline{C})$ be the events that G is connected or is disconnected respectively. It can be seen that the probability of $E(C)$ is less than the sum of all q_i (i.e., the sum of disconnect probability of all the lower level blocks), where $1 \leq i \leq k$, thus we have

$$P(E(\overline{C})) \leq \sum_{i=1}^{k} q_i \leq \sum_{i=1}^{k} [1 - (1 - (q^{\frac{n}{2}})^{2^{i-1}})] \frac{m^2}{4^{i-1}} P(E(A_{i-1})) \qquad (21.6)$$

where

$$P(E(A_i)) = P\left(\bigcap_j E(B_i^j)\right) = \prod_j P(E(B_i^j))$$

Therefore, the CP in the entire system $G(A, m, n, R)$ is derived as

$$P(E(C)) \geq 1 - \sum_{i=1}^{j} [1 - (1 - (q^{\frac{n}{2}})^{2^{i-1}})^{\frac{m^2}{4^{i-1}}}] P(E(A_{i-1}))$$

$$\geq 1 - \sum_{i=1}^{k} [1 - (1 - (q^{\frac{n}{2}})^{2^{i-1}})^{\frac{m^2}{4^{i-1}}}] \qquad (21.7)$$

where $k = \log m$.

21.4.4 Simulation and Performance Analysis

The simulation system is programmed with C++. We will look at CP of entire network in line with different number of nodes distributed in each block and R/L ratio. At the same time, we will also examine the relationship between the CP and transmission range of nodes.

For a network $G(A, m, n, R)$, we set $A = 3200$, thus G is a compose of 3200×3200 graphs. Initially, n is chosen from 1 to 10 nodes and distributed uniformly at one B_0 block in the whole area G. We gathered the data by creating 10,000 graphs for measuring the average CPs. We first consider the performance observation for two level 0 blocks and then for the overall network.

21.4.5 CP of Two Adjacent Blocks

In this section, we compare the mathematical derivation of the disconnectivity probability with our simulation results. Figure 21.10 shows the simulation data of the disconnectivity probabilities with cell value of $q \times 100$. The values are arranged in the number nodes and the ratio R/L. The disconnectivity probability in thick line are nearly the same as the analytical results (with error 0.01%). This is because that the transmission range R should be no less than $\sqrt{2}L$ to ensure that the nodes within block level 0 are connected.

Figure 21.6 shows the comparison between the analytical and the simulation results with different ratio R/L, where only one node resides in each block. The analytical results are compared using *function p(r)*. In this figure, CP is computed using a geometrical probability method that is almost the same as the simulation results. Figure 21.7 gives the comparison between the analytical results and the simulation data with different node number n, where the ratio R/L of the

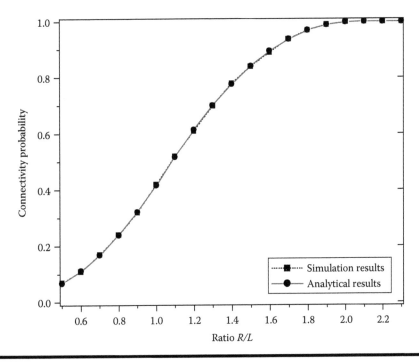

Figure 21.6 **The connectivity probability with different ratio R/L between two adjacent blocks and only one node residing in each block.**

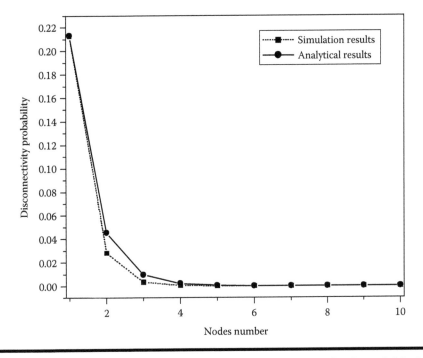

Figure 21.7 **The disconnectivity probability with different node number in each block and the transmission range R of each node is $\sqrt{2}L$.**

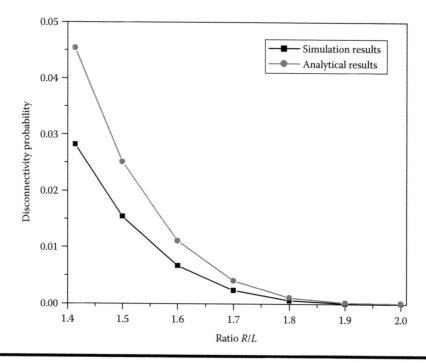

Figure 21.8 **The disconnectivity probability with different ratio R/L, where two nodes reside in each block, having $CP > 97.7\%$.**

transmission range to the side length of each block is $\sqrt{2}$. The analytical results are computed by (21.1) and it can be seen that the disconnectivity probability derived from geometrical probability is the lower bound and is very close to the simulation results. Both results shown from Figures 21.6 and 21.7 reflect the preciseness of our derivation basis for the CP in the network G.

Figure 21.8 provides the comparison between the analytical results and the simulation results with different ratio R/L, where two nodes reside in each block. The analytical results are also computed by (21.1). As shown in the Figure 21.8, the connectivity probability, which is computed in geometrical probability methods, is the lower bound and is close to the simulation results in the range of 1%.

21.4.6 Performance Observation and Analysis

Assume that the network G is divided into 16×16 blocks and there are n nodes that are uniformly distributed in a block. Figure 21.9a and b show the simulation results for the CP with different number of nodes and transmission power. From the graph, we have the following 15 observations. To achieve the desired CP, say $= 0.9$ with $n = 2, 4, 6, 8, 10$ respectively, the corresponding transmission power must be set at least as 230, 170, 150, 130, and 120 units respectively. Clearly, there is a trade-off between the number of nodes allocated to a block and the transmission power of each node. However, our analytical results provide specific indications for allocating the nodes in each block in relation with the power transmission if high CP is desired.

To compare the CP for nodes with local uniform distribution with that global (uniform) distribution, simulation about the global distribution of nodes are also collected. The simulation

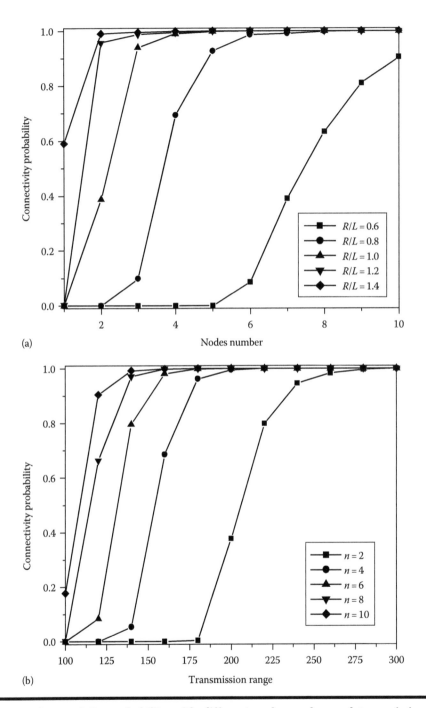

(a)

(b)

Figure 21.9 Connectivity probability with different node number and transmission power r, where $A = 3200$, $m^2 = 256$, and $N = nm^2$. (a) Analytical CP as against the theoretical transmission range ratios. (b) Simulation results of CP as against different node number and transmission range.

n / R/L	1	2	3	4	5	6	7	8	9	10
0.5	93	97	96	93	87	78	67	54	42	31
0.6	89	90	83	72	55	39	23	16	10	6
0.7	83	78	62	41	24	13	6.6	3.4	1.7	0.85
0.8	76	61	34	18	8	3.4	1.4	0.57	0.23	0.1
0.9	68	43	19	6.9	2.4	0.83	0.28	0.11	0.035	0.012
1.0	58	27	9	2.6	0.78	0.23	0.07	0.02	0.007	0.001
1.1	48	16	4.2	1.1	0.27	0.06	0.01	0.003	0.002	0
1.2	39	9.6	2.1	0.4	0.08	0.01	0.002	0	0	0
1.3	30	5.6	0.9	0.14	0.02	0.002	0	0	0	0
1.414	22	3.1	0.4	0.042	0.004	0	0	0	0	0
1.5	16	1.5	0.14	0.009	0	0	0	0	0	0
1.6	11	0.67	0.04	0.001	0	0	0	0	0	0
1.7	6.3	0.24	0.007	0	0	0	0	0	0	0
1.8	3.2	0.054	0.003	0	0	0	0	0	0	0
1.9	1.2	0.006	0	0	0	0	0	0	0	0
2.0	0.22	0.001	0	0	0	0	0	0	0	0

Figure 21.10   The disconnectivity probability q in a block which is composed of two adjacent B_0 blocks with different node number and ratio R/L, data in the table scale to $q \times 100$.

results are shown in Figure 21.11a and b. It is interesting to observe that the data demonstrated are similar to that shown in Figure 21.9. This fact also demonstrates the feasibility of our approach. The following figures show the computation results of lower bound CP based on (21.2) and values of q given by Figure 21.10. Figure 21.12 shows the CP with different node number n and the ratio R/L. To achieve CP ≥ 0.95, if we assign $R/L = 0.8, 1.0, 1.2$, and 1.4 respectively, then the number of nodes in each block B_0 should be at least as 8, 5, 4, and 3 respectively, with the number of level 0 blocks $m^2 = 256$. Figures 21.13 and 21.14 illustrate that the CP decreases as number of blocks increases under the same n and R/L. Figure 21.13 shows CPs under different R/L when $n = 4$. It can be seen that if the desired CP is set as 9.5 for $G(3200, 16, 4, R)$, then the ratio R/L should be greater than 1.1. Figure 21.14 shows the change of CP with the number of nodes. In a network $G(3200, 16, 4, R)$, we have CP $> 97.7\%$.

21.5 Conclusion

We have investigated the relationships among the number of nodes, transmission power, and CP for sensor networks. Our approach is novel in the sense that we use a geometrical probability approach to compute the connectivity probability of two adjacent blocks and the precise result can be calculated directly when $n = 1$. Based on the result, we derive the lower bound for connectivity probability when $n > 1$, which is close to the simulation results.

For overall network we provided a progressive approach to determine the lower bound of CP. The results presented are easy to understand and do not require the infinity of number of nodes for the probabilistic analysis. This is different from the previous asymptotic approach in which only the probability of connectedness converging as $n \rightarrow \infty$. Our approach can be further extended into the research for probabilistic QoS in the *ad hoc* network while connectivity, real-time bandwidth are on demand. The connectivity must be maintained while the required bandwidth is also guaranteed.

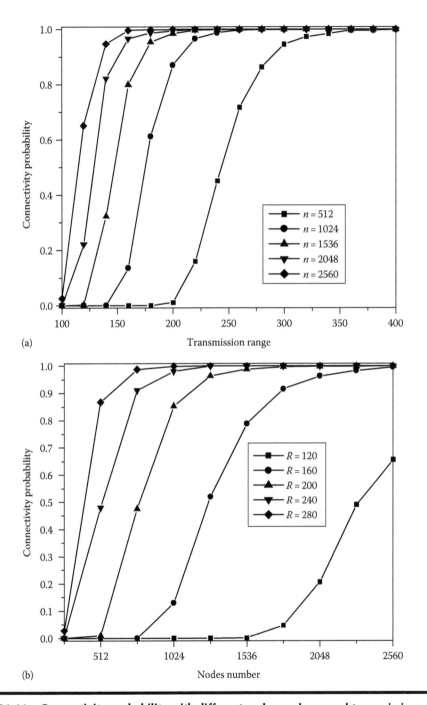

(a)

(b)

Figure 21.11 **Connectivity probability with different node number n and transmission range R, where $A = 3200$. (a) CP for different transmission ranges as against the various number of nodes. (b) CP for different number of nodes as against the transmission power.**

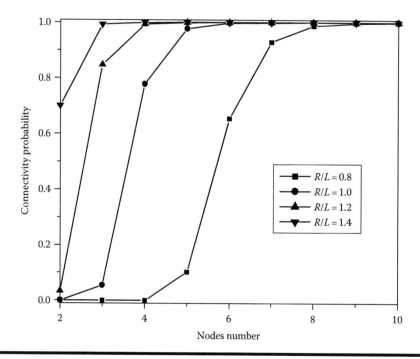

Figure 21.12 Connectivity probability with different node number *n* and the ratio *R/L*, where *m* = 16 and *q* is given in Figure 21.10.

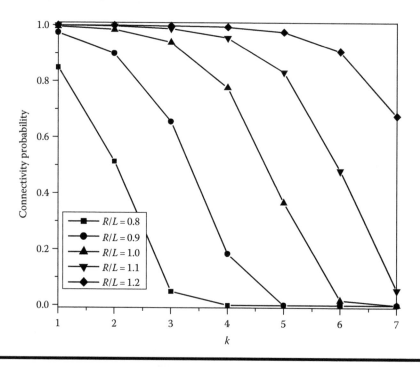

Figure 21.13 Connectivity probability with different blocks number *m* = 2^*k*^ and the ratio *R/L*, where *n* = 4 and *q* is given in Figure 21.10.

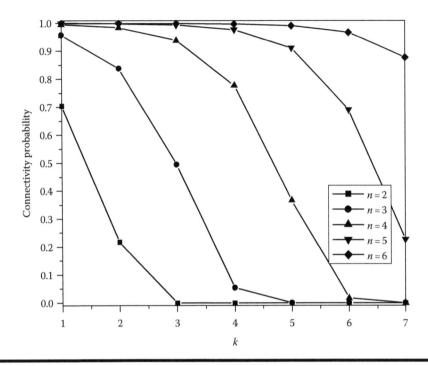

Figure 21.14   Connectivity probability with different block number $m = 2^k$ and the node number n, where $R/L = 1.0$ and q is given in Figure 21.10.

Acknowledgments

The work is supported in part by the National Science Foundation of China under Grant Nos. 61173169, 61103203, 70921001, the Program for New Century Excellent Talents in University under Grant No. NCET-10-0798.

References

1. E.M. Royer, A review of current routing protocols for ad hoc mobile wireless networks, *IEEE Personal Communications*, 1999, 4(2), 46–55.
2. S. Chakrabarti and A. Mishra, QoS issues in ad hoc wireless networks, *IEEE Communications Magazine*, 2001, 39(2), 142–148.
3. S. Basagni, D. Turgut, and S.K. Das, Mobility-adaptive protocols for managing large ad hoc networks, *Proceedings of IEEE International Conference on Communications (ICC 2001)*, Helsinki, Finland, 2001, Vol. 5, pp. 1539–1543.
4. M. Sanchez, P. Manzoni, and Z.J. Haas, Determination of critical transmission range in ad-hoc networks, *Proceeding of Multiaccess Mobility and Teletraffic for Wireless Communications 1999 Workshop (MMT 99)*, San Diego, CA, 1999.
5. L.M. Kirousis, E. Kranakis, D. Krizanc, and A. Pelc, Power consumption in packet radio networks, *Theoretical Computer Science*, 2000, 243, 289–305.
6. P. Santi, D.M. Blough, and F. Vainstein, A probabilistic analysis for the range assignment problem in ad hoc networks, *Proceedings of ACM Symposium on Mobile Ad Hoc Networking and Computing (MobiHoc)*, Long Beach, CA, 2001, pp. 212–220.

7. L. Li and J. Halpern, Minimum energy mobile wireless networks revisited, *Proceeding of IEEE International Conference on Communication (ICC 2003)*, Anchorage, AK, 2001, Vol. 1, pp. 278–283.

8. X. Wang, G. Xing, and Y. Zhang, Integrated coverage and connectivity configuration in wireless sensor networks, *Proceeding of the First ACM Conference on Embedded Networked Sensor Systems (SenSys)*, Los Angeles, CA, 2003.

9. H. Zhang and J.C. Hou, Maintaining sensing coverage and connectivity in large sensor networks, *Proceedings of NSF International Workshop on Theoretical and Algorithmic Aspect in Sensor, Ad Hoc Wireless and Peer-to-Peer Networks*, Chicago, IL, 2004.

10. M. Cardei and J. Wu, Energy-efficient coverage problems in wireless ad hoc sensor networks, *Computer Communications*, 2005, 29(4),413–420.

11. S. Slijepcevic and M. Potkonjak, Power efficient organization of wireless sensor networks, *Proceedings of IEEE International Conference on Communications (ICC 2001)*, 2001, Helsinki, Finland, Vol. 2, pp. 472–476.

12. S. Meguerdichian and M.Potkonjak, Low power 0/1 coverage and scheduling techniques in sensor networks, UCLA Technical Reports 030001, 2003.

13. C.F. Hsin and M. Liu, Network coverage using low duty-cycle sensors: Random & coordinated sleep algorithm, *Proceeding of the Third International Symposium on Information Processing in Sensor Networks*, Berkeley, CA, 2004, pp. 433–442.

14. F. Ye, G. Zhong, S. Lu, and L. Zhang, Energy efficient robust sensing coverage in large sensor networks, Technical Report, http://www.cs.ucla.edu/ yefan/coveragetech-report.ps, 2002.

15. D. Tian and N.D. Georganas, A coverage-preserving node scheduling scheme for large wireless sensor networks, *Processings of ACM Wireless Sensor Network and Application Workshop 2002*, Atlanta, GA, 2002.

16. T. Yan, T. He, and J.A. Stankovic, Differentiated surveillance for sensor networks. *Proceedings of the First ACM Conference on Embedded Networked Sensor Systems (SenSys 2003)*, Los Angeles, CA, 2003, pp. 51–62.

17. L. Wang and S.S. Kulkarni, pCover: partial coverage for long-lived surveillance sensor networks, Dept. Computer Sci. Eng., Michigan State Univ., Tech. Rep. MSC-CSE-0530, 2005.

18. Y. Liu and W. Liang, Approximate coverage in wireless sensor networks, *Proceedings of IEEE Conference on Local Computer Networks*, Sydney, New South Wales, Australia, 2005.

19. S. Ren, Q. Li, H. Wang, X. Chen, and X. Zhung, A study on object tracking quality under probabilistic coverage in sensor networks, *Mobile Computing and Communications Review*, 2005, 9, 73–76.

20. Y. Zou and K. Chakrabarty, Sensor deployment and target localization in distributed sensor networks, *ACM Transactions on Embedded Computing Systems*, 2003, 3, 1293–1303.

21. N. Ahmed, S. Kanhere, and S. Jha, Probabilistic coverage in wireless sensor networks, *Proceedings of the 30th Conference on Local Computer Networks, IEEE*, Los Alamitos, NJ, 2005.

22. G. Wang, G. Cao, and T.L. Porta, Movement-assisted sensor deployment, *Proceedings of INFOCOM*, Hong Kong, China, 2004, pp. 2313–2324.

23. D.W. Gage, Command control for many-robot systems, *Proceedings of AUVS-92*, Huntsville, AL, 1992.

24. S. Hynes, Multi-agent simulations (MAS) for assessing massive sensor coverage and deployment, Technical report, Master's thesis, Naval Postgraduate School, Monterey, CA, 2003.

25. S. Kumar, T. Lai, and A. Arora, Barrier coverage with sensors, *Proceedings of MobiCom*, Cologne, Germany, 2005, pp. 284–298.

26. A. Howard, M.J. Mataric, and G.S. Sukhatme, Mobile sensor networks deployment using potential fields: A distributed, scalable solution to the area coverage problem, *Proceedings of the 6th International Symposium on Distributed Autonomous Robotics Systems*, Fukuoka, Japan, June 2002.

27. C. Bettstetter, On the minimum node degree and connectivity of a wireless multihop networks, *Proceedings of ACM MobiHoc*, Lausanner, Switzerland, 2002, pp. 7–14.

28. J. Carle and D. Simplot, Energy efficient area monitoring by sensor networks, *IEEE Computer*, 2004, 37(2), 40–46.

29. H. Gupta, S.R. Das, and Q. Gu, Connected sensor cover: Self-organization of sensor networks for efficient query execution. *Proceedings of ACM MobiHoc*, Annapolis, MD, 2003, pp. 189–200.

30. Z. Zhou, S. Das, and H. Gupta, Connected k-coverage problem in sensor networks, *Proceedings of International Conference on Computer Communications and Networks (ICCCN)*, Chicago, IL, 2004, pp. 373–378.

31. S. Kumar, T. Lai, and J. Balogh, On k-coverage in mostly sleeping sensor networks, *Proceedings of MobiCom*, Philadelphia, PA, 2004, pp. 144–158.

32. F. Xue and P.R. Kumar, The number of neighbors needed for connectivity of wireless networks, *Wireless Network*, 2004, 10(2), pp. 169–181.

33. S. Shakkottai, R. Srikant, and N. Shroff, Unreliable sensor grids: Coverage, connectivity and diameter, *Proceedings of INFOCOM*, San Francisco, CA, 2003, pp. 1073–1083.

Chapter 22

Weighted Localized Clustering

A Coverage-Aware Reader Collision Arbitration Protocol in RFID Networks

Joongheon Kim, Eunkyo Kim, Wonjun Lee, Dongshin Kim, Jihoon Choi, Jaewon Jung, and Christian K. Shin

Contents

22.1 Introduction

In the wireless networking and mobile computing environment, because mobile devices generally are energy-constraints, the power management of mobile devices is an important issue with regard to network lifetime. Due to the limited power source of mobile devices, energy consumption has been considered as the most critical factor in designing network protocols. Facing these challenge and research issues, several approaches to prolong the lifetime of the wireless networking and mobile computing, including clustering schemes and structured schemes with a two-tiered hierarchy, have been investigated [1–5]. The clustering technology facilitates the distribution of control over the network and enables locality of communications [2]. The two-tiered hierarchical structuring method is an energy-efficient scheme for wireless networking [6]. It consists of the upper tier for communicating among cluster heads (CHs) and the lower tier for acquired events and transmitting them to CHs. However, in the clustering scheme and the two-tiered hierarchical structuring scheme, if the cluster range is larger than optimal, a CH consumes more energy than required. On the other hand, a smaller-than-necessary range results in a shortage of covering the entire network field. Therefore, we propose a novel clustering-based algorithm that aims to minimize the energy consumption of CHs under the hierarchical structure [7]. Our proposed clustering scheme, *low-energy localized clustering (LLC)*, is able to regulate the cluster radius by communicating with CHs for energy savings. We extend our basic scheme to *weighted low-energy localized clustering (w-LLC)* to cope with the case that events occur more frequently in a certain area of the sensor network field. Also when the CHs have different computing power, we need to assign weight factors to each CH. In these cases, *w*-LLC, therefore, is better than LLC in practical environment to apply the algorithm. The major application areas of *w*-LLC are "wireless sensor networks" and "RFID networks." In wireless sensor networks, sensors are deployed over the network sensing fields and perform the specific tasks of processing, sensing, and communicating capacities [8,9]. Because of the limited power source of sensors, energy consumption has been considered as the most critical factor in designing sensor network protocols. To achieve energy-efficiency in wireless sensor networks, clustering schemes and hierarchically structured schemes are proposed [1–6]. RFID network systems, also, have two-tiered hierarchical structure. In the upper tier, there are RFID readers to receive the signals from the RFID tags. In the lower tier, there are RFID tags. In the hierarchical clustering-based two-tiered network architecture, the larger the overlapping areas of clusters that RFID readers form, the higher the collision probability among the readers. We propose *weighted localized clustering for RFID networks (w-LCR)* as an application area of *w*-LLC, which minimizes the overlapping areas among clusters by regulating an RFID reader's cluster radius dynamically to minimize the RFID reader collisions. The basic feature of RFID reader collision arbitration protocol is proposed in [10]. In [10], however, we do not consider energy efficiency but guarantee perfect coverage. This chapter shows energy efficiency via performance evaluations and perfect coverage via theoretical analysis. The remainder of this chapter is organized as follows. In Section 22.2, we investigate previous work on the clustering scheme and the hierarchical structure scheme in wireless sensor networks and RFID networks. In Section 22.3, we propose *w*-LLC, a weighted dynamic localized scheme designed for hierarchical clustering protocols. We evaluate the effectiveness of w-LLC with simulations and theorem proving in Section 22.4. In Section 22.5, we apply w-LLC to RFID reader collision arbitration algorithm and, in Section 22.6, we show the performance of the coverage-aware reader collision arbitration. Section 22.7 concludes this chapter and presents the direction of future work.

22.2 Related Work

Several clustering and hierarchical structure schemes aiming to improve energy efficiency in wireless networked systems have been proposed. In the case of the wireless sensor network environment, because the energy conservation of each sensor is important in designing the wireless sensor network protocol, the architecture proposed in this chapter is better than the other network environments. Also numerous clustering schemes and hierarchical schemes are developed in wireless sensor networks such as ∼ LEACH (low energy adaptive clustering hierarchy) [3], a protocol architecture for sensor networks that combines the ideas of energy-efficient cluster-based routing with application-specific data aggregation to achieve good performance with regard to system lifetime, latency, and application-perceived quality. It preserves limited amount of energy by selecting a CH at random among sensors. By doing this, LEACH must have the energy constraint in each CH, whereas w-LLC has less energy constraint than LEACH. In [4], Gupta and Younis propose a two-tiered hierarchical clustering scheme in which a CH with less energy constraint is selected among sensors. Based on this method, it pursues an energy-efficient network management. However, a CH is selected among the sensors under restricted assumptions as assumed in [3]. Similar to LEACH, the CH of proposed scheme in [4] has more energy constraint than the CH of w-LLC, because the CH in [4] is a sensor with low energy and computing power while the CH of w-LLC has more computing power than regular sensors. FLOC (fast local clustering) [5] provides an adjustable method of cluster radius, but it is limited to one-hop or two-hop distances. The concept of 'regulating the radius of cluster' on FLOC is similar to that of w-LLC. However, w-LLC does not consider the hop count when it regulates the radius of cluster. Also in FLOC, the CHs are selected among the regular sensors in the sensor network field. Therefore, a CH in FLOC has more energy constraint than one in w-LLC. The clustering-based topology control scheme [6] consists of two tiers: (1) an upper tier for communicating between CHs and (2) a lower tier for sensing, processing, and transmitting by sensors. It has similarity on hierarchical structuring concept like w-LLC. However, its performance depends on the radius of each cluster; as a cluster is increased covering the whole sensor network area, the energy consumption can be increased. On the other hand, w-LLC can regulate the radius of each cluster to minimize the energy consumption of each CH and consider the local properties of the region by assigning weight functions to each CH. To cite another example, the RFID network system also has a two-tiered clustering-based hierarchical network architecture. RFID systems consist of RFID readers as an upper layer and RFID tags as a lower layer in a clustering-based logical hierarchical model. After clustering tags in the lower layer, RFID readers recognize RFID tags in the cluster of RFID reader and send the information stored or delivered in RFID tags to a server by using multi-hop routing among RFID readers. RFID tags have one-hop communication with RFID readers, which allows us to exploit general clustering schemes. Therefore, the RFID system has a similar architecture to the hierarchical clustering-based two-tiered sensor network architecture proposed in [6].

22.3 *w*-LLC: Weighted Low-Energy Localized Clustering

The *w*-LLC aims to minimize overlapping areas of clusters by regulating the cluster range of each cluster. If the cluster range is larger than optimal, a CH consumes more energy than required. On the other hand, a smaller cluster range than optimal results in the entire wireless network field of lower tier not being covered. The basic system model is shown in Figure 22.1.

In *w*-LLC, the whole wireless network area of low tier is totally covered by CHs and the CHs in network field of upper tier consider their weights assigned by *penalty functions* and *reward functions*. For achieving more energy efficiency, a server computes equations presented later such that CHs can consume energy as less as possible. Energy-efficient radii for each cluster are calculated based on the objective functions given by *w*-LLC. *w*-LLC consists of two phases and one policy. Following are what we assume in *w*-LLC:

- The proposed architecture has a two-tiered hierarchical structure. The upper tier consists of CHs and the lower tier consists of mobile devices.
- A server knows the position of each CH.

22.3.1 Initial Phase

In the initial phase, the CHs deployed at random construct a triangle to determine a *cluster radius decision point* (*CRDP*) that is able to minimize overlapping areas of clusters as shown in Figure 22.1. The distance between CRDP and each point can be estimated as the radius of each cluster. *Delaunay triangulation* [11,12], which guarantees the construction of an approximate equilateral triangle, is used for constructing a triangle consisting of three CHs. The construction of equilateral triangles

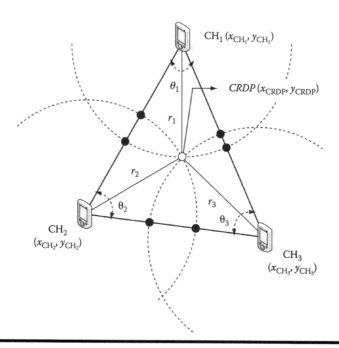

Figure 22.1   System model for *w*-LLC.

leads to load-balanced energy consumption of each CH. By the concept of load-balancing, the prolonging of network lifetime can be achieved [13].

22.3.2 Weighted Localized Clustering Phase

The cluster radius of three points, including the construction of a triangle, can be dynamically controlled by using a CRDP as a pivot. In LLC, our previous work [7], the goal is to determine the CDRPs that minimize the overlapping areas of clusters by finding optimal cluster radii. However, in some cases where a subset of CHs are assigned specific tasks or where the computing power of each CH is not the same, they need to assign weight factors to each CH. Therefore, we suggest an extended algorithm of LLC, *w-LLC*, by assigning priorities to the CHs using weight functions.

22.3.2.1 NLP-Based Approach for w-LLC

In LLC, by using a NLP-based approach, a CRDP is determined by an energy-constrained objective function as Equation 22.1:

$$min : f(\Omega) = \frac{1}{2}\sum_{k=1}^{3}\theta_k \cdot r_k^2 \cdot \frac{E_k}{\frac{1}{3}\sum_{j=1}^{3}E_j} - S_{triangle} \qquad (22.1)$$

s.t.

$$r_i^2 = (x_{CRDP} - x_i)^2 + (y_{CRDP} - y_i)^2$$

$$\Omega = \begin{bmatrix} r_1 & \theta_1 & E_1 \\ r_2 & \theta_2 & E_2 \\ r_3 & \theta_3 & E_3 \end{bmatrix}$$

In Equation 22.1, θ_k denotes the angle value of CH_k, r_k denotes the distance between CRDP and CH_k, and E_k denotes the energy state of CH_k. Also $S_{triangle}$ is the triangle area by *Delaunay* triangulation. Figure 22.2 shows the conceptual diagram for Equation 22.1. The purpose of Equation 22.1 is to minimize the overlapped cluster coverage by considering the residual energy of each CH. If the overlapping area becomes larger, CHs may consume more and more energy. Therefore, maintaining a minimized overlapping area can save considerable energy of each CH. To compute the position of a CRDP that can minimize the overlapping areas, we obtain the areas of three sectors that have the angles of CHs and the area of $S_{triangle}$. For the purpose of computing the area of $S_{triangle}$, Heron's formula is used. A server calculates the area of $S_{triangle}$ using this formula. As for an NLP method to solve Equation 22.1, we use a *limited memory Broyden Fletcher Goldfarb-Shanno (L-BFGS) method* [14–16], one of the most efficient NLP methods for solving unconstrained optimizationed problems. In [14], the theoretical concept of non-linear programming is presented. In [15] and [16], the L-BFGS algorithm is presented. The other values except the angular values can be transmitted to the server; however, all the CHs should be aware of these values. Taking communication overheads into account, it is desirable to compute these at the server rather than CHs since a server is less energy constrained than CHs or mobile devices. The server eventually obtains the energy state and the positions of each CH. Each angular value is computed by the second law of cosine in the server. As shown in Equation 22.1, the CH of

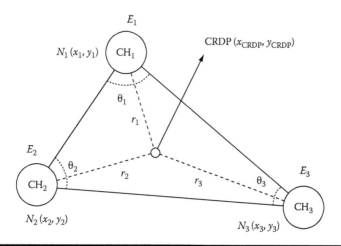

Figure 22.2   Notations for NLP-based approach for *w*-LLC.

a NLP-based approach has the same priorities. However, in certain cases, we need to assign a different weight to each CH. If the CH in certain areas is more important than that in other CHs, we need to assign higher priority to the CH. Also if events occur in some endemic-specific area, the CH in that area must be assigned with a higher priority. Therefore, the consideration of weight functions is necessary. Finally, in the case of the CHs having different computing power, different weight factors must be considered. If one CH has more computing power than the other CHs, the CH needs to enlarge its cluster range to preserve the energy of the neighbor CHs to achieve a load-balancing effect. Therefore, we also need to assign, weight functions to each CH with respect to computing power of each CH. We can consider *penalty functions* and *reward functions* for weight functions. If the penalty function of a CH has a large value, the CH must reduce its cluster radius. If the reward function of a CH has a large value, the CH must enlarge its cluster radius. In other words, a smaller penalty function value means a higher priority for a CH, while a smaller reward function value indicates a lower priority. Equation 22.2 shows the objective function of a NLP-based approach for *w*-LLC:

$$min : f(\Omega) = \frac{1}{2}\sum_{k=1}^{3}\theta_k \cdot r_k^2 \cdot \Phi \cdot \Psi - S_{triangle} \qquad (22.2)$$

s.t.

$$r_i^2 = (x_{CRDP} - x_i)^2 + (y_{CRDP} - y_i)^2$$

$$\Phi = 3\prod_{l=1}^{m}\frac{\phi_{l,k}(\vec{x})}{\sum_{i=1}^{3}\phi_{l,i}(\vec{x})}$$

$$\Psi = 3\prod_{g=1}^{n}\frac{\frac{1}{\psi_{g,k}(\vec{x})}}{\sum_{i=1}^{3}\frac{1}{\psi_{g,i}(\vec{x})}}$$

$$\Omega = \begin{bmatrix} r_1 & \theta_1 & \phi_{1,1}(\overrightarrow{x})\ldots\phi_{m,1}(\overrightarrow{x}) & \psi_{1,1}(\overrightarrow{x})\ldots\psi_{n,1}(\overrightarrow{x}) \\ r_2 & \theta_2 & \phi_{1,2}(\overrightarrow{x})\ldots\phi_{m,2}(\overrightarrow{x}) & \psi_{1,2}(\overrightarrow{x})\ldots\psi_{n,2}(\overrightarrow{x}) \\ r_3 & \theta_3 & \phi_{1,3}(\overrightarrow{x})\ldots\phi_{m,3}(\overrightarrow{x}) & \psi_{1,3}(\overrightarrow{x})\ldots\psi_{n,3}(\overrightarrow{x}) \end{bmatrix}$$

The notations of Equation 22.2 are the same as the notations of Figure 22.1 and Equation 22.1. In Equation 22.2, w-LLC assigns a weight function to each CH, where $\phi_{l,k}(\overrightarrow{x})$ and $\psi_{l,k}(\overrightarrow{x})$ represent a penalty function and a reward function, respectively. This objective function, Equation 22.2, has m penalty functions and n reward functions. The example of penalty function and reward function is "residual energy" and "priority," respectively. If residual energy is quite small in a CH, the CH must preserve its residual energy. In this case, the CH becomes more important than the other CHs with respect to energy conservation. The CH needs to reduce its cluster radius for preserving its energy for load balancing. Therefore, the smaller the residual energy of a CH, the higher the weight of the CH (because this objective function tries to conserve the energy of CH which has less residual energy than the other CHs). Therefore, the "residual energy" is considered as the example of "penalty function" (because one variable is higher, the depended variables become lower). If the priority of a CH is higher, the importance of CH also becomes higher. Therefore, the priorities can be an example of reward functions.

22.3.2.2 VC-Based Approach for w-LLC

In an NLP-based approach, additional computation overheads may be generated due to an iterative NLP method for solving the objective function in Equation 22.2. We thus consider another method based on vector computation to reduce the computation overheads to find the optimal position of CRDP. The notations of a VC-based approach for determining a CRDP is depicted in Figure 22.3. The factors of vector values for calculating the vector coordination is added. It initially executes an NLP computation. Weight factors do not need to be considered in the initial NLP computation. The basic notations and the objective function for obtaining minimized energy consumption in the VC-based method are the same as those in the NLP-based method, except that the former

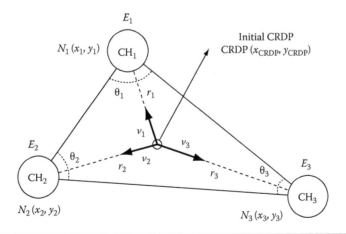

Figure 22.3   Notations for VC-based approach for *w*-LLC.

does not consider any energy constraints and weight factors for simplifying the NLP computation. The equation to obtain the initial position of CDRP is described as an objective function Equation 22.3:

$$min : f(\Omega) = \frac{1}{2} \sum_{k=1}^{3} \theta_k \cdot r_k^2 - S_{triangle} \tag{22.3}$$

s.t.

$$r_i^2 = (x_{\text{CRDP}} - x_i)^2 + (y_{\text{CRDP}} - y_i)^2$$

$$\Omega = \begin{bmatrix} r_1 & \theta_1 \\ r_2 & \theta_2 \\ r_3 & \theta_3 \end{bmatrix}$$

As time passes, however, the objective function may move a CRDP toward the CH that has consumed the most amount of energy. In the objective function of a VC-based approach, we do not consider the energy factor. In the iterative procedure, we need to consider the energy factor. Therefore, the next position of CRDP is determined by using Equation 22.4 under the iteration policy:

$$\text{CRDP}_{i+1} = \text{CRDP}_i - \sum_{k=1}^{3} \frac{E_k}{\frac{1}{3} \sum_{j=1}^{3} E_j} \cdot \frac{v_k}{\|v_k\|} \tag{22.4}$$

$$\text{s.t. CRDP}_k = (x_{\text{CRDP}_k}, y_{\text{CRDP}_k})$$

Equation 22.3, the objective function of "VC-based approach," does not consider the energy state of each CH when it determines the position of CRDP. Therefore, by using Equation 22.4, we can consider the energy state of each CH. By Equation 22.4, the CH that has more energy consumption is assigned with higher priority, whereas the CH that has less energy consumption is assigned lower priority. By this operation, we can obtain load-balanced energy consumption and the optimal position of a CRDP. Using this VC-based approach, we can preserve the energy of CHs as much as the NLP-based approach so that we can reduce computation overheads. In the VC-based approach, we have to update the coordination of CRDP given previously. In the NLP-based approach, however, recomputation to find the optimal solution to the objective function using the NLP method can be overburdened. Therefore, the repeated vector computation is much simpler than the NLP-based computation with respect to algorithm complexity. To apply the concept of assigning the weight factors in CHs, we can update Equation 22.4 as

$$\text{CRDP}_{i+1} = \text{CRDP}_i - \sum_{k=1}^{3} \Phi \cdot \Psi \cdot \frac{v_k}{\|v_k\|} \tag{22.5}$$

s.t.

$$\text{CRDP}_k = (x_{\text{CRDP}_k}, y_{\text{CRDP}_k})$$

$$\Phi = 3 \prod_{l=1}^{m} \frac{\phi_{l,k}(\vec{x})}{\sum_{l=1}^{3} \phi_{l,i}(\vec{x})}$$

$$\Psi = 3 \prod_{g=1}^{n} \frac{\frac{1}{\psi_{g,k}(\vec{x})}}{\sum_{i=1}^{3} \frac{1}{\psi_{g,i}(\vec{x})}}$$

by assigning weight functions. The notations of Equation 22.5 are the same as Equation 22.2 and Equation 22.4. By remodeling Equation 22.4 as Equation 22.5, we can consider the many weight factors. As shown in Equation 22.5, we can consider m penalty functions and n reward functions for finding the coordination of CRDP. The coordination of CRDP can provide minimized energy consumption in each CH.

22.3.3 Iterative Policy

Events can ∼ occur most frequently in some specific areas where a certain CH consumes its energy a lot. In this situation, the CHs in the target region will consume more energy and the operation will be relatively more important than the other CHs in the lower tier of the network. Then a server has to change the radius of clusters to balance energy consumption of each CH and to preserve the energy of the CH which has higher priority than the other CH. Moreover, since the server has an iteration timer, if no events occur until the timer is expired, the server requests energy information to all the CHs and starts the *w*-LLC algorithm. The server collects data including energy information of CHs and executes *w*-LLC periodically. Based on this design rationale, we are able to make the "iteration policy of *w*-LLC," as shown in Figure 22.4.

As shown in Figure 22.4, the server uses the *expire time* and *threshold*. By using these values the *w*-LLC can estimate the computation time of the objective functions of the NLP-based approach and the VC-based approach.

22.4 Effectiveness of w-LLC

Effectiveness of *w*-LLC is shown through "performance evaluation" and "guarantee of perfect coverage." In performance evaluation, Section 22.4.1, we show the energy efficiency of *w*-LLC. Also in guarantee of perfect coverage, Section 22.4.2, we show the proof that the hierarchical sensor network filed is being covered totally.

22.4.1 Performance Evaluation

In this simulation result section, we evaluate and analyze the performance of *w*-LLC through simulation-based performance evaluation. The notations used in simulation are described in Table 22.1.

As a simulation environment, we consider the wireless sensor network environment. To consider the wireless sensor network environment with the network architecture proposed in this chapter,

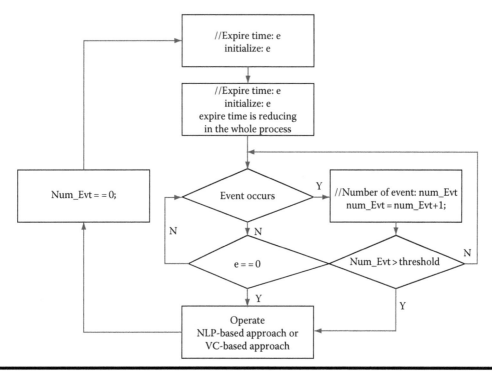

Figure 22.4 Flowchart for "iteration policy of *w*-LLC."

Table 22.1 List of Symbols Used in Simulation-Based Performance Evaluation

E_k	Residual energy of CH_k
N	Number of CHs
r_k	Distance between CRDP and CH_k
$\phi_{i,j}(\vec{x})$	i th penalty function of CH_j
$\psi_{i,j}(\vec{x})$	i th reward function of CH_j

the wireless network system consists of the upper tier for communicating among CHs and the lower tier for sensing events and transmitting them to CHs. There are "clustering scheme with fixed radius (FR)" and LLC for the comparison studies with *w*-LLC. We consider two kinds of performance evaluation metrics. We show the "comparison of the variance and average of the energy status in each CH of FR, LLC, and *w*-LLC" and "comparison of the residual energy in each CH of FR, LLC, and *w*-LLC." Tables 22.2 and 22.3 present the average lifetime and variance of the five CHs evaluated by FR, LLC, and *w*-LLC. We generate the events in some specific areas and assign weight functions to the CH. In the case of FR where cluster radii are fixed, the variance is constant. *w*-LLC shows less variance than LLC, which denotes the fairness of energy consumption. As shown in Figure 22.5, FR shows the worst performance among others in terms of the average lifetime of five CHs. When events occur in certain specific areas more frequently, *w*-LLC shows

Table 22.2   Comparison of Var. among FR, LLC, and *w*-LLC

Scheme	FR	LLC	w-LLC
Round #1	0.9151	9.1334	6.6106
Round #2	0.9151	3.7465	1.3900
Round #3	0.9151	1.6010	1.6251
Round #4	0.9151	0.6715	0.3749
Round #5	0.9151	3.8771	0.7259

Table 22.3   Comparison of Avg. among FR, LLC, and *w*-LLC

Scheme	FR	LLC	w-LLC
Round #1	13.9311	34.2577	35.5435
Round #2	13.9311	34.0341	35.0131
Round #3	13.9311	33.5493	35.2474
Round #4	13.9311	33.8866	34.6837
Round #5	13.9311	33.9239	35.1925

better performance than LLC. We also consider two different scenarios: (1) sensing events occur around certain hot spots, named focused sensing, and (2) events occur evenly across the sensor network, named fair sensing. As shown in Tables 22.2 and 22.3, residual energies in FR are further quickly consumed than LLC and *w*-LLC. FR shows a longer lifetime in fair sensing than focused sensing. The CHs in the hot spots of focused sensing exhaust their own energies rapidly. In focused sensing, *w*-LLC is the most energy efficient, since it uses weight functions that reflect territorial characteristics. In fair sensing, the weights of all CHs in *w*-LLC are mostly equal.

We evaluate the performance of a scenario where sensing events intensively occur in a certain area of a network. In such an environment, the weight factors of *w*-LLC can deal with the situation very well. As shown in Figure 22.6, *w*-LLC outperforms the earlier version, LLC. The range of overlapping area per total area of LLC is between 0.30 and 0.34. However, the range of overlapping area per total area of *w*-LLC is between 0.26 and 0.31. Therefore, *w*-LLC is more energy efficient than LLC. Furthermore, as time passes, the gap of between LLC and *w*-LLC gets larger and larger. This means that the effectiveness of load-balancing disappears as time passes. Therefore, as the time passes, *w*-LLC can become more energy-efficient than LLC.

If the CHs in the system achieve load balancing, the lifetime of the system can increase [13]. As shown in Figure 22.7, the standard deviation of CHs in LLC is 3.171049598 and the average lifetime of CHs in LLC is 18.5 min. On the other hand, the standard deviation of CHs in *w*-LLC is 2.67498702 and the average lifetime of CHs in *w*-LLC is 20.6 min. By the result of this simulation, we show that *w*-LLC is more energy efficient than LLC with regard to lifetime of CHs.

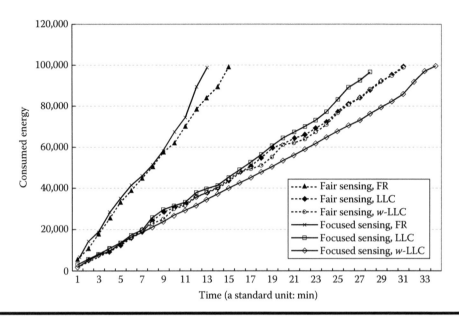

Figure 22.5   Comparison of the residual energy.

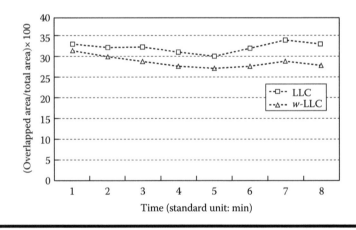

Figure 22.6   Percentage of overlapped area/total area (events intensively occur in certain areas).

22.4.2 Guarantee of Perfect Coverage

The main contributions of *w*-LLC are (1) *energy efficiency* and (2) *perfect coverage*. Energy efficiency has been discussed in a previous section. Perfect coverage will be shown in this section by a theoretical analysis.

Theorem 22.1   By using *w*-LLC, perfect coverage can be achieved. That is, the entire sensor network can be covered by clusters in the upper layer.

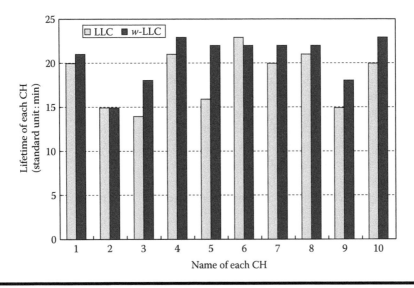

Figure 22.7   Comparison of LLC and *w*-LLC with regard to load balancing on CHs.

Proof

By Corollaries 22.1 through 22.3, Theorem 22.1 can be proven. The proofs of Corollaries 22.1 through 22.3 are described in Appendices 22.A through 22.C, respectively.

Corollary 22.1   In sector OAA' in Figure 22.8, $\overline{OQ'}$, which is constructed by central point O and any point, the Q' in Figure 22.8, on the arc AA', meets $\overline{QA}$ or $\overline{QA'}$, necessarily.

Corollary 22.2   Sector OAA' in Figure 22.8 covers rectangle $OAQA'$, perfectly.

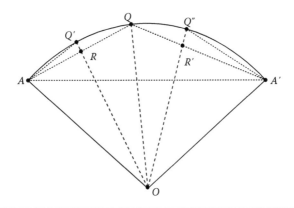

Figure 22.8   Notations for Corollary 1.

Corollary 22.3 The triangle constructed by *Delaunay* triangulation in an initial phase is covered by three rectangles, perfectly.

By Corollaries 22.1 through 22.3, the `rectangle` $OA_1N_1B_1$ is covered by `sector` $A_1N_1B_1$. The `rectangle` $OA_2N_2B_2$ is covered by `sector` $A_2N_2B_2$. The `rectangle` $OA_3N_3B_3$ is covered by `sector` $A_3N_3B_3$. Also, `triangle` $N_1N_2N_3$ = `rectangle` $OA_1N_1B_1$ + `rectangle` $OA_2N_2B_2$ + `rectangle` $OA_3N_3B_3$. Then, by the `triangle` $N_1N_2N_3$, the *Delaunay* triangle which is constructed in an initial phase, is covered by three rectangles, the `triangle` $N_1N_2N_3$ is covered by the three sectors, perfectly. That is,

$$\text{\texttt{triangle}} \ N_1N_2N_3 \subset \text{\texttt{three rectangles}} \subset \text{\texttt{three sectors}}$$

Therefore, based on these three corollaries, Theorem 22.1 is proved.

22.5 Adaptability for RFID Reader Collision Arbitration

Radio frequency IDentification (RFID) is the next generation wireless communication technology applicable to various fields such as distribution, circulation, transportation, etc. RFID is a non contact technology that identifies objects attached with tags. Tags consist of a microchip and antenna. RFID readers obtain the information of objects and surroundings through communication with tag antennas [17]. Minimizing collisions among RFID reader and tag signals has a substantial effect on performance because tag recognition rate effectively determines RFID system performance. However, the RFID reader and tag collision problem has not received much attention before we proposed the basic feature of coverage-based RFID reader anticollision algorithm in [10]. In [10], however, we did not consider energy efficiency and show "guarantee of perfect coverage." Therefore, as one of the promising application areas of *w*-LLC, RFID networks can be considered to develop an RFID reader collision arbitration protocol. RFID systems consist of RFID readers that are fixed a priori as an upper layer and RFID tags as a lower layer in a two-tiered logical hierarchical model. After clustering tags in the lower layer, RFID readers recognize tags in the cluster and send information stored in tags to a server by using multi-hop routing among RFID readers. RFID tags have one-hop communication with RFID readers, which allows us to exploit general clustering schemes. In other words, this RFID system has a similar architecture to *w*-LLC. Based on the *w*-LLC-like system architecture, we can reduce the overlapping area of RFID reader clusters. By reducing the overlapping area, we can reduce the collision of signals from RFID readers.

22.5.1 Classification on RFID Reader Collision Arbitration

Research efforts for the RFID collision arbitration problem [17] can be classified into two groups: RFID reader collision arbitration protocols [18] and RFID tag collision arbitration protocols [19]. The main focus of this chapter will be to develop an RFID reader collision arbitration protocol based on *w*-LLC. The approaches to RFID reader collision arbitration protocols are further divided into a scheduling-based approach, which prevents RFID readers from simultaneously transmitting signal to an RFID tag and coverage-based approach. A widely known scheduling-based protocol is the *Colorwave* proposed in [18]. The *Colorwave* performs scheduling instructed by RFID readers using distributed color selection (DCS) or enhanced DCS, variable-maximum distributed color selection (VDCS), after it divides medium into time slots. The other reader collision arbitration approach is the coverage-based approach, which minimizes collision possibility by optimizing the overlapping areas of clusters that RFID readers have to cover up. Under the concept of a coverage-aware RFID reader collision arbitration mechanism to minimize the overlapping area of the cluster, we can apply w-LLC to overcome reader collision problems occurring among RFID readers that have different computing power.

22.5.2 Coverage-Aware RFID Reader Collision Arbitration: w-LCR

In the hierarchical clustering-based two-tiered network architecture, the larger the overlapping areas of clusters that RFID readers form, the higher the collision probability among the readers. We propose an algorithm that minimizes overlapping areas among clusters by regulating RFID readers cluster radii dynamically to minimize reader collisions. Our initial version of the dynamic cluster coverage algorithm, named LLC, is presented in [7]. Also, each RFID reader in the RFID network has different computing powers. Therefore, it is hard to apply LLC to the RFID networks directly. By this constraint, we can use w-LLC for considering the different computing powers of each RFID reader. The w-LLC is to minimize the energy consumption of each cluster head by dynamically adjusting cluster radius. Based on this, this chapter proposes the (w-LCR) scheme to minimize the overlapping areas and then minimizes RFID reader collisions.

22.6 Performance Evaluation of w-LCR

A simulation study is conducted to evaluate the performance of w-LCR. RFID tags were randomly deployed and RFID reader was placed according to each simulation metric. Our simulations were designed to evaluate the effect of (1) probability of RFID reader collision, and (2) energy consumption. We compare w-LCR against the method that has a fixed cluster radius. The cluster radius is fixed but it has to be extended to the distance to allow communication among all RFID readers. In Figure 22.1, the cluster radius is determined by the longer distance between one RFID reader (applying RFID network to Figure 22.1, the CH means RFID reader) and other RFID readers.

22.6.1 Possibility of RFID Reader Collision

Figure 22.9 shows the possibility of collision in FR and w-LCR in RFID networks. w-LCR have much lower possibility of collision than FR methods because w-LCR algorithm regulates cluster radius dynamically and minimizes the overlapping (tag-colliding) areas in RFID networks. As shown in Figure 22.10, the possibility of collision of w-LCR is between 0.09 and 0.11. However, the possibility of collision of FR is between 0.25 and 0.35. Therefore, w-LCR has better performance than FR (almost three times). Furthermore, as shown the shape of graph, FR has more variance than w-LCR. The variance of FR is 0.1 and that of w-LCR is 0.02. FR has more variance than w-LCR (almost five times). The load-balancing concept based on the weight functions of w-LCR can make the variance smaller than FR as shown the Figure 22.9. Also, as shown the shape of the FR graph, the possibility of collision increases continuously. On the other hand, the shape of w-LCR graph does not increase continuously. This situation also occurs under the effect of load-balancing concept of weight functions. Figure 22.10 shows that the possibility of collision increases continuously as the number of RFID reader increases. Here, we show the rate of overlapped areas

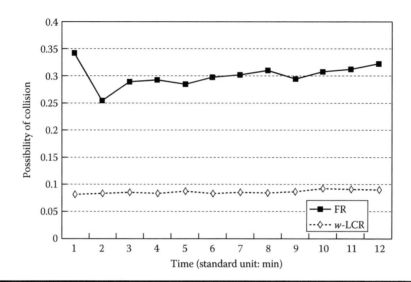

Figure 22.9   Possibility of collision in RFID networks.

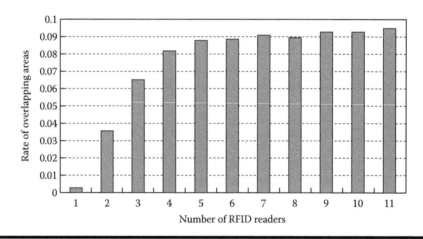

Figure 22.10   Possibility of collision on the number of RFID readers.

on the given number of RFID readers when we use the *w*-LCR. Of course, RFID reader collision will not occur if there is only one RFID reader in the RFID networks. However, only one RFID reader may not cover the whole area of a big RFID network because of the limited transmission ability. Trivially, it will be true that the more the number of RFID readers, the more collisions occur (but the rate of collisions linearly does not increase). Therefore, the number of RFID readers needs to be appropriately handled according to the size of the RFID networks as well as hardware specifications of RFID readers.

22.6.2 Energy Consumption

w-LCR consider the residual energy state of RFID readers as another important metric. We compare the performance of proposed schemes in terms of energy consumption in RFID networks.

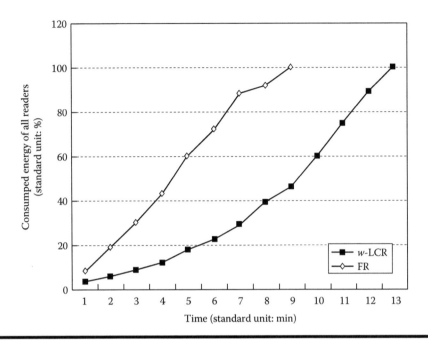

Figure 22.11   Energy consumption in RFID networks.

Figure 22.11 shows the results of simulation. As shown in Figure 22.11, *w*-LCR can achieve more energy efficiency than FR. FR consumes the entire energy of all RFID readers at 8.7 min. However, *w*-LCR consumes the entire energy of all RFID readers at 12.6 min. Therefore, *w*-LCR is more energy efficient than FR (1.45 times). Furthermore, FR consumes half the energy of all RFID readers at 4.1 min. However, *w*-LCR consumes half the energy of all RFID readers at 9.4 min. Therefore, *w*-LCR is more energy efficient than FR-based algorithms (2.29 times). As shown until now, at the beginning of the operating protocol, *w*-LCR is more energy efficient than the terminal stage. As shown in the shape of *w*-LCR graph, the increment at the beginning is smaller. However, as time passes, the increment of the graph is higher. As time passes, some RFID readers consume all energy and become "battery-drained RFID reader" until recharge. Then the neighbor RFID readers must cover the area controlled by the "battery-drained RFID reader." Therefore, as time passes, the increment becomes higher. In FR, the cluster radius is not controlled. Therefore, RFID reader, which uses the FR, consumes a fixed amount of energy. Hence, the shape of the FR graph has a linear form.

22.7   Conclusions and Future Work

We extended our previous research, LLC [7], to an extended scheme, *w-LLC*, to cope with the situation where events more frequently occur in a certain area of a wireless network field or where the computing powers of each CH are not equal. For improving our previous work, we apply the concept of weight functions to the LLC. In LLC, as a weight factor, we only consider the residual energy of each CH. As an extension of LLC, we modify the LLC by considering multiple weight factors. Equations 22.2 through 22.5 represented in this chapter have $m + n$ weight factors. Based on simulation-based performance evaluation, we observe that *w*-LLC achieves better throughput than LLC. As an application area, we consider RFID networks to solve RFID reader collision problem. We develop *the w-LCR* scheme based on the concept of *w-LLC*. By reducing the overlapping areas of clusters, we can reduce the possibility of collision of signals. The proposed RFID reader collision arbitration protocol in this chapter is a coverage-aware reader collision arbitration protocol. As a future research direction, first, we will make more sensitive iteration policy with regard to energy efficiency. If we can consider the local information, we can design more energy-efficient iteration policy. Next, we can design more efficient RFID reader collision arbitration with the concept of "hybrid RFID reader anticollision algorithm" which is based on the concept of scheduling-based RFID reader anticollision algorithm used by *Colorwave*, based on *w*-LCR.

22.A Appendix A: Proof of Corollary 22.1

Corollary 22.1 I n sector OAA' in Figure 22.8, $\overline{OQ'}$, which is constructed by central point O and any point, the Q' in Figure 22.9, on the arc AA', meets $\overline{QA}$ or $\overline{QA'}$, necessarily.

Proof

The sector OAA' is divided into two parts as sector OAQ and sector $OA'Q$ with Q (CRDP in *w*-LLC) considered as a reference point.

Part 1 Q' **lies on arc** QA

When Q' lies on the arc QA as shown in Figure 22.12, by the property of sector ($\overline{OA} = \overline{OQ} = \overline{OQ'}$), $\triangle OAQ'$ is an isosceles triangle.

If we consider that $\angle AOQ' = \varphi$ and $\angle AOQ = \alpha$, then $\overline{AQ'} = 2 \cdot \overline{OA} \cdot \sin \frac{\varphi}{2}$.

By *the law of sines*,

$$\frac{\overline{AQ'}}{\sin \angle ARQ'} = \frac{\overline{Q'R}}{\sin \angle RAQ'}$$

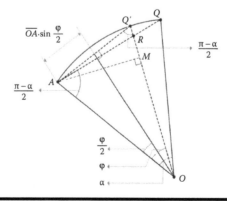

Figure 22.12   Detailed description of the notations for Corollary 22.1.

Then,

$$\overline{Q'R}$$

$$= \frac{2 \cdot \overline{OA} \cdot \sin \frac{\varphi}{2} \cdot \sin(\angle OAQ' + \angle OAQ)}{\sin(\angle RAO + \angle AOR)} = \frac{2 \cdot \overline{OA} \cdot \sin \frac{\varphi}{2} \cdot \sin(\frac{\pi - \varphi}{2} - \frac{\pi - \alpha}{2})}{\sin(\frac{\pi - \alpha}{2} + \varphi)}$$

$$= \frac{2 \cdot \overline{OA} \cdot \sin \frac{\varphi}{2} \cdot \sin \frac{\alpha - \varphi}{2}}{\sin(\frac{\pi - \alpha}{2} + \varphi)}.$$

Based on the condition of $0 < \varphi < \alpha < \pi$, $\sin \frac{\varphi}{2} > 0$, $\sin \frac{\alpha - \varphi}{2} > 0$, and $\sin(\frac{\pi - \alpha}{2} + \varphi) > 0$. Therefore, $\overline{Q'R} > 0$. Then $\overline{OR} < \overline{OQ'}$. Therefore, R is within the sector. R lies on $\overline{QA}$ while it lies on $\overline{OQ'}$ too. Therefore, there exists a point where $\overline{OQ'}$ meets $\overline{QA}$.

End of Proof: Part 1

Part 2: Q'' **lies on arc** QA'
　　Part 2 can be proved in the same way as Part 1 because they are dual.
End of Proof: Part 2

End of Proof

22.B Appendix B: Proof of Corollary 22.2

Corollary 22.2

Sector OAA' in Figure 22.8 covers rectangle $OAQA'$ perfectly.

Proof

If we choose any Q (CRDP in w-LLC) in the arc AA', (1) an isosceles triangle AOQ is covered by a sector AOQ (by Corollary 22.1), (2) an isosceles triangle $A'OQ$ is covered by a sector $A'OQ$ (by Corollary 22.1), (3) a sector OAA' = a sector AOQ + a sector $A'OQ$, and (4) a rectangle $OAQA'$ = an isosceles triangle AOQ + an isosceles triangle $A'OQ$.
By (1), (2), (3), and (4), a sector OAA' covers a rectangle $OAQA'$ perfectly.

End of Proof

22.C Appendix C: Proof of Corollary 22.3

Corollary 22.3

The triangle constructed by *Delaunay* triangulation in an initial phase is covered by three rectangles perfectly.

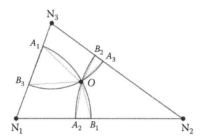

Figure 22.13   Notations for Corollary 22.3.

Proof

As shown in Figure 22.13, when the three rectangles are united, the overlapped areas are generated while the triangle is covered perfectly. Overlapped areas are `triangle` OA_1B_3, `triangle` OA_2B_1, and `triangle` OA_3B_2. Therefore, we prove that the triangle generated in an initial phase is covered by three rectangles.

End of Proof

Acknowledgments

This research was jointly sponsored by MEST, Korea, under WCU (R33-2008-000-10044-0); MEST, Korea, under Basic Science Research Program (2011-0012216); MKE/KEIT, Korea, under the IT R&D program [KI001810041244, SmartTV 2.0 Software Platform]; and MKE, Korea, under ITRC NIPA-2012-(H0301-12-3001).

References

1. V. Mhatre and C. Rosenberg, Design guidelines for wireless sensor networks: Communication, clustering and aggregation, *Elsevier Ad Hoc Networks Journal*, 2(1), 45–63, 2004.
2. O. Younis and S. Fahmy, HEED: A hybrid, energy-efficient, distributed clustering approach for ad hoc sensor networks, *IEEE Transactions on Mobile Computing*, 3(1), pp. 366–379, 2004.
3. W. B. Heinzelman, A. P. Chandrakasan, and H. Balakrishnan, An application-specific protocol architecture for wireless microsensor networks, *IEEE Transactions on Wireless Communications*, 1(4), 660–670, 2002.
4. G. Gupta and M. Younis, Load-balanced clustering for wireless sensor networks, In *Proceedings of IEEE International Conference on Communications (ICC)*, Vol. 3, pp. 1848–1852, Anchorage, AK, May 2003.
5. M. Demirbas, A. Arora, V. Mittal, and V. Kulathumani, Design and analysis of a fast local clustering service for wireless sensor networks, In *Proceedings of IEEE International Conference on Broad Networks (BROADNETS)*, San Jose, CA, Oct 2004.
6. J. Pan, Y. T. Hou, L. Cai, Y. Shi, and S. X. Shen, Topology control for wireless sensor networks, In *Proceedings of ACM International Conference on Mobile Computing and Networking (MobiCom)*, San Diego, CA, Sep 2003.

7. J. Kim, E. Kim, S. Kim, D. Kim, and W. Lee, Low-energy localized clustering: An adaptive cluster radius configuration scheme for topology control in wireless sensor networks, In *Proceedings of IEEE Vehicular Technology Conference (VTC)*, Stockholm, Sweden, May 2005.

8. I. F. Akyildiz, W. L. Su, Y. Sankarasubramaniam, and E. Cayirci, Wireless sensor networks: A survey, *Elsevier Computer Networks*, 38(4), 393–422, Mar 2002.

9. D. Estrin, R. Govindan, J. Heidemann, and S. Kumar, Next century challenges: Scalable coordination in sensor networks, In *Proceedings of ACM International Conference on Mobile Computing and Networking (MobiCom)*, Seattle, WA, Aug 1999.

10. J. Kim, W. Lee, J. Yu, J. Myung, E. Kim, and C. Lee, Effect of localized optimal clustering for reader anti-collision in RFID networks: Fairness aspect to the readers, In *Proceedings of IEEE International Conference on Computer Communications and Networks (ICCCN)*, San Diego, CA, Oct 2005.

11. M. de Berg, M. van Kreveld, M. Overmars, and O. Schwarzkopf, *Computational Geometry: Algorithms and Applications*, 2nd edn., Springer-Verlag, Berlin, Germany, 2000.

12. F. Aurenhammer, Voronoi diagrams — A survey of a fundamental geometric data structure, *ACM Computing Surveys*, 23 (3), 345–405, Sep 1991.

13. S. Yin and X. Lin, Adaptive load balancing in mobile ad hoc networks, In *Proceedings of IEEE Wireless Communications and Networking Conference (WCNC)*, New Orleans, LA, Ma. 2005.

14. M. S. Bazaraa, H. D. Sherali, and C. M. Shetty, *Nonlinear Programming: Theory and Algorithms*, 2nd edn., Wiley, New York, 1993.

15. J. Nocedal, Updating quasi-Newton matrices with limited storage, *Mathematics of Computation*, 35, 773–782, 1980.

16. D. C. Liu and J. Nocedal, On the limited memory BFGS method for large scale optimization, *ACM Mathematical Programming*, 45, 503–528, Dec 1989.

17. F. Zhou, C. Chen, D. Jin, C. Huang, and H. Min, Evaluating and optimizing power consumption of anti-collision protocols for applications in RFID systems, In *Proceedings of ACM International Symposium on Low Power Electronics and Design (ISLPED)*, Newport Beach, CA, 2004.

18. J. Waldrop, D. W. Engels, and S. E. Sarma, Colorwave: An anticollision algorithm for the reader collision problem, In *Proceedings of IEEE International Conference on Communications (ICC)*, Anchorage, AK, May 2003.

19. C. Law, K. Lee, and K.-Y. Siu, Efficient memoryless protocol for tag identification, In *Proceedings of ACM International Workshop on Discrete Algorithms and Methods for Mobile Computing and Communications (DIALM)*, Boston, MA, 2000.

Chapter 23

Coupled Simulation/Emulation for Cross-Layer Design of Mobile Robotics and Vehicular Network Applications

Sven Hanemann, Bernd Freisleben, and Matthew Smith

Contents

23.1 Introduction

The main goal of ubiquitous and embedded computing is to bring computers seamlessly into our environment and make computers adapt to us and not force us to adapt to them. Examples range from smart houses to wearable computers. While some applications of ubiquitous computing are based on a fixed infrastructure and operate in a static environment, many consist of a number of mobile nodes operating in a particular area without using any pre-existing infrastructure. To fully realize their potential, smart applications communicate with other devices to create a wireless ad hoc network helping to fulfill their goal.

Such mobile wireless ad hoc networks can only operate if nodes offer their forwarding capabilities to other nodes. The special conditions of wireless ad hoc networks, such as limited transmission ranges of the nodes, limited bandwidth, possible interference, limited computational and energy resources, require special care during system development, since interdependencies between the different conditions can cause negative side effects when optimizing only a single criteria. For instance, increasing transmission range also increases energy consumption and internode interference [7,40]. Developers must therefore take a unified approach to platform and application design to ensure seamless integration of the disparate components. This encompasses everything from application and operating system concerns right down to routing strategies and the physical communication medium.

Furthermore, since the movement of the nodes has a significant effect on network integrity, the mobility of the network needs to be dealt with. Most current mobile ad hoc network research is done with simulators and deals with mobility in a static way, viewing it as something which happens to the network in a predefined scenario-based way. This form of mobility cannot be influenced during simulation time and reduces mobility to something the network has to handle as an outside influence. In the case of ubiquitous computing, this approach is not sufficient since embedded devices have a task or goal to fulfill, so they will not move around randomly while avoiding obstacles but will follow a path which lets them fulfill their task. Furthermore, as the network is used for the task at hand, availability and quality of service of the network can actively influence the movement of the network participant. Thus, mobility is not something which solely happens to a network but is something which can be affected by the network state and controlled by an application. When used correctly, mobility can even be beneficial for the network. If, for instance, a mobile node moves outside the communication range of all other nodes, it might be advisable to change course to integrate itself back into the network. Or if the task at hand does not allow this, the node needs a way to find back into the network after its task has been completed. This form of behavior can only be implemented by processing information from the network layer and then adjusting the node behavior in the application layer accordingly.

Traditional design separates the system in layers each dealing with a specific aspect of the system. The typical layered structure as defined in OSI is comprised of the following layers: Application, Presentation, Session, Transport, Network, Data Link, and Physical Layer each with well-defined static interfaces to the layer above and below. This enables the functional components in each layer to be developed and tested separately and thus simplifies the development process.

However, since ad hoc networks operate in a dynamic and usually interactive environment, it is not always possible to predetermine the optimal setting of the functional components in each separate layer. For example, high quality of service in the application layer might be important to the users in which case increased energy consumption caused by increased activity in the transport and network layers is acceptable, at other times battery life is at a premium and should be conserved wherever possible.

A purely layered approach isolates functional components such as a routing algorithm from the application layer. This hides vital interdependencies needed for efficient system design [12]. A cross-layer approach adds custom interfaces between neighbor and non-neighbor layers to facilitate information exchange between the layers, thereby enabling a global optimization of the system behavior at runtime. Enabling this cross-layer approach leads to interdependent development cycles of the functional components in the separate layers, making an efficient test bed vital to the production process. There is currently no environment with which to study such cross-layer interaction in a truly dynamic way.

Although hardware prices are continuously dropping, buying a large number of mobile nodes equipped with wireless communication devices to test the system in the wild is usually beyond the scope of most projects. Instead, simulation and emulation environments are used, since a large number of nodes can be simulated at low cost in a repeatable manner. The downside of simulation and emulation is that the scope and level of detail of the simulation/emulation is limited to a certain extent.

In this chapter, we present a new approach to develop applications for embedded and ubiquitous computing based on wireless ad hoc networks, which makes use of the benefits of cross-layer system design. The approach is based on the combination of a network emulator with a virtual world simulator. The benefits of our approach to ubiquitous application design are

- There is a virtual surrounding in which the network nodes are embedded and can fulfill a task. This allows the application designer to test the primary software for an application.
- A wireless network is emulated to enable the nodes within the virtual world to communicate in a realistic manner. This allows the application to be designed in the lab while at the same time taking the real-world problems of wireless communication into account.
- Cross-layer interaction between the application and network layer is possible. This is particularly important because ubiquitous and embedded applications have stringent requirements to efficiently adapt to their environment.
- Mobility can be influenced by the application based on cross-layer interaction, allowing the application design to be tested in a feedback-enabled unified environment.

To demonstrate the feasibility of our proposal, the following two scenarios have been implemented, each showing cross-layer interaction between the application and the network layer but in different directions:

- The main scenario is from the area of embedded computing. A number of robots equipped with embedded Linux operating systems and wireless communication devices cooperatively explore an unknown environment searching for minefields. The number of robots is relatively small compared to the area to be covered, thus leading to a sparsely populated network which due to limited communication range will be partitioned on a regular basis. Two network self-healing strategies are demonstrated where the network layer affects the application layer in a cross-layer fashion.

◼ The second scenario is from the area of ubiquitous computing. Cars equipped with a driver assistance system and a wireless communication device dynamically set speed limits on a multilane highway depending on current traffic. This application demonstrates the benefits of cross-layer information flow from the application to the network layer.

The main focus of this chapter is on the first scenario, the second scenario will only be introduced briefly to show how different application environments are integrated into the emulation environment and show that cross-layer interaction can be beneficial in both directions.

This chapter is organized as follows. First, related work for ad hoc wireless networks in the fields of cross-layer design, mobility patterns, and simulation/emulation is discussed. Then, a novel coupling of simulation and emulation systems for unified cross-layer research for mobile wireless ad hoc networks and embedded system design is introduced, and a cross-layer method for utilizing the mobility of the network is demonstrated. In the following sections, a cross-layer self-healing mechanism for mobile ad hoc networks is introduced, and an experimental setup demonstrating how five robots cooperatively clear two minefields and cars set speed limits on a multilane highway using cross-layer techniques is shown. The final section concludes this chapter and outlines areas for future research.

23.2 Related Work

Since this chapter covers a number of topics, the related work is divided into several sections. First, related work in the field of cross-layer design is examined. Then, research in the field of mobility patterns is described and an overview of available simulators and emulators which have bearing on cross-layer design in mobile ad hoc networks is given. Finally, the special conditions of vehicular ad hoc networks will be described.

23.2.1 Cross-Layer Design

A wireless multi-hop ad hoc network with autonomous mobile nodes is a collection of mobile nodes with wireless communication facilities forming a temporary network without relying on any existing infrastructure. Limited available bandwidth and highly dynamic network topology due to continuously moving nodes require dedicated networking protocols to achieve reliable communication.

In an ubiquitous computing environment, these factors vary greatly between applications and within applications over time. Furthermore, there are interdependencies between these factors: For example, increasing the transmission range of a device has an adverse effect on its energy supply and possibly limits the overall network bandwidth since it now interferes with the medium access of nodes previously not in range.

However, flexibility in link, access, network, and application layer protocols can be exploited to compensate for and even take advantage of the dynamics of wireless ad hoc networks. Many applications can benefit from a deviation of the strictly layered protocol stack implementations originally proposed by the ISO OSI reference model [34]. The isolation of functional components such as the routing algorithm and the application layer hides information and interdependencies that could be used to improve the efficiency of the node or the network as a whole. If, for instance, a node moves out of the communication range of the other nodes, it might be advised to change course to integrate itself back into the network. This behavior can be implemented by processing information from the network layer and then adjusting the node behavior in the application layer accordingly. Autonomous mobile nodes—for example, mobile robots equipped with a wireless communication device—have a specific task to complete which may also govern their movement pattern. The availability of network services, such as the propagation of terrain or environmental conditions provided by other nodes, can actively influence the movement of a network participant. This leads to changes in an application's control flow based on the network state as one example of cross-layer concerns.

A cross-layer approach adds custom interfaces between neighbor and non-neighbor layers to facilitate information exchange between the layers, thereby enabling a global optimization of the system behavior at runtime. Cross-layer design for wireless networks has been studied in a number of papers [21,34,35,40,43]. These papers mainly address the two fundamental questions in cross layer design: as shown in Figure 23.1: what kind of information should be exchanged across

Extended topology information	
Layer	Topology information
Application layer	Topology control algorithm Server location Network map
Transport layer	Congestion window Timeout clock Packet losses rate
Network layer	Routing affinity Route lifetime Multiple routing Network connectivity
Mac/link layer	Link bandwidth Link quality Mac packet delay Node neighborhood
Physical layer	Movement patterns Radio transmission range SNR information

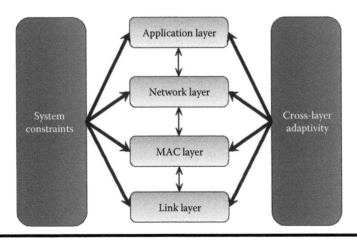

Figure 23.1   Cross-layer design architecture.

protocol layers and how to adapt to them while factoring global system constraints and characteristics into the protocol design. The channel and network dynamics of ad hoc networks with multi-hop routing make it also difficult to support, e.g., multimedia requirements of high speed and low delay. However, flexibility in link, access, network, and application layer protocols can be exploited to compensate for and even take advantage of the dynamics of wireless ad hoc networks. Thus, the focus of these papers is on interaction between the lower levels, such as the network and data link layers. They concentrate on the design problems concerning power control, queuing discipline, the choice of routing, media access, and their interaction.

To the best of our knowledge, there is no research specifically handling cross-layer interaction between the application layer and the lower layers to influence the mobility of nodes.

23.2.2 Mobility Patterns

In a mobile ad hoc network environment, nodes move in and out of communication range frequently, thus constantly changing the network structure. The problem of routing messages in such mobile networks is discussed in several papers [6,20,24,27,32]. In all of these papers, mobility is usually scenario or random waypoint based in order to isolate the impact of mobility on the communicating traffic patterns of ad hoc networks. The communicating traffic pattern consists of randomly chosen source—destination pairs with long enough session times. In the mostly used network simulator (NS2) distribution, the implementation of this mobility model is as follows: At every instance, a node randomly chooses a destination and moves towards it with a velocity chosen uniformly randomly from $[0, V_{\max}]$, where $V_{\max}$ is the maximum allowable velocity for every node.

Current ad hoc network scenarios like battlefield applications, rescue operations or mine sweeping have complex node mobility and connectivity dynamics. Wildly varying mobility characteristics have a significant impact on the performance of the routing protocol. The random waypoint model is insufficient to capture the following characteristics:

1. Spatial dependence of movement among nodes
2. Temporal dependence of movement of a node over time
3. Existence of obstacles constraining mobility

Jardosh et al. [19] argue that this approach does not adequately model the problem faced by mobile ad hoc networks. The simulations on which the above papers are based use flat, unobstructed areas in which the nodes move. In reality, mobile nodes — be they people with PDAs or robots with wireless communication facilities — usually follow certain well-defined paths and have to avoid obstacles in their way, which leads to certain areas having a high density of nodes, while others are only sparsely populated or completely empty. To remedy this discrepancy between simulation and real world, Jardosh et al. introduce obstacles into the environment and use Voronoi diagrams to construct movement paths which more adequately model real life moment patterns, interactive control of the nodes is not possible though.

Yoon et al. [44] even argue that the random way point model may generate misleading or incorrect results due to the fact that the average nodal speed consistently decreases over time.

Bai et al. [1] propose a framework to systematically analyze the impact of mobility on the performance of routing protocols for wireless ad hoc networks in order to relieve some of deficiencies of the random way point model.

Currently, there is no research where the mobility model of the network is based on dynamic scenarios. To enable research in this area, simulators or emulators which allow the node movement to be influenced dynamically during the simulation are required.

23.2.3 Simulation/Emulation

Currently, applications and communication protocols for ad hoc wireless network are usually evaluated in either simulations or small-scale live deployments. Large-scale deployment is typically costly and difficult to perform under controlled circumstances. Simulation allows researchers to vary system configurations.

Network simulators like NS2 [23] and GloMoSim [2,11] are usually used to develop solutions for the transport layer and below, i.e., routing algorithms, link layer error correction, and so forth, but are not very suitable, for instance, to develop power aware applications which need to run on a mobile node such as a PDA or robot. Simulation requires the duplication of application, operating system, and network behavior within the simulator. Many assumptions in simulation models are hidden; they must be carefully examined to ensure that the model behaves as expected [1,44]. A further problem is that code developed for the simulator usually does not run on real systems without modification, since the simulator does not offer a complete model of the system.

A network emulator such as MobiEmu [45], MobiNet [22] or m-ORBIT [28], on the other hand, can emulate system behavior on all levels, thus allowing more realistic development for a given system such as Linux. Such an emulator allows the layers of a standard Linux PC to be accessed in a realistic manner. In emulation systems, the communication of unmodified applications running on stock operating systems is subject to real-time emulation of a user-specified wireless network environment. The infrastructure of emulation systems is extensible, facilitating the development and evaluation of new MAC layers, routing protocols, mobility, and traffic models.

To develop code for specific mobile hardware such as mobile phones, PDAs, or robots, additional emulators such as the J2ME Wireless Toolkit [38] or Player [30] can be used.

All of the above solutions only cover part of the problem domain. And none allow the system to be embedded in a virtual environment which would allow realistic development of application layer code. Simulators like Stage [10,31], OneSAF [37] or RoboCup Rescue [29] offer such functionality. To efficiently develop and test systems under near real-world conditions, an overall research environment is needed. To this end, the separate simulators and emulators need to be combined in such a way that all aspects of mobile ad hoc development can be studied in a unified manner.

23.2.4 Vehicular Ad Hoc Networks

Currently available cost-effective wireless technology is mostly based on the IEEE 802.11b [15] standard and its follow-on, 802.11g [16], which both operate in the free ISM band at a frequency of 2.4 GHz. Wireless hardware based on 802.11b has been used in the HOCMAN mobile ad hoc communication system [15] for motorcyclists. These experiments, however, consisted mostly of the exchange of small HTML and graphics documents between moving vehicles. In addition, if a moving object exceeded the speeds of around 90 km/h, success rates for transmission dropped significantly (in the worst case to about 20%). This indicates that the common 802.11b/g-based WLAN protocols are not adequate for Roadcasting applications. This problem was also recognized by the IEEE wireless network task groups. Currently, the IEEE 802.11p working group is defining a standard for Wireless Access for the Vehicular Environment (WAVE) [17], which is specially designed for data exchange between high-speed vehicles and between these vehicles and a roadside infrastructure. IEEE 802.11p will use the licensed ITS band at a frequency of 5.9 GHz; the distance between nodes of a 802.11p network may be up to 1000 m with relative speeds of the moving vehicles up to 200 km/h. In this scenario, 802.11p is expected to reach data rates of up to 6 MB/s. Currently, there is no device which meets the 802.11p specifications with which to actually build a system; however it will soon be possible to at least simulate some of the more important features of 802.11p as they become known.

Furthermore, [40] shows how cross-layer optimization can benefit mobile ad hoc networks, while [5] crucially goes on to show that traditional peer-to-peer software like gnutella or pastry running on mobile ad hoc networks perform poorly if node mobility is high. Partitioning and reintegration of the networked nodes is particularly critical. Multimedia applications like Roadcasting (http://roadcasting.org/) are very data intensive and are only usable if a high quality of service can be guaranteed [36]. If the peer-to-peer software creates a high amount of overhead solely because it is being applied in a mobile ad hoc scenario, it is not possible to meet the quality of service demand required by the Roadcasting application.

Also [5] shows how cross-layer techniques can be utilized to significantly improve the performance of a peer-to-peer protocol by accessing cross-layer information supplied by the routing daemon. Overall performance gains of 40% were achieved in a random mobility scenario compared to the pure peer-to-peer system in the same scenario. Such developments can only be done in an environment which offers a realistic view of the host operating system in which Roadcasting applications can be embedded. A network simulator like NS2 might be used to develop an efficient routing protocol for vehicular ad hoc networks, but it is not possible to run a peer-to-peer framework and the application based thereon transparently as well, since the simulation system does not model the host operating system. Thus, cross-layer optimization, which is a vital requirement for Roadcasting applications can not be fully researched in such an environment. A simulation/emulation environment is needed that is geared towards cross-layer development of both routing strategies as well as mobile ad hoc applications which offers a full operating system for each mobile node, so the routing strategies and application can be developed and tested in a realistic environment taking full advantage of possible cross-layer optimization techniques.

23.3 An Approach to Mobile Cross-Layer Design Based on Coupled Simulation/Emulation

To enable realistic and efficient cross-layer research in embedded and ubiquitous computing using mobile ad hoc wireless networks, we introduce a coupled simulation/emulation environment which encompasses the vital aspect of mobility, range-limited communication and realistic interaction with a virtual world. To this end, a number of existing solutions were extended and combined, so that cross-layer studies can be undertaken within a single environment. In the following subsections, a brief overview of the simulator and emulators utilized in this chapter and our extension to them is given. Then, our approach of coupling these simulators is presented and the architecture of the overall system is examined. In the final subsection we examine the control structures needed to utilize mobility in a cross-layer way. To illustrate this approach, a novel mechanism for self-healing networks is introduced using the system and mechanisms introduced in this section.

The combined simulation/emulation environment needs to encapsulate the following aspects to offer a unified research environment: a network simulation/emulation, a hardware simulation/emulation, and a virtual world with mobility support.

23.3.1 Network Emulation

As stated above, network simulation does not offer the necessary depth to allow cross-layer studies to be done in a realistic manner. To enable cross-layer design, the network emulation system must offer an operating system environment which closely models a real operating system. Our emulation system MarNet [3] is based on MobiEmu [45] and utilizes XenoLinux [2] as its operating system base to fulfill this requirement. By utilizing XenoLinux, MarNet offers the full capability and detail of a Linux environment for use with the emulation. XenoLinux is a modified Linux kernel which runs as an executable binary on a host Linux machine. Virtual resources, including a root file system and swap space can be assigned to the Xen kernel and can have a hardware configuration entirely separated from that of the host. This allows multiple Xen instances to be run on a single real machine, without interfering with each other. The Xen Virtual Node is the host operating system in which the cross-layer solution (including the hardware emulator) is embedded.

Each Xen Node thus has its own ethernet sockets with a unique MAC and assigned IP address and can communicate realistically with other Xen Nodes on its own or on remote machines. This is done transparently so there is no difference if a Xen Node communicates with another Xen Node or a real machine. This feature is used by MarNet to emulate a wireless ad hoc network based on the wired network used by the Xen Nodes.

MarNet uses a Manager/Worker architecture for emulation. The Xen Nodes represent the mobile nodes in the ad hoc network. The manager of MarNet loads a scenario file (NS2 compatible) with which the topology of the network for a given point in time can be calculated. Based on the topology information it is possible to calculate which nodes would be able to communicate with each other in the wireless world. The controller then sets the IP-Tables of the Xen Nodes accordingly to block all traffic between Xen Nodes which are too far apart to communicate.

Figure 23.2 shows the basic Manager/Worker architecture of MarNet. The manager receives a scenario file from the user and then forwards it to all workers. Since all participants have the scenario information, the manager then only needs to send time stamps to all workers. Based on the current emulation time and the scenario file, the workers can block all communication from Xen Nodes which are unreachable in the virtual world. The routing daemon on the workers thus have a realistic view of their part of the network and can build their routing tables accordingly. By analyzing the entire scenario file, it is possible to calculate the optimal route to each node. This information can then be used to compare the routing daemon results with the optimal solution.

In the field of ubiquitous and embedded computing, the environment in which an application operates can be subject to frequent change, which can adversely affect the routing module and vice versa. For instance, while addressing energy preservation issues, some algorithms reduce the number of network transmissions to a minimum, thereby becoming unsuitable in a scenario of rapidly moving nodes. The `marnetd` allows the adaptation to changing environments through its modular design: Applications or users may change routing strategies on demand by hot-swapping new routing modules into the system at run time.

The `marnetd` is registered as a system service which defines an API for the development of such routing modules and provides an interface between data link, network, and application layers.

Figure 23.3 shows the developer view of the `marnetd` system service and outlines its integration into the operating system: It acts as a mediator between operating system and algorithm researchers

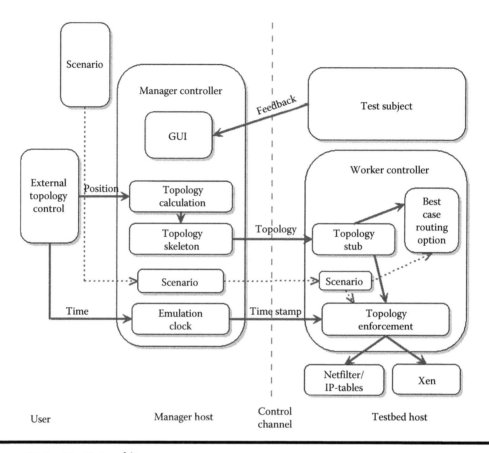

Figure 23.2 MarNet architecture.

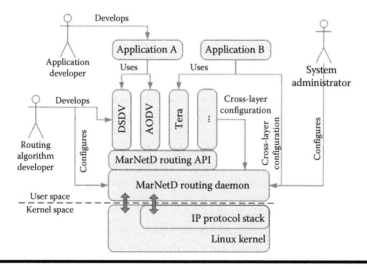

Figure 23.3 The MarNet daemon.

and developers who create or use modules such as DSDV [24], AODV [25] or TERA [14], mobile ad hoc network enabled application developers who make use of `marnetd` as a provider of routing services and network state information (through cross-layer facilities) and normal users who simply wish to set up a working mobile ad hoc network.

By analyzing the entire scenario file, it is possible to calculate the optimal route to each node. This information can then be used to compare the routing daemon results with the optimal solution.

MobiEmu [45], which is the base of MarNet, only supports scenario file-based mobility. Since we require the ability to modify the movement of nodes at any time during the simulation, we extended the functionality of MobiEmu such that arbitrary position information for all nodes can be entered into the system from an outside source. This allows us to write custom code to control node movement within the emulation system or plug in a virtual world where nodes move to rules completely unknown to the network emulation environment. Based on exchangeable topology calculation metrics, the manager calculates which nodes are capable of communication based on the position information it receives. This extended topology information is then partitioned and sent to the Xen workers as required. There, based on the information sent by the manager, the IP-Tables of the Xen Nodes are set, so communication between virtually unconnected nodes is not accepted. The extension can be run as an alternative to the standard scenario based mobility structures. This allows comparative studies to be held within the same system based on existing scenarios. In Figure 23.2, the dotted lines show the legacy information flow.

Once running, the daemon provides a local pipe interface, which appears as a file in the host system's file system, and can be used to load and unload modules. The local terminal interface (LTI) is designed to be an extensible command processor which allows runtime read and write access to any of the daemon's state variables. Furthermore, module programmers can also unlock any of the module specific parameters for access through this interface and register custom commands.

For our second example scenario, a traffic density algorithm is used to determine which routing strategy is best suited for the current traffic situation for a specific application. In [4], several ad hoc routing strategies are compared showing that routing algorithms that perform well at low node mobility levels do not perform as well with high node mobility and vice versa. Based on that, the onboard sensors of the car which measure speed and traffic density are used by the application to switch routing strategies depending on which algorithm has the best performance metric for the given situation. To accurately design and test Roadcasting frameworks, a realistic traffic model is required to ensure that the application is tested in the environment it will later run in.

Since node mobility and node density have such large effects on the routing strategy and both of these attributes change drastically in the vehicular scenario, it is unlikely that a single routing algorithm will be able to produce near-optimal results in all situations which can face vehicles on the road (i.e., free motorway and traffic jams). To be able to swap routing strategies to dynamically adapt to the current situation, we utilize the modular routing architecture, which we previously introduced in [18], which offers a programming API to facilitate the development of cross-layer enabled routing modules and allows the hot-swapping of routing strategies at runtime. Figure 23.4 shows the modular routing architecture used by the Roadcasting application. Three routing algorithms are plugged into the system DSDV (pro-active), AODV (re-active) and Tera [18], an optimized routing algorithm previously developed to achieve fast convergence, avoid the count to infinity problem, and eliminate routing loops quickly [13,14,18]. The local terminal interface (LTI) shown in Figure 23.4 is the component which exports the cross-layer information from the application and from the routing algorithms and can be used to switch routing algorithms on-demand. Using the LTI, it is now possible to create peer-to-peer infrastructures which not only utilize cross-layer optimizations using a single routing strategy (as in [5]) but with a number of

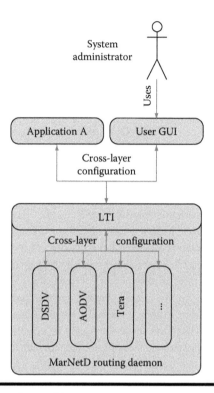

Figure 23.4   The modular routing architecture.

different routing strategies. This presents a major benefit to the Roadcasting environment since all the work done by researches in the field of ad hoc routing strategies can be seamlessly integrated into peer-to-peer overlay networks based on mobile nodes. This enables the convergence of mobile ad hoc networks and peer-to-peer computing in general and specifically offers great potential for Roadcasting applications.

23.3.2 Hardware Emulation

For the two scenarios described in the introduction we require two hardware emulation systems which model the target platform on which the application will be deployed.

The target hardware platforms for the first scenario described in the introduction are mobile robots. To develop code which can be used during the simulation and emulation phase but also run on the actual target system, either an emulator for the target hardware is required or a common API must be used to ensure portability. Player [30] provides a network interface for a variety of mobile robot and sensor hardware. Player's client/server model allows control programs to be written in any programming language and run on any computer which has a network connection to the robot. It supports multiple concurrent client connections to devices, creating new possibilities for distributed and collaborative sensing and control. Support is offered for a variety of robot hardware. The original Player platform is the ActivMedia Pioneer 2 family, but several other robots and many common sensors are supported [30]. Using the player interface we can develop code which will run on real mobile robots but we can test it on a desktop machine without incurring the cost of buying real robots. An added bonus is that with Stage [31] there is a virtual world in which multiple player robots can interact with a virtual world and each other (see the next section). Player clients can be implemented in C++, Tcl, Java, and Python. This diverse choice will need to be dealt with during the cross-layer design phase of the client (discussed below).

Unlike the robot scenario where a specialized hardware platform is targeted, the second scenario requires the software to run on board of smart cars, which means we do not have any particular hardware constraints since future cars will have sufficient computational power to allow the use the Java JVM to write the application software.

23.3.3 Virtual World

To successfully develop ubiquitous applications, it is vital that the application can be tested in a realistic environment. This environment is application specific, thus we supply two simulation systems for our two applications.

The first scenario requires an environment which allows mobile robots to be tested. Stage is a simulator which offers such an environment. A population of mobile robots/sensor nodes are simulated and can move within a two-dimensional bitmapped environment. Various sensor models are provided, including sonar, laser rangefinder, pan-tilt-zoom camera with color blob detection, and odometry.

Stage devices present a standard Player interface, such that few or no changes are required in order to move between simulation and hardware. Several controllers designed in Stage have been demonstrated to work on real robots [31].

Stage is designed to simulate a large number of robots/sensor nodes at low fidelity. This matches well with our aims, since the interaction of a large number of mobile nodes is of central interest. The simulator is not very resource intensive, since it only simulates in 2D. Player clients can be run on remote PCs, using ready made communication clients bundled with Stage. This allows processing requirements to be distributed across a network of PCs.

For the second scenario we require an environment in which traffic on a motorway is simulated, allowing different traffic situations to be examined. There are several commercial traffic simulation systems available [8,26,39,41], but since they are not available in source code it is not possible to interact with the cars in a cross-layer manner, furthermore the sensors required by the cars to sense the traffic environment are not available in these simulators. TrafficSim [33], a proof-of-concept implementation of a multi-lane motorway traffic simulator, has been developed to show the benefits of cross-layer interaction for ad hoc network routing in motorway traffic. Each car has sensors measuring the distance to the car directly in front and behind it and which lane it is on. A configurable driver-agent controls the cars based on an extension of the driver model described in [42]. Scenario files can be used to reproducibly create different traffic situations, like accidents and traffic jams.

23.3.4 Coupled Simulation/Emulation

Each of the simulation/emulation systems introduced above deals with one aspect of mobile wireless ad hoc networks. To enable mobility-based cross-layer research, the simulation/emulation systems must now be connected. Figure 23.5 shows the abstract information flow for the first scenario between the simulators/emulators. The central box represents Stage, our virtual world simulator. Within the virtual world, there are several robots which are controlled from Player clients at the top of the diagram. The Player clients receive realistic sensor information from the virtual world and can influence the world using virtual actuators. There can be any number of virtual

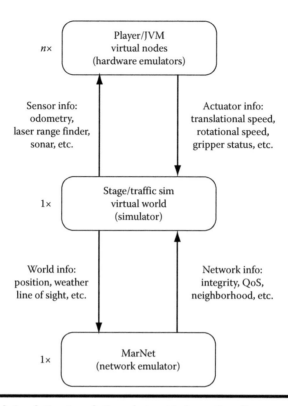

Figure 23.5   Coupled simulation/emulation abstract information flow.

robots limited only by the available processing power of the hardware emulation system. At the bottom of the figure, the network emulator is shown. Here, the topology information is gathered from the virtual world and the network state is modified accordingly. This in turn has an effect on how the nodes within the virtual world perceive their environment. In the case of the traffic scenario, the Stage simulator is replaced by the TrafficSim simulator, and the Player clients are replaced by the Java programs controlling the cars.

Figure 23.6 gives a more detailed view of the system. Each Player client is hosted on a separate Xen Node system. All Xen Nodes are connected to each other, the Manager Controller of the MarNet and the Stage Simulator via a wired network. There are two wired control networks and one emulated wireless ad hoc network to be examined. Solid lines represent control network connections which are not interrupted during the simulation. The first control network is used to exchange sensor and actuator control information between Player clients on the Xen Nodes and the Stage virtual world simulator. These connections are represented by the bidirectional arrows between the Xen Nodes and the Stage simulator in Figure 23.6. The second control network allows the network emulator to control the Xen Nodes. This is illustrated by the lines between the emulator and the Xen Nodes annotated with "Set IP-Tables." The third network is the emulated wireless network of the Xen Nodes. It is based on the same wires as the two control networks but its connections are controlled by the network emulator. These emulated connections are represented by the broken lines in the figure. As mentioned in Section 23.3.1, MarNet emulates the wireless ad hoc network by connection/traffic shaping between the Xen Nodes to create the illusion of a wireless ad hoc network. The network emulator calculates which connections to enable/disable

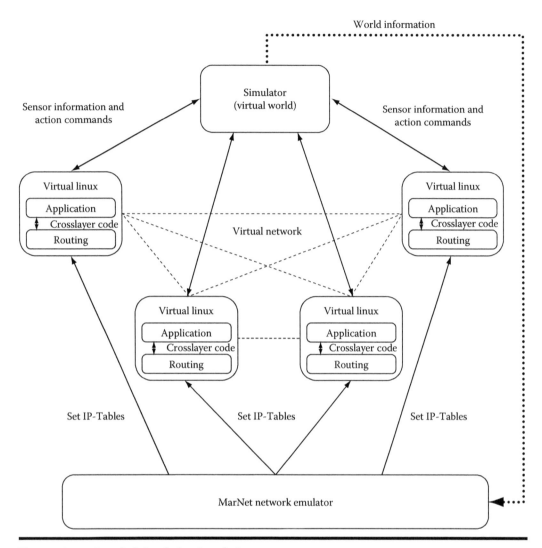

Figure 23.6   Coupled simulation/emulation.

based on information gathered via the dotted line from the virtual world. It is advisable to host the network emulator manager, the virtual world, and the Xen Nodes on separate machines because network congestion can otherwise slow the simulation down. The more Xen Nodes can be hosted on one machine, the less load actually has to be carried by the real network since the network emulator workers are capable of realistically swapping messages on one real machine without going through the wired network.

23.3.5 *Cross-Layer Coupling*

The cross-layer design environment as described above is based on three separate components: network emulator, hardware emulator, and virtual world. From the application developers' point of view, the virtual world is the surrounding in which the application operates; the hardware emulator offers the physical resources available and the network emulator in combination with Xen is the operating system and communication facility. Since the host operating system is Linux, applications can be written in any number of programming languages. The core operating system is implemented in C and C++, which means that most lower layer codes like routing daemons are also written in C and C++. To enable cross-layer design, code running in different layers must be able to communicate with each other. Usually, this is done via direct API calls, virtual devices, or even shared memory. This approach requires that development of code in the cross-coupled layers must be tightly linked, since mutual APIs need to be used during compilation or shared memory access needs to be coordinated. This produces efficient code but it also creates tightly interwoven program structures and thus highly interdependent components. Should one component change or be replaced, all dependent components need to be changed as well. Furthermore, the programming language must be compatible to C or C++. To a certain extent, this is a problem of all cross-layer development. A case-by-case decision must be made where cross-layer benefits outweighs the increased code complexity. As mentioned above, our application can be implemented in C++, Tcl, Java, or Python. This also poses a difficulty for standard cross-layer techniques because Java or Python cannot directly share APIs or memory with C and C++ code. To enable multi-language cross-layer design and to reduce component interdependencies at the same time, a loose cross-layer coupling can be used. This loose coupling is achieved by using socket communication between the components on the separate layers. A language-independent communication protocol can then be used to exchange messages between the components.

Since UDP is less resource intensive than TCP and has all the required features, datagram sockets are used here for cross-layer coupling. During initializing, *localhost* and an arbitrary free port are set as the `rttAddress/rttSocket` communication end point. To send information to a component in a different layer, a datagram packet is constructed containing the information to be sent. For instance, the x and y position, the speed and orientation of a mobile robot gathered in the application layer can be sent to the routing layer to help with routing choices.

The use of sockets for inter-component communication is less efficient than direct API calls or shared memory but is much more robust and interoperable. If APIs change or the memory mapping is altered, components relying on set standards will crash. Error recovery or prevention when using incorrect API calls or wrong shared memory addresses is next to impossible, since machine code when calling a non-existing method will crash instantly. The loose coupling of the cross-layer interaction achieved through socket communication makes the separate components less interdependent, the worst that can happen if a component changes is that no messages or incorrect messages are received. An absent or erroneous message in itself does not cause a system to crash, so error handling algorithms can handle these situations. Since interaction is purely message based, error handling can easily be included without specialized knowledge of the other components in the system.

23.3.6 Cross-Layer Mobility Control

For the robot scenario, a major feature of the coupled mobile ad hoc network simulation environment is that the mobility can be influenced during simulation time. This allows feedback effects between node movement and network connectivity to be researched and later utilized to benefit the application.

To enable cross-layer mobility control, a simple control structure is used for node movement. It should be noted that this control structure is a proof-of-concept implementation to show the benefits of cross-layer interaction for a mobile ad hoc network for ubiquitous computing and is not intended for production grade robotics systems.

All aspects of a node's physical behavior are separated into simple modules called "motivations" which can then be combined freely to form complex behaviors. Each motivation handles one specific type of situation. For instance, the ability to avoid obstacles based on laser range data was coded into a motivation called `AvoidMotivation`. A different motivation called `DestinationWalkMotivation` guides the node straight toward a destination, irrespective of obstacles. The combination of these two motivations allows the node to head toward a destination, while avoiding obstacles on the way. The `AvoidMotivation` has a higher priority than the `DestinationWalkMotivation` to ensure that obstacles in the nodes path are circumvented. But the latter is not completely subsumed. Through a weighted merging of all motivations, a safe best compromise is reached. This is achieved by gathering action suggestions from each motivation and not giving the motivation direct control over the node. A suggestion is a vector of acceptance values for each possible action. A behavioral arbitrator then merges all suggestions paying greater attention to critical motivation suggestions. Based on the merging of all the different acceptance levels, one action as the best compromise is chosen. Figure 23.7 shows how different suggestions cast their vote for a set of possible actions. Solid lines represent a positive and broken lines a negative stance toward a given action. The thickness of the line shows the level of acceptance.

Any number of motivations can be combined in this way and new motivations can be added during runtime. The priority of suggestions can also be altered to adapt to new situations. Through this mechanism, a behavioral module in the form of a `Motivation` can be added to influence the overall behavior of the node based on cross-layer information. The amount of influence this module has varies depending on the importance of its goal and the state of the rest of the motivation architecture. Figure 23.8 illustrates this approach. Motivations A and B base their suggestion solely on information gathered from sensors integrated in the node. Motivations C and D get information from both sensors and from cross-layer sources. Motivation E bases its suggestion using only information gathered from cross-layer sources. The suggestions are weighted according to their priority of the motivation and then merged by the behavioral arbitrator.

The next section shows an example how cross-layer information can be woven into the decision-making process of a node via two cross-layer motivations.

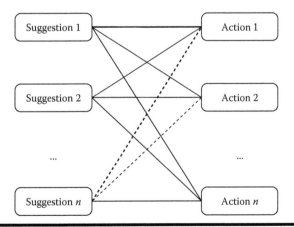

Figure 23.7 Suggestion architecture.

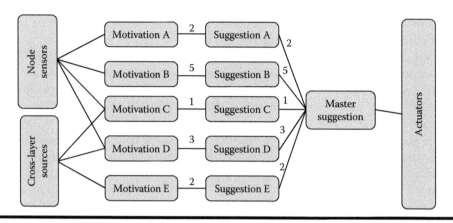

Figure 23.8   Control structure.

23.4 Cross-Layer Examples

23.4.1 Utilizing Node Mobility for Self-Healing Networks

For our first example scenario, a proof-of-concept cross-layer system utilizing node mobility was implemented to test the cross-layer research environment introduced in this chapter. One problem faced by mobile ad hoc wireless networks is that due to node movement the node density can become unevenly distributed over the area, and the network can even become partitioned so that some nodes lose contact to the rest of the network. Depending on the primary task of the network, this can be unavoidable but in some cases the primary task is not affected adversely by a minor position change. In these cases, it is desirable that the nodes move to a position in which the network density is spread as evenly as possible in its region. More significant position changes are required when the primary task requires a communication channel to a certain node which is currently not available. In such a case, the primary task is suspended and the node tries to move to a position in which communication is restored. To enable this behavior, the application controlling the nodes utilizes information gathered from the routing daemon located in the network layer. This repair mechanism works without outside supervision or predefined static rules. Since the decisions made by this mechanism influence the network status, there is a constant feedback loop which automatically regulates the amount of influence the self-healing mechanisms have on the overall system.

This adaptive self-healing capability is implemented in two separate motivations:

1. Passive network preservation motivation
2. Active network preservation motivation

The passive network preservation motivation is a low-priority motivation which constantly tries to move the node, so that it stays within communication range of as many other nodes as possible. It also tries to ensure that equal distance is kept to all neighbors. The information on which nodes are within communication range and the connection strength is gathered from the network layer. Since this motivation has a lower priority than the DestinationWalkMotivation, it does not interfere with the execution of the main task of the node as the primary movement is never completely overridden. Through the improved subsumption architecture, it will however influence the movement of the node, so that within the bounds of the task at hand an optimal position is

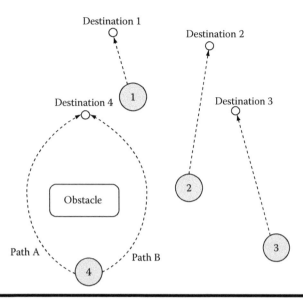

Figure 23.9 Passive network preservation.

adopted. In Figure 23.9, four nodes are shown, each heading to its own destination. Node 4 must avoid an obstacle either via path A or path B. Through the passive network preservation motivation the node decides to take path B because network connectivity is higher than with path A. Passive network preservation is particularly useful in densely populated networks. This motivation will also try and keep the signal strength of all neighboring nodes at a roughly equal level. Thus, if no other task is given to a network an even distribution of nodes will emerge.

The active network preservation motivation has a higher priority and is only used if a certain network connection is essential to the nodes operation. The active network preservation motivation has a higher priority than the normal task of the node and thus can subsume that behavior. The active network preservation notes where its neighbors are headed and shorty before losing connection to them sends a query, asking where those nodes are headed. To be able to do this, the routing daemon receives position direction and speed information from the application layer. This information is then passed to other nodes by the routing daemon. If later those nodes need to be contacted but are not reachable due to network partitioning, an educated guess based on last known position and destination can be made as to where they might be found. The active network preservation motivation then steers the node in that direction. If possible, other tasks are taken into account, but until network connectivity is reestablished the node will continue searching for the lost connection.

A scenario where active network preservation is needed is shown in the following figures. Figure 23.10 shows four nodes heading towards their respective destinations. The large circles around node 1 and 2 show the communication range of the nodes. Node 1 has two destinations which it will drive to sequentially. Since the passive network preservation motivation has a lower priority than the node tasks, it does not stop node 1 from moving out of communication range of the other nodes. Before communication is lost, node 1 queries its neighbors where they are headed and when they expect to be where.

In Figure 23.11 node 1 reaches its first destination. If the node finds something which needs to be handled by a different node, it cannot call them because it has no connection to the network anymore. In this example, node 3 needs to be contacted so it can examine the site.

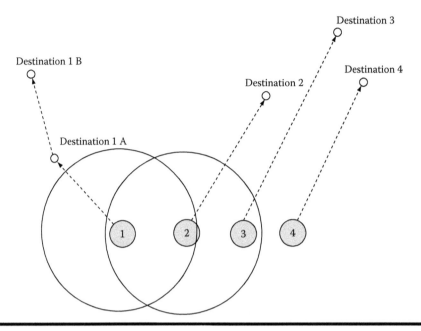

Figure 23.10   Active network preservation 1.

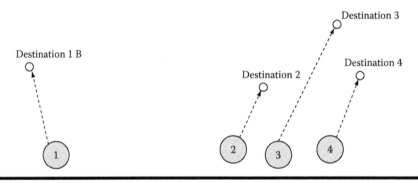

Figure 23.11   Active network preservation 2.

Based on the knowledge gathered before leaving the network, node 1 heads toward where node 3 wanted to go, in the hope of finding it there. This is shown in Figure 23.12.

In Figure 23.13, node 1 enters the communication range of node 2. Since node 2 can also communicate with node 3, the message from node 1 can now be sent to node 3 with two hops. Node 1 now does not need to go to the destination point of node 3 anymore since the message has been sent and received.

In Figure 23.14, node 1 is heading back to its destination and node 3 has been informed of a new destination. Depending on the priority the different tasks have, it must decide where to go first. If both destinations have the same priority, the closest one will be picked first.

This self-healing behavior is greatly facilitated by the cross-layer information exchange between the application and the network layer. Through "loose" coupling the separate components remain

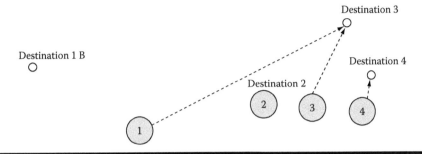

Figure 23.12 Active network preservation 3.

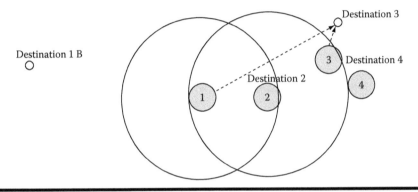

Figure 23.13 Active network preservation 4.

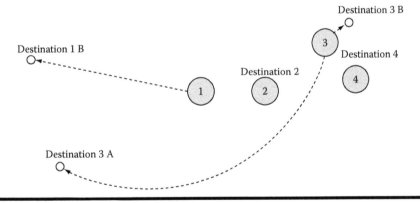

Figure 23.14 Active network preservation 5.

functional even when the cross-layer interface breaks down. Only the novel coupling of different simulators and emulators and dynamic mobility patterns enabled the development of this self-healing mobile ad hoc network. Without the ability to access all different aspects of the entire system, best guesses would have been needed to develop each component separately, and there would have been no way to test the complete system without resorting to expensive real hardware tests.

23.5 Experimental Results

To illustrate the cross-layer mobile ad hoc network environment introduced in this chapter, the following two scenarios were implemented to show some of the key benefits of the system. The self-healing network strategy for the mobile robots scenario is based upon a cross-layer information flow from the network to the application layer. The routing optimization for motorway ad hoc networks is based upon cross-layer information flow from the application to the network layer.

23.5.1 Self-Healing Mobile Robot Ad hoc Network

A number of mobile robots are charged to clear an area of mines. All robots can detect mines but only one robot is equipped with a mine clearing device. To complete their task in an efficient manner, the robots must cooperate since only one of them is capable of clearing the mines. The robots are connected via range-limited wireless communication, thus forming a mobile wireless ad hoc network. In this example, the benefits of mobility control through cross-layer interaction between the application layer and the network layer are shown. Due to the complex interaction between events in the virtual world, actions of the mobile nodes and the network state, the following scenario could not have been investigated in any traditional network simulation environment.

The experimental setup is as follows: Five robots are located at the bottom of the Stage playing field littered with obstacles. In the top left hand and top right hand corner, there are two separate minefields. The mines are represented by small black dots. Each robot is equipped with a laser range scanner, a position device, and a blob detecting camera. The mine clearing robot D is additionally equipped with a gripper device for defusing mines. The task of robots A through C and robot E is to locate the mines and report the position of them to robot D. The mine clearing robot then must drive to the reported minefield and clear it.

The robots are controlled by Player clients each running in a separate Xen system which in turn is part of the MarNet network emulator. The mine searching and clearing behavior is divided into separate motivations listed below. Lower priority motivation are listed first.

Mine searching:

■ Drive to destination on opposite side of field
■ Head toward blobs detected by camera
■ If a minefield is identified, inform mine clearing robot

Mine clearing:

■ If informed about a minefield head there
■ Head toward blobs detected by camera
■ Clear mines

The communication range of the robots is limited, thus not all robots will be able to communicate with each other. A multi-hop routing daemon installed on the Xen nodes routes messages through intermediate robots wherever possible. Due to the fact that the robots have a task to fulfill, the network will become partitioned during the course of the experiment. To ensure that nonetheless communication is possible, the active and passive network preservation motivations introduced are added to the robots' motivations. This gives the network the necessary self-healing capabilities which are needed to fulfill the task at hand.

The screenshots in Figures 23.15a through 23.17b were taken from the Stage simulator and annotated to illustrate some of the key benefits of the approach taken in this chapter.

■ Figure 23.15a: To begin with, all the robots are set up at the bottom of the playing field. The four mine searching robots A,B,C, and E will head out across the field in search of mines. The mine clearing robot D does not yet have any task to fulfill so it stays at its starting position.

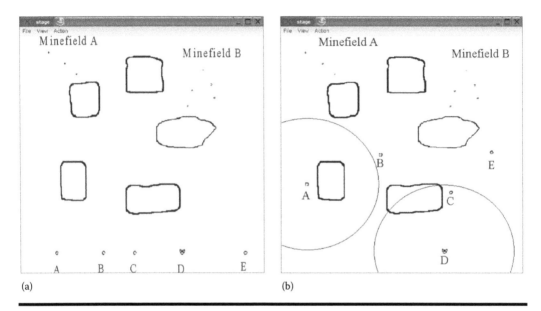

Figure 23.15 Self healing ad hoc network. (a) Stage screenshot 1. (b) Stage screenshot 2.

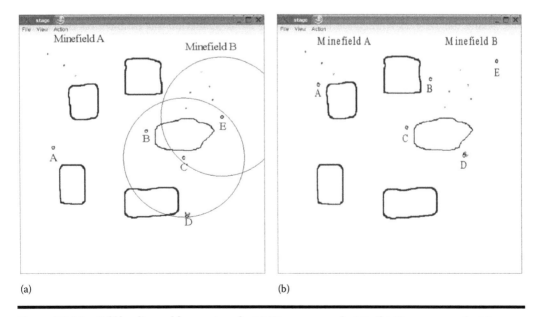

Figure 23.16 Self healing ad hoc network. (a) Stage screenshot 3. (b) Stage screenshot 4.

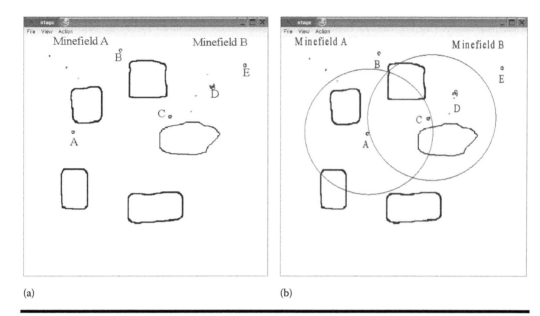

(a)   (b)

Figure 23.17   Self Healing ad hoc network. (a) Stage screenshot 5. (b) Stage screenshot 6.

■ Figure 23.15b: The mine clearing robot D is informed by the network layer that it is losing the connection to the network. The large circles around robot A and D show the communication range of the wireless communication device carried by each robot. The passive network preservation motivation kicks in as the mine clearing robot does not currently have any task; it will ensure that it stays within communication range of the bulk of the network. This action is only possible through the flexible coupled simulation and emulation system described in this chapter. In a scenario-based simulation, robot D would be aware that it is losing the network connection but it can not move to alleviate the problem because it can not alter the scenario file. In a strictly layered approach, robot D would notice that it lost the rest of the network only when it is too late. The cross-layer information gathered from the network layer informs the application before the fact so the application can take measures to prevent the loss of connection. Mine searching robot A is also aware that it has lost its connection to the network but since it has already detected the blobs from minefield A, it continues on its way. The passive network preservation motivation does not override the robots, primary task. Before robot A left the network, the active network preservation code queried the network layer in which direction the other nodes are headed and sent messages to the other nodes asking for their destinations, so it can attempt to rejoin the network later.

■ Figure 23.16a: Mine searching robot E has found minefield B. The message is transmitted to the mine clearing robot D by multi-hop routing via robot C. Due to the network emulation system, the routing daemon is under the illusion that only the multi-hop route to robot D is possible. The second motivation of robot D will now override the passive network preservation and lead the robot to the minefield.

■ Figure 23.16b: Mine searching robot A has found mine field A. Since it is not connected to the network it cannot inform the mine clearing robot of its find. To regain entry, it calls up the information gathered before it left the network and then tries to rejoin the network by moving toward where the other nodes were headed.

- Figure 23.17a: The mine clearing robot is defusing the mines of minefield B.
- Figure 23.17b: The mine searching robot A has rejoined the network and can relay its message to the mine clearing robot D via robot C. The mine clearing robot will finish clearing field B and then head to minefield A.

23.5.2 Routing Optimization for Motorway Ad Hoc Networks

Roadcasting (http://roadcasting.org/) is a novel technology that emerged from the adaptation of methods for personalized audio transmissions over the Internet (Podcasting) to the area of mobile ad hoc networks, in particular car-to-car networks. The original idea of Roadcasting was restricted to a system of mobile peer-to-peer radio stations. The technology behind Roadcasting, however, allows for a vastly larger range of applications, some of which are detailed below:

- Driver routing assistance: Navigating through an unknown area in dense traffic with a convoy of cars can prove difficult since single cars may get lost. Roadcasting can improve this situation by providing a walkie-talkie like service for a group of related vehicles.
- Distribution of emergency information: Mobile Roadcasting peers can be installed by the police at the roadside to warn drivers about traffic jams, accidents, and similar dangerous situations. This can be done using natural voice messages which is a more human form of warning than an the ever-growing number of warning icons or fixed audio messages, used by traditional navigation systems.
- Parking information system: When entering an area, drivers can be informed about the number of free parking places. Interconnected systems at various parking locations can, in addition, provide guidance to the next available space.
- Touristic information: Drivers can be informed of landmarks and other items of touristic interest by Roadcasting peers.
- Roadside advertisements: Gas stations, motels, etc., can send information on current prices and special offers to interested Roadcasting peers. This service must be implemented as an Opt-In service since the privacy and the peace of mind of road users must be protected.

To be able to research the effect of routing algorithms on Roadcasting applications, the following traffic simulator was created. Each car within the simulation is equipped with a wireless communication ranging from 802.11b to 802.11p and is tasked to exchange information on its current speed, lane, and distance to the car in front and behind with its neighboring cars. The communication device currently only affects the broadcast radius. Figure 23.18 shows a sixed-laned motorway (three lanes heading in each direction) with an average traffic level. From this information, a snapshot of the current traffic environment can be gathered. Depending on which routing strategies are available to the system, thresholds can be defined at which the routing algorithms are reconfigured to adapt to the current situation. Figure 23.19 shows the same motorway but two lorries are blocking the two lower lanes creating a traffic jam. The traffic density is higher and the average car speed much lower. By switching routing strategies, unnecessary routing overhead can be avoided based on this information gathered on the application layer.

23.6 Conclusions

The simulation/emulation system introduced in this chapter allows research into cross-layer design for mobile ad hoc networks to be undertaken in a unified way. The novel coupling of a virtual

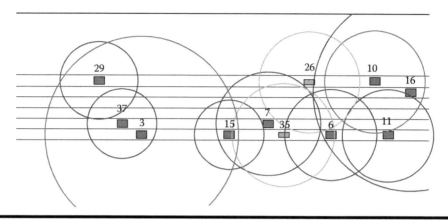

Figure 23.18 TrafficSim screenshot 1.

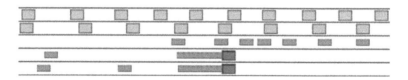

Figure 23.19 TrafficSim screenshot 2.

world simulator with a hardware and a network emulator allows the research to achieve new depth of realism. A unique feature is the ability to simulate the feedback loop between network state and application state (i.e., mobility). Changes made at runtime on one layer which affect a different layer can be fed back to the layer of origin, allowing a much more accurate study of mobile ad hoc networks and cross-layer concerns of ubiquitous and embedded applications to be made. Using this mechanism, cross-layer code can do more than make existing code more efficient. Spatial dependence of movements among nodes, the existence of barriers or areas of constraint mobility and the temporal dependence of movement of an node over time can be factored into mobility characteristics of the emulation in real time. As one example of the new possibilities, a novel self-healing mechanism based on cross-layer interaction was presented. The self-healing ability is an emergent behavior which is created out of the dynamic merging of the primary task of the network and behaviors basing their suggestions on the current network status. The coupled simulation/emulation approach allows researchers to vary system configuration through its modular design without any changes to the application or the used simulation system. New MAC layer protocols, ad hoc routing protocols, and mobility or traffic models can be evaluated. Thus, the emulation infrastructure can be easily extented and facilitates the development of new cross-layer solutions and interdependencies.

Future work will include a more detailed study of the routing optimization of the motorway scenario and of further cross-layer optimization possibilities tied to the state of the virtual world and mobility in particular. The routing daemon will be extended to utilize the available position information from the nodes to implement a geographic multi-hop routing algorithm. The network traffic shaping metric by which the network emulator determines the network topology will be extended by features such as: interference from jammers or bad weather conditions, obstacles obstructing line of site and different transmission power levels. As details of the 802.11p hardware become available, the traffic simulation will be improved to better model the real world environment.

Furthermore, different routing strategies will be tested in different traffic scenarios to quantify the benefits gained by cross-layer optimization. In addition, research in specialized routing algorithms as well as adapted lightweight service discovery [9] mechanisms for ad hoc networks can be used to improve the traffic simulation.

References

1. F. Bai, N. Sadagopan, and A. Helmy. IMPORTANT: A framework to systematically analyze the impact of mobility on the performance of routing protocols for ad hoc networks. In *Proceedings of the IEEE INFOCOM 2003 (The 22nd Annual Joint Conference of the IEEE Computer and Communications Societies. IEEE)*, pp. 825–835. San Francisco, CA, IEEE, 2003.

2. P. Barham, B. Dragovic, K. Fraser, S. Hand, T. Harris, A. Ho, R. Neugebauer, I. Pratt, and A. Warfield. Xen and the art of virtualization. In *SOSP'03: Proceedings of the Nineteenth ACM Symposium on Operating Systems Principles*, pp. 164–177. Bolton Landing, New York, ACM Press, 2003.

3. O. Battenfeld, M. Smith, P. Reinhardt, T. Friese, and B. Freisleben. A modular routing architecture for hot swappable mobile ad hoc routing algorithms. In *Proceedings of the Second International Conference on Embedded Software and Systems*, pp. 359–366. Xian, China, Springer-Verlag, 2005.

4. J. Broch, D. A. Maltz, D. B. Johnson, Y.-C. Hu, and J. Jetcheva. A performance comparison of multi-hop wireless ad hoc network routing protocols. In *MobiCom'98: Proceedings of the 4th Annual ACM/IEEE International Conference on Mobile Computing and Networking*, pp. 85–97. New York, ACM Press, 1998.

5. M. Conti, E. Gregori, and G. Turi. A cross-layer optimization of gnutella for mobile ad hoc networks. In *Proceedings of the 6th ACM International Symposium on Mobile ad hoc Networking and Computing*, pp. 343–354. Urbana-Champaign, IL, ACM Press, 2005.

6. S. Das, R. Castaneda, and J. Yan. Simulation based performance evaluation of mobile, ad hoc network routing protocols. In *Proceedings of the 7th International Conference on Computer Communication and Networks (IC3N)*, pp. 1–24. Lafayette, LA, 1998.

7. D. S. J. De Couto, D. Aguayo, B. A. Chambers, and R. Morris. Effects of loss rate on ad hoc wireless routing. Technical report MIT-LCS-TR-836. Cambridge, MA, MIT Laboratory for Computer Science, 2002.

8. Dynalogic. http://www.dynalogic.fr

9. M. Engel and B. Freisleben. A lightweight communication infrastructure for spontaneously networked devices with limited resources. In *Proceedings of the 2002 International Conference on Objects, Components, Architectures, Services, and Applications for a Networked World*, pp. 22–40, LNCS 2591. Erfurt, Germany, Springer-Verlag, 2002.

10. B. Gerkey, R. Vaughan, and A. Howard. The player/stage project: tools for multi-robot and distributed sensor systems. In *Proceedings of the International Conference on Advanced Robotics (ICAR 2003),* June 30–July 3, 2003, pp. 317–323. Coimbra, Portugal, 2003.

11. GloMoSim: A Scalable Network Simulation Environment. Los Angeles, CA, UCLA Computer Science Department. http://pcl.cs.ucla.edu/projects/glomosim/

12. A. Goldsmith and S. Wicker. design challenges for energy-constrained ad hoc wireless networks. *IEEE Wireless Communication Magazine*, 9 no. 4, pp. 8–27, 2002.

13. S. Hanemann, R. Jansen, and B. Freisleben. Reducing packet transmissions in ad hoc routing protocols by adaptive neighbor discovery. In *Proceedings of the International Conference on Wireless Networks (ICWN 2003)*, NV, pp. 369–375. Las Vegas, CSREA Press, 2003.

14. S. Hanemann, R. Jansen, and B. Freisleben. Update message delay: An approach for improving distance vector routing in wireless ad hoc networks. In *Proceedings of the 10th Symposium on Communications and Vehicular Technology (SCVT 2003)*, pp. 4–11. Eindhoven, the Netherlands, IEEE Press, 2003.

15. IEEE. IEEE Std 802.11b-1999 R2003. In *IEEE Standard for Information Technology Std 802.11b-1999 R2003*. New York, IEEE Press, 1999.

16. IEEE. IEEE Std 802.11g-2003. In *IEEE Standard for Information Technology Std 802.11g-2003*. New York, IEEE, 2003.

17. IEEE. IEEE 802.11p. In *IEEE Draft 802.11p D1.4*. New York, IEEE 802.11 Group, 2005.

18. R. Jansen and B. Freisleben. Bandwidth efficient distance vector routing for ad hoc networks. In *Proceedings of the 2001 Wireless and Optical Communications Conference*, pp. 117–122. Banff, Albeita, Canada, ACTA Press, 2001.

19. A. Jardosh, E. Belding-Royer, K. Almeroth, and S. Suri. Towards realistic mobility models for mobile ad hoc networks. In *ACM MobiCom'03: The 9th Annual International Conference on Mobile Computing and Networking*, pp. 217–229. San Diego, CA, ACM, 2003.

20. D. B. Johnson and D. A. Maltz. Dynamic source routing in ad hoc wireless networks. In T.Imielinski and H.Korth, eds., *Mobile Computing*, Vol. 353, pp. 1–18. Amsterdam, the Netherlands Kluwer Academic Publishers, 1996.

21. J. MacDonald and Y. Fang. Cross layer performance effects of path coupling in wireless ad hoc networks: Power throughput implications of IEEE 802.11 Mac. In *Proceedings of IEEE International Performance, Computing and Communications Conference*, pp. 281–290. Phoenix, AZ, IEEE, 2003, http://ieeexplore.ieee.org/stamp/stamp.jsp?arnumber=01201985

22. P. Mahadevan, A. Rodriguez, D. Becker, and A. Vahdat. MobiNet: A scalable emulation infrastructure for ad hoc wireless Networks. In *Proceedings of the International Conference On Mobile Systems, Applications and Services*, pp. 7–12. Barkely, CA, USENIX Association, 2005.

23. NS2. Network simulator NS-2 Project Group. A collaboration between researchers at UC Berkeley, LBL, USC/ISI, Xerox PARC and DAPRA, The Network Simulator NS-2, July 2012, http://www.isi.edu/nsnam/ns/

24. C. Perkins and P. Bhagwat. Highly dynamic destination-sequenced distance-vector routing (DSDV) for mobile computers. In *ACM SIGCOMM'94 Conference on Communications Architectures, Protocols and Applications*, pp. 234–244. London, U.K., ACM, 1994.

25. C. E. Perkins and E. M. Royer. Ad hoc on-demand distance vector routing. In *Proceedings of the 2nd IEEE Workshop on Mobile Computing Systems and Applications*, pp. 90–100. New Orleans, LA, IEEE, 1999, http://www.hotmobile.org/1999/

26. PTV Planung Transport Verkehr AG. Homepage. http://www.ptv.de/cgibin/traffic/traf_referenzen.pl

27. J. Raju and J. Garcia-Luna-Aceves. A comparison of on-demand and table driven routing for ad-hoc wireless networks. In *Proceedings of IEEE International Conference on Communication*, pp. 1–5. New Orleans, LA, ICC 2000, http://www.icc00.org/

28. K. Ramachandran, S. Kaul, S. M. M. Gruteser, and I. Seskar. Towards large-scale mobile network emulation through spatial switching on a wireless grid. In *Proceedings of the 2005 ACM SIGCOMM Workshop on Experimental Approaches to Wireless Network Design and Analysis*, pp. 46–51. New York, ACM Press, 2005.

29. Rescue Simulation Project. RoboCup Rescue, Jun 2012, http://roborescue.sourceforge.net/

30. B. Gerkey. robotics.usc.edu. Player robot device interface, Jun 2012, http://playerstage.sourceforge.net/index.php?src=playe

31. R. T. Vaughan.robotics.usc.edu. Stage multiple robot simulator, Jun 2012, http://playerstage.sourceforge.net/index.php?src=stage

32. I. Rubin and A. Behzad. Cross-layer routing and multiple-access protocol for power-controlled wireless access nets. *IEEE CAS Workshop on Wireless Communications and Networking*, pp. 1–7. Pasadena, CA, IEEE, 2002.

33. C. Schlosser. Verkehrssimulation zur Analyse drahtlos kommunizierender fahrerassistenzsysteme in mobilen ad-hoc netzwerken. Diploma thesis, Department of Mathematics and Computer Science, University of Marburg, Marburg, Germany, 2005.

34. S. Shakkottai, T. S. Rappaport, and P. Karlson. Cross layer design for wireless networks. *IEEE Communications Magazin*, 41 issue 10, 1–14, 2003.

35. A. K. Singh, V. Raisinghani, and S. Iyer. Improving TCP performance over mobile wireless environments using cross layer feedback. In *Proceedings on the IEEE International Conference on Personal Wireless Communications*, pp. 81–85. New Delhi, India, IEEE, 2002.

36. M. Smith, M. Engel, S. Hanemann, and B. Freisleben. Towards a roadcasting communications infrastructure. In *Proceedings of the First International Conference on Mobile Communications and Learning*, pp. (213– 213). Moka, Mauritius, IEEE Press, 2006.

37. STRICOM. OneSAF, Jun 2012, http://www.onesaf.net/community/

38. Sun Microsystems. J2ME Wireless Toolkit, Jun 2012, http://www.oracle.com/technetwork/java/download-135801.html

39. The Federal Highway Administration, U.S. Department of Transportation. CORSIM User Manual, Jun 2012, http://sites.poli.usp.br/ptr/lemt//CORSIM/CORSIMUsersGuide.pdf

40. S. Toumpis and A. Goldsmith. Performance, optimization, and cross-layer design of media access protocols for wireless ad hoc networks. In *International Conference on Communication (ICC) 2003*, pp. 2234–2240. Anchorage, AK, IEEE, May 2003.

41. Trafficware Corporation. Homepage. http://www.trafficware.com

42. M. Treiber and D. Helbing. Realistische Mikrosimulation von Strassenverkehr mit einem einfachen Modell. *Simulationstechnik ASIM*, pp. 514–520. Rostock, Germany, 2002.

43. H. L. W. Yuen and T. Andersen. A simple and effective cross layer networking system for mobile ad hoc networks. In *Proceedings of PIMRC 2002*, pp. 1–5. Lisbon, Portugal, 2002.

44. J. Yoon, M. Liu, and B. Nobel. Random Waypoint Considered Harmful. In *Proceedings of the IEEE 22nd Annual Joint Conference of the IEEE Computer and Communications Societies*, pp. 1312–1321. San Francisco, CA, IEEE, 2003.

45. Y. Zhang and W. Li. An Integrated Environment for Testing Mobile Ad Hoc Networks. In *Proceedings of the 3rd ACM International Symposiun on Mobile Ad Hoc Networking and Computing*, pp. 104–111. Lausanne, Switzerland, MobiHoc 2002, http://www.sigmobile.org/mobihoc/2002/

Chapter 24

Dynamic Routing and Wavelength Assignment in Optical WDM Networks with Ant-Based Mobile Agents

Son Hong Ngo, Xiaohong Jiang, Susumu Horiguchi, and Minyi Guo

Contents

24.1 Introduction

All-optical networks using wavelength-division multiplexing (WDM) technology are promising infrastructure for the future Internet [1]. In such networks, data are switched and routed in all-optical domain via lightpaths. The Routing and Wavelength Assignment (RWA) problem is one of the most important issues in optical WDM networks; it involves determining routes and wavelengths to establish lightpaths for connection requests. In optical networks without wavelength converters, the same wavelength must be assigned on every link of a lightpath; this referred to as the wavelength-continuity constraint. The RWA problem can be generally classified into two types: static and dynamic RWA. In a static RWA problem, network topology and connection-establishing requirements are given in advance, and the problem is to find a solution that satisfies certain optimizing conditions, such as minimizing the number of wavelengths or maximizing the total number of lightpaths that can be established. A dynamic RWA problem involves the on-line establishment of lightpaths for connection requests that arrive dynamically. The static RWA problem is proved as a NP-hard problem [2]. The dynamic RWA problem is much more difficult to solve, therefore heuristics algorithms are usually employed to solve it.

We focus on the dynamic RWA problem in this chapter. Many approaches have been proposed for solving the dynamic RWA problem and these approaches can be generally classified into three basic classes: fixed routing, alternate routing, and adaptive routing. In the fixed routing approach, a fixed shortest path is statically computed for every pair of network nodes. Whenever a connection request arrives between a node pair, that fixed path is always attempted for wavelength assignment. This method is simple but it tends to introduce a high blocking probability. In alternate routing, a node pair has a set of precomputed shortest paths and all of them will be attempted for lightpath establishment upon the arrival of a connection request. In the adaptive routing approach, the path for a connection request is computed upon its arrival time based on the current network state, thus it provides the lowest blocking probability. However, adaptive routing usually requires higher computation complexity and thus introduces a longer setup delay than alternate routing. Moreover, adaptive routing requires higher control overhead, including special support from control protocols to keep track of the up-to-date global network state, which makes it practically infeasible for large and highly dynamic networks.

The adaptive routing approach can be further divided in two categories: the routing with precise information and the routing with imprecise information [3–5]. In a network with a ideal condition, a global routing information can be stored with precision and we can apply the optimal RWA. In a large network, however, maintaining precise global network state information is almost impossible [6]. Many factors such as propagation delay and out-of-order state update due to control overhead can affect the precision of global network state information. Several researches indicate that optimal adaptive routing with imprecise information performs even worse than a simple fixed routing approach [7]. Thus, an advanced routing approach without requiring global routing information is much more preferable in practical networks.

In this chapter, we are interested in adaptive routing approach for dynamic RWA with the lack of global routing information and propose a novel algorithm for it based on the principle of Ant-Colony Optimization [8,9]. Inspired by the behavior of natural ant systems, a new class of ant-based algorithms for routing in communication networks are currently being developed. Previous work has shown the potential success of ant-based routing for both packet switching networks (e.g., AntNet [10,11]) and circuit switching telephone networks (e.g., ABC [12]). However, few results are available for applying the ant-based approach to the challenging RWA problem (especially the dynamic RWA problem) in WDM networks. We first introduce an adaptive ant-based routing (ABR) algorithm with the wavelength continuity constraint. In this algorithm, each network node is equipped with a probabilistic routing table (*pheromone table*). The path for a connection request is determined promptly based on the current states of these routing tables, which are continuously updated by an adjustable number of cooperative ants. The ABR algorithm is attractive in the sense that it does not require global and centralized routing information for route selection as other adaptive RWA routing algorithms, therefore, the path for a connection request can be determined promptly by simply looking up routing tables in a small computation time. An extensive simulation study shows that the proposed ABR algorithm performs better than the fixed-alternated routing algorithm in terms of blocking probability.

We further improve the performance of ant-based routing approach by proposing a hybrid ant-based routing algorithm. Actually, it is difficult for the ABR algorithm to search more than one path for a connection request, that is because the routing tables are distributed in every nodes of the network. Once a connection request arrives, a control packet is sent to every node in the network to gather information about possible routes. Therefore, it is complicated for the ABR algorithm to gathering information about more than two routes for lightpath establishment. On other hand, the alternate routing method is a good candidate for multipath routing because its routing table always contains many alternate routes at each node, thus it is easier to attempt multiple routes for setup establishment. From the above observations, we then propose a hybrid ant-based routing algorithm (HABR) for dynamic RWA by combining the mobile agent approach and the alternate routing methods. To enable the new HABR algorithm, we adopt a novel twin routing table structure on each network node that comprises a P-route table for connection setup and a pheromone table for ants' foraging. The P-route table contains a set of P feasible routes between a source-destination pair, which are dynamically updated by ant-based mobile agents based on current network congestion information. By keeping a suitable number of ants in a network to continually update the twin routing tables in the network, the candidate alternate routes are available at the arrival of a connection request, thus a small setup delay is guaranteed in our new algorithm. Extensive simulation results indicate that the HABR algorithm significantly outperforms the alternate routing approaches with a suitable number of ants and a small value of P.

The rest of this chapter is organized as follows. Section 24.2 summarizes some conventional RWA approaches. Section 24.3 describes our proposed adaptive ant-based routing algorithm. Section 24.4 presents the hybrid ant-based routing algorithm. Section 24.5 investigates the complexity and control overhead of ant-based approaches. Section 24.6 shows the simulation results. Finally, Section 24.7 concludes this chapter.

24.2 Overview of Conventional RWA Algorithms

In the following part, we summarize some basic algorithms for RWA problem: (a) The most simple fixed routing using shortest path (SP) algorithm summarize; (b) The alternate routing approaches such as alternate shortest path (ASP) routing and Fixed Path Least Congested (FPLC) routing and (c) The adaptive unconstrained routing algorithm (AUR). We will later compare the performance of our proposed algorithm with the shortest path and alternate routing approaches because they do not require global routing information like the ant-based approach. It is not meaningful to compare an ant-based algorithm with an optimal adaptive routing algorithm that requires global routing information, because it certainly outperforms any other algorithms which uses non-global routing information. However, we will describe the adaptive unconstrained routing algorithm here as a reference for a qualitative comparison.

24.2.1 Fixed Shortest Path Routing

In a WDM network where the RWA uses the fixed shortest path (SP) routing algorithm. The lightpath is always established on the shortest path regardless of the availability of wavelength. On each network node, the Dijkstra's algorithm or Bellman–Ford's algorithm [13] is used off-line to compute the shortest path (in terms of number of hops) for every source-destination node pairs. The list of the shortest path to all destination is stored in a fixed routing table. When a connection request arrives at a source node, the fixed shortest path between the source node and its destination is taken from the routing table to establish the lightpath. This kind of table is the explicit routing table because it describes in detail every node of a route from the source node to the destination node. An example of the routing table is shown in Figure 24.1a.

For the wavelength assignment, any scheme as described in [14] can be used. If the First-Fit wavelength assignment scheme is adopted (as in this work) then the first available wavelength is selected for lightpath establishment. In case that no available wavelength is found on that shortest path, the connection request is then blocked. This algorithm is very simple and the route is almost ready by the time that a connection request arrives. However, the performance is low and it is rarely used in practice. We just use it as a baseline algorithm for the comparison purpose. An implementation of the shortest-path algorithm is available in NS-2. [15].

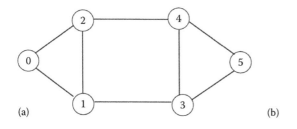

Destination	Shortest route
0	2, 0
1	2, 1
...	...
5	2, 4, 5

(a)   (b)

Destination	Shortest routes pairs
0	[2,0] ; [2,1,0]
1	[2,1] ; [2,0.1]
...	...
5	[2,4,5] ; [2,1,3,5]

(c)

Figure 24.1 SimpleNet network topology and its routing tables on node 2. (a) SimpleNet topology. (b) Routing table for shortest path routing. (c) Routing table for alternate shortest path routing.

24.2.2 Alternate Routing

24.2.2.1 Alternate Shortest Path Routing

Fixed alternate routing or so-called Alternate Shortest Path routing (ASP) has been studied extensively in WDM mesh optical networks at the wavelength level. An ASP algorithm predetermines a number of link-disjoint paths for each pair of nodes and selects one of these paths according to the network state information at run time. Alternate shortest routing requires that each access node in the network have a routing table, which contains an ordered list of a limited number of fixed routes to each destination node. When a connection request arrives, the source node attempts routes in sequence from the routing table, until a route with a valid wavelength assignment is found (the wavelength assignment algorithm is specified in previous chapter). If no available route is found from the list of alternate routes, then the connection request is blocked and lost.

ASP routing provides benefits such as (1) simplicity of control to setup and teardown lightpaths, and (2) fault tolerance upon link failures. A direct route between a source node and a destination node is defined as the first route in the list of routes to in the routing table. An alternate route between is any route other than the first route in the list of routes to in the routing table. The term alternate routes is also employed to describe all routes (including the direct route) from a source node to a destination node. As an example, Figure 24.1b illustrates the routing table at node 2 for the network shown beside. In this example, each source maintains one direct route and one alternate route, for a total of two alternate routes, to each destination node. For the networks considered here, the routing tables at each node are ordered by the hop distance to the destination. Therefore, the shortest-hop path to the destination is the first route in the routing table. When there are ties in the hop distance between different routes, the ordering among them in the routing table is random.

Extensive empirical results in the literature indicate that the performance of fixed-alternate routing with a small number of alternate routes asymptotically approaches that of adaptive routing in reduced blocking probability. For this reason, we will use two alternate routes for testing in our experiments: the first route is the shortest path; the alternate route is the shortest among all other paths.

An example of the implementation of fixed-alternate routing is available in the OWNS simulator [16]. In this module, the two alternate routes are computed using a modified Dijsktra's algorithm. At first, the direct route is computed directly as the first shortest path in terms of number of hops. Next, to obtain the second route that is link-disjoint with, the cost of every link that belongs to the direct route is assigned to infinity. The Dijkstra's algorithm is then applied again to obtained an alternate route which is the second shortest path.

24.2.2.2 Fixed-Paths Least Congestion Routing

The Fixed-Paths Least Congestion routing (FPLC) algorithm is a combination of Lead Loaded Routing (LLR) and alternate routing approaches [17]. Least-loaded routing (LLR) algorithms are introduced as dynamic routing methods for fully connected networks in [18,19]. The main idea of LLR is borrowed from telephone network. If a connection cannot be set up along the direct route, a two-hop alternate route with the largest number of free wavelengths is chosen. Least-loaded routing methods can provide a significantly low blocking probability among dynamic routing approach. The main problems with these dynamic-routing methods are longer setup delays and higher control overheads, including the introduction of a central control node that keeps track of the network's global state. Therefore, the alternate routing approach is much more preferred on practice.

In [20], Li et al. has considered alternate dynamic routing algorithms in all-optical networks and introduced a dynamic routing algorithm, called fixed-paths least-congestion (FPLC). In FPLC routing, a statical set of routes is computed which is later used for each source-destination pair in a network. These routes are stored at each source node. Upon arrival of a connection setup request, the source node searches the available number of wavelengths on the routes in parallel by sending needle packets requesting a path setup toward the destination node. (The needle packets can be sent out through a control network, either out-of-band or in-band, as proposed in [21]. At the destination node, a route with the maximum number of idle wavelengths is selected to set up the connection (least congested route). If no wavelength is available on any of the routes, the request is blocked.

Normally, two link disjoint shortest paths are used for FPLC but this principle can be applied for the case of any number of routes. These routes are referred as the first and second route. The length of the first route is less than or equal to that of the second route. If two routes have the same number of idle wavelengths, the first route is selected to set up the request. It is shown that we should restrict the number of preferred routes to two because network resources cannot be used efficiently if many longer routes are allowed in the network. One reason that the two routes are required to be edge-disjoint is that we try to search two paths in parallel. Another consideration is fault tolerance. If one path fails, the connection can be rerouted to another. Numerical results show that the FPLC routing with the first-fit wavelength assignment method significantly improves network performance compared to the alternate paths routing algorithms. The reason is that more wavelengths are left free on a network when the FPLC with the first-fit wavelength assignment method is used.

The FPLC still has higher setup delay and higher control overhead. To overcome these shortcomings, a new routing method based on FPLC that uses neighborhood information is also proposed by the same author (FPLC(k)). In this method, for each source-destination pair, a set of preferred paths are precomputed and stored at the source. Moreover, instead of searching all the links on the preferred routes for availability of free wavelengths, only the first k links on each path are searched. A route is selected based on the availability of free wavelengths on the first k links on the preferred paths. If several free wavelengths are available on the selected route, a wavelength is selected according to a prespecified wavelength assignment algorithm. If no free wavelengths are available on the first k links of all the preferred routes, the request is blocked. It is shown that a value of $k = 2$ is generally enough to ensure good network performance of two typical networks, such as a 4×4 mesh-torus network and the NSFnet T1 backbone network.

24.2.3 Adaptive Unconstrained Routing

The Adaptive Unconstrained Routing algorithm (AUR) is proposed by Mokhtar et al. [22]. This is called unconstrained routing scheme because it considers all paths between the source and the destination in the routing decision. The routing solution is accomplished by executing a dynamic shortest path algorithm with link costs obtained from network state information at the time of connection request.

In AUR algorithm, the RWA problem is formulated as follows. In a network with K links and W wavelengths, the state of a link $i, 0 \leq i \leq K - 1$, at time t can be specified by a column vector $\sigma_t^i = (\sigma_t^i(0), \sigma_t^i(1), \ldots, \sigma_t^i(W - 1))^T$ where $\sigma_t^i(j) = 1$ if wavelength λ_j is utilized by some connection at time and $\sigma_t^i(j) = 0$ otherwise. The state of the network at time t is then described by the matrix $\sigma_t = (\sigma_t^{(0)}, \ldots, \sigma_t^{(K-1)})$. Given a connection request that arrives at time t, the RWA algorithm searches for a path $P = (i_1, i_2, \ldots, i_l)$ from the source of the request to its destination such that $\sigma_t^{i_l}(j) = 0$ for all $k = 1, 2, \ldots, l$ and some j.

The AUR is not limited to a set of predetermined paths or search sequences. It makes use of the network state at the time of connection establishment to improve the blocking performance fixed routing techniques. Since the search for a route and a wavelength assignment may be viewed as a search over the rows and columns of the network state matrix, there are many ways in which an RWA algorithm may proceed. The following approach is adopted: For a connection request that arrives at time, the rows are searched in an adaptively varying order. Each row specifies the available topology at the corresponding wavelength; therefore, our approach is to search sequentially over the wavelength set until an available path is found (a standard shortest path algorithm is used to find a path on the effective topology). If no path is found after exhausting the wavelength set, the connection request is blocked.

Five adaptive AUR RWA algorithms is considered with different sorting mechanisms of the wavelength set. Among them the AUR with exhaustive (AUR-E) wavelength search performs the best. In this scheme, all of the wavelengths are searched for the shortest available path and the shortest path among them is selected.

The use of unconstrained routing yields significant improvements in the call-blocking performance over traditional constrained routing techniques. In particular, AUR outperforms fixed routing by a significant margin. AUR also performs better than alternate routing; however, as the number of alternate routes increases, the performance approaches that of AUR. Alternate routing is found to be a good trade-off between fixed routing and AUR.

24.3 Adaptive Ant-Based Routing Algorithm

24.3.1 Network Model and Assumptions

We focus on optical wide area networks with an irregular mesh topology. A network composes of a number of optical switching nodes connected by optical fiber links. Each network node is considered as an optical router that can routes a lightpath from an input port to an output port. The fiber links may span across a large area, e.g., many cities or countries.

Each network node has two separate planes: the data plane and the control plane. User data are carried in the optical domain, which refers to the lightpaths. The control plain carried the control signal and control data in order to set up and tear down lightpaths. We suppose that the control plane of a WDM optical network is carried out in a packet switching network that has the same topology as the optical network. This assumption is based on the fact that control data can be transported in several ways, such as through an out-of-band electronic network or through a dedicated wavelength in the optical domain [23].

We assume that the lightpaths are undirected and symmetric. This assumption assures that the same route and wavelength are assigned for both directions with the same cost. We also focus on the wavelength continuity constraint in which a same wavelength must be used for all the links of a lightpath. With the above assumptions, a network can be represented by an undirected graph with N nodes and some bi-directional links. Each link has the same capacity of W wavelengths and no nodes have wavelength conversion capability (wavelength continuity constraint).

Our purpose is to build a distributed routing based on ant-based mobile agents, thus we suppose that there is no link-state update mechanism [3] on every network node. In contrast, the control tasks are performed by a number of artificial ant-based mobile agents that works in the separate control plane. We suppose that the number of ants and also their sizes are small enough to not make the control plane become congested.

24.3.2 New Routing Table Structure

To support dynamic route selection, a node i with k_i neighbors has a routing table $R_i = [r_{n,d}^i]_{k_i, N-1}$ with $N - 1$ rows and k_i columns (N is the number of network nodes). Each row corresponds to a destination node and each column corresponds to a neighbor node. Like the routing table structure proposed by the ABC algorithm [12], the value $r_{n,d}^i$ expresses the goodness of choosing n as the next hop in establishing a lightpath. It is also used as the selection probability of neighbor node n when an ant moves toward its destination node d. For each destination, the sum of all neighbor's selection probabilities must be 1 to satisfy the normalized condition:

$$\sum_{n \in Nb_k} r_{n,d}^i = 1 \ \forall i, d \in [1, N] \tag{24.1}$$

where Nb_k is the set of all the neighbor nodes of the node k.

An example of the routing table is shown in Figure 24.2. When a connection request occurs between the source node 3 and the destination node 0, node 1 will be selected as the next hop because $r_{1,0}^0 > r_{4,0}^0 > r_{5,0}^0$.

24.3.3 Ant's Foraging and Routing Table Updating

The main idea of the ant-based algorithm is adopting the cooperative behavior of ant colonies to adaptively discover a route for lightpath establishment. To achieve that goal, ants are launched from each node with a given probability ρ to a randomly selected destination every T time units. Here ρ and T are design parameters. Each ant is considered as a mobile agent: It collects data on its trip, performs routing table updating on visited nodes and continues to move forward as illustrated in Figure 24.3.

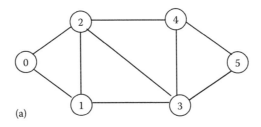

	Neighbor		
Destination	1	4	5
0	$r^3_{1,0} = 0.6$	$r^3_{4,0} = 0.3$	$r^3_{5,0} = 0.1$
1	$r^3_{1,1} = 0.8$	$r^3_{4,1} = 0.2$	$r^3_{5,1} = 0.0$
...	...	...	...
5	$r^3_{1,5} = 0.0$	$r^3_{4,5} = 0.2$	$r^3_{5,5} = 0.8$

(b)

Figure 24.2 A simple network and its routing table of node 3 (a) SimpleNet topology. (b) Pheromone routing table.

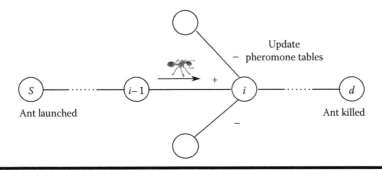

Figure 24.3 Ant's moving and updating task.

24.3.3.1 Data Collected by Ants

Ants collect two kinds of data along their trips: path length and congestion information. Each ant carries a binary mask M_{ant} that indicates the available wavelengths on its path; this mask has W bits corresponding to the number of wavelengths. The bit value 1 corresponds to a free wavelength while the bit value 0 corresponds to a busy wavelength. Under the wavelength continuity constraint, a wavelength can be assigned only if it is free on all links of the path. Thus, at each node, the wavelength mask is updated as follows:

$$M_{ant} = M_{ant} \text{ AND } M_{link} \qquad (24.2)$$

where

M_{ant} is the actual mask carried by the ant
M_{link} is the mask for available wavelengths on next selected link

24.3.3.2 Routing Table Updating

Whenever an ant visits a node, it updates the routing table (pheromone updating). Suppose an ant moves from the source node s to destination node d following the route $(s, \ldots, i-1, i, \ldots, d)$, it will update the entry corresponding to the source node s in the routing table of node i as follows: the probability of selecting neighbor $i-1$ is increased while the probabilities of selecting other neighbors are decreased (Figure 24.3).

More formally, suppose that an ant visits node i at time t, so the values for routing entry in time $t+1$ are determined by the following formula (remember that the sum of selecting probabilities for all neighbors is always 1):

For the previous visited node:

$$r^i_{i-1,s}(t+1) = \frac{r^i_{i-1,s}(t) + \delta r}{1 + \delta r} \tag{24.3}$$

For the other nodes that is different from the previous visited node:

$$r^i_{n,s}(t+1) = \frac{r^i_{n,s}(t)}{1 + \delta r}, \forall n \neq i-1 \tag{24.4}$$

Here, δr is the reinforcement parameter and is derived from the data collected by the ant. In the basic ant-based routing algorithm for WDM networks, this parameter (amount of trailing pheromone) is computed by the following principle: the amount of trailing pheromone decreases with increasing path length and increases with an increased number of available wavelengths.

Let δl be the amount of pheromone corresponding to the path length (refer to the number of hops) and δw be the amount of pheromone corresponding to the percent of free wavelengths of the path that the ant has moved along. We introduce here a scalar parameter such that we can adjust the weight between δl and δw in δr as follows:

$$\delta r = \alpha \cdot \delta l + (1 - \alpha) \cdot \delta w, \ 0 \leq \alpha \leq 1 \tag{24.5}$$

The factor δl is derived from the length of the path that the ant has moved along. The shorter the path length, the bigger the δl value, and vice versa. Note that the length l of a path between a source-destination node pair is always greater than or equal to the length of the shortest path between the source and the destination (denoted by l_{min} hereafter), and as a reinforcement parameter, δl must be small and $0 < \delta l < 1$. Thus, we compute δl as follows:

$$\delta l = e^{-\beta \cdot \Delta l}, \Delta l = l - l_{min} \tag{24.6}$$

where β is a control parameter. Since the absolute value of path length varies significantly in a large network, using the length difference instead of the absolute length in determining the reinforcement parameter δl enables the pheromone updating process for all node pairs to be controlled by the same parameter β.

The factor δw is derived from the percent of free wavelengths in the path that the ant has moved along. The path with more free wavelengths has a larger value of δw. Similar to δl, δw must be a small value and $0 < \delta w < 1$. We can compute dw as follows:

$$\delta w = e^{\gamma \cdot w} - 1 \tag{24.7}$$

where γ is another control parameter. Here α, β, and γ are design parameters and can be adjusted to get the best system performance.

24.3.3.3 Ant Movement

When an ant moves from a source to a destination, its next hop is determined stochastically: a neighbor is selected according to its selection probabilities in the routing table. This is the basic principle of ant colony optimization [8,9,24,25]. As a result, an ant colony tends to discover the better path between a node pair in terms of path length and the degree of congestion along this path.

In ABR algorithm, an ant is killed when it reaches its destination node or when it cannot find a path with a free wavelength to which to move. An ant is also killed if its lifetime exceeds a predefined value TTL (time-to-live), or if it detects a loop on its path (by searching the visited nodes in the ant's stack)

Ant-based algorithms usually suffer from *stagnation*, in which an optimal path is found by ants so the pheromone for this path is recursively increased [26]. In this case, too many ants concentrate on this optimal path, which prevents them from discovering other, better paths when the network state changes. To avoid this "frozen" situation, a random scheme with an exploration factor P_{noise} is introduced: at each node; the ant selects its next hop randomly with an exploiting probability P_{noise} and selects its next hop according to the routing table with probability $1 - P_{noise}$. The using of P_{noise} allows ants to keep exploring for a better solution to a lightpath request.

24.3.4 Smart Updating

As described in [27], agents supplemented with dynamic programming capacity, or smart agents, are efficient in improving the performance of ant-based routing systems. With the idea of smart agents, the pheromone updating affects not only the entry corresponding to the source node, but also all the entries corresponding to previous nodes the ant has visited. In order to facilitate the smart updating, an ant must push into its stack the node identification and a binary mask that determines the states of wavelengths on all the links it has traversed. Under the wavelength continuity constraint, this wavelength mask is determined in a same way as described in Section 24.3.3.1. This stack also serves for loop detection and backtracking, to ensure that ants will not move forever on the network.

24.3.5 Selection of Path and Wavelength

With support from the routing table and the ant's foraging, path selection can be performed in a straightforward manner as described in all other algorithms on ant-based algorithms: When a connection request arrives at the source node, the next hop will be determined by the node with the highest selection probability among all its neighboring entries. The visited nodes will never be selected as the next hop. This principle is applied from the source node to the destination node. We call this the First-Highest scheme. With this scheme, the route is already determined upon the arrival of a connection request.

After determining the route for a connection request, any wavelength assignment scheme can be applied. Our work in this thesis does not focus on ant-based wavelength assignment heuristics, so we use First-Fit [14]—a simple yet efficient heuristic for wavelength assignment. In the First Fit heuristic, all wavelengths are considered and the first available one is selected.

As explained in the previous section, ant-based algorithms adopting the First-Highest scheme have a stagnation state problem. When the network state changes, e.g., when a lightpath is established or released, ants may not be able to quickly find a better path for a new request. This phenomenon greatly affects the performance of ant-based dynamic RWA algorithms with a wavelength continuity constraint, since the state change of a wavelength on one link may affect the overall RWA solution. Here, we introduce here a new path selection scheme, called the Second-Highest scheme, to enlarge the search space for RWA solution based on the ants's searching. In the Second-Highest scheme, we will try the second route as follows if there is no free wavelength available on the first route, as obtained by the First-Highest scheme: At the source node, the next hop is determined by the node with the second highest selection probability, after this hop, the First-Highest scheme is used again until the destination node is reached. The wavelength assignment for the second route is also based on the First-Fit heuristic.

The pseudo-code of the ant-based RWA algorithm with the Second-Highest scheme and First-Fit wavelength assignment is summarized as follows:

{Ant generation}
Do
    For each node in network
        Select a random destination
        Launch ants to it with a probability ρ
    End for
    Increase time by a time-step for ants' generation
Until (end of simulation)

{Ant foraging}
For each ant from source s to destination d do (in parallel)
    While current node $i \neq d$ and TTL > 0
        Smart-update routing table elements
        Push trip's state into stack
        If (found a next hop)
          Move to next hop and decrease TTL
        Else
          Kill ant
        End if
    End while
End for

{Routing and Wavelength Assignment}
For each connection request do (in parallel)
    Select a path based on first-highest-probability-lookup
    Select the first available wavelength on path
    If (found)
        Setup a lightpath
    Else
        Select another based on second highest probability
        Select the first available wavelength on path
        If (found)
          Setup a lightpath

```
        Else
            Consider a blocking case
        End if
    End if
End for
```

24.4 Alternate Ant-Based Routing Algorithm

24.4.1 Motivation of Ant-Based Alternate Routing Approach

It is notable that the current alternate routing algorithms are all based on a set of fixed and static paths computed in advance, so these algorithms have little adaptability to network traffic variations and may significantly limit the routing performance in terms of blocking probability. In previous chapter, we have proposed the adaptive ant-based routing algorithm that is more adaptive to the dynamic traffic than the alternate routing method. However, one of drawback of ABR algorithm is that it is difficult to search more than one paths for a connection request in order to further improve the blocking performance. This is because the ABR use next-hop routing pheromone tables (pheromone). These tables are distributed in every node of the network. Once a node gets a route, it must send a control packet to every node in the network to gather information about possible routes. Therefore, it is complicated for ABR algorithm in trying more than two routes for lighpath establishment. On other hand, the alternate routing method is a good candidate for multi paths routing because the k-routes routing table is located at each node and it is easier to try multiple routes for setup establishment.

Inheriting from the originality of alternate routing method and the ant colony optimization, we propose a hybrid ant-based routing algorithm (HABR) for dynamic RWA by combining the best of mobile agent approach's good adaptability [8] and the alternate routing method's simplicity while avoiding their shortcomings. The HABR is presented in the following section.

24.4.2 Twin Routing Table Structure

In HABR algorithm, a network node i with k_i neighbors is equipped with a probabilistic pheromone table (as the pheromone table of ABR algorithm, Figure 24.2), $R_i = [r^i_{n,d}]_{k_i, N-1}$ with $N-1$ rows (N is the number of network nodes) and k_i columns, and a P-route table with $N-1$ rows, as illustrated in Figure 24.4. In the pheromone table, each row corresponds to a destination node and each column corresponds to a neighbor node. The value $r^i_{n,d}$ is used as the selection probability of neighbor node n when an ant is moving toward its destination node d. In the P-route table, each row corresponds to a destination and contains a list of P routes to the destination. Each route

Destination	List of P routes
0	3.1.0; 3.4.2.0; ...
1	3.1; 3.4.2.1; 3.4.2.0.1; ...
...	...
5	...

Figure 24.4   *P*-route table on the node 3 of the SimpleNet topology.

is assigned a value that represents the goodness of this route based on its length and the number of idle wavelengths along it. The bigger the goodness of a route, the better path is considered for connection setup.

24.4.3 Ant's Foraging and Routing Tables Updating

Ant's moving and the pheromone table updating is similar to the ABR algorithm. On each network node and for every t time units (s), an ant is launched with a given probability ρ to a randomly selected destination; here ρ and t are design parameters. Each ant is considered to be a mobile agent: It collects information on its trip, performs routing tables updating on visited nodes, and continues to move forward as illustrated in Figure 24.3.

24.4.3.1 P-Route Table Updating

Besides updating the pheromone table, ant also updates the P-route table. As mentioned above, each route is associated with a goodness value that is calculated by the following formula:

$$dr = \varphi \times \frac{1}{dl + 1} + (1 - \varphi) \times dw \qquad (24.8)$$

Here, φ is a scalar parameter used to adjust the emphasis between the route length and the percentage of free wavelengths. The bigger value of φ, the larger goodness value the shorter route has. The smaller value of φ, the larger goodness value the route with more idle wavelengths has. The parameter φ should be chosen such that the shorter route has a larger goodness value, while for two routes with the same length, the route with a larger number of idle wavelengths gets a larger goodness value. The selection of φ is discussed in [28].

When an ant reports its trip to a P-route table, it updates only the goodness value if the reported route already exists in the table. Otherwise, ant inserts the new route in the table or replaces the route that has the smallest goodness value by the new one.

24.4.4 Connection Setup

Upon the arrival of a connection request, the k best routes that have the biggest goodness values in the P-route table are selected as the routes candidates for lightpath setup. The lightpath setup is similar to other alternate routing methods: The source node sends in parallel some needle packets [23] along these k routes toward destination to request a path setup. At the destination node, the route with the highest goodness value is selected for connection setup. If no free wavelength is available for all these k routes, the connection request will be blocked. We use in our algorithm two alternate routes ($k = 2$) as suggested in previous works [17,20] that alternate routing methods using too many alternate routes do not significantly improve network performance.

To select a route for the connection setup in HABR algorithm, not only the number of idle wavelengths but also the network congestion is considered. However, the network congestion information is continuously updated by ant-based mobile agents. Thus, the setup delay depends only on the time that the needle packets get the results along these k alternate routes. It is noticed that only the number of idle wavelengths are searched by needle packets. This is also similar to the FPLC algorithm where the maximum number of idle wavelengths is considered. Moreover, it is noticed that the time complexity to select the k best routes from the P-route table is only $O(k \times P)$. This

search is conducted locally on the source node based on the goodness value of each route. For the above reasons, the setup delay of HABR is similar to FPLC with the same number of alternate routes.

For the wavelength assignment, any method in [14] can be applied. We adopt the First Fit heuristic because it is simple but still can result in a good performance. In the First Fit heuristic, all wavelengths are considered and the first available one is selected.

The pseudo-code of the main steps in the HABR algorithm can be summarized as follows:

{Ant generation}
Do
    For each node in network
        Select a random destination;
        Launch ants to this destination
        with a probability ρ;
    End for
    Increase time by a time-step for ants' generation;
Until (End of simulation)

{Ant foraging}
For each ant from source s to destination d do (in parallel)
    While current node $i <> d$
        Update pheromone table elements;
        Update P-route table;
        Push trip's state into stack;
        If (found a next hop)
          Move ant to next hop ;
        Else
          Kill ant;
        End if
    End while
End for
{Connection setup}
Select k best route from the P-route table;
If (no wavelength available)
  Consider a blocking case;
Else
  Setup lightpath on highest goodness route;

24.5 Numerical Results and Discussions

In this section, we verify the performance of our proposed algorithms based on simulation in a circuit-switched routing module for WDM network in the *ns-2* Network Simulator [15]. For comparison, we also conduct simulation for the fixed routing algorithm (SP) and the alternate routing algorithms (ASP and FPLC) that adopt two alternate routes. To simplify, we use the FPLC algorithm where not neighborhood information but global information is used (information about wavelength availability is considered whole along the path, not only 2 links as in FPLC(2)). The network topologies used in our simulation are illustrated in Figure 24.5.

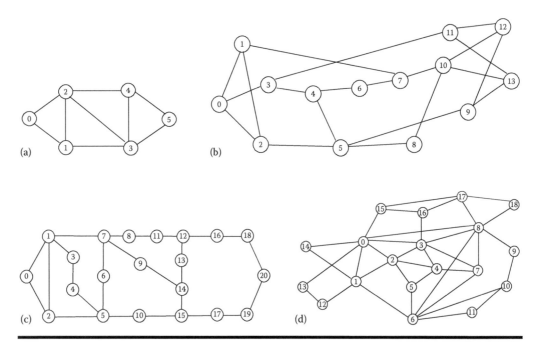

Figure 24.5 Network topologies used in simulation. (a) SimpleNet, 6 nodes, 8 links. (b) NSF, 14 nodes, 21 links. (c) ARPA-2, 21 nodes, 26 links. (d) EON, 19 nodes, 35 links.

We adopt the general traffic model widely used in performance analysis of data communication networks [29]. There are totally T arriving traffic sessions; arriving sessions are distributed randomly over the network. Each traffic session has many connection requests: For each traffic session, connection requests arrive according to a Poisson process with an arrival rate λ (call/s). Connection holding time is exponentially distributed with mean $1/\mu(s)$. The total network load is measured by $T \times \lambda/\mu$ (Erlang).

The average holding time is set as $1/\mu = 10$ s, and two values of the number of wavelengths, $W = 8$ and $W = 16$, are considered here. For each case, the number of traffic sessions, the arrival rate are selected to have a reasonable range of traffic load such that the total blocking probability is about 5%—a practical value for WDM networks. The time step for ant's generation is set as $t = 0.1$ s. To get a stable result, each experiment is run in 2000 s and it is repeated five times to get the average value of blocking probability.

As explained in the previous section, we keep the number of alternate routes $k = 2$. ASP and FPLC also use two alternate routes for two reasons. Firstly, alternate routing methods like ASP and FPLC using more than two routes do not significantly improve performance [17,20,30]. Secondly, the more alternate routes are searched by needle packets over the network, the higher the cost for the connection setup process is. Thus we should apply the same number of alternate routes in all ASP, FPLC, and HABR algorithms for a fair performance comparison.

Each experiment is conducted in an initial time without the traffic load to get an initial list of P routes. We found that ants can report a list of P shortest paths because ants only consider the path length at initial period. This confirms the results in previous work [8] that ant-based mobile agents tend to find the shortest paths if only the path length is considered in path selection process.

24.5.1 Simulation Results for ABR Algorithm

24.5.1.1 Parameters Setting and Tuning

24.5.1.1.1 Setting Parameters α, β, γ

In any ant-based algorithms, the space of possible parameter settings is huge. The most common way for parameter setting is resolved through experiments. The values used in the simulations reported here were those found to be best according to our experience. The way of setting the pheromone parameters α, β, and γ can affect the blocking performance of ABR algorithm. Here, β is the parameter to adjust the emphasize of path length on the amount of pheromone (Equation 24.6). γ is the parameter to adjust the emphasis of wavelength availability on the amount of pheromone (Equation 24.7) and α is a scalar parameter to adjust the importance of path length versus wavelength availability in the amount of increased pheromone (Equation 24.5).

The bigger value of α, the more emphasis of path length on the increased amount of pheromone δr. The smaller value of α, the more emphasis of wavelength availability on the increased amount of pheromone δr. For example, if $\alpha = 1$, the path with shorter length is considered as better solution regardless of its wavelength availability. If $\alpha = 0$, the path with higher wavelength availability is considered as better routing solution regardless of its length. Intuitively, $\alpha \approx 1$ is a better solution because a RWA solution is good with shortest available path. Moreover, for two paths with lengths nearly equal, a RWA solution may select the path with the larger number of available wavelength. In our experiments, we reported that the value $\alpha = 0.8$ or 0.6 are good scalar parameters for ABR algorithm as shown in Table 24.1. For example, with the load of 0.5 Erlang on the NSF network, we can observe the blocking probability versus scalar parameter α (Figure 24.6). It is reported that $\alpha = 0.8$ is a good parameter value.

The selection of the values β and γ depends on the setting of value δr—the amount of adjusted pheromone. In fact, δr will affect the performance and the stability of ant-based algorithm. With a high value of δr, the pheromone value of a route may change too fast thus causing an instability on blocking performance. In contrast, a small value of δr makes the pheromone value of a route change gradually, but we may need a large number of ants to update and increase the value of pheromone on a good route so that this route will become a good RWA solution. Experiments show that the value of δr should range between 0.05 and 0.2. For lower and higher value, the blocking performance may be degraded.

The values of β and γ should be selected so that δr ranges between 0.05 and 0.2. Because δl and δw are summed with scale factor α and $1 - \alpha$ (Equation 24.5), we will select β and γ so that δl and δw take the value around 0.2. According to Equation 24.6 and Equation 24.7, β should be set around 2 and γ should be set around 0.2. In our experiments, we change the values of β

Table 24.1 Parameter Values Used in Simulations

Network	(W_1, T_1)	(W_2, T_2)	α, β, γ	P_{noise}
SimpleNet	(8, 60)	(16, 120)	0.8, 1.75, 0.2	0.05
NSFNet	(8, 100)	(16, 200)	0.8, 1.75, 0.2	0.06
ARPA-2	(8, 75)	(16, 150)	0.6, 1.75, 0.2	0.06
EON	(8, 150)	(16, 300)	0.6, 1.75, 0.2	0.06

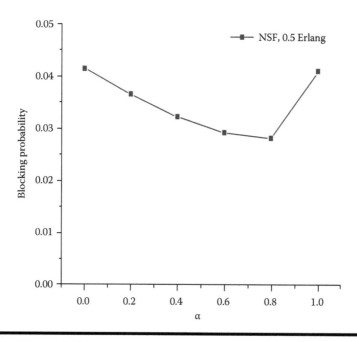

Figure 24.6   Blocking probability versus scalar parameter α.

and γ around these values and found that $\beta = 1.75$ and $\gamma = 0.2$ are "good" parameters for ABR algorithm.

To determine the good pheromone parameters β and γ, we repeat the experiments with their values in the expected range. We set $\beta \in [0.5 : 5]$ and $\gamma \in [0 : 0.5]$. Figure 24.7 shows the results in blocking probability versus different values of pair (β and γ) when the scalar parameter $\alpha = 0.8$.

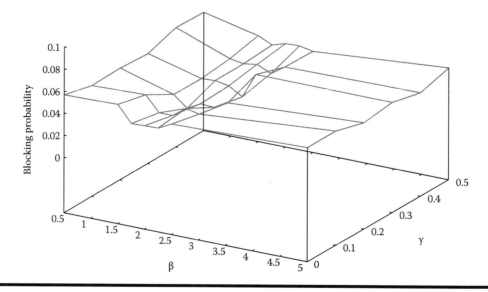

Figure 24.7   Blocking probability of ABR in NSF network with different values of pheromone parameters (normalized load $\nu = 0.5$).

The experiments are conducted on NSF network with the normalized load $v = 0.5$. On observing the results, we found that there exist "good" ranges of value β (from 1.5 to 2.5) and γ (from 0.1 to 0.3) in that the ABR can perform much better than other cases. We decide to select $\beta = 1.75$ and $\gamma = 0.2$ as the good pheromone parameters to use later in our experiments. Note that these values are not expected to be the optimal choice.

24.5.1.1.2 Tuning Number of Ants

In order to get the routing tables into a ready state, the algorithm is first allocated an initial predefined running time with a fixed number of ants and without traffic load, in which the routing tables are initialized with equal probability for each neighbor node. During this initial period, only path length information is taken into account when ants update the routing tables. At the end of the initial period, our simulation results show that the routing tables indicate the shortest path between each node pair. This result has also been demonstrated in previous work [12]. After the initial period, the traffic is then loaded into the network. We first keep $\rho = 0$, so the performance of ant-based routing is the same as the fixed routing algorithm because there is no ants available to explore the network state. We then tune the number of ants by increasing ρ to get a lower blocking probability. Figure 24.8 illustrates the relation between the blocking probability and the ants' launching probability in the SimpleNet topology.

We can observe from Figure 24.8 that the blocking probability decreased significantly when ρ was increased from 0 to 0.1. However, when we increase the value of ρ further, from 0.1 to 0.9, to have an increased numbers of ants exploring the network, we are not able to significantly reduce the blocking probability any further. In fact, too many ants in the network ($\rho \approx 1$) will cause a high control overhead. It is interesting to note in Figure 24.8 that a good performance in terms of

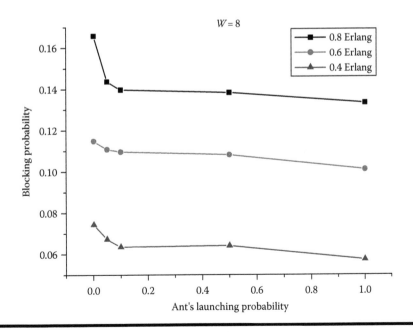

Figure 24.8 **Blocking probability versus ρ on SimpleNet.**

low blocking probability can be obtained within a large range of ρ. Thus, the ABR algorithm has good robustness to the variation of parameter ρ.

24.5.1.2 Results and Analysis

24.5.1.2.1 Blocking Probability Performance

Extensive simulation has been conducted upon the three networks shown in Figure 24.5, and the corresponding comparison results among shortest path (SP), alternate shortest path (ASP), and the ant-based routing (ABR) algorithms are summarized in Figure 24.9 through 24.11.

Figure 24.9 show that in terms of blocking probability for different networks, the new ant-based algorithm is much better than the SP routing algorithm and also outperforms the ASP routing algorithm. It is interesting to note in Fig. 24.9 that in a small network such as SimpleNet, the results of the ABR algorithm are only slightly better than the ASP algorithm. The reason is that

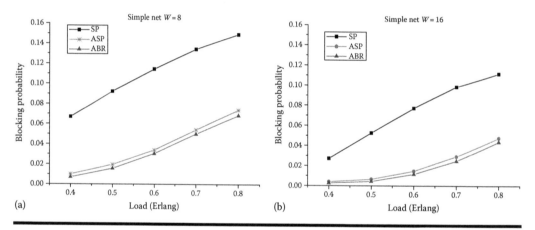

Figure 24.9 **Comparisons between ABR and other routing algorithms on SimpleNet. (a) The number of wavelengths W = 8. (b) The number of wavelengths W = 16.**

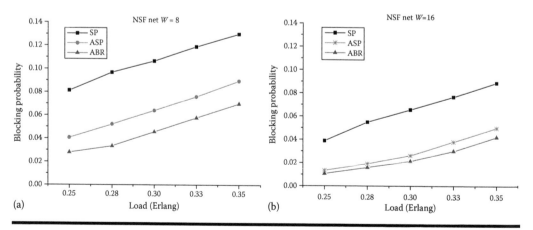

Figure 24.10 **Comparisons between ABR and other routing algorithms on NSFNet. (a) The number of wavelengths W = 8. (b) The number of wavelengths W = 16.**

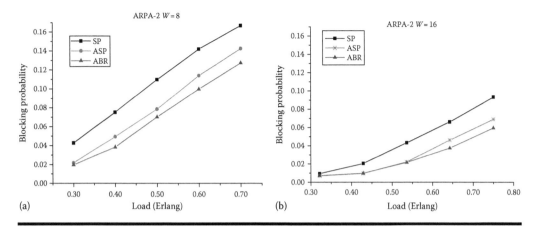

Figure 24.11 **Comparisons between ABR and other routing algorithms on ARPA-2. (a) The number of wavelengths W = 8. (b) The number of wavelengths W = 16.**

for the small network with a small average node degree of 2.7, there are only a few alternate routes between any two nodes.

The results in Figure 24.10 indicate clearly that for a larger network with a bigger average node degree (such as NSFNet backbone), the ABR algorithm significantly outperforms both the SP and ASP algorithms. This is because the NSFNet network has a bigger average node degree of 3.0, so the ABR algorithm has more chances to find a good path among more alternate routes, and thus can be more adaptive.

The comparisons in Figure 24.11 show that for a large network with a relatively small average node degree (such as the APRA-2 network, with an average node degree of 2.4), the ABR algorithm can still outperform the ASP algorithm, but not very significantly. One reason is that there are only a few alternate routes (about two) between two nodes. Another reason is that the size of the network is relatively larger, so ants must take a longer time on their trips; this also prevents the Ant-based algorithm from achieving a better performance.

In all the tested networks, the ABR algorithm slightly outperforms the ASP algorithms when the traffic load is small. This is because there is not much difference between these routing algorithms when the network is slightly loaded. When the network is highly loaded, the same result is observed because ants cannot update the routing tables quickly enough when the network states change too fast. Overall, the ABR is especially better than ASP algorithms in the reasonable region of traffic load (the region in which the network is moderately loaded).

24.5.1.2.2 Number of Required Wavelengths

To explain more about the advantages of ABR algorithm versus SP and ASP algorithms, we tune the number of wavelengths for each network to obtain a blocking probability that is less than a threshold value. Figure 24.12 shows the number of required wavelengths versus traffic load to guarantee a blocking probability that is lower than 2%. It is observed that for all of cases, ASP algorithm requires less number of wavelengths than SP to guarantee 2% blocking threshold. On SimpleNet, the ABR require the same number of wavelengths as the ASP algorithm. This is because the blocking probability of ASP and ABR is almost similar on a small network such as SimpleNet (Figure 24.9). On the NSFNET and ARPA-2 where the blocking probability of ABR is smaller

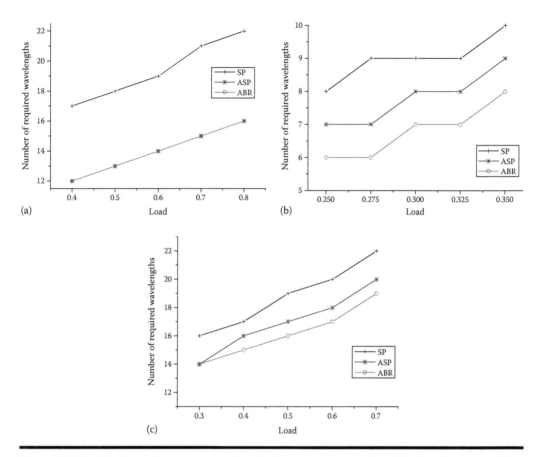

Figure 24.12 Number of required wavelengths to guarantee less than 2% blocking probability. (a) SimpleNet. (b) NSF. (c) ARPA-2.

than that of SP and ASP, the number of required wavelengths of ABR is smaller than that of ASP and it much smaller than that of SP algorithm. These results conform with the blocking probability performance as shown in Figures 24.10 and 24.11.

In summary, the above extensive simulation study based on different network topologies and various traffic loads has demonstrated that the ABR algorithm has a very good adaptability, and that it consistently achieves a better performance in terms of low blocking probability when used for dynamic routing in WDM networks.

24.5.2 Simulation Results for HABR Algorithm

24.5.2.1 Setting of the Number of Alternate Routes

To find a suitable value of P for simulation, we compare the performance of HABR with ASP by keeping ant's launching probability at $\rho = 1$. Figure 24.13 shows the network blocking probability versus the total traffic load with $P = 4, 6, 8$, and 16 on the NSF network topology. When P is small, HABR slightly outperforms ASP, the difference is more significant when $P = 6$ or 8 but there is no much improvement when $P = 16$. This can be explained intuitively that when P is small, the

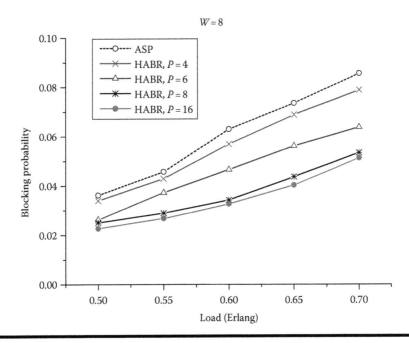

Figure 24.13 Performance of HABR on difference value of *P* when *k* = 2.

P-route table is not large enough to contain many feasible routes reported by ants, which will reduce the chance to find a good route when *k* best routes are selected. Moreover, a small value of *P* may cause routes to be replaced too frequently so a good route may be replaced by another route that could become a bad route later. When *P* increases, ants can report a larger number of feasible routes into *P*-route table, so the blocking probability will be decreased. Too large value of *P* does not much increase performance significantly, because the number of good routes between two nodes is not large in reality. In fact, a too high value of *P* will increase the complexity of HABR, so taking $P = 6$ or 8 is good enough for HABR to significantly outperform ASP in terms of blocking probability.

24.5.2.2 Tuning Number of Ants

Figure 24.14 illustrates the performance of the HABR algorithm on the NSF network topology when $P = 8$ with different values of ant's launching probability ρ. We observe that the blocking probability decrease significantly as ρ increases. When the value of ρ is big enough to have a suitable numbers of ants exploring the network, the blocking probability does not decreases significantly anymore. In fact, too many ants in the networks ($\rho \approx 1$) will cause a high control overhead. However, our simulation results (Figure 24.14) show that a good performance can be obtained within a large range value of ρ, as we can see in the figure when $\rho \in [0.5, 1]$. We will select $\rho = 0.75$ for our experiments because with this value, the blocking probability is significantly decreased in comparison with ASP algorithm.

24.5.2.3 Blocking Probability Performance

The blocking probability versus the traffic load of the routing algorithms ASP, FPLC, and HABR with different numbers of wavelengths are shown in Figures 24.15 and 24.16 for the NSF network

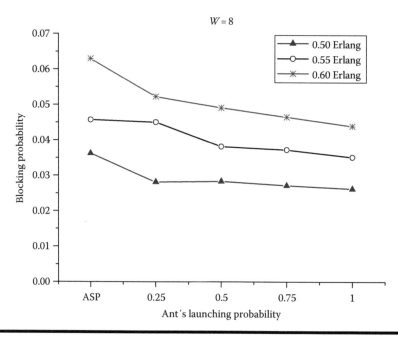

Figure 24.14   Performance gain of HABR with difference value of ant's launching probability in compare with ASP on the NSF network.

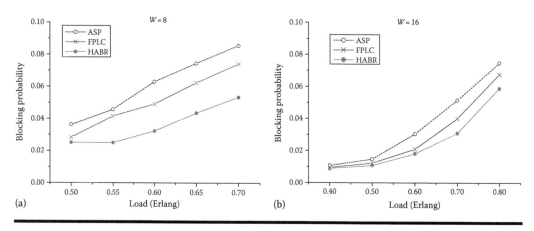

Figure 24.15   Comparisons between HABR and other algorithms on NSF network with $P = 6, k = 2, \rho = 0.75$. (a) The number of wavelengths $W = 8$. (b) The number of wavelengths $W = 16$.

and the EON network, respectively. All of the routing algorithms use $k = 2$ alternate routes. From previous experiments, we select $P = 6$ and $\rho = 0.75$ because HABR can perform a good network performance with these values. These figures show clearly that the HABR always outperform much better than both ASP and FPLC algorithms in terms of blocking probability under different traffic load conditions.

The above results can be explained as follows. The HABR algorithm can outperform the ASP algorithm because the k alternate routes used for connection setup in HABR are continually updated

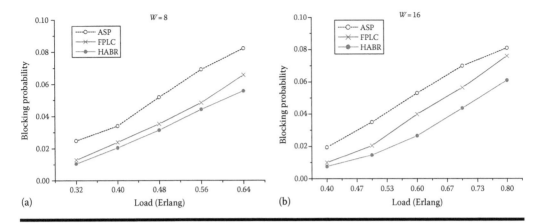

Figure 24.16 **Comparisons between HABR and other algorithms on EON network with** $P = 6, k = 2, \rho = 0.75$. **(a) The number of wavelengths W = 8. (b) The number of wavelengths W = 16.**

by ant-based mobile agents, thus these k routes are much better than the fixed routes used by ASP. Moreover, each route in HABR is assigned with a goodness value as in Equation 24.8; this goodness value allows ants to introduce into P-route table the routes with smaller path length and bigger number of idle wavelengths. As a result, HABR will select the least congested route in a more adaptive manner to the traffic variation than FPLC does. By consequence, the HABR algorithm can outperform the FPLC algorithm in terms of blocking probability.

We can also observe from Figures 24.15 and 24.16 that when the traffic load is small, the HABR algorithm just slightly outperforms the FPLC and ASP algorithms. In contrast, the difference is much more significant when the traffic load is high. This is because the ASP algorithm is a fixed routing algorithm, while the FPLC algorithm is a dynamic routing but it is not flexible enough to adapt to the traffic variation. The FPLC algorithm is dynamic in terms of using only the information of idle wavelengths on a set of precomputed routes. In contrast, the HABR algorithm is much more dynamic in terms of using information about networks congestion that is explored by mobile agent. That is why the ABR algorithm can much more outperform the ASP and FPLC algorithm under different value of traffic load, even when the traffic load is very high.

24.5.2.4 Number of Required Wavelengths

Figure 24.17a shows the number of required wavelengths versus traffic load to guarantee a blocking probability level om NSFNet. In this figure, the blocking probability is expected to be less than 2%, the number of traffic is 200 source-destination pairs. We can observe that with a same traffic load, the HABR requires a smaller number of wavelengths than that of ASP and FPLC. For example at load 0.5, the HABR requires only 8 wavelength while FPLC and ASP require 9 and 10 wavelengths, respectively. The same result can be found on the Figure 24.17b on EON network with 300 source-destination pairs. These results show that using HABR algorithm allows a network to save more resource than other alternate routing algorithms.

In summary, the HABR algorithm can always outperform significantly the ASP and the promising FPLC algorithms in terms of blocking probability while using the same alternate routes. The setup process of HABR is similar to other alternate methods as it takes only k alternate routes for connection setup. Moreover, the ant-based mobile agents always explore and update the P-route

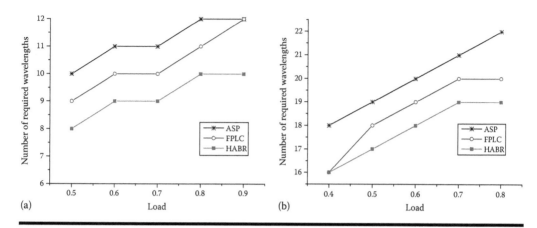

Figure 24.17   Number of required wavelengths to guarantee less than 2% blocking probability. (a) NSF. (b) EON.

tables before connection requests arrive, so the HABR algorithm does not increase the setup time in comparison with other alternate routing methods.

24.6 Complexity Analysis

24.6.1 ABR Algorithm

24.6.1.1 Computation Times for Route Setup

The setup delay includes many factors such as computation time for determining a selected route, time for searching an available wavelength, delay for sending and receiving the setup message. It is notable that the delay of setup message mainly depends on propagation delay. The time for searching an available wavelength depends on wavelength assignment method. Thus we will mainly discuss the route computation time, i.e., the time consumed by a routing algorithm to determine a route to establish the lightpath.

The time complexity of the fixed shortest path (SP) routing algorithm is $O(1)$ because the route is fixed and determined. The time complexity of the ASP algorithm is $O(k)$ where k is the number of alternate routes. For the adaptive unconstrained routing algorithm ABR [22], the route is determined by applying the shortest path finding Dijkstra's algorithm for all wavelengths, thus the complexity of AUR is $O(WN^2)$ where W is the number of wavelength and N is the number of nodes.

In the ABR algorithm, the route is determined by selecting the highest blocking probability at each node, for a route, if l is the route length and d is the average node degree, thus the complexity is $O(l(d-1)) = O(ld)$. Because the route length l is the number of node N in the worse case and the average path length h in the average case, thus the complexity of ABR will be $O(Nd)$ in the average case and be $O(hd)$ is the average case.

In a real network, d is a small value ($d \approx 3$) and the average path length is very small in compared with the number of nodes ($h << N$). We can see that $O(1) < O(k) < O(hd) << O(WN^2)$. For this reason, the time complexity of ABR algorithm is bigger than fixed routing such as SP or ASP but is much smaller than the AUR algorithm. Thus, the ant-based routing algorithm can reduce the setup delay in comparison with other algorithms that use adaptive routing method.

The flowing tables show the average computation time of a route for a connection request on three experimental networks with different values of normalized load $\nu = \lambda/\mu$ as described in the previous section. To give a reference, we also show the computation time of the AUR algorithm which applies an exhaustive search on global link state information for shortest available path. It is clear that the computation time of ABR algorithm is very small and can be comparable to the ASP algorithm (note that in the ASP algorithm, the route is precomputed and the route is almost available). For three networks as in Table 24.2a through c, the computation time of ABR is just two times that of ASP algorithm.

24.6.1.2 Size of Routing Tables

The size of routing tables is defined as the amount of memory required to store the routing information on each network node. For the SP and ASP algorithms, they need a routing table with $N - 1$ entries and $k(N - 1)$ entries, respectively, where each entry contains a route from a source node to a destination node with average path length h. Therefore, the memory required of SP and

Table 24.2 Average Computation Time (μs) of a Route for a Connection Request

	\multicolumn{6}{c	}{a. SimpleNet}
	\multicolumn{3}{c	}{$W = 8$}	\multicolumn{3}{c	}{$W = 16$}
Algorithm	$\nu = 0.4$	$\nu = 0.6$	$\nu = 0.8$	$\nu = 0.4$	$\nu = 0.6$	$\nu = 0.8$
ASP	1.44	1.46	1.47	1.64	1.66	1.86
ABR	2.04	2.18	2.26	2.06	2.27	2.35
AUR	10.38	11.63	11.32	16.90	17.00	18.42

	\multicolumn{6}{c	}{b. NSF}
	\multicolumn{3}{c	}{$W = 8$}	\multicolumn{3}{c	}{$W = 16$}
Algorithm	$\nu = 0.25$	$\nu = 0.30$	$\nu = 0.35$	$\nu = 0.25$	$\nu = 0.30$	$\nu = 0.35$
ASP	1.90	2.02	2.00	1.97	2.03	2.14
ABR	3.19	3.71	3.58	3.92	4.13	4.10
AUR	33.65	33.83	34.93	56.57	58.72	62.54

	\multicolumn{6}{c	}{c. APRA2}
	\multicolumn{3}{c	}{$W = 8$}	\multicolumn{3}{c	}{$W = 16$}
Algorithm	$\nu = 0.30$	$\nu = 0.50$	$\nu = 0.70$	$\nu = 0.30$	$\nu = 0.50$	$\nu = 0.70$
ASP	2.33	2.33	2.34	2.24	2.35	2.36
ABR	5.21	5.36	5.56	5.25	5.49	5.67
AUR	43.70	49.15	62.91	100.05	107.75	144.73

All simulations are run on a DELL PC Intel 2.66 GHz, RAM 256 MB.

ASP is $O(Nh)$ and $O(kNh)$, respectively. The AUR algorithm need the global routing information that contains the link state of every network links, i.e., the availability of any wavelength on any link must be known, thus the memory required is $O(Wl)$ or $O(WNd)$.

The ABR algorithm works in a distributed manner and does not require global routing information as AUR. All routing information are stored in the pheromone routing table. According to the pheromone routing table description, each pheromone table on each network node has $d(N-1)$ entries in average where d is the average node degree that corresponds to the number of neighbor nodes. Thus, the memory complexity of ABR algorithm is $O(Nd)$. For common networks, $O(Nd) \approx O(Nh) < O(kNh) << O(WNd)$. Therefore, it is notable that in terms of memory consumed, the complexity of ABR is as same as SP algorithm and is much more smaller than those of the ASP and AUR algorithms.

24.6.1.3 Control Overhead

Control overhead of a routing algorithm is defined as the ratio between the bandwidth occupied by the control messages and the total available network bandwidth. The control messages may be divided into two types: routing information packets and setup establishment packets. For any RWA algorithm, the amount of setup establishment packets are very small and can be negligible. The main control overhead come from routing information packets. Moreover, the fixed routing approaches such as SP and ASP do not have routing information packets because the routes are statically precomputed. Therefore, it is worthy to discuss about the control overhead of adaptive routing approaches such as ABR algorithm (that uses ant-based mobile agents to update the routing tables) and AUR algorithm (that uses a link-state updating protocol to distribute the routing information).

Assume that a network uses the link state advertisement (LSA) flooding for updating routing information the AUR-E routing algorithm, as in OSPF specification [31]. The LSA flooding is summarized as follows. Each node periodically (in a time interval t) distributes a LSA packet containing the state information of every neighbor links to every node in the network. In case of optimization flooding where the number of flooding packets is optimized, the LSA packets may followed a flooding tree as in Figure 24.18. In this case, the number of exchanged messages for

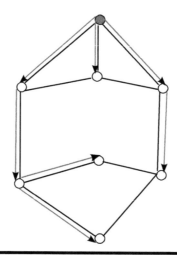

Figure 24.18   LSA flooding tree in a network: The number of exchanged LSA packets is reduced.

advertising a LSA packet is about N, as same as the number of nodes. As a result, the total number of LSA-exchanged packets for N nodes in a time unit is N^2/t. If the average size of a LSA packet is $size_{LSA}$, so the amount of control overhead can be roughly estimated as $N^2 \times size_{LSA}/t$.

In ant-based routing algorithm ABR, each node also periodically send an ant to a random destination. Assume that an ant will follow a route with the average length $\bar{h}$, the number of ants in the network will be $\rho N\bar{h}/t$ where ρ is the ant launching probability. If the average size of an ant is $size_{ANT}$, so the amount of control overhead can be roughly estimated as $\rho N\bar{h} \times size_{ANT}/t$.

In practice, it is very difficult to evaluate the parameters in the above overhead formulas. Instead, those values are estimated through experimental works. Recent extensive simulation results in [7] has shown that for the adaptive routing using periodical LSA flooding, the value t must be very small in order to keep a low overall blocking probability. In these works, t is set to 0.1 or 1 s. These works even show that with $t = 1$ s, the blocking probability can be only improved for the case of low traffic load. This results shows a disadvantage of adaptive routing algorithm using global routing information: The AUR is good only if the update interval t must be very small, which can cause a high control overhead in the networks.

The average length of route that ant moves in each networks is obtained from simulation (Table 24.3). It is noticed that the average length is very small in compared with the number of node in the network $\bar{h} << N$. This is because an ant does not move randomly but follow the route indicated by pheromone tables. This allows to reduce the control overhead in comparison with mobile agents with random walk.

One way to evaluate the control overhead of ABR algorithm is to compare it with the well-known LSA flooding of OSPF routing method [31]. Let θ be the ratio of the overhead of ABR in comparison with LSA-flooding method (see Figure 24.19). We have:

$$\theta = \rho\bar{h}/N \times \frac{t_{LSA}}{t_{ANT}} \times \frac{size_{ANT}}{size_{LSA}} \tag{24.9}$$

To evaluate the size of ant, let us consider a possible way to implement an ant-based mobile agent. The data that an ant carries have the hop count, a stack for route trace, and a wavelength available mask, as in Figure 24.20.

In this case, the size of data carried by an ant will be $1 + n + 4 = \bar{h} + 5$ B. Assume that each ant is encapsulated in an IP packet with 20 B header, the average size of an ant will be $\bar{h} + 25$ B. Note that the size of LSA packet is $64 + 8N_i$ where N_i is the number of neighbors of node i [25]. On average, $size_{LSA} = 64 + 8d$ where d is the average node degree.

Table 24.3   Ratio of overhead of ABR versus AUR (LSA-Flooding Method)

	SimpleNet	NSFNet	ARPA-2
$\bar{h}$	1.79	2.61	3.99
d	2.67	3.00	2.48
$size_{ANT}$	26.79	27.61	28.99
$size_{LSA}$	78.32	84.88	95.92
θ	0.76	0.45	0.36

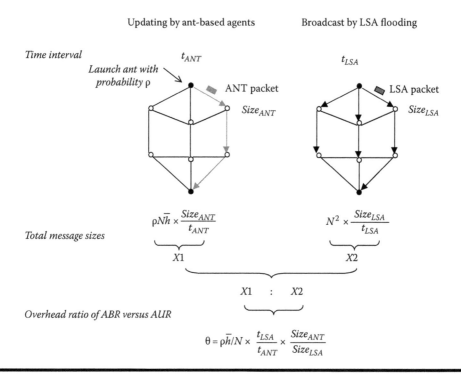

Figure 24.19 Routing overhead for updating routing information and the overhead ratio of ant-based ABR algorithm versus AUR algorithm that uses LSA flooding method.

```
struct hdr_ant {
    byte n;        // Hop count
    byte *data;    // Stack for route trace
    long mask;     // 64 bits (4 bytes) wavelength mask
    ....
}
```

Figure 24.20 A possible implementation of ant header data structure.

For ABR algorithm, the good parameter setting is: $\rho = 0.75$ and $t_{ANT} = 1$ ms. For adaptive routing algorithm using LSA flooding method, recent researches [7] show that t_{LSA} should be 10 ms to avoid the effect of inaccurate routing information. In this case we have:

$$\theta = 0.75 \times 10/1 \times \bar{h}/N \times \frac{size_{ANT}}{size_{LSA}} = 7.5 \frac{\bar{h}}{N} \times \frac{size_{ANT}}{size_{LSA}} \tag{24.10}$$

The overload ratio θ for each network in our experiment is shown in the following table. It is observed that the overload of ABR is about one-half of that of AUR-E algorithm which use the LSA-flooding method. Because LSA-flooding method is a high overhead updating mechanism, the results in the table shows that ant-based algorithm still has a high overhead. This is a drawback of all ant-based algorithms. Note that the overhead of AntNet algorithm is about 15 times bigger than that of OSPF routing in a packet switching network.

24.6.1.4 Summary

We summarize some complexity and overhead parameters estimation in Table 24.4.

24.6.2 HABR Algorithm

24.6.2.1 Computation Time

The computation time of a route for a connection request is very small because HABR is based on alternate routing methods. The time complexity to select the k best routes from the P-route table is $O(kP)$. This is a local search on the source node based on the goodness value of each route. For the above reasons, the setup delay of HABR is similar to FPLC with the same number of alternate routes. It is notable from the previous section that the time complexity of ASP is $O(k)$. The FPLC has a little more complexity because it searches for idle wavelengths for the least congested path among k routes. To find the number of idle wavelengths, the complexity is $O(W)$ thus the total complexity will be $O(kW)$. In practice, the value of parameters k, P, W is small so the previous mentioned algorithms can be considered to have low complexity (Table 24.4 and 24.5).

The average computation time of a route for a connection request with different values of normalized load $v = \lambda/\mu$ are shown in the Table 24.6. We can see that both FPLC and HABR have

Table 24.4   Complexity and Overhead Estimation of RWA Algorithms

Overhead	Algorithms			
	SP	ASP	ABR	AUR
Comp. time	$O(1)$	$O(k)$	$O(hd)$	$O(WN^2)$
Size of tables	$O(Nh)$	$O(kNh)$	$O(Nd)$	$O(WNd)$
Messages	—	—	$\rho N\bar{h} \times size_{ANT}/t$	$N^2 \times size_{LSA}/t$

Table 24.5   Notation

N	Number of nodes.
d	Average node degree (number of links $= N \times d/2$).
W	Number of wavelengths.
h	Average route length in number of hops.
k	Number of alternate routes in FAR.
$\bar{h}$	Average length of route that an ant moves on.
ρ	Ant launching probability.
t	Time interval to send a LSA or ant packet.
$size_{ANT}$	Average size of an ant-based mobile agent.
$size_{LSA}$	Average size of a link-state packet.

Table 24.6   Average Computation Time of a Route (μs)

a. NSF						
	W = 8			W = 16		
Algorithm	$\gamma = 0.5$	$\gamma = 0.6$	$\gamma = 0.7$	$\gamma = 0.5$	$\gamma = 0.6$	$\gamma = 0.7$
ASP	2.09	2.12	2.20	2.27	2.33	2.44
FPLC	2.38	2.40	2.42	3.04	3.10	3.16
HABR	2.42	2.54	2.46	2.45	2.53	2.55

b. EON						
	W = 8			W = 16		
Algorithm	$\gamma = 0.4$	$\gamma = 0.6$	$\gamma = 0.8$	$\gamma = 0.4$	$\gamma = 0.6$	$\gamma = 0.8$
ASP	2.17	2.25	2.36	2.57	2.63	2.76
FPLC	2.42	2.58	2.56	3.36	3.18	3.21
HABR	2.95	3.03	3.12	3.23	3.20	3.47

All simulations are run on a DELL PC Intel 2.66 GHz, RAM 256 MB.

almost the same computation time as ASP algorithm. This means the HABR can take advantage from alternate routing methods to keep a very slow setup time while reducing significantly the blocking probability performance.

24.6.2.2 Routing Tables' Size and Overhead

In HABR algorithm, the routing overhead is mainly caused by the amount of ants exploring the network state. The amount of exchanged ant-based agents is similar to the basic ABR algorithm because we use the same pheromone table and pheromone parameters, that means the routing overhead of HABR is $\rho N \bar{h} \times size_{ANT}/t$. As explained before, this is about one-half of the routing overhead of the OSPF routing method.

The size of routing tables is the sum of both type of tables: pheromone table and k-routes table. As explained before, the size of pheromone table ($O(Nd)$) is much more smaller than that of k-routes table ($O(kNh)$). Therefore, the P-routes tables take the main part in the total size of routing tables, that means the size of routing tables of HABR is about P/k times larger than that of other alternate routing methods. In our experiment, a good blocking probability performance is obtained when $P = 8$, which means HABR need four times of memory consumed in compared with ASP or FPLC with $k = 2$ alternate routes.

It is noticed that we can apply more advanced techniques of alternate routing methods to improve furthermore the performance of the HABR algorithm, such as routing using neighborhood information [20], or another techniques to evaluate the goodness of a route to introduce the load balancing as in [32]. However, these problems are out of scope of this chapter. In this work, we use the simple function to evaluate the goodness of a route as in Equation 24.8 in order to emphasize the advantages when using ant-based approach in combination with alternate routing method to improve the network performance. The problem of how to find the ant's launching probability and the number of routes P to reduce the control overhead still remain open for further research.

24.7 Concluding Remarks

24.7.1 Conclusion

In this chapter, we proposed an adaptive ant-based routing (ABR) algorithm and a hybrid alternate ant-based routing (HABR) algorithm for the dynamic Routing and Wavelength Assignment problem in WDM networks with wavelength continuity constraint. In the ABR algorithm, the ant-based agents continuously update the routing tables according to the network state. To enhance the performance of our new algorithm, we proposed an efficient scheme for scoring a path based on both the number of available wavelengths and the path length, and also a scheme for path searching based on the second-highest value from the routing table entries. These new schemes can help an ant-based RWA algorithm overcome the stagnation problem, and thus effectively improve its performance in terms of blocking probability. In the HABR algorithm, we proposed to use a novel twin routing table structure that consists of a *P*-route table for routing and a pheromone table for mobile agents' foraging. The HABR can take advantages of both ant-based and alternate routing methods while avoiding their shortcomings. A significant advantage of our new algorithms is that the path can be determined immediately based on the routing tables, significantly reducing setup delay. Moreover, the new algorithms are flexible in the sense that to achieve a good performance, the number of ants can be adjusted by their launching probability. Extensive simulation results indicate that the ABR consistently outperforms the promising fixed-alternate routing algorithm, while the HABR algorithm can always outperform the promising alternate shortest-path routing algorithm and the fixed-path least congestion routing algorithm in terms of blocking probability while guaranteeing a small setup time in comparison with other adaptive routing methods.

24.7.2 Future Works

In this work, we have shown that ant-based mobile agents, with the behavior of natural ant colonies, can be an efficient method in building fully distributed algorithms for solving difficult problems such as dynamic routing and wavelength assignment. Most of our work in the thesis are based on the simulation methods. The behavior of our algorithms are verified by extensive experiments. In the future, we are investigate an analytical model for investigating the behaviors of this approach. Although there have been some analytical models for mobile agents network routing such as in [33], these models are so simple in the sense that the agents are assumed to not be ant-based agents with complex cooperating tasks. A method for setting up these parameters such as the number of agents or the value of pheromone parameters will also be investigated to further improve the performance and stability of these algorithms in case of optical WDM networks. Another work is the evaluation of this ant-based approach in various network situations and its applicability to real optical fiber networks.

References

1. B. Mukherjee, *IEEE J. Select. Areas Commun.* **18**, 1810 (2000).
2. R. G. Michael and D. S. Johnson, *Computers and Intractability: A Guide to the Theory of NP Completeness* (W. H. Freeman and Company, San Francisco, CA, 1979).
3. J. Zhou and X. Yuan, in *Proceedings of the International Conference on Parallel Processing Workshops, ICPPW'02* (IEEE Computer Society, Los Alamitos, CA, Vancouver British Columbia, Canada, 2002), pp. 207–213.

4. X. Masip-Bruin, S. Sanchez-Lopez, J. Sole-Pareta, J. Domingo-Pascual, and D. Colle, in *Proceedings of the 2003 IEEE Global Telecommunications Conference, GLOBECOM'03*, San Francisco, CA (New York, 2003), vol. 5, pp. 2575–2579.

5. X. Masip-Bruin, R. Munoz, S. Sanchez-Lopez, J. Sole-Pareta, J. Domingo-Pascual, and G. Junyent, in *Proceedings of the Optical Network Design and Modeling, ONDM'03*, Budapest, Hungary, 2003, pp. 333–349.

6. R. A. Guerin and A. Orda, *IEEE/ACM Trans.* Netw. **7**, 350 (1999).

7. S. Shen, G. Xiao, and T. H. Cheng, in *Proceedings of the IEEE International Conference on Communications, ICC'05*, Seoul, Korea, 2005, vol. 3, pp. 1782–1786.

8. M. Dorigo and V. Maniezzo, *IEEE Trans. Syst. Man Cybernetics, Part B* **26**, 29 (1996).

9. E. Bonabeau, M. Dorigo, and G. Theraulaz, *From Natural to Artificial Swarm Intelligence* (Oxford University Press, New York, 1999).

10. G. Di Caro and M. Dorigo, in *Proceedings of the Thirty-First Annual Hawaii International Conference on System Sciences, HICSS'98*, Mauna Lani, HI (IEEE Computer Society, Washington, DC, 1998), p. 74.

11. M. Di Caro and M. Dorigo, J. Artif. Intell. Res. **9**, 317–365 (1998).

12. R. Schoonderwoerd, O. Holland, and J. Bruten, in *Proceedings of the First International Conference on Autonomous Agents*, Marina Del Rey, CA (ACM Press, New York, 1997), pp. 209–216.

13. D. Bertsekas and R. Gallager, *Data Networks* (Prentice-Hall, Inc., Upper Saddle River, NJ, 1987).

14. H. Zang, J. P. Jue, and B. Mukherjee, *Optical Networks Magazine* **1**, 47 (2000).

15. The VINT project, The network simulator, NS-2 (2003), accessed on July 2012, http://www.isi.edu/nsnam/ns/index.html

16. B. Wen, N. M. Bhide, R. K. Shenai, and K. M. Sivalingam, Special Issue on Simulation, CAD and Measurement of Optical Networks *SPIE Optical Networks Magazine*, 16–26 (2001).

17. S. Ramamurthy and B. Mukherjee, *IEEE/ACM Trans. Net.* **10**, 351 (2002).

18. K. Chan and T. P. Yum, in *Proceedings of the EEE INFOCOM'94*, Toronto, ontario, Canada, 1994, vol. 2, pp. 962–969.

19. A. Birman, *IEEE J. Select. Areas Commun.* **14**, 852 (1996).

20. L. Li and A. Somani, *IEEE/ACM Trans. Net.* **7**, 779 (1999).

21. H. Zang, J. Jue, L. Sahasrabuddhe, R. Ramamurthy, and B. Mukherjee, *IEEE Commun. Mag.* **39**, 100 (2001).

22. H. Mokhtar and M. Azizoglu, *IEEE/ACM Trans. Net.* **6**, 197 (1998).

23. R. Ramaswami and A. Segall, in *Proceedings of IEEE INFOCOM'96*, San Francisco, CA, 1996, pp. 138–147.

24. M. Dorigo, E. Bonabeau, and G. Theraulaz, *Fut. Gener. Comput. Syst.* **16**, 851 (2000).

25. M. Dorigo and T. Stutzle, *Ant Colony Optimization* (Bradford Book, Cambridge, MA, 2004).

26. K. Sim and W. Sun, *IEEE Trans. Syst. Man Cybernetics, Part A* **33**, 560 (2003).

27. E. Bonabeau, F. Henaux, S. Gurin, D. Snyers, P. Kuntz, and G. Thraulaz, in *Proceedings of the Second International Workshop on Agents in Telecommunications Applications, IATA'98* (Springer Verlag, Paris, France, 1998), vol. 1437 of *Lectures Notes in AI*, pp. 60–71.

28. V. Le, X. Jiang, S. Ngo, and S. Horiguchi, Special Issue on Software Agent and its Applications *IEICE Transactions on Information System*, **E88-D**, 2067 (2005).

29. A. Girard, *Routing and Dimensioning in Circuit-Switched Networks* (Addison-Wesley, Reading, MA, 1990).

30. E. Karasan and E. Ayanoglu, *IEEE/ACM Trans. Net.* **6**, 186 (1998).

31. J. Moy, Internet request for comments, RFC 1247 (1991).

32. R. M. Garlick and R. S. Barr, in *Proceedings of the Third International Workshop on Ant Algorithms, ANTS'02* (Springer, Brussels, Belgium, 2002), vol. 2463 of *Lecture Notes in Computer Science*, p. 250.

33. J. Sum, H. Shen, C. S. Leung, and G. Young, *IEEE Trans. Parallel Distrib. Syst.* **14**, 193 (2003).

Chapter 25

A Modular Software Architecture for Adaptable Mobile Ad Hoc Routing Strategies

Oliver Battenfeld, Patrick Reinhardt,
Matthew Smith, and Bernd Freisleben

Contents

25.1 Introduction

Mobile ad hoc networks (MANETs) are spontaneously formed networks of mobile devices utilizing wireless communication technology to interconnect with each other or to other networks. An essential property of such networks is their ability to self-organize in terms of both handling the inference of available communication paths from broadcast or unicast messages and relaying data destined for remote nodes. In other words, a MANET does not rely on any pre-existing infrastructure such as access points or relays, but rather requires the participating nodes to act as routers for other devices in their range, forwarding data as needed. However, finding optimal paths in a distributed fashion is complicated by several factors [15]:

- Most scenarios allow the participants to move around freely, which may result in significant topology changes even in networks with a relatively low "nomadicity" due to different effective transmission ranges. The latter are the result of heterogenous hardware characteristics such as transmission power levels and antennae features or environmental conditions such as obstructions or interferences.
- Wireless links can be subject to significant bandwidth constraints, not only in comparison to their wired counterparts but also in comparison to their nominal (i.e., ideal) data rates. The available bandwidth varies over time, making the redirection of traffic flows based on load ratios very difficult.
- Mobile devices also have exhaustible energy supplies which potentially gives rise to conflicts between the nodes' original task, such as the collection of sensor data and its services to the network, both of which require energy and thus some kind of governor to allocate it.

There are interdependencies between these factors. Increasing the transmission range of a device, for example, has an adverse effect on its energy supply and possibly limits the overall network bandwidth since it now interferes with the medium access of nodes previously not in range [16,36]. It is therefore impossible to develop a universally optimal routing algorithm for modern portable devices such as PDAs, smartphones, notebooks, which are not dedicated to a specific task in the same way less complex embedded devices are. Context information, e.g., location, current speed or heading, network traffic, running applications or user profile settings, should be taken into account when making routing decisions. The mobility patterns of people exploring a museum using wireless PDAs differ greatly from an ad hoc network of cars traveling on an empty freeway, which again differs from cars stuck in a traffic jam. Furthermore, there is currently no support for ad hoc networking in widely used operating systems such as the Windows family or Unix derivatives like Linux or the BSD family.

To address these issues, we present a modular software architecture for mobile ad hoc networks in this chapter which allows dynamic routing algorithm switches, adaptive parameter changes at runtime while at the same time providing a simple and concise API for developers. The compiled modules can be run without modifications on both our scalable MANET emulation testbed MarNET [13] and on real Linux 2.6 based computer systems, including portable devices such as PDAs and notebooks. Moreover, it is possible to design and integrate routing algorithms and application software using the so-called cross-layer approach [34].

This chapter is organized as follows: Section 25.2 presents the design and implementation of the proposed modular routing framework. Section 25.3 deals with our scalable distributed emulation environment for the examination and verification of MANET application software. In Section 25.4, related work is discussed. Section 25.5 concludes this chapter and outlines areas for future research.

25.2 A Novel Modular Routing Framework

In this section, we introduce a modular routing framework which offers hot-swappable routing modules, cross-layer programming facilities, and an API to ad hoc routing algorithm developers, application developers, and system administrators. Section 25.2.1 introduces the basic architecture of the framework, and Section 25.2.2 outlines implementation details of the routing framework and the routing modules.

25.2.1 Routing Architecture

The development of routing algorithms can roughly be divided into three stages: Once a hypothesis or theoretical concept has been established, a researcher would normally try to verify his/her assumptions with a simulation software, optimizing the developed approach step by step. Finally, the algorithm is ported to the target platform and must be reexamined under real conditions. Both simulator and real system have their own API, both of which must be carefully studied. What at first seemed to be an implementation detail, may turn out to be a time and effort consuming obstacle: Having to deal with the realities and intricacies of systems and network programming requires massive work on details such as synchronization issues, binary formats of network packets, interoperability issues with other hardware platforms, thread scheduling or performance optimizations. Bridging the gap between specification and implementation is a tedious task: Since, e.g., Linux does not (explicitly) support ad hoc networks, the kernel cannot maintain custom route attributes such as expiration times, service indicators, or network load information. It is also incapable of enqueuing outgoing packets while a route search is in progress—an indispensable feature needed for some reactive ad hoc routing algorithms.

We propose an alternative approach to extend the kernel or other parts of the operating system, a task which requires expertise in an area entirely different from the development of routing algorithms: The software framework `marnetd` (MarNET Daemon; the name stems from its roots as part of the MANET emulation testbed MarNET, see Section 25.3) currently runs on all Linux-based systems with a version 2.6 series kernel.

Figure 25.1 presents an overview and outlines its integration into the operating system. It acts as a mediator between operating system and algorithm researchers and developers who create or use modules based on DSDV [32], AODV [33], TERA [23] or any other, possibly their own, algorithm, application developers of MANET-enabled software who make use of `marnetd` as a provider of routing services and network state information (through cross-layer facilities) and the average user who simply wishes to set up a working mobile ad hoc network.

`marnetd` provides a single concise API to implement routing algorithms as so-called modules which can be "hot-plugged" (i.e., switched) at runtime. In the field of ubiquitous and embedded computing, for example, the environment in which an application operates can be subject to frequent change, which can adversely affect the routing module and vice versa. For instance, while

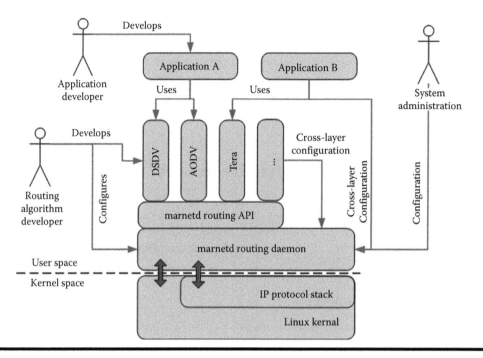

Figure 25.1   The MarNET daemon.

addressing energy preservation issues, some algorithms reduce the number of network transmissions to a minimum, thereby becoming unsuitable in a scenario of rapidly moving nodes. marnetd allows the adaptation to changing environments through its modular design: Applications, users or even another marnetd module may change routing algorithms on demand by hot-swapping new routing modules into the system at runtime.

Once running, the marnetd daemon provides a local pipe interface, which appears as a file in the host system's file system: The local terminal interface (LTI) is designed to be an extensible command processor. Using a defined protocol, it is possible to execute registered "commands" and access their return values. marnetd comes with a number of built-in commands to load or unload modules or to allow runtime read and write access to any of the daemon's state variables. Furthermore, programmers can have their modules register additional commands thus allowing the exposure of module parameters and cross-layer information exchange.

Figure 25.2 illustrates, from the user's perspective, how LTI provides external access to the configuration of both the daemon and any loaded modules. LTI provides applications with an access point to the routing system enabling them to switch between routing strategies or exchange data to facilitate cross-layer information propagation. An interactive menu-driven software (User GUI) is provided to allow human configuration of the routing service as well as an application-driven approach.

25.2.2 Marnetd API

Figure 25.3 shows a screenshot of the client software implementation. If marnetd authorizes the connection (LTI access can be limited to certain users or user groups which can be configured in the daemon's configuration file), the client requests information about loaded modules. It then

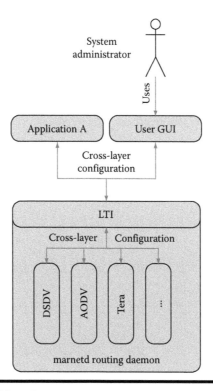

Figure 25.2   Local terminal interface (LTI).

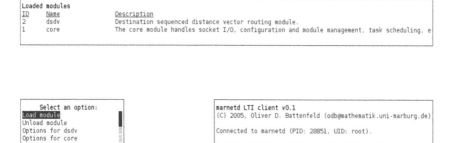

Figure 25.3   A console LTI client.

```
#include <marnetd_api.h>
[...]

char *module_name = "Skeleton module";
const char *module_description = "Skeleton module for illustration purposes";
const char *module_configfile = NULL;
const int module_version = 1;
const int api_version = 1;
const int module_type = MARNETD_MODULE_ALGORITHM;

const marnetd_option_t module_options[] = {
    MARNETD_OPTION_END()
};

int marnetd_module_init(void) {
  /* Initialization procedure: Automatically
     invoked when this module is loaded. */
  read_configuration();
  register_listening_port();
  [...]
}

int marnetd_module_fini(void) {
  /* Cleanup procedure: Automatically
     invoked on module unload requests. */
  unregister_tasks();
  free_memory();
  [...]
}
```

Figure 25.4 A skeleton marnetd module.

presents the user with a menu which can be used to load or eject a currently loaded module or access the options/parameters provided by either `marnetd` itself or by a loaded module. If the routing module developer configures the module to allow runtime modifications, the software can be used to alter the values during runtime.

To illustrate the functionality of `marnetd`, we now present an overview of its API. Modules are best created with the C programming language. The C code fragment in Figure 25.4 represents the most basic `marnetd` module.

It shows an empty module skeleton containing the required symbols which are checked during the module loading procedure, namely, an array of options exposed through LTI, module type, and versioning information and the initialization functions `marnetd_module_init` and `marnetd_module_fini` which must be implemented; they are automatically executed by `marnetd`.

With the "Makefiles" provided by the source code distribution archive, it is possible to have GNU *make* generate the binary module automatically. The use of the C compiler included with the GNU compiler collection (GCC) is recommended due to its capability to execute the linker transparently with options to create position-independent code, which is required to build the modules from the object files.

25.2.2.1 Network I/O Management

On POSIX systems, communication with other processes (mostly) happens through so-called sockets. The actual transmission protocol and medium (if any) is handled by the operating system, thus the initialization and maintenance can be cumbersome. While other routing frameworks provide a set of "building blocks" to construct the final algorithm, we have refrained from using this approach since it takes away the possibility to define the wire format of the network packets. This is an important field of optimization, a good example being algorithm-specific flooding techniques such as OLSR's MPR (RFC 3626) flooding. In cases where compliance with existing standards such as IETF RFCs is required, it is even necessary for compatibility and interoperability. Thus, the routing framework must offer a flexible and extensible mechanism for plugging in these different approaches into the same system.

To enable this, modules request a listening handler on a network port of a specific network interface. The interface's properties such as address configuration, maximum transmission unit size, hardware type, wireless support, can be queried by the `marnetd` API. Incoming data is then processed either through a traditional approach using a `Unix select()` wrapper or by a dynamic number of pooled handler threads. Figure 25.5 illustrates this behavior: Available network interfaces capable of IP transport can be used as devices for incoming or outgoing network traffic. Upon arrival of incoming data, `marnetd` determines the appropriate actions and delivers the data to a handler function. Figure 25.6 shows a stub of such an ingress handler (i.e., a certain function type as outlined in the API), its actual functionality is implemented by the module programmer who does not have to care about handling packet reception or UDP header data. The procedure simply handles a buffer which already contains the routing protocol data. Figure 25.7 shows how such an ingress handler procedure can be registered with `marnetd`'s multiplexer.

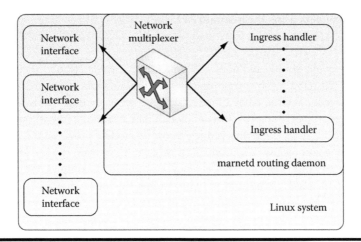

Figure 25.5 Network I/O multiplexer

```
void ingress_handler(size_t rcv_size, void *rcvbuf, size_t rcvbuf_len, struct ↩
    sockaddr *frombuf) {
    /* rcvbuf is our network message readily prepared to be processed */
    if (((struct protocol_hdr*) rcvbuf)->msg_type == NEIGHBOR_ANNOUNCE) {
        handle_new_neighbor();
    }
    [...]
}
```

Figure 25.6 A basic ingress handler.

```
char *ifname;
int socket_id;
int port = 520;

/*Retrieve interface name defined in configuration text file*/
ifname = marnetd_get_optval(this_module, "interface");
socket_id = marnetd_register_sock_dgram(ifname, port, BUF_SIZ, ingress_handler, \
    0);
```

Figure 25.7   Ingress handler registration.

```
marnetd_addr_t dest_addr;
marnetd_if_t if;
int port;

/* Construct a network message */
struct protocol_msg msg;
struct timeval now;
gettimeofday(&now, NULL);
msg.type = NEIGHBOR_ANNOUNCE;
memcpy(msg.sendtime, now, sizeof(struct timeval);

/* Add more protocol data and send it */
[...]
if (0 > marnetd_send_dgram(&msg, sizeof(msg), dest_addr, if, port)) {
  // An error has occured
}
```

Figure 25.8   Sending data using the marnetd API.

Sending data is equally simple and does not require any consideration of socket programming details. Figure 25.8 illustrates this by means of a hypothetical network message defined in a routing protocol. The message is constructed by filling it with the necessary data, the buffer is then passed to the `marnetd_send_dgram` API function which only needs to know where to send the data to. System programming aspects are thus hidden from the module programmer.

With an increasing complexity of a protocol of message exchanges, the implementation of the "state machine" that processes the data flow becomes more and more complex. One approach would have been to provide an ever-growing array of basic routines to support the development effort, such as automatic neighbor detection, message flooding approaches, etc. The downside of this approach is not only an unstable API due to the constant changes to the daemon's core functionality, it also means the loss of control of the wire format of the network messages. A key element of `marnetd`, however, is its versatility with respect to the deployment environment. In order to create RFC-compliant implementations of AODV or OLSR (e.g., to study the interoperability with other devices or implementations), it is necessary to follow the specification of the message format as outlined in the respective documents.

Recent developments in the open-source community have made it possible to use a standardized approach previously reserved for commercial environments: ASN.1, an International Telecommunications Union (ITU) standard, defines a portable specification language which can be used as an input to a compiler system [3]. Its output can be integrated immediately into the module code.

Figure 25.9 shows a simple ASN.1 module definition which only contains a single message type. It describes a data structure containing three integer values. Using the ASN.1 language, it

```
TestProtocol DEFINITIONS ::=
BEGIN

Beacon ::= SEQUENCE {
   messagetype      INTEGER,   -- Message type
   sourceip         INTEGER,   -- Source IP address
   sequencenumber   INTEGER    -- Sequence number
}
END
```

Figure 25.9   A simple ASN.1 definition.

```
RequestResponse ::=
   CHOICE { request CHOICE {
               request1 GetRoutetable,
               request2 GetPowerstatusInfo,
               request3 GetPositionInfo,
               request4 NeighbourlinkDown
               ... },
         response CHOICE {
               response1 SimpleACK,
               response2 SimpleNACK,
               response3 Routetable,
               response4 PositionInfo,
               ... }}
```

Figure 25.10   ASN.1 protocol definition.

is possible to define a protocol by defining the possible network messages in the above manner and then specifying possible message exchange scenarios as a graph or tree of request-reply packets (s. Figure 25.10).

The ASN.1 compiler then generates platform-independent source code which greatly simplifies protocol tests with heterogenous hardware. It contains decoder (deserializer) and encoder (serializer) functions which operate on a buffer. These functions can be integrated with the module's ingress handlers, as shown in Figure 25.6, or packet sending functions, as shown in Figure 25.8. The decoding and encoding functions automatically handle endianess and type size issues.

Using this mechanism, `marnetd` leaves the wire format fully in the hands of the routing module developer while still providing a useful framework on which to base new routing modules.

25.2.2.2 Packet Forwarding and Source Routing

Although routing and packet forwarding are intrinsically tied to each other, it is necessary to differentiate between the two as a programmer. Routing is a process which determines the communication paths to be used when communicating with remote systems. Packet forwarding, on the other hand, acts on routing information which is generated dynamically in MANETs using a routing system service. For example, the Linux kernel maintains a kernel space routing table which is queried by the packet forwarding routines. Separating routing and forwarding allows for high-performance kernel space forwarding routines based on simple table lookups. A disadvantage of this construct is, as far as MANETs are concerned, the relatively fixed nature of the table entries, since there is no way to add additional attributes or to have the kernel delete routes after these have not been used for some amount of time, apart, of course, from changing the kernel code.

`marnetd` still lets the kernel take care of packet forwarding while at the same time allowing any kind of custom route attribute to be stored with route table entries. This feature also works with any standard Linux 2.6 kernel provided by current distributions. Adding a route is fairly simple, as Figure 25.11 illustrates:

`marnetd` stores custom attributes, expiration times, etc., in an internal data structure called RIB (Routing Information Base). The kernel table is synchronized transparently with the RIB: `marnetd` automatically determines whether a new route must be reported to the kernel which then does the actual forwarding. The module programmer simply uses the functions provided by the daemon API, as illustrated in Figure 25.11, which outlines the situation when the routing algorithm has determined that a new node was discovered. The type `marnetd_route_t` represents a route to a given destination, a variable of this type is filled with the information about the new node and then passed to the API's `marnetd_add_route` function.

The listing in Figure 25.12 again uses the `marnetd_route_t` type, only this time it is used as a value-result parameter supplied to a lookup function which is perhaps the most often used function in an algorithm implementation: `marnetd_get_route` will place the properties of a route to the given destination in the supplied memory area. Custom route attributes can be added using the `marnetd_add_route_info` function, see Figure 25.13. In this example, a timer is added as an attribute indicating the time the route was added to the RIB.

`marnetd` maintains at most one custom route attribute per route per module. `marnetd_add_route_info`, however, accepts any (pointer to a) data type; custom attributes

```
marnetd_route_t route;
marnetd_addr_t dest, nexthop;
int distance;
[...]

/* A new node was discovered: */
route.kroute.dest = dest; // 10.1.0.42
route.kroute.gw = nexthop; // 10.0.0.5
route.kroute.i_metric = distance;
[...]
marnetd_add_route(&route);
```

Figure 25.11 Adding routes using the marnetd API.

```
marnetd_route_t route;
marnetd_addr_t dest;

dest = /* some destination we are
          interested in, e.g. 10.1.0.42 */
[...]

if (marnetd_get_route(&dest, &route)) {
    /* route is present in our database,
       its properties have been placed
       in the 'route' variable. */
    [...]
}
```

Figure 25.12 Routing table lookup using the marnetd API.

```
/* Module ID, usually filled upon
   module initialization */
int module_id;
marnetd_route_t route;
marnetd_addr_t some_destination;
struct timeval *route_age;

[...]

if (marnetd_get_route(some_destination, &route)) {
  /* route is known, add attribute */
  marnetd_add_route_info(&route, module_id, route_age);
}
else {
  /* route unknown */
}
```

Figure 25.13 Adding custom attributes to routes.

may thus be realized as a `struct` defined by the module programmer which then may contain arbitrary data.

Beyond that, route table management includes a configurable "route arbitration" mechanism to resolve conflicts created by the execution of multiple protocol modules on a single node (on so-called "bridge nodes" which forward data between disjoint routing subnets).

25.2.2.3 On-Demand Routing

Reactive routing protocols such as AODV [33] only start to search for valid or efficient routes when needed, i.e., when a local process wants to send data to a remote system. The separation between routing and packet forwarding poses a problem in this scenario: The kernel returns an error (on POSIX.1 systems EHOSTUNREACH, i.e., "Host is unreachable") when a process tries to send data to a remote system or network that is not listed in the routing table. This behavior is undesired because at the time of the send request it is undetermined whether the receiver is actually unreachable or not yet discovered. The current POSIX API used on Linux, BSD systems, and even Windows does not include any functionality to tie a user program's packet flow to the routing state (see [25] for a detailed review).

`marnetd` explicitly supports the development of reactive protocols with its ODH subsystem ("On-Demand Handler"). It intercepts outgoing packets, caches them, and notifies routing modules about their destination address, protocol type, etc. As most other `marnetd` components, ODH works in an event-driven manner: A routing module registers its interest in being notified about outgoing packets destined either for any destination or for certain address ranges or IP networks only.

Figure 25.14 shows an example of how to use the ODH subsystem. Upon initialization, the module registers a callback function which is called when a packet to a host in the specified network is enqueued. It is possible to set both network address and netmask to zero, thereby activating the callback for every outgoing packet. The last parameter (ODH_ALL or ODH_NEW_DEST) determines whether the callback function is called only for packets to destinations that are not in the routing table and for which no route search is in progress (ODH_NEW_DEST) or for every single outgoing packet (ODH_ALL). The latter should only be used for debugging purposes to avoid unnecessary performance degradation in the packet forwarding process.

```
/* Call back function for ODH */
void odh(marnetd_packet_t *packet) {
  /* Start route search */
}

int marnetd_module_init(void) {
  [...]
  marnetd_addr_t network, netmask;
  network.ipv4 = 0; netmask.ipv4 = 0;
  marnetd_register_odh(odh, &network, &netmask, ODH_ALL);
  [...]
}
```

Figure 25.14   Registering an ODH callback.

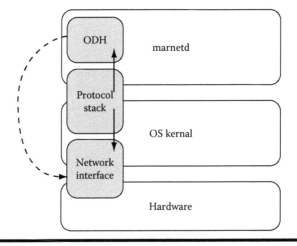

Figure 25.15   ODH architecture.

Once the algorithm has determined that the route search has either succeeded or failed, the module must call `marnetd_odh_route_ack()` or `marnetd_odh_route_nack()` with the respective destination address as an argument in order to have `marnetd` either drop the cached packets or reinject them into the local forwarding chain.

Figure 25.15 illustrates the subsystem architecture. Upon registration of an ODH callback, `marnetd` sets a new network route to the specified network which forces the kernel to divert traffic through a special network tunnel device. Packets sent through the tunnel are cached by `marnetd` which then notifies the routing module, delivering packet header information such as destination address, payload protocol type, etc. When the cached packets are rescheduled for network transfer, a raw socket is used to directly enqueue them in the buffer of the network device. No further preparation steps are needed in this case because the packet encapsulation was already done by the kernel when the packets were originally scheduled.

25.2.2.4 Task Management

Upon module initialization (or at runtime), it is possible to register "tasks" with the `marnetd` scheduling system thus avoiding, for example, the overhead of creating and managing additional threads.

```
/* A function to schedule */
struct timeval *cleanup(void *p) {
 static struct timeval next_run;
 /* check RIB for expired routes */
 return &next_run;
}

int marnetd_module_init(void) {
  int task_id; struct timeval interval;
  interval.tv_sec = 2; interval.tv_usec = 0;
  [...]
  task_id = marnetd_register_task("Route cleanup", &interval, cleanup, NULL);
  if (task_id < 0) {
    marnetd_log(LOGLEVEL_NORMAL, "Test module", MESSAGE_LOG, "Route cleanup ↩
       task failure");
    return -1;
  }
}
```

Figure 25.16   Task registration.

A task is essentially a function that is to be executed in regular intervals which may also be altered dynamically. Such a task might, for example, be the broadcasting of HELLO announce messages or the deletion of expired routing table entries. Tasks are attributed with a function pointer, a function argument of arbitrary type and a timer. When a module is unloaded, all its registered functions are automatically removed from the scheduler queue. Figure 25.16 shows a function stub for a task which removes expired routes from the routing information base. During the registration process, the programmer can determine whether the tasks should run in fixed intervals. The function may also determine the (earliest) time of its next run, thereby allowing adaptive rescheduling of certain regular tasks such as sending out beaconing messages.

25.2.2.5 Utility Functions

Many of the daemon's internal functions and most algorithm modules require the means to create, search, and alter data storage structures. Neither the C programming language nor the GNU C library (which is essentially the foundation of the Linux system programming API since it wraps around the "real" system calls) offer support for such constructs. As a remedy, implementations of various types of dynamic data structures such as linked lists, stacks, and hash tables with support for keys of fixed and arbitrary length are provided. Both structures support the storage of any data type. Moreover, they are fully thread-safe, since concurrent access and synchronization issues are handled internally.

Beyond that, it is often useful to be able to read a set of variables from an external source. During the development and test phase of a routing module, many of the algorithm's parameters are subject to frequent changes, e.g., in order to fine-tune certain aspects such as, for example, the time interval between broadcast messages, the size of the routing table, etc. Having to recompile the module each time, copying it back into place perhaps even on another computer or PDA can quickly become a nuisance. The marnetd API addresses this by providing a convenient approach to define a set of variables which will be read from a text configuration file during the module's initialization procedure. By gearing parameter handling code toward this API feature, it is also possible to integrate the module with the LTI system (see Section 25.2.2.6). Valid variable types are, for example, integers, floating point numbers, string or lists of such types.

```
static marnetd_option_t opt[] = {
  OPTION_STRING_AS_INT("loglevel", "Sets verbosity of logging messages.", ↵
      OPTION_DEFAULT, "verbose", translator),
  OPTION_STRING("autoload", "Autoload module on startup", OPTION_DEFAULT, NULL),
  OPTION_BOOL("enable_ip_fw", "Enable kernel IP forwarding during startup.", ↵
      OPTION_READ_ONLY, true),
  OPTION_INT("route_validity", "Number of seconds after which a route is ↵
      considered expired.", OPTION_DEFAULT),
  OPTION_INT("infinity", "Maximum route length.", OPTION_DEFAULT),
  OPTION_END()
};
```

Figure 25.17 Defining algorithm parameters.

As shown in Figure 25.17, an option can be attributed with a textual description for human readability, access flags, and a type translation matrix. The latter allows a user, for example, to set a variable "logging level" to "quiet," "normal," or "verbose" while the daemon internally treats them as integers 0, 1, and 2.

With the variable definitions of Figure 25.17, a text file containing the keywords shown in Figure 25.18 may be placed in the module directory. This file can then be used as a configuration file to be edited when needed. `marnetd` automatically parses the contents of the configuration file prior to the module initialization procedure. If it encounters syntax or type translation errors, the initialization procedure is aborted, leaving an error message in the log files.

Figure 25.19 illustrates how to retrieve the value of such an option. Setting option values is equally simple as is writing back values to the configuration file.

25.2.2.6 Local Terminal Interface

With the Local Terminal Interface (LTI), `marnetd` provides an extensible command processor to both module programmers and users. Using the API, it is possible to register "commands" with

```
# This is a comment
loglevel = normal
autoload = aodv.so
route_validity = 5
infinity = 10
enable_ip_fw = true
```

Figure 25.18 An excerpt from a marnetd configuration file.

```
marnetd_option_t *interface_option;
marnetd_module_t *this_module;
char *interface;

this_module = marnetd_get_module_by_name ("foo");

interface_option = marnetd_get_option (this_module, "interface");

interface = marnetd_option_get_string (interface_option);
```

Figure 25.19 Querying options data.

the LTI subsystem which can both read arguments from and return results to the interface when using a defined protocol. There are a number of built-in commands to allow module loading and unloading, to access routing data or MAC layer information such as neighbors in range, SNR, signal strength, etc., or perform administrative tasks such as log facility selection, etc. A special command provides a wrapper around the options described in Section 25.2.25: An application connected through LTI may query the name, description, data type, and current values of any option registered with `marnetd` unless the module programmer specifically denied access (by setting `OPTION_NO_LTI` when defining the option). The option wrapper command also accepts new values for options. The LTI client software introduced in Section 25.2.1 makes use of this feature, as shown in Figure 25.20.

Module programmers may extend LTI with their own commands thus enabling cross-layer information exchange through a defined interface: This way, applications may not only extract and inject routing data such as route tables or switch modules but may also, read GPS positioning information of other nodes, request that their data be routed through routes considered stable or query about services offered by other nodes. The cross-layer approach, which relies on interactions between various layers of the networks, has been subject of increased interest in the past [27,34,37]. A vivid example of a cross-layer design implementation was presented in [35]: A network of autonomous robots communicating over a wireless ad hoc network was shown to benefit from the distribution of geographical location information through the routing layer, allowing the node to adjust their movement patterns to maintain connectivity while at the same time reducing the amount of network traffic in comparison to other approaches.

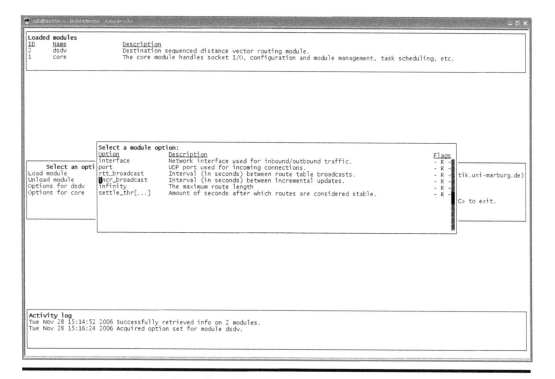

Figure 25.20   Accessing option values through LTI.

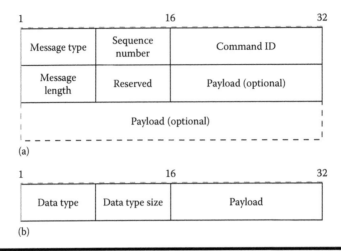

Figure 25.21   LTI protocol messages. (a) LTI protocol header. (b) LTI payload unit.

The default settings configure LTI as a local socket (also known as a Unix socket), which, as the name implies, allows local processes to connect to `marnetd` using a special node in the filesystem (usually `/tmp/marnetd-lti`). Privilege checks are then conducted by the operating system kernel. It is also possible to connect to remote nodes using an encrypted SSL/TLS channel, but this requires the user to set up an SSL certificate authority.

The LTI protocol is designed as a request/response protocol, the binary layout of the packets is shown in Figure 25.21.

The command ID numbers for built-in commands remain fixed and documented in the API documentation. Dynamically registered commands of running modules can be queried at runtime. Any such query will always return a unique ID, data types, and number of arguments of the command (if any), possible return values or data types and description strings for human readability.

The documented source code of the GUI client included with the `marnetd` source code distribution archive can be used as an example for other applications. Core functionality such as connection setup or authentication can even be copied as a whole which is supported by the GNU GPL license. The client currently uses the portable "ncurses" library to generate graphical menu elements in a text mode console. It thus does not require X11 or any additional widget libraries such as GTK+ or Qt, making it possible to run the program on PDAs with a framebuffer GUI (or none at all).

25.2.2.7 Additional Supporting Functionality

Other utility functions include memory management helpers, string construction functions, and a flexible logging system which can print messages on a console, deliver them to a syslog daemon, or write its own log file. Thus, most modules will not have to rely on any external libraries, making binary redistribution simple and effortless.

Module programmers using the API introduced can greatly reduce the effort involved in developing an implementation of their routing algorithm. Using our approach, we were able to create a DSDV [32] algorithm implementation with 180 lines of module code whereas our previously existing implementation as a monolithic routing daemon required 1700 lines of C code. Our current software includes are a simple demonstration implementation of a RIP-based distance vector algorithm as well as DSDV [32], AODV [33], and TERA [23].

25.3 Marnet: Marburg Ad Hoc Network Emulation Testbed

In order to evaluate the suitability and performance of a specific algorithm in varying environments, it has become common practice to utilize network simulators which deal with mobility statically by means of predefined scenarios. The reproducibility of test results is an inherent property of most simulation architectures. Combined with the global view and clock control of the simulators, it is possible to easily gather statistics and observe the behavior of the network, its traffic volume on any link or the routing table of an individual node at any point in (simulation) time. The implementation of an algorithm, however, requires the programming interface of the specific simulator to be used. The transition to a real application platform thus requires a re-implementation of the algorithm which very often raises new issues with respect to the algorithm's design (as a piece of software).

To address these issues, we have developed MarNET, the Marburg Ad Hoc Network Emulation Testbed. Using MarNET in conjunction with the `marnetd` API does not require any transitional steps. It provides an integrated and highly scalable emulation environment to test and evaluate the initial hypotheses while at the same time supporting the creation of software which can be executed both in the test environment as well as on real machines. It offers a virtually unmodified Linux instance for each of the emulated nodes which can be executed in parallel on a single or multiple computers. The actual requirements depend, of course, on the available hardware resources, scenario size, and the requirements of test subject software. However, the employed virtualization technology called XenoLinux is capable of running up to 100 virtual machine instances on a single machine made of commodity hardware (s. Section 25.3.2). Because each of the emulated nodes can provide a fully featured Linux execution environment (right down to system calls), MarNET is especially useful for the evaluation of cross-layer algorithms and application software. This design approach often leads to interdependent development cycles of the functional components in the separate layers and unintended side effects (s., e.g., [24]), requiring an efficient test environment as an integral part of the development process.

The overall architecture of MarNET is illustrated in Figure 25.22, the components shown are discussed in the subsequent sections.

25.3.1 Topology Manager

A wireless network link is characterized by its communication parameters, i.e., the maximum bandwidth, the round trip time, and the packet loss ratio. These values are primarily influenced by the type of network hardware, the physical layer protocol, the node locations, and the nodes' environment.

By governing the communication parameters of a set of network nodes, a wireless network can be emulated on a wired one. This is the main functionality of the Topology Manager as the principal component of MarNET.

As illustrated in Figure 25.23, the Topology Manager consists of two components, the *Controller* and the *Drone*. The Controller permits a global view on the emulated network (as shown in Figure 25.24) and provides the global emulation time to the connected Drones. A Drone is started on every system that is used as a network node. The Controller determines the scenario file to be used

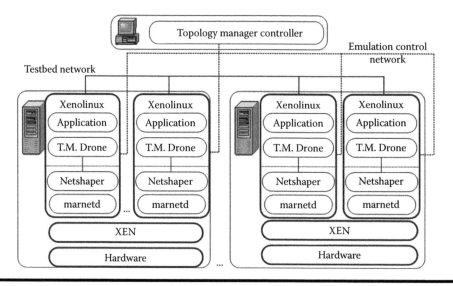

Figure 25.22   MarNET architecture.

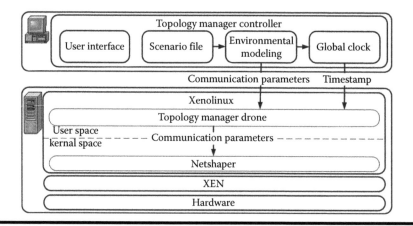

Figure 25.23   Topology manager.

which contains the locations and movement patterns of the network nodes. The communication parameters are calculated and announced to the Drones. Triggered by regular timestamps, the communication parameters are switched, thus enabling MarNET to model the highly fluctuating transmission quality of MANETs caused by high node mobility and the limited range of wireless devices [15,30].

The calculation of the communication parameters is executed by the Environmental Engine. Individual emulation models can be implemented using the API provided by this component. In contrast to other emulation models recreating communication parameters with binary states, i.e., *reachable* and *not reachable*, MarNET supports the development of models achieving a finer granularity by introducing packet loss ratios between 0% and 100%, thus enabling the developer to analyze the impact of lost and corrupted packets on the stability and performance of the software under test.

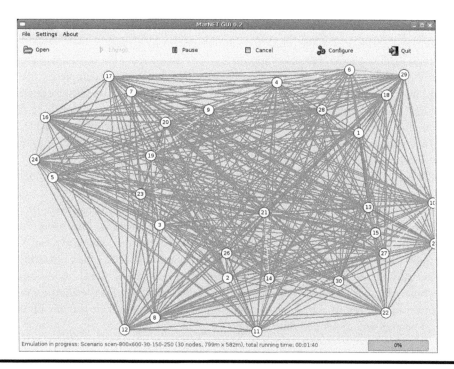

Figure 25.24   Topology manager visualization.

In the current version of MarNET, two models have already been implemented:

1. The *Pisa Model* which is based on measurements of the University of Pisa and the Istituto IIT [11].
2. The *Marburg Model* which is the result of our experiments conducted at the University of Marburg.

The measurements leading to the Marburg Model were conducted using two notebooks running Debian GNU/Linux. One of the notebooks was set up to send groups of 1000 UDP broadcast packets of 1500 byte size. Since packet loss in broadcast UDP packages is not handled, it was possible to count the received packages and to calculate the loss ratio. Since distance is only one of the possible values influencing packet loss, these measurements are possibly affected by side effects, such as (a) interferences of other devices especially 802.11 networks and bluetooth appliances emitting electromagnetic waves, and (b) the multi-path problem where deferred data fragments arrive at the receiver and possibly destroy data. For a detailed discussion of these side effects see [9].

When using the Marburg Model, the Environmental Engine calculates the communication parameters as follows: After parsing the scenario file, the positions of all nodes and the distances between all node pairs are calculated at regular timestamps. Based on both distances and the signal attenuation caused by obstructions on the playing field, the signal levels for each node pair are computed. Finally, the packet loss ratio is derived from the signal levels.

25.3.2 Xen

During the verification phase of software for mobile ad hoc networks, it is vital to conduct large-scale examinations to be able to analyze the scalability and stability of the software under test.

MarNET makes use of the system resource virtualization software Xen [12] which allows us to run multiple emulated nodes on a single physical computer system in parallel.

Xen uses a so-called paravirtualization approach which differs from the traditional approach of full virtualization. In full virtualization solutions, the virtual hardware is completely identical to an existing physical system or a specification thereof. Hence, the guest operating system can be run unmodified on the virtual machine. The drawback of this approach, however, is performance degradation, since complete hardware emulation typically causes a significant overhead [12], especially when software techniques such as dynamic recompilation are employed which may recompile code segments during execution to adapt to the runtime environment.

Using the paravirtualization approach, the virtual machines are similar but not identical to the physical one. Hence, the guest operating system needs to be slightly modified to run. This enables the virtual machines to directly communicate with the physical hardware of the host system. Consequently, the virtualization overhead is reduced significantly, improving the overall performance. Measurements conducted at the Cambridge Computer Laboratory and Clarkson University show that the performance of XenoLinux (the paravirtualized Linux operating system kernel) is virtually identical to the one of a standard Linux System [12,14]. Additionally, the application binary interface does not require to be changed. This allows for the effortless migration from emulation to field test or actual use since changes to the program code become unnecessary. *Live code* can be used and developed much earlier.

25.3.3 Network Topology

Figure 25.22 also gives an overview of the MarNET network topology. The testbed consists of two separate networks: The control network is mainly used as the Topology Manager control channel. Furthermore, the virtual nodes access a dedicated file server containing, among other programs, the software under test and debugging output. The testbed network—being the actual emulated network—is exclusively utilized by the test subjects. Its communication parameters are adjusted during the emulation run as described in Section 25.3.1. Both networks are implemented as follows: In each virtual node, Xen creates two ethernet devices which are tunnel endpoints of virtual ethernet devices on the host machine. Each of the MarNET networks uses a Linux software bridge containing one virtual ethernet device per virtual node and a physical device on the host machine to expand the testbed over multiple hosts. This solution benefits from the fact that the testbed size is easily extensible by adding additional (physical) machines, thus leading to a high scalability of the testbed.

25.3.4 Discussion

In contrast to simulation environments, MarNET offers the possibility to evaluate unmodified Linux applications. Hence, after finishing the verification step, the software under test can be used on real devices such as notebooks or PDAs without effort. Additionally, existing applications can be evaluated on the testbed enabling the developer to compare different algorithms or implementations regarding scalability and stability. The distributed emulation approach of MarNET as well as the node virtualization allow for a large-scale testbed by running multiple commodity hardware computer systems each of which provides multiple network nodes.

We conducted a series of experiments to evaluate the scalability of the testbed. A 2 GHz AMD Opteron 246 with two Broadcom NetXtreme BCM5704 Gigabit Ethernet adapters and 2 GBs of main memory was used as a test environment. A second test system was installed on an Intel

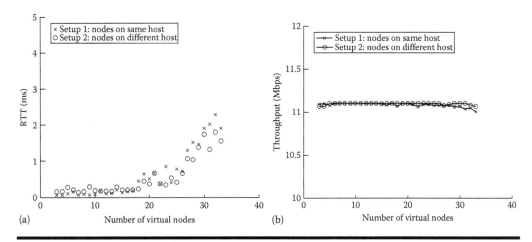

Figure 25.25 **(a) Round trip time and (b) throughput versus the number of emulated nodes.**

Pentium IV 3 GHz with Intel 82547EI Gigabit Ethernet controllers and 1 GB of main memory. Both systems were interconnected through a Cisco Catalyst 3550-12T Gigabit switch.

We consider the round trip time (RTT) to be an important performance criterion when analyzing the delay of packets through the emulated network. Hence, we conducted measurements using two setups: In the first one, the sending and the receiving emulated node were located on the same physical host. In the second setup, both emulated nodes were located on different hosts. Figure 25.25a shows the RTT as a function of the number of emulated nodes. In both setups, the RTT is rising with the number of virtual nodes, as expected. The maximum value reached is 2.298 ms which is below the typical range for wireless LANs of 3–100 ms [21]. The difference between the two setups averages at 0.098 ms which amounts to roughly 4% of the maximum value. We conclude that the influence the wired network between the physical nodes has on the RTT is negligible.

Apart from the RTT, we consider the achievable throughput between two emulated nodes to be a major performance concern. Thus, we conducted measurements using Iperf [6], a tool for measuring TCP/UDP bandwidth performance. Figure 25.25b shows the maximum achievable throughput as a function of the number of emulated nodes: Since the bandwidth was limited to 11 Mbps during the test run (which is the nominal data rate for 802.11b networks), the determined values are near this maximum. The difference between the two test setups averages at 0.010 Mbps, which is 0.09% of the maximum value. We conclude that the influence of the wired network on the achievable throughput is marginal.

The results of both experiments show that testbed sizes of hundreds of nodes are achievable. However, we consider the ethernet network connecting the physical computer systems to be the limiting factor in the testbed size. A high medium usage in the emulation network will have influence on the round trip times and therefore falsify the evaluation results. This drawback can be addressed by adding multiple network interfaces per emulated node as well as dynamically creating virtual LANs and thus collision domains according to the current topology.

The currently implemented environmental models rely on measurements made in a series of outdoor experiments. However, the drawback of these facsimiles is the missing shared medium of the wireless channel, qualifying it for large-scale application evaluations, but making it rather unsuitable for quality of service verifications at the current development stage. Consequently, the development of a shared medium model by creating a centralized control instance dispatching the medium access is required. By introducing a clustering algorithm which dynamically recreates the wireless collision domains on the wire at runtime, we are currently developing an approach to decentralize the implementation and thus maintain the current scalability. Additionally, we are currently evaluating the Xen source code using the system-wide code profiler Xenoprof [29] in an effort to improve the network virtualization performance in our execution environment.

25.4 Related Work

25.4.1 Routing APIs and System Services

There are several routing algorithm implementations for wired networks based on protocol specifications such as OSPF, RIP, or BGP. Apart from closed-source carrier-grade implementations based on Cisco's IOS or Lucent's TOAS operating systems, there are alternatives for commodity hardware, most notably the GNU project's Zebra [5], a collection of multiple routing daemons (i.e., protocol implementations) interoperating with a centralized service instance (the Zebra core). Zebra aims at providing a stable router software for large wired networks; it was, however, not designed to be an extensible routing framework for mobile networks (or wired networks, for that matter). The integration of the multiple routing daemons primarily happens through an administration interface in the core system.

There are several routing algorithm implementations designed specifically for mobile ad hoc networks. For example, OLSRd [8] is an open source community OLSR (RFC 3626) implementation based on an initial release from the University of Oslo. It features a plug-in system allowing the extension of the original OLSR protocol as specified by the RFC which led to the creation of numerous add-on programs, e.g., to support new metrics based on link quality or energy resources. AODV-NIST (US National Institute for Standards and Technology, [1]) and AODV-UU (University of Uppsala, [2]) implement the AODV specification (RFC 3561). The former is designed as an extension to the Linux kernel, the latter runs under Linux 2.4 and can be integrated with the ns-2 simulation system. All of these implementations concentrate on the implementation and extension of a specific protocol. Neither provides the means to coordinate with other locally running implementations or can be made aware of the existence of other implementations, apart from major source code overhauls.

The Click [26] software architecture is a tool to create configurable routers from basic "elements." An element represents a conceptually simple computation, such as decrementing an IP packet's time-to-live field. Using a declarative configuration language, a directed graph of elements is composed, its edges represent the packet flow. The fine-grained elements are a key to Click's flexibility, one may even create an Ethernet switch or Network Address Translation (NAT) setups using the Click approach. Our `marnetd` routing daemon, however, provides a concise API tailored to the use and research of mobile ad hoc routing algorithms. Ease of use and ease of programming are central aspects since many users want to avoid the often cumbersome details of programming a Unix system or creating network messages. He and Raghavendra [22] describe programmable routing services for sensor networks; `marnetd` does not provide a predefined set of routing services (which also define a wire format for the network messages) but rather gives its users the ability to implement such services using its API. He and Raghavendra [22] also focus on the characteristic properties of sensor networks and do not address cross-layer issues.

The Ad Hoc Service Library [25] aims at providing API functions to augment the routing functionality of current operating systems to avoid having to make kernel code changes which, according to the authors, is not only nontrivial but may also lead to unstable systems and unde-ployable solutions. We explicitly support this assessment, but, as already outlined in Section 25.2, also see the need to address usage scenario changes which may also require switching the algorithm. Additionally, `marnetd` explicitly supports information exchange with the application layer, thereby enabling the development of software using the cross-layer design approach. The same is true for the user level framework presented by Allard et al. [10]: The solution presented in this paper, however, also limits access to the ad hoc network by forcing applications to make use of a SOCKS5 proxy interface. Although many widely used network services such as WWW, FTP, or IRC can be channeled through a SOCKS proxy, newer multimedia or streaming applications would have to be modified for this scenario. Microsoft Research has recently provided a driver module [17] named "Mesh Connectivity Layer" (MCL) which provides a virtual network interface on Windows systems. This interface can act as a gateway to a mesh ad hoc network using a modified version of the DSR routing protocol called "Link Quality Source Routing" protocol (LQSR). It is intended to be used in community networks allowing connected homes access to the Internet over some gateway. Although the authors state that, in principle, MCL can be extended with other protocols, there is no documentation except the source code for the driver itself which is provided with a non-free license.

25.4.2 Network Simulators

Network simulators like ns-2 [7] and GloMoSim [4] are usually used to develop solutions for the transport layer and below, i.e., routing algorithms, link layer error correction and so forth, but are not very suitable to develop power-aware applications which need to run on a mobile node such as a PDA or a robot. A further problem is that code developed for the simulator usually does not run on real systems without modification, since the simulator does not offer a complete model of the system.

A network emulator such as MobiEmu [38], on the other hand, can emulate system behavior on all levels, thus allowing more realistic development for a given system such as Linux. The main drawback of MobiEmu is the two-state-emulation approach, i.e., the communication between two nodes is either possible without limitations or it is completely impossible depending on the distance between the communication partners. MobiEmu does not consider physical effects such as packet loss or data corruption.

Fall [20] has presented an emulation facility for the NS Simulator. Traffic originating from physical nodes is piped through the simulator. Hence, this offers the possibility to evaluate code developed as a NS module using real-world traffic. Mahrenholz and Ivanov [28] have examined the real-time behavior of this emulator, which showed an incorrect scheduling behavior leading to false simulation results in the IEEE 802.11 protocol. An extension to the emulator by a time correction feature was introduced to correct this error. However, scalability is one of the main drawbacks of this solution since the centralized NS simulator is likely to become a bottleneck concerning performance.

Ni and Zheng have proposed the EMPOWER [31] emulation system. The basic components in this test environment are so-called emulator nodes equipped with several four-port network devices connected to a hardware switch. Additionally, virtual devices are used to interfere with packet loss and introduce transmission delays.

Engel et al. [19] have developed an emulation testbed using several Linux instance on a L4 microkernel. In this emulation environment, a multiplexer component is used to imitate wireless network connections and offer the possibility to inject transmission errors.

All of the above solutions only cover part of the problem domain, and none allow the system to be embedded in a virtual environment which would allow realistic development of application layer code. To efficiently develop and test systems under near real-world conditions, an overall research environment is needed. To this end, the separate simulators and emulators need to be combined in such a way that all aspects of mobile ad hoc development can be studied in a unified manner.

25.5 Conclusions

In this chapter, we have presented a modular software architecture (called `marnetd`) for mobile ad hoc networks, which allows dynamic routing algorithm switches, adaptive parameter changes at runtime while at the same time providing a simple and concise API for developers. The compiled modules can be run without modifications on both the presented scalable MANET emulation testbed MarNET and on real Linux 2.6-based computer systems, including portable devices such as PDAs and notebooks. Our proposal allows to develop and integrate routing algorithms and application software using cross-layer design principles.

`marnetd` has been extensively tested with our routing modules (see Section 25.2.1) which previously existed as monolithic implementations. These older versions are self-contained (i.e., they did not use any external libraries apart from glibc) and unaware of other routing services (or even information from other layers) which in turn allowed for the creation of fast straightforward code with very small memory footprints. As such, they are well suited for a comparison with their modularized counterparts written for the `marnetd` API which, internally, heavily rely on the library functions provided by the framework. On top of that, there is an additional overhead generated by module management mechanisms and interprocess communication.

In order to evaluate code performance and code behavior we have conducted a test series with the GNU execution profiler software "gprof" which produces call graph profile data at program run time. Its output thus gives information about the number of individual function calls and CPU time spent by those functions. Besides giving an insight as to what parts of the code may need optimization, the data also provides us with the possibility to compare CPU time usage of modular vs. non-modular algorithm implementations in detail.

They have shown that the module implementations use between 1% and 6% additional total CPU time during test runs in our MarNET environment with a total length of 15 h using 10 randomly generated scenarios. Considering that the CPU utilization ranges between less than 1.2% when run on a WiFi router (Linksys WRT54GSv3) with a 200 MHz MIPS CPU and approximately 0.001% on a 2 GHz dual CPU AMD Opteron system, the use of the `marnetd` framework does not cause any significant performance penalty.

Nevertheless, `marnetd` needs additional work both in terms of technical and design aspects. For example, as we have already described in Section 25.2.2.3, the ODH subsystem responsible for the provision of reactive routing support mechanisms uses a Linux-specific kernel extension which provides a virtual network interface. This interface essentially tunnels outgoing packets from the protocol stack to `marnetd`, thus allowing the algorithm implementation to search for viable routes while the packets are "on hold." Unfortunately, this part of the source code currently makes it impossible to compile `marnetd` on BSD-based systems. In addition, the current implementation copies cached packets from kernel space to user space and, in case the search was successful, back again. We therefore intend to find a portable solution which would make it possible to run `marnetd` on BSD systems like FreeBSD, NetBSD, or even MacOS and perhaps Microsoft Windows systems while at the same time avoid expensive copies between kernel and user space. This will also involve studies on how to integrate such an implementation in modern OS kernels in an efficient manner, especially concerning the memory usage required by the reactive routing approach. Engel and Freisleben [18] have introduced new ways of improving kernel code using the aspect-oriented programming paradigm which we will try to adapt for our architecture.

A major next step in the development process of `marnetd` will be the construction of an administrative module to allow the specification of routing strategies: With the term "routing strategy," we refer to a strategy that determines when a particular routing algorithm will be used in a scenario and what the values of the algorithm-specific parameters will be under those circumstances. Switches between modules must be coordinated in a distributed manner. Obviously, this cannot be achieved by extending the routing modules; such "forced extensions" would make interoperation with devices not running `marnetd` impossible. Thus, we intend to specify and implement an out-of-band distributed coordination protocol which allows sets of MANET nodes to vote for the use of a specific algorithm module on the basis of locally perceived network and environmental conditions. With a working coordinator module, we will then perform studies on its impact on network performance and efficiency which will finally allow us to devise and analyze routing strategies for various user roles and scenarios.

25.6 Acknowledgment

This work is financially supported by the Deutsche Forschungsgemeinschaft (DFG, SSP 1140, FR-791/7-2).

References

1. National Institute of Standards and Technology. Kernel AODV, 2004. December 11, 2006. http://w3.antd.nist.gov/wctg/aodv_kernel/ (accessed on July 20, 2012).
2. E. Nordström. AODV-UU, 2011. http://sourceforge.net/projects/aodvuu/ (accessed on July 20, 2012).
3. L. Walkin. ASN.1 compiler, 2007. http://lionet.info/asn1c/download.html (accessed on July 20, 2012).
4. UCLA Parallel Computing Laboratory. GloMoSim, 2000. http://pcl.cs.ucla.edu/projects/glomosim/ (accessed on July 20, 2012).
5. K. Ishiguro. GNU Zebra, 1999. http://www.gnu.org/software/zebra/ (accessed on July 20, 2012). http://www.zebra.org
6. J. Dugan, M. Kutzko. Iperf bandwidth measurement tool, 2011. http://sourceforge.net/projects/iperf/ (accessed on July 20, 2012).
7. K. Fall, K. Varadhan. Network simulator NS-2, 2005, 2012. http://nsnam.isi.edu/nsnam/index.php/User_Information (accessed on July 20, 2012).
8. A. Tonnesen. OLSRd, 2012. http://www.olsr.org (accessed on July 20, 2012).
9. D. Aguayo, J. Bicket, S. Biswas, G. Judd, and R. Morris. Link-level measurements from an 802.11b mesh network. *SIGCOMM Computer Communication Review*, 34(4):121–132, 2004.
10. J. Allard, P. Gonin, M. Singh, and G. G. Richard III. A user level framework for ad hoc routing. In *Proceedings of the IEEE International Conference on Local Computer Networks (LCN 2002)*, pp. 13–19, Tampa, FL, 2002.
11. G. Anastasi, E. Borgia, M. Conti, and E. Gregori. IEEE 802.11 ad hoc networks: Performance measurements. *Cluster Computing Journal, Special Issue on Ad Hoc Networks*, 8:135–145, 2005.

12. P. Barham, B. Dragovic, K. Fraser, S. Hand, T. Harris, A. Ho, R. Neugebauer, I. Pratt, and A. Warfield. Xen and the art of virtualization. In *SOSP'03: Proceedings of the Nineteenth ACM Symposium on Operating Systems Principles*, pp. 164–177, New York, ACM Press, 2003.

13. O. Battenfeld, M. Smith, P. Reinhardt, T. Friese, and B. Freisleben. A modular routing architecture for hot swappable mobile ad hoc routing algorithms. In *Proceedings of the Second International Conference on Embedded Software and Systems*, pp. 359–366, Xian, China, Springer, 2005.

14. B. Clark, T. Deshane, E. Dow, S. Evanchik, M. Finlayson, J. Herne, and J. N. Matthews. Xen and the art of repeated research. In *Proceedings of the Usenix Annual Technical Conference, Freenix Track*, pp. 135–144, Boston, MA, 2004.

15. S. Corson and J. Macker. Mobile ad-hoc networking (MANET): Routing protocol performance issues and evaluation considerations, IETF Request for Comment 2501, Fremont, CA, U.S.A., 1999.

16. D. De Couto, D. Aguayo, B. A. Chambers, and R. Morris. Effects of loss rate on ad hoc wireless routing. Technical report MIT-LCS-TR-836, MIT Laboratory for Computer Science, Cambridge, Massachusetts, U.S.A., 2002.

17. R. Draves, J. Padhye, and B. Zill. Link quality source routing, 2005. http://research.microsoft.com/mesh/ (accessed on July 20, 2012).

18. M. Engel and B. Freisleben. Supporting autonomic computing functionality via dynamic operating system kernel aspects. In *Proceedings of the 4th International Conference on Aspect-Oriented Software Development, AOSD 2005*, pp. 51–62, Chicago, Illinois, U.S.A., 2005.

19. M. Engel, M. Smith, S. Hanemann, and B. Freisleben. Wireless ad-hoc network emulation using microkernel-based virtual linux systems. In *Proceedings of the 5th EUROSIM Congress on Modeling and Simulation*, pp. 198–203, Marne la Vallee, France, EUROSIM Publishers, 2004.

20. K. Fall. Network emulation in the VINT/NS simulator. *Proceedings of the Fourth IEEE Symposium on Computers and Communications*, pp. 244–250, Sharm EL Sheikh Red sea, Egypt, 1999.

21. A. Gurtov and S. Floyd. Modeling wireless links for transport protocols. *SIGCOMM Computer Communication Review*, 34(2):85–96, 2004.

22. Y. He and C. S. Raghavendra. Building programmable routing service for sensor networks. *Computer Communications*, 28(6):664–675, 2005.

23. R. Jansen, S. Hanemann, and B. Freisleben. Proactive distance-vector multipath routing for wireless ad hoc networks. In *Proceedings of the IASTED International Conference on Communication Systems and Networks 2003 (ICCSN2003)*, pp. 1–6, Benalmadena, Spain, ACTA Press, 2003.

24. V. Kawadia and P. Kuma. A cautionary perspective on cross-layer design. In *IEEE Wireless Communications*, 12(1): 3-11, 2005, IEEE. IEEE, 2005.

25. V. Kawadia, Y. Zhang, and B. Gupta. system services for implementing ad-hoc routing: Architecture, implementation and experiences. In *MobiSys 2003: The First International Conference on Mobile Systems, Applications, and Services*, San Francisco, CA, 2003.

26. E. Kohler, R. Morris, B. Chen, J. Janotti, and M. F. Kaashoek. The click modular router. *ACM Transactions on Computer Systems*, 18: 263-297, 2000.

27. J. MacDonald and Y. Fang. Cross layer performance effects of path coupling in wireless ad hoc networks: Power throughput implications of IEEE 802.11 Mac. In *Proceedings of IEEE International Performance , Computing and Communications Conference*, pp. 281–290, Phoenix, AZ, 2003.

28. D. Mahrenholz and S. Ivanov. Real-time network emulation with NS-2. In *Proceedings of IEEE International Symposium on Distributed Simulation and Real-Time Applications*, pp. 29–36, Budapest, Hungary, 2004.

29. A. Menon, J. R. Santos, Y. Turner, G. J. Janakiraman, and W. Zwaenepoel. Diagnosing performance overheads in the xen virtual machine environment. In *Proceedings of the 1st ACM/USENIX International Conference on Virtual Execution Environments*, pp. 13–23, New York, ACM Press, 2005.

30. P. Mohapatra and S. Krishnamurthy. *Ad Hoc Networks - Technologies and Protocols*, Boston, MA, Springer, 2005.
31. L. Ni and P. Zheng. EMPOWER: A network emulator for wireline and wireless networks. In *IEEE InfoCom 2003*, pp. 1933–1942, San Francisco, CA, 2003.
32. C. Perkins and P. Bhagwat. Highly dynamic destination-sequenced distance-vector routing (DSDV) for mobile computers. In *ACM SIGCOMM'94 Conference on Communications Architectures, Protocols and Applications*, pp. 234–244, London, U.K., 1994.
33. C. E. Perkins and E. M. Royer. Ad hoc on-demand distance vector routing. In *Proceedings of the 2nd IEEE Workshop on Mobile Computing Systems and Applications*, pp. 90–100, New Orleans, LA, 1999.
34. S. Shakkottai, T. S. Rappaport, and P. Karlson. Cross layer design for wireless networks. *IEEE Communications Magazin*, 41(10): 74–80, 2003.
35. M. Smith, S. Hanemann, and B. Freisleben. Coupled simulation/emulation for cross-layer enabled mobile wireless cxomputing. In *Proceedings of the Second International Conference on Embedded Software and Systems*, pp. 375–381, Xian, China, 2005.
36. S. Toumpis and A. Goldsmith. Performance, optimization, and cross-layer design of media access protocols for wireless ad hoc networks. In *International Conference on Communication (ICC) 2003*, pp. 2234–2240, Anchorage, AK, IEEE, May 2003.
37. H. L. W. Yuen and T. Andersen. A simple and effective cross layer networking system for mobile ad hoc networks. *Proceedings of PIMRC 2002*, pp. 1-5, Lisbon, Portugal, 2002.
38. Y. Zhang and W. Li. An integrated environment for testing mobile ad hoc networks. In *Proceedings of the 3rd ACM International Symposium on Mobile Ad Hoc Networking & Computing*, pp. 104–111, Lausanne, Switzerland, ACM Press, 2002.

Chapter 26

Efficient Mobility Management Schemes

Distance and Direction-Based Location Update and Sectional Ring Paging

Seung-Yeon Kim, Woo-Jae Kim, Young-Joo Suh

Contents

26.1 Introduction

In recent years, we have witnessed growing demands for personal communication services. In personal communication networks, location management is an important component for scarce resource utilization. Mobile hosts move from cell to cell, and thus the network should keep track of the location of each mobile host in order to deliver incoming calls successfully. Two basic

639

components for tracking location of each mobile host are location update and paging. Location update is performed by each mobile host, which notifies the network of its current location. Paging is the process in which the network searches the mobile host by sending polling signals to cells close to the last reported location of the mobile host. Personal communication networks are partitioned into several location areas (LAs), and each LA consists of one or more cells. A mobile host sends its location information each time it enters into a new LA. When a call for the mobile host arrives at the network, all cells in the LA are paged.

Location update schemes are classified into static algorithms and dynamic algorithms. In static algorithms, mobile hosts transmit location update messages when they enter into a predetermined cell. In dynamic algorithms, mobile hosts transmit location update messages based on their movement patterns. Dynamic algorithms include time-based, movement-based, and distance-based schemes, where location updates are performed based on the elapsed time, the number of cell boundary crossings, and the distance mobile hosts moved, respectively [1–5].

In this chapter, we propose a location update scheme, that considers both the moving direction and distance information of a mobile host. In the proposed scheme, the LA of a mobile host is determined by both moving direction and distance information, and the mobile host updates its location each time it enters into a new LA. In addition, we propose a paging scheme. When a call arrives for a mobile host, the network performs the sectional ring paging in the LA until the host is found.

One of the main issues of location management schemes is to decrease the waste of wireless bandwidth by reducing the active traffic between the network and the mobile host. Generally, if the size of an LA increases, the location update load decreases, while the paging load increases due to the increased number of cells for paging. Thus, a trade-off between location update cost and paging cost is required. Our goal is to propose a new location update and paging scheme so that the total cost (i.e., location update cost and paging cost) required managing location of mobile hosts is reduced.

26.2 Background and Related Works

In this section, we present general features of location management schemes. We also give a brief summary of the previous research results for location management schemes.

26.2.1 Location Management Scheme

Location management is an essential function in order to provide personal communication services to mobile hosts. In the location management scheme, a mobile host performs a location update to explicitly notify the network of its new access point, and the network stores the changes to its location profile. Then, when an incoming call arrives, the network performs paging by sending the paging message to all cells where the called mobile host is likely to exist. The location update schemes can be classified into static and dynamic.

In a static algorithm, a mobile host updates its location at a predetermined set of cells regardless of its mobility. In the location management, the service coverage area is partitioned into LAs, and each LA consists of multiple contiguous cells. The base station of each cell broadcasts the identification of LA to which the cell belongs. A mobile host knows which LA it is in. A mobile host will update its location whenever it moves into a cell which belongs to a new LA. When an incoming call arrives for the mobile host, the network will page all cells of the LA that were last reported by the mobile host. The static algorithm is generally not efficient since the size of the LA is usually static even though the mobile host's mobility and traffic characteristics change dynamically.

In a dynamic algorithm, mobile hosts transmit location update messages based on their movement patterns. Dynamic algorithms are further divided into time-based schemes [1,2], movement-based schemes [1,3], and distance-based schemes [1]. The time-based scheme is considered the simplest strategy in which each mobile host updates its location periodically at every predefined unit of time. When there is an incoming call for a mobile host, the network will first search the cell the mobile host last reported. If it is not found there, the network will search the next ring of cells, and the process continues until the mobile host is located. In the movement-based scheme, each mobile host counts the number of cell boundary crossing after the last location update. When this counter exceeds a predefined threshold, the mobile host transmits a location update message. When an incoming call arrives for the mobile host, the network only needs to page all the cells within a distance of a predefined threshold from the last reported cell. In the distance-based location update scheme, each mobile host tracks the distance between the current located cell and the last reported cell. When the distance exceeds a predefined threshold, the mobile host transmits a location update message. When an incoming call arrives for a mobile host, the network pages all the cells within a distance (predefined threshold) from the last reported cell. These three kinds of dynamic algorithms can reduce location management costs if each optimal threshold value is properly selected, depending on a mobile host's mobility and call arrival pattern. However, each location management cost itself is largely diverse according to the mobile host's mobility pattern [1–3].

26.2.2 Related Works

26.2.2.1 Location Update Scheme

Recent research results on dynamic location update schemes consider several factors to determine whether location updates should be sent or not [6–8]. In a basic distance-based scheme such as proposed in [1,5], the update threshold is determined by measuring the distance between the mobile host's current location and its last updated location. Because it only considers the traveled distance in the update threshold, this scheme is insufficient to adapt the characteristics of a mobile host which has more directive pattern. It is important that update threshold depend not only on traveling distance but also on the moving direction.

In [6], Hwang et al. proposed a direction-based location update (DBLU) scheme, where a mobile host performs a location update when its moving direction changes. In DBLU, a mobile host registers its location after crossing each direction-changing point and informs the network of its new moving direction. Thus, the network can always keep track of the mobile host's moving direction, and the LA is determined by the line as a moving direction of the mobile host. Under the random walk model, however, the DBLU scheme generates more location update messages than those of the distance-based scheme [1,5]. The authors also proposed a paging scheme called line-paging (LP), in which, when a call arrives for a mobile host, the network sequentially pages the LAs on the target mobile host's moving line until the mobile host is found.

26.2.2.2 Paging Scheme

For paging, the ring-structured paging scheme has been widely used in various location management studies [1,3,9,5]. Figure 26.1 shows an example of cells and rings for ring-structure paging. When a call arrives, the ring-structured paging scheme pages the last updated cell first (Ring 0 in Figure 26.1). If the mobile host is not found there, it pages all cells within the innermost ring which surrounds the last updated cell (Ring 1 in Figure 26.1). If still not be found, the

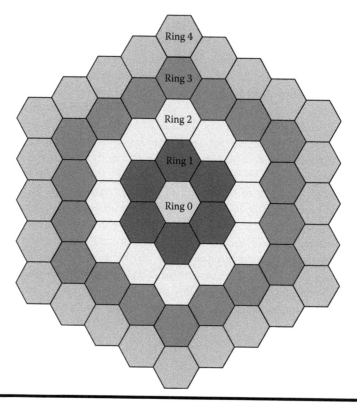

Figure 26.1   The ring-structure scheme for ring paging.

scheme pages all cells within the second ring (Ring 2 in Figure 26.1). This process repeats ring by ring from the innermost ring to the outermost one until it finds the mobile host. The ring-structured paging scheme does not consider a mobile host's moving direction and thus may page too many cells.

In [10], Lee and Hwang proposed a predictive paging scheme based on the movement direction, called movement direction paging (MDP) scheme, where a paging area is selected as cells predicted by the future movement direction of a mobile host. In MDP scheme, a mobile host updates its location with a movement-based scheme in [3]. When a location registration is performed, the history of movement direction for a mobile host is also recorded. Using this history of the movement direction, the future movement direction of the mobile host can be predicted. When a call comes for a mobile host, the paging area is selected as cells predicted by the future movement direction. The total cost of the MDP scheme can be reduced compared with the basic velocity paging scheme in [11], since the MDP scheme may have a smaller paging area. However, the unit registration cost is a little bit higher due to the additional operation of the mobile switching center (MSC) and the extra management of the movement direction history. Moreover, if the information of prediction is not correct, the total cost may be increased.

26.3 Proposed Location Update and Paging Schemes

In this section, we describe the proposed location update scheme called distance and direction-based location update (DDLU) and the proposed paging scheme called sectional ring paging (SRP).

26.3.1 Distance and Direction-Based Location Update Scheme

In this section, we propose the DDLU scheme. We assume that the network is homogeneous and that all cells are hexagonal and that the initial movement direction of each mobile host is known. In addition, each mobile host is assumed to move toward each of the six directions—*u* (up) *ur* (up-right), *ul* (up-left), *d* (down), *dr* (down-right), and *dl* (down-left)—in equal probability. We also assume that the initial movement direction of each mobile host at each LA is *u*, and after the initial movement, subsequent movement directions of mobile hosts are determined by the relative direction with the initial movement direction. Figure 26.2a shows the movement directions of mobile hosts.

The proposed DDLU scheme uses the distance-based location update scheme, where "distance" means that the ring distance between the current located ring and the last located cell, just as the basic ring-structured system. But DDLU uses additional information on a mobile host's movement direction to construct the LA. When a mobile host moves, the LA is determined by the initial movement direction of the mobile host. First draw a line (*l*) of length *Th* from the current cell to the mobile host's movement direction, where *Th* is the distance threshold, and then build two equilateral triangles that share a side of the line *l*. The constructed LA includes the cells on and inside of the rhombus made by the two equilateral triangles. Figure 26.2b shows an LA assuming that the initial movement direction is *u* and the distance threshold value *Th* is 3. We call cells on the border of an LA (i.e., cells on the rhombus) as boundary cells. In Figure 26.2b, cells marked as "b" are boundary cells.

In the proposed scheme, the location update is performed each time a mobile host leaves the current LA. It is checked by two ways—by distance only and by both distance and direction. The first case is very simple. Each time a mobile host changes cells, it checks the current ring ID (CRID). If it is larger than the distance threshold value *Th*, it means that the mobile host has entered into the new LA. This is the case, as shown in Figure 26.3, where a mobile lost enters into one of the cells in ring ID 4 (cells 1–11) from a boundary cell in ring ID 3. Since *Th* = 3, if the CRID is larger than 3, then it means that the mobile host is out of the LA. Thus, a location update message is sent and a new LA is constructed.

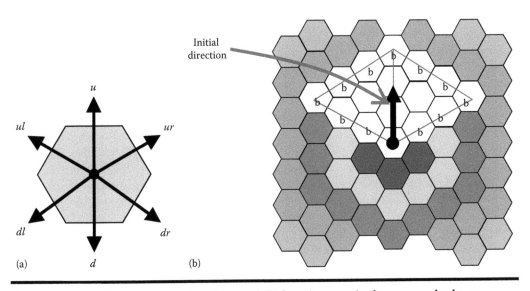

Figure 26.2 **(a) The movement directions and (b) location area in the proposed scheme.**

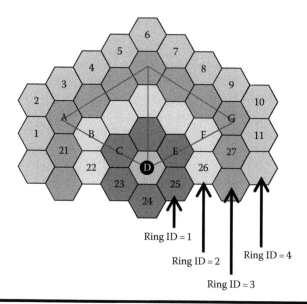

Figure 26.3   Overview of location update.

The second case is more difficult, when a mobile host leaves the LA by entering into cells 21–27 via boundary cells located in the lower two sides of the LA rhombus, i.e., cells A, B, C, D, E, F, and G in Figure 26.3. In this case, it is impossible to know whether the mobile host leaves the LA by distance. So, direction information is needed in this case. When a mobile host changes cells, it checks the CRID. If it is equal to or smaller than the distance threshold value *Th*, then the mobile host compares the CRID with the previous ring ID (PRID). If the CRID is equal to or larger than the PRID, the mobile host can make a decision on whether it updates its location or not. We used a variable called *bound_cnt*, which is the number of right-side movements (i.e., *ur* or *dr*) minus the number of left-side movements (i.e., *ul* or *dl*). If the absolute value of *bound_cnt* equals to the previous ring ID and the current movement direction of the mobile host is *d*, *dl*, or *dr*, then it means that the previous cell is one of the boundary cells of A, B, C, D, E, F, and G in Figure 26.3. Thus, the mobile host transmits a location update message. Otherwise, the mobile host is located inside of the LA, and thus there is no location update. Figures 26.4 and 26.5 summarize the proposed location update scheme in an algorithmic description and a flowchart, respectively.

Figure 26.6a shows an example of the proposed location update scheme; Figure 26.6b shows the PRID, CRID, and *bound_cnt* values in each step. Initially, *bound_cnt* = 0. When a mobile host moves *u* (1 in Figure 26.6a), then CRID = 2, PRID = 1, and *bound_cnt* = 0. Although CRID > PRID, the absolute value of *bound_cnt* (abs(*bound_cnt*)) does not equal to PRID, and thus there is no location update. After the movement 2, CRID equals to PRID, but abs(*bound_cnt*) (0) is not equal to PRID (1), and thus there is no location update. Since the direction of movement 2 is *dl*, abs(*bound_cnt*) becomes 1. When the mobile host performs movement 3, CRID becomes 2. CRID is larger than PRID, but abs(*bound_cnt*) (1) is not equal to PRID (2), and thus there is no location update. Since the direction of movement 3 is *ul*, abs(*bound_cnt*) becomes 2. Now the mobile host performs movement 4. CRID becomes 3 which is larger than PRID, abs(*bound_cnt*) is equal to PRID, and the current movement direction is *dl*. Since the mobile host moves out of the LA, the mobile host transmits a location update message.

Algorithm : DDLU(Distance and Direction based-Location Update)
Input : CRID(Current Ring ID), PRID(Previous Ring ID)
     CMD(Current Moving Direction), bound_cnt(boundary counter)
Output : To perform Location Update or Not
Begin
  If CRID exceeds the threshold Then execute location update
  Else
    If CRID is greater than or equal to PRID
      If abs(bound_cnt) equals to PRID && CMD equals to d, dl or dr
        Then execute Location Update
      Else save CRID /*no update action*/
    Else save CRID /*no update action*/
  If CMD equals to ur or dr Then increase bound_cnt
  Else if CMD equals to ul or dl Then decrease bound_cnt
End

*abs(): absolute value

Figure 26.4 Algorithmic description of the location update scheme.

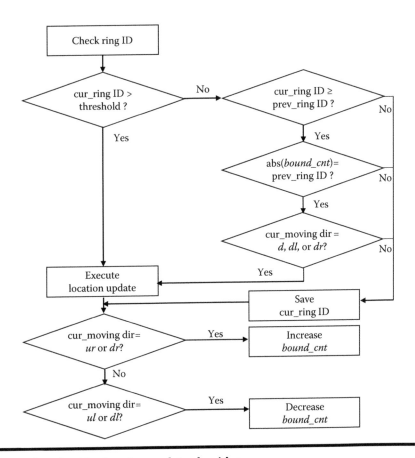

Figure 26.5 Flowchart for location update algorithm.

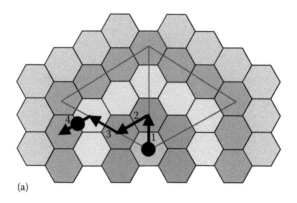

(a)

Step	PRID	CRID	bound_cnt
Init.	0	0	0
1	0	1	0 (CMD = u)
2	1	1	−1 (CMD = dl)
3	1	2	−2 (CMD = ul)
4	2	3	Update (CMD = dl)

(b)

Figure 26.6 Example operation. (a) Movement of a mobile host. (b) PRID, CRID, bound_cnt values.

26.3.2 Sectional Ring Paging Scheme

In this section, we propose a paging scheme based on ring-structured paging. In the proposed sectional ring paging (SRP) scheme, when a call arrives, it pages the last updated cell first. If the mobile host is not found, it pages cells in the innermost ring and within the LA. If still not found, SRP pages cells in the second ring and within the LA. This process repeats until it finds the mobile host. Figure 26.7 shows an example. Assume that, a mobile host has performed a location update at "cell 1" and it is now located in "cell 15." If a call to the mobile host arrives, SRP first pages the last updated cell (e.g., cell 1). If the mobile host cannot be found, then SRP pages the three cells in the first ring within the LA (e.g., cells 2–4). If still not found, SRP locates the mobile host by paging the five cells in the second ring within the LA (e.g., cells 5–9). This process repeats until it locates the mobile host.

26.4 Performance Evaluation

26.4.1 Simulation Environment

We evaluated the performance of the proposed schemes using a discrete-event simulation. We assumed that 400 (20 × 20) cells are located on the $x - y$ coordinate system and mobile hosts are located with the x and y coordinates chosen at random.

We assume that each mobile host moves with a speed ranging from 0 to 1 cell/unit time. We used the unit time as a relative speed of host movements, which is a virtual time tick in the simulation model. We studied the performance of the proposed schemes by varying several parameters,

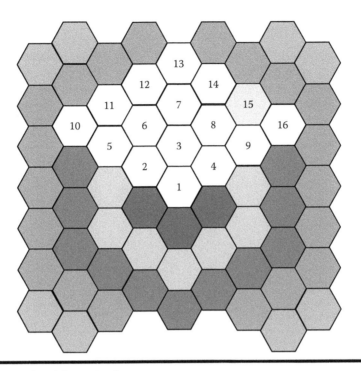

Figure 26.7   Example of the SRP scheme.

Table 26.1   Simulation Parameters

Parameter	Description	Values
N	Network size (# of cells)	20 × 20 (400)
MH	Number of mobile hosts	20 100
C	Call arrival interval	30, 50, 70
MR	Host mobility rate	20, 40, 60
Th	Distance-threshold	3, 4, 5
Time	Total simulation time	600 unit times

including the number of mobile hosts, call arrival interval, and host mobility rate. Table 26.1 summarizes the parameters used in our simulation study. In Table 26.1, mobility rate means handoff probability, and mobility rate of 40 means the probability that the mobile host moves to another cell at each time tick is 40%. The call arrival interval (C) is the time gap between incoming calls. We performed the simulation five times and obtained the average value.

Our simulation study compares the performance of the proposed distance and direction-based location update with sectional ring paging (DDLU-SRP) scheme with the distance-based location update scheme [1] with ring paging (distance-ring) and the DBLU scheme with line paging (DBLU-LP) [6]. The main features considered are location update cost, paging cost, and the total cost in various situations.

26.5 Simulation Results

Figures 26.8 and 26.9 show the location update cost (i.e., the number of location update signals per unit time) and the paging cost (i.e., the number of paging signals per unit time), respectively, as a function of the number of mobile hosts, when C is 50, MR is 40, and *Th* is 3. In general, larger sized LA leads to fewer number of location update signals since fewer number of mobile hosts

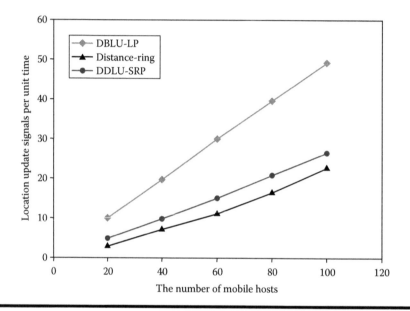

Figure 26.8   Location update cost (MR = 40, C = 50, *Th* = 3).

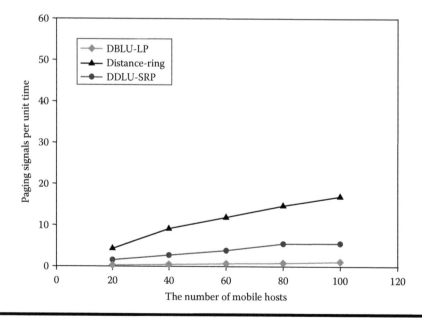

Figure 26.9   Paging cost (MR = 40, C = 50, *Th* = 3).

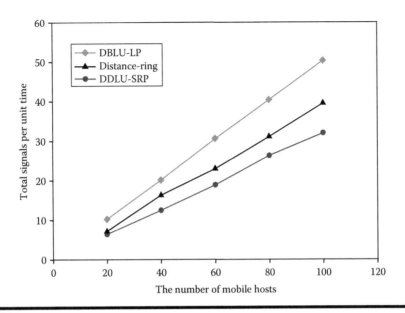

Figure 26.10   Total cost (MR = 40, C = 50, *Th* = 3).

cross LA boundaries. However, the paging load increases when the LA size becomes larger due to the increased number of cells for paging. Generally, the size of LA in the distance-based scheme is larger than that in the proposed scheme, while the size of LA in the DBLU scheme is the smallest since mobile hosts performs location updates whenever they change moving directions. Thus, the distance-based location update scheme with ring paging (distance-ring) shows the smallest location update cost and the largest signaling cost. On the other hand, DBLU scheme with line paging (DBLU-LP) shows the largest location update cost and the smallest signaling cost. The proposed DDLU-SRP scheme shows intermediate performance in all of the costs. Figure 26.10 shows the total cost, which is the sum of the location update cost and paging cost, as a function of the number of mobile hosts. As shown in Figure 26.10, the proposed DDLU-SRP shows the lowest total cost performance.

We also performed our simulation study by varying the distance threshold value. Figures 26.11 and 26.12 show the total cost as a function of the number of mobile hosts when C is 50, MR is 40, and *Th* are 4 (Figure 26.11) and 5 (Figure 26.12). From these figures, we can see that DDLU-SRP shows the lowest total cost performance regardless of the distance threshold values. Therefore, we present the rest of our simulation results when the distance threshold value is 3.

Figures 26.13 and 26.14 show the total costs of the three schemes when the call arrival interval (C) values are 70 (Figure 26.13) and 30 (Figure 26.14) and when MR is 40. The call arrival interval is the frequency of incoming calls. Thus, larger call arrival interval indicates less incoming calls. As shown in Figure 26.13, when the call arrival interval is relatively long (C = 70), distance-ring and DDLU-SRP show very comparable performances. On the other hand, when the call arrival interval is relatively short (C = 30), distance-ring and DBLU-UP show very comparable performance. But, in both of the cases, the proposed DDLU-SRP shows the best performance.

Figures 26.15 and 26.16 show the total costs when the mobility rate (MR) is 20 and 60, respectively, and when C is 50. Higher mobility rate means more frequent handoffs. When the mobility rate is high, the total cost is largely affected by location update cost due to more frequent

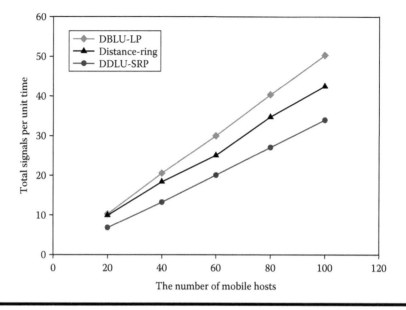

Figure 26.11   Total cost (MR = 40, C = 50, *Th* = 4).

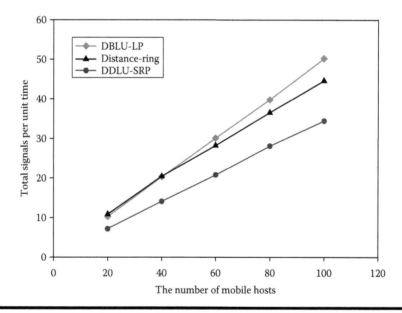

Figure 26.12   Total cost (MR = 40, C = 50, *Th* = 5).

handoffs. As shown in Figure 26.15, when the mobility rate is relatively low, the proposed DDLU-SRP scheme shows very good performance. When the mobility rate is relatively high, DDLU-SRP and distance-ring show very comparable performances, as shown in Figure 26.16. But in both of the cases, the proposed DDLU-SRP scheme shows the best performance. From our simulation results, the proposed scheme shows much better performance when calls arrive more frequently and/or the mobility rate is relatively low.

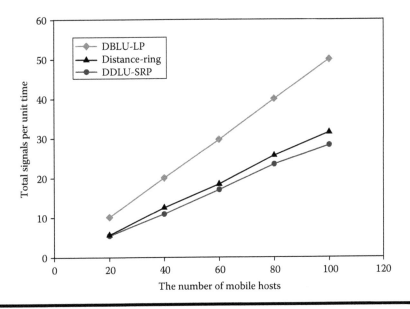

Figure 26.13 Total cost (MR = 40, C = 70, *Th* = 3).

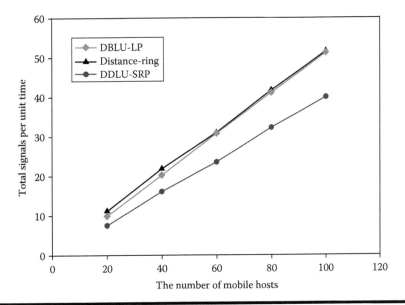

Figure 26.14 Total cost (MR = 40, C = 30, *Th* = 3).

From our simulation study, we can see that the proposed scheme reduces the total cost required for location management. Furthermore, the size of an LA can be reduced by reflecting on the mobile host's characteristics, which are the distance and moving direction. Moreover, the proposed scheme can reduce the total cost successfully regardless of call arrival interval, mobility rate, and the number of mobile hosts.

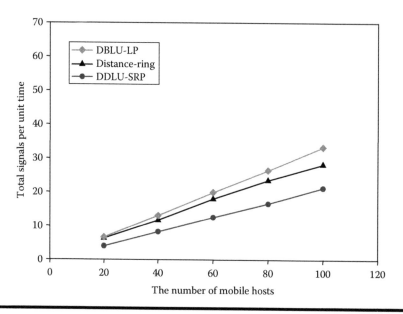

Figure 26.15   Total cost (MR = 20, C = 50, *Th* = 3).

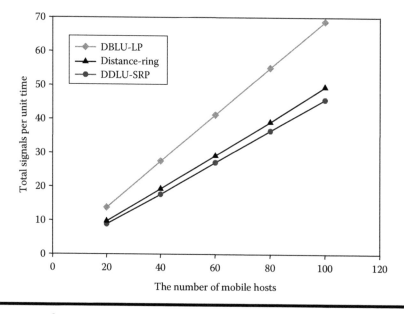

Figure 26.16   Total cost (MR = 60, C = 50, *Th* = 3).

26.6 Conclusion

In this chapter, we propose an efficient location update scheme based on both the movement direction and distance information of mobile hosts, and we also propose a paging scheme. In the proposed scheme, the location area of a mobile host is determined by both moving direction and distance information, which reduces the total cost required for location management of mobile hosts. According to our simulation study, the proposed scheme shows an improved cost performance compared with other schemes.

References

1. A. Bar-Noy, I. Kessler, and M. Sidi, Mobile users: To update or not to update?, *Proceedings of ACM Wireless Networks*, 1, (2), 175–185, 1995.
2. C. Rose, Minimizing the average Cost of paging and registration: A timer-based method, *Proceedings of ACM Wireless Networks*, 2, (2), 109–116, 1996.
3. I. F. Akyildiz, J. S. M. Ho, and Y.-B. Lin, Movement-based location update and selective paging for PCS networks, *Proceedings of IEEE/ACM Transaction on Networking*, 4, (4), 629–638, 1996.
4. C. Rose and R. Yates, Minimizing the average Cost of paging under delay constraints, *Proceedings of ACM/Baltzer Wireless Networks*, 1, (2), 211–219, 1995.
5. J. S. M. Ho and I. F. Akyildiz, Mobile user location update and paging under delay constraints, *Proceedings of ACM/Baltzer Wireless Networks*, 1, (4), 413–425, 1995.
6. H.-W. Hwang, M.-F. Chang, and C.-C. Tseng, A direction-based location update scheme with a line-paging strategy for PCS networks, *Proceedings of IEEE Communications Letters*, 4, (5), 149, May 2000.
7. T. Tung and A. Jamalipour, Adaptive location management strategy combining distance-based update and sectional paging techniques, *Proceedings of HICSS*, Big Island, HT, Jan 2003.
8. Z. Naor and H. Levy, Minimizing the wireless cost of tracking mobile users: An adaptive threshold scheme, *Proceedings of IEEE INFOCOM*, San Francisco, CA, 1998.
9. T. Tung and A. Jamalipour, A novel sectional paging strategy for PCS networks, *Proceedings of Globecom*, Taipei, Taiwan, 2002.
10. B.-K. Lee and C.-S. Hwang, A predictive paging scheme based on the movement direction of a mobile host, *Proceedings of VTC*, Amsterdam, Holland, 1999.
11. G. Wan and E. Lin, A dynamic paging scheme for wireless communication systems, *Proceedings of ACM MobiCom*, Budapest, Hungary, 1997.

Chapter 27

Enhancing TCP Performance and Connectivity in Ad Hoc Networks

Vinh Dien Hoang, Yan Zhang, Lili Zhang, and Maode Ma

Contents

27.1 Introduction

The telecommunication industry and wireless technology have developed rapidly during the last decade and have resulted in several wireless networks such as GPRS, PHS, UMTS, IEEE 802.11x, ad hoc networks, etc. Certainly there is need for interconnecting these wireless networks with other existing networks. The de facto standard for internetworking between these networks is TCP/IP. However, there are currently two main problems for widely used Reno TCP. The first one is how to initiate and maintain connections when the mobile host in the wireless network moves frequently and changes its IP address. The second problem is how to provide a good performance TCP services in the wireless environment with lousy links and frequent handoffs.

To solve the first problem and provide connectivity between the mobile host (MH) and the fixed host (FH), mobile IP had been proposed [1]. Using this scheme, an FH/MH can initiate the connection to other MHs using the home agent (HA) and the foreign agent (FA). Details will be presented in Section 27.3.

As for the second problem, it is because during the development of the TCP protocol, timeout event occurs mainly due to congestion at the intermediate routers. So the TCP protocol had been optimized for the reliable transmission network where losses are due to congestion. This as resulted in Reno TCP [2], which is widely deployed nowadays. When a timeout occurs, Reno TCP assumes that this is due to the congestion at the intermediate nodes and activates its slow start algorithm. Although this algorithm works fine in the wired network, it shows a lot of problems when applied to wireless networks, where losses are also due to lousy wireless links and frequent movement of MHs.

In this chapter, a solution for this problem, TCPMobile, is presented. Basically, TCPMobile separates the TCP connection between the wired part and the wireless part at the FA. In addition, important modifications on the TCP protocol for the connection in the wireless part are introduced to improve TCP performance. This chapter also proposes a new method — MARs — for MHs to access the Internet by using multiple BSs at the same time. TCPMobile and MARs are the complete solution for the MH to effectively communicate with the fixed host. TCPMobile operates in layer 4 whereas MARs is in layer 3.

This chapter is organized as follows. Related works are presented in Section 27.2. Mobile IP and the proposed solution (TCPMobile) are presented in Section 27.3. In Section 27.4, the proposal of using multiple BSs at the same time—MARs is discussed. Section 27.5 evaluates performance of the proposed solution based on the NS2 network simulator [3]. The simulation is used to evaluate the proposed solution over Reno TCP and M-TCP [4] as well as MARs. Finally, Section 27.6 concludes this chapter.

27.2 Related Works

Many solutions have been proposed during the last 10 years to tackle the aforementioned problem. They can be classified into two major categories keeps the end-to-end TCP connection: or separates the wired and wireless part of TCP connection (split TCP connection).

27.2.1 End-to-End TCP Connection

Proposals in this scheme either try to hide the losses due to the lousy wireless link and the frequent mobile handoff from the TCP sender (hence, requiring changes on the receiver side) or use some algorithms at the sender side to differentiate between congestion losses and random losses (hence, requiring changes in the sender side).

Freeze-TCP [5], based on the fact that the MH can somehow (e.g., from signal strength, etc.) predict the disconnection with the wired network beforehand so that it could advertise a zero window size ACK to the TCP sender to stop it from sending more packets or shrinking its congestion window. When the wireless connection resumes, the MH advertises the same ACK with a suitable window size to the sender, and the transmission continues. This solution maintains the end-to-end TCP connection and requires changes only at the MH.

In TCP-MD&R [6], an MH bases on TCP-MD (moving detection algorithm) to predict handoff and TCP-R (registration) to freeze sender from sending more packets during the handoff. This solution improves the performance of TCP during handoff, especially in frequent handoff.

However, when losses happen due to the lousy wireless link, the performance of Freeze-TCP and TCP-MD&R is degraded significantly since each time a packet is lost (easily happens in the wireless environment), the senders congestion window is shrunk to one segment window size.

TCP k-SACK[7] is a little different from TCP SACK in its congestion detection and avoidance algorithms. The TCP sender does not use the single packet loss as an indication of the congestion. Instead, it uses a new parameter—loss window. If more than k packets transmitted in the loss window are lost, then it is due to the congestion, otherwise it is due to the lousy link. TCP k-SACK can achieve better performance compared with normal Reno TCP since it does not shrink its congestion window when less than k packets are lost. This solution requires changes at both sender and receiver sides.

Fast retransmission [8,9] uses triple ACKs to activate fast recovery algorithm of Reno TCP in the sender side. When detecting a packet loss, the MH sends triple ACKs to the sender. The sender upon receiving triple ACKs will resend the lost packet and keep its congestion window unchanged. This solution is simple and requires changes only in the MH side. However, each time fast recovery is activated using Fast retransmission, the TCP senderąřs threshold window is shrunk by half, which will reduce the connection performance later on.

TCP Veno [10] tries to distinguish the random losses caused by the lousy links with losses caused by the congestion and adjusts the slow start threshold to a suitable value (not 1 as in Reno TCP). TCP Veno requires change in the sender only.

However, TCP Veno, TCP k-SACK, and Fast retransmission cannot effectively deal with the frequent handoff of the MH, which is expected in the wireless network. And this will degrade the performance of the connection during and after handoff.

27.2.2 Split TCP Connection

These proposals here break each TCP connection between the sender and the receiver into two separate connections: one TCP connection in the wired network and one TCP connection in the wireless network. The break point is usually the base station (BS). I-TCP [11] at belongs to this category. In the wired network between the FH and the BS, a traditional TCP implementation (Reno TCP) is used. In the wireless network between the BS and the MH, a customized TCP protocol for wireless is used. BS will act as a bridge to forward packets between the two connections. This solution can solve all the problems since a new, customized TCP protocol is used in the wireless network. However, it breaks the end-to-end TCP semantics.

M-TCP [4] is a very special solution in this group because it breaks the TCP connection at the BS but manages to keep the end-to-end TCP semantics. The original TCP connection from the FH to the MH will be split at the BS. When the BS receives a TCP segment from the FH, it sends the segment to the MH and waits for the MH's ACK. During that waiting time, the BS does not acknowledge the segment with the FH. Only when the BS receives the ACK from the MH does it acknowledge the segment with the FH. By doing so, the TCP end-to-end semantic is maintained. However, the most interesting feature of M-TCP is that the BS always keeps the last byte unacknowledged. This means that if the MH had acknowledged with the BS k bytes, the BS only acknowledged with the FH $k - 1$ bytes. When the BS senses that the MH was disconnected (e.g., during handoff), it will send an ACK packet with zero-window size to the FH for the last byte C the kth byte. This ACK will freeze the TCP sender at the FH and prevent it from shrinking its transmission windows. When the BS-MH connection is restored, the BS will send the same ACK with suitable window size to the FH to reactivate its TCP sender and continue with the data transmission.

It is observed from the M-TCP operations that the BS does not acknowledge the TCP segment it received immediately. It has to wait for the ACK from the MH first. This will delay the connection in the FH-BS part. Moreover, when the MH finishes handoff and restores the BS-MH connection, data transmission also does not resume immediately because it has to wait for the BS to wake up the TCP sender at the FH. M-TCP also did not make any significant changes on the TCP protocol at the wireless connection except on handling handoffs. These factors will reduce the overall performances of M-TCP connections.

Snoop [12] can be considered as belonging to the first category. However, it needs information from transport protocol (TCP) to operate and it is sometimes considered as belonging to the link-layer protocol category. In fact, snoop can be classified as somewhere between the two earlier categories. By using an agent residing in the intermediate node (usually the BS) to read the TCP header, packets are cached and retransmitted to the MH if they are lost. However, like Fast retransmission, TCP k-SACK, Snoop is unable to effectively deal with frequent handoffs. Its performance will suffer after a longtime disconnection.

These analyses are the reasons for us to find a TCP solution that can maintain a good performance in the lousy wireless link environment with frequent handoffs. This is the main objective of this chapter.

27.3 TCPMobile

To provide complete TCP services for the MH, we have to ensure that the MH can establish and maintain the connection with other hosts and vice versa when the MH connects to the network. But due to the frequent movement of the MH, it can be expected that the IP address of the MH will change as MH moves to a new place. With this new IP address, the MH still can send data to the other hosts but the other hosts cannot do so because there is no mechanism to inform them about the MH's new IP address. Mobile IP [1] is the best solution to date to tackle this problem.

Using Mobile IP, the MH is assigned to a home network that is represented by a home agent (HA). Each MH has a static IP address belonging to the home network. Other hosts will communicate with the MH using this static IP address. When the MH moves to other networks, it will obtain another IP address called IP care-of address (CoA) from the FA in that network. The FA in that network in turn will inform the MH's HA about the presence of its MH in the foreign network. If the foreign network does not have a FA or all FAs are busy, the MH will act as the FA and obtain its so-called co-located CoA (using DHCP, etc).

Figure 27.1 illustrates how mobile IPs work. When an FH wants to send data to an MH, it sends the IP packet to the MH's home IP address (1). This packet will be routed to the MHs home network and intercepted by the HA. The HA will encapsulate this entire datagram and forward it to the MH's CoA (or co-located CoA). This encapsulated packet will eventually reach the FA (2) where the MH resides. The FA will then decapsulate it and send the original IP datagram to the destination MH (3).

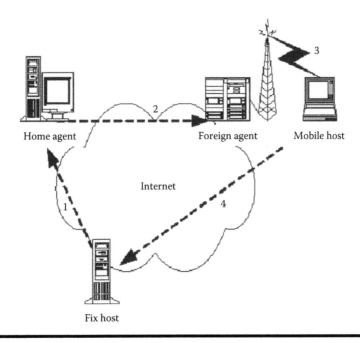

Figure 27.1   Mobile IP operations.

If the MH wants to send a packet to other hosts (could be FH or MH), it will send this packet to the FH's IP address or MH's home IP address. This packet will then be routed by foreign network routers to the FH's or MH's home network (4).

Since the MH moves frequently, there should be a mechanism for the MH to detect its movement. The MH can do so by listening to the agent advertisement messages periodically broadcasted by the FA. Based on the information on these messages, the MH can determine whether it has moved to a new foreign network or not.

Using mobile IP, a TCP connection can be established between a host and an MH without being afraid of IP address changes of the MH. However, this TCP connection's performance will be affected because of the packet losses in the wireless network and during MH handoff.

It can be seen from the Mobile IP operations that the FA has to strip off the HA's encapsulation to recover the original datagram from the FH and forward it to the MH through the foreign network, which may have different MTU (maximum transmission unit), etc. Moreover, all packets that arrive at the MH are forwarded by the FA. This is true even with the use of binding update to eliminate the triangle routing problem in Mobile IP. In addition, FA belongs to the foreign network (in most cases, FA is BS) which is usually only one or two hops away from the MH. That is the reason we propose to break the connection between the FH and the MH at the FA.

In the proposed solution (TCPMobile), separated connection idea used in I-TCP is adapted and integrated with Mobile IP at the FA. Whenever the FA receives a datagram from the FH through the HA, it will send an ACK for this datagram on MH's behalf. This datagram will be stored in the FA's buffer and eventually sent to the MH through a separate connection. So only the MH and the FA are aware of this separation. It is true that the FA acts as a proxy for the MH. In the wireless connection between the FA and the MH, a modified TCP implementation is used to improve the performance of the connection in the wireless network. Moreover, an effective handoff mechanism is introduced. Using this handoff mechanism, the FA will actively freeze the FH's TCP sender whenever it detects a handoff in progress. Following are the detailed operations of TCPMobile.

27.3.1 Connection Establishment and Data Transfer

If the FH initiates the connection, it will send a TCP segment with the SYN bit on and the ACK bit off to the MH through the HA. When this segment reaches FA, FA will issue an ACK to the FH on behalf of the MH, and the connection between FA and FH is established. The FA then establishes a separate connection to the MH. The connections are now ready for data transfer.

If the MH initiates the connection with another host (FH or another MH), it will send a TCP segment with the SYN bit on and the ACK bit off to the FH through the foreign network router (may not be the FA). This connection request segment will eventually arrive at the FH and an ACK is issued to the MH through the HA and FA. Upon receiving this ACK, the FA will reply to the FH on MH's behalf and the connection between FA and FH is established. The FA then establishes a separate connection with the MH. After this connection has been established, the FA will act as a bridge and forward data between the two connections.

To mitigate the lousy wireless links, two simple but important modifications are applied at the Reno TCP for the connection between the FA and MH:

1. When the FA receives two duplicate ACKs from the MH, it will assume that the packet has been lost and retransmit it. It will not enter fast recovery (Reno TCP enters fast recovery when it receives three duplicate ACKs). When the number of duplicate ACKs seen at the FA reaches 5, TCP sender at the FA enters fast recovery. This modification ensures that the lost packet is retransmitted in a timely manner (after two duplicate ACKs). This modification also ensures that if congestion happens, the TCP sender in the FA would reduce its congestion window and enter congestion avoidance phase (after five duplicate ACKs).
2. When timeout happens, the TCP sender at the FA retransmits the timeout packet. It does not enter the slow start phase as Reno TCP does. During that time, if another timeout happens and no new ACK come, the TCP sender will enter the slow start phase to avoid congestion at the intermediate nodes.

27.3.2 Handoff Management

The FA detects handoff in progress after a timeout period from the MH (currently we choose $4 \times$ FA-MH round trip delay) or after receiving a handoff request message from another FA. When detecting that handoff is in progress (Figure 27.2), the FA will restore the TCP sockets associated with the MH to its previous state, send a zero window size ACK to the FH to freeze FHs socket, and wait for the handoff request message from the new FA. When the handoff request message comes (2), the old FA will copy all of the socket's status associated with the MH involved to the new FA (3). The new FA then issues an ACK with nonzero window size to the FH and reactivates the connection. The FH is not aware of this handoff process.

When the MH moves to a new cell, it could detect its movement based on agent advertisement messages broadcasted by the new FA or it can issue its own solicitation message to explicitly ask for advertisement messages after a timeout period. Upon receiving these messages, the MH could detect its movement and liaise with the new FA to complete its Mobile IP registration process. The MH then informs the new FA about its previous connections and the old FA's address (1). The new FA based on that information will liaise with the old FA to create exactly the same TCP

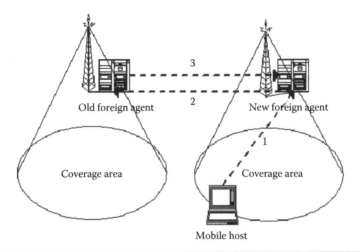

Figure 27.2   Handoff.

sockets associated with the MH in the old FA (2, 3). After handoff finishes, the old FA will delete all TCP sockets associated with the old MH.

Unlike [6], in our TCPMobile solution, the FA will send a zero window size ACK to freeze FH's TCP sender after a timeout period. If the MH sends a zero window size ACK to the FH after detecting its movements as in [6], it is too late since it usually takes more than a round trip time to detect the MH movement, and the FH as already shrunk its congestion windows.

27.4 MARs for Internet Connection

TCPMobile works well in layer 4 and improve the TCP connections. However, if routing protocol in layer 3 does not work well in the wireless network, TCPMobile will not improve the connection throughput much. In other words, single layer solution cannot solve the problem completely. In this section, layer 3 problems for Internet connection will be discussed, and MARs will be proposed to tackle these problems.

Internet services for MHs are provided by attachment points called Internet gateways, BS, or access routers (AR). Nodes in wireless network requiring connection to the Internet have to register with BSs and obtain a global routable address. This is done through address autoconfiguration [13], gateway discovery [14,15], and registration processes. Depending on what type of ad hoc routing protocol is used in wireless network, these processes could be manual or proactive or integrated with the ad hoc routing protocol.

However, it is not that simple in reality because of the dynamic nature of wireless networks, where nodes can join and leave the network at any time. It becomes even worse when more than one BS and/or large and dynamic nodes movement are involved. A number of solutions [14–16] have been proposed so far to tackle these problems. In these existing solutions, an MH registers with a BS (usually the shortest C in term of hops) and sticks to that BS until it is disconnected or finds a better one. This is the motivation for us to propose a solution in which an MH registers with two or more BSs simultaneously for Internet connectivity. Whenever an MH wants to send a packet to the Internet, it will send the packet to one of its registered BSs. So the load on the connection will be divided between the registered BSs. In addition, when an inter-BS handoff occurs, data will not be lost because the MH can still use the remaining BSs during handoff.

27.4.1 HMIP

As present in Section 27.3, Mobile IP requires the mobile host to update its HA and all of its CNs about its new CoA whenever handoff occurs. If the HA CNs are far away or handoff happens frequently, network performance will be severely reduced. HMIP [17,18] is proposed to solve this problem.

HMIP (Figure 27.3) could be considered as the extension of Mobile IP with a new functional node C mobility anchor point (MAP). MAP locates in the foreign network at a higher level compared to BSs. Each MAP serves a domain consisting of several BSs. The main idea of HMIP is when a mobile host visits a BS under MAP it will register with this BS to obtain an on-link CoA (LCoA) and with MAP to obtain a regional CoA (RCoA). The mobile host then uses this RCoA for registration with HA and communication with CNs. When the mobile host moves to a new BS within MAP domain, it only has to register for a new LCoA and inform MAP about its new LCoA. HA/CNs do not need to know anything about this movement and still keep connected with the mobile host using the existing RCoA. When MAP receives a packet destined for a mobile host, it will look at its mapping table to find the correspondent LCoA and forward this packet to the correct BS and then to the mobile host. So MAP is actually a local HA that serves a lot of MHs from different home networks. Using MAP will reduce overhead packets and registration time and enhance the performance of mobile IP, not only in signaling but also in handoff.

27.4.2 The Proposed Multiple ARS Solution: MARs

To use multiple BSs (in other words—multiple access routers [MARs]), the MH has to register with multiple BSs and obtain an LCoA for each registered BS. With these multiple LCoAs, it is very

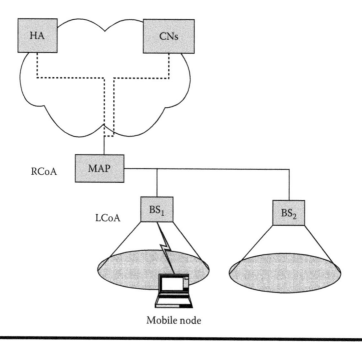

Figure 27.3 HMIP.

difficult for the MH to communicate with CNs. So the main idea here is to use HMIP and utilize the RCoA managed by MAP. All packets sent by MH set the source address to the RCoA assigned by MAP. Packets from CN sent to this RCoA source address will reach MAP. MAP will choose which Bs this packet should be forwarded to and eventually this packet will reach the destination MH through the chosen BS. Details are discussed is the following.

27.4.2.1 Address Configuration and Gateway Discovery

Initial address configuration is quite straitforward [13]. If an MH does not have address when it joins the wireless network, it needs to build two IPv6 addresses. The first one is the MANET address based on MANET_PREFIX ($fec0 :: ffff/64$) and its EUI 64 bits global ID. The second one is a temporary address to check the uniqueness of the MANET address through the duplicate address detection (DAD) operations. The MANET address will be used for further protocol operations and has to ensure that it is unique in the whole wireless network. The temporary address is used for this objective and will be discarded when DAD is over.

This temporary address is built from MANET_INITIAL_PREFIX ($fec0 :: ffff/96$) and another 32 random bits. It is used for a very short time and will be discarded so that the chance of having another node using the same address at the same time is very rare. The MH uses the temporary address broadcast and its chosen MANET address in an AREQ (address request message) in the network. If another node is already used, this chosen MANET address, an AREP (address reply message), will be sent back to the source node. If this happens, the MH has to choose another MANET address either manually or automatically. After a timeout period, if no AREP is received, the MH could assume that its chosen MANET address is unique and could use it for communication with other nodes in the network.

The MH has then to discover the BS to obtain information such as global network prefixes, address lifetime, MAP information, etc., for registration and routing packet to the Internet. Basically, there are two ways for the MH to discover the BSs [14]. In both ways, ad hoc routing protocols or a modified Neighbor Discovery Protocol (NDP) could be used.

In the first way, the MH discovers the BSs by listening to the gateway advertisement messages (GWADV) periodically broadcasted by BSs as part of the ad hoc routing protocol or NDP. If a proactive ad hoc routing protocol is used, GWADV could be piggybacked to any broadcast routing message such as HELLO message. If a reactive ad hoc routing protocol is used, GWADV could be stored in the sent route reply/notify messages. If NDP is used, GWADV message could be sent in router advertisement message of NDP. In all cases, the GWADV message will be broadcast to all nodes in the wireless network.

In the second way, the MH will explicitly ask for BSs to send it the GWADV information. Again, both ad hoc routing protocols and NDP could be used. If a reactive routing protocol is used in wireless network, MH could use the route request packet (RREQ), which is destined for all BSs as the gateway solicitation message (GWSOL). Source address used in this RREQ message could be the chosen MANET address or the MANET home address. BSs will reply by the route reply message (RREP) as the GWADV message. If NDP is used, MH could send the GWSOL message to Internet gateway multicast group (IGW_MCAST). BSs should belong to this group so they could be reached and reply with the GWADV message to the requesting MH.

With these GWADV messages, the MH will have enough information to process the next step C registration for data communication with outside networks.

27.4.2.2 Registration

To use multiple BSs, the MH has to register with multiple BSs. In this chapter, the MH will register and use two BSs. However, more than two BSs could also be used using the same principle.

The MH, based on the information advertised by BSs in GWADV messages, will know how many BSs are available, mobility anchor point (MAP) addresses (used in HMIP [17]), their IP prefixes, and other metric such as number of hops to reach BSs, etc. It then chooses the two best BSs (usually the shortest BS in terms of hops) from the same MAP as its candidate BSs to register with.

First, the MH creates two local LCoA and RCoA by appending its MH's 64 bit EUI to the 64 bit IPv6 prefixes of the two candidate BSs and the MAP. It then sends REG (registration) packets containing information about the selected MAP, the newly created LCoA, and the RCoA to all candidate BSs. These BSs will update their list of registered MHs and forward the REG packet to the selected MAP to update MAPs routing table and the list of registered nodes. Only registered nodes can use BSs, MAP for communication with outside networks. The MH is now ready to receive and send data from and to the Internet.

27.4.2.3 Routing and Data Delivering

MH should create a default route in its routing table so that connection to global CNs can be made. The default route is as shown in Table 27.1, depending on which type of ad hoc routing protocol is used.

Note that the BS address in Table 27.1 could be either the first registered BS, or the second registered BS depending on the load balancing algorithm at the MH. So this address will change from time to time, from the first BS to the second BS and back to the first BS and so on.

Routing inside wireless networks is quite straightforward using the ad hoc routing protocol. Routing outside wireless networks is done through registered BSs and MAP. When the MH wants to send a packet to the Internet, it will choose one of the two registered BSs to send this packet to, based on the MH load balancing function. This load balancing function may be as simple as the function returned the BS which received lesser packets so far. Or it could be other types of complex functions of hops, delay, bandwidth, network configuration, etc., which is out of the scope of this chapter.

When the MH knows which BS it should send the packet to, it sends the packet to the chosen BS using the routing header option as in Figure 27.4. The source address and the destination address in this packet IPv6 header are RCoA and the chosen BS address, respectively. The next header field in the IPv6 header has the value of 63, which will force the BS to examine the next header C in the routing header [19]. The routing type field in the routing header is 0, and the fields next to reserved field are the 128 bit LCoA and 128 bit CN destination address, respectively. This routing header will inform BS that this packet has come from LCoA and to the destination CN address. BS could check to determine whether this packet has come from a registered MH or not based on this LCoA. Because the routing type in the packet routing header is 0, the BS will

Table 27.1 Routing Table

Destination/Prefix Length	Next Hop Gateway
Default route/0	<BS address>
<BS address> /128	<next-hop address>

IPv6 header	Routing header	Data
Src addr : RCoA Dst addr : AR	Addr 1: LCoA Addr 2: CN's addr	. . . . . .

(a)

IPv6 header	Routing header	Data
Src addr : RCoA Dst addr : CN's addr	Addr 1: LCoA Addr 2: AR	. . . . . .

(b)

Figure 27.4   Packet's headers: (a) Original packet headers. (b) After being processed by the selected AR.

then exchange the BS address in the IPv6 header with the CN address in the routing header. This means that the packet is sent from RCoA node to the CN node through the LCoA node and the BS. The destination of the packet is now the CN address.

The BS will now check the destination CN address in the packet header. If this destination is inside the wireless network, an ICMP error packet is sent back to the source to force it to use the route within the wireless network for data communication. If the destination is a host in the Internet, BS will forward the packet normally using the Internet routing protocol. All nodes outside wireless networks will only know the RCoA of the MH. Packets sent by CNs/HA to this RCoA will eventually reach MAP. Based on the load balancing function, MAP will choose which BS it should use to forward the packet to the MH. BSs then route the packet to the destination MH using the ad hoc routing protocol.

27.4.2.4 Handoff

Within MAP, handoff to a new BS could follow [16]. This means if the MH finds a new BS (under the same MAP), which is two hops nearer one of its registered BSs, it will handoff to this new BS. The MH sends a REG packet to the new BS containing the MH's new LCoA. The new BS will update and insert a new entry for the newly registered MH. This REG packet is then forwarded to MAP so MAP could also update its registered tables. During this time, MH still maintains normal operation with CNs/HA using the remaining BS. MAP then sends an acknowledgment packet to the MH through the new registered BS. The MH sends the De_REG packet to the old BS to remove it from the registered node at the old BS. The handoff is completed. The new BS could be used for data transfer normally, as described in previous section.

When an MH finds two new BSs of a new MAP, both of these BSs are two hops nearer its current registered BSs. Handoff to a new MAP will be carried out.

The handoff process between MAP domains is as follows: the MH sends the REG packet to the two new BSs and in turn to the new MAP. In this packet, the address of the old MAP is also included. The new MAP will liaise with the old MAP for buffered packets to be forwarded to the new MAP and then to MH through two new BSs. After handoff finishes, MH will inform HA/CNs about its new RCoA. This new RCoA will be used for all subsequent packets.

The MH handoffs to the new MAP only when both BSs of the new MAP are two hops nearer its current registered BSs. This ensures that handoff between MAP is minimized and hence reduces the effect of handoff over the existing connections.

27.5 Performance Evaluation

Simulations on the NS2 simulator [3] are used to evaluate the effectiveness of the proposed TCPMobile solution. To achieve a fair evaluation, Reno TCP and M-TCP will be studied together with TCPMobile in two different scenarios.

The following metrics are used for performance comparison:

1. *Throughput*: The number of packets received at the destination over the simulation duration.
2. *End-to-end packet delay*: The average delivery delay of the packet from the source to the destination. This delay is calculated from the time the packet is created by the application at the source to the time the packet is received by the application at the destination. This includes buffering time, transmission time, processing time at the source, intermediate nodes, and the destination.

In the first scenario, TCPMobile and Reno TCP are simulated in the model shown on Figure 27.5. This model consists of an FH, a router, a BS, and an MH. In this model, the FH and the BS also act as the HA and FA, respectively. TCP segments are sent from the HA (FH) to the FA (also BS). Upon receiving them, the FA will buffer and send ACKs back on behalf of the MH and be responsible for delivering these TCP segments to the MH through another separate connection.

The connection from the HA to the FA is in the wired network with 5 Mbps bandwidth. The connection from the FA to the MH is in the wireless network with 1 Mbps bandwidth. The time delays between the HA and the router and between the router and the FA are 2 ms.

Packets transmitted through the wireless link will be lost with the probability of p. In the wireless environment, packet loss can happen at any time. So by using loss probability p, the lousy wireless link could be modeled. By varying p, effects of the lousy wireless link on the TCP's performance can be studied. In addition, the wireless link will be disconnected twice, each time 2 during the simulation. This is used to study the protocol ability to recover from the long-time disconnection. All simulations last 550s.

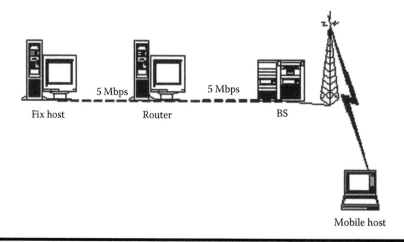

Figure 27.5 Reno TCP simulation model.

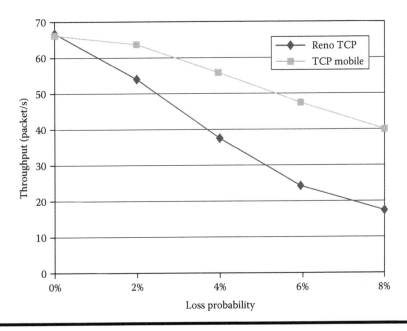

Figure 27.6   Reno TCP-TCPMobile throughput vs loss probability.

TCP parameters:

Initial windows size 20
Initial threshold size 20
Packet size 1040 bytes
ACK packet size 40 bytes
TCP buffer 50 packets

Figure 27.6 shows the throughput of Reno TCP and TCPMobile when loss probability in the wireless network increased. This simulation proved that Reno TCP throughput will suffer especially in the high-loss environment. Each time the packet is lost, Reno TCP reduces its threshold and congestion window by half and slows its data transmission, in this case, unnecessarily. TCPMobile, on the other hand, retransmits the lost packet and slowly adjusts its threshold and congestion window to mitigate the lousy link, and as a result, impressively improves the throughput, especially in the high packet loss scenario.

The second simulation is used to compare the performance of the proposed solution with a prominent protocol, and M-TCP has been chosen because it has many advantages and uses the same philosophy as TCPMobile. In this scenario, the difference and the behavior of M-TCP and TCPMobile will be studied in the model shown in Figure 27.7. This model only consists of the wireless part because the most significant differences between TCPMobile and M-TCP are at this part. Doing so also simplifies the simulation. Using this model, in a wireless network that consists of a BS, two MHs will be simulated. TCP segments will be transmitted from the source (the BS) to the destination (the MH two hops away) using M-TCP and TCPMobile. All other parameters are the same as in the previous scenario.

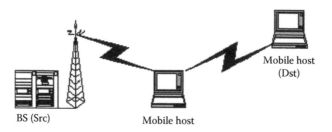

Figure 27.7   M-TCP simulation model.

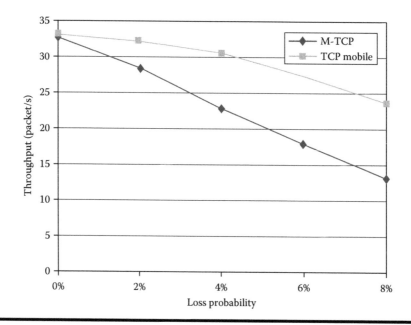

Figure 27.8   M-TCP-TCPMobile throughput vs loss probability.

The throughput of M-TCP and TCPMobile is shown in Figure 27.8. It can be seen that both M-TCP and TCPMobile are affected by packet losses. However, when the losses increase, M-TCP throughput reduces much faster than that of TCPMobile. When loss probability reaches 8%, there is quite a big difference between throughput of TCPMobile and M-TCP as shown in Figure 27.8. This is because TCPMobile, with two modifications on handling timeouts and duplicate ACKs, retransmits the loss packet more timely and adjusts the congestion window more precisely with the current network situation. These two modifications effectively improve TCPMobile throughput even in a high-loss network. The throughput of TCPMobile is almost double that of M-TCP when losses increase to 8%.

However, the higher throughput of TCPMobile comes with a price, that is, the higher end-to-end delay. Figure 27.9 shows the comparison of the end-to-end delay between the TCPMobile and M-TCP. TCPMobile suffers higher end-to-end delay even in the low-loss network. There are two contributing factors to this higher delay. The first factor is that the throughput of TCPMobile is higher. This higher throughput adds more delay to the transmission time and processing time at the source, destination, and intermediate nodes. The second factor is the way TCPMobile

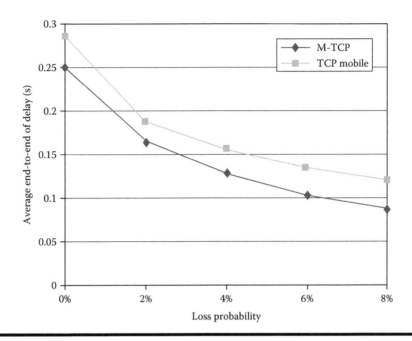

Figure 27.9 M-TCP-TCPMobile average end-to-end delay.

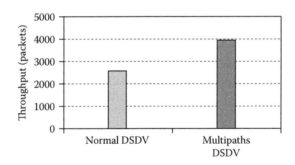

Figure 27.10 Throughput of TCP connection when MARs is used in proactive DSDV routing protocol.

sends its TCP segments. After spending some time adjusting the congestion window and threshold window, TCPMobile will usually send the TCP segments in burst. This burst adds more stress to the intermediate nodes, wireless bandwidth resource and causes higher end-to-end delay.

These simulation results clearly show that TCPMobile's performance is very good compared with Reno TCP and M-TCP, especially in the high-loss wireless network.

Figures 27.10 and 27.11 plot the throughput of a TCP connection when MARs is applied. In Figure 27.10, Normal DSDV is the TCP throughput when using normal DSDV routing protocol. Multipath DSDV is the TCP throughput when using two paths between the source node and destination node. The load is shared equally between the two paths. Figure 27.11 uses the same notions, except that AODV is used in the simulation. Figure 27.11 shows that using the

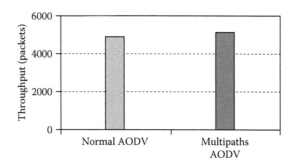

Figure 27.11 Throughput of TCP connection when MARs is used in reactive AODV routing protocol.

proposed MARs solution with TCP in both proactive DSDV and reactive AODV routing protocol significantly increased the connection throughput.

27.6 Conclusion

Mobile IP and different TCP schemes used for the wireless network have been studied in this chapter. Because of the different nature of the packet loss in wired and the wireless networks, the best solution for improving TCP performance in the wireless network is that it should have different TCP implementations. The FA is the most suitable point to split the TCP connection so that a different TCP protocol for the connection from the FA to MH can be implemented. Details of the proposed solution—TCPMobile—are then introduced, which include all connection establishment, data transfer, and handoff management phases. This solution also uses a modified TCP implementation for the wireless part, which differs with Reno TCP in the way it handles duplicate ACK packets and timeout events. A novel solution for internetworking between wireless networks and the Internet, MARs has also been proposed to provide a complete solution in both layers 3 and 4. The main idea of MAR is to use more than one BS's simultaneously for Internet connectivity. By doing so, the MH can maintain its connection even during handoff. Simulations on the NS2 simulator are used to evaluate the proposed solutions. The simulation proves that the proposed TCPMobile solution achieves better performance than Reno TCP and M-TCP do in terms of throughput. However, more research should be carried out on the connection between the FA and MH to reduce its higher packet end-to-end delay. Simulations also show that using MARs improves throughput, especially in small-sized networks.

References

1. C. Perkins, IP mobility support, RFC 2002, 1996.
2. W.R. Stevens, TCP slow start, congestion avoidance, fast retransmission, and fast recovery algorithms, IETF, RFC 2001, Jan 1997.
3. K. Fall and K. Varadhan, NS Manual, The VINT project, http://www.isi.edu/nsnam/ns/doc/
4. K. Brown and S. Singh, M-TCP: TCP for mobile cellular network, *ACM Computer Communication Review*, 27(5), 19–42, 1997.
5. T. Goff, J. Moronski, and D. S. Phatak, Freeze-TCP: A true end to end TCP enhancement mechanism for mobile environments, *IEEE Proceedings of 19th Annual Joint Conference of the IEEE Computer and Communications Societies*, Tel Aviv, Israel, Vol. 3, pp. 1537–1545, 2000.

6. J. W. Kwon, H. D. Park, and Y. Z. Cho, An efficient mechanism for mobile IP handoffs, *IEICE Transactions on Communications*, E85-B (4), 796–801, 2002.

7. A. Chrungoo, V. Gupta, H. Saran, and R. Shorey: TCP k-SACK: A simple protocol to improve performance over lossy links, *IEEE Global Telecommunications Conference*, San Antonio, TX, Vol. 3, pp. 1713–1717, 2001.

8. R. Caceres and L. Iftode, Improving the performance of reliable transport protocols in mobile computing environments, *IEEE Journal on Selected Areas in Communications*, 13, 850–857, 1995.

9. J. Nakanishi, K. Tsukamoto, and S. Komaki, Proposal of continuous-FRT method for TCP/IP access in spot type DERC system, *Proceedings of the 3rd International Workshop on ITS Telecommunications*, Seoul, South Korea, 269–273, 2002.

10. C. P. Fu and S. C. Liew, TCP veno: TCP enhancement for transmission over wireless access networks, *IEEE Journal on Selected Areas in Communications*, 21(2), 216–228, 2003.

11. A. Bakre and B. R. Badrinath, I-TCP: Indirect TCP for mobile hosts, *IEEE Proceeding of 15th International Conference Distributed Computing Systems (ICDCS)*, Vancouver, British Columbian Canada, 136–143, 1995.

12. H. Balakrishnan, S. Seshan, and R. H. Katz, Improving reliable transport and handoff performance in cellular wireless networks, *ACM Wireless Networks*, 1, 469–481, 1995.

13. C. E. Perkins, J. T. Malinen, R. Wakikawa, E. M. Belding-Royer, and Y. Sun, Ad hoc address autoconfiguration, Internet Draft, draft-ietf-manet-autoconf-01.txt, Nov. 2001.

14. R. Wakikawa, J. T. Malinen, C. E. Perkins, A. Nilsson, and A. J. Tuominen, Global connectivity for IPv6 mobile ad hoc networks, Internet Draft, draft-wakikawa-manet-globalv6-02.txt, Nov. 2002.

15. Y. Sun, E. M. Belding-Royer, and C. E. Perkins, Internet connectivity for ad hoc mobile networks, *International Journal of Wireless Information Networks Special Issue on MANETs: Standards, Research, Applications*, 9(2), 75–88, Apr 2002.

16. U. Jonsson, F. Alriksson, T. Larsson, P. Johansson, and G. Maguire, Jr., MIPMANET—Mobile IP for mobile ad hoc networks, *Proceedings of the MobiHOC'00*, Boston, MA, pp. 75–85, Aug 2000.

17. H. Soliman, C. Castelluccia, K. El-Malki, and L. Bellier, Hierarchical mobile IPv6 mobility management (HMIPv6), Internet Draft, draft-ietf-mobileip-hmipv6-08.txt, Jun. 2003 (work in progress).

18. N. Montavont and T. Noel, Handover management for mobile nodes in IPv6 networks, *IEEE Communications Magazine*, 40, 38–43, Aug 2002.

19. S. Deering and R. Hinden, Internet protocol, Version 6 (IPv6) specification, RFC 2460, pp. 12–17, Dec 1998.

Index

9 781439 848111